$f(t)$	$\mathscr{L}\{f(t)\} = F(s)$
39. $\dfrac{e^{bt} - e^{at}}{t}$	$\ln \dfrac{s - a}{s - b}$
40. $\dfrac{2(1 - \cos kt)}{t}$	$\ln \dfrac{s^2 + k^2}{s^2}$
41. $\dfrac{2(1 - \cosh kt)}{t}$	$\ln \dfrac{s^2 - k^2}{s^2}$
42. $\dfrac{\sin at}{t}$	$\arctan\left(\dfrac{a}{s}\right)$
43. $\dfrac{\sin at \cos bt}{t}$	$\dfrac{1}{2}\arctan\dfrac{a + b}{s} + \dfrac{1}{2}\arctan\dfrac{a - b}{s}$
44. $\dfrac{1}{\sqrt{\pi t}}\, e^{-a^2/4t}$	$\dfrac{e^{-a\sqrt{s}}}{\sqrt{s}}$
45. $\dfrac{a}{2\sqrt{\pi t^3}}e^{-a^2/4t}$	$e^{-a\sqrt{s}}$
46. $\operatorname{erfc}\left(\dfrac{a}{2\sqrt{t}}\right)$	$\dfrac{e^{-a\sqrt{s}}}{s}$
47. $2\sqrt{\dfrac{t}{\pi}}\, e^{-a^2/4t} - a\operatorname{erfc}\left(\dfrac{a}{2\sqrt{t}}\right)$	$\dfrac{e^{-a\sqrt{s}}}{s\sqrt{s}}$
48. $e^{ab}e^{b^2 t}\operatorname{erfc}\left(b\sqrt{t} + \dfrac{a}{2\sqrt{t}}\right)$	$\dfrac{e^{-a\sqrt{s}}}{\sqrt{s}(\sqrt{s} + b)}$
49. $-e^{ab}e^{b^2 t}\operatorname{erfc}\left(b\sqrt{t} + \dfrac{a}{2\sqrt{t}}\right)$ $\quad + \operatorname{erfc}\left(\dfrac{a}{2\sqrt{t}}\right)$	$\dfrac{be^{-a\sqrt{s}}}{s(\sqrt{s} + b)}$
50. $\delta(t)$	1
51. $\delta(t - t_0)$	e^{-st_0}
52. $e^{at}f(t)$	$F(s - a)$
53. $f(t - a)\mathscr{U}(t - a)$	$e^{-as}F(s)$
54. $\mathscr{U}(t - a)$	$\dfrac{e^{-as}}{s}$
55. $f^{(n)}(t)$	$s^n F(s) - s^{(n-1)}f(0) - \cdots - f^{(n-1)}(0)$
56. $t^n f(t)$	$(-1)^n \dfrac{d^n}{ds^n}F(s)$
57. $\displaystyle\int_0^t f(\tau)g(t - \tau)\, d\tau$	$F(s)\,G(s)$

DIFFERENTIAL EQUATIONS
WITH BOUNDARY-VALUE
PROBLEMS

5TH EDITION

DENNIS G. ZILL
Loyola Marymount University

MICHAEL R. CULLEN

BROOKS/COLE

THOMSON LEARNING

Australia • Canada • Mexico • Singapore • Spain • United Kingdom • United States

BROOKS/COLE

THOMSON LEARNING ™

Publisher: *Gary Ostedt*
Marketing Team: *Karin Sandberg, Samantha Cabaluna*
Marketing Associate: *Beth Kroenke*
Assistant Editor: *Carol Benedict*
Editorial Assistant: *Daniel Thiem*
Production Editor: *Keith Faivre*
Production Service: *Lifland et al., Bookmakers*
Manuscript Editor: *Gail Magin/Lifland*

Cover Design: *Roy R. Neuhaus*
Cover Photo: *The Stock Market/Cameron Davidson*
Interior Illustration: *Network Graphics*
Photo Researcher: *Gail Magin/Lifland*
Print Buyer: *Vena Dyer*
Typesetting: *The PRD Group*
Cover Printing: *Phoenix Color Corporation*
Printing and Binding: *R.R. Donnelley–Willard*

Photo Credits: p. 104, Fig. 3.8: www.corbis.com/Archivo Iconografico, S.A.; Fig. 3.9: www.corbis.com/Bettmann;
p. 135: © John K.B. Ford/*Innerspace Visions–Ursus* WS-1E; p. 263: AP/WIDE WORLD PHOTOS; p. 406:
© Roger Ressmeyer/CORBIS.

Library of Congress Cataloging-in-Publication Data

Zill, Dennis G. [date]

    Differential equations with boundary-value problems / Dennis G. Zill, Michael R.
Cullen.—5th ed.
      p.   cm.
    Includes index.
    ISBN 0-534-38002-6 (alk. paper)
    1. Differential equations. 2. Boundary value problems. I. Cullen, Michael R. II. Title.

QA371.Z55 2000
515′.35—dc21

00-057921

This edition is dedicated to my coauthor, colleague,
and friend of 27 years, as well as an award-winning teacher,
Michael R. Cullen, who passed away during its production.

Michael Cullen
1943–1999

CONTENTS

4 HIGHER-ORDER DIFFERENTIAL EQUATIONS 138

5 MODELING WITH HIGHER-ORDER DIFFERENTIAL EQUATIONS 215

6 SERIES SOLUTIONS OF LINEAR EQUATIONS 267

7 THE LAPLACE TRANSFORM 306

8 SYSTEMS OF LINEAR FIRST-ORDER DIFFERENTIAL EQUATIONS 364

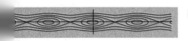

9 NUMERICAL SOLUTIONS OF ORDINARY DIFFERENTIAL EQUATIONS 410

10 PLANE AUTONOMOUS SYSTEMS AND STABILITY 439

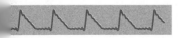

11 ORTHOGONAL FUNCTIONS AND FOURIER SERIES 483

12 PARTIAL DIFFERENTIAL EQUATIONS AND BOUNDARY-VALUE PROBLEMS IN RECTANGULAR COORDINATES   521

13 BOUNDARY-VALUE PROBLEMS IN OTHER COORDINATE SYSTEMS   561

14 INTEGRAL TRANSFORM METHOD   581

15 NUMERICAL SOLUTIONS OF PARTIAL DIFFERENTIAL EQUATIONS   610

APPENDIXES APP-1

SELECTED ANSWERS FOR ODD-NUMBERED PROBLEMS AN-1

PREFACE

In a revision of a textbook, the author is pulled by many influences in many directions. The author is expected to satisfy the requests from reviewers and editors to add things and to change things so that the text stays both current and competitive. At the same time the author is also expected not to increase the bulk of the book (a request to delete something is rare) and is certainly expected not to do anything that might upset the users of the previous edition. So yes, I have tried to meet, or at least strike a compromise with, all these demands. But no, I have not (in my opinion) deviated from the principle I have stated in prefaces over the many years that this text has been in print—that is, an undergraduate text should be written with the student's understanding kept firmly in mind, which means to me that the material should be presented in a straightforward, readable, and helpful manner, while the level of theory is kept consistent with the notion of a "first course."

In recent years, partially due to evolving technology and its expanding importance in pedagogy and to the reform movement in calculus, changes have occurred in the course called "Differential Equations." Instructors are questioning aspects of both the traditional teaching methods used and the traditional content of the course. This healthy introspection is important in making the subject matter not only more interesting for students but also more relevant to the world in which they live.

WHAT IS NEW TO THIS EDITION?

- This edition has a clearer delineation of the three major approaches to differential equations: analytical, qualitative, and numerical. Before launching into the search for analytic solutions of first-order differential equations, Chapter 2 starts with a section that is new to this book. In this section the qualitative behavior of solutions of first-order differential equations is examined using direction fields and phase-line analysis. There is no doubt in my mind that a reasonable amount of qualitative analysis should be, and will be, a permanent part of a typical first course.
- Since this is the 5th edition, an effort was made to liven up the exercise sets through the addition of new kinds of problems. Some of the problems that call for the use of a computer algebra system are new to this edition. With the realization that many colleges still lack the computational resources to incorporate problems of this sort into their curricula, I have placed the

majority of these problems at the end of the exercise sets under the heading "Computer Lab Assignments." This was done so that these kinds of problems would not "get in the way"; that is, they do not have to be "weeded out" of the more standard types of problems if an instructor wants to skip them entirely or postpone their coverage. Problems that require simpler technology such as a graphics calculator or graphing software have been marked by an icon in the margin. Finally, many conceptual and discussion problems have been added throughout most exercise sets. In some instances I have *slightly* reduced the number of routine "drill problems"—problems that usually require formal manipulation of some solution method—to accommodate these new problems. Like the Computer Lab Assignments, problems that emphasize understanding of concepts and problems that are suited to class or group discussion are, for convenience and recognition, placed near the end of the exercise sets.

- Three new *Project Modules,* all composed by Prof. Gilbert N. Lewis, appear after Chapters 3, 5, and 8. These modules explore mathematical models relating to the conservation of natural resources, the destruction of the Tacoma Narrows Bridge, and the modes of vibration of a multistory building during an earthquake. These modules are more than essays because each contains a set of questions that can be used as a computer lab assignment if desired.

WHAT HAS CHANGED IN THIS EDITION?

Although there are changes in almost all chapters, the greatest number of changes occur in Chapters 2, 6, and 7.

Chapter 2: *First-Order Differential Equations*

The topics in this chapter have been rearranged somewhat, and one new section has been added.

- Before the chase begins for solutions of first-order differential equations, some qualitative analysis is presented in Section 2.1, called "Solution Curves Without the Solution." How the behavior of solutions and the shape of solution curves can be discerned is discussed through the use of direction fields and phase-line analysis. Approximately half of the material in this section is new to this text; the direction field material appeared in Chapter 9 in the 4th edition. In testing the new material on autonomous first-order equations for several years in my own classes while using my alternative text *Differential Equations with Computer Lab Experiments,* I found that students actually enjoyed the brief excursion into the qualitative aspects of autonomous first-order differential equations because it did not involve complicated procedures and simply built upon the concept of interpreting a derivative—a concept already familiar to them from their study of differential calculus. This brief introduction to the notions of critical points, equilibrium solutions, and stability of classification of critical points as attractors, repellers, and semistable is carried out in a nontheoretical and nonthreatening manner.

- In this edition I have moved the discussion of linear equations (Section 2.3) before that on exact equations (Section 2.4). In Section 2.3 I have retained the "Property, Procedure, and Variation of Parameters" format, even though this approach to linear first-order differential equations did not meet with universal acclaim from the users of the 4th edition. Some texts present linear first-order operations (for example, obtaining the general form of the integrating factor from an appeal to exactness) as if linear first-order equations were somehow different from linear higher-order equations. Bear in mind that I am trying to get across early on the point that the same procedures presented in Chapter 4 for linear higher-order differential equations are applicable to linear first-order equations.
- In the discussion of exact equations in the 4th edition, the concept of an integrating factor was relegated to an exercise set. The procedure whereby an integrating factor can be determined for certain kinds of nonexact equations has now been added to Section 2.4.
- The intuitive Euler's method has been moved from Chapter 9 in the 4th edition to Section 2.6. This was done for two reasons: to strike a better balance among the analytical, qualitative, and numerical approaches to differential equations and to better illustrate the graphical aspects of computer software known generically as numerical solvers. Moreover, the presentation of Euler's method in Chapter 2 is consistent with the early consideration of this topic in most current calculus texts.

Chapter 6: *Series Solutions of Linear Equations*

- The lengthy discussion of the various cases of solving linear differential equations with variable coefficients using the method of Frobenius (Section 6.2) has been abbreviated to the essentials.

Chapter 7: *The Laplace Transform*

Chapter 7 has been completely reorganized in order to get to the point of the chapter sooner than in all previous editions—namely, that the Laplace transform is a useful tool for solving certain kinds of equations.

- The Laplace transform of derivatives, and how this result is used in solution of simple linear initial-value problems, is now introduced in Section 7.2.
- In the subsequent sections of this chapter initial-value problems of increasing difficulty, as well as the solution of other kinds of equations, are examined in conjunction with the unfolding of the various operational properties of the transform. In earlier editions all applications were considered in one, I admit, rather formidable section.

Chapter 8: *Systems of Linear First-Order Differential Equations*

- More figures and graphs were added to Chapter 8. The graphs of solutions of plane systems are given in the tx-plane, ty-plane, and xy-plane. Although the words "phase plane" are introduced, no qualitative analysis of autonomous second-order equations or of autonomous plane systems is presented at this point in the text.
- A discussion of how the Laplace transform can be used to determine a matrix exponential has been added to Section 8.4.

WHAT REMAINS THE SAME?

The chapter lineup by topics and the basic underlying philosophy remain the same as in the previous edition. Like its predecessors, this edition contains a generous supply of examples, exercises, and applications. Nuances such as presenting applications of differential equations in separate chapters also continue in this revision. For every reviewer who states "incorporate applications with the discussion of solutions of differential equations" there is another reviewer who makes the plea to keep applications separate from the methods of solution. I mostly agree with the latter viewpoint. I feel that presenting applications of first- and higher-order ordinary differential equations in separate chapters not only gives the greatest flexibility to the text but also allows the user to get his or her "feet on the ground" by focusing on fewer concepts at the beginning of the course. From the student's viewpoint the mixture of applications with methods of solution can be overwhelming, and from the instructor's viewpoint it makes the applied topics difficult to skip if they are not part of the course syllabus. But this philosophy is not carved in stone; I have now interwoven applications into the various sections of Chapter 7 on the Laplace transform. I was convinced to do this for two reasons: students have attained a certain comfort level with applied problems by the time Chapter 7 is usually covered, and, as noted above, the alternative (namely, lumping all applications into one section) is unpalatable. In conclusion, I have resisted the recommendations to present the solution of linear second-order differential equations and the solution of linear equations of order higher than two in separate sections.

SUPPLEMENTS AVAILABLE

For Students
Student Solutions Manual, by Warren S. Wright and Carol D. Wright (ISBN 0-534-38003-4), provides the solution to every third problem in each exercise set, with the exception of the Discussion Problems and the Computer Lab Assignments.

For Instructors
Complete Solutions Manual, by Warren S. Wright and Carol D. Wright (ISBN 0-534-38004-2), provides worked-out solutions to all problems in the text.

ACKNOWLEDGMENTS

A large measure of gratitude is owed to the following persons, who contributed to this revision through their help, suggestions, and criticisms:

Zaven Margosian, *Lawrence Technological University*
Brian M. O'Connor, *Tennessee Technological University*
Mohsen Razzaghi, *Mississippi State University*

A special acknowledgment is reserved for Gilbert Lewis, Michigan Technological University, who generously took time out of a busy schedule to craft the three new *Project Modules* in this edition. I also want to thank Barbara Lovenvirth for coordinating the work on these modules. Finally, I would like to express my sincere appreciation to my multitalented colleague Michael Berg for rendering the cartoon on page 31.

The task of compiling a text such as this one is time-consuming and difficult. Undoubtedly some errors have sifted through, as hundreds of manuscript pages have passed through many hands. I apologize for this in advance. To expedite the correction of an error please send it directly to my editor, Gary Ostedt, at Brooks/Cole.

Dennis G. Zill
Los Angeles

DIFFERENTIAL EQUATIONS
WITH BOUNDARY-VALUE
PROBLEMS

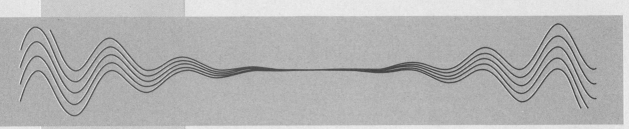

A family of solutions of a DE; see page 8.

1 INTRODUCTION TO DIFFERENTIAL EQUATIONS

INTRODUCTION   The words *differential* and *equations* certainly suggest solving some kind of equation that contains derivatives. Analogous to a course in algebra and trigonometry, where a good amount of time is spent solving equations such as $x^2 + 5x + 4 = 0$ for the unknown variable x, in this course one of our tasks will be to solve differential equations such as $y'' + 2y' + y = 0$ for the unknown function $y = \phi(x)$.

The first paragraph tells something, but not the complete story, about the course you are about to begin. As the course unfolds, you will see that there is more to the study of differential equations than just mastering methods someone has devised to solve them. But first, in order to read, study, and be conversant in a specialized subject, one first has to learn the terminology, the jargon, of that discipline.

1

1.1 DEFINITIONS AND TERMINOLOGY

• Ordinary and partial differential equations   • Order of a DE   • Linear and nonlinear DEs   • Normal form   • Solution of an ODE   • Explicit and implicit solutions   • Trivial solution   • Family of solutions   • Singular solution

In calculus you learned that the derivative dy/dx of a function $y = \phi(x)$ is itself another function $\phi'(x)$ found by an appropriate rule. The function $y = e^{0.1x^2}$ is differentiable on the interval $(-\infty, \infty)$, and its derivative is $dy/dx = 0.2xe^{0.1x^2}$. If we replace $e^{0.1x^2}$ on the right-hand side of the derivative by the symbol y, we obtain

$$\frac{dy}{dx} = 0.2xy. \tag{1}$$

Now imagine that a friend of yours simply hands you equation (1)—you have no idea how it was constructed—and asks: What is the function represented by the symbol y? You are now face to face with one of the basic problems in this course: How do you solve such an equation for the unknown function $y = \phi(x)$? The problem is loosely equivalent to the familiar reverse problem of differential calculus: Given a derivative, find an antiderivative.

Differential Equation   The equation that we made up in (1) is called a **differential equation.** Before proceeding any further, let us consider a more precise definition of this concept.

> ### DEFINITION 1.1   Differential Equation
>
> An equation containing the derivatives of one or more dependent variables with respect to one or more independent variables is said to be a **differential equation (DE).**

In order to talk about them, we shall classify differential equations by **type, order,** and **linearity.**

Classification by Type   If an equation contains only ordinary derivatives of one or more dependent variables with respect to a single independent variable, it is said to be an **ordinary differential equation (ODE).** For example,

$$\frac{dy}{dx} + 5y = e^x, \quad \frac{d^2y}{dx^2} - \frac{dy}{dx} + 6y = 0, \quad \text{and} \quad \frac{dx}{dt} + \frac{dy}{dt} = 2x + y \tag{2}$$

are ordinary differential equations. An equation involving the partial derivatives of one or more dependent variables of two or more independent variables is called a **partial differential equation (PDE).** For example,

$$\frac{\partial^2 u}{\partial x^2} + \frac{\partial^2 u}{\partial y^2} = 0, \quad \frac{\partial^2 u}{\partial x^2} = \frac{\partial^2 u}{\partial t^2} - 2\frac{\partial u}{\partial t}, \quad \text{and} \quad \frac{\partial u}{\partial y} = -\frac{\partial v}{\partial x} \tag{3}$$

are partial differential equations.

Ordinary derivatives throughout this text will be written using either the **Leibniz notation** dy/dx, d^2y/dx^2, d^3y/dx^3, ... or the **prime notation** y', y'', y''', ... Using the latter notation, we can write the first two differential equations in (2) a little more compactly as $y' + 5y = e^x$ and $y'' - y' + 6y = 0$. Actually the prime notation is used to denote only the first three derivatives; the fourth derivative is written $y^{(4)}$ instead of y''''. In general, the nth derivative is written d^ny/dx^n or $y^{(n)}$. Although less convenient to write and to typeset, the Leibniz notation has an advantage over the prime notation in that it clearly displays both dependent and independent variables. For example, in the equation $d^2x/dt^2 + 16x = 0$ it is immediately seen that the symbol x now represents a dependent variable whereas the independent variable is t. You should also be aware that in physical sciences and engineering **Newton's dot notation** (derogatively referred to by some as the "flyspeck" notation) is sometimes used to denote derivatives with respect to time t. Thus the differential equation $d^2s/dt^2 = -32$ becomes $\ddot{s} = -32$. Partial derivatives are often denoted by a **subscript notation** indicating the independent variables. For example, with the subscript notation the second equation in (3) becomes $u_{xx} = u_{tt} - 2u_t$.

Classification by Order   The **order of a differential equation** (either ODE or PDE) is the order of the highest derivative in the equation. For example,

second-order $\downarrow$ $\downarrow$ first-order

$$\frac{d^2y}{dx^2} + 5\left(\frac{dy}{dx}\right)^3 - 4y = e^x$$

is a second-order ordinary differential equation. First-order ordinary differential equations are occasionally written in differential form $M(x, y)\, dy + N(x, y)\, dx = 0$. For example, if we assume that y denotes the dependent variable in $(y - x)\, dx + 4x\, dy = 0$, then $y' = dy/dx$, and so by dividing by the differential dx we get the alternative form $4xy' + y = x$. See the Remarks at the end of this section.

In symbols, we can express an nth-order ordinary differential equation in one dependent variable by the general form

$$F(x, y, y', \ldots, y^{(n)}) = 0, \tag{4}$$

where F is a real-valued function of $n + 2$ variables, $x, y, y', \ldots, y^{(n)}$, and where $y^{(n)} = d^ny/dx^n$. For both practical and theoretical reasons, we shall also make the assumption hereafter that it is possible to solve an ordinary differential equation in the form (4) uniquely for the highest derivative $y^{(n)}$ in terms of the remaining $n + 1$ variables. The differential equation

$$\frac{d^ny}{dx^n} = f(x, y, y', \ldots, y^{(n-1)}), \tag{5}$$

where f is a real-valued continuous function, is referred to as the **normal form** of (4). Thus, when it suits our purposes, we shall use the normal forms

$$\frac{dy}{dx} = f(x, y) \quad \text{and} \quad \frac{d^2y}{dx^2} = f(x, y, y')$$

to represent general first- and second-order ordinary differential equations. For example, the normal form of the first-order equation $4xy' + y = x$ is $y' = (x - y)/4x$. See the Remarks.

Classification by Linearity   An nth-order ordinary differential equation (4) is said to be **linear** if F is linear in $y, y', \ldots, y^{(n-1)}$. This means that an nth-order ODE is linear when (4) is $a_n(x)y^{(n)} + a_{n-1}(x)y^{(n-1)} + \cdots + a_1(x)y' + a_0(x)y - g(x) = 0$ or

$$a_n(x)\frac{d^ny}{dx^n} + a_{n-1}(x)\frac{d^{n-1}y}{dx^{n-1}} + \cdots + a_1(x)\frac{dy}{dx} + a_0(x)y = g(x). \tag{6}$$

From (6) we see the characteristic two properties of a linear differential equation: First, the dependent variable and all its derivatives are of the first degree—that is, the power of each term involving y is 1. Second, each coefficient depends at most on the independent variable x. The equations

$$(y - x)\,dx + 4x\,dy = 0, \quad y'' - 2y' + y = 0, \quad \text{and} \quad \frac{d^3y}{dx^3} + x\frac{dy}{dx} - 5y = e^x$$

are, in turn, linear first-, second-, and third-order ordinary differential equations. We have just demonstrated that the first equation is linear in the variable y by writing it in the alternative form $4xy' + y = x$. A **nonlinear** ordinary differential equation is simply one that is not linear. Nonlinear functions of the dependent variable or its derivatives, such as $\sin y$ or $e^{y'}$, cannot appear in a linear equation. Therefore,

$$\begin{array}{ccc}
\text{nonlinear term:} & \text{nonlinear term:} & \text{nonlinear term:} \\
\text{coefficient depends on } y & \text{nonlinear function of } y & \text{power not 1} \\
\downarrow & \downarrow & \downarrow
\end{array}$$

$$(1 - y)y' + 2y = e^x, \quad \frac{d^2y}{dx^2} + \sin y = 0, \quad \text{and} \quad \frac{d^4y}{dx^4} + y^2 = 0$$

are examples of nonlinear first-, second-, and fourth-order ordinary differential equations, respectively.

Solutions   As stated before, one of the goals in this course is to solve, or find solutions of, differential equations. In the next definition we consider the concept of a solution of an ordinary differenial equation.

DEFINITION 1.2   Solution of an ODE

Any function ϕ, defined on an interval I and possessing at least n derivatives that are continuous on I, which when substituted into an nth-order ordinary differential equation reduces the equation to an identity, is said to be a **solution** of the equation on the interval.

In other words, a solution of an nth-order ordinary differential equation (4) is a function ϕ that possesses at least n derivatives and for which

$$F(x, \phi(x), \phi'(x), \ldots, \phi^{(n)}(x)) = 0 \quad \text{for all } x \text{ in } I.$$

We say that ϕ *satisfies* the differential equation on I. For our purposes, we shall also assume that a solution ϕ is a real-valued function. In our introductory discussion we saw that $y = e^{0.1x^2}$ is a solution of $dy/dx = 0.2xy$ on the interval $(-\infty, \infty)$.

Occasionally it will be convenient to denote a solution by the alternative symbol $y(x)$.

Interval of Definition   You cannot think *solution* of an ordinary differential equation without simultaneously thinking *interval*. The interval I in Definition 1.2 is variously called the **interval of definition,** the **interval of existence,** the **interval of validity,** or the **domain** of the solution and can be an open interval (a, b), a closed interval $[a, b]$, an infinite interval (a, ∞), and so on.

EXAMPLE 1   **Verification of a Solution**

Verify that the indicated function is a solution of the given differential equation on the interval $(-\infty, \infty)$.

(a) $dy/dx = xy^{1/2}$; $y = \frac{1}{16}x^4$ **(b)** $y'' - 2y' + y = 0$; $y = xe^x$

Solution   One way of verifying that the given function is a solution is to see, after substituting, whether each side of the equation is the same for every x in the interval.

(a) From

$$\text{left-hand side:} \quad \frac{dy}{dx} = \frac{1}{16}(4 \cdot x^3) = \frac{1}{4}x^3$$

$$\text{right-hand side:} \quad xy^{1/2} = x \cdot \left(\frac{1}{16}x^4\right)^{1/2} = x \cdot \left(\frac{1}{4}x^2\right) = \frac{1}{4}x^3$$

we see that each side of the equation is the same for every real number x. Note that $y^{1/2} = \frac{1}{4}x^2$ is, by definition, the nonnegative square root of $\frac{1}{16}x^4$.

(b) From the derivatives $y' = xe^x + e^x$ and $y'' = xe^x + 2e^x$ we have, for every real number x,

$$\text{left-hand side:} \quad y'' - 2y' + y = (xe^x + 2e^x) - 2(xe^x + e^x) + xe^x = 0$$
$$\text{right-hand side:} \quad 0. \qquad\qquad ■$$

Note, too, that in Example 1 each differential equation possesses the constant solution $y = 0$, $-\infty < x < \infty$. A solution of a differential equation that is identically zero on an interval I is said to be a **trivial solution.**

Solution Curve   The graph of a solution ϕ of an ODE is called a **solution curve.** Since ϕ is a differentiable function, it is continuous on its interval I of definition. Thus there may be a difference between the graph of the *function* ϕ and the graph of the *solution* ϕ. Put another way, the domain of the function ϕ need not be the same as the interval I of definition (or domain) of the solution ϕ. Example 2 illustrates the difference.

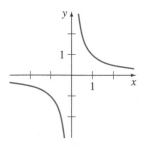

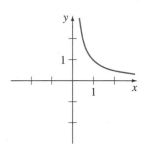

(a) function $y = 1/x$, $x \neq 0$

(b) solution $y = 1/x$, $(0, \infty)$

Figure 1.1

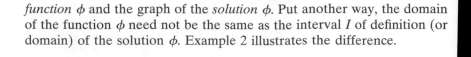

EXAMPLE 2   **Domain vs. Interval I of Definition**

The domain of $y = 1/x$, considered simply as a *function*, is the set of all real numbers x except 0. When we graph $y = 1/x$, we plot points in the xy-plane corresponding to a judicious sampling of numbers taken from its domain. The rational function $y = 1/x$ is discontinuous at 0, and its graph, in a neighborhood of the origin, is given in Figure 1.1(a). The function $y = 1/x$ is not differentiable at $x = 0$, since the y-axis (whose equation is $x = 0$) is a vertical asymptote of the graph.

Now $y = 1/x$ is also a solution of the linear first-order differential equation $xy' + y = 0$. (Verify.) But when we say $y = 1/x$ is a *solution* of this DE, we mean it is a function defined on an interval I on which it is differentiable and satisfies the equation. In other words, $y = 1/x$ is a solution of the DE on *any* interval not containing 0, such as $(-3, -1)$, $(\frac{1}{2}, 10)$, $(-\infty, 0)$, or $(0, \infty)$. Because the solution curves defined by $y = 1/x$ on the intervals $-3 < x < -1$ and $\frac{1}{2} < x < 10$ are simply segments, or pieces, of the solution curves defined by $y = 1/x$ on $-\infty < x < 0$ and $0 < x < \infty$, respectively, it makes sense to take the interval I to be as large as possible. Thus we take I to be either $(-\infty, 0)$ or $(0, \infty)$. The solution curve on $(0, \infty)$ is shown in Figure 1.1(b). ∎

Explicit and Implicit Solutions   You should be familiar with the terms *explicit* and *implicit functions* from your study of calculus. A solution in which the dependent variable is expressed solely in terms of the independent variable and constants is said to be an **explicit solution.** For our purposes, let us think of an explicit solution as an explicit formula $y = \phi(x)$ that we can manipulate, evaluate, and differentiate using the standard rules. We have just seen in the last two examples that $y = \frac{1}{16}x^4$, $y = xe^x$, and $y = 1/x$ are, in turn, explicit solutions of $dy/dx = xy^{1/2}$, $y'' - 2y' + y = 0$, and $xy' + y = 0$. Moreover, the trivial solution $y = 0$ is an explicit solution of all three equations. When we get down to the business of actually solving some ordinary differential equations, you will see that methods of solution do not always lead directly to an explicit solution $y = \phi(x)$. This is particularly true when we attempt to solve nonlinear first-order differential equations. Often we have to be content with a relation or expression $G(x, y) = 0$ that defines a solution ϕ implicitly.

DEFINITION 1.3   **Implicit Solution of an ODE**

A relation $G(x, y) = 0$ is said to be an **implicit solution** of an ordinary differential equation (4) on an interval I, provided there exists at least one function ϕ that satisfies the relation as well as the differential equation on I.

It is beyond the scope of this course to investigate the conditions under which a relation $G(x, y) = 0$ defines a differentiable function ϕ. So we shall assume that if the formal implementation of a method of solution leads to a relation $G(x, y) = 0$, then there exists at least one function ϕ that satisfies both the relation (that is, $G(x, \phi(x)) = 0$) and the differential equation on an interval I. If the implicit solution $G(x, y) = 0$ is fairly simple, we may be able to solve for y in terms of x and obtain one or more explicit solutions. See the Remarks.

EXAMPLE 3   Verification of an Implicit Solution

The relation $x^2 + y^2 = 25$ is an implicit solution of the differential equation

$$\frac{dy}{dx} = -\frac{x}{y} \tag{7}$$

on the interval $-5 < x < 5$. By implicit differentiation we obtain

$$\frac{d}{dx}x^2 + \frac{d}{dx}y^2 = \frac{d}{dx}25 \quad \text{or} \quad 2x + 2y\frac{dy}{dx} = 0.$$

Solving the last equation for the symbol dy/dx gives (7). Moreover, solving $x^2 + y^2 = 25$ for y in terms of x yields $y = \pm\sqrt{25 - x^2}$. The two functions $y = \phi_1(x) = \sqrt{25 - x^2}$ and $y = \phi_2(x) = -\sqrt{25 - x^2}$ satisfy the relation (that is, $x^2 + \phi_1^2 = 25$ and $x^2 + \phi_2^2 = 25$) and are explicit solutions defined on the interval $-5 < x < 5$. The solution curves given in Figures 1.2(b) and (c) are segments of the graph of the implicit solution in Figure 1.2(a). ▬

Any relation of the form $x^2 + y^2 - c = 0$ *formally* satisfies (7) for any constant c. However, it is understood that the relation should always make sense in the real number system; thus, for example, if $c = -25$ we cannot say that $x^2 + y^2 + 25 = 0$ is an implicit solution of the equation. (Why not?)

Because the distinction between an explicit solution and an implicit solution should be intuitively clear, we will not belabor the issue by always saying "Here is an explicit (implicit) solution."

Families of Solutions   The study of differential equations is similar to that of integral calculus. In some texts, a solution ϕ is sometimes referred to as an **integral** of the equation, and its graph is called an **integral curve.** When evaluating an antiderivative or indefinite integral in calculus, we use a single constant c of integration. Analogously, when solving a first-order differential equation $F(x, y, y') = 0$, we *usually* obtain a solution containing a single arbitrary constant or parameter c. A solution containing an arbitrary constant represents a set $G(x, y, c) = 0$ of solutions called a **one-parameter family of solutions.** When solving an nth-order differential equation $F(x, y, y', \ldots, y^{(n)}) = 0$, we seek an **$n$-parameter family of solutions** $G(x, y, c_1, c_2, \ldots, c_n) = 0$. This means that a single differential equation can possess an infinite number of solutions corresponding to

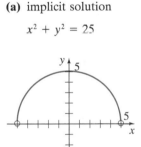

(a) implicit solution

$x^2 + y^2 = 25$

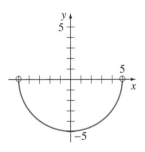

(b) explicit solution

$y_1 = \sqrt{25 - x^2}, \; -5 < x < 5$

(c) explicit solution

$y_2 = -\sqrt{25 - x^2}, \; -5 < x < 5$

Figure 1.2 An implicit and two explicit solutions of $y' = -x/y$

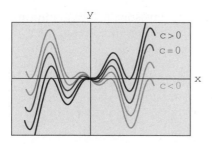

Figure 1.3 Some solutions of $xy' - y = x^2 \sin x$

the unlimited number of choices for the parameter(s). A solution of a differential equation that is free of arbitrary parameters is called a **particular solution.** For example, the one-parameter family $y = cx - x \cos x$ is an explicit solution of the linear first-order equation $xy' - y = x^2 \sin x$ on the interval $(-\infty, \infty)$. (Verify.) Figure 1.3, obtained using graphing software, shows the graphs of some of the solutions in this family. The solution $y = -x \cos x$, the colored curve in the figure, is a particular solution corresponding to $c = 0$. Similarly, on the interval $(-\infty, \infty)$, $y = c_1 e^x + c_2 x e^x$ is a two-parameter family of solutions of the linear second-order equation $y'' - 2y' + y = 0$ in Example 1. (Verify.) Some particular solutions of the equation are the trivial solution $y = 0$ $(c_1 = c_2 = 0)$, $y = xe^x$ $(c_1 = 0, c_2 = 1)$, $y = 5e^x - 2xe^x$ $(c_1 = 5, c_2 = -2)$, and so on.

Sometimes a differential equation possesses a solution that is not a member of a family of solutions of the equation—that is, a solution that cannot be obtained by specializing *any* of the parameters in the family of solutions. Such an extra solution is called a **singular solution.** For example, we have seen that $y = \frac{1}{16}x^4$ and $y = 0$ are solutions of the differential equation $dy/dx = xy^{1/2}$ on $(-\infty, \infty)$. In Section 2.2 we shall demonstrate, by actually solving it, that the differential equation $dy/dx = xy^{1/2}$ possesses the one-parameter family of solutions $y = (\frac{1}{4}x^2 + c)^2$. When $c = 0$, the resulting particular solution is $y = \frac{1}{16}x^4$. But notice that the trivial solution $y = 0$ is a singular solution since it is not a member of the family $y = (\frac{1}{4}x^2 + c)^2$; there is no way of assigning a value to the constant c to obtain $y = 0$.

In all the preceding examples we have used x and y to denote the independent and dependent variables, respectively. But you should become accustomed to seeing and working with other symbols to denote these variables. For example, we could denote the independent variable by t and the dependent variable by x.

EXAMPLE 4 **Using Different Symbols**

The functions $x = c_1 \cos 4t$ and $x = c_2 \sin 4t$, where c_1 and c_2 are arbitrary constants or parameters, are both solutions of the linear differential equation

$$x'' + 16x = 0.$$

For $x = c_1 \cos 4t$ the first two derivatives with respect to t are $x' = -4c_1 \sin 4t$ and $x'' = -16c_1 \cos 4t$. Substituting x'' and x then gives

$$x'' + 16x = -16c_1 \cos 4t + 16(c_1 \cos 4t) = 0.$$

In like manner, for $x = c_2 \sin 4t$ we have $x'' = -16c_2 \sin 4t$, and so

$$x'' + 16x = -16c_2 \sin 4t + 16(c_2 \sin 4t) = 0.$$

Finally, it is straightforward to verify that the linear combination of solutions, or the two-parameter family $x = c_1 \cos 4t + c_2 \sin 4t$, is also a solution of the differential equation.

The next example shows that a solution of a differential equation can be a piecewise-defined function.

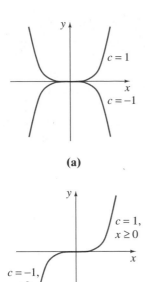

(a)

(b)

Figure 1.4 Some solutions of $xy' - 4y = 0$

EXAMPLE 5 **A Piecewise-Defined Solution**

You should verify that the one-parameter family $y = cx^4$ is a one-parameter family of solutions of the differential equation $xy' - 4y = 0$ on the inverval $(-\infty, \infty)$. See Figure 1.4(a). The piecewise-defined differentiable function

$$y = \begin{cases} -x^4, & x < 0 \\ x^4, & x \geq 0 \end{cases}$$

is a particular solution of the equation but cannot be obtained from the family $y = cx^4$ by a single choice of c; the solution is constructed from the family by choosing $c = -1$ for $x < 0$ and $c = 1$ for $x \geq 0$. See Figure 1.4(b). ∎

Systems of Differential Equations Up to this point we have been discussing single differential equations containing one unknown function. But often in theory, as well as in many applications, we must deal with systems of differential equations. A **system of ordinary differential equations** is two or more equations involving the derivatives of two or more unknown functions of a single independent variable. For example, if x and y denote dependent variables and t denotes the independent variable, then a system of two first-order differential equations is given by

$$\frac{dx}{dt} = f(t, x, y)$$
$$\frac{dy}{dt} = g(t, x, y). \tag{8}$$

A **solution** of a system such as (8) is a pair of differentiable functions $x = \phi_1(t)$, $y = \phi_2(t)$, defined on a common interval I, that satisfy each equation of the system on this interval.

Remarks (*i*) A few last words about implicit solutions of differential equations are in order. In Example 3 we were able to solve the relation $x^2 + y^2 = 25$ for y in terms of x to get two explicit solutions of the differential equation $dy/dx = -x/y$, $\phi_1(x) = \sqrt{25 - x^2}$ and $\phi_2(x) = -\sqrt{25 - x^2}$. But don't read too much into this one example. Unless it is easy, obvious, convenient, important, or you are instructed to, there is usually no need to try to solve an implicit solution $G(x, y) = 0$ for y explicitly in terms of x. Also do not misinterpret the second sentence following Definition 1.3. An implicit solution $G(x, y) = 0$ can define a perfectly good differentiable function ϕ that is a solution of a DE, yet we may not be able to solve $G(x, y) = 0$ using analytical methods such as algebra. The solution curve of ϕ may be a segment, or piece, of the graph of $G(x, y) = 0$. See Problems 41 and 42 in Exercises 1.1. Also, read the discussion following Example 4 in Section 2.2.

(*ii*) In the differential form $M(x, y)\,dx + N(x, y)\,dy = 0$ it may not be obvious whether a first-order ODE is linear or nonlinear, since there is nothing in this form that tells us which symbol denotes the dependent variable. See Problems 3 and 4 in Exercises 1.1.

(*iii*) It may not seem like a big deal to assume $F(x, y, y', \ldots, y^{(n)}) = 0$ can be solved for $y^{(n)}$, but you should be a little bit careful here. There are exceptions and there certainly are some problems connected with this assumption. See Problems 36 and 37 in Exercises 1.1.

(*iv*) You may run across the words "closed-form solutions" in DE texts or in lectures by instructors of courses in differential equations. Translated, this phrase usually refers to explicit solutions that are expressible in terms of *elementary* (or familiar) *functions*: finite combinations of integer powers of x, roots, exponential and logarithmic functions, and trigonometric and inverse trigonometric functions.

(*v*) If *every* solution of an nth-order ordinary differential equation $F(x, y, y', \ldots, y^{(n)}) = 0$ on an interval I can be obtained from an n-parameter family $G(x, y, c_1, c_2, \ldots, c_n) = 0$ by appropriate choices of the parameters c_i, $i = 1, 2, \ldots, n$, we then say that the family is the **general solution** of the differential equation. In solving linear differential equations we shall impose relatively simple restrictions on the coefficients of the equation; with these restrictions we can be assured not only that a solution exists on an interval but also that a family of solutions yields all possible solutions. Nonlinear equations, with the exception of some first-order equations, are usually difficult or impossible to solve in terms of elementary functions. Furthermore, if we happen to obtain a family of solutions for a nonlinear equation, it is not obvious whether this family contains all solutions. On a practical level, then, the designation "general solution" is applied only to linear differential equations. This concept will be important in Section 2.3 and in Chapters 4–6.

EXERCISES 1.1

Answers to odd-numbered problems begin on page AN-1.

In Problems 1–10 give the order of each differential equation. State whether the equation is linear or nonlinear.

1. $(1 - x)y'' - 4xy' + 5y = \cos x$

2. $x\dfrac{d^3y}{dx^3} - \left(\dfrac{dy}{dx}\right)^4 + y = 0$

3. $(y^2 - 1)\,dx + x\,dy = 0$

4. $u\,dv + (v + uv - ue^u)\,du = 0$

5. $t^5 y^{(4)} - t^3 y'' + 6y = 0$

6. $\dfrac{d^2u}{dr^2} + \dfrac{du}{dr} + u = \cos(r + u)$

7. $\dfrac{d^2y}{dx^2} = \sqrt{1 + \left(\dfrac{dy}{dx}\right)^2}$

8. $\dfrac{d^2R}{dt^2} = -\dfrac{k}{R^2}$

9. $(\sin\theta)y''' - (\cos\theta)y' = 2$

10. $\ddot{x} - \left(1 - \dfrac{\dot{x}^2}{3}\right)\dot{x} + x = 0$

In Problems 11–14 verify that the indicated function is an explicit solution of the given differential equation. Assume an appropriate interval I of definition.

11. $2y' + y = 0;$ $y = e^{-x/2}$

12. $\dfrac{dy}{dt} + 20y = 24;$ $y = \frac{6}{5} - \frac{6}{5}e^{-20t}$

13. $y'' - 6y' + 13y = 0;$ $y = e^{3x}\cos 2x$

14. $y'' + y = \tan x;$ $y = -(\cos x)\ln(\sec x + \tan x)$

In Problems 15 and 16 verify that the indicated expression is an implicit solution of the given differential equation. Find at least one explicit solution in each case. Use a graphing utility to obtain the graphs of the explicit solutions. Give the interval I of definition of each solution ϕ.

15. $\dfrac{dX}{dt} = (X - 1)(1 - 2X);$ $\ln\left(\dfrac{2X - 1}{X - 1}\right) = t$

16. $2xy\,dx + (x^2 - y)\,dy = 0;$ $-2x^2y + y^2 = 1$

In Problems 17–20 verify that the indicated family of functions is a solution of the given differential equation. Assume an appropriate interval I of definition of each solution.

17. $P' = P(1 - P);$ $P = \dfrac{c_1 e^t}{1 + c_1 e^t}$

18. $y' + 2xy = 1;$ $y = e^{-x^2}\displaystyle\int_0^x e^{t^2}\,dt + c_1 e^{-x^2}$

19. $\dfrac{d^2y}{dx^2} - 4\dfrac{dy}{dx} + 4y = 0;$ $y = c_1 e^{2x} + c_2 x e^{2x}$

20. $x^3\dfrac{d^3y}{dx^3} + 2x^2\dfrac{d^2y}{dx^2} - x\dfrac{dy}{dx} + y = 12x^2;$ $y = c_1 x^{-1} + c_2 x + c_3 x \ln x + 4x^2$

21. (a) Verify that $y = \phi_1(x) = x^2$ and $y = \phi_2(x) = -x^2$ are solutions of the differential equation $xy' - 2y = 0$ on the interval $(-\infty, \infty)$.

  (b) Verify that the piecewise-defined function
$$y = \begin{cases} -x^2, & x < 0 \\ x^2, & x \geq 0 \end{cases}$$
is also a solution of $xy' - 2y = 0$ on the interval $(-\infty, \infty)$.

22. In Example 3 we saw that $y = \phi_1(x) = \sqrt{25 - x^2}$ and $y = \phi_2(x) = -\sqrt{25 - x^2}$ are solutions of the differential equation $dy/dx = -x/y$ on the interval $(-5, 5)$. Explain why the piecewise-defined function
$$y = \begin{cases} \sqrt{25 - x^2}, & -5 < x < 0 \\ -\sqrt{25 - x^2}, & 0 \leq x < 5 \end{cases}$$
is not a solution of the differential equation on the interval $(-5, 5)$.

23. (a) Give the domain of the function $y = x + 2\sqrt{x + 3}$.

  (b) Verify that the function in part (a) is a solution of the differential equation $(y - x)y' = y - x + 2$ on some interval I. Give the largest interval I of definition of this solution.

24. The indicated function is a solution of the given differential equation. Find at least one interval I of definition of each solution.

  (a) $y' = 25 + y^2;$ $y = \tan 5x$

  (b) $2y' = y^3\cos x;$ $y = (1 - \sin x)^{-1/2}$

25. Find values of m so that the function $y = e^{mx}$ is a solution of the given differential equation. Explain your reasoning.

(a) $y' + 2y = 0$ (b) $y'' - 5y' + 6y = 0$

26. Find values of m so that the function $y = x^m$ is a solution of the given differential equation. Explain your reasoning.

(a) $xy'' + 2y' = 0$ (b) $x^2y'' - 7xy' + 15y = 0$

In Problems 27 and 28 verify that the given pair of functions is a solution of the indicated system of differential equations on the interval $(-\infty, \infty)$.

27. $x = e^{-2t} + 3e^{6t}, \quad y = -e^{-2t} + 5e^{6t};$

$$\frac{dx}{dt} = x + 3y$$

$$\frac{dy}{dt} = 5x + 3y$$

28. $x = \cos 2t + \sin 2t + \frac{1}{5}e^t, \quad y = -\cos 2t - \sin 2t - \frac{1}{5}e^t;$

$$\frac{d^2x}{dt^2} = 4y + e^t$$

$$\frac{d^2y}{dt^2} = 4x - e^t$$

Discussion Problems

29. Make up a differential equation that does not possess any real solutions.

30. Make up a differential equation that you feel confident possesses only the trivial solution $y = 0$. Explain your reasoning.

31. What function do you know from calculus is such that its first derivative is itself? Its first derivative is a constant multiple k of itself? Write each answer in the form of a first-order differential equation with a solution.

32. What function (or functions) do you know from calculus is such that its second derivative is itself? Its second derivative is the negative of itself? Write each answer in the form of a second-order differential equation with a solution.

33. Suppose $y = \phi(x)$ is a solution of an nth-order differential equation $F(x, y, y', \ldots, y^{(n)}) = 0$ on an interval I. Explain why $\phi, \phi', \ldots, \phi^{(n-1)}$ are necessarily continuous on I.

34. Suppose the two-parameter family $y(x) = c_1y_1(x) + c_2y_2(x)$ is a solution of a linear second-order differential equation on an interval I. If $x = 0$ is in I, find a particular solution in the family that satisfies both conditions: $y(0) = 2, y'(0) = 0$. State any assumptions that you must make.

35. Discuss, and illustrate with examples, how to solve differential equations of the forms $dy/dx = f(x)$ and $d^2y/dx^2 = f(x)$.

36. The differential equation $x(y')^2 - 4y' - 12x^3 = 0$ has the general form given in (4). Determine whether the equation can be put in the normal form $dy/dx = f(x, y)$.

37. The normal form (5) of an nth-order differential equation is equivalent to (4) whenever both forms have exactly the same solutions. Make up a first-order differential equation in which $F(x, y, y') = 0$ is not equivalent to the normal form $dy/dx = f(x, y)$.

38. Find a linear second-order differential equation $F(x, y, y', y'') = 0$ for which $y = c_1 x + c_2 x^3$ is a two-parameter family of solutions. Make sure your equation is free of the arbitrary parameters c_1 and c_2.

39. Examine form (6) of a linear differential equation. Discuss under what conditions $y = 0$ is a solution of a linear equation.

40. We know that $y' = 0$ if and only if y is a constant function. Use this idea to determine whether the given differential equation possesses constant solutions.

   (a) $3xy' + 5y = 10$ **(b)** $y' = y^2 + 2y - 3$
   (c) $(y - 1)y' = 1$ **(d)** $y'' + 4y' + 6y = 10$

In Problems 41 and 42 the given figure represents the graph of an implicit solution $G(x, y) = 0$ of a differential equation $dy/dx = f(x, y)$. In each case, the relation $G(x, y) = 0$ implicitly defines several solutions of the DE. Carefully reproduce each figure on a piece of paper. Use different colored pencils to mark off segments, or pieces, on each graph that correspond to graphs of solutions. Keep in mind that a solution ϕ must be a function and differentiable. Use the solution curve to estimate the interval I of definition of each solution ϕ.

41.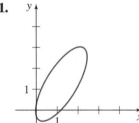

42.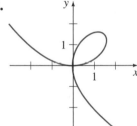

Figure 1.5   Figure 1.6

43. The graphs of the members of the one-parameter family $x^3 + y^3 = 3cxy$ are called **folia of Descartes.** Verify that this family is an implicit solution of the first-order differential equation

$$\frac{dy}{dx} = \frac{y(y^3 - 2x^3)}{x(2y^3 - x^3)}.$$

44. The graph in Figure 1.6 is the member of the family of folia in Problem 43 corresponding to $c = 1$. Discuss how the differential equation in Problem 43 can help in finding points on the graph of $x^3 + y^3 = 3xy$ where the tangent line is vertical. How does knowing where a tangent line is vertical help in determining an interval I of definition of a solution ϕ of the DE? Carry out your ideas and compare with your estimates of the intervals in Problem 42.

Qualitative information about a solution of a differential equation can often be obtained from the equation itself. Before working Problems 45–48, review your calculus text on the geometric significance of the derivatives dy/dx and d^2y/dx^2.

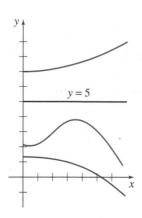

Figure 1.7

45. Suppose $y = \phi(x)$ is a solution of $y' = y^2 + 4$ defined on an interval I. Describe the graph of $y = \phi(x)$ on I. For example, can a solution curve have any relative extrema?

46. **(a)** Verify that $y = 5$, $-\infty < x < \infty$ is a solution of $y' = 5 - y$. The graph of this constant solution is the horizontal black line in Figure 1.7.

(b) Why aren't the curves shown in color in Figure 1.7 plausible solution curves?

(c) Sketch several plausible solution curves in the regions defined by $y < 5$ and $y > 5$.

47. Suppose $y = \phi(x)$ is a solution of the differential equation $dy/dx = y(a - by)$, where a and b are positive constants.

(a) By inspection find two constant solutions of the equation.

(b) Using only the differential equation, find intervals on the y-axis on which a nonconstant solution $y = \phi(x)$ is increasing. On which $y = \phi(x)$ is decreasing.

(c) Using only the differential equation, explain why $y = a/2b$ is the y-coordinate of a point of inflection of the graph of a nonconstant solution $y = \phi(x)$.

(d) On the same coordinate axes, sketch the graphs of the two constant solutions found in part (a). These constant solutions partition the xy-plane into three regions. In each region, sketch the graph of a nonconstant solution $y = \phi(x)$ whose shape is suggested by the results in parts (b) and (c).

48. Each indicated one-parameter family is an implicit solution of the given first-order differential equation:

$$y\,dx - 2x\,dy = 0,\ y^2 = c_1 x \quad \text{and} \quad 2x\,dx + y\,dy = 0,\ 2x^2 + y^2 = c_2^2.$$

Carefully plot three curves in the first family by choosing $c_1 > 0$ and three more curves by choosing $c_1 < 0$. Carefully plot at least three curves in the second family. What is true at a point where a solution curve from one family intersects a solution curve from the other family? Using only the differential equations, can you demonstrate that this property must hold at a point of intersection?

Computer Lab Assignments

In Problems 49 and 50 use a CAS to compute all derivatives and to carry out the simplifications needed to verify that the indicated function is a particular solution of the given differential equation.

49. $y^{(4)} - 20y''' + 158y'' - 580y' + 841y = 0$; $\quad y = xe^{5x}\cos 2x$

50. $x^3 y''' + 2x^2 y'' + 20xy' - 78y = 0$; $\quad y = 20\dfrac{\cos(5\ln x)}{x} - 3\dfrac{\sin(5\ln x)}{x}$

1.2   INITIAL-VALUE PROBLEMS

• nth-order initial-value problem   • Initial conditions   • Existence and uniqueness of a solution   • Interval of existence   • Interval of existence and uniqueness

We are often interested in solving a differential equation subject to prescribed side conditions—conditions that are imposed on the unknown solution $y = y(x)$ or its derivatives. On some interval I containing x_0, the problem

$$\text{Solve:} \quad \frac{d^n y}{dx^n} = f(x, y, y', \ldots, y^{(n-1)})$$

$$\text{Subject to:} \quad y(x_0) = y_0, \quad y'(x_0) = y_1, \quad \ldots, \quad y^{(n-1)}(x_0) = y_{n-1}$$

$$\tag{1}$$

where $y_0, y_1, \ldots, y_{n-1}$ are arbitrarily specified real constants, is called an **initial-value problem (IVP).** The values of $y(x)$ and its first $n - 1$ derivatives at a single point x_0: $y(x_0) = y_0, y'(x_0) = y_1, \ldots, y^{(n-1)}(x_0) = y_{n-1}$ are called **initial conditions.**

First- and Second-Order IVPs   The problem given in (1) is also called an ***n*th-order initial-value problem.** For example,

$$\text{Solve:} \quad \frac{dy}{dx} = f(x, y)$$

$$\text{Subject to:} \quad y(x_0) = y_0$$

$$\tag{2}$$

and

$$\text{Solve:} \quad \frac{d^2 y}{dx^2} = f(x, y, y')$$

$$\text{Subject to:} \quad y(x_0) = y_0, \quad y'(x_0) = y_1$$

$$\tag{3}$$

are first- and second-order initial-value problems, respectively. These two problems are easy to interpret in geometric terms. For (2) we are seeking a solution of the differential equation on an interval I containing x_0 so that a solution curve passes through the prescribed point (x_0, y_0). See Figure 1.8. For (3) we want to find a solution of the differential equation whose graph not only passes through (x_0, y_0) but passes through so that the slope of the curve at this point is y_1. See Figure 1.9. The term *initial condition* derives from physical systems where the independent variable is time t and where $y(t_0) = y_0$ and $y'(t_0) = y_1$ represent, respectively, the position and velocity of an object at some beginning, or initial, time t_0.

Solving an nth-order initial-value problem frequently entails using an n-parameter family of solutions of the given differential equation to find n specialized constants so that the resulting particular solution of the equation also "fits"—that is, satisfies—the n initial conditions.

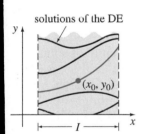

Figure 1.8 First-order IVP

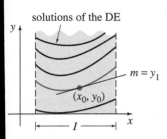

Figure 1.9 Second-order IVP

EXAMPLE 1   **First-Order IVPs**

It is readily verified that $y = ce^x$ is a one-parameter family of solutions of the simple first-order equation $y' = y$ on the interval $(-\infty, \infty)$. If we

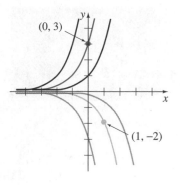

Figure 1.10 Solutions of IVPs

specify an initial condition, say, $y(0) = 3$, then substituting $x = 0$, $y = 3$ in the family determines the constant $3 = ce^0 = c$. Thus the function $y = 3e^x$ is a solution of the initial-value problem

$$y' = y, \quad y(0) = 3.$$

Now if we demand that a solution of the differential equation pass through the point $(1, -2)$ rather than $(0, 3)$, then $y(1) = -2$ will yield $-2 = ce$ or $c = -2e^{-1}$. The function $y = -2e^{x-1}$ is a solution of the initial-value problem

$$y' = y, \quad y(1) = -2.$$

The graphs of these two functions are shown in color in Figure 1.10. ∎

EXAMPLE 2   **Second-Order IVP**

In Example 4 of Section 1.1 we saw that $x = c_1 \cos 4t + c_2 \sin 4t$ is a two-parameter family of solutions of $x'' + 16x = 0$. Find a solution of the initial-value problem

$$x'' + 16x = 0, \quad x\left(\frac{\pi}{2}\right) = -2, \quad x'\left(\frac{\pi}{2}\right) = 1. \tag{4}$$

Solution   We first apply $x(\pi/2) = -2$ to the given family of solutions: $c_1 \cos 2\pi + c_2 \sin 2\pi = -2$. Since $\cos 2\pi = 1$ and $\sin 2\pi = 0$, we find that $c_1 = -2$. We next apply $x'(\pi/2) = 1$ to the one-parameter family $x(t) = -2 \cos 4t + c_2 \sin 4t$. Differentiating and then setting $t = \pi/2$ and $x' = 1$ gives $8 \sin 2\pi + 4c_2 \cos 2\pi = 1$, from which we see that $c_2 = \frac{1}{4}$. Hence $x = -2 \cos 4t + \frac{1}{4} \sin 4t$ is a solution of (4). ∎

Existence and Uniqueness   Two fundamental questions arise in considering an initial-value problem:

> *Does a solution of the problem exist?*
> *If a solution exists, is it unique?*

For an initial-value problem such as (2), we ask:

Existence $\begin{cases} \textit{Does the differential equation } dy/dx = f(x, y) \textit{ possess solutions?} \\ \textit{Do any of the solution curves pass through the point } (x_0, y_0)? \end{cases}$

Uniqueness $\begin{cases} \textit{When can we be certain that there is precisely one solution} \\ \textit{curve passing through the point } (x_0, y_0)? \end{cases}$

Note that in Examples 1 and 2 the phrase "*a* solution" is used rather than "*the* solution" of the problem. The indefinite article "a" is used deliberately to suggest the possibility that other solutions may exist. At this point it has not been demonstrated that there is a single solution of each problem. The next example illustrates an initial-value problem with two solutions.

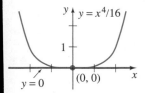

Figure 1.11 Two solutions of the same IVP

| EXAMPLE 3 | **An IVP Can Have Several Solutions** |

Each of the functions $y = 0$ and $y = \frac{1}{16}x^4$ satisfies the differential equation $dy/dx = xy^{1/2}$ and the initial condition $y(0) = 0$, and so the initial-value problem

$$\frac{dy}{dx} = xy^{1/2}, \quad y(0) = 0$$

has at least two solutions. As illustrated in Figure 1.11, the graphs of both functions pass through the same point $(0, 0)$. ∎

Within the safe confines of a formal course in differential equations one can be fairly confident that *most* differential equations will have solutions and that solutions of initial-value problems will *probably* be unique. Real life, however, is not so idyllic. Thus it is desirable to know in advance of trying to solve an initial-value problem whether a solution exists and, when it does, whether it is the only solution of the problem. Since we are going to consider first-order differential equations in the next two chapters, we state here without proof a straightforward theorem that gives conditions that are sufficient to guarantee the existence and uniqueness of a solution of a first-order initial-value problem of the form given in (2). We shall wait until Chapter 4 to address the question of existence and uniqueness of a second-order initial-value problem.

| THEOREM 1.1 | **Existence of a Unique Solution** |

Let R be a rectangular region in the xy-plane defined by $a \leq x \leq b$, $c \leq y \leq d$ that contains the point (x_0, y_0) in its interior. If $f(x, y)$ and $\partial f/\partial y$ are continuous on R, then there exist some interval I_0: $x_0 - h < x < x_0 + h$, $h > 0$, contained in $a \leq x \leq b$, and a unique function $y(x)$, defined on I_0, that is a solution of the initial-value problem (2).

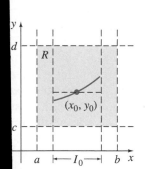

Figure 1.12 Rectangular region R

The foregoing result is one of the most popular existence and uniqueness theorems for first-order differential equations because the criteria of continuity of $f(x, y)$ and $\partial f/\partial y$ are relatively easy to check. The geometry of Theorem 1.1 is illustrated in Figure 1.12.

| EXAMPLE 4 | **Example 3 Revisited** |

We saw in Example 3 that the differential equation $dy/dx = xy^{1/2}$ possesses at least two solutions whose graphs pass through $(0, 0)$. Inspection of the functions

$$f(x, y) = xy^{1/2} \quad \text{and} \quad \frac{\partial f}{\partial y} = \frac{x}{2y^{1/2}}$$

shows that they are continuous in the upper half-plane defined by $y > 0$. Hence Theorem 1.1 enables us to conclude that through any point (x_0, y_0), $y_0 > 0$ in the upper half-plane there is some interval centered at x_0 on which the given differential equation has a unique solution. Thus, for example, even without solving it we know that there exists some interval centered at 2 on which the initial-value problem $dy/dx = xy^{1/2}$, $y(2) = 1$ has a unique solution. ∎

In Example 1, Theorem 1.1 guarantees that there are no other solutions of the initial-value problems $y' = y$, $y(0) = 3$ and $y' = y$, $y(1) = -2$ other than $y = 3e^x$ and $y = -2e^{x-1}$, respectively. This follows from the fact that $f(x, y) = y$ and $\partial f/\partial y = 1$ are continuous throughout the entire xy-plane. It can be further shown that the interval I on which each solution is defined is $(-\infty, \infty)$.

Interval of Existence/Uniqueness   Suppose $y(x)$ represents a solution of the initial-value problem (2). The following three sets on the real x-axis may not be the same: the domain of the function $y(x)$, the interval I over which the solution $y(x)$ is defined or exists, and the interval I_0 of existence *and* uniqueness. Example 2 of Section 1.1 illustrated the difference between the domain of a function and the interval I of definition. Now suppose (x_0, y_0) is a point in the interior of the rectangular region R in Theorem 1.1. It turns out that the continuity of the function $f(x, y)$ on R by itself is sufficient to guarantee the existence of at least one solution of $dy/dx = f(x, y)$, $y(x_0) = y_0$, defined on some interval I. The interval I of definition for this initial-value problem is usually taken to be the largest interval containing x_0 over which the solution $y(x)$ is defined and differentiable. The interval I depends on both $f(x, y)$ and the initial condition $y(x_0) = y_0$. See Problems 27 and 28 in Exercises 1.2. The extra condition of continuity of the first partial derivative $\partial f/\partial y$ on R enables us to say that not only does a solution exist on some interval I_0 containing x_0, but it is the *only* solution satisfying $y(x_0) = y_0$. However, Theorem 1.1 does not give any indication of the sizes of intervals I and I_0; *the interval I of definition need not be as wide as the region R, and the interval I_0 of existence and uniqueness may not be as large as I.* The number $h > 0$ that defines the interval I_0: $x_0 - h < x < x_0 + h$ could be very small, and so it is best to think that the solution $y(x)$ is *unique in a local sense*—that is, a solution defined near the point (x_0, y_0). See Problem 34 in Exercises 1.2.

Remarks   (*i*) The conditions in Theorem 1.1 are sufficient but not necessary. When $f(x, y)$ and $\partial f/\partial y$ are continuous on a rectangular region R, it must always follow that a solution of (2) exists and is unique whenever (x_0, y_0) is a point interior to R. However, if the conditions stated in the hypothesis of Theorem 1.1 do not hold, then anything could happen: Problem (2) *may* still have a solution and this solution *may* be unique, or it may have several solutions, or it may have no solution at all. A rereading of Example 4 reveals that the hypotheses of Theorem 1.1 do not hold on the line $y = 0$ for the differential equation $dy/dx = xy^{1/2}$, and so it is not surprising, as we saw in Example 3 of this section, that there are two solutions defined on a common interval $-h < x < h$ satisfying $y(0) = 0$. On the other hand, the hypotheses of Theorem 1.1 do not hold on the line $y = 1$ for the differential equation $dy/dx = |y - 1|$. Nevertheless it can be proved that the solution of the initial-value problem $dy/dx = |y - 1|$, $y(0) = 1$ is unique. Can you guess this solution? (*ii*) You are encouraged to read, think about, work, and then keep in mind Problem 33 in Exercises 1.2.

EXERCISES 1.2

Answers to odd-numbered problems begin on page AN-1.

In Problems 1 and 2 use the fact that $y = 1/(1 + c_1e^{-x})$ is a one-parameter family of solutions of $y' = y - y^2$ to find a solution of the initial-value problem consisting of the differential equation and the given initial condition.

1. $y(0) = -\dfrac{1}{3}$ **2.** $y(-1) = 2$

In Problems 3–6 use the fact that $x = c_1 \cos t + c_2 \sin t$ is a two-parameter family of solutions of $x'' + x = 0$ to find a solution of the initial-value problem consisting of the differential equation and the given initial conditions.

3. $x(0) = -1, x'(0) = 8$ **4.** $x(\pi/2) = 0, x'(\pi/2) = 1$
5. $x(\pi/6) = 1/2, x'(\pi/6) = 0$ **6.** $x(\pi/4) = \sqrt{2}, x'(\pi/4) = 2\sqrt{2}$

In Problems 7–10 use the fact that $y = c_1e^x + c_2e^{-x}$ is a two-parameter family of solutions of $y'' - y = 0$ to find a solution of the initial-value problem consisting of the differential equation and the given initial conditions.

7. $y(0) = 1, y'(0) = 2$ **8.** $y(1) = 0, y'(1) = e$
9. $y(-1) = 5, y'(-1) = -5$ **10.** $y(0) = 0, y'(0) = 0$

In Problems 11 and 12 determine by inspection at least two solutions of the given initial-value problem.

11. $y' = 3y^{2/3}, \quad y(0) = 0$ **12.** $xy' = 2y, \quad y(0) = 0$

In Problems 13–20 determine a region of the xy-plane for which the given differential equation would have a unique solution whose graph passes through a point (x_0, y_0) in the region.

13. $\dfrac{dy}{dx} = y^{2/3}$ **14.** $\dfrac{dy}{dx} = \sqrt{xy}$

15. $x\dfrac{dy}{dx} = y$ **16.** $\dfrac{dy}{dx} - y = x$

17. $(4 - y^2)y' = x^2$ **18.** $(1 + y^3)y' = x^2$
19. $(x^2 + y^2)y' = y^2$ **20.** $(y - x)y' = y + x$

In Problems 21–24 determine whether Theorem 1.1 guarantees that the differential equation $y' = \sqrt{y^2 - 9}$ possesses a unique solution through the given point.

21. $(1, 4)$ **22.** $(5, 3)$

23. $(2, -3)$ **24.** $(-1, 1)$

25. (a) By inspection find a one-parameter family of solutions of the differential equation $xy' = y$. Verify that each member of the family is a solution of the initial-value problem $xy' = y$, $y(0) = 0$.

    (b) Explain part (a) by determining a region R in the xy-plane for which the differential equation $xy' = y$ would have a unique solution through a point (x_0, y_0) in R.

    (c) Verify that the piecewise-defined function

$$y = \begin{cases} 0, & x < 0 \\ x, & x \geq 0 \end{cases}$$

satisfies the condition $y(0) = 0$. Determine whether this function is also a solution of the initial-value problem in part (a).

26. (a) Verify that $y = \tan(x + c)$ is a one-parameter family of solutions of the differential equation $y' = 1 + y^2$.

    (b) Since $f(x, y) = 1 + y^2$ and $\partial f/\partial y = 2y$ are continuous everywhere, the region R in Theorem 1.1 can be taken to be the entire xy-plane. Use the family of solutions in part (a) to find an explicit solution of the initial-value problem $y' = 1 + y^2$, $y(0) = 0$. Even though $x_0 = 0$ is in the interval $-2 < x < 2$, explain why the solution is not defined on this interval.

    (c) Determine the largest interval I of definition for the solution of the initial-value problem in part (b).

27. (a) Verify that $y = -1/(x + c)$ is a one-parameter family of solutions of the differential equation $y' = y^2$.

    (b) Since $f(x, y) = y^2$ and $\partial f/\partial y = 2y$ are continuous everywhere, the region R in Theorem 1.1 can be taken to be the entire xy-plane. Find a solution from the family in part (a) that satisfies $y(0) = 1$. Find a solution from the family in part (a) that satisfies $y(0) = -1$. Determine the largest interval I of definition for the solution of each initial-value problem.

    (c) Find a solution from the family in part (a) that satisfies $y' = y^2$, $y(0) = y_0$, $y_0 \neq 0$. Explain why the largest interval I of definition for this solution is either $-\infty < x < 1/y_0$ or $1/y_0 < x < \infty$.

    (d) Determine the largest interval I of definition for the solution of the initial-value problem $y' = y^2$, $y(0) = 0$.

28. (a) Verify that $3x^2 - y^2 = c$ is a one-parameter family of solutions of the differential equation $y \, dy/dx = 3x$.

    (b) By hand, sketch the graph of the implicit solution $3x^2 - y^2 = 3$. Find all explicit solutions $y = \phi(x)$ of the DE in part (a) defined by this relation. Give the interval I of definition of each explicit solution. The point $(-2, 3)$ is on the graph of $3x^2 - y^2 = 3$, but which of the explicit solutions satisfies $y(-2) = 3$?

    (c) Use the family of solutions in part (a) to find an implicit solution of the initial-value problem $y \, dy/dx = 3x$, $y(2) = -4$. Then, by hand, sketch the graph of the explicit solution of this problem and give its interval I of definition.

(d) Are there any explicit solutions of $y\,dy/dx = 3x$ that pass through the origin?

Discussion Problems

29. Consider the initial-value problem $y' = x - 2y$, $y(0) = \frac{1}{2}$. Using sound mathematics, explain which of the two curves shown in Figure 1.13 is the only plausible solution curve.

30. Explain why a first-order initial-value problem $dy/dx = f(x, y)$, $y'(x_0) = y_1$, where y_1 is an arbitrary real number, does not make sense.

31. Determine a plausible value of x_0 for which the graph of the solution of the initial-value problem $y' + 2y = 3x - 6$, $y(x_0) = 0$ is tangent to the x-axis at $(x_0, 0)$. Explain your reasoning.

32. Figure 1.14 shows the graphs of four members of a family of solutions of a second-order differential equation $d^2y/dx^2 = f(x, y, y')$. Match each solution curve with appropriate initial conditions.
  (a) $y(1) = 1$, $y'(1) = -2$ **(b)** $y(-1) = 0$, $y'(-1) = -4$
  (c) $y(1) = 1$, $y'(1) = 2$ **(d)** $y(0) = -1$, $y'(0) = 2$
  (e) $y(0) = -1$, $y'(0) = 0$ **(f)** $y(0) = -4$, $y'(0) = -2$

33. Suppose that the first-order differential equation $dy/dx = f(x, y)$ possesses a one-parameter family of solutions and that $f(x, y)$ satisfies the hypotheses of Theorem 1.1 in some rectangular region R of the xy-plane. Discuss why two different solution curves cannot intersect or be tangent to each other at a point (x_0, y_0) in R.

34. The functions

$$y(x) = \tfrac{1}{16}x^4, \quad -\infty < x < \infty \quad \text{and} \quad y(x) = \begin{cases} 0, & x < 0 \\ \tfrac{1}{16}x^4, & x \geq 0 \end{cases}$$

shown in Figures 1.15(a) and (b), respectively, are clearly different. Both functions are solutions of the initial-value problem $dy/dx = xy^{1/2}$, $y(2) = 1$ on the interval $(-\infty, \infty)$. Resolve the apparent contradiction between this fact and the last sentence in Example 4.

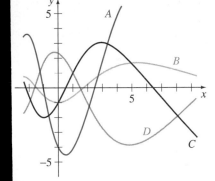

Figure 1.13

Figure 1.14

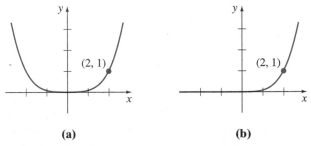

Figure 1.15 Two solutions of $dy/dx = xy^{1/2}$, $y(2) = 1$

35. Consider the differential equation $y' = y - x + 1$. If we differentiate the differential equation and resubstitute y', then $y'' = y' - 1 = (y - x + 1) - 1 = y - x$. Discuss the geometric significance of the following information:

$$y' = 0 \text{ for } y = x - 1, \quad y'' = 0 \text{ for } y = x;$$
$$y' > 0 \text{ for } y > x - 1, \quad y'' > 0 \text{ for } y > x;$$
$$y' < 0 \text{ for } y < x - 1, \quad y'' < 0 \text{ for } y < x.$$

Use this information as an aid in sketching a family of solution curves for the differential equation. Also reread Problem 33.

1.3 DIFFERENTIAL EQUATIONS AS MATHEMATICAL MODELS

• Mathematical model   • Level of resolution   • Population dynamics   • Radioactive decay and carbon dating   • Spread of disease   • Newton's law of cooling/warming   • Mixtures • Torricelli's law   • Kirchhoff's second law   • Newton's second law of motion   • Viscous damping   • Dynamical system

In this section we focus only on the construction of differential equations as mathematical models. Once we have examined some methods for solving differential equations in Chapters 2 and 4, we return to, and solve, some of these models in Chapters 3 and 5.

Mathematical Models   It is often desirable to describe the behavior of some real-life system or phenomenon, whether physical, sociological, or even economic, in mathematical terms. The mathematical description of a system or a phenomenon is called a **mathematical model** and is constructed with certain goals in mind. For example, we may wish to understand the mechanisms of a certain ecosystem by studying the growth of animal populations in that system, or we may wish to date fossils by analyzing the decay of a radioactive substance either in the fossil or in the stratum in which it was discovered.

Construction of a mathematical model of a system starts with

(*i*) *identification of the variables that are responsible for changing the system. We may choose not to incorporate all these variables into the model at first. In this step we are specifying the **level of resolution** of the model.*

Next

(*ii*) *we make a set of reasonable assumptions, or hypotheses, about the system we are trying to describe. These assumptions will also include any empirical laws that may be applicable to the system.*

For some purposes it may be perfectly within reason to be content with low-resolution models. For example, you may already be aware that in beginning physics courses, the retarding force of air friction is sometimes ignored, in modeling the motion of a body falling near the surface of the earth, but if you are a scientist whose job it is to accurately predict the flight path of a long-range projectile, you have to take into account air resistance and other factors such as the curvature of the earth.

Since the assumptions made about a system frequently involve *a rate of change* of one or more of the variables, the mathematical depiction of all these assumptions may be one or more equations involving *derivatives*. In other words, the mathematical model may be a differential equation or a system of differential equations.

Once we have formulated a mathematical model that is either a differential equation or a system of a differential equations, we are faced with the not insignificant problem of trying to solve it. *If* we can solve it, then we deem the model to be reasonable if its solution is consistent with either experimental data or known facts about the behavior of the system. But if the predictions produced by the solution are poor, we can either increase the level of resolution of the model or make alternative assumptions about the mechanisms for change in the system. The steps of the modeling process are then repeated, as shown in the following diagram:

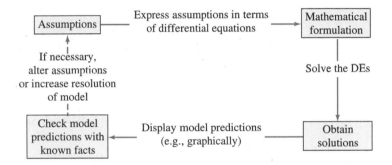

Of course, by increasing the resolution we add to the complexity of the mathematical model and increase the likelihood that we cannot obtain an explicit solution.

A mathematical model of a physical system will often involve the variable time t. A solution of the model then gives the **state of the system;** in other words, the values of the dependent variable (or variables) for appropriate values of t describe the system in the past, present, and future.

Population Dynamics   One of the earliest attempts to model human **population growth** by means of mathematics was by the English economist Thomas Malthus in 1798. Basically, the idea behind the Malthusian model is the assumption that the rate at which the population of a country grows at a certain time is proportional* to the total population of the country at that time. In other words, the more people there are at time t, the more there are going to be in the future. In mathematical terms, if $P(t)$ denotes the total population at time t, then this assumption can be expressed as

$$\frac{dP}{dt} \propto P \quad \text{or} \quad \frac{dP}{dt} = kP, \tag{1}$$

---

*If two quantities u and v are proportional, we write $u \propto v$. This means one quantity is a constant multiple of the other: $u = kv$.

where k is a constant of proportionality. This simple model, which fails to take into account many factors that can influence human populations to either grow or decline (immigration and emigration, for example), nevertheless turned out to be fairly accurate in predicting the population of the United States during the years 1790–1860. Populations that grow at a rate described by (1) are rare; nevertheless, (1) is still used to model *growth of small populations over short intervals of time* (bacteria growing in a petri dish, for example).

Radioactive Decay   The nucleus of an atom consists of combinations of protons and neutrons. Many of these combinations of protons and neutrons are unstable—that is, the atoms decay or transmute into atoms of another substance. Such nuclei are said to be radioactive. For example, over time the highly radioactive radium, Ra-226, transmutes into the radioactive gas radon, Rn-222. To model the phenomenon of **radioactive decay,** it is assumed that the rate dA/dt at which the nuclei of a substance decay is proportional to the amount (more precisely, the number of nuclei) $A(t)$ of the substance remaining at time t:

$$\frac{dA}{dt} \propto A \quad \text{or} \quad \frac{dA}{dt} = kA. \tag{2}$$

Of course equations (1) and (2) are exactly the same; the difference is only in the interpretation of the symbols and the constants of proportionality. For growth, as we expect in (1), $k > 0$, and for decay, as in (2), $k < 0$.

The model (1) for growth can also be seen as the equation $dS/dt = rS$, which describes the growth of capital S when an annual rate of interest r is compounded continuously. The model (2) for decay also occurs in biological applications such as determining the half-life of a drug—the time that it takes for 50% of a drug to be eliminated from a body by excretion or metabolism. In chemistry the decay model (2) appears in the mathematical description of a first-order chemical reaction. The point is this:

A single differential equation can serve as a mathematical model for many different phenomena.

Mathematical models are often accompanied by certain side conditions. For example, in (1) and (2) we would expect to know, in turn, the initial population P_0 and the initial amount of radioactive substance A_0 on hand. If the initial point in time is taken to be $t = 0$, then we know that $P(0) = P_0$ and $A(0) = A_0$. In other words, a mathematical model can consist of either an initial-value problem or, as we shall see later on in Section 5.2, a boundary-value problem.

Newton's Law of Cooling/Warming   According to Newton's empirical law of cooling/warming, the rate at which the temperature of a body changes is proportional to the difference between the temperature of the body and the temperature of the surrounding medium, the so-called ambient temperature. If $T(t)$ represents the temperature of a body at time t, T_m the temperature of the surrounding medium, and dT/dt the rate at which the temperature of the body changes, then Newton's law of cooling/warming translates into the mathematical statement

$$\frac{dT}{dt} \propto T - T_m \quad \text{or} \quad \frac{dT}{dt} = k(T - T_m), \tag{3}$$

where k is a constant of proportionality. In either case, cooling or warming, if T_m is a constant it stands to reason that $k < 0$.

Spread of a Disease   A contagious disease—for example, a flu virus—is spread throughout a community by people coming into contact with other people. Let $x(t)$ denote the number of people who have contracted the disease and $y(t)$ denote the number of people who have not yet been exposed. It seems reasonable to assume that the rate dx/dt at which the disease spreads is proportional to the number of encounters, or *interactions,* between these two groups of people. If we assume that the number of interactions is jointly proportional to $x(t)$ and $y(t)$—that is, proportional to the product xy—then

$$\frac{dx}{dt} = kxy, \tag{4}$$

where k is the usual constant of proportionality. Suppose a small community has a fixed population of n people. If one infected person is introduced into this community, then it could be argued that $x(t)$ and $y(t)$ are related by $x + y = n + 1$. Using this last equation to eliminate y in (4) gives us the model

$$\frac{dx}{dt} = kx(n + 1 - x). \tag{5}$$

An obvious initial condition accompanying equation (5) is $x(0) = 1$.

Chemical Reactions   The disintegration of a radioactive substance, governed by the differential equation (1), is said to be a **first-order reaction.** In chemistry a few reactions follow this same empirical law: If the molecules of substance A decompose into smaller molecules, it is a natural assumption that the rate at which this decomposition takes place is proportional to the amount of the first substance that has not undergone conversion; that is, if $X(t)$ is the amount of substance A remaining at any time, then $dX/dt = kX$, where k is negative constant since X is decreasing. An example of a first-order chemical reaction is the conversion of t-butyl chloride, $(CH_3)_3CCl$, into t-butyl alcohol, $(CH_3)_3COH$:

$$(CH_3)_3CCl + NaOH \rightarrow (CH_3)_3COH + NaCl.$$

Only the concentration of the t-butyl chloride controls the rate of reaction. But in the reaction

$$CH_3Cl + NaOH \rightarrow CH_3OH + NaCl$$

one molecule of sodium hydroxide, NaOH, is consumed for every molecule of methyl chloride, CH_3Cl, thus forming one molecule of methyl alcohol, CH_3OH, and one molecule of sodium chloride, NaCl. In this case the rate at which the reaction proceeds is proportional to the product of the remaining concentrations of CH_3Cl and NaOH. To describe this second reaction in general, let us suppose *one* molecule of a substance A combines with *one* molecule of a substance B to form *one* molecule of a substance C. If X denotes the amount of chemical C formed at time t and if α and β are, in turn, the amounts of the two chemicals A and B at $t = 0$ (the initial amounts), then the instantaneous amounts of A and B not converted to chemical C are $\alpha - X$ and $\beta - X$, respectively. Hence the rate of formation of C is given by

$$\frac{dX}{dt} = k(\alpha - X)(\beta - X), \tag{6}$$

where k is a constant of proportionality. A reaction whose model is equation (6) is said to be a **second-order reaction.**

Mixtures   The mixing of two salt solutions of differing concentrations gives rise to a first-order differential equation for the amount of salt contained in the mixture. Let us suppose that a large mixing tank holds 300 gallons of brine (that is, water in which a certain number of pounds of salt has been dissolved). Another brine solution is pumped into the large tank at a rate of 3 gallons per minute; the concentration of the salt in this inflow is 2 pounds per gallon. When the solution in the tank is well stirred, it is pumped out at the same rate as the entering solution. See Figure 1.16. If $A(t)$ denotes the amount of salt (measured in pounds) in the tank at time t, then the rate at which $A(t)$ changes is a net rate:

$$\frac{dA}{dt} = \left(\begin{array}{c} \text{input rate} \\ \text{of salt} \end{array}\right) - \left(\begin{array}{c} \text{output rate} \\ \text{of salt} \end{array}\right) = R_i - R_o. \tag{7}$$

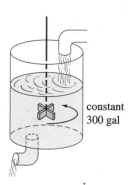

Figure 1.16

constant 300 gal

The input rate R_i of salt (in pounds per minute) is

$$
\begin{array}{ccc}
\text{input rate} & \overset{\text{concentration}}{\underset{\text{of salt}}{\text{in inflow}}} & \text{input rate} \\
\text{of brine} & & \text{of salt} \\
\downarrow & \downarrow & \downarrow
\end{array}
$$

$$R_i = (3 \text{ gal/min}) \cdot (2 \text{ lb/gal}) = 6 \text{ lb/min}.$$

Now since solution is being pumped out of the tank at the same rate that it is being pumped in, the number of gallons of brine in the tank at time t is a constant 300 gallons. Hence the concentration of salt in the tank and in the outflow is $A(t)/300$ lb/gal, and so the output rate R_o of salt is

$$
\begin{array}{ccc}
\text{output rate} & \overset{\text{concentration}}{\underset{\text{in outflow}}{\text{of salt}}} & \text{output rate} \\
\text{of brine} & & \text{of salt} \\
\downarrow & \downarrow & \downarrow
\end{array}
$$

$$R_o = (3 \text{ gal/min}) \cdot \left(\frac{A}{300} \text{ lb/gal}\right) = \frac{A}{100} \text{ lb/min}.$$

Equation (7) then becomes

$$\frac{dA}{dt} = 6 - \frac{A}{100}. \tag{8}$$

Draining a Tank   In hydrodynamics, **Torricelli's law** states that the speed v of efflux of water through a sharp-edged hole at the bottom of a tank filled to a depth h is the same as the speed that a body (in this case, a drop of water) would acquire in falling freely from a height h—that

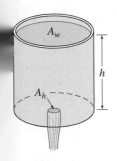

Figure 1.17

is, $v = \sqrt{2gh}$, where g is the acceleration due to gravity. This last expression comes from equating the kinetic energy $\frac{1}{2}mv^2$ with the potential energy mgh and solving for v. Suppose a tank filled with water is allowed to drain through a hole under the influence of gravity. We would like to find the depth h of water remaining in the tank at any time t. Consider the tank shown in Figure 1.17. If the area of the hole is A_h (in ft^2) and the speed of the water leaving the tank is $v = \sqrt{2gh}$ (in ft/s), then the volume of water leaving the tank per second is $A_h\sqrt{2gh}$ (in ft^3/s). Thus if $V(t)$ denotes the volume of water in the tank at time t, then

$$\frac{dV}{dt} = -A_h\sqrt{2gh}, \qquad (9)$$

where the minus sign indicates that V is decreasing. Note here that we are ignoring the possibility of friction at the hole that might cause a reduction of the rate of flow there. Now if the tank is such that the volume of water in it at any time t can be written $V(t) = A_w h$, where A_w (in ft^2) is the *constant* area of the upper surface of the water (see Figure 1.17), then $dV/dt = A_w\, dh/dt$. Substituting this last expression into (9) gives us the desired differential equation for the height of the water at time t:

$$\frac{dh}{dt} = -\frac{A_h}{A_w}\sqrt{2gh}. \qquad (10)$$

It is interesting to note that (10) remains valid even when A_w is not constant. In this case we must express the upper surface area of the water as a function of h—that is, $A_w = A(h)$. See Problem 12 in Exercises 1.3.

Series Circuits   Consider the single-loop series circuit shown in Figure 1.18(a), containing an inductor, resistor, and capacitor. The current in a circuit after a switch is closed is denoted by $i(t)$; the charge on a capacitor at time t is denoted by $q(t)$. The letters L, C, and R are known as inductance, capacitance, and resistance, respectively, and are generally constants. Now according to **Kirchhoff's second law**, the impressed voltage $E(t)$ on a closed loop must equal the sum of the voltage drops in the loop. Figure 1.18(b) shows the symbols and the formulas for the respective voltage drops across an inductor, a capacitor, and a resistor. Since current $i(t)$ is related to charge $q(t)$ on the capacitor by $i = dq/dt$, adding the three voltage drops

$$\underset{\text{inductor}}{L\frac{di}{dt} = L\frac{d^2q}{dt^2}}, \quad \underset{\text{resistor}}{iR = R\frac{dq}{dt}}, \quad \text{and} \quad \underset{\text{capacitor}}{\frac{1}{C}q}$$

and equating the sum to the impressed voltage yields a second-order differential equation

$$L\frac{d^2q}{dt^2} + R\frac{dq}{dt} + \frac{1}{C}q = E(t). \qquad (11)$$

We will examine a differential equation analogous to (11) in great detail in Section 5.1.

Falling Bodies   To construct a mathematical model of the motion of a body moving in a force field, one often starts with Newton's second law

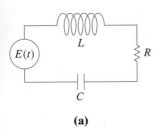

(a)

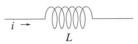

Inductor
inductance L: henrys (h)
voltage drop across: $L\frac{di}{dt}$

$i \rightarrow$ L

Resistor
resistance R: ohms (Ω)
voltage drop across: iR

$i \rightarrow$ R

Capacitor
capacitance C: farads (f)
voltage drop across: $\frac{1}{C}q$

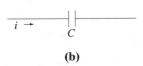

$i \rightarrow$ C

(b)

Figure 1.18

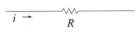

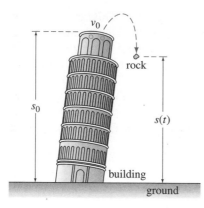

Figure 1.19

of motion. Recall from elementary physics that Newton's **first law of motion** states that a body either will remain at rest or will continue to move with a constant velocity unless acted upon by an external force. In each case this is equivalent to saying that when the sum of the forces $F = \Sigma F_k$—that is, the *net* or resultant force—acting on the body is zero, then the acceleration a of the body is zero. **Newton's second law of motion** indicates that when the net force acting on a body is not zero, then the net force is proportional to its acceleration a or, more precisely, $F = ma$, where m is the mass of the body.

Now suppose a rock is tossed upward from the roof of a building as illustrated in Figure 1.19. What is the position $s(t)$ of the rock relative to the ground at time t? The acceleration of the rock is the second derivative d^2s/dt^2. If we assume that the upward direction is positive and that no force acts on the rock other than the force of gravity, then Newton's second law gives

$$m \frac{d^2s}{dt^2} = -mg \quad \text{or} \quad \frac{d^2s}{dt^2} = -g. \tag{12}$$

In other words, the net force is simply the weight $F = F_1 = -W$ of the rock near the surface of the earth. Recall that the magnitude of the weight is $W = mg$, where m is the mass of the body and g is the acceleration due to gravity. The minus sign in (12) is used because the weight of the rock is a force directed downward, which is opposite to the positive direction. If the height of the building is s_0 and the initial velocity of the rock is v_0, then s is determined from the second-order initial-value problem

$$\frac{d^2s}{dt^2} = -g, \quad s(0) = s_0, \quad s'(0) = v_0. \tag{13}$$

Although we have not been stressing solutions of the equations we have constructed, note that (13) can be solved by integrating the constant $-g$ twice with respect to t. The initial conditions determine the two constants of integration. From elementary physics, you might recognize the solution of (13) as the formula $s(t) = -\frac{1}{2}gt^2 + v_0 t + s_0$.

Falling Bodies and Air Resistance Prior to Galileo's famous experiment from the leaning tower of Pisa, it was generally believed that heavier objects in free fall, such as a cannonball, fell with a greater acceleration than lighter objects, such as a feather. Obviously a cannonball and a feather when dropped simultaneously from the same height *do* fall at different rates, but it is not because a cannonball is heavier. The difference in rates is due to air resistance. The resistive force of air was ignored in the model given in (13). Under some circumstances a falling body of mass m, such as a feather with low density and irregular shape, encounters air resistance proportional to its instantaneous velocity v. If we take, in this circumstance, the positive direction to be oriented downward, then the net force acting on the mass is given by $F = F_1 + F_2 = mg - kv$, where the weight $F_1 = mg$ of the body is force acting in the positive direction and air resistance $F_2 = -kv$ is a force, called **viscous damping**, acting in the opposite or upward direction. See Figure 1.20. Now since v is related to acceleration a by $a = dv/dt$, Newton's second law becomes $F = ma = m \, dv/dt$. By equating the net force to this form of Newton's

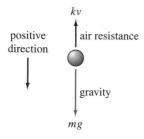

Figure 1.20 Falling body of mass m

second law, we obtain a first-order differential equation for the velocity $v(t)$ of the body at time t,

$$m\frac{dv}{dt} = mg - kv. \tag{14}$$

Here k is a positive constant of proportionality. If $s(t)$ is the distance the body falls in time t from its initial point of release, then $v = ds/dt$ and $a = dv/dt = d^2s/dt^2$. In terms of s, (14) is a second-order differential equation

$$m\frac{d^2s}{dt^2} = mg - k\frac{ds}{dt} \quad \text{or} \quad m\frac{d^2s}{dt^2} + k\frac{ds}{dt} = mg. \tag{15}$$

A Slipping Chain   Suppose a uniform chain of length L feet is draped over a metal peg anchored into a wall high above ground level. Let us assume that the peg is frictionless and that the chain weighs ρ lb/ft. Figure 1.21(a) illustrates the position of the chain where it hangs in equilibrium; if displaced a little to either the right or the left, the chain would slip off the peg. Suppose the positive direction is taken to be downward, and suppose $x(t)$ denotes the distance that the right end of the chain would fall in time t. The equilibrium position corresponds to $x = 0$. In Figure 1.21(b) the chain is displaced an amount x_0 feet and is held on the peg until it is released at an initial time that is designated as $t = 0$. For the chain in motion, as shown in Figure 1.21(c), we have the following quantities:

weight of the chain: $W = (L \text{ ft}) \cdot (\rho \text{ lb/ft}) = L\rho,$

mass of the chain: $m = W/g = L\rho/32$

net force: $F = \left(\dfrac{L}{2} + x\right)\rho - \left(\dfrac{L}{2} - x\right)\rho = 2x\rho.$

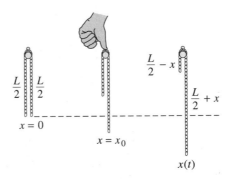

(a) equilibrium   **(b)** chain held   **(c)** motion
                     until $t = 0$ for $t > 0$

Figure 1.21

Since $a = d^2x/dt^2$, $ma = F$ becomes

$$\frac{L\rho}{32}\frac{d^2x}{dt^2} = 2\rho x \quad \text{or} \quad \frac{d^2x}{dt^2} - \frac{64}{L}x = 0. \tag{16}$$

Remarks Each example in this section has described a dynamical system—a system that changes or evolves with the flow of time t. Since the study of dynamical systems is a branch of mathematics currently in vogue, we shall occasionally relate the terminology of that field to the discussion at hand.

In more precise terms, a **dynamical system** consists of a set of time-dependent variables, called **state variables,** together with a rule that enables us to determine (without ambiguity) the state of the system (this may be a past, present, or future state) in terms of a state prescribed at some time t_0. Dynamical systems are classified as either discrete-time systems or continuous-time systems. In this course we shall be concerned only with continuous-time dynamical systems—systems in which *all* variables are defined over a continuous range of time. The rule, or mathematical model, in a continuous-time dynamical system is a differential equation or a system of differential equations. The **state of the system** at a time t is the value of the state variables at that time; the specified state of the system at a time t_0 is simply the initial conditions that accompany the mathematical model. The solution of the initial-value problem is referred to as the **response of the system.** For example, in the case of radioactive decay, the rule is $dA/dt = kA$. Now if the quantity of a radioactive substance at some time t_0 is known, say $A(t_0) = A_0$, then by solving the rule we find that the response of the system for $t \geq t_0$ is $A(t) = A_0 e^{(t-t_0)}$ (see Section 3.1). The response $A(t)$ is the single state variable for this system. In the case of the rock tossed from the roof of a building, the response of the system—the solution of the differential equation $d^2s/dt^2 = -g$, subject to the initial state $s(0) = s_0$, $s'(0) = v_0$—is the function $s(t) = -\frac{1}{2}gt^2 + v_0 t + s_0$, $0 \leq t \leq T$, where T represents the time when the rock hits the ground. The state variables are $s(t)$ and $s'(t)$, which are, respectively, the vertical position of the rock above ground and its velocity at time t. The acceleration $s''(t)$ is *not* a state variable since we have to know only any initial position and initial velocity at a time t_0 to uniquely determine the rock's position $s(t)$ and velocity $s'(t) = v(t)$ for any time in the interval $t_0 \leq t \leq T$. The acceleration $s''(t) = a(t)$ is, of course, given by the differential equation $s''(t) = -g$, $0 < t < T$.

One last point: Not every system studied in this text is a dynamical system. We shall also examine some static systems in which the model is a differential equation.

What Lies Ahead Throughout this text you will see three different types of approaches to, or analyses of, differential equations. Over the centuries, differential equations would often spring from the efforts of a scientist/engineer to describe some physical phenomenon or to translate an empirical or experimental law into mathematical terms. As a consequence, a scientist/engineer/mathematician would often spend many years of his/her life trying to find the solutions of a DE. With a solution in hand, the study of its properties then followed. This quest for solutions is called by some the *analytical approach* to differential equations. Once they realized that explicit solutions are at best difficult to obtain and at worst impossible to obtain, mathematicians learned that a differential equation itself could be a font of valuable information. It is possible, in some instances, to glean directly from the differential equation answers to questions such as, Does the DE actually have solutions? If a solution of the DE exists and satisfies an initial condition, is it the only such solution? What are some of the properties of the unknown solutions? What can we say about the geometry of the solution curves? Such an approach is *qualitative analysis*. Finally, if a differential equation cannot be solved by analytical methods, yet we can prove that a solution exists, the next logical query is, Can we somehow approximate the values of an unknown solution? Here we enter the realm of *numerical analysis*. An affirmative answer to the last question stems from the fact that a differential equation can be used as a cornerstone for constructing very accurate approximation algorithms. In Chapter 2 we start with qualitative considerations of first-order ODEs, then examine analytical stratagems for solving some special first-order equations, and conclude with an introduction to an elementary numerical method. See Figure 1.22.

(a) analytical   **(b)** qualitative   **(c)** numerical

Figure 1.22 Different approaches to the study of differential equations

EXERCISES 1.3

Answers to odd-numbered problems begin on page AN-1.

Population Dynamics

1. Under the same assumptions underlying the model in (1), determine a differential equation governing the growing population $P(t)$ of a country when individuals are allowed to immigrate into the country at a constant rate r. What is the differential equation when individuals are allowed to emigrate at a constant rate r?

2. The model given in (1) fails to take death into consideration; the growth rate equals the birth rate. In another model of the changing population of a community, it is assumed that the rate at which the

population changes is a *net* rate—that is, the difference between the rate of births and the rate of deaths in the community. Determine differential equation governing the population $P(t)$ if the birth rate and the death rate are both proportional to the population present at time t.

3. Using the concept of a net rate introduced in Problem 2, determine a differential equation governing a population $P(t)$ if the birth rate is proportional to the population present at time t but the death rate is proportional to the square of the population present at time t.

4. The number of field mice in a certain pasture is given by the function $200 - 10t$, where t is measured in years. Determine a differential equation governing a population of owls that feed on the mice if the rate at which the owl population grows is proportional to the difference between the number of owls at time t and the number of field mice at time t.

Newton's Law of Cooling/Warming

5. A cup of coffee cools according to Newton's law of cooling (3). Use data from the graph of temperature $T(t)$ in Figure 1.23 to estimate the constants T_m, T_0, and k in a model of the form of a first-order initial-value problem: $dT/dt = k(T - T_m)$, $T(0) = T_0$.

6. The ambient temperature T_m in (3) could be a function of time t. Suppose that in an artificially controlled environment, $T_m(t)$ is periodic with a 24-hour period, as illustrated in Figure 1.24. Devise a mathematical model for the temperature $T(t)$ of a body within this environment.

Spread of Disease/Technology

7. Suppose a student carrying a flu virus returns to an isolated college campus of 1000 students. Determine a differential equation governing the number of people $x(t)$ who have contracted the flu if the rate at which the disease spreads is a proportional to the number of interactions between the number of students with the flu and the number of students who have not yet been exposed to it.

8. At a time $t = 0$ a technological innovation is introduced into a community with a fixed population of n people. Determine a differential equation governing the number of people $x(t)$ who have adopted the innovation at time t if it is assumed that the rate at which the innovation spreads through the community is jointly proportional to the number of people who have adopted it and the number of people who have not adopted it.

Mixtures

9. Suppose that a large mixing tank initially holds 300 gallons of water in which 50 pounds of salt has been dissolved. Pure water is pumped into the tank at a rate of 3 gal/min, and then when the solution is well stirred, it is pumped out at the same rate. Determine a differential equation for the amount $A(t)$ of salt in the tank at time t.

10. Suppose that a large mixing tank initially holds 300 gallons of water in which 50 pounds of salt has been dissolved. Another brine solution is pumped into the tank at a rate of 3 gal/min, and then when the

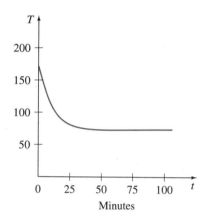

Figure 1.23

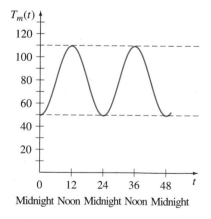

Midnight Noon Midnight Noon Midnight

Figure 1.24

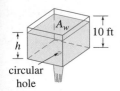

Figure 1.25

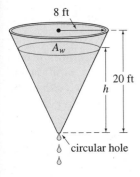

Figure 1.26

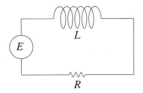

Figure 1.27

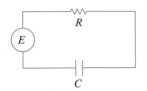

Figure 1.28

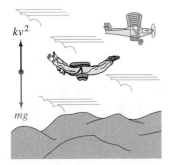

Figure 1.29

solution is well stirred, it is pumped out at a *slower* rate of 2 gal/min. If the concentration of the solution entering is 2 lb/gal, determine a differential equation for the amount $A(t)$ of salt in the tank at time t.

Draining a Tank

11. Suppose water is leaking from a tank through a circular hole of area A_h at its bottom. When water leaks through a hole, friction and contraction of the water stream near the hole reduce the volume of water leaving the tank per second to $cA_h\sqrt{2gh}$, where c ($0 < c < 1$) is an empirical constant. Determine a differential equation for the height h of water at time t for the cubical tank in Figure 1.25. The radius of the hole is 2 in., and $g = 32$ ft/s².

12. The right-circular conical tank shown in Figure 1.26 loses water out of a circular hole at its bottom. Determine a differential equation for the height of the water h at time t. The radius of the hole is 2 in., $g = 32$ ft/s², and the friction/contraction factor introduced in Problem 11 is $c = 0.6$.

Series Circuits

13. A series circuit contains a resistor and an inductor as shown in Figure 1.27. Determine a differential equation for the current $i(t)$ if the resistance is R, the inductance is L, and the impressed voltage is $E(t)$.

14. A series circuit contains a resistor and a capacitor as shown in Figure 1.28. Determine a differential equation for the charge $q(t)$ on the capacitor if the resistance is R, the capacitance is C, and the impressed voltage is $E(t)$.

Falling Bodies and Air Resistance

15. For high-speed motion through the air—such as the skydiver shown in Figure 1.29, falling before the parachute is opened—air resistance is closer to a power of the instantaneous velocity. Determine a differential equation for the velocity $v(t)$ of a falling body of mass m if air resistance is proportional to the square of the instantaneous velocity.

Newton's Second Law and Archimedes' Principle

16. A cylindrical barrel s feet in diameter of weight w lb is floating in water. After an initial depression, the barrel exhibits an up-and-down bobbing motion along a vertical line. Using Figure 1.30(b), determine

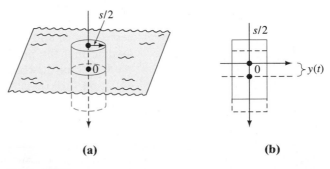

(a)   (b)

Figure 1.30

a differential equation for the vertical displacement $y(t)$ if the origin is taken to be on the vertical axis at the surface of the water when the barrel is at rest. Use **Archimedes' principle**: Buoyancy, or upward force of the water on the barrel, is equal to the weight of the water displaced. Assume that the downward direction is positive, that the density of water is 62.4 lb/ft³, and that there is no resistance between the barrel and the water.

Newton's Second Law and Hooke's Law

17. After a mass m is attached to a spring, the spring stretches s units and then hangs at rest in the equilibrium position shown in Figure 1.31(b). After the spring/mass system has been set in motion, let $x(t)$ denote the directed distance of the mass beyond the equilibrium position. Assume that the downward direction is positive, that the motion takes place in a vertical straight line through the center of gravity of the mass, and that the only forces acting on the system are the weight of the mass and the restoring force of the stretched spring. Use **Hooke's law**: The restoring force of a spring is proportional to its total elongation. Determine a differential equation for the displacement $x(t)$ at time t.

18. In Problem 17, what is a differential equation for the displacement $x(t)$ if the motion takes place in a medium that imparts a damping force on the spring/mass system that is proportional to the instantaneous velocity of the mass and acts in a direction opposite to that of motion?

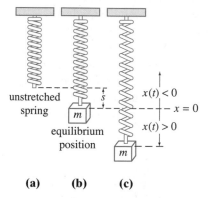

unstretched spring

equilibrium position

$x(t) < 0$

$x = 0$

$x(t) > 0$

(a)   **(b)**   **(c)**

Figure 1.31

Newton's Second Law and Variable Mass

When the mass m of a body moving through a force field is variable, **Newton's second law** takes the following form: When the net force acting on a body is not zero, then the net force F is equal to the time rate of change of momentum of the body—that is, $F = \dfrac{d}{dt}(mv),$* where mv is momentum. Use this formulation of Newton's second law in Problems 19 and 20.

19. A uniform 10-foot chain is coiled loosely on the ground. As shown in Figure 1.32, one end of the chain is pulled vertically upward by means of a constant force of 5 lb. The chain weighs 1 lb/ft. Determine a differential equation for the height $x(t)$ of the end above ground level at time t. Assume that the positive direction is upward.

20. A uniform chain of length L, measured in feet, is held vertically so that the lower end just touches the floor, as shown in Figure 1.33. The chain weighs 2 lb/ft. The upper end that is held is released from rest at $t = 0$, and the chain falls straight down. Ignore air resistance, assume that the positive direction is downward, and let $x(t)$ denote the length of the chain on the floor at time t. Show that a differential equation for $x(t)$ is

$$(L - x)\frac{d^2x}{dt^2} - \left(\frac{dx}{dt}\right)^2 = Lg.$$

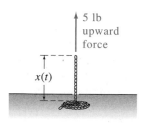

5 lb upward force

$x(t)$

Figure 1.32

L

Figure 1.33

---

*Note that when m is constant, this is the same as $F = ma$.

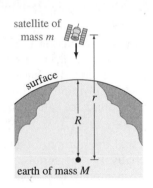

satellite of mass m

surface

r

R

earth of mass M

Figure 1.34

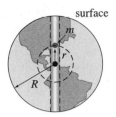

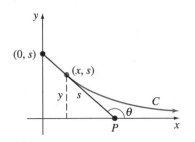

surface

m

r

R

Figure 1.35

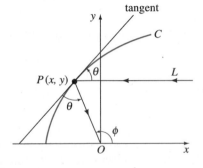

y

$(0, s)$

(x, s)

y s

θ

C

P

x

Figure 1.36

y

tangent

C

θ

$P(x, y)$

L

θ

ϕ

O

x

Figure 1.37

Newton's Second Law and The Law of Universal Gravitation

21. By **Newton's universal law of gravitation,** the free-fall acceleration a of a body, such as the satellite shown in Figure 1.34, falling a great distance to the surface is not the constant g. Rather, the acceleration a is inversely proportional to the square of the distance from the center of the earth: $a = k/r^2$, where k is the constant of proportionality. Using the fact that at the surface of the earth $r = R$ and $a = g$, determine k. Assuming that the positive direction is upward, use Newton's second law and his universal law of gravitation to find a differential equation for the distance r.

22. Suppose a hole is drilled through the center of the earth and a bowling ball of mass m is dropped into the hole, as shown in Figure 1.35. Construct a mathematical model that describes the motion of the ball. Let r denote the distance from the center of the earth to the mass at time t, M denote the mass of the earth, M_r denote the mass of that portion of the earth within a sphere of radius r, and δ denote the constant density of the earth.

Miscellaneous Mathematical Models

23. In the theory of learning, the rate at which a subject is memorized is assumed to be proportional to the amount that is left to be memorized. Suppose M denotes the total amount of a subject to be memorized and $A(t)$ is the amount memorized in time t. Determine a differential equation for the amount $A(t)$.

24. In Problem 23 assume that the rate at which material is *forgotten* is proportional to the amount memorized in time t. Determine a differential equation for $A(t)$ when forgetfulness is taken into account.

25. A drug is infused into a patient's bloodstream at a constant rate of r grams per second. Simultaneously, the drug is removed at a rate proportional to the amount $x(t)$ of the drug present at time t. Determine a differential equation governing the amount $x(t)$.

26. A person P, starting at the origin, moves in the direction of the positive x-axis, pulling a weight along the curve C, called a **tractrix,** as shown in Figure 1.36. The weight, initially located on the y-axis at $(0, s)$, is pulled by a rope of constant length s, which is kept taut throughout the motion. Determine a differential equation of the path of motion. Assume that the rope is always tangent to C.

27. As illustrated in Figure 1.37, light rays strike a plane curve C in such a manner that all rays L parallel to the x-axis are reflected to a single point O. Assuming that the angle of incidence is equal to the angle of reflection, determine a differential equation that describes the shape of the curve C. [*Hint:* Inspection of Figure 1.37 shows that we can write $\phi = 2\theta$. Why? Now use an appropriate trigonometric identity.]

Discussion Problems

28. Reread Problem 31 in Exercises 1.1 and then give an explicit solution $P(t)$ for equation (1). Find a one-parameter family of solutions of (1).

29. Reread the sentence following equation (3). If we assume that T_m is a positive constant, supply the reasons why we would expect $k < 0$

in (3) in cases of both cooling and warming. You might start by interpreting, say, $T(t) > T_m$ in a graphical manner.

30. Reread the discussion leading up to equation (8). If we assume that initially the tank holds, say, 50 lb of salt, it stands to reason, since salt is being added to the tank continuously for $t > 0$, that $A(t)$ should be an increasing function. Discuss how you might determine from the DE, without actually solving it, the number of pounds of salt in the tank after a long period of time.

31. Generalize model (8) under the assumptions that the number of gallons of brine in the large mixing tank is initially V_0; the input and output rates (in gal/min) of brine are r_i and r_o, respectively; the concentrations of salt in the inflow and the outflow are c_i and c_o, respectively; and $A(t)$ is the amount of salt (in pounds) in the tank at time t.

32. The differential equation $\dfrac{dP}{dt} = (k \cos t)P$, where k is a positive constant, is a model of the human population $P(t)$ of a certain community. Discuss an interpretation for the solution of this equation. In other words, what kind of population do you think the differential equation describes?

33. Reread Section 1.3 and classify each mathematical model as linear or nonlinear.

34. In Problem 21, suppose $r = R + s$, where s is the distance from the surface of the earth to the falling body. What does the differential equation obtained in Problem 21 become when s is very small compared to R?

35. In meteorology the term *virga* refers to falling raindrops or ice particles that evaporate before they reach the ground. Assume that a typical raindrop is spherical in shape. Starting at some time, which we can designate as $t = 0$, the raindrop of radius r_0 falls from rest from a cloud and begins to evaporate. If it is assumed that a raindrop evaporates in such a manner that its shape remains spherical, then it also makes sense to assume that the rate at which the raindrop evaporates (that is, the rate at which it loses mass) is proportional to its surface area. Show that this latter assumption implies that the rate at which the radius r of the raindrop decreases is a constant. Find $r(t)$. [*Hint:* See Problem 35 in Exercises 1.1.] If the positive direction is downward, construct a mathematical model for the velocity v of the falling raindrop at time t. Ignore air resitance. [*Hint:* See the introduction to Problems 19 and 20.]

36. The "snowplow problem" is a classic. It appears in many differential equations texts, but was probably made famous by Ralph Palmer Agnew.

One day it started snowing at a heavy and steady rate. A snowplow started out at noon, going 2 miles the first hour and 1 mile the second hour. What time did it start snowing?

Find the text *Differential Equations,* by Ralph Palmer Agnew, McGraw-Hill Book Co., and then discuss the construction and solution of the mathematical model.

CHAPTER 1 IN REVIEW

Answers to odd-numbered problems begin on page AN-1.

In Problems 1 and 2 fill in the blank and then write this result as a linear first-order differential equation that is free of the symbol c_1 and has the form $dy/dx = f(x, y)$. The symbols c_1 and k represent constants.

1. $\dfrac{d}{dx} c_1 e^{kx} = $ _____

2. $\dfrac{d}{dx}(5 + c_1 e^{-2x}) = $ _____

In Problems 3 and 4 fill in the blank and then write this result as a linear second-order differential equation that is free of the symbols c_1 and c_2 and has the form $F(y, y'') = 0$. The symbols c_1, c_2, and k represent constants.

3. $\dfrac{d^2}{dx^2}(c_1 \cos kx + c_2 \sin kx) = $ _____

4. $\dfrac{d^2}{dx^2}(c_1 \cosh kx + c_2 \sinh kx) = $ _____

In Problems 5 and 6 compute y' and y'' and then combine these results as a linear second-order differential equation that is free of the symbols c_1 and c_2 and has the form $F(y, y', y'') = 0$. The symbols c_1 and c_2 represent constants.

5. $y = c_1 e^x + c_2 x e^x$ **6.** $y = c_1 e^x \cos x + c_2 e^x \sin x$

In Problems 7–12 match each of the given differential equations with one or more of these solutions: **(a)** $y = 0$, **(b)** $y = 2$, **(c)** $y = 2x$, **(d)** $y = 2x^2$.

7. $xy' = 2y$ **8.** $y' = 2$

9. $y' = 2y - 4$ **10.** $xy' = y$

11. $y'' + 9y = 18$ **12.** $xy'' - y' = 0$

In Problems 13 and 14 determine by inspection at least one solution of the given differential equation.

13. $y'' = y'$ **14.** $y' = y(y - 3)$

In Problems 15 and 16 interpret each statement as a differential equation.

15. On the graph of $y = \phi(x)$, the slope of the tangent line at a point P is the square of the distance from P to the origin.

16. On the graph of $y = \phi(x)$, the rate at which the slope changes with respect to x at a point P is the negative of the slope of the tangent line at P.

17. (a) Give the domain of the function $y = x^{2/3}$.
    (b) Give the interval I of definition over which $y = x^{2/3}$ is a solution of $3xy' - 2y = 0$.

18. (a) Verify that the one-parameter family $y^2 - 2y = x^2 - x + c$ is an implicit solution of the differential equation $(2y - 2)y' = 2x - 1$.
    (b) Use part (a) to find an implicit solution of the DE that satisfies the initial condition $y(0) = 1$.
    (c) Use part (b) to find an explicit *function* $y = \phi(x)$ that satisfies $y(0) = 1$. Give the domain of ϕ. Is the function $y = \phi(x)$ a *solution*

of the initial-value problem? If so, give its interval I of definition; if not, explain.

19. A differential equation may possess more than one family of solutions.
    (a) Plot different members of the families $y = \phi_1(x) = x^2 + c_1$ and $y = \phi_2(x) = -x^2 + c_2$.
    (b) Verify that $y = \phi_1(x)$ and $y = \phi_2(x)$ are two solutions of the nonlinear first-order differential equation $(y')^2 = 4x^2$.
    (c) Construct a piecewise-defined function that is a solution of the nonlinear DE in part (b) but is not a member of either family of solutions in part (a).

20. What is the slope of the tangent line to the graph of the solution of $y' = 6\sqrt{y} + 5x^3$ that passes through $(-1, 4)$?

In Problems 21 and 22 verify that the indicated function is a particular solution of the given differential equation. Give an interval of definition I for each solution.

21. $x^2 y'' + xy' + y = 0; \quad y = \sin(\ln x)$

22. $x^2 y'' + xy' + y = \sec(\ln x); \quad y = \cos(\ln x)\ln(\cos(\ln x)) + (\ln x)\sin(\ln x)$

23. The graph of a solution of a second-order initial-value problem $d^2y/dx^2 = f(x, y, y')$, $y(2) = y_0$, $y'(2) = y_1$ is given in Figure 1.38. Use the graph to estimate the values of y_0 and y_1.

24. A tank in the form of a right circular cylinder of radius 2 feet and height 10 feet is standing on end. If the tank is initially full of water and water leaks from a circular hole of radius $\frac{1}{2}$ inch at its bottom, determine a differential equation for the height h of the water at time t. Ignore friction and contraction of water at the hole.

25. A box weighing 96 pounds slides down an incline that makes a 30° angle with the horizontal. If the coefficient of sliding friction is μ, determine a differential equation for the velocity $v(t)$ of the box at time t. Use the fact that the force of friction opposing the motion is μN, where N is the normal component of the weight. See Figure 1.39.

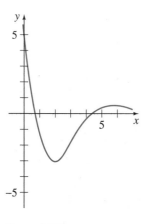

Figure 1.38

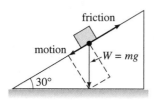

Figure 1.39

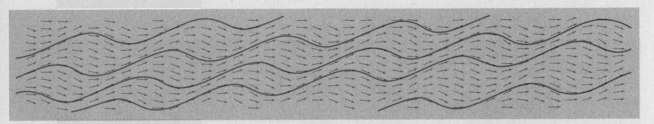

Solution curves following the flow of a direction field; see page 42.

2 FIRST-ORDER DIFFERENTIAL EQUATIONS

INTRODUCTION We are now in a position to solve some differential equations. We begin with first-order equations. The ability to find exact solutions usually depends on the ability to recognize what *kind* of DE it is and then apply an equation-specific method of solution. In other words, what works for one kind of first-order equation does not necessarily apply to another kind. Before solving anything, however, we shall see in the first section of this chapter that a first-order DE can often yield enough information about how its solutions "behave" to enable us to sketch rough renditions of solution curves. In the last section of the chapter we shall again use the differential equation itself, this time to construct a numerical solution. This last concept extends the notion of the solution of a DE, since, strictly speaking, a numerical solution is not a solution at all but a formula-generated list of numerical data that approximate the values of an unknown solution.

2.1 SOLUTION CURVES WITHOUT THE SOLUTION

• Direction field • Autonomous DEs • Critical points
• Asymptotically stable • Attractor • Repeller • Semi-stable

Some differential equations do not possess *any* solutions. For example, there is no real function that can satisfy $(y')^2 + 1 = 0$. (Why?) Some differential equations possess solutions that can be found in explicit or implicit form by implementing an equation-specific method of solution (applying certain manipulations and analytical procedures such as integration to the DE). Some differential equations possess solutions but cannot be solved analytically. In other words, when we say that a solution of a DE *exists,* we do not mean that there necessarily exists a method of solution that will produce an explicit or an implicit solution. Over the course of time mathematicians have devised ingenious procedures to solve some very specialized equations, so there are, not surprisingly, a large number of differential equations that *can* be solved analytically. We will study some of these solution methods for *first-order equations* in the subsequent sections of this chapter. But first, we examine how we can get information about a solution from a first-order DE without actually solving the equation.

What a First-Order DE Can Tell Us   Let us imagine for the moment that we have in front of us a first-order differential equation in normal form $dy/dx = f(x, y)$, and let us further imagine that we can neither find nor invent a method for solving it analytically. This is not as bad a predicament as one might think, since it is often possible to glean useful information about the nature of solutions directly from the differential equation itself. For example, we have just seen that whenever $f(x, y)$ and $\partial f/\partial y$ satisfy certain continuity conditions, the qualitative questions about the existence and uniqueness of solutions can be answered. We shall see in this section that other qualitative questions about properties of solutions—How does a solution behave near a certain point? How does a solution behave as $x \to \infty$?—can often be answered when the function f depends solely on the variable y. We begin, however, with a simple concept from calculus: A derivative dy/dx of a differential function $y = y(x)$ gives slopes of tangent lines at points on its graph.

Slope   Because a solution $y = y(x)$ of a first-order differential equation $dy/dx = f(x, y)$ is necessarily a differentiable function on its interval I of definition, it must also be continuous on I. Thus the corresponding solution curve on I must have no breaks and must possess a tangent line at each point $(x, y(x))$. The slope of the tangent line at $(x, y(x))$ on a solution curve is the value of the first derivative dy/dx at this point, and this we know from the differential equation: $f(x, y(x))$. Now suppose that (x, y) represents any point in a region of the xy-plane over which the function f is defined. The value $f(x, y)$ that the function f assigns to the point represents the slope of a line or, as we shall envision it, a line segment called a **lineal element.** For example, consider the equation $dy/dx = 0.2xy$,

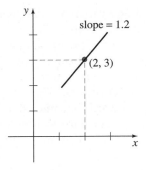

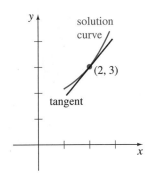

(a) lineal element at a point

(b) lineal element is tangent to solution curve that passes through the point

Figure 2.1

where $f(x, y) = 0.2xy$. At, say, the point $(2, 3)$, the slope of a lineal element is $f(2, 3) = 1.2$. Figure 2.1(a) shows a line segment with positive slope passing through $(2, 3)$. As shown in Figure 2.1(b), *if* a solution curve also passes through $(2, 3)$, it does so tangent to this line segment; in other words, the lineal element is a miniature tangent line at that point.

Direction Fields   If we systematically evaluate *f* over a rectangular grid of points in the *xy*-plane and draw a lineal element at each point (x, y) of the grid with slope $f(x, y)$, then the collection of all these lineal elements is called a **direction field** or a **slope field** of the differential equation $dy/dx = f(x, y)$. Visually the direction field suggests the appearance or shape of a family of solution curves of the differential equation, and consequently it may be possible to see at a glance certain qualitative aspects of the solutions—regions in the plane, for example, in which a solution exhibits an unusual behavior. A single solution curve that wends its way through a direction field must follow the flow pattern of the field; it is tangent to a lineal element when it intersects a point in the grid.

Sketching a direction field by hand is straightforward but time consuming. It is one of those tasks about which an argument can be made for doing it once or twice in a lifetime, but overall it is most efficiently carried out by means of computer software.

EXAMPLE 1   **Direction Field**

Figure 2.2(a) was obtained using the direction field application in one of the software supplements to this text with $dy/dx = 0.2xy$ and a 5×5 grid with points defined by (mh, nh), $h = 1$, *m* and *n* integers, $-5 \leq m \leq 5$, $-5 \leq n \leq 5$. In Figure 2.2(a), notice that at any point along the *x*-axis $(y = 0)$ and the *y*-axis $(x = 0)$ the slopes are $f(x, 0) = 0$ and $f(0, y) = 0$, respectively, so the lineal elements are horizontal. Moreover, observe in the first quadrant that for a fixed value of *x* the values of $f(x, y) = 0.2xy$ increase as *y* increases; similarly, for a fixed *y* the values of $f(x, y) = 0.2xy$

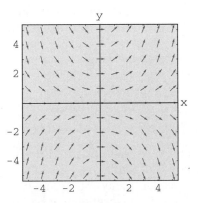

(a) direction field for $dy/dx = 0.2xy$

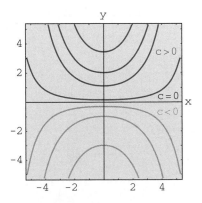

(b) some solution curves in the family $y = ce^{0.1x^2}$

Figure 2.2

increase as x increases. This means that as both x and y increase the lineal elements almost become vertical and have positive slope ($f(x, y) = 0.2xy > 0$ for $x > 0$, $y > 0$). In the second quadrant, $|f(x, y)|$ increases as $|x|$ and y increase, and so the lineal elements again become almost vertical but this time have negative slope ($f(x, y) = 0.2xy < 0$ for $x < 0$, $y > 0$). Reading left to right, imagine a solution curve that starts at a point in the second quadrant, moves steeply downward, becomes flat at it passes through the y-axis, and then moves steeply upward as it enters the first quadrant—in other words, has a concave upward shape similar to that of a horseshoe. From this it could be surmised that $y \to \infty$ as $x \to \pm\infty$. Now in the third and fourth quadrants, since $f(x, y) = 0.2xy > 0$ and $f(x, y) = 0.2xy < 0$, respectively, the situation is reversed; a solution curve increases and then decreases as we move from left to right. We saw in (1) of Section 1.1 that $y = e^{0.1x^2}$ is an explicit solution of the differential equation $dy/dx = 0.2xy$; you should verify that the one-parameter family of solutions of the same equation is given by $y = ce^{0.1x^2}$. For purposes of comparison with Figure 2.2(a), some representative graphs of members of this family are shown in Figure 2.2(b). ■

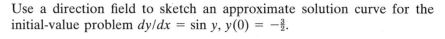

EXAMPLE 2 Direction Field

Use a direction field to sketch an approximate solution curve for the initial-value problem $dy/dx = \sin y$, $y(0) = -\frac{3}{2}$.

Solution Before proceeding, recall that, from the continuity of $f(x, y) = \sin y$ and $\partial f/\partial y = \cos y$, Theorem 2.1 guarantees the existence of a unique solution curve passing through any specified point (x_0, y_0) in the plane. Now we again set our computer software for a 5×5 rectangular region and specify (because of the initial condition) points in that region with vertical and horizontal separation of $\frac{1}{2}$ unit—that is, points defined by (mh, nh), $h = \frac{1}{2}$, m and n integers such that $-10 \le m \le 10$, $-10 \le n \le 10$. The result is shown in Figure 2.3. Since the right-hand side of $dy/dx = \sin y$ is 0 at $y = 0$ and at $y = -\pi$, the lineal elements are horizontal at all points whose second coordinates are $y = 0$ or $y = -\pi$. It makes sense then that a solution curve passing through the initial point $(0, -\frac{3}{2})$ has the shape shown in color in the figure. ■

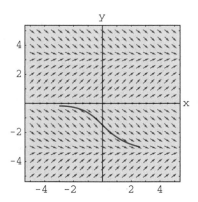

Figure 2.3 Direction field for $dy/dx = \sin y$

Increasing/Decreasing Interpretation of the derivative dy/dx as a function that gives slope plays the key role in the construction of a direction field. Another telling property of the first derivative will be used next; namely, if $dy/dx > 0$ (or $dy/dx < 0$) for all x in an interval I, then a differentiable function $y = y(x)$ is increasing (or decreasing) on I.

Autonomous First-Order DEs In Section 1.1 we divided the class of ordinary differential equations into two types: linear and nonlinear. We now consider briefly another kind of classification of ordinary differential equations, a classification that is of particular importance in the qualitative investigation of differential equations. A differential equation in which the independent variable does not appear explicitly is said to be **autonomous.** If x is the independent variable, then an autonomous first-order differential equation can be written as $F(y, y') = 0$ or in normal form as

$$\frac{dy}{dx} = f(y).$$ **(1)**

We shall assume throughout that the function f in (1) and its derivative f' are continuous functions of y on some interval I. The first-order equations

$$\underset{\substack{\downarrow \\ f(y)}}{\frac{dy}{dx} = 1 + y^2} \quad \text{and} \quad \underset{\substack{\downarrow \\ f(x, y)}}{\frac{dy}{dx} = 0.2xy}$$

are autonomous and nonautonomous, respectively.

Many differential equations encountered in applications, as equations that are models of physical laws that do not change over time, are autonomous. As we have already seen in Section 1.3, in applied context, symbols other than y and x are routinely used to represent the dependent and independent variables. For example, if t represents time, then inspection of

$$\frac{dA}{dt} = kA, \quad \frac{dx}{dt} = kx(n + 1 - x), \quad \frac{dT}{dt} = k(T - T_m), \quad \text{and} \quad \frac{dA}{dt} = 6 - \frac{1}{100}A,$$

where k, n, and T_m are constants, shows that each equation is time-independent. Indeed, *all* of the first-order differential equations introduced in Section 1.3 are time-independent and so are autonomous.

Critical Points The zeros of the function f in (1) are of special importance. We say that a real number c is a **critical point** of the autonomous differential equation (1) if it is a zero of f—that is, if $f(c) = 0$. A critical point is also called an **equilibrium point** or **stationary point.** Now observe that if we substitute $y(x) = c$ into (1), then both sides of the equation equal zero. This means:

> If c is a critical point of (1), then $y(x) = c$ is a constant solution of the autonomous equation.

A constant solution $y(x) = c$ of (1) is called an **equilibrium solution;** equilibria are the *only* constant solutions of (1).

As mentioned previously, we can tell when a nonconstant solution $y = y(x)$ of (1) is increasing or decreasing by determining the algebraic sign of the derivative dy/dx; we do this by identifying the intervals over which $f(y)$ is positive or negative.

EXAMPLE 3   **An Autonomous Differential Equation**

The differential equation

$$\frac{dP}{dt} = P(a - bP), \quad a > 0, b > 0$$

has the form $dP/dt = f(P)$, which is (1) with t and P playing the parts of x and y, respectively, and hence is autonomous. From $f(P) = P(a - bP) = 0$ we see that 0 and a/b are critical points of the equation

Figure 2.4

and so the equilibrium solutions are $P(t) = 0$ and $P(t) = a/b$. By putting the critical points on a vertical line, we divide the line into three intervals: $-\infty < P < 0$, $0 < P < a/b$, and $a/b < P < \infty$. The arrows on the line shown in Figure 2.4 indicate the algebraic sign of $f(P) = P(a - bP)$ on these intervals and whether a nonconstant solution $P(t)$ is increasing or decreasing on the interval. The following table explains the figure.

Interval	Sign of $f(P)$	$P(t)$	Arrow
$(-\infty, 0)$	minus	decreasing	points down
$(0, a/b)$	plus	increasing	points up
$(a/b, \infty)$	minus	decreasing	points down

Figure 2.4 is called a **one-dimensional phase portrait**, or simply a **phase portrait**, of the differential equation $dP/dt = P(a - bP)$. The vertical line is called a **phase line**.

Solution Curves   Without solving an autonomous differential equation we can usually say a great deal about its solution curves. Since the function f in (1) is independent of the variable x, we may consider f defined for $-\infty < x < \infty$ or for $0 \le x < \infty$. Also, since f and its derivative f' are continuous functions of x on some interval I, the fundamental results of Theorem 1.1 hold in some horizontal strip or region R in the xy-plane corresponding to I, and so through any point (x_0, y_0) in R there passes only one solution curve of (1). See Figure 2.5(a). For the sake of discussion let us suppose that (1) possesses exactly two critical points c_1 and c_2 and that $c_1 < c_2$. The graphs of the equilibrium solutions $y(x) = c_1$ and $y(x) = c_2$ are horizontal lines, and these lines partition the region R into three subregions R_1, R_2, and R_3, as illustrated in Figure 2.5(b). Without proof, here are some conclusions that we can draw about a nonconstant solution $y(x)$ of (1):

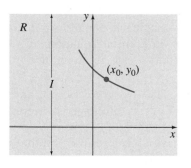

(a) region R

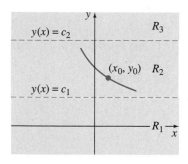

(b) subregions R_1, R_2, and R_3 of R

Figure 2.5

- If (x_0, y_0) is in a subregion R_i, $i = 1, 2, 3$ and $y(x)$ is a solution whose graph passes through this point, then $y(x)$ remains in the subregion for all x. As illustrated in Figure 2.5(b), the solution $y(x)$ in R_2 is bounded below by c_1 and above by c_2; that is, $c_1 < y(x) < c_2$ for all x. The solution curve stays within R_2 for all x because the graph of a nonconstant solution of (1) cannot cross the graph of either equilibrium solution $y(x) = c_1$ or $y(x) = c_2$. See Problem 33 in Exercises 2.1.
- By the continuity of f we must then have either $f(y) > 0$ or $f(y) < 0$ for all x in a subregion R_i, $i = 1, 2, 3$. In other words, $f(y)$ cannot change signs in a subregion. See Problem 33 in Exercises 2.1.
- Since $dy/dx = f(y(x))$ is either positive or negative in a subregion, a solution $y(x)$ is strictly monotonic—that is, $y(x)$ is either increasing or decreasing in a subregion R_i, $i = 1, 2, 3$. Therefore $y(x)$ cannot be oscillatory nor can it have a relative extremum (maximum or minimum). See Problem 33 in Exercises 2.1.

- If $y(x)$ is bounded above by a critical point (as in subregion R_1) or is bounded below by a critical point (as in subregion R_3), then $y(x)$ must approach this point either as $x \to \infty$ or as $x \to -\infty$. If $y(x)$ is bounded—that is, bounded above and below by two consecutive critical points (as in subregion R_2)—then $y(x)$ must approach both critical points, one as $x \to \infty$ and the other as $x \to -\infty$. See Problem 34 in Exercises 2.1.

With the foregoing facts in mind let us reexamine the differential equation in Example 2.

EXAMPLE 4 **Example 3 Revisited**

The three intervals determined on the P-axis or phase line by the critical points 0 and a/b now correspond in the tP-plane to three subregions:

$$R_1: -\infty < P < 0, \quad R_2: 0 < P < a/b, \quad \text{and} \quad R_3: a/b < P < \infty,$$

where $-\infty < x < \infty$. The phase portrait in Figure 2.4 tells us that $P(t)$ is decreasing in R_1, increasing in R_2, and decreasing in R_3. If $P(0) = P_0$ is an initial value, then in R_1, R_2, and R_3 we have, respectively, the following:

(i) For $P_0 < 0$, $P(t)$ is bounded above. Since $P(t)$ is decreasing, $P(t)$ decreases without bound for increasing t and $P(t) \to 0$ as $t \to -\infty$. This means that the negative t-axis, $P = 0$, is a horizontal asymptote for a solution curve.

(ii) For $0 < P_0 < a/b$, $P(t)$ is bounded. Since $P(t)$ is increasing, $P(t) \to a/b$ as $t \to \infty$ and $P(t) \to 0$ as $t \to -\infty$. The two lines $P = 0$ and $P = a/b$ are horizontal asymptotes for any solution curve starting in this subregion.

(iii) For $P_0 > a/b$, $P(t)$ is bounded below. Since $P(t)$ is decreasing, $P(t) \to a/b$ as $t \to \infty$. This means that $P = a/b$ is a horizontal asymptote for a solution curve.

In Figure 2.6 the original phase portrait is reproduced to the left of the tP-plane in which the subregions R_1, R_2, and R_3 are shaded. The graphs of the equilibrium solutions $P = a/b$ and $P = 0$ (the t-axis) are shown in the figure as colored dashed lines; the solid graphs represent typical graphs of $P(t)$ illustrating the three cases just discussed. ■

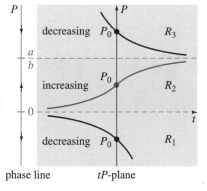

phase line tP-plane

Figure 2.6

In a subregion such as R_1 in Example 4, where $P(t)$ is decreasing and unbounded below, we must necessarily have $P(t) \to -\infty$. Do *not* interpret this last statement to mean $P(t) \to -\infty$ as $t \to \infty$; we could have $P(t) \to -\infty$ as $t \to T$, where $T > 0$ is a finite number that depends on the initial condition $P(t_0) = P_0$. In dynamic terms, $P(t)$ could "blow up" in finite time; graphically, $P(t)$ could have a vertical asymptote at $t = T > 0$. A similar remark holds for the subregion R_3.

The differential equation $dy/dx = \sin y$ in Example 2 is autonomous and has an infinite number of critical points, since $\sin y = 0$ at $y = n\pi$, n an integer. Moreover, we now know that because the solution $y(x)$ that passes through $(0, -\frac{3}{2})$ is bounded above and below by two consecutive critical points $(-\pi < y(x) < 0)$ and is decreasing ($\sin y < 0$ for $-\pi < y < 0$), $y(x)$ must approach the horizontal asymptotes $y = -\pi$ and $y = 0$ as $x \to \infty$ and as $x \to -\infty$, respectively.

EXAMPLE 5   **Solution Curves of an Autonomous DE**

The autonomous equation $dy/dx = (y - 1)^2$ possesses the single critical point 1. From the phase portrait in Figure 2.7(a) we conclude that a solution $y(x)$ is an increasing function in the subregions of the xy-plane defined by the inequalities $-\infty < y < 1$ and $1 < y < \infty$. For an initial condition $y(0) = y_0 < 1$, a solution $y(x)$ is increasing and bounded above by 1 and so $y(x) \to 1$ as $x \to \infty$; for $y(0) = y_0 > 1$, a solution $y(x)$ is increasing and unbounded.

Now $y(x) = 1 - 1/(x + c)$ is a one-parameter family of solutions of the differential equation. (See Problem 4 in Exercises 2.2.) A given initial condition determines a value for c. For the initial conditions, say, $y(0) = -1 < 1$ and $y(0) = 2 > 1$, we find, in turn, $y(x) = 1 - 1/(x + \frac{1}{2})$ and $y(x) = 1 - 1/(x - 1)$. As shown in Figures 2.7(b) and (c), the graph of each of these *functions* possesses a vertical asymptote. But bear in mind that the *solutions* of the initial-value problems

$$\frac{dy}{dx} = (y - 1)^2, \quad y(0) = -1 \quad \text{and} \quad \frac{dy}{dx} = (y - 1)^2, \quad y(0) = 2$$

are the functions defined on special intervals. They are, respectively,

$$y(x) = 1 - \frac{1}{x + \frac{1}{2}}, \quad -\frac{1}{2} < x < \infty \quad \text{and} \quad y(x) = 1 - \frac{1}{x - 1}, \quad -\infty < x < 1.$$

The solution curves are the portions of the graphs in Figures 2.7(b) and (c) shown in solid color. As predicted by the phase portrait, for the solution in Figure 2.7(b), $y(x) \to 1$ as $x \to \infty$; for the solution in Figure 2.7(c), $y(x) \to \infty$ as $x \to 1$ from the left.

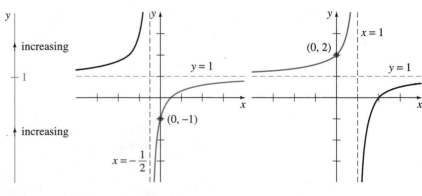

(a) phase line

(b) xy-plane
$y(0) < 1$

(c) xy-plane
$y(0) > 1$

Figure 2.7

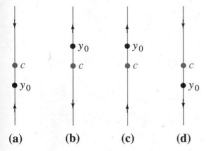

(a)   (b)   (c)   (d)

Figure 2.8
(a) asymptotically
    stable; attractor
(b) unstable; repeller
(c) semi-stable
(d) semi-stable

Attractors and Repellers   There are basically three types of behavior of a solution of (1) near a critical point c. When the arrowheads on both sides of c point toward c, as in Figure 2.8(a), all solutions $y(x)$ of (1) that start from an initial point (x_0, y_0) sufficiently near c exhibit the asymptotic behavior $\lim_{x \to \infty} y(x) = c$. For this reason c is said to be **asymptotically stable.** Using a physical analogy, a solution that starts near c can be compared to a charged particle that, over time, is drawn to the particle c of opposite charge, and so c is also referred to as an **attractor.** The arrowheads in Figure 2.8(b) pointing away from c indicate that all solutions that start near c move away from c as x increases. In this case c is said to be an **unstable** critical point. An unstable critical point is also called a **repeller,** for obvious reasons. The critical point c illustrated in Figures 2.8(c) and (d) is neither an attractor nor a repeller. But since c exhibits characteristics of both an attractor and a repeller—that is, a solution starting from an initial point (x_0, y_0) sufficiently near c is attracted to c from one side and repelled from the other side—we say that c is **semi-stable.** In Example 3 the critical points a/b and 0 are asymptotically stable (an attractor) and unstable (a repeller), respectively. The critical point 1 in Example 5 is semi-stable.

EXERCISES 2.1 ———————————————————

Answers to odd-numbered problems begin on page AN-1.

In Problems 1–4 use the given computer-generated direction field to sketch, by hand, an approximate solution curve that passes through the indicated points.

1. $\dfrac{dy}{dx} = \dfrac{x}{y}$

   (a) $y(0) = 5$ **(b)** $y(3) = 3$
   (c) $y(4) = 2$ **(d)** $y(-5) = -3$

2. $\dfrac{dy}{dx} = e^{-0.01xy^2}$

   (a) $y(-6) = 0$ **(b)** $y(0) = 1$
   (c) $y(0) = -4$ **(d)** $y(8) = -4$

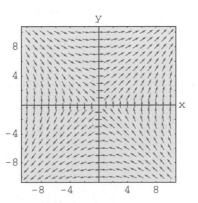

Figure 2.9

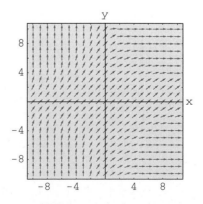

Figure 2.10

3. $\dfrac{dy}{dx} = 1 - xy$ **4.** $\dfrac{dy}{dx} = (\sin x)\cos y$

     (a) $y(0) = 0$ **(b)** $y(-1) = 0$ **(a)** $y(0) = 1$ **(b)** $y(1) = 0$

     (c) $y(2) = 2$ **(d)** $y(0) = -4$ **(c)** $y(3) = 3$ **(d)** $y(0) = -\frac{5}{2}$

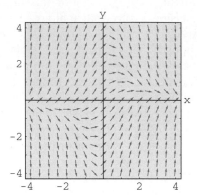

Figure 2.11   **Figure 2.12**

In Problems 5–12 use computer software to obtain a direction field for the given differential equation. By hand, sketch an approximate solution curve passing through each of the given points.

5. $y' = x$

     (a) $y(0) = 0$ **(b)** $y(0) = -3$

6. $y' = x + y$

     (a) $y(-2) = 2$ **(b)** $y(1) = -3$

7. $y\dfrac{dy}{dx} = -x$

     (a) $y(1) = 1$ **(b)** $y(0) = 4$

8. $\dfrac{dy}{dx} = \dfrac{1}{y}$

     (a) $y(0) = 1$ **(b)** $y(-2) = -1$

9. $\dfrac{dy}{dx} = 0.2x^2 + y$

     (a) $y(0) = \frac{1}{2}$ **(b)** $y(2) = -1$

10. $\dfrac{dy}{dx} = xe^y$

     (a) $y(0) = -2$ **(b)** $y(1) = 2.5$

11. $y' = y - \cos\dfrac{\pi}{2}x$

     (a) $y(2) = 2$ **(b)** $y(-1) = 0$

12. $\dfrac{dy}{dx} = 1 - \dfrac{y}{x}$

     (a) $y(-\frac{1}{2}) = 2$ **(b)** $y(\frac{3}{2}) = 0$

In Problems 13 and 14 the given figure represents the graph of $f(y)$ and $f(x)$, respectively. By hand, sketch a direction field over an appropriate grid for $dy/dx = f(y)$ (Problem 13) and then for $dy/dx = f(x)$ (Problem 14).

13.   **14.**

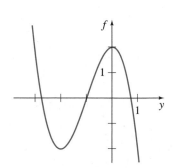

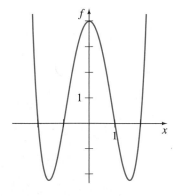

Figure 2.13   **Figure 2.14**

15. Consider the autonomous first-order differential equation $dy/dx = y - y^3$ and the initial condition $y(0) = y_0$. By hand, sketch the graph of a typical solution $y(x)$ when y_0 has the given values.

(a) $y_0 > 1$ (b) $0 < y_0 < 1$ (c) $-1 < y_0 < 0$ (d) $y_0 < -1$

16. Consider the autonomous first-order differential equation $dy/dx = y^2 - y^4$ and the initial condition $y(0) = y_0$. By hand, sketch the graph of a typical solution $y(x)$ when y_0 has the given values.

(a) $y_0 > 1$ (b) $0 < y_0 < 1$ (c) $-1 < y_0 < 0$ (d) $y_0 < -1$

In Problems 17–24 find the critical points and phase portrait of the given autonomous first-order differential equation. Classify each critical point as asymptotically stable, unstable, or semi-stable. By hand, sketch typical solution curves in the regions in the xy-plane determined by the graphs of the equilibrium solutions.

17. $\dfrac{dy}{dx} = y^2 - 3y$

18. $\dfrac{dy}{dx} = y^2 - y^3$

19. $\dfrac{dy}{dx} = (y - 2)^4$

20. $\dfrac{dy}{dx} = 10 + 3y - y^2$

21. $\dfrac{dy}{dx} = y^2(4 - y^2)$

22. $\dfrac{dy}{dx} = y(2 - y)(4 - y)$

23. $\dfrac{dy}{dx} = y \ln(y + 2)$

24. $\dfrac{dy}{dx} = \dfrac{ye^y - 9y}{e^y}$

25. The autonomous differential equation

$$m \frac{dv}{dt} = mg - kv,$$

where k is a positive constant of proportionality and g is the acceleration due to gravity, governs the velocity v of a body of mass m falling under the influence of gravity. Because the term $-kv$ represents air resistance, the velocity of a body falling from a great height does not increase without bound as time t increases.

(a) Use a phase portrait of the differential equation to find the limiting, or terminal, velocity of the body. Explain your reasoning.

(b) Find the terminal velocity of the body if air resistance is proportional to v^2. See pages 28 and 33.

26. When certain kinds of chemicals are combined, the rate at which a new compound is formed is governed by the autonomous differential equation

$$\frac{dX}{dt} = k(\alpha - X)(\beta - X),$$

where $k > 0$ is a constant of proportionality and $\beta > \alpha > 0$. Here $X(t)$ denotes the number of grams of the new compound formed in time t. See pages 25–26.

(a) From the phase portrait of the differential equation, can you predict the behavior of X as $t \to \infty$?

(b) Consider the case $\alpha = \beta$. What is the behavior of X as $t \to \infty$ if $X(0) < \alpha$? From the phase portrait of the differential equation, can you predict the behavior of X as $t \to \infty$ if $X(0) > \alpha$?

(c) Verify that an explicit solution of the differential equation in the case when $k = 1$ and $\alpha = \beta$ is $X(t) = \alpha - 1/(t + c)$. Find a solution satisfying $X(0) = \alpha/2$. Find a solution satisfying $X(0) = 2\alpha$. Graph these two solutions. Does the behavior of the solutions agree with your answers to part (b)?

In Problems 27 and 28 consider the autonomous differential equation $dy/dx = f(y)$, where the graph of f is given. Use the graph to locate the critical points of each differential equation. Sketch a phase portrait of each differential equation. By hand, sketch typical solution curves in the subregions in the xy-plane determined by the graphs of the equilibrium solutions.

27.

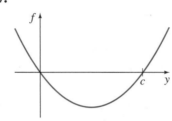

Figure 2.15

28.

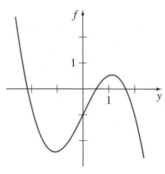

Figure 2.16

Discussion Problems

29. Consider the differential equation $dy/dx = x^2 + y^2$. What can you say about the lineal elements at the points (x, y) on the graph of the circles $x^2 + y^2 = c^2$, $c > 0$?

30. (a) Consider the direction field of the differential equation $dy/dx = x(y - 4)^2 - 2$, but do not use technology to obtain it. Describe the slopes of the lineal elements on the lines $x = 0$, $y = 3$, $y = 4$, and $y = 5$.

(b) Consider the initial-value problem $dy/dx = x(y - 4)^2 - 2$, $y(0) = y_0$, where $y_0 < 4$. Can a solution $y(x) \to \infty$ as $x \to \infty$? Based on the information in part (a), discuss.

 31. For the first-order differential equation $dy/dx = f(x, y)$ a curve in the plane defined by $f(x, y) = 0$ is called a **nullcline,** since a lineal element at a point on the curve has zero slope. Use computer software to obtain a direction field over a rectangular grid of points for $dy/dx = x^2 - 2y$ and then superimpose the graph of the nullcline $y = \frac{1}{2}x^2$ over the direction field. Discuss the behavior of solution curves in regions defined by $y < \frac{1}{2}x^2$ and by $y > \frac{1}{2}x^2$. Sketch some approximate solution curves. Try to generalize your observations.

32. A critical point c of an autonomous first-order differential equation is said to be **isolated** if there exists some open interval that contains c but no other critical point. Can there exist an autonomous equation of the form given in (1) for which *every* critical point is non-isolated? Discuss; do not think profound thoughts.

33. Suppose that $y(x)$ is a nonconstant solution of the autonomous differential equation (1). If c is a critical point of (1), why can't the graph of $y(x)$ cross the graph of the equilibirum solution $y = c$? Discuss. Why can't $f(y)$ change signs in one of the subregions discussed on page 44? Why can't $y(x)$ be oscillatory or have a relative extremum (maximum or minimum)?

34. Suppose that $y(x)$ is a solution of the autonomous equation $dy/dx = f(y)$ and is bounded above and below by two consecutive critical points $c_1 < c_2$, as in subregion R_2 of Figure 2.5(b). If $f(y) > 0$ in the region, then $\lim_{x \to \infty} y(x) = c_2$. Discuss why there cannot exist a number $L < c_2$ such that $\lim_{x \to \infty} y(x) = L$. As part of your discussion, consider what happens to $y'(x)$ as $x \to \infty$.

35. (a) Using the autonomous equation (1), discuss how it is possible to obtain information about the location of **points of inflection** of a solution curve.

(b) Consider the differential equation $dy/dx = y^2 - y - 6$. Use your ideas from part (a) to find intervals on the y-axis for which solution curves are concave up and intervals for which solution curves are concave down. Discuss why *each* solution curve of an initial-value problem of the form $dy/dx = y^2 - y - 6$, $y(0) = y_0$, where $-2 < y_0 < 3$, has a point of inflection with the same y-coordinate. What is that y-coordinate? Carefully sketch the solution curve for which $y(0) = -1$. Repeat for $y(2) = 2$.

2.2 SEPARABLE VARIABLES

• Solution by integration   • Definition of a separable DE   • Method of solution   • Losing a solution   • Alternative forms of a solution

In solving differential equations you will often have to integrate, and this may require some special technique of integration. It will be worth a few minutes of your time to review basic integration formulas from a calculus text or, if you use a CAS (computer algebra system), the command syntax for integration. Two important items to include in your review are integration by parts and the partial fraction decomposition of a rational function.

Solution by Integration   We begin our study of the methodology of solving first-order equations $dy/dx = f(x, y)$ with the simplest of all differential equations. When f is independent of the variable y—that is, $f(x, y) = g(x)$—the differential equation

$$\frac{dy}{dx} = g(x) \tag{1}$$

can be solved by integration. If $g(x)$ is a continuous function, then integrating both sides of (1) gives the solution $y = \int g(x)\, dx = G(x) + c$, where $G(x)$ is an antiderivative (indefinite integral) of $g(x)$. For example, if $dy/dx = 1 + e^{2x}$, then $y = \int (1 + e^{2x})\, dx$; that is, $y = x + \frac{1}{2}e^{2x} + c$.

Equation (1), as well as its method of solution, is just a special case when f in $dy/dx = f(x, y)$ is a product of a function of x and a function of y.

DEFINITION 2.1　Separable Equation

A first-order differential equation of the form

$$\frac{dy}{dx} = g(x)h(y)$$

is said to be **separable** or to have **separable variables.**

For example, the equations

$$\frac{dy}{dx} = y^2xe^{3x+4y} \quad \text{and} \quad \frac{dy}{dx} = y + \sin x$$

are separable and nonseparable, respectively. In the first equation we can factor $f(x, y) = y^2xe^{3x+4y}$ as

$$f(x, y) = y^2xe^{3x+4y} = \overset{\overset{g(x)}{\downarrow}}{(xe^{3x})}\,\overset{\overset{h(y)}{\downarrow}}{(y^2e^{4y})},$$

but in the second equation there is no way of expressing $y + \sin x$ as a product of a function of x times a function of y.

Observe that by dividing by the function $h(y)$ we can write a separable equation $dy/dx = g(x)h(y)$ as

$$p(y)\frac{dy}{dx} = g(x), \tag{2}$$

where, for convenience, we have denoted $1/h(y)$ by $p(y)$. From this last form we can see immediately that (2) reduces to (1) when $h(y) = 1$.

Now if $y = \phi(x)$ represents a solution of (2), we must have $p(\phi(x))\phi'(x) = g(x)$, and therefore

$$\int p(\phi(x))\phi'(x)\,dx = \int g(x)\,dx. \tag{3}$$

But $dy = \phi'(x)\,dx$, and so (3) is the same as

$$\int p(y)\,dy = \int g(x)\,dx \quad \text{or} \quad H(y) = G(x) + c, \tag{4}$$

where $H(y)$ and $G(x)$ are antiderivatives of $p(y) = 1/h(y)$ and $g(x)$, respectively.

Method of Solution　Equation (4) indicates the procedure for solving separable equations. A one-parameter family of solutions, usually given implicitly, is obtained by integrating both sides of $p(y)\,dy = g(x)\,dx$.

Note　There is no need to use two constants in the integration of a separable equation, because if we write $H(y) + c_1 = G(x) + c_2$ then the difference $c_2 - c_1$ can be replaced by a single constant c, as in (4). In many instances throughout the chapters that follow, we will relabel constants in a manner convenient to a given equation. For example, multiples of constants or combinations of constants can sometimes be replaced by a single constant.

EXAMPLE 1 Solving a Separable DE

Solve $(1 + x)\, dy - y\, dx = 0$.

Solution   Dividing by $(1 + x)y$, we can write $dy/y = dx/(1 + x)$, from which it follows that

$$\int \frac{dy}{y} = \int \frac{dx}{1 + x}$$

$$\ln|y| = \ln|1 + x| + c_1$$

$$y = e^{\ln|1+x|+c_1} = e^{\ln|1+x|} \cdot e^{c_1} \qquad \leftarrow \text{laws of exponents}$$

$$= |1 + x|e^{c_1}$$

$$= \pm\, e^{c_1}(1 + x). \qquad \leftarrow \begin{cases} |1 + x| = 1 + x, & x \geq -1 \\ |1 + x| = -(1 + x), & x < -1 \end{cases}$$

Relabeling $\pm e^{c_1}$ as c then gives $y = c(1 + x)$.

Alternative Solution   Since each integral results in a logarithm, a judicious choice for the constant of integration is $\ln|c|$ rather than c. Rewriting the second line of the solution as $\ln|y| = \ln|1 + x| + \ln|c|$ enables us to combine the terms on the right-hand side by the properties of logarithms. From $\ln|y| = \ln|c(1 + x)|$ we immediately get $y = c(1 + x)$. Even if the indefinite integrals are not *all* logarithms, it may still be advantageous to use $\ln|c|$. However, no firm rule can be given.

In Section 1.1 we saw that a solution curve may be only a segment or an arc of the graph of an implicit solution $G(x, y) = 0$.

EXAMPLE 2 Solution Curve

Solve the initial-value problem $\dfrac{dy}{dx} = -\dfrac{x}{y}$, $y(4) = -3$.

Solution   Rewriting the equation as $y\, dy = -x\, dx$, we get

$$\int y\, dy = -\int x\, dx \quad \text{and} \quad \frac{y^2}{2} = -\frac{x^2}{2} + c_1.$$

We can write the result of the integration as $x^2 + y^2 = c^2$ by replacing the constant $2c_1$ by c^2. This solution of the differential equation represents a family of concentric circles centered at the origin.

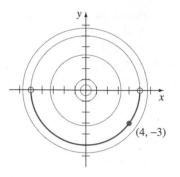

Figure 2.17

Now when $x = 4$, $y = -3$, so $16 + 9 = 25 = c^2$. Thus the initial-value problem determines the circle $x^2 + y^2 = 25$ with radius 5. Because of its simplicity we can solve this implicit solution for an explicit solution that satisfies the initial condition. We have seen this solution as $y = \phi_2(x)$ or $y = -\sqrt{25 - x^2}$, $-5 < x < 5$ in Example 3 of Section 1.1. A solution curve is the graph of a differentiable *function*. In this case the solution curve is the lower semi-circle, shown in color in Figure 2.17, containing the point $(4, -3)$. ∎

Losing a Solution   Some care should be exercised when separating variables, since the variable divisors could be zero at a point. Specifically, if r is a zero of the function $h(y)$, then substituting $y = r$ into $dy/dx = g(x)h(y)$ makes both sides zero; in other words, $y = r$ is a constant solution of the differential equation. But after variables are separated, the left-hand side of $\dfrac{dy}{h(y)} = g(x)\, dx$ is undefined at r. As a consequence, $y = r$ may not show up in the family of solutions obtained after integration and simplification. Recall that such a solution is called a singular solution.

EXAMPLE 3   **Losing a Solution**

Solve $\dfrac{dy}{dx} = y^2 - 4$.

Solution   We put the equation in the form

$$\frac{dy}{y^2 - 4} = dx \quad \text{or} \quad \left[\frac{\frac{1}{4}}{y - 2} - \frac{\frac{1}{4}}{y + 2}\right] dy = dx. \tag{5}$$

The second equation in (5) is the result of using partial fractions on the left-hand side of the first equation. Integrating and using the laws of logarithms gives

$$\frac{1}{4}\ln|y - 2| - \frac{1}{4}\ln|y + 2| = x + c_1 \quad \text{or} \quad \ln\left|\frac{y - 2}{y + 2}\right| = 4x + c_2 \quad \text{or} \quad \frac{y - 2}{y + 2} = \pm e^{4x + c_2}.$$

Here we have replaced $4c_1$ by c_2. Finally, after replacing $\pm e^{c_2}$ by c and solving the last equation for y, we get the one-parameter family of solutions

$$y = 2\frac{1 + ce^{4x}}{1 - ce^{4x}}. \tag{6}$$

Now if we factor the right-hand side of the differential equation as $dy/dx = (y - 2)(y + 2)$, we know from the discussion of critical points in Section 2.1 that $y = 2$ and $y = -2$ are two constant (equilibrium) solutions. The solution $y = 2$ is a member of the family of solutions defined by (6) corresponding to the value $c = 0$. However, $y = -2$ is a singular solution; it cannot be obtained from (6) for any choice of the parameter c. This latter solution was lost early on in the solution process. Inspection of (5) clearly indicates that we must preclude $y = \pm 2$ in these steps. ∎

EXAMPLE 4 **An Initial-Value Problem**

Solve $(e^{2y} - y) \cos x \dfrac{dy}{dx} = e^y \sin 2x$, $y(0) = 0$.

Solution Dividing the equation by $e^y \cos x$ gives

$$\frac{e^{2y} - y}{e^y} dy = \frac{\sin 2x}{\cos x} dx.$$

Before integrating we use termwise division on the left-hand side and the trigonometric identity $\sin 2x = 2 \sin x \cos x$ on the right-hand side. Then

integration by parts → $\quad \displaystyle\int (e^y - ye^{-y}) \, dy = 2 \int \sin x \, dx$

yields $\qquad\qquad e^y + ye^{-y} + e^{-y} = -2 \cos x + c.$ **(7)**

The initial condition $y = 0$ when $x = 0$ implies $c = 4$. Thus a solution of the initial-value problem is

$$e^y + ye^{-y} + e^{-y} = 4 - 2 \cos x. \qquad \textbf{(8)} \quad \blacksquare$$

Use of Computers The Remarks at the end of Section 1.1 mentioned that it may be difficult to use an implicit solution $G(x, y) = 0$ to find an explicit solution $y = \phi(x)$. Equation (8) shows that the task of solving for y in terms of x may present more problems than just the drudgery of symbol pushing—sometimes it simply cannot be done! Implicit solutions such as (8) are somewhat frustrating; neither the graph of the equation nor an interval over which a solution satisfying $y(0) = 0$ is defined is apparent. The problem of "seeing" what an implicit solution looks like can be overcome in some cases by means of technology. One way* of proceeding is to use the contour plot application of a CAS. Recall from multivariate calculus that for a function of two variables $z = G(x, y)$ the *two-dimensional* curves defined by $G(x, y) = c$, where c is constant, are called the *level curves* of the function. With the aid of a CAS, some of the level curves of the function $G(x, y) = e^y + ye^{-y} + e^{-y} + 2 \cos x$ have been reproduced in Figure 2.18. The family of solutions defined by (7) is the level curves $G(x, y) = c$. Figure 2.19 illustrates the level curve $G(x, y) = 4$, which is the particular solution (8), in solid color. The other curve in Figure 2.19 is the level curve $G(x, y) = 2$, which is the member of the family $G(x, y) = c$ that satisfies $y(\pi/2) = 0$.

If an initial condition leads to a particular solution by yielding a specific value of the parameter c in a family of solutions for a first-order differential equation, there is a natural inclination for most students (and instructors) to relax and be content. However, a solution of an initial-value problem may not be unique. We saw in Example 3 of Section 1.2

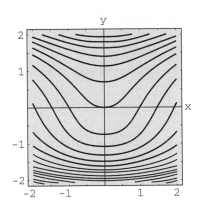

Figure 2.18 Level curves $G(x, y) = c$, where $G(x, y) = e^y + ye^{-y} + e^{-y} + 2 \cos x$

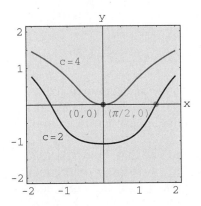

Figure 2.19 Level curves $c = 2$ and $c = 4$

---

*In Section 2.6 we will discuss several other ways of proceeding that are based on the concept of a numerical solver.

that the initial-value problem

$$\frac{dy}{dx} = xy^{1/2}, \quad y(0) = 0 \tag{9}$$

has at least two solutions, $y = 0$ and $y = \frac{1}{16}x^4$. We are now in a position to solve the equation. Separating variables and integrating $y^{-1/2}\,dy = x\,dx$ gives

$$2y^{1/2} = \frac{x^2}{2} + c_1 \quad \text{or} \quad y = \left(\frac{x^2}{4} + c\right)^2.$$

When $x = 0$, then $y = 0$, and so necessarily $c = 0$. Therefore $y = \frac{1}{16}x^4$. The trivial solution $y = 0$ was lost by dividing by $y^{1/2}$. In addition, the initial-value problem (9) possesses infinitely many more solutions, since for any choice of the parameter $a \geq 0$ the piecewise-defined function

$$y = \begin{cases} 0, & x < a \\ \dfrac{(x^2 - a^2)^2}{16}, & x \geq a \end{cases}$$

satisfies both the differential equation and the initial condition. See Figure 2.20.

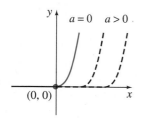

Figure 2.20

Remarks   (i) If g is a function continuous on an interval I containing x_0, then from the fundamental theorem of calculus we have

$$\frac{d}{dx}\int_{x_0}^{x} g(t)\,dt = g(x).$$

In other words, $\int_{x_0}^{x} g(t)\,dt$ is an antiderivative of the function g.

There are times when this form is convenient. For example, if g is continuous on an interval I containing x_0, then a solution of the simple initial-value problem $dy/dx = g(x)$, $y(x_0) = y_0$ that is defined on I is given by

$$y(x) = y_0 + \int_{x_0}^{x} g(t)\,dt.$$

You should verify this. Since an antiderivative of a continuous function cannot always be expressed in terms of elementary functions, this may be the best we can do in obtaining an explicit solution of an IVP.

(ii) In some of the preceding examples we saw that the constant in the one-parameter family of solutions for a first-order differential equation can be relabeled when convenient. Also, it can easily happen that two individuals, solving the same equation correctly, arrive at dissimilar expressions for their answers. For example, by separation of variables we can show that one-parameter families of solutions for the DE $(1 + y^2)\,dx + (1 + x^2)\,dy = 0$ are

$$\arctan x + \arctan y = c \quad \text{or} \quad \frac{x + y}{1 - xy} = c.$$

As you work your way through the next several sections, keep in mind that families of solutions may be equivalent in the sense that one family may be obtained from another by either relabeling the constant or applying algebra and trigonometry. See Problems 27 and 28 in Exercises 2.2.

EXERCISES 2.2

Answers to odd-numbered problems begin on page AN-1.

In Problems 1–22 solve the given differential equation by separation of variables.

1. $\dfrac{dy}{dx} = \sin 5x$
2. $\dfrac{dy}{dx} = (x + 1)^2$

3. $dx + e^{3x}\, dy = 0$
4. $dy - (y - 1)^2\, dx = 0$

5. $x\dfrac{dy}{dx} = 4y$
6. $\dfrac{dy}{dx} + 2xy = 0$

7. $\dfrac{dy}{dx} = e^{3x+2y}$
8. $e^x y\dfrac{dy}{dx} = e^{-y} + e^{-2x-y}$

9. $y \ln x \dfrac{dx}{dy} = \left(\dfrac{y + 1}{x}\right)^2$
10. $\dfrac{dy}{dx} = \left(\dfrac{2y + 3}{4x + 5}\right)^2$

11. $\sec^2 x\, dy + \csc y\, dx = 0$
12. $\sin 3x\, dx + 2y \cos^3 3x\, dy = 0$

13. $(e^y + 1)^2 e^{-y}\, dx + (e^x + 1)^3 e^{-x}\, dy = 0$

14. $y\, dy = x(1 + x^2)^{-1/2}(1 + y^2)^{1/2}\, dx$

15. $\dfrac{dS}{dr} = kS$
16. $\dfrac{dQ}{dt} = k(Q - 70)$

17. $\dfrac{dP}{dt} = P - P^2$
18. $\dfrac{dN}{dt} + N = Nte^{t+2}$

19. $\dfrac{dy}{dx} = \dfrac{xy + 3x - y - 3}{xy - 2x + 4y - 8}$
20. $\dfrac{dy}{dx} = \dfrac{xy + 2y - x - 2}{xy - 3y + x - 3}$

21. $\dfrac{dy}{dx} = x\sqrt{1 - y^2}$
22. $(e^x + e^{-x})\dfrac{dy}{dx} = y^2$

In Problems 23–26 solve the given initial-value problem.

23. $\dfrac{dx}{dt} = 4(x^2 + 1), \quad x(\pi/4) = 1$
24. $\dfrac{dy}{dx} = \dfrac{y^2 - 1}{x^2 - 1}, \quad y(2) = 2$

25. $x^2\dfrac{dy}{dx} = y - xy, \quad y(-1) = -1$
26. $\dfrac{dy}{dt} + 2y = 1, \quad y(0) = \frac{5}{2}$

In Problems 27 and 28 find an implicit and an explicit solution of the given initial-value problem.

27. $\sqrt{1 - y^2}\, dx - \sqrt{1 - x^2}\, dy = 0, \; y(0) = \dfrac{\sqrt{3}}{2}$

28. $(1 + x^4)\, dy + x(1 + 4y^2)\, dx = 0, \; y(1) = 0$

29. (a) Find a solution of the initial-value problem consisting of the differential equation in Example 3 and the initial conditions $y(0) = 2$, $y(0) = -2$, and $y(\frac{1}{4}) = 1$.

(b) Find the solution of the differential equation in Example 4 when $\ln c_1$ is used as the constant of integration on the *left-hand* side in the solution and $4 \ln c_1$ is replaced by $\ln c$. Then solve the same initial-value problems in part (a).

30. Find a solution of $x\dfrac{dy}{dx} = y^2 - y$ that passes through the indicated points.

(a) $(0, 1)$ **(b)** $(0, 0)$ **(c)** $(\frac{1}{2}, \frac{1}{2})$ **(d)** $(2, \frac{1}{4})$

31. Find a singular solution of Problem 21. Of Problem 22.

32. Show that an implicit solution of

$$2x \sin^2 y \, dx - (x^2 + 10) \cos y \, dy = 0$$

is given by $\ln(x^2 + 10) + \csc y = c$. Find the constant solutions, if any, that were lost in the solution of the differential equation.

Often a radical change in the form of the solution of a differential equation corresponds to a very small change in either the initial condition or the equation itself. In Problems 33–36 find an explicit solution of the given initial-value problem. Use a graphing utility to plot the graph of each solution. Compare each solution curve in a neighborhood of $(0, 1)$.

33. $\dfrac{dy}{dx} = (y - 1)^2$, $y(0) = 1$ **34.** $\dfrac{dy}{dx} = (y - 1)^2$, $y(0) = 1.01$

35. $\dfrac{dy}{dx} = (y - 1)^2 + 0.01$, $y(0) = 1$ **36.** $\dfrac{dy}{dx} = (y - 1)^2 - 0.01$, $y(0) = 1$

37. Every autonomous first-order equation $dy/dx = f(y)$ is separable. Find explicit solutions $y_1(x)$, $y_2(x)$, $y_3(x)$, and $y_4(x)$ of the differential equation $dy/dx = y - y^3$ that satisfy, in turn, the initial conditions $y_1(0) = 2$, $y_2(0) = \frac{1}{2}$, $y_3(0) = -\frac{1}{2}$, and $y_4(0) = -2$. Use a graphing utility to plot the graphs of each solution. Compare these graphs with those predicted in Problem 15 of Exercises 2.1. Give the exact interval of definition for each solution.

38. (a) The autonomous differential equation $dy/dx = 1/(y - 3)$ has no critical points. Nevertheless, place 3 on the phase line and obtain a phase portrait of the equation. Compute d^2y/dx^2 to determine where solution curves are concave up and where they are concave down (see Problem 35 in Exercises 2.1). Use the phase portrait and concavity to sketch, by hand, some typical solution curves.

(b) Find explicit solutions $y_1(x)$, $y_2(x)$, $y_3(x)$, and $y_4(x)$ of the differential equation in part (a) that satisfy, in turn, the initial conditions $y_1(0) = 4$, $y_2(0) = 2$, $y_3(1) = 2$, and $y_4(-1) = 4$. Graph each solution and compare with your sketches in part (a). Give the exact interval of definition for each solution.

39. (a) Find an explicit solution of the initial-value problem

$$\frac{dy}{dx} = \frac{2x + 1}{2y}, \quad y(-2) = -1.$$

(b) Use a graphing utility to plot the graph of the solution in part (a). Use the graph to estimate the interval I of definition of the solution.

(c) Determine the exact interval I of definition by analytical methods.

40. Repeat parts (a)–(c) of Problem 39 for the initial-value problem consisting of the differential equation in Problem 7 and the condition $y(0) = 0$.

Discussion Problems

41. (a) Explain why the interval of definition of the explicit solution $y = \phi_2(x)$ of the initial-value problem in Example 2 is the *open* interval $-5 < x < 5$.

(b) Can any solution of the differential equation cross the x-axis? Do you think that $x^2 + y^2 = 1$ is an implicit solution of the initial-value problem $dy/dx = -x/y$, $y(1) = 0$?

42. (a) If $a > 0$, discuss the differences, if any, between the solutions of the initial-value problems consisting of the differential equation $dy/dx = x/y$ and each of the initial conditions $y(a) = a$, $y(a) = -a$, $y(-a) = a$, and $y(-a) = -a$.

(b) Does the initial-value problem $dy/dx = x/y$, $y(0) = 0$ have a solution?

(c) Solve $dy/dx = x/y$, $y(1) = 2$ and give the exact interval I of definition of its solution.

43. In Problems 37 and 38 we saw that every autonomous first-order differential equation $dy/dx = f(y)$ is separable. Does this fact help in solution of the initial-value problem $\dfrac{dy}{dx} = \sqrt{1 + y^2} \sin^2 y$, $y(0) = \frac{1}{2}$?

Discuss. Sketch, by hand, a plausible solution curve of the problem.

44. Without the use of technology, how would you solve

$$(\sqrt{x} + x)\frac{dy}{dx} = (\sqrt{y} + y)?$$

Carry out your ideas.

45. Find a function whose square plus the square of its derivative is 1.

46. (a) The differential equation in Problem 27 is equivalent to the normal form

$$\frac{dy}{dx} = \sqrt{\frac{1 - y^2}{1 - x^2}}$$

in the square region in the xy-plane defined by $|x| < 1$, $|y| < 1$. But the quantity under the radical is nonnegative also in the regions defined by $|x| > 1$, $|y| > 1$. Sketch all regions in the xy-plane for which this differential equation possesses real solutions.

(b) Solve the DE in part (a) in the regions defined by $|x| > 1$, $|y| > 1$. Then find an implicit and an explicit solution of the differential equation subject to $y(2) = 2$.

Computer Lab Assignments

47. (a) Use a CAS and the concept of level curves to plot representative graphs of members of the family of solutions of the differential equation $\dfrac{dy}{dx} = -\dfrac{8x + 5}{3y^2 + 1}$. Experiment with different numbers of level curves as well as various rectangular regions defined by $a \le x \le b$, $c \le y \le d$.

(b) On separate coordinate axes plot the graphs of the particular solutions corresponding to the initial conditions $y(0) = -1$, $y(0) = 2$, $y(-1) = 4$, and $y(-1) = -3$.

48. (a) Use a CAS and the concept of level curves to plot representative graphs of members of the family of solutions of the differential equation $\dfrac{dy}{dx} = \dfrac{x(1-x)}{y(-2+y)}$. Experiment with different numbers of level curves as well as various rectangular regions in the xy-plane until your result resembles Figure 2.21.

(b) On separate coordinate axes plot the graph of the implicit solution corresponding to the initial condition $y(0) = \frac{3}{2}$. Use a colored pencil to mark off that segment of the graph that corresponds to the solution curve of a solution ϕ that satisfies the initial condition. With the aid of a root-finding application, approximate the largest interval I of definition of the solution ϕ. [*Hint:* First find the points on the curve in part (a) where the tangent is vertical.]

(c) Repeat part (b) for the initial condition $y(0) = -2$.

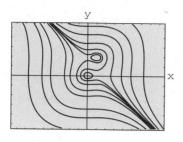

Figure 2.21

2.3   LINEAR EQUATIONS

- *Definition of a linear DE* • *Standard form of a linear DE*
- *Variation of parameters* • *Method of solution* • *Integrating factor* • *General solution* • *Error function* • *Transient term*

We continue our quest for solutions of first-order DEs by next examining linear equations. Linear differential equations are an especially "friendly" family of differential equations in that, given a linear equation, whether first order or a higher-order kin, there is always a good possibility that we can find some sort of solution of the equation that we can look at.

Linear First-Order DE   Recall from Section 1.1 that a differential equation is said to be linear when it is of the first degree in the dependent variable and all its derivatives. When $n = 1$ in (6) of Section 1.1, we obtain a linear first-order differential equation.

DEFINITION 2.2   Linear Equation

A first-order differential equation of the form

$$a_1(x)\frac{dy}{dx} + a_0(x)y = g(x) \tag{1}$$

is said to be a **linear equation**.

When $g(x) = 0$, the linear equation is said to be **homogeneous;** otherwise, it is **nonhomogeneous.**

Standard Form   By dividing both sides of (1) by the lead coefficient $a_1(x)$, we obtain a more useful form, the **standard form,** of a linear equation:

$$\frac{dy}{dx} + P(x)y = f(x). \tag{2}$$

We seek a solution of (2) on an interval I for which both functions P and f are continuous.

In the discussion that follows we illustrate a property and a procedure and end up with a formula representing the form that every solution of (2) must have. But more than the formula, the property and the procedure are important, because these two concepts carry over to linear equations of higher order.

The Property   The differential equation (2) has the property that its solution is the **sum** of the two solutions: $y = y_c + y_p$, where y_c is a solution of the associated homogeneous equation

$$\frac{dy}{dx} + P(x)y = 0 \tag{3}$$

and y_p is a particular solution of the nonhomogeneous equation (2). To see this, observe that

$$\frac{d}{dx}[y_c + y_p] + P(x)[y_c + y_p] = \underbrace{\left[\frac{dy_c}{dx} + P(x)y_c\right]}_{0} + \underbrace{\left[\frac{dy_p}{dx} + P(x)y_p\right]}_{f(x)} = f(x).$$

Now the homogeneous equation (3) is also separable. This fact enables us to find y_c by writing (3) as

$$\frac{dy}{y} + P(x)\, dx = 0$$

and integrating. Solving for y gives $y_c = ce^{-\int P(x)dx}$. For convenience let us write $y_c = cy_1(x)$, where $y_1 = e^{-\int P(x)dx}$. The fact that $dy_1/dx + P(x)y_1 = 0$ will be used next to determine y_p.

The Procedure   We can now find a particular solution of equation (2) by a procedure known as **variation of parameters.** The basic idea here is to find a function u so that $y_p = u(x)y_1(x) = u(x)e^{-\int P(x)dx}$ is a solution of (2). In other words, our assumption for y_p is the same as $y_c = cy_1(x)$ except that c is replaced by the "variable parameter" u. Substituting $y_p = uy_1$ into (2) gives

$$\overset{\substack{\text{Product Rule}\\\downarrow}}{u\frac{dy_1}{dx}} + y_1\frac{du}{dx} + P(x)uy_1 = f(x) \quad\text{or}\quad u\left[\overset{\substack{\text{zero}\\\downarrow}}{\frac{dy_1}{dx} + P(x)y_1}\right] + y_1\frac{du}{dx} = f(x)$$

so

$$y_1\frac{du}{dx} = f(x).$$

Separating variables and integrating then gives

$$du = \frac{f(x)}{y_1(x)}\, dx \quad \text{and} \quad u = \int \frac{f(x)}{y_1(x)}\, dx.$$

Since $y_1(x) = e^{-\int P(x)\,dx}$, we see that $1/y_1(x) = e^{\int P(x)\,dx}$. Therefore

$$y_p = uy_1 = \left(\int \frac{f(x)}{y_1(x)}\, dx\right) e^{-\int P(x)\,dx} = e^{-\int P(x)\,dx} \int e^{\int P(x)\,dx}\, f(x)\, dx$$

and

$$y = \underbrace{ce^{-\int P(x)\,dx}}_{y_c} + \underbrace{e^{-\int P(x)\,dx} \int e^{\int P(x)\,dx} f(x)\, dx}_{y_p}. \tag{4}$$

Hence if (2) has a solution, it must be of form (4). Conversely, it is a straightforward exercise in differentiation to verify that (4) constitutes a one-parameter family of solutions of equation (2).

You should not memorize the formula given in (4). *However,* you should remember the special term

$$e^{\int P(x)\,dx} \tag{5}$$

because it is used in an equivalent but easier way of solving (2). If equation (4) is multiplied by (5),

$$e^{\int P(x)\,dx} y = c + \int e^{\int P(x)\,dx} f(x)\, dx, \tag{6}$$

and then (6) is differentiated,

$$\frac{d}{dx}\left[e^{\int P(x)\,dx} y\right] = e^{\int P(x)\,dx} f(x), \tag{7}$$

we get

$$e^{\int P(x)\,dx} \frac{dy}{dx} + P(x)e^{\int P(x)\,dx} y = e^{\int P(x)\,dx} f(x). \tag{8}$$

Dividing the last result by $e^{\int P(x)\,dx}$ gives (2).

Method of Solution　The recommended method of solving (2) actually consists of (6)–(8) worked in reverse order. In other words, if (2) is multiplied by (5), we get (8). The left-hand side of (8) is recognized as the derivative of the product of $e^{\int P(x)\,dx}$ and y. This gets us to (7). We then integrate both sides of (7) to get the solution (6). Because we can solve (2) by integration after multiplication by $e^{\int P(x)\,dx}$, we call this function an **integrating factor** for the differential equation.* For convenience we summarize these results. We again emphasize that you should not memorize formula (4) but work through the following procedure each time.

---

*This integrating factor can be derived by an alternative procedure discussed in Section 2.4.

Solving a Linear First-Order Equation

(*i*) Put a linear equation of form (1) into the standard form (2).

(*ii*) From the standard form identify $P(x)$ and then find the integrating factor $e^{\int P(x)dx}$.

(*iii*) Multiply the standard form of the equation by the integrating factor. The left-hand side of the resulting equation is automatically the derivative of the integrating factor and y:

$$\frac{d}{dx}[e^{\int P(x)dx}y] = e^{\int P(x)dx}f(x).$$

(*iv*) Integrate both sides of this last equation.

EXAMPLE 1   **Solving a Homogeneous Linear DE**

Solve $\dfrac{dy}{dx} - 3y = 0$.

Solution   This linear equation can be solved by separation of variables. Alternatively, since the equation is already in the standard form (2), we see that $P(x) = -3$ and so the integrating factor is $e^{\int(-3)dx} = e^{-3x}$. We multiply the equation by this factor and recognize that

$$e^{-3x}\frac{dy}{dx} - 3e^{-3x}y = 0 \quad \text{is the same as} \quad \frac{d}{dx}[e^{-3x}y] = 0.$$

Integrating both sides of the last equation gives $e^{-3x}y = c$. Solving for y gives us the explicit solution $y = ce^{3x}$, $-\infty < x < \infty$.   ■

EXAMPLE 2   **Solving a Nonhomogeneous Linear DE**

Solve $\dfrac{dy}{dx} - 3y = 6$.

Solution   The associated homogeneous equation for this DE was solved in Example 1. Again the equation is already in the standard form (2), and the integrating factor is still $e^{\int(-3)dx} = e^{-3x}$. This time multiplying the given equation by this factor gives

$$e^{-3x}\frac{dy}{dx} - 3e^{-3x}y = 6e^{-3x}, \quad \text{which is the same as} \quad \frac{d}{dx}[e^{-3x}y] = 6e^{-3x}.$$

Integrating both sides of the last equation gives $e^{-3x}y = -2e^{-3x} + c$ or $y = -2 + ce^{3x}$, $-\infty < x < \infty$.   ■

The final solution in Example 2 is the sum of two solutions: $y = y_c + y_p$, where $y_c = ce^{3x}$ is the solution of the homogeneous equation in Example 1 and $y_p = -2$ is a particular solution of the nonhomogeneous equation $y' - 3y = 6$. You need not be concerned about whether a linear first-order equation is homogeneous or nonhomogeneous; when you follow the solution procedure outlined above, a solution of a nonhomogeneous equation necessarily turns out to be $y = y_c + y_p$. However, the distinction between solving a homogeneous DE and solving a nonhomogeneous DE becomes more important in Chapter 4, where we solve linear higher-order equations.

When a_1, a_0, and g in (1) are constants, the differential equation is autonomous. In Example 2 you can verify from the normal form $dy/dx = 3(y + 2)$ that -2 is a critical point and that it is unstable (a repeller). Thus a solution curve with an initial point either above or below the graph of the equilibrium solution $y = -2$ pushes away from this horizontal line as x increases. Figure 2.22, obtained with the aid of a graphing utility, shows the graph of $y = -2$ along with some additional solution curves.

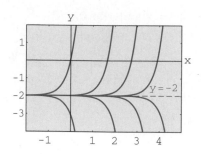

Figure 2.22 Some solutions of $y' - 3y = 6$

Constant of Integration Notice that in the general discussion and in Examples 1 and 2 we disregarded a constant of integration in the evaluation of the indefinite integral in the exponent of $e^{\int P(x)\,dx}$. If you think about the laws of exponents and the fact that the integrating factor multiplies both sides of the differential equation, you should be able to explain why writing $\int P(x)\,dx + c$ is unnecessary. See Problem 44 in Exercises 2.3.

General Solution Suppose again that the functions P and f in (2) are continuous on a common interval I. In the steps leading to (4) we showed that *if* (2) has a solution on I, then it must be of the form given in (4). Conversely, it is a straightforward exercise in differentiation to verify that any function of the form given in (4) is a solution of the differential equation (2) on I. In other words, (4) is a one-parameter family of solutions of equation (2) and *every solution of (2) defined on I is a member of this family.* Therefore we call (4) the **general solution** of the differential equation on the interval I. (See the Remarks at the end of Section 1.1.) Now by writing (2) in the normal form $y' = F(x, y)$, we can identify $F(x, y) = -P(x)y + f(x)$ and $\partial F/\partial y = -P(x)$. From the continuity of P and f on the interval I we see that F and $\partial F/\partial y$ are also continuous on I. With Theorem 1.1 as our justification, we conclude that there exists one and only one solution of the initial-value problem

$$\frac{dy}{dx} + P(x)y = f(x), \quad y(x_0) = y_0 \tag{9}$$

defined on *some* interval I_0 containing x_0. But when x_0 is in I, finding a solution of (9) is just a matter of finding an appropriate value of c in (4)—that is, to each x_0 in I there corresponds a distinct c. In other words, the interval I_0 of existence and uniqueness in Theorem 1.1 for the initial-value problem (9) is the entire interval I.

EXAMPLE 3   **General Solution**

Solve $x\dfrac{dy}{dx} - 4y = x^6 e^x$.

Solution Dividing by x, we get the standard form

$$\frac{dy}{dx} - \frac{4}{x}y = x^5 e^x. \tag{10}$$

From this form we identify $P(x) = -4/x$ and $f(x) = x^5 e^x$ and further observe that P and f are continuous on $(0, \infty)$. Hence the integrating factor is

we can use $\ln x$ instead of $\ln |x|$ since $x > 0$
$\downarrow$

$$e^{-4\int dx/x} = e^{-4\ln x} = e^{\ln x^{-4}} = x^{-4}.$$

Here we have used the basic identity $b^{\log_b N} = N,\ N > 0$. Now we multiply (10) by x^{-4} and rewrite

$$x^{-4}\frac{dy}{dx} - 4x^{-5}y = xe^x \quad \text{as} \quad \frac{d}{dx}[x^{-4}y] = xe^x.$$

It follows from integration by parts that the general solution defined on the interval $(0, \infty)$ is $x^{-4}y = xe^x - e^x + c$ or $y = x^5 e^x - x^4 e^x + cx^4$. ∎

Except in the case when the lead coefficient is 1, the recasting of equation (1) into the standard form (2) requires division by $a_1(x)$. Values of x for which $a_1(x) = 0$ are called **singular points** of the equation. Singular points are potentially troublesome. Specifically, in (2), if $P(x)$ (formed by dividing $a_0(x)$ by $a_1(x)$) is discontinuous at a point, the discontinuity may carry over to solutions of the differential equation.

EXAMPLE 4 **General Solution**

Find the general solution of $(x^2 - 9)\dfrac{dy}{dx} + xy = 0$.

Solution We write the differential equation in standard form

$$\frac{dy}{dx} + \frac{x}{x^2 - 9}y = 0 \tag{11}$$

and identify $P(x) = x/(x^2 - 9)$. Although P is continuous on $(-\infty, -3)$, $(-3, 3)$, and $(3, \infty)$, we shall solve the equation on the first and third intervals. On these intervals the integrating factor is

$$e^{\int x\,dx/(x^2-9)} = e^{\frac{1}{2}\int 2x\,dx/(x^2-9)} = e^{\frac{1}{2}\ln|x^2-9|} = \sqrt{x^2 - 9}.$$

After multiplying the standard form (11) by this factor, we get

$$\frac{d}{dx}\left[\sqrt{x^2 - 9}\,y\right] = 0.$$

Integrating both sides of the last equation gives $\sqrt{x^2 - 9}\,y = c$. Thus for either $x > 3$ or $x < -3$ the general solution of the equation is

$$y = \frac{c}{\sqrt{x^2 - 9}}. \quad ∎$$

Notice in the preceding example that $x = 3$ and $x = -3$ are singular points of the equation and that every function in the general solution $y = c/\sqrt{x^2 - 9}$ is discontinuous at these points. On the other hand, $x = 0$ is a singular point of the differential equation in Example 3, but the general solution $y = x^5 e^x - x^4 e^x + c x^4$ is noteworthy in that every function in this one-parameter family is continuous at $x = 0$ and is defined on the interval $(-\infty, \infty)$ and not just on $(0, \infty)$, as stated in the solution. However, the family $y = x^5 e^x - x^4 e^x + c x^4$ defined on $(-\infty, \infty)$ cannot be considered the general solution of the DE, since the singular point $x = 0$ still causes a problem. See Problem 39 in Exercises 2.3.

EXAMPLE 5 An Initial-Value Problem

Solve $\dfrac{dy}{dx} + y = x$, $\quad y(0) = 4$.

Solution The equation is in standard form, and $P(x) = 1$ and $f(x) = x$ are continuous on $(-\infty, \infty)$. The integrating factor is $e^{\int dx} = e^x$, and so integrating

$$\frac{d}{dx}[e^x y] = x e^x$$

gives $e^x y = x e^x - e^x + c$. Solving this last equation for y yields the general solution $y = x - 1 + c e^{-x}$. But from the initial condition we know that $y = 4$ when $x = 0$. Substituting these values into the general solution implies $c = 5$. Hence the solution of the problem is

$$y = x - 1 + 5 e^{-x}, \quad -\infty < x < \infty. \tag{12}$$ ■

Figure 2.23, obtained with the aid of a graphing utility, shows the graph of (12) in solid color, along with the graphs of other representative solutions in the one-parameter family $y = x - 1 + c e^{-x}$. In this general solution we identify $y_c = c e^{-x}$ and $y_p = x - 1$. It is interesting to observe that as x increases the graphs of *all* members of the family are close to the graph of the particular solution $y_p = x - 1$, which is shown in solid black in Figure 2.23. This is because the contribution of $y_c = c e^{-x}$ to the values of a solution becomes negligible for increasing values of x. We say that $y_c = c e^{-x}$ is a **transient term,** since $y_c \to 0$ as $x \to \infty$. While this behavior is not a characteristic of all general solutions of linear equations (see Example 3), the notion of a transient is often important in applied problems.

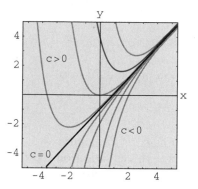

Figure 2.23 Some solutions of $y' + y = x$

EXAMPLE 6 A Discontinuous $f(x)$

Find a continuous solution satisfying

$$\frac{dy}{dx} + y = f(x), \quad \text{where} \quad f(x) = \begin{cases} 1, & 0 \le x \le 1 \\ 0, & x > 1 \end{cases}$$

and the initial condition $y(0) = 0$.

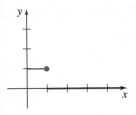

Figure 2.24

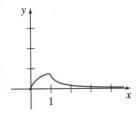

Figure 2.25

Solution From Figure 2.24 we see that f is piecewise continuous with a discontinuity at $x = 1$. Consequently, we solve the problem in two parts corresponding to the two intervals over which f is defined. For $0 \leq x \leq 1$ we have

$$\frac{dy}{dx} + y = 1 \quad \text{or, equivalently,} \quad \frac{d}{dx}[e^x y] = e^x.$$

Integrating this last equation and solving for y gives $y = 1 + c_1 e^{-x}$. Since $y(0) = 0$, we must have $c_1 = -1$ and therefore $y = 1 - e^{-x}, 0 \leq x \leq 1$. Then for $x > 1$, the equation

$$\frac{dy}{dx} + y = 0$$

leads to $y = c_2 e^{-x}$. Hence we can write

$$y = \begin{cases} 1 - e^{-x}, & 0 \leq x \leq 1 \\ c_2 e^{-x}, & x > 1. \end{cases}$$

Now in order for y to be a continuous function we certainly want $\lim_{x \to 1^+} y(x) = y(1)$. This latter requirement is equivalent to $c_2 e^{-1} = 1 - e^{-1}$ or $c_2 = e - 1$. As Figure 2.25 shows, the function

$$y = \begin{cases} 1 - e^{-x}, & 0 \leq x \leq 1 \\ (e - 1)e^{-x}, & x > 1 \end{cases} \tag{13}$$

is continuous on $[0, \infty)$. See Problem 42 in Exercises 2.3. ■

Functions Defined by Integrals Some simple functions do not possess antiderivatives that are elementary functions, and integrals of these kinds of functions are called **nonelementary.** For example, you may have seen in calculus that $\int e^{x^2} dx$ and $\int \sin x^2 dx$ are nonelementary integrals. In applied mathematics some important functions are *defined* in terms of nonelementary integrals. Two such functions are the **error function** and the **complementary error function:**

$$\text{erf}(x) = \frac{2}{\sqrt{\pi}} \int_0^x e^{-t^2} dt \quad \text{and} \quad \text{erfc}(x) = \frac{2}{\sqrt{\pi}} \int_x^\infty e^{-t^2} dt. \tag{14}$$

Since $(2/\sqrt{\pi}) \int_0^\infty e^{-t^2} dt = 1$, it is seen from (14) that the error function $\text{erf}(x)$ and the complementary error function $\text{erfc}(x)$ are related by $\text{erf}(x) + \text{erfc}(x) = 1$. Because of its importance in areas such as probability and statistics, the error function has been extensively tabulated. Note that $\text{erf}(0) = 0$ is one obvious functional value. Values of $\text{erf}(x)$ can also be found using a CAS. Before working through the next example, you are urged to reread (*i*) of the Remarks at the end of Section 2.2.

EXAMPLE 7 **The Error Function**

Solve the initial-value problem $\dfrac{dy}{dx} - 2xy = 2, \; y(0) = 1.$

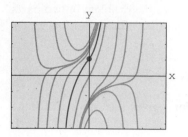

Figure 2.26 Some solutions of $y' - 2xy = 2$

Solution   Since the equation is already in standard form, we see that the integrating factor is e^{-x^2}, and so from

$$\frac{d}{dx}[e^{-x^2}] = 2e^{-x^2} \quad \text{we get} \quad y = 2e^{x^2} \int_0^x e^{-t^2}\, dt + ce^{x^2}. \qquad \textbf{(15)}$$

Applying $y(0) = 1$ to the last expression then gives $c = 1$. Hence the solution to the problem is

$$y = 2e^{x^2} \int_0^x e^{-t^2}\, dt + e^{x^2} = e^{x^2}[1 + \sqrt{\pi}\,\text{erf}(x)].$$

The graph of this solution on $(-\infty, \infty)$, shown in color in Figure 2.26 among other members of the family defined by (15), was obtained with the aid of a computer algebra system. ∎

Use of Computers   Some computer algebra systems are capable of producing explicit solutions for some kinds of differential equations. For example, solving the equation $y' + 2y = x$ with the input commands

$$\textbf{DSolve[y' [x] + 2 y[x] == x, y[x], x]} \qquad \text{(in \textit{Mathematica})}$$

and   **dsolve(diff(y(x), x) + 2*y = x, y(x));**   (in *Maple*)

yields, in turn, the outputs

$$\textbf{y[x] } - > - \left(\frac{1}{4}\right) + \frac{\text{x}}{2} + \frac{\text{C[1]}}{\text{E}^{2\,\text{x}}}$$

and   **y(x) = 1/2 x − 1/4 + exp(−2 x)_C1**

Translating to standard symbols, we have $y = -\frac{1}{4} + \frac{1}{2}x + ce^{-2x}$.

Remarks   (*i*) Occasionally a first-order differential equation is not linear in one variable but is linear in the other variable. For example, the differential equation

$$\frac{dy}{dx} = \frac{1}{x + y^2}$$

is not linear in the variable y. But its reciprocal

$$\frac{dx}{dy} = x + y^2 \quad \text{or} \quad \frac{dx}{dy} - x = y^2$$

is recognized as linear in the variable x. You should verify that the integrating factor $e^{\int(-1)dy} = e^{-y}$ and integration by parts yields an implicit solution of the first equation: $x = -y^2 - 2y - 2 + ce^y$.

(*ii*) Mathematicians have "adopted" as their own certain words from engineering, which they found appropriately descriptive. The word *transient*, used earlier, is one of these terms. In future discussions the words *input* and *output* will occasionally pop up. The function f in (2) is called the **input** or **driving function;** a solution of the differential equation for a given input is called the **output** or **response.**

EXERCISES 2.3 ——————————————————————————

Answers to odd-numbered problems begin on page AN-1.

In Problems 1–24 find the general solution of the given differential equation. Give the largest interval over which the general solution is defined. Determine whether there are any transient terms in the general solution.

1. $\dfrac{dy}{dx} = 5y$

2. $\dfrac{dy}{dx} + 2y = 0$

3. $\dfrac{dy}{dx} + y = e^{3x}$

4. $3\dfrac{dy}{dx} + 12y = 4$

5. $y' + 3x^2 y = x^2$

6. $y' + 2xy = x^3$

7. $x^2 y' + xy = 1$

8. $y' = 2y + x^2 + 5$

9. $x\dfrac{dy}{dx} - y = x^2 \sin x$

10. $x\dfrac{dy}{dx} + 2y = 3$

11. $x\dfrac{dy}{dx} + 4y = x^3 - x$

12. $(1 + x)\dfrac{dy}{dx} - xy = x + x^2$

13. $x^2 y' + x(x + 2)y = e^x$

14. $xy' + (1 + x)y = e^{-x} \sin 2x$

15. $y\, dx - 4(x + y^6)\, dy = 0$

16. $y\, dx = (ye^y - 2x)\, dy$

17. $\cos x\, \dfrac{dy}{dx} + (\sin x)y = 1$

18. $\cos^2 x \sin x\, dy + (y \cos^3 x - 1)\, dx = 0$

19. $(x + 1)\dfrac{dy}{dx} + (x + 2)y = 2xe^{-x}$

20. $(x + 2)^2 \dfrac{dy}{dx} = 5 - 8y - 4xy$

21. $\dfrac{dr}{d\theta} + r \sec \theta = \cos \theta$

22. $\dfrac{dP}{dt} + 2tP = P + 4t - 2$

23. $x\dfrac{dy}{dx} + (3x + 1)y = e^{-3x}$

24. $(x^2 - 1)\dfrac{dy}{dx} + 2y = (x + 1)^2$

In Problems 25–30 solve the given initial-value problem. Give the largest interval over which the solution is defined.

25. $xy' + y = e^x, \quad y(1) = 2$

26. $y\dfrac{dx}{dy} - x = 2y^2, \quad y(1) = 5$

27. $L\dfrac{di}{dt} + Ri = E, \quad i(0) = i_0;$
$L, R, E,$ and i_0 constants

28. $\dfrac{dT}{dt} = k(T - T_m), \quad T(0) = T_0;$
$k, T_m,$ and T_0 constants

29. $(x + 1)\dfrac{dy}{dx} + y = \ln x,$
$y(1) = 10$

30. $y' + (\tan x)y = \cos^2 x,$
$y(0) = -1$

In Problems 31–34 find a continuous solution satisfying the given differential equation and the indicated initial condition. Use a graphing utility to plot the solution curve for the IVP.

31. $\dfrac{dy}{dx} + 2y = f(x), \quad f(x) = \begin{cases} 1, & 0 \le x \le 3 \\ 0, & x > 3 \end{cases}, \quad y(0) = 0$

32. $\dfrac{dy}{dx} + y = f(x)$, $f(x) = \begin{cases} 1, & 0 \le x \le 1 \\ -1, & x > 1 \end{cases}$, $y(0) = 1$

33. $\dfrac{dy}{dx} + 2xy = f(x)$, $f(x) = \begin{cases} x, & 0 \le x < 1 \\ 0, & x \ge 1 \end{cases}$, $y(0) = 2$

34. $(1 + x^2)\dfrac{dy}{dx} + 2xy = f(x)$, $f(x) = \begin{cases} x, & 0 \le x < 1 \\ -x, & x \ge 1 \end{cases}$, $y(0) = 0$

35. Find a continuous solution of $y' + P(x)y = 4x$, $y(0) = 3$, where

$$P(x) = \begin{cases} 2, & 0 \le x \le 1 \\ -2/x, & x > 1. \end{cases}$$

Use a graphing utility to plot the solution curve for the IVP.

36. Express the solution of the initial-value problem $y' - 2xy = 1$, $y(1) = 1$ in terms of erf(x).

37. Consider the initial-value problem $y' + e^x y = f(x)$, $y(0) = 1$. Express the solution of the IVP for $x \ge 0$ as a nonelementary integral when $f(x) = 1$. What is the solution when $f(x) = 0$? When $f(x) = e^x$?

Discussion Problems

38. Reread the second paragraph following Example 2. Construct a linear first-order differential equation for which all nonconstant solutions approach the horizontal asymptote $y = 4$ as $x \to \infty$.

39. Reread Example 3 and then discuss, with reference to Theorem 1.1, the existence and uniqueness of a solution of the initial-value problem consisting of $xy' - 4y = x^6 e^x$ and the given initial condition.
   (a) $y(0) = 0$ **(b)** $y(0) = y_0$, $y_0 \ne 0$ **(c)** $y(x_0) = y_0$, $x_0 \ne 0$, $y_0 \ne 0$

40. Find the general solution of the differential equation in Example 4 on the interval $(-3, 3)$.

41. Reread the discussion following Example 5. Construct a linear first-order differential equation for which all solutions are asymptotic to the line $y = 3x - 5$ as $x \to \infty$.

42. Reread Example 6 and then discuss why it is technically incorrect to say that the function in (13) is a solution of the IVP on the interval $[0, \infty)$.

43. **(a)** Construct a linear first-order differential equation of the form $xy' + a_0(x)y = g(x)$ for which $y_c = c/x^3$ and $y_p = x^3$. Give an interval on which $y = x^3 + c/x^3$ is the general solution of the DE.
   (b) Give an initial condition $y(x_0) = y_0$ for the DE found in part (a) so that the solution of the IVP is $y = x^3 - 1/x^3$. Repeat for the solution $y = x^3 + 2/x^3$. Give an interval I of definition of each of these solutions. Graph the solution curves. Is there an initial-value problem whose solution is defined on $-\infty < x < \infty$?
   (c) Is each IVP found in part (b) unique? That is, can there be more than one IVP for which, say, $y = x^3 - 1/x^3$, x in some interval I, is the solution?

44. In determining the integrating factor (5) we did not use a constant of integration in the evaluation of $\int P(x)\, dx$. Explain why using $\int P(x)\, dx + c$ has no effect on the solution of (2).

45. Suppose $P(x)$ is continuous on some interval I and a is a number in I. What can be said about the solution of the initial-value problem $y' + P(x)y = 0$, $y(a) = 0$?

46. The following system of differential equations is encountered in the study of a special type of radioactive series of elements:

$$\frac{dx}{dt} = -\lambda_1 x$$

$$\frac{dy}{dt} = \lambda_1 x - \lambda_2 y,$$

where λ_1 and λ_2 are constants. Discuss how to solve the system subject to $x(0) = x_0$, $y(0) = y_0$.

47. In the two parts of this problem assume α and β are constants; $P(x)$, $f(x)$, $f_1(x)$, and $f_2(x)$ are continuous on an interval I; and x_0 is any point in I.
   (a) Suppose that y_1 is a solution of the initial-value problem $y' + P(x)y = 0$, $y(x_0) = \alpha$ and that y_2 is a solution of $y' + P(x)y = f(x)$, $y(x_0) = 0$. Find a solution of $y' + P(x)y = f(x)$, $y(x_0) = \alpha$. Prove your assertion.
   (b) Suppose that y_1 is a solution of $y' + P(x)y = f_1(x)$, $y(x_0) = \alpha$ and that y_2 is a solution of $y' + P(x)y = f_2(x)$, $y(x_0) = \beta$. If y is a solution of $y' + P(x)y = f_1(x) + f_2(x)$, what is the value $y(x_0)$? Prove your assertion. If y is a solution of $y' + P(x)y = c_1 f_1(x) + c_2 f_2(x)$, where c_1 and c_2 are arbitrarily specified constants, what is the value $y(x_0)$? Prove your assertion.

Computer Lab Assignments

48. (a) Express the solution of the initial-value problem $y' - 2xy = -1$, $y(0) = \sqrt{\pi}/2$ in terms of erfc(x).
   (b) Use tables or a computer to calculate $y(2)$. Use a CAS to obtain the solution curve for the IVP on $(-\infty, \infty)$.

49. (a) The **sine integral function** is defined by $\text{Si}(x) = \int_0^x \frac{\sin t}{t}\, dt$, where the integrand is defined to be 1 at $t = 0$. Express the solution $y(x)$ of the initial-value problem $x^3 y' + 2x^2 y = 10 \sin x$, $y(1) = 0$ in terms of Si(x).
   (b) Use a CAS to plot the solution curve for the IVP for $x > 0$.
   (c) Use a CAS to find the value of the absolute maximum of the solution $y(x)$ for $x > 0$.

50. (a) The **Fresnel sine integral** is defined by $S(x) = \int_0^x \sin\left(\frac{\pi}{2} t^2\right) dt$.
   Express the solution $y(x)$ of the initial-value problem $y' - (\sin x^2)y = 0$, $y(0) = 5$ in terms of $S(x)$.
   (b) Use a CAS to plot the solution curve for the IVP on $(-\infty, \infty)$.
   (c) It is known that $S(x) \to \frac{1}{2}$ as $x \to \infty$ and $S(x) \to -\frac{1}{2}$ as $x \to -\infty$. What does the solution $y(x)$ approach as $x \to \infty$? As $x \to -\infty$?
   (d) Use a CAS to find the values of the absolute maximum and the absolute minimum of the solution $y(x)$.

2.4 EXACT EQUATIONS

• Differential • Exact differential • Definition of an exact
DE • Criterion for an exact differential • Method of solution

Although the simple equation $y\,dx + x\,dy = 0$ is separable, we can solve the equation in an alternative manner by recognizing that the left-hand side is equivalent to the differential of the product of x and y; that is,

$$y\,dx + x\,dy = d(xy) = 0.$$

By integrating both sides of $d(xy) = 0$ we immediately obtain the implicit solution $xy = c$. The DE $y\,dx + x\,dy = 0$ is an example of an **exact** first-order equation, which we study next.

Differential of a Function of Two Variables   If $z = f(x, y)$ is a function of two variables with continuous first partial derivatives in a region R of the xy-plane, then its differential (also called the total differential) is

$$dz = \frac{\partial f}{\partial x}\,dx + \frac{\partial f}{\partial y}\,dy. \tag{1}$$

Now if $f(x, y) = c$, it follows from (1) that

$$\frac{\partial f}{\partial x}\,dx + \frac{\partial f}{\partial y}\,dy = 0. \tag{2}$$

In other words, given a one-parameter family of curves $f(x, y) = c$, we can generate a first-order differential equation by computing the differential. For example, if $x^2 - 5xy + y^3 = c$, then (2) gives

$$(2x - 5y)\,dx + (-5x + 3y^2)\,dy = 0. \tag{3}$$

For our purposes it is more important to turn the problem around; namely, given a first-order DE such as (3), can we recognize that it is equivalent to the differential $d(x^2 - 5xy + y^3) = 0$?

DEFINITION 2.3   Exact Equation

A differential expression $M(x, y)\,dx + N(x, y)\,dy$ is an **exact differential** in a region R of the xy-plane if it corresponds to the differential of some function $f(x, y)$. A first-order differential equation of the form

$$M(x, y)\,dx + N(x, y)\,dy = 0$$

is said to be an **exact equation** if the expression on the left-hand side is an exact differential.

For example, $x^2y^3\,dx + x^3y^2\,dy = 0$ is an exact equation, because the left-hand side of the equation is an exact differential:

$$d(\tfrac{1}{3}x^3y^3) = x^2y^3\,dx + x^3y^2\,dy.$$

Notice that if $M(x, y) = x^2y^3$ and $N(x, y) = x^3y^2$, then $\partial M/\partial y = 3x^2y^3 = \partial N/\partial x$. Theorem 2.1 shows that the equality of these partial derivatives is no coincidence.

THEOREM 2.1   **Criterion for an Exact Differential**

Let $M(x, y)$ and $N(x, y)$ be continuous and have continuous first partial derivatives in rectangular region R defined by $a < x < b, c < y < d$. Then a necessary and sufficient condition that $M(x, y)\, dx + N(x, y)\, dy$ be an exact differential is

$$\frac{\partial M}{\partial y} = \frac{\partial N}{\partial x}. \tag{4}$$

Proof of the Necessity   For simplicity let us assume that $M(x, y)$ and $N(x, y)$ have continuous first partial derivatives for all (x, y). Now if the expression $M(x, y)\, dx + N(x, y)\, dy$ is exact, there exists some function f such that for all x in R,

$$M(x, y)\, dx + N(x, y)\, dy = \frac{\partial f}{\partial x}\, dx + \frac{\partial f}{\partial y}\, dy.$$

Therefore $$M(x, y) = \frac{\partial f}{\partial x}, \qquad N(x, y) = \frac{\partial f}{\partial y},$$

and $$\frac{\partial M}{\partial y} = \frac{\partial}{\partial y}\left(\frac{\partial f}{\partial x}\right) = \frac{\partial^2 f}{\partial y\, \partial x} = \frac{\partial}{\partial x}\left(\frac{\partial f}{\partial y}\right) = \frac{\partial N}{\partial x}.$$

The equality of the mixed partials is a consequence of the continuity of the first partial derivatives of $M(x, y)$ and $N(x, y)$. ∎

The sufficiency part of Theorem 2.1 consists of showing that there exists a function f for which $\partial f/\partial x = M(x, y)$ and $\partial f/\partial y = N(x, y)$ whenever (4) holds. The construction of the function f actually reflects a basic procedure for solving exact equations.

Method of Solution   Given an equation in the differential form $M(x, y)\, dx + N(x, y)\, dy = 0$, determine whether the equality in (4) holds. If it does, then there exists a function f for which

$$\frac{\partial f}{\partial x} = M(x, y).$$

We can find f by integrating $M(x, y)$ with respect to x, while holding y constant:

$$f(x, y) = \int M(x, y)\, dx + g(y), \tag{5}$$

where the arbitrary function $g(y)$ is the "constant" of integration. Now differentiate (5) with respect to y and assume $\partial f/\partial y = N(x, y)$:

$$\frac{\partial f}{\partial y} = \frac{\partial}{\partial y}\int M(x, y)\, dx + g'(y) = N(x, y).$$

This gives
$$g'(y) = N(x, y) - \frac{\partial}{\partial y} \int M(x, y)\, dx. \tag{6}$$

Finally, integrate (6) with respect to y and substitute the result in (5). The implicit solution of the equation is $f(x, y) = c$.

Some observations are in order. First, it is important to realize that the expression $N(x, y) - (\partial/\partial y) \int M(x, y)\, dx$ in (6) is independent of x, because

$$\frac{\partial}{\partial x}\left[N(x, y) - \frac{\partial}{\partial y} \int M(x, y)\, dx \right] = \frac{\partial N}{\partial x} - \frac{\partial}{\partial y}\left(\frac{\partial}{\partial x} \int M(x, y)\, dx \right)$$

$$= \frac{\partial N}{\partial x} - \frac{\partial M}{\partial y} = 0.$$

Second, we could just as well start the foregoing procedure with the assumption that $\partial f/\partial y = N(x, y)$. After integrating N with respect to y and then differentiating that result, we would find the analogues of (5) and (6) to be, respectively,

$$f(x, y) = \int N(x, y)\, dy + h(x) \quad \text{and} \quad h'(x) = M(x, y) - \frac{\partial}{\partial x} \int N(x, y)\, dy.$$

In either case *none of these formulas should be memorized.*

EXAMPLE 1   **Solving an Exact DE**

Solve $2xy\, dx + (x^2 - 1)\, dy = 0$.

Solution   With $M(x, y) = 2xy$ and $N(x, y) = x^2 - 1$ we have

$$\frac{\partial M}{\partial y} = 2x = \frac{\partial N}{\partial x}.$$

Thus the equation is exact, and so, by Theorem 2.1, there exists a function $f(x, y)$ such that

$$\frac{\partial f}{\partial x} = 2xy \quad \text{and} \quad \frac{\partial f}{\partial y} = x^2 - 1.$$

From the first of these equations we obtain, after integrating,

$$f(x, y) = x^2 y + g(y).$$

Taking the partial derivative of the last expression with respect to y and setting the result equal to $N(x, y)$ gives

$$\frac{\partial f}{\partial y} = x^2 + g'(y) = x^2 - 1. \qquad \leftarrow N(x, y)$$

It follows that $g'(y) = -1$ and $g(y) = -y$. Hence $f(x, y) = x^2 y - y$, and so the solution of the equation in implicit form is $x^2 y - y = c$. The explicit form of the solution is easily seen to be $y = c/(1 - x^2)$ and is defined on any interval not containing either $x = 1$ or $x = -1$. ∎

Note   The solution of the DE in Example 1 is *not* $f(x, y) = x^2 - y$. Rather, it is $f(x, y) = c$; if a constant is used in the integration of $g'(y)$, we can then write the solution as $f(x, y) = 0$. Note, too, that the equation could be solved by separation of variables.

| EXAMPLE 2 | **Solving an Exact DE** |

Solve $(e^{2y} - y \cos xy) \, dx + (2xe^{2y} - x \cos xy + 2y) \, dy = 0$.

Solution   The equation is exact because

$$\frac{\partial M}{\partial y} = 2e^{2y} + xy \sin xy - \cos xy = \frac{\partial N}{\partial x}.$$

Hence a function $f(x, y)$ exists for which

$$M(x, y) = \frac{\partial f}{\partial x} \quad \text{and} \quad N(x, y) = \frac{\partial f}{\partial y}.$$

Now for variety we shall start with the assumption that $\partial f / \partial y = N(x, y)$; that is,

$$\frac{\partial f}{\partial y} = 2xe^{2y} - x \cos xy + 2y$$

$$f(x, y) = 2x \int e^{2y} \, dy - x \int \cos xy \, dy + 2 \int y \, dy.$$

Remember, the reason x can come out in front of the symbol $\int$ is that in the integration with respect to y, x is treated as an ordinary constant. It follows that

$$f(x, y) = xe^{2y} - \sin xy + y^2 + h(x)$$

$$\frac{\partial f}{\partial x} = e^{2y} - y \cos xy + h'(x) = e^{2y} - y \cos xy, \qquad \leftarrow M(x, y)$$

and so $h'(x) = 0$ or $h(x) = c$. Hence a family of solutions is

$$xe^{2y} - \sin xy + y^2 + c = 0. \qquad \blacksquare$$

| EXAMPLE 3 | **An Initial-Value Problem** |

Solve $\dfrac{dy}{dx} = \dfrac{xy^2 - \cos x \sin x}{y(1 - x^2)}, \, y(0) = 2$.

Solution   By writing the differential equation in the form

$$(\cos x \sin x - xy^2) \, dx + y(1 - x^2) \, dy = 0$$

we recognize that the equation is exact because

$$\frac{\partial M}{\partial y} = -2xy = \frac{\partial N}{\partial x}.$$

Now $$\frac{\partial f}{\partial y} = y(1 - x^2)$$

$$f(x, y) = \frac{y^2}{2}(1 - x^2) + h(x)$$

$$\frac{\partial f}{\partial x} = -xy^2 + h'(x) = \cos x \sin x - xy^2.$$

The last equation implies that $h'(x) = \cos x \sin x$. Integrating gives

$$h(x) = -\int (\cos x)(-\sin x \, dx) = -\frac{1}{2}\cos^2 x.$$

Thus $\qquad \dfrac{y^2}{2}(1 - x^2) - \dfrac{1}{2}\cos^2 x = c_1 \quad$ or $\quad y^2(1 - x^2) - \cos^2 x = c, \qquad$ **(7)**

where $2c_1$ has been replaced by c. The initial condition $y = 2$ when $x = 0$ demands that $4(1) - \cos^2(0) = c$, and so $c = 3$. An implicit solution of the problem is then $y^2(1 - x^2) - \cos^2 x = 3$.

The solution curve of the IVP is the curve drawn in color in Figure 2.27; it is part of an interesting family of curves. The graphs of the members of the one-parameter family of solutions given in (7) can be obtained in several ways, two of which are using software to graph level curves (as discussed in the last section) and using a graphing utility to carefully graph the explicit functions obtained for various values of c by solving $y^2 = (c + \cos^2 x)/(1 - x^2)$ for y. ∎

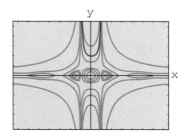

Figure 2.27 Some graphs of members of the family $y^2(1 - x^2) - \cos^2 x = c$

Integrating Factors Recall from the last section that the left-hand side of the linear equation $y' + P(x)y = f(x)$ can be transformed into a derivative when we multiply the equation by an integrating factor. The same basic idea sometimes works for a nonexact differential equation $M(x, y) \, dx + N(x, y) \, dy = 0$. That is, it is sometimes possible to find an **integrating factor** $\mu(x, y)$ so that, after multiplying, the left-hand side of

$$\mu(x, y)M(x, y) \, dx + \mu(x, y)N(x, y) \, dy = 0 \qquad \textbf{(8)}$$

is an exact differential. In an attempt to find μ, we turn to the criterion (4) for exactness. Equation (8) is exact if and only if $(\mu M)_y = (\mu N)_x$, where the subscripts denote partial derivatives. By the Product Rule of differentiation the last equation is the same as $\mu M_y + \mu_y M = \mu N_x + \mu_x N$ or

$$\mu_x N - \mu_y M = (M_y - N_x)\mu. \qquad \textbf{(9)}$$

Although M, N, M_y, and N_x are known functions of x and y, the difficulty here in determining the unknown $\mu(x, y)$ from (9) is that we must solve a partial differential equation. Since we are not prepared to do that, we make a simplifying assumption. Suppose μ is a function of one variable; for example, say that μ depends only on x. In this case, $\mu_x = du/dx$ and (9) can be written as

$$\frac{d\mu}{dx} = \frac{M_y - N_x}{N}\mu. \qquad \textbf{(10)}$$

We are still at an impasse if the quotient $(M_y - N_x)/N$ depends on both x and y. However, if after all obvious algebraic simplifications are made the quotient $(M_y - N_x)/N$ turns out to depend solely on the variable x, then (10) is a first-order ordinary differential equation. We can finally determine μ because (10) is *separable* as well as *linear*. It follows from either Section 2.2 or Section 2.3 that $\mu(x) = e^{\int ((M_y - N_x)/N)dx}$. In like manner, it follows from (9) that if μ depends only on the variable y, then

$$\frac{d\mu}{dy} = \frac{N_x - M_y}{M}\mu. \tag{11}$$

In this case, if $(N_x - M_y)/M$ is a function of y only, then we can solve (11) for μ.

We summarize the results for the differential equation

$$M(x, y)\,dx + N(x, y)\,dy = 0. \tag{12}$$

- If $(M_y - N_x)/N$ is a function of x alone, then an integrating factor for equation (12) is

$$\mu(x) = e^{\int \frac{M_y - N_x}{N} dx}. \tag{13}$$

- If $(N_x - M_y)/M$ is a function of y alone, then an integrating factor for equation (12) is

$$\mu(y) = e^{\int \frac{N_x - M_y}{M} dy}. \tag{14}$$

EXAMPLE 4   **A Nonexact DE Made Exact**

The nonlinear first-order differential equation

$$xy\,dx + (2x^2 + 3y^2 - 20)\,dy = 0$$

is not exact. With the identifications $M = xy$, $N = 2x^2 + 3y^2 - 20$, we find the partial derivatives $M_y = x$ and $N_x = 4x$. The first quotient from (13) gets us nowhere, since

$$\frac{M_y - N_x}{N} = \frac{x - 4x}{2x^2 + 3y^2 - 20} = \frac{-3x}{2x^2 + 3y^2 - 20}$$

depends on x and y. However, (14) yields a quotient that depends only on y:

$$\frac{N_x - M_y}{M} = \frac{4x - x}{xy} = \frac{3x}{xy} = \frac{3}{y}.$$

The integrating factor is then $e^{\int 3\,dy/y} = e^{3\ln y} = e^{\ln y^3} = y^3$. After we multiply the given DE by $\mu(y) = y^3$, the resulting equation is

$$xy^4\,dx + (2x^2y^3 + 3y^5 - 20y^3)\,dy = 0.$$

You should verify that the last equation is now exact as well as show, using the method of this section, that a family of solutions is $\frac{1}{2}x^2y^4 + \frac{3}{7}y^3 - 5y^4 = c$. ∎

Remarks (*i*) When testing an equation for exactness, make sure it is of the precise form $M(x, y) \, dx + N(x, y) \, dy = 0$. Sometimes a differential equation is written $G(x, y) \, dx = H(x, y) \, dy$. In this case, first rewrite it as $G(x, y) \, dx - H(x, y) \, dy = 0$ and then identify $M(x, y) = G(x, y)$ and $N(x, y) = -H(x, y)$ before using (4).
(*ii*) In some texts on differential equations the study of exact equations precedes that of linear DEs. Then the method for finding integrating factors just discussed can be used to derive an integrating factor for $y' + P(x)y = f(x)$. By rewriting the last equation in the differential form $(P(x)y - f(x)) \, dx + dy = 0$ we see that

$$\frac{M_y - N_x}{N} = P(x).$$

From (13) we arrive at the already familiar integrating factor $e^{\int P(x) \, dx}$, used in Section 2.3.

EXERCISES 2.4

Answers to odd-numbered problems begin on page AN-2.

In Problems 1–20 determine whether the given differential equation is exact. If it is exact, solve it.

1. $(2x - 1) \, dx + (3y + 7) \, dy = 0$

2. $(2x + y) \, dx - (x + 6y) \, dy = 0$

3. $(5x + 4y) \, dx + (4x - 8y^3) \, dy = 0$

4. $(\sin y - y \sin x) \, dx + (\cos x + x \cos y - y) \, dy = 0$

5. $(2xy^2 - 3) \, dx + (2x^2y + 4) \, dy = 0$

6. $\left(2y - \dfrac{1}{x} + \cos 3x\right)\dfrac{dy}{dx} + \dfrac{y}{x^2} - 4x^3 + 3y \sin 3x = 0$

7. $(x^2 - y^2) \, dx + (x^2 - 2xy) \, dy = 0$

8. $\left(1 + \ln x + \dfrac{y}{x}\right) dx = (1 - \ln x) \, dy$

9. $(x - y^3 + y^2 \sin x) \, dx = (3xy^2 + 2y \cos x) \, dy$

10. $(x^3 + y^3) \, dx + 3xy^2 \, dy = 0$

11. $(y \ln y - e^{-xy}) \, dx + \left(\dfrac{1}{y} + x \ln y\right) dy = 0$

12. $(3x^2y + e^y) \, dx + (x^3 + xe^y - 2y) \, dy = 0$

13. $x\dfrac{dy}{dx} = 2xe^x - y + 6x^2$

14. $\left(1 - \dfrac{3}{y} + x\right)\dfrac{dy}{dx} + y = \dfrac{3}{x} - 1$

15. $\left(x^2y^3 - \dfrac{1}{1 + 9x^2}\right)\dfrac{dx}{dy} + x^3y^2 = 0$

16. $(5y - 2x)y' - 2y = 0$

17. $(\tan x - \sin x \sin y) \, dx + \cos x \cos y \, dy = 0$

18. $(2y \sin x \cos x - y + 2y^2 e^{xy^2}) \, dx = (x - \sin^2 x - 4xy e^{xy^2}) \, dy$

19. $(4t^3y - 15t^2 - y) \, dt + (t^4 + 3y^2 - t) \, dy = 0$

20. $\left(\dfrac{1}{t} + \dfrac{1}{t^2} - \dfrac{y}{t^2 + y^2} \right) dt + \left(ye^y + \dfrac{t}{t^2 + y^2} \right) dy = 0$

In Problems 21–26 solve the given initial-value problem.

21. $(x + y)^2 \, dx + (2xy + x^2 - 1) \, dy = 0, \quad y(1) = 1$

22. $(e^x + y) \, dx + (2 + x + ye^y) \, dy = 0, \quad y(0) = 1$

23. $(4y + 2t - 5) \, dt + (6y + 4t - 1) \, dy = 0, \quad y(-1) = 2$

24. $\left(\dfrac{3y^2 - t^2}{y^5} \right) \dfrac{dy}{dt} + \dfrac{t}{2y^4} = 0, \quad y(1) = 1$

25. $(y^2 \cos x - 3x^2y - 2x) \, dx + (2y \sin x - x^3 + \ln y) \, dy = 0, \quad y(0) = e$

26. $\left(\dfrac{1}{1 + y^2} + \cos x - 2xy \right) \dfrac{dy}{dx} = y(y + \sin x), \quad y(0) = 1$

In Problems 27 and 28 find the value of k so that the given differential equation is exact.

27. $(y^3 + kxy^4 - 2x) \, dx + (3xy^2 + 20x^2y^3) \, dy = 0$

28. $(6xy^3 + \cos y) \, dx + (2kx^2y^2 - x \sin y) \, dy = 0$

In Problems 29 and 30 verify that the given differential equation is not exact. Multiply the given differential equation by the indicated integrating factor $\mu(x, y)$ and verify that the new equation is exact. Solve.

29. $(-xy \sin x + 2y \cos x) \, dx + 2x \cos x \, dy = 0; \quad \mu(x, y) = xy$

30. $(x^2 + 2xy - y^2) \, dx + (y^2 + 2xy - x^2) \, dy = 0; \quad \mu(x, y) = (x + y)^{-2}$

In Problems 31–36 solve the given differential equation by finding, as in Example 4, an appropriate integrating factor.

31. $(2y^2 + 3x) \, dx + 2xy \, dy = 0$

32. $y(x + y + 1) \, dx + (x + 2y) \, dy = 0$

33. $6xy \, dx + (4y + 9x^2) \, dy = 0$

34. $\cos x \, dx + \left(1 + \dfrac{2}{y} \right) \sin x \, dy = 0$

35. $(10 - 6y + e^{-3x}) \, dx - 2 \, dy = 0$

36. $(y^2 + xy^3) \, dx + (5y^2 - xy + y^3 \sin y) \, dy = 0$

In Problems 37 and 38 solve the given initial-value problem by finding, as in Example 4, an appropriate integrating factor.

37. $x \, dx + (x^2y + 4y) \, dy = 0, \quad y(4) = 0$

38. $(x^2 + y^2 - 5) \, dx = (y + xy) \, dy, \quad y(0) = 1$

39. (a) Show that a one-parameter family of solutions of the equation

$$(4xy + 3x^2) \, dx + (2y + 2x^2) \, dy = 0$$

is $x^3 + 2x^2y + y^2 = c$.

(b) Show that the initial conditions $y(0) = -2$ and $y(1) = 1$ determine the same implicit solution.

(c) Find explicit solutions $y_1(x)$ and $y_2(x)$ of the differential equation in part (a) such that $y_1(0) = -2$ and $y_2(1) = 1$. Use a graphing utility to plot the graphs of $y_1(x)$ and $y_2(x)$.

Discussion Problems

40. Consider the concept of an integrating factor, introduced in Problems 29–38. Are the two equations $M \, dx + N \, dy = 0$ and $\mu M \, dx + \mu N \, dy = 0$ necessarily equivalent in the sense that a solution of one is also a solution of the other? Discuss.

41. Reread Example 3 and then discuss why we can conclude that the interval of definition of the explicit solution of the IVP (the colored curve in Figure 2.27) is $(-1, 1)$.

42. Discuss how the functions $M(x, y)$ and $N(x, y)$ can be found so that each differential equation is exact. Carry out your ideas.

(a) $M(x, y) \, dx + \left(xe^{xy} + 2xy + \dfrac{1}{x} \right) dy = 0$

(b) $\left(x^{-1/2}y^{1/2} + \dfrac{x}{x^2 + y} \right) dx + N(x, y) \, dy = 0$

43. Differential equations are sometimes solved by having a clever idea. Here is a little exercise in cleverness: Although the differential equation $\left(x - \sqrt{x^2 + y^2} \right) dx + y \, dy = 0$ is not exact, show how the rearrangement $(x \, dx + y \, dy)/\sqrt{x^2 + y^2} = dx$ and the observation $\frac{1}{2}d(x^2 + y^2) = x \, dx + y \, dy$ can lead to a solution.

44. True or False: Every separable first-order equation $dy/dx = g(x)h(y)$ is exact.

2.5 SOLUTIONS BY SUBSTITUTIONS

• *Substitution in a DE*   • *Homogeneous function*   • *Homogeneous DEs*   • *Bernoulli's equation*

We solve a differential equation by first recognizing it as a certain kind of equation (say, separable) and then carrying out a procedure, consisting of equation-specific mathematical steps, that yields a sufficiently differentiable function that satisfies the equation. Often this solution procedure begins with transforming a given differential equation into another differential equation by means of a **substitution.** For example, suppose we wish to transform the first-order equation $dy/dx = f(x, y)$ by the substitution $y = g(x, u)$, where u is regarded as a function of the variable x. If g possesses first-partial derivatives, then the Chain Rule gives

$$\frac{dy}{dx} = g_x(x, u) + g_u(x, u)\frac{du}{dx}.$$

If we replace dy/dx by the foregoing derivative and y in $f(x, y)$ by $g(x, u)$, the DE $dy/dx = f(x, y)$ becomes $g_x(x, u) + g_u(x, u)\dfrac{du}{dx} = f(x, g(x, u))$, which, solved for du/dx, has the form $\dfrac{du}{dx} = F(x, u)$. If we can determine a solution $u = \phi(x)$ of this second equation, then a solution of the original differential equation is $y = g(x, \phi(x))$.

Homogeneous Equations If a function f possesses the property $f(tx, ty) = t^{\alpha}f(x, y)$ for some real number α, then f is said to be a **homogeneous function** of degree α. For example, $f(x, y) = x^3 + y^3$ is a homogeneous function of degree 3 since

$$f(tx, ty) = (tx)^3 + (ty)^3 = t^3(x^3 + y^3) = t^3 f(x, y),$$

whereas $f(x, y) = x^3 + y^3 + 1$ is not homogeneous. A first-order DE in differential form

$$M(x, y)\, dx + N(x, y)\, dy = 0 \tag{1}$$

is said to be **homogeneous*** if both coefficients M and N are homogeneous functions of the *same* degree. In other words, (1) is homogeneous if

$$M(tx, ty) = t^{\alpha}M(x, y) \quad \text{and} \quad N(tx, ty) = t^{\alpha}N(x, y).$$

In addition, if M and N are homogeneous functions of degree α, we can also write

$$M(x, y) = x^{\alpha}M(1, u) \quad \text{and} \quad N(x, y) = x^{\alpha}N(1, u), \quad \text{where } u = y/x, \tag{2}$$

and

$$M(x, y) = y^{\alpha}M(v, 1) \quad \text{and} \quad N(x, y) = y^{\alpha}N(v, 1), \quad \text{where } v = x/y. \tag{3}$$

See Problem 31 in Exercises 2.5. Properties (2) and (3) suggest the substitutions that can be used to solve a homogeneous differential equation. Specifically, *either* of the substitutions $y = ux$ or $x = vy$, where u and v are new dependent variables, will reduce a homogeneous equation to a *separable* first-order differential equation. To show this, observe that as a consequence of (2) a homogeneous equation $M(x, y)\, dx + N(x, y)\, dy = 0$ can be rewritten as

$$x^{\alpha}M(1, u)\, dx + x^{\alpha}N(1, u)\, dy = 0 \quad \text{or} \quad M(1, u)\, dx + N(1, u)\, dy = 0,$$

where $u = y/x$ or $y = ux$. By substituting the differential $dy = u\, dx + x\, du$ into the last equation and gathering terms, we obtain a separable DE in the variables u and x:

$$M(1, u)\, dx + N(1, u)[u\, dx + x\, du] = 0$$
$$[M(1, u) + uN(1, u)]\, dx + xN(1, u)\, du = 0$$

or

$$\frac{dx}{x} + \frac{N(1, u)\, du}{M(1, u) + uN(1, u)} = 0.$$

At this point we offer the same advice as in the preceding sections: Do not memorize anything here (especially the last formula); rather, *work through the procedure each time.* The proof that the substitutions $x = vy$ and $dx = v\, dy + y\, dv$ also lead to a separable equation follows in an analogous manner from (3).

---

*Here the word "homogeneous" does not mean the same as it did in Section 2.3. Recall that a linear first-order equation $a_1(x)y' + a_0(x)y = g(x)$ is homogeneous when $g(x) = 0$.

EXAMPLE 1 **Solving a Homogeneous DE**

Solve $(x^2 + y^2)\, dx + (x^2 - xy)\, dy = 0$.

Solution   Inspection of $M(x, y) = x^2 + y^2$ and $N(x, y) = x^2 - xy$ shows that these coefficients are homogeneous functions of degree 2. If we let $y = ux$, then $dy = u\, dx + x\, du$, so, after substituting, the given equation becomes

$$(x^2 + u^2 x^2)\, dx + (x^2 - ux^2)[u\, dx + x\, du] = 0$$
$$x^2(1 + u)\, dx + x^3(1 - u)\, du = 0$$
$$\frac{1 - u}{1 + u}\, du + \frac{dx}{x} = 0$$
$$\left[-1 + \frac{2}{1 + u}\right] du + \frac{dx}{x} = 0. \qquad \leftarrow \text{long division}$$

After integration the last line gives

$$-u + 2 \ln|1 + u| + \ln|x| = \ln|c|$$
$$-\frac{y}{x} + 2 \ln \left|1 + \frac{y}{x}\right| + \ln|x| = \ln|c|. \qquad \leftarrow \text{resubstituting } u = y/x$$

Using the properties of logarithms, we can write the preceding solution as

$$\ln \left|\frac{(x + y)^2}{cx}\right| = \frac{y}{x} \quad \text{or} \quad (x + y)^2 = cxe^{y/x}. \qquad \blacksquare$$

Although either of the indicated substitutions can be used for every homogeneous differential equation, in practice we try $x = vy$ whenever the function $M(x, y)$ is simpler than $N(x, y)$. Also it could happen that after using one substitution, we may encounter integrals that are difficult or impossible to evaluate in closed form; switching substitutions may result in an easier problem.

Bernoulli's Equation   The differential equation

$$\frac{dy}{dx} + P(x)y = f(x)y^n, \qquad \textbf{(4)}$$

where n is any real number, is called **Bernoulli's equation.** Note that for $n = 0$ and $n = 1$ equation (4) is linear. For $n \neq 0$ and $n \neq 1$ the substitution $u = y^{1-n}$ reduces any equation of form (4) to a linear equation.

EXAMPLE 2 **Solving a Bernoulli DE**

Solve $x\dfrac{dy}{dx} + y = x^2 y^2$.

Solution   We first rewrite the equation as

$$\frac{dy}{dx} + \frac{1}{x}y = xy^2$$

by dividing by x. With $n = 2$, we next substitute $y = u^{-1}$ and

$$\frac{dy}{dx} = -u^{-2}\frac{du}{dx} \qquad \leftarrow \text{Chain Rule}$$

into the given equation and simplify. The result is

$$\frac{du}{dx} - \frac{1}{x}u = -x.$$

The integrating factor for this linear equation on, say, $(0, \infty)$ is

$$e^{-\int dx/x} = e^{-\ln x} = e^{\ln x^{-1}} = x^{-1}.$$

Integrating $$\frac{d}{dx}[x^{-1}u] = -1$$

gives $x^{-1}u = -x + c$ or $u = -x^2 + cx$. Since $u = y^{-1}$, we have $y = 1/u$, and so a solution of the given equation is $y = 1/(-x^2 + cx)$.   ▄

Note that we have not obtained the general solution of the original nonlinear differential equation in Example 2, since $y = 0$ is a singular solution of the equation.

Reduction to Separation of Variables   A differential equation of the form

$$\frac{dy}{dx} = f(Ax + By + C) \tag{5}$$

can always be reduced to an equation with separable variables by means of the substitution $u = Ax + By + C$, $B \neq 0$. Example 3 illustrates the technique.

EXAMPLE 3   **An Initial-Value Problem**

Solve $\dfrac{dy}{dx} = (-2x + y)^2 - 7, \quad y(0) = 0.$

Solution   If we let $u = -2x + y$, then $du/dx = -2 + dy/dx$, and so the differential equation is transformed into

$$\frac{du}{dx} + 2 = u^2 - 7 \quad \text{or} \quad \frac{du}{dx} = u^2 - 9.$$

The last equation is separable. Using partial fractions

$$\frac{du}{(u - 3)(u + 3)} = dx \quad \text{or} \quad \frac{1}{6}\left[\frac{1}{u - 3} - \frac{1}{u + 3}\right]du = dx$$

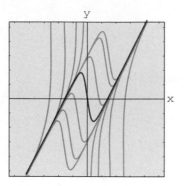

Figure 2.28 Some solutions of $y' = (-2x + y)^2 - 7$

and then integrating yields

$$\frac{1}{6} \ln \left| \frac{u - 3}{u + 3} \right| = x + c_1 \quad \text{or} \quad \frac{u - 3}{u + 3} = e^{6x + 6c_1} = ce^{6x}. \qquad \leftarrow \text{replace } e^{6c_1} \text{ by } c$$

Solving the last equation for u and then resubstituting gives the solution

$$u = \frac{3(1 + ce^{6x})}{1 - ce^{6x}} \quad \text{or} \quad y = 2x + \frac{3(1 + ce^{6x})}{1 - ce^{6x}}. \qquad \textbf{(6)}$$

Finally, applying the initial condition $y(0) = 0$ to the last equation in (6) gives $c = -1$. Figure 2.28, obtained with the aid of a graphing utility, shows the graph of the particular solution

$$y = 2x + \frac{3(1 - e^{6x})}{1 + e^{6x}}$$

in color, along with the graphs of some other members of the family of solutions (6). ∎

EXERCISES 2.5

Answers to odd-numbered problems begin on page AN-2.

In Problems 1–10 solve the given homogeneous equation by using an appropriate substitution.

1. $(x - y)\,dx + x\,dy = 0$ **2.** $(x + y)\,dx + x\,dy = 0$

3. $x\,dx + (y - 2x)\,dy = 0$ **4.** $y\,dx = 2(x + y)\,dy$

5. $(y^2 + yx)\,dx - x^2\,dy = 0$ **6.** $(y^2 + yx)\,dx + x^2\,dy = 0$

7. $\dfrac{dy}{dx} = \dfrac{y - x}{y + x}$ **8.** $\dfrac{dy}{dx} = \dfrac{x + 3y}{3x + y}$

9. $-y\,dx + (x + \sqrt{xy})\,dy = 0$ **10.** $x\dfrac{dy}{dx} - y = \sqrt{x^2 + y^2}$

In Problems 11–14 solve the given initial-value problem.

11. $xy^2 \dfrac{dy}{dx} = y^3 - x^3, \quad y(1) = 2$

12. $(x^2 + 2y^2)\dfrac{dx}{dy} = xy, \quad y(-1) = 1$

13. $(x + ye^{y/x})\,dx - xe^{y/x}\,dy = 0, \quad y(1) = 0$

14. $y\,dx + x(\ln x - \ln y - 1)\,dy = 0, \quad y(1) = e$

In Problems 15–20 solve the given Bernoulli equation by using an appropriate substitution.

15. $x\dfrac{dy}{dx} + y = \dfrac{1}{y^2}$ **16.** $\dfrac{dy}{dx} - y = e^x y^2$

17. $\dfrac{dy}{dx} = y(xy^3 - 1)$ **18.** $x\dfrac{dy}{dx} - (1 + x)y = xy^2$

19. $t^2 \dfrac{dy}{dt} + y^2 = ty$

20. $3(1 + t^2) \dfrac{dy}{dt} = 2ty(y^3 - 1)$

In Problems 21 and 22 solve the given initial-value problem.

21. $x^2 \dfrac{dy}{dx} - 2xy = 3y^4, \quad y(1) = \dfrac{1}{2}$

22. $y^{1/2} \dfrac{dy}{dx} + y^{3/2} = 1, \quad y(0) = 4$

In Problems 23–28 use the procedure illustrated in Example 3 to solve the given differential equation.

23. $\dfrac{dy}{dx} = (x + y + 1)^2$

24. $\dfrac{dy}{dx} = \dfrac{1 - x - y}{x + y}$

25. $\dfrac{dy}{dx} = \tan^2(x + y)$

26. $\dfrac{dy}{dx} = \sin(x + y)$

27. $\dfrac{dy}{dx} = 2 + \sqrt{y - 2x + 3}$

28. $\dfrac{dy}{dx} = 1 + e^{y-x+5}$

In Problems 29 and 30 solve the given initial-value problem.

29. $\dfrac{dy}{dx} = \cos(x + y), \quad y(0) = \pi/4$

30. $\dfrac{dy}{dx} = \dfrac{3x + 2y}{3x + 2y + 2}, \quad y(-1) = -1$

Discussion Problems

31. Explain why it is always possible to express any homogeneous differential equation $M(x, y) \, dx + N(x, y) \, dy = 0$ in the form

$$\dfrac{dy}{dx} = F\left(\dfrac{y}{x}\right) \quad \text{or} \quad \dfrac{dy}{dx} = G\left(\dfrac{x}{y}\right).$$

You might start by proving that

$$M(x, y) = x^\alpha M(1, y/x) \quad \text{and} \quad N(x, y) = x^\alpha N(1, y/x).$$

32. In Example 3 the solution $y(x)$ becomes unbounded as $x \to \pm\infty$. Nevertheless $y(x)$ is asymptotic to a curve as $x \to -\infty$ and to a different curve as $x \to \infty$. What are the equations of these curves?

33. The differential equation $dy/dx = P(x) + Q(x)y + R(x)y^2$ is known as **Ricatti's equation.**

    (a) A Ricatti equation can be solved by a succession of two substitutions *provided* we know a particular solution y_1 of the equation. First use the substitution $y = y_1 + u$ and then discuss what to do next.

    (b) Find a one-parameter family of solutions for the differential equation

$$\dfrac{dy}{dx} = -\dfrac{4}{x^2} - \dfrac{1}{x}y + y^2,$$

    where $y_1 = 2/x$ is a known solution of the equation.

34. Devise an appropriate substitution to solve $xy' = y \ln(xy)$.

2.6 A NUMERICAL SOLUTION

• Linearization   • Euler's method   • Step size   • Numerical solver

So far we have examined first-order DEs qualitatively (Section 2.1) and analytically (Sections 2.2–2.5). To round out the picture of the different types of analyses of differential equations, we conclude this chapter with one of the simplest ways of approximating solutions of a differential equation. The numerical procedure known as Euler's method utilizes the idea that a tangent line can be used to approximate the values of a function in a small neighborhood of the point of tangency. We shall use the geometry of Euler's method as a means of explaining the graphical component of some of the software packages referred to generically hereafter as numerical solvers. A more extensive treatment of approximation techniques for differential equations is found in Chapter 9.

Using the Tangent Line   One way to approximate a solution of the initial-value problem

$$y' = f(x, y), \quad y(x_0) = y_0 \tag{1}$$

is to use tangent lines. For example, let $y(x)$ denote the unknown solution of the initial-value problem $y' = 0.1\sqrt{y} + 0.4x^2$, $y(2) = 4$. The nonlinear differential equation cannot be solved directly by the methods considered in Sections 2.2–2.4, but we can still find approximate numerical values of the unknown $y(x)$. Specifically, suppose we wish to know the value of $y(2.5)$. The IVP has a solution and, as the flow of the direction field in Figure 2.29(a) suggests, a solution curve must have a shape similar to the curve shown in color.

The direction field was generated with lineal elements passing through points in a grid with integer coordinates. As the solution curve passes through the initial point $(2, 4)$, the lineal element at this point is a tangent line with slope given by $f(2, 4) = 0.1\sqrt{4} + 0.4(2)^2 = 1.8$. As is obvious in Figure 2.29(a) and the "zoom in" in Figure 2.29(b), when x is close to

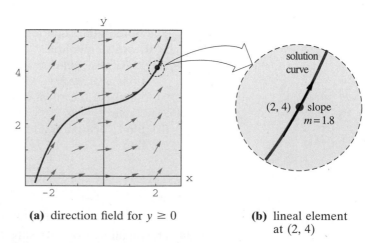

(a) direction field for $y \geq 0$

(b) lineal element at $(2, 4)$

Figure 2.29

2, the points on the solution curve are close to the points on the tangent line (the lineal element). Using the point $(2, 4)$, the slope $f(2, 4) = 1.8$, and the point-slope form of a line, we find that an equation of the tangent line is $y = L(x)$, where $L(x) = 1.8x + 0.4$. This last equation, called a **linearization** of $y(x)$ at $x = 2$, can be used to approximate $y(x)$ within a small neighborhood of $x = 2$. If $y_1 = L(x_1)$ denotes the value of the y-coordinate on the tangent line and $y(x_1)$ is the y-coordinate on the solution curve corresponding to an x-coordinate x_1 that is close to $x = 2$, then $y(x_1) \approx y_1$. If we choose, say, $x_1 = 2.1$, then $y_1 = L(2.1) = 1.8(2.1) + 0.4 = 4.18$ and so $y(2.1) \approx 4.18$.

Euler's Method   To generalize the procedure just illustrated, we use the linearization of the unknown solution $y(x)$ of (1) at $x = x_0$:

$$L(x) = y_0 + f(x_0, y_0)(x - x_0). \tag{2}$$

The graph of this linearization is a straight line tangent to the graph of $y = y(x)$ at the point (x_0, y_0). We now let h be a positive increment of the x-axis, as shown in Figure 2.30. Then by replacing x by $x_1 = x_0 + h$ in (2), we get

$$L(x_1) = y_0 + f(x_0, y_0)(x_0 + h - x_0) \quad \text{or} \quad y_1 = y_0 + hf(x_0, y_0),$$

where $y_1 = L(x_1)$. The point (x_1, y_1) on the tangent line is an approximation to the point $(x_1, y(x_1))$ on the solution curve. Of course the accuracy of the approximation $L(x_1) \approx y(x_1)$ or $y_1 \approx y(x_1)$ depends heavily on the size of the increment h. Usually we must choose this **step size** to be "reasonably small." We now repeat the process using a second "tangent line" at (x_1, y_1).* By identifying the new starting point as (x_1, y_1) with (x_0, y_0) in the above discussion, we obtain an approximation $y_2 \approx y(x_2)$ corresponding to two steps of length h from x_0; that is, $x_2 = x_1 + h = x_0 + 2h$ and

$$y(x_2) = y(x_0 + 2h) = y(x_1 + h) \approx y_2 = y_1 + hf(x_1, y_1).$$

Continuing in this manner, we see that $y_1, y_2, y_3, \ldots$ can be defined recursively by the general formula

$$y_{n+1} = y_n + hf(x_n, y_n), \tag{3}$$

where $x_n = x_0 + nh$, $n = 0, 1, 2, \ldots$. This procedure of using successive "tangent lines" is called **Euler's method.**

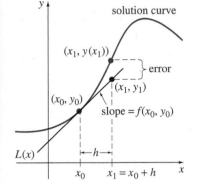

Figure 2.30

EXAMPLE 1   **Euler's Method**

Consider the initial-value problem $y' = 0.1\sqrt{y} + 0.4x^2$, $y(2) = 4$. Use Euler's method to obtain an approximation to $y(2.5)$ using first $h = 0.1$ and then $h = 0.05$.

---

*This is not an actual tangent line, since (x_1, y_1) lies on the first tangent and not on the solution curve.

Solution   With the identification $f(x, y) = 0.1\sqrt{y} + 0.4x^2$, (3) becomes

$$y_{n+1} = y_n + h(0.1\sqrt{y_n} + 0.4x_n^2).$$

Then for $h = 0.1$, $x_0 = 2$, $y_0 = 4$, and $n = 0$, we find

$$y_1 = y_0 + h(0.1\sqrt{y_0} + 0.4x_0^2) = 4 + 0.1(0.1\sqrt{4} + 0.4(2)^2) = 4.18$$

which, as we have already seen, is an estimate to the value of $y(2.1)$. However, if we use the smaller step size $h = 0.05$, it takes two steps to reach $x = 2.1$. From

$$y_1 = 4 + 0.05(0.1\sqrt{4} + 0.4(2)^2) = 4.09$$
$$y_2 = 4.09 + 0.05(0.1\sqrt{4.09} + 0.4(2.05)^2) = 4.18416187$$

we have $y_1 \approx y(2.05)$ and $y_2 \approx y(2.1)$. The remainder of the calculations were carried out using software; the results are summarized in Tables 2.1 and 2.2. We see in Tables 2.1 and 2.2 that it takes five steps with $h = 0.1$ and ten steps with $h = 0.05$, respectively, to get to $x = 2.5$. Each entry has been rounded to four decimal places.

TABLE 2.1 $h = 0.1$

x_n	y_n
2.00	4.0000
2.10	4.1800
2.20	4.3768
2.30	4.5914
2.40	4.8244
2.50	5.0768

TABLE 2.2 $h = 0.05$

x_n	y_n
2.00	4.0000
2.05	4.0900
2.10	4.1842
2.15	4.2826
2.20	4.3854
2.25	4.4927
2.30	4.6045
2.35	4.7210
2.40	4.8423
2.45	4.9686
2.50	5.0997

In Example 2 we apply Euler's method to a differential equation for which we have already found a solution. We do this to compare the values of the approximations y_n at each step with the true values of the solution $y(x_n)$ of the initial-value problem.

EXAMPLE 2   **Comparison of Approximate and Exact Values**

Consider the initial-value problem $y' = 0.2xy$, $y(1) = 1$. Use Euler's method to obtain an approximation to $y(1.5)$ using first $h = 0.1$ and then $h = 0.05$.

Solution   With the identification $f(x, y) = 0.2xy$, (3) becomes

$$y_{n+1} = y_n + h(0.2x_n y_n),$$

where $x_0 = 1$ and $y_0 = 1$. Again with the aid of computer software, we obtain the values in Tables 2.3 and 2.4.

TABLE 2.3 $h = 0.1$

x_n	y_n	True Value	Abs. Error	% Rel. Error
1.00	1.0000	1.0000	0.0000	0.00
1.10	1.0200	1.0212	0.0012	0.12
1.20	1.0424	1.0450	0.0025	0.24
1.30	1.0675	1.0714	0.0040	0.37
1.40	1.0952	1.1008	0.0055	0.50
1.50	1.1259	1.1331	0.0073	0.64

TABLE 2.4 $h = 0.05$

x_n	y_n	True Value	Abs. Error	% Rel. Error
1.00	1.0000	1.0000	0.0000	0.00
1.05	1.0100	1.0103	0.0003	0.03
1.10	1.0206	1.0212	0.0006	0.06
1.15	1.0318	1.0328	0.0009	0.09
1.20	1.0437	1.0450	0.0013	0.12
1.25	1.0562	1.0579	0.0016	0.16
1.30	1.0694	1.0714	0.0020	0.19
1.35	1.0833	1.0857	0.0024	0.22
1.40	1.0980	1.1008	0.0028	0.25
1.45	1.1133	1.1166	0.0032	0.29
1.50	1.1295	1.1331	0.0037	0.32

In Example 1 the true values were calculated from the known solution $y = e^{0.1(x^2 - 1)}$. (Verify.) The **absolute error** is defined to be

$$|true\ value - approximation|.$$

The **relative error** and **percentage relative error** are, in turn,

$$\frac{absolute\ error}{|true\ value|} \quad \text{and} \quad \frac{absolute\ error}{|true\ value|} \times 100.$$

It is apparent from Tables 2.3 and 2.4 that the accuracy of the approximations improves as the step size h decreases. Also, we see that even though the percentage relative error is growing with each step, it does not appear to be that bad. But you should not be deceived by one example. If we simply change the coefficient of the right side of the DE in Example 2 from 0.2 to 2, then at $x_n = 1.5$ the percentage relative errors increase dramatically. See Problem 4 in Exercises 2.6.

A Caveat Euler's method is just one of many different ways a solution of a differential equation can be approximated. Although attractive for its simplicity, *Euler's method is seldom used in serious calculations.* It was introduced here simply to give you a first taste of numerical methods. We will go into greater detail in discussing numerical methods that give significantly greater accuracy, notably the **fourth-order Runge-Kutta method,** in Chapter 9.

Numerical Solvers Regardless of whether we can actually find an explicit or implicit solution, if a solution of a differential equation exists, it represents a smooth curve in the Cartesian plane. The basic idea behind *any* numerical method for first-order ordinary differential equations is to somehow approximate the y-values of a solution for preselected values of x. We start at a specified initial point (x_0, y_0) on a solution curve and proceed to calculate in a step-by-step fashion a sequence of points $(x_1, y_1), (x_2, y_2), \ldots, (x_n, y_n)$ whose y-coordinates y_i approximate the y-coordinates $y(x_i)$ of points $(x_1, y(x_1)), (x_2, y(x_2)), \ldots, (x_n, y(x_n))$ that lie on the graph of the usually unknown solution $y(x)$. By taking the x-coordinates close together (that is, for small values of h) and by joining the points $(x_1, y_1), (x_2, y_2), \ldots, (x_n, y_n)$ with short line segments, we obtain a polygonal curve whose qualitative characteristics we hope are close to those of an actual solution curve. Drawing curves is something well-suited to a computer. A computer program written to either implement a numerical method or render a visual representation of an approximate solution curve fitting the numerical data produced by this method is referred to as a **numerical solver** or an **ODE solver.** There are many different numerical solvers commercially available, either embedded in a larger software package, such as a computer algebra system, or provided as a stand-alone package. Some software packages simply plot the generated numerical approximations, whereas others generate hard numerical data as well as the corresponding approximate or numerical solution curves. By way of illustration of the connect-the-dots nature of the graphs produced by a numerical solver, the two black polygonal graphs in Figure 2.31 are the numerical solution curves for the initial-value problem $y' = 0.2xy, y(0) = 1$ on the interval $[0, 4]$ obtained from Euler's method and the Runge-Kutta method using the step size $h = 1$. The colored smooth curve is the graph of the exact solution $y = e^{0.1x^2}$ of the IVP. Notice in Figure 2.31 that, even with the ridiculously large step size of $h = 1$, the Runge-Kutta method produces the more believable "solution curve." The numerical solution curve obtained from the Runge-Kutta method is indistinguishable from the actual solution curve on the interval $[0, 4]$ when a more typical step size of $h = 0.1$ is used.

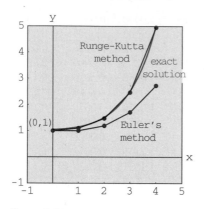

Figure 2.31 Comparison of numerical methods

Using a Numerical Solver   Knowledge of the various numerical methods is not necessary in order to use a numerical solver. A solver usually requires that the differential equation be expressed in normal form $dy/dx = f(x, y)$. Numerical solvers that generate only curves usually require that you supply $f(x, y)$ and the initial data x_0 and y_0 and specify the desired numerical method. If the idea is to approximate the numerical value of $y(a)$, then a solver may additionally require that you state a value for h or, equivalently, give the number of steps that you want to take to get from $x = x_0$ to $x = a$. For example, if we wanted to approximate $y(4)$ for the IVP illustrated in Figure 2.31, then, starting at $x = 0$, it would take four steps to reach $x = 4$ with a step size of $h = 1$; forty steps is equivalent to a step size of $h = 0.1$. Although we will not delve here into the many problems that one can encounter when attempting to approximate mathematical quantities, you should at least be aware of the fact that a numerical solver may break down near certain points or give an incomplete or misleading picture when applied to some first-order differential equations in the normal form. Figure 2.32 illustrates the graph obtained by applying Euler's method to a certain first-order initial-value problem $dy/dx = f(x, y), y(0) = 1$. Equivalent results were obtained using three

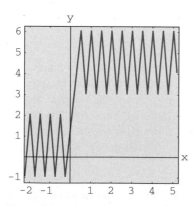

Figure 2.32

different commercial numerical solvers, yet the graph is hardly a plausible solution curve. (Why?) There are several avenues of recourse when a numerical solver has difficulties; three of the more obvious are decrease the step size, use another numerical method, and try a different numerical solver.

EXERCISES 2.6

Answers to odd-numbered problems begin on page AN-2.

In Problems 1 and 2 use Euler's method to obtain a four-decimal approximation of the indicated value. Carry out the recursion of (3) by hand, first using $h = 0.1$ and then using $h = 0.05$.

1. $y' = 2x - 3y + 1$, $y(1) = 5$; $y(1.2)$

2. $y' = x + y^2$, $y(0) = 0$; $y(0.2)$

In Problems 3 and 4 use Euler's method to obtain a four-decimal approximation of the indicated value. First use $h = 0.1$ and then use $h = 0.05$. Find an explicit solution for each initial-value problem and then construct tables similar to Tables 2.3 and 2.4.

3. $y' = y$, $y(0) = 1$; $y(1.0)$ **4.** $y' = 2xy$, $y(1) = 1$; $y(1.5)$

In Problems 5–10 use a numerical solver and Euler's method to obtain a four-decimal approximation to the indicated value. First use $h = 0.1$ and then use $h = 0.05$.

5. $y' = e^{-y}$, $y(0) = 0$; $y(0.5)$

6. $y' = x^2 + y^2$, $y(0) = 1$; $y(0.5)$

7. $y' = (x - y)^2$, $y(0) = 0.5$; $y(0.5)$

8. $y' = xy + \sqrt{y}$, $y(0) = 1$; $y(0.5)$

9. $y' = xy^2 - \dfrac{y}{x}$, $y(1) = 1$; $y(1.5)$

10. $y' = y - y^2$, $y(0) = 0.5$; $y(0.5)$

In Problems 11 and 12 use a numerical solver to graph the solution curve for the given initial-value problem. First use Euler's method and then use the fourth-order Runge-Kutta method. Use $h = 0.25$ in each case. Superimpose both solution curves on the same coordinate axes. If possible, use a different color for each curve. Repeat using $h = 0.1$ and $h = 0.05$.

11. $y' = 2(\cos x)y$, $y(0) = 1$ **12.** $y' = y(10 - 2y)$, $y(0) = 1$

Discussion Problems

13. Use a numerical solver and Euler's method to approximate $y(1.0)$, where $y(x)$ is the solution to $y' = 2xy^2$, $y(0) = 1$. First use $h = 0.1$ and then use $h = 0.05$. Repeat using the fourth-order Runge-Kutta method. Discuss what might cause the approximations to $y(1.0)$ to differ so greatly.

Computer Lab Assignments

14. **(a)** Use a numerical solver and the fourth-order Runge-Kutta method to graph the solution curve for the initial-value problem $y' = -2xy + 1$, $y(0) = 0$.

(b) Solve the initial-value problem by one of the analytic procedures developed earlier in this chapter.

(c) Use the analytic solution $y(x)$ found in part (b) and a CAS to find the coordinates of all relative extrema.

CHAPTER 2 IN REVIEW

Answers to odd-numbered problems begin on page AN-2.

In Problems 1 and 2 fill in the blanks.

1. The DE $y' - ky = A$, where k and A are constants, is autonomous. The critical point _____ of the equation is a(n) _____ (attractor or repeller) for $k > 0$ and a(n) _____ (attractor or repeller) for $k < 0$.

2. The initial-value problem $x\dfrac{dy}{dx} - 4y = 0$, $y(0) = k$ has an infinite number of solutions for $k = $ _____ and no solution for $k = $ _____.

In Problems 3 and 4 construct an autonomous first-order differential $dy/dx = f(y)$ whose phase portrait is consistent with the given figure.

3.

Figure 2.33

4.

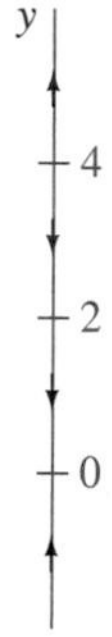

Figure 2.34

5. The number 0 is a critical point of the autonomous differential equation $dx/dt = x^n$, where n is a positive integer. For what values of n is 0 asymptotically stable? Semi-stable? Unstable? Repeat for the differential equation $dx/dt = -x^n$.

6. Consider the differential equation

$$\frac{dP}{dt} = f(P), \quad \text{where } f(P) = -0.5P^3 - 1.7P + 3.4.$$

The function $f(P)$ has one real zero, as shown in Figure 2.35. Without attempting to solve the differential equation, estimate the value of $\lim_{t \to \infty} P(t)$.

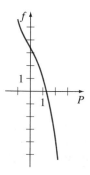

Figure 2.35

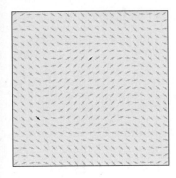

Figure 2.36

7. Figure 2.36 is a portion of the direction field of a differential equation $dy/dx = f(x, y)$. By hand, sketch two different solution curves—one that is tangent to the lineal element shown in black and one that is tangent to the lineal element shown in color.

8. Classify each differential equation as separable, exact, linear, homogeneous, or Bernoulli. Some equations may be more than one kind. Do not solve.

(a) $\dfrac{dy}{dx} = \dfrac{x - y}{x}$ (b) $\dfrac{dy}{dx} = \dfrac{1}{y - x}$

(c) $(x + 1)\dfrac{dy}{dx} = -y + 10$ (d) $\dfrac{dy}{dx} = \dfrac{1}{x(x - y)}$

(e) $\dfrac{dy}{dx} = \dfrac{y^2 + y}{x^2 + x}$ (f) $\dfrac{dy}{dx} = 5y + y^2$

(g) $y\,dx = (y - xy^2)\,dy$ (h) $x\dfrac{dy}{dx} = ye^{x/y} - x$

(i) $xy\,y' + y^2 = 2x$ (j) $2xy\,y' + y^2 = 2x^2$

(k) $y\,dx + x\,dy = 0$ (l) $\left(x^2 + \dfrac{2y}{x}\right)dx = (3 - \ln x^2)\,dy$

(m) $\dfrac{dy}{dx} = \dfrac{x}{y} + \dfrac{y}{x} + 1$ (n) $\dfrac{y}{x^2}\dfrac{dy}{dx} + e^{2x^3 + y^2} = 0$

In Problems 9–16 solve the given differential equation.

9. $(y^2 + 1)\,dx = y\sec^2 x\,dy$

10. $y(\ln x - \ln y)\,dx = (x \ln x - x \ln y - y)\,dy$

11. $(6x + 1)y^2\dfrac{dy}{dx} + 3x^2 + 2y^3 = 0$ 12. $\dfrac{dx}{dy} = -\dfrac{4y^2 + 6xy}{3y^2 + 2x}$

13. $t\dfrac{dQ}{dt} + Q = t^4 \ln t$ 14. $(2x + y + 1)y' = 1$

15. $(x^2 + 4)\,dy = (2x - 8xy)\,dx$

16. $(2r^2 \cos\theta \sin\theta + r\cos\theta)\,d\theta + (4r + \sin\theta - 2r\cos^2\theta)\,dr = 0$

In Problems 17 and 18 solve the given initial-value problem and give the largest interval I on which the solution is defined.

17. $\sin x\dfrac{dy}{dx} + (\cos x)y = 0, \quad y\left(\dfrac{7\pi}{6}\right) = -2$

18. $\dfrac{dy}{dt} + 2(t + 1)y^2 = 0, \quad y(0) = -\tfrac{1}{8}$

19. (a) Without solving, explain why the initial-value problem

$$\dfrac{dy}{dx} = \sqrt{y}, \quad y(x_0) = y_0$$

has no solution for $y_0 < 0$.

(b) Solve the initial-value problem in part (a) for $y_0 > 0$ and find the largest interval I on which the solution is defined.

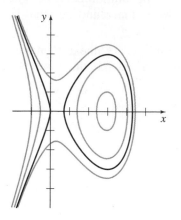

Figure 2.37

20. (a) Find an implicit solution of the initial-value problem

$$\frac{dy}{dx} = \frac{y^2 - x^2}{xy}, \quad y(1) = -\sqrt{2}.$$

(b) Find an explicit solution of the problem in part (a) and give the largest interval I over which the solution is defined. A graphing utility may be helpful here.

21. Graphs of some members of a family of solutions for a first-order differential equation $dy/dx = f(x, y)$ are shown in Figure 2.37. The graphs of two implicit solutions, one that passes through the point $(1, -1)$ and one that passes through $(-1, 3)$, are shown in black. Reproduce the figure on a piece of paper. With colored pencils trace out the solution curves for the solutions $y = y_1(x)$ and $y = y_2(x)$ defined by the implicit solutions such that $y_1(1) = -1$ and $y_2(-1) = 3$, respectively. Estimate the intervals on which the solutions $y = y_1(x)$ and $y = y_2(x)$ are defined.

22. Use Euler's method with step size $h = 0.1$ to approximate $y(1.2)$, where $y(x)$ is a solution of the initial-value problem $y' = 1 + x\sqrt{y}$, $y(1) = 9$.

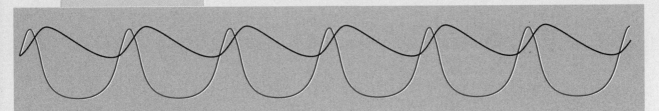

Solution of a predator-prey model; see pages 124–125.

3

MODELING WITH FIRST-ORDER DIFFERENTIAL EQUATIONS

INTRODUCTION　We saw in Section 1.3 that mathematical models for population growth, radioactive decay, continuous compound interest, chemical reactions, cooling of bodies, liquid draining through a hole in a tank, velocity of a falling body, rate of memorization, and current in a series circuit are often first-order differential equations. With the methods developed in Chapter 2, we can now solve some of the linear and nonlinear differential equations that commonly arise in applications. The chapter concludes with the natural next step—systems of first-order differential equations as mathematical models.

3.1 LINEAR EQUATIONS

- *Exponential growth and decay* • *Half-life* • *Carbon dating*
- *Newton's law of cooling* • *Mixtures* • *Series circuits*
- *Transient term* • *Steady-state term*

In this section we are going to solve some of the mathematical models introduced in Section 1.3 that are linear first-order differential equations. Recall from Section 2.3 that multiplying the standard form of a linear equation,

$$\frac{dy}{dt} + P(t)y = f(t),$$

by the integrating factor $e^{\int P(t)\,dt}$ enables us to write the DE as

$$\frac{d}{dt}\left[e^{\int P(t)\,dt}y\right] = e^{\int P(t)\,dt}f(t).$$

The equation can be solved by integrating both sides of this last form.

Growth and Decay The initial-value problem

$$\frac{dx}{dt} = kx, \qquad x(t_0) = x_0, \tag{1}$$

where k is a constant of proportionality, serves as a model for diverse phenomena involving either **growth** or **decay.** We saw in Section 1.3 that, in biological applications, the rate of growth of certain populations (bacteria, small animals) over short periods of time is proportional to the population present at time t. Knowing the population at some arbitrary initial time t_0, we can then use the solution of (1) to predict the population in the future—that is, at times $t > t_0$. The constant of proportionality k in (1) can be determined from the solution of the initial-value problem, using a subsequent measurement of x at a time $t_1 > t_0$. In physics and chemistry, (1) is seen in the form of a *first-order reaction*—that is, a reaction whose rate, or velocity, dx/dt is directly proportional to the amount x of a substance that is unconverted or remaining at time t. The decomposition, or decay, of U-238 (uranium) by radioactivity into Th-234 (thorium) is a first-order reaction.

EXAMPLE 1 **Bacterial Growth**

A culture initially has P_0 number of bacteria. At $t = 1$ h, the number of bacteria is measured to be $\frac{3}{2}P_0$. If the rate of growth is proportional to the number of bacteria $P(t)$ present at time t, determine the time necessary for the number of bacteria to triple.

Solution We first solve the differential equation in (1), with the symbol x replaced by P. With $t_0 = 0$, the initial condition is $P(0) = P_0$. We then use the empirical observation that $P(1) = \frac{3}{2}P_0$ to determine the constant of proportionality k.

Notice that the differential equation $dP/dt = kP$ is both separable and linear. When it is put in the standard form of a linear first-order DE,

$$\frac{dP}{dt} - kP = 0,$$

we can see by inspection that the integrating factor is e^{-kt}. Multiplying both sides of the equation by this term and integrating gives, in turn,

$$\frac{d}{dt}[e^{-kt}P] = 0 \quad \text{and} \quad e^{-kt}P = c.$$

Therefore $P(t) = ce^{kt}$. At $t = 0$ it follows that $P_0 = ce^0 = c$, and so $P(t) = P_0 e^{kt}$. At $t = 1$ we have $\frac{3}{2}P_0 = P_0 e^k$ or $e^k = \frac{3}{2}$. From the last equation we get $k = \ln \frac{3}{2} = 0.4055$, and so $P(t) = P_0 e^{0.4055t}$. To find the time at which the number of bacteria has tripled, we solve $3P_0 = P_0 e^{0.4055t}$ for t. It follows that $0.4055t = \ln 3$, or

$$t = \frac{\ln 3}{0.4055} \approx 2.71 \text{ h.}$$

See Figure 3.1.

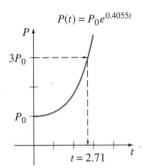

$P(t) = P_0 e^{0.4055t}$

Figure 3.1

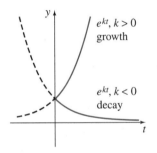

$e^{kt}, k > 0$ growth

$e^{kt}, k < 0$ decay

Figure 3.2

Notice in Example 1 that the actual number P_0 of bacteria present at time $t = 0$ played no part in determining the time required for the number in the culture to triple. The time necessary for an initial population of, say, 100 or 1,000,000 bacteria to triple is still approximately 2.71 hours.

As shown in Figure 3.2, the exponential function e^{kt} increases as t increases for $k > 0$ and decreases as t increases for $k < 0$. Thus problems describing growth (whether of populations, bacteria, or even capital) are characterized by a positive value of k, whereas problems involving decay (as in radioactive disintegration) yield a negative k value. Accordingly, we say that k is either a **growth constant** $(k > 0)$ or a **decay constant** $(k < 0)$.

Half-Life   In physics the **half-life** is a measure of the stability of a radioactive substance. The half-life is simply the time it takes for one-half of the atoms in an initial amount A_0 to disintegrate, or transmute, into the atoms of another element. The longer the half-life of a substance, the more stable it is. For example, the half-life of highly radioactive radium, Ra-226, is about 1700 years. In 1700 years one-half of a given quantity of Ra-226 is transmuted into radon, Rn-222. The most commonly occurring uranium isotope, U-238, has a half-life of approximately 4,500,000,000 years. In about 4.5 billion years, one-half of a quantity of U-238 is transmuted into lead, Pb-206.

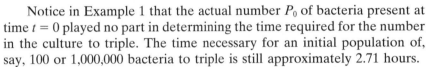

EXAMPLE 2   **Half-Life of Plutonium**

A breeder reactor converts relatively stable uranium 238 into the isotope plutonium 239. After 15 years it is determined that 0.043% of the initial amount A_0 of plutonium has disintegrated. Find the half-life of this isotope if the rate of disintegration is proportional to the amount remaining.

Solution   Let $A(t)$ denote the amount of plutonium remaining at any time. As in Example 1, the solution of the initial-value problem

$$\frac{dA}{dt} = kA, \qquad A(0) = A_0$$

is $A(t) = A_0 e^{kt}$. If 0.043% of the atoms of A_0 have disintegrated, then 99.957% of the substance remains. To find the decay constant k, we use $0.99957A_0 = A(15)$—that is, $0.99957A_0 = A_0 e^{15k}$. Solving for k then gives $k = \frac{1}{15} \ln 0.99957 = -0.00002867$. Hence $A(t) = A_0 e^{-0.00002867t}$. Now the half-life is the corresponding value of time at which $A(t) = \frac{1}{2}A_0$. Solving for t gives $\frac{1}{2}A_0 = A_0 e^{-0.00002867t}$ or $\frac{1}{2} = e^{-0.00002867t}$. The last equation yields

$$t = \frac{\ln 2}{0.00002867} \approx 24{,}180 \text{ yr.} \qquad \blacksquare$$

Carbon Dating   About 1950 the chemist Willard Libby devised a method of using radioactive carbon as a means of determining the approximate ages of fossils. The theory of **carbon dating** is based on the fact that the isotope carbon-14 is produced in the atmosphere by the action of cosmic radiation on nitrogen. The ratio of the amount of C-14 to ordinary carbon in the atmosphere appears to be a constant, and as a consequence the proportionate amount of the isotope present in all living organisms is the same as that in the atmosphere. When an organism dies, the absorption of C-14, by either breathing or eating, ceases. Thus by comparing the proportionate amount of C-14 present, say, in a fossil with the constant ratio found in the atmosphere, it is possible to obtain a reasonable estimation of the fossil's age. The method is based on the knowledge that the half-life of radioactive C-14 is approximately 5600 years. For his work Libby won the Nobel Prize for chemistry in 1960. Libby's method has been used to date wooden furniture in Egyptian tombs, the woven flax wrappings of the Dead Sea scrolls, and the cloth of the enigmatic shroud of Turin.

EXAMPLE 3   Age of a Fossil

A fossilized bone is found to contain one-thousandth the original amount of C-14. Determine the age of the fossil.

Solution   The starting point is again $A(t) = A_0 e^{kt}$. To determine the value of the decay constant k, we use the fact that $A_0/2 = A(5600)$ or $\frac{1}{2}A_0 = A_0 e^{5600k}$. From $5600k = \ln \frac{1}{2} = -\ln 2$ we then get $k = -(\ln \ 2)/5600 = -0.00012378$. Therefore $A(t) = A_0 e^{-0.00012378t}$. With $A(t) = \frac{1}{1000}A_0$ we have $\frac{1}{1000}A_0 = A_0 e^{-0.00012378t}$, so $-0.00012378t = \ln \frac{1}{1000} = -\ln 1000$. Thus

$$t = \frac{\ln 1000}{0.00012378} \approx 55{,}800 \text{ yr.} \qquad \blacksquare$$

The age found in Example 3 is really at the border of accuracy for this method. The usual carbon-14 technique is limited to about 9 half-lives of the isotope, or about 50,000 years. One reason for this limitation is that the chemical analysis needed to obtain an accurate measurement of the remaining C-14 becomes somewhat formidable around the point of $\frac{1}{1000} A_0$. Also, this analysis demands the destruction of a rather large sample of the specimen. If this measurement is accomplished indirectly, based on the actual radioactivity of the specimen, then it is very difficult to distinguish between the radiation from the fossil and the normal background radiation.* But recently the use of a particle accelerator has enabled scientists to separate C-14 from stable C-12 directly. When the precise value of the ratio of C-14 to C-12 is computed, the accuracy of this method can be extended to 70,000–100,000 years. Other isotopic techniques such as using potassium 40 and argon 40 can give ages of several million years.** Nonisotopic methods based on the use of amino acids are also sometimes possible.

Newton's Law of Cooling   In equation (3) of Section 1.3 we saw that the mathematical formulation of Newton's empirical law of cooling of an object is given by the linear first-order differential equation

$$\frac{dT}{dt} = k(T - T_m), \tag{2}$$

where k is a constant of proportionality, $T(t)$ is the temperature of the object for $t > 0$, and T_m is the ambient temperature—that is, the temperature of the medium around the object. In Example 4 we assume that T_m is constant.

EXAMPLE 4   **Cooling of a Cake**

When a cake is removed from an oven, its temperature is measured at 300°F. Three minutes later its temperature is 200°F. How long will it take for the cake to cool off to a room temperature of 70°F?

Solution   In (2) we make the identification $T_m = 70$. We must then solve the initial-value problem

$$\frac{dT}{dt} = k(T - 70), \quad T(0) = 300 \tag{3}$$

and determine the value of k so that $T(3) = 200$.

Equation (3) is both linear and separable. Separating variables,

$$\frac{dT}{T - 70} = k \, dt,$$

---

*The number of disintegrations per minute per gram of carbon is recorded using a Geiger counter. The lower level of detectability is about 0.1 disintegrations per minute per gram.
**Potassium-argon dating is used in dating terrestrial materials such as minerals, rocks, and lava and extraterrestrial materials such as meteorites and lunar rocks. The age of a fossil can be estimated by determining the age of the rock strata in which it was found.

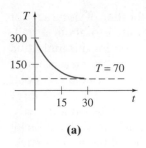

(a)

$T(t)$	t (min)
75°	20.1
74°	21.3
73°	22.8
72°	24.9
71°	28.6
70.5°	32.3

(b)

Figure 3.3

yields $\ln|T - 70| = kt + c_1$, and so $T = 70 + c_2 e^{kt}$. When $t = 0$, $T = 300$, so $300 = 70 + c_2$ gives $c_2 = 230$ and, therefore, $T = 70 + 230e^{kt}$. Finally, the measurement $T(3) = 200$ leads to $e^{3k} = \frac{13}{23}$ or $k = \frac{1}{3} \ln \frac{13}{23} = -0.19018$. Thus

$$T(t) = 70 + 230e^{-0.19018t}. \tag{4}$$

We note that (4) furnishes no finite solution to $T(t) = 70$ since $\lim_{t \to \infty} T(t) = 70$. Yet intuitively we expect the cake to reach room temperature after a reasonably long period of time. How long is "long"? Of course, we should not be disturbed by the fact that the model (3) does not quite live up to our physical intuition. Parts (a) and (b) of Figure 3.3 clearly show that the cake will be approximately at room temperature in about one-half hour.   ■

Mixtures   The mixing of two fluids sometimes gives rise to a linear first-order differential equation. When we discussed the mixing of two brine solutions in Section 1.3, we assumed that the rate $A'(t)$ at which the amount of salt in the mixing tank changes was a net rate:

$$\frac{dA}{dt} = \left(\begin{matrix} \text{input rate} \\ \text{of salt} \end{matrix} \right) - \left(\begin{matrix} \text{output rate} \\ \text{of salt} \end{matrix} \right) = R_i - R_o. \tag{5}$$

In Example 5 we solve equation (8) of Section 1.3.

EXAMPLE 5   Mixture of Two Salt Solutions

Recall that the large tank considered in Section 1.3 initially held 300 gallons of brine solution. Salt was entering and leaving the tank; a brine solution was being pumped into the tank at the rate of 3 gal/min, mixed with the solution there, and then pumped out of the tank at the rate of 3 gal/min. The concentration of the entering solution was 2 lb/gal, and so salt was entering the tank at the rate $R_i = (2 \text{ lb/gal}) \cdot (3 \text{ gal/min}) = 6 \text{ lb/min}$ and leaving the tank at the rate $R_o = (3 \text{ gal/min}) \cdot (A/300 \text{ lb/gal}) = A/100 \text{ lb/min}$. From these data and (5), we get equation (8) of Section 1.3. Let us pose this question: If 50 pounds of salt was dissolved in the initial 300 gallons, how much salt would be in the tank after a long time?

Solution   To find $A(t)$ we solve the initial-value problem

$$\frac{dA}{dt} = 6 - \frac{A}{100}, \quad A(0) = 50.$$

Note here that the side condition is the initial amount of salt, $A(0) = 50$, and not the initial amount of liquid. Now since the integrating factor of the linear differential equation is $e^{t/100}$, we can write the equation as

$$\frac{d}{dt}[e^{t/100} A] = 6e^{t/100}.$$

Integrating the last equation and solving for A gives the general solution $A = 600 + ce^{-t/100}$. When $t = 0$, $A = 50$, so we find that $c = -550$. Thus

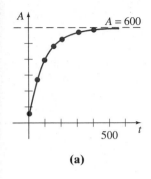

(a)

t (min)	A (lb)
50	266.41
100	397.67
150	477.27
200	525.57
300	572.62
400	589.93

(b)

Figure 3.4

the amount of salt in the tank at any time t is given by

$$A(t) = 600 - 550e^{-t/100}. \qquad (6)$$

The solution (6) was used to construct the table in Figure 3.4(b). Also, it can be seen from (6) and Figure 3.4 that $A \to 600$ as $t \to \infty$. Of course, this is what we would expect in this case; over a long time the number of pounds of salt in the solution must be (300 gal)(2 lb/gal) = 600 lb.

In Example 5 we assumed that the rate at which the solution was pumped in was the same as the rate at which the solution was pumped out. However, this need not be the situation; the mixed brine solution could be pumped out at a rate faster or slower than the rate at which the other solution was pumped in. For example, if the well-stirred solution in Example 5 is pumped out at a slower rate of, say, 2 gal/min, liquid will accumulate in the tank at the rate of (3 − 2) gal/min = 1 gal/min. After t minutes the tank will contain $300 + t$ gallons of brine. The rate at which the salt is leaving is then

$$R_o = (2 \text{ gal/min}) \left(\frac{A}{300 + t} \text{ lb/gal} \right).$$

Hence equation (5) becomes

$$\frac{dA}{dt} = 6 - \frac{2A}{300 + t} \quad \text{or} \quad \frac{dA}{dt} + \frac{2}{300 + t}A = 6.$$

You should verify that the solution of the last equation, subject to $A(0) = 50$, is $A(t) = 600 + 2t - (4.95 \times 10^7)(300 + t)^{-2}$.

Series Circuits   For a series circuit containing only a resistor and an inductor, Kirchhoff's second law states that the sum of the voltage drop across the inductor $(L(di/dt))$ and the voltage drop across the resistor (iR) is the same as the impressed voltage $(E(t))$ on the circuit. See Figure 3.5.

Thus we obtain the linear differential equation for the current $i(t)$,

$$L\frac{di}{dt} + Ri = E(t), \qquad (7)$$

where L and R are constants known as the inductance and the resistance, respectively. The current $i(t)$ is also called the **response** of the system.

The voltage drop across a capacitor with capacitance C is given by $q(t)/C$, where q is the charge on the capacitor. Hence, for the series circuit shown in Figure 3.6, Kirchhoff's second law gives

$$Ri + \frac{1}{C}q = E(t). \qquad (8)$$

But current i and charge q are related by $i = dq/dt$, so (8) becomes the linear differential equation

$$R\frac{dq}{dt} + \frac{1}{C}q = E(t). \qquad (9)$$

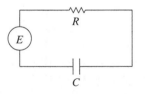

Figure 3.5 *LR* series circuit

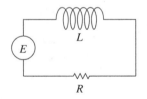

Figure 3.6 *RC* series circuit

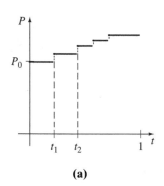

(a)

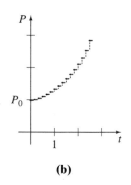

(b)

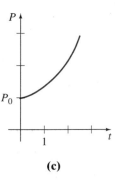

(c)

Figure 3.7

| EXAMPLE 6 | **Series Circuit** |

A 12-volt battery is connected to a series circuit in which the inductance is $\frac{1}{2}$ henry and the resistance is 10 ohms. Determine the current i if the initial current is zero.

Solution   From (7) we see that we must solve

$$\frac{1}{2}\frac{di}{dt} + 10i = 12,$$

subject to $i(0) = 0$. First, we multiply the differential equation by 2 and read off the integrating factor e^{20t}. We then obtain

$$\frac{d}{dt}[e^{20t}i] = 24e^{20t}.$$

Integrating each side of the last equation and solving for i gives $i(t) = \frac{6}{5} + ce^{-20t}$. Now $i(0) = 0$ implies $0 = \frac{6}{5} + c$ or $c = -\frac{6}{5}$. Therefore the response is $i(t) = \frac{6}{5} - \frac{6}{5}e^{-20t}$.   ■

From (4) of Section 2.3 we can write a general solution of (7):

$$i(t) = \frac{e^{-(R/L)t}}{L}\int e^{(R/L)t}E(t)\,dt + ce^{-(R/L)t}. \qquad \textbf{(10)}$$

In particular, when $E(t) = E_0$ is a constant, (10) becomes

$$i(t) = \frac{E_0}{R} + ce^{-(R/L)t}. \qquad \textbf{(11)}$$

Note that as $t \to \infty$, the second term in equation (11) approaches zero. Such a term is usually called a **transient term;** any remaining terms are called the **steady-state** part of the solution. In this case E_0/R is also called the **steady-state current;** for large values of time it appears that the current in the circuit is simply governed by Ohm's law ($E = iR$).

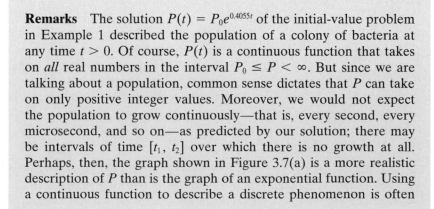

Remarks   The solution $P(t) = P_0e^{0.4055t}$ of the initial-value problem in Example 1 described the population of a colony of bacteria at any time $t > 0$. Of course, $P(t)$ is a continuous function that takes on *all* real numbers in the interval $P_0 \le P < \infty$. But since we are talking about a population, common sense dictates that P can take on only positive integer values. Moreover, we would not expect the population to grow continuously—that is, every second, every microsecond, and so on—as predicted by our solution; there may be intervals of time $[t_1, t_2]$ over which there is no growth at all. Perhaps, then, the graph shown in Figure 3.7(a) is a more realistic description of P than is the graph of an exponential function. Using a continuous function to describe a discrete phenomenon is often more a matter of convenience than of accuracy. However, for some purposes we may be satisfied if our model describes the system fairly closely when viewed macroscopically in time, as in Figures 3.7(b) and (c), rather than microscopically, as in Figure 3.7(a).

EXERCISES 3.1

Answers to odd-numbered problems begin on page AN-2.

Growth and Decay

1. The population of a community is known to increase at a rate proportional to the number of people present at time t. If the population has doubled in 5 years, how long will it take to triple? To quadruple?

2. Suppose it is known that the population of the community in Problem 1 is 10,000 after 3 years. What was the initial population? What will be the population in 10 years?

3. The population of a town grows at a rate proportional to the population present at time t. The initial population of 500 increases by 15% in 10 years. What will be the population in 30 years?

4. The population of bacteria in a culture grows at a rate proportional to the number of bacteria present at time t. After 3 hours it is observed that there are 400 bacteria present. After 10 hours there are 2000 bacteria present. What was the initial number of bacteria?

5. When a vertical beam of light passes through a transparent medium, the rate at which its intensity I decreases is proportional to $I(t)$, where t represents the thickness of the medium (in feet). In clear seawater, the intensity 3 feet below the surface is 25% of the initial intensity I_0 of the incident beam. What is the intensity of the beam 15 feet below the surface?

6. When interest is compounded continuously, the amount of money increases at a rate proportional to the amount S present at time t—that is, $dS/dt = rS$, where r is the annual rate of interest.
   (a) Find the amount of money accrued at the end of 5 years when $5000 is deposited in a savings account drawing $5\frac{3}{4}$% annual interest compounded continuously.
   (b) In how many years will the initial amount deposited have doubled?
   (c) Use a calculator to compare the amount obtained in part (a) with the amount $S = 5000(1 + \frac{1}{4}(0.0575))^{5(4)}$ that is accrued when interest is compounded quarterly.

7. The radioactive isotope of lead, Pb-209, decays at a rate proportional to the amount present at time t and has a half-life of 3.3 hours. If 1 gram of lead is present initially, how long will it take for 90% of the lead to decay?

8. Initially 100 milligrams of a radioactive substance was present. After 6 hours the mass had decreased by 3%. If the rate of decay is proportional to the amount of substance present at time t, find the amount remaining after 24 hours.

9. Determine the half-life of the radioactive substance described in Problem 8.

10. (a) Consider the initial-value problem $dA/dt = kA$, $A(0) = A_0$ as the model for the decay of a radioactive substance. Show that, in general, the half-life T of the substance is $T = -(\ln 2)/k$.

   (b) Show that the solution of the initial-value problem in part (a) can be written $A(t) = A_0 2^{-t/T}$.

   (c) If a radioactive substance has the half-life T given in part (a), how long will it take an initial amount A_0 of the substance to decay to $\frac{1}{8}A_0$?

Carbon Dating

11. Archaeologists used pieces of burned wood, or charcoal, found at the site to date prehistoric paintings and drawing on the walls and ceilings of a cave in Lascaux, France. See Figure 3.8. Use the information in Example 3 to determine the approximate age of a piece of wood, if it was found that 85.5% of the C-14 had decayed.

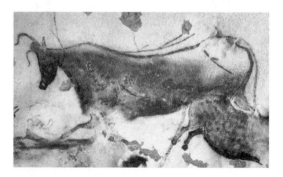

Figure 3.8 Lascaux cave painting

Figure 3.9 Shroud of Turin

12. The shroud of Turin, which shows the negative image of a body of a crucified man, is believed by many to be the burial shroud of Jesus of Nazareth. In 1988 the Vatican granted permission to have the shroud carbon dated. Three independent scientific laboratories analyzed the cloth and concluded that the shroud was approximately 660 years old,* an age consistent with its historical appearance. Using this age, determine what percentage of the original amount of C-14 remained in the cloth as of 1988.

Newton's Law of Cooling/Warming

13. A thermometer is removed from a room where the air temperature is 70°F and is taken outside, where the temperature is 10°F. After

---

*Some scholars have disagreed with this finding. For more information on this fascinating mystery see the Shroud of Turin Website Home Page at http://www.shroud.com/.

$\frac{1}{2}$ minute the thermometer reads 50°F. What is the reading of the thermometer at $t = 1$ min? How long will it take for the thermometer to reach 15°F?

14. A thermometer is taken from an inside room to the outside, where the air temperature is 5°F. After 1 minute the thermometer reads 55°F, and after 5 minutes it reads 30°F. What was the initial temperature of the room?

15. A small metal bar, whose initial temperature is 20°C, is dropped into a container of boiling water. How long will it take the bar to reach 90°C if its temperature increases 2° in 1 second? How long will it take the bar to reach 98°C?

16. A thermometer reading 70°F is placed in an oven preheated to a constant temperature. Through a glass window in the oven door, an observer records that the thermometer reads 110°F after $\frac{1}{2}$ minute and 145°F after 1 minute. How hot is the oven?

Mixtures

17. A tank contains 200 liters of fluid in which 30 grams of salt is dissolved. Brine containing 1 gram of salt per liter is then pumped into the tank at a rate of 4 L/min; the well-mixed solution is pumped out at the same rate. Find the number $A(t)$ of grams of salt in the tank at time t.

18. Solve Problem 17 assuming that pure water is pumped into the tank.

19. A large tank is filled with 500 gallons of pure water. Brine containing 2 pounds of salt per gallon is pumped into the tank at a rate of 5 gal/min. The well-mixed solution is pumped out at the same rate. Find the number $A(t)$ of pounds of salt in the tank at time t. What is the concentration of the solution in the tank at $t = 5$ min?

20. Solve Problem 19 under the assumption that the solution is pumped out at a faster rate of 10 gal/min. When is the tank empty?

21. A large tank is partially filled with 100 gallons of fluid in which 10 pounds of salt is dissolved. Brine containing $\frac{1}{2}$ pound of salt per gallon is pumped into the tank at a rate of 6 gal/min. The well-mixed solution is then pumped out at a slower rate of 4 gal/min. Find the number of pounds of salt in the tank after 30 minutes.

22. In Example 5 the size of the tank containing the salt mixture was not given. Suppose, as discussed on page 101, that the rate at which brine is pumped into the tank is 6 lb/min, but the mixed solution is pumped out at a rate of 2 gal/min. It stands to reason that, since brine is accumulating in the tank at the rate of 1 gal/min, any finite tank must eventually overflow. Now suppose that the tank has an open top and has a total capacity of 400 gallons.
   (a) When will the tank overflow?
   (b) What will be the number of pounds of salt in the tank at the instant it overflows?
   (c) Assume that, although the tank is overflowing, brine solution continues to be pumped in at a rate of 3 gal/min and the well-stirred solution continues to be pumped out at a rate of 2 gal/min. Devise a method for determining the number of pounds of salt in the tank at $t = 150$ min.

(d) Determine the number of pounds of salt in the tank as $t \to \infty$. Does your answer agree with your intuition?

(e) Use a graphing utility to obtain the graph $A(t)$ on the interval $[0, 500)$.

Series Circuits

23. A 30-volt electromotive force is applied to an LR series circuit in which the inductance is 0.1 henry and the resistance is 50 ohms. Find the curve $i(t)$ if $i(0) = 0$. Determine the current as $t \to \infty$.

24. Solve equation (7) under the assumption that $E(t) = E_0 \sin \omega t$ and $i(0) = i_0$.

25. A 100-volt electromotive force is applied to an RC series circuit in which the resistance is 200 ohms and the capacitance is 10^{-4} farad. Find the charge $q(t)$ on the capacitor if $q(0) = 0$. Find the current $i(t)$.

26. A 200-volt electromotive force is applied to an RC series circuit in which the resistance is 1000 ohms and the capacitance is 5×10^{-6} farad. Find the charge $q(t)$ on the capacitor if $i(0) = 0.4$. Determine the charge and current at $t = 0.005$ s. Determine the charge as $t \to \infty$.

27. An electromotive force

$$E(t) = \begin{cases} 120, & 0 \le t \le 20 \\ 0, & t > 20 \end{cases}$$

is applied to an LR series circuit in which the inductance is 20 henries and the resistance is 2 ohms. Find the current $i(t)$ if $i(0) = 0$.

28. Suppose an RC series circuit has a variable resistor. If the resistance at time t is given by $R = k_1 + k_2 t$ and k_1 and k_2 are known positive constants, then (9) becomes

$$(k_1 + k_2 t)\frac{dq}{dt} + \frac{1}{C}q = E(t).$$

If $E(t) = E_0$ and $q(0) = q_0$, show that

$$q(t) = E_0 C + (q_0 - E_0 C)\left(\frac{k_1}{k_1 + k_2 t}\right)^{1/Ck_2}.$$

Miscellaneous Linear Models

29. In (14) of Section 1.3 we saw that a differential equation governing the velocity v of a falling mass subject to air resistance proportional to the instantaneous velocity is

$$m\frac{dv}{dt} = mg - kv,$$

where $k > 0$ is a constant of proportionality. The positive direction is downward.

(a) Solve the equation subject to the initial condition $v(0) = v_0$.

(b) Use the solution in part (a) to determine the limiting, or terminal, velocity of the mass. We saw how to determine the terminal velocity without solving the DE in Problem 25 in Exercises 2.1.

(c) If the distance s, measured from the point where the mass is released above ground, is related to velocity by $ds/dt = v$, find an explicit expression for s if $s(0) = 0$.

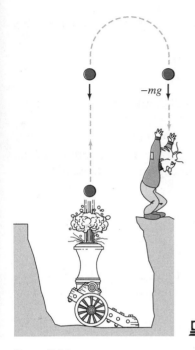

Figure 3.10

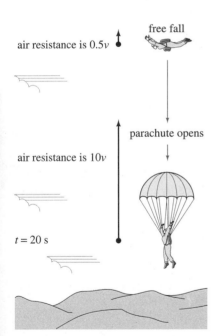

Figure 3.11

30. Suppose a small cannonball weighing 16 pounds is shot vertically upward, as shown in Figure 3.10, with an initial velocity $v_0 = 300$ ft/s. The answer to the question "How high does the cannonball go?" depends on whether we take air resistance into account.

(a) Suppose air resistance is ignored. If the positive direction is upward, then a model governing the state of the cannonball is given by $d^2s/dt^2 = -g$ (equation (12) of Section 1.3). Integrate this equation once to show that velocity is $v(t) = -32t + 300$, where we take $g = 32$ ft/s².

(b) Use the fact that $ds/dt = v(t)$ to determine the height $s(t)$ of the cannonball measured from ground level. Find the maximum height attained by the cannonball.

31. Repeat Problem 30, but this time assume that air resistance is proportional to instantaneous velocity. It stands to reason that the maximum height attained by the cannonball must be *less* than that obtained in part (b) of Problem 30. Show this by supposing that the constant of proportionality is $k = 0.0025$. [*Hint:* You will have to slightly modify the DE in Problem 29.]

32. A skydiver weighs 125 pounds, and her parachute and equipment combined weigh another 35 pounds. After exiting from a plane at an altitude of 15,000 feet, she waits 15 seconds and opens her parachute. Assume the constant of proportionality in the model in Problem 29 has the value $k = 0.5$ during free fall and $k = 10$ after the parachute is opened. Assume her initial velocity upon leaving the plane is zero. What is her velocity and how far has she traveled 20 seconds after leaving the plane? See Figure 3.11. How does her velocity at 20 seconds compare with her terminal velocity? How long does it take her to reach the ground? [*Hint:* Think in terms of two distinct IVPs.]

33. As a raindrop falls, it evaporates while retaining its spherical form. If the further assumptions are made that the rate at which the raindrop evaporates is proportional to its surface area and that air resistance is negligible, then the velocity $v(t)$ of the raindrop is given by

$$\frac{dv}{dt} + \frac{3(k/\rho)}{(k/\rho)t + r_0} v = g.$$

Here ρ is the density of water, r_0 is the radius of the raindrop at $t = 0$, $k < 0$ is the constant of proportionality, and the downward direction is taken to be the positive direction.

(a) Solve for $v(t)$ if the raindrop falls from rest.

(b) Reread Problem 35 of Exercises 1.3 and then show that the radius of the raindrop at time t is $r(t) = (k/\rho)t + r_0$.

(c) If $r_0 = 0.01$ ft and if $r = 0.007$ ft 10 seconds after the raindrop falls from a cloud, determine the time at which the raindrop has evaporated completely.

34. The differential equation $dP/dt = (k \cos t)P$, where k is a positive constant, is a mathematical model for a population $P(t)$ that undergoes yearly seasonal fluctuations. Solve the equation subject to $P(0) = P_0$. Use a graphing utility to obtain the graph of the solution for different choices of P_0.

35. In one model of the changing population $P(t)$ of a community, it is assumed that

$$\frac{dP}{dt} = \frac{dB}{dt} - \frac{dD}{dt},$$

where dB/dt and dD/dt are the birth and death rates, respectively.
(a) Solve for $P(t)$ if $dB/dt = k_1 P$ and $dD/dt = k_2 P$.
(b) Analyze the cases $k_1 > k_2$, $k_1 = k_2$, and $k_1 < k_2$.

36. The following system of differential equations is encountered in the study of a series of elements that are decaying by radioactivity:

$$\frac{dx}{dt} = -\lambda_1 x$$

$$\frac{dy}{dt} = \lambda_1 x - \lambda_2 y,$$

where λ_1 and λ_2 are constants. It does not take any specialized knowledge of systems to solve this particular system. Solve for $x(t)$ and $y(t)$, subject to $x(0) = x_0$ and $y(0) = y_0$.

37. When forgetfulness is taken into account, the rate of memorization of a subject is given by

$$\frac{dA}{dt} = k_1(M - A) - k_2 A,$$

where $k_1 > 0$, $k_2 > 0$, $A(t)$ is the amount to be memorized in time t, M is the total amount to be memorized, and $M - A$ is the amount remaining to be memorized. (See Problems 23 and 24 in Exercises 1.3.)
(a) Since the DE is autonomous, use the phase portrait concept of Section 2.1 to find the limiting value of $A(t)$ as $t \to \infty$. Interpret the result.
(b) Solve for $A(t)$, subject to $A(0) = 0$. Sketch the graph of $A(t)$ and verify your prediction in part (a).

38. The rate at which a drug disseminates into the bloodstream is governed by $dx/dt = r - kx$, where r and k are positive constants. The function $x(t)$ describes the concentration of the drug in the bloodstream at time t.
(a) Since the DE is autonomous, use the phase portrait concept of Section 2.1 to find the limiting value of $x(t)$ as $t \to \infty$.
(b) Solve the DE subject to $x(0) = 0$. Sketch the graph of $x(t)$ and verify your prediction in part (a). At what time is the concentration one-half this limiting value?

Discussion Problems

39. Suppose a medical examiner arrives at the scene of a homicide and finds that the temperature of the corpse is 82°F. Make up additional, but plausible, data necessary to determine an approximate time of death of the victim using Newton's law of cooling (2).

40. Mr. Jones puts two cups of coffee on the breakfast table at the same time. He immediately pours cream into his coffee from a pitcher that was sitting on the table for a long time. He then reads the morning paper for 5 minutes before taking his first sip. Mrs. Jones arrives at the table 5 minutes after the cups were set down, adds cream to her coffee, and takes a sip. Assume that Mr. and Mrs. Jones add exactly the same amount of cream to their cups. Discuss who drinks the hotter cup of coffee. Defend your assertion with sound mathematics.

41. There are several ways of determining the value of the constant of proportionality k in the falling body model in Problem 29. Discuss how to devise a simple physical experimental and a corresponding algebraic equation for determining k.

Computer Lab Assignments

42. It is well known that the model in which air resistance is ignored (part (a) of Problem 30) predicts that the time t_a it takes the cannonball to attain its maximum height is the same as the time t_d it takes the cannonball to fall from its maximum height to the ground. Moreover, the magnitude of the impact velocity v_i will be the same as the initial velocity v_0 of the cannonball. Verify these results. Then, using the model in Problem 31 that takes air resistance into account, compare the value of t_a with t_d and the value of the magnitude of v_i with v_0. A root-finding application of a CAS (or a graphic calculator) may be useful here.

3.2 NONLINEAR EQUATIONS

• Population models   • Relative growth rate   • Density-dependent hypothesis   • Logistic differential equation   • Logistic function   • Second-order chemical reaction   • Variable mass

We turn now to the solution of nonlinear first-order mathematical models. Most— but not all—of the differential equations in this section can be solved by separation of variables. You should review the nonlinear models in Section 1.3, Section 2.2, and, if necessary, techniques of integration.

Population Dynamics   If $P(t)$ denotes the size of a population at time t, the model for exponential growth begins with the assumption that $dP/dt = kP$ for some $k > 0$. In this model the **relative,** or **specific, growth rate** defined by

$$\frac{dP/dt}{P} \tag{1}$$

is assumed to be a constant k. True cases of exponential growth over long periods of time are hard to find, because the limited resources of the environment will at some time exert restrictions on the growth of a population. Thus (1) can be expected to decrease as P increases in size.

The assumption that the rate at which a population grows (or declines) is dependent only on the number present and not on any time-dependent mechanisms such as seasonal phenomena (see Problem 32 in Exercises 1.3) can be stated as

$$\frac{dP/dt}{P} = f(P) \quad \text{or} \quad \frac{dP}{dt} = Pf(P). \tag{2}$$

The differential equation in (2), which is widely assumed in models of animal populations, is called the **density-dependent hypothesis.**

Logistic Equation   Suppose an environment is capable of sustaining no more than a fixed number K of individuals in its population. The quantity K is called the **carrying capacity** of the environment. Hence for the function f in (2) we have $f(K) = 0$, and we simply let $f(0) = r$. Figure 3.12 shows three functions f that satisfy these two conditions. The simplest assumption that we can make is that $f(P)$ is linear—that is, $f(P) = c_1P + c_2$. If we use the conditions $f(0) = r$ and $f(K) = 0$, we find, in turn, $c_2 = r$ and $c_1 = -r/K$, and so f takes on the form $f(P) = r - (r/K)P$. Equation (2) becomes

$$\frac{dP}{dt} = P\left(r - \frac{r}{K}P\right). \tag{3}$$

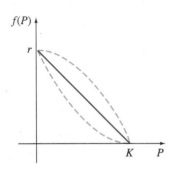

Figure 3.12

With constants relabeled, the nonlinear equation (3) is the same as

$$\frac{dP}{dt} = P(a - bP). \tag{4}$$

Around 1840 the Belgian mathematician-biologist P. F. Verhulst was concerned with mathematical models for predicting the human populations of various countries. One of the equations he studied was (4), where $a > 0$ and $b > 0$. Equation (4) came to be known as the **logistic equation,** and its solution is called the **logistic function.** The graph of a logistic function is called a **logistic curve.**

The linear differential equation $dP/dt = kP$ does not provide a very accurate model for population when the population itself is very large. Overcrowded conditions, with the resulting detrimental effects on the environment such as pollution and excessive and competitive demands for food and fuel, can have an inhibiting effect on population growth. As we shall now see, the solution of (4) is bounded as $t \to \infty$. If we rewrite (4) as $dP/dt = aP - bP^2$, the nonlinear term $-bP^2$, $b > 0$ can be interpreted as an "inhibition" or "competition" term. Also, in most applications the positive constant a is much larger than the constant b.

Logistic curves have proved to be quite accurate in predicting the growth patterns, in a limited space, of certain types of bacteria, protozoa, water fleas (*Daphnia*), and fruit flies (*Drosophila*).

Solution of the Logistic Equation   One method of solving (4) is separation of variables. Decomposing the left side of $dP/P(a - bP) = dt$ into partial fractions and integrating gives

$$\left(\frac{1/a}{P} + \frac{b/a}{a - bP}\right) dP = dt$$

$$\frac{1}{a}\ln|P| - \frac{1}{a}\ln|a - bP| = t + c$$

$$\ln\left|\frac{P}{a - bP}\right| = at + ac$$

$$\frac{P}{a - bP} = c_1 e^{at}.$$

It follows from the last equation that

$$P(t) = \frac{ac_1 e^{at}}{1 + bc_1 e^{at}} = \frac{ac_1}{bc_1 + e^{-at}}.$$

If $P(0) = P_0$, $P_0 \neq a/b$, we find $c_1 = P_0/(a - bP_0)$, and so, after substituting and simplifying, the solution becomes

$$P(t) = \frac{aP_0}{bP_0 + (a - bP_0)e^{-at}}. \tag{5}$$

Graphs of $P(t)$ The basic shape of the graph of the logistic function $P(t)$ can be obtained without too much effort. Although the variable t usually represents time and we are seldom concerned with applications in which $t < 0$, it is nonetheless of some interest to include this interval when displaying the various graphs of P. From (5) we see that

$$P(t) \to \frac{aP_0}{bP_0} = \frac{a}{b} \text{ as } t \to \infty \quad \text{and} \quad P(t) \to 0 \text{ as } t \to -\infty.$$

The dashed line $P = a/2b$ shown in Figure 3.13 corresponds to the ordinate of a point of inflection of the logistic curve. To show this, we differentiate (4) by the product rule:

$$\frac{d^2P}{dt^2} = P\left(-b\frac{dP}{dt}\right) + (a - bP)\frac{dP}{dt} = \frac{dP}{dt}(a - 2bP)$$

$$= P(a - bP)(a - 2bP)$$

$$= 2b^2 P\left(P - \frac{a}{b}\right)\left(P - \frac{a}{2b}\right).$$

From calculus recall that the points where $d^2P/dt^2 = 0$ are possible points of inflection, but $P = 0$ and $P = a/b$ can obviously be ruled out. Hence $P = a/2b$ is the only possible ordinate value at which the concavity of the graph can change. For $0 < P < a/2b$ it follows that $P'' > 0$, and $a/2b < P < a/b$ implies $P'' < 0$. Thus, as we read from left to right, the graph changes from concave up to concave down at the point corresponding to $P = a/2b$. When the initial value satisfies $0 < P_0 < a/2b$, the graph of $P(t)$ assumes the shape of an S, as we see in Figure 3.13(a). For $a/2b < P_0 < a/b$ the graph is still S-shaped, but the point of inflection occurs at a negative value of t, as shown in Figure 3.13(b).

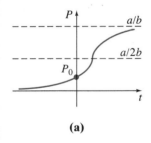

(a)

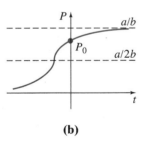

(b)

Figure 3.13

We have already seen equation (4) in (5) of Section 1.3 in the form $dx/dt = kx(n + 1 - x)$, $k > 0$. This differential equation provides a reasonable model for describing the spread of an epidemic brought about initially by introducing an infected individual into a static population. The solution $x(t)$ represents the number of individuals infected with the disease at time t.

EXAMPLE 1   Logistic Growth

Suppose a student carrying a flu virus returns to an isolated college campus of 1000 students. If it is assumed that the rate at which the virus spreads is proportional not only to the number x of infected students but also to the number of students not infected, determine the number of infected students after 6 days if it is further observed that after 4 days $x(4) = 50$.

Solution   Assuming that no one leaves the campus throughout the duration of the disease, we must solve the initial-value problem

$$\frac{dx}{dt} = kx(1000 - x), \qquad x(0) = 1.$$

By making the identifications $a = 1000k$ and $b = k$, we have immediately from (5) that

$$x(t) = \frac{1000k}{k + 999ke^{-1000kt}} = \frac{1000}{1 + 999e^{-1000kt}}.$$

Now, using the information $x(4) = 50$, we determine k from

$$50 = \frac{1000}{1 + 999e^{-4000k}}.$$

We find $-1000k = \frac{1}{4} \ln \frac{19}{999} = -0.9906$. Thus

$$x(t) = \frac{1000}{1 + 999e^{-0.9906t}}.$$

Finally $x(6) = \dfrac{1000}{1 + 999e^{-5.9436}} = 276$ students.

Additional calculated values of $x(t)$ are given in the table in Figure 3.14(b).

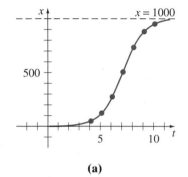

(a)

t (days)	x (number infected)
4	50 (observed)
5	124
6	276
7	507
8	735
9	882
10	953

(b)

Figure 3.14

Modifications of the Logistic Equation   There are many variations of the logistic equation. For example, the differential equations

$$\frac{dP}{dt} = P(a - bP) - h \quad \text{and} \quad \frac{dP}{dt} = P(a - bP) + h \qquad \textbf{(6)}$$

could serve, in turn, as models for the population in a fishery where fish are **harvested** or are **restocked** at rate h. When $h > 0$ is a constant, the DEs in (6) can be readily analyzed qualitatively or solved analytically by separation of variables. The equations in (6) could also serve as models of the human population decreased by emigration or increased by immi-gration, respectively. The rate h in (6) could be a function of time t or could be population dependent; for example, harvesting might be done periodically over time or might be done at a rate proportional to the population P at time t. In the latter instance, the model would look like $P' = P(a - bP) - cP, c > 0$. See the *Project Module* at the end of this chapter. The human population of a community might change because of immigration in such a manner that the contribution due to immigration was large when the population P of the community was itself small but small when P was large; a reasonable model for the population of the community would then be $P' = P(a - bP) + ce^{-kP}, c > 0, k > 0$. See Problem 20 in Exercises 3.2. Another equation of the form given in (2),

$$\frac{dP}{dt} = P(a - b \ln P),\tag{7}$$

is a modification of the logistic equation known as the **Gompertz differential equation.** This DE is sometimes used as a model in the study of the growth or decline of populations, the growth of solid tumors, and certain kinds of actuarial predictions. See Problems 5 and 6 in Exercises 3.2.

Chemical Reactions Suppose that a grams of chemical A is combined with b grams of chemical B. If there are M parts of A and N parts of B formed in the compound and $X(t)$ is the number of grams of chemical C formed, then the number of grams of chemical A and the number of grams of chemical B remaining at time t are, respectively,

$$a - \frac{M}{M + N}X \quad \text{and} \quad b - \frac{N}{M + N}X.$$

The law of mass action states that when no temperature change is involved, the rate at which the two substances react is proportional to the product of the amounts of A and B that are untransformed (remaining) at time t:

$$\frac{dX}{dt} \propto \left(a - \frac{M}{M + N}X\right)\left(b - \frac{N}{M + N}X\right).\tag{8}$$

If we factor out $M/(M + N)$ from the first factor and $N/(M + N)$ from the second and introduce a constant proportionality $k > 0$, (8) has the form

$$\frac{dX}{dt} = k(\alpha - X)(\beta - X),\tag{9}$$

where $\alpha = a(M + N)/M$ and $\beta = b(M + N)/N$. Recall from (6) of Section 1.3 that a chemical reaction governed by the nonlinear differential equation (9) is said to be a **second-order reaction.**

EXAMPLE 2 **Second-Order Chemical Reaction**

A compound C is formed when two chemicals A and B are combined. The resulting reaction between the two chemicals is such that for each gram of A, 4 grams of B is used. It is observed that 30 grams of the compound C is formed in 10 minutes. Determine the amount of C at any time if the rate of the reaction is proportional to the amounts of A and B remaining and if initially there are 50 grams of A and 32 grams of B. How much of the compound C is present at 15 minutes? Interpret the solution as $t \to \infty$.

Solution   Let $X(t)$ denote the number of grams of the compound C present at any time t. Clearly $X(0) = 0$ g and $X(10) = 30$ g.

If, for example, 2 grams of compound C is present, we must have used, say, a grams of A and b grams of B, so $a + b = 2$ and $b = 4a$. Thus we must use $a = \frac{2}{5} = 2(\frac{1}{5})$ g of chemical A and $b = \frac{8}{5} = 2(\frac{4}{5})$ g of B. In general, for X grams of C we must use

$$\frac{X}{5} \text{ grams of } A \quad \text{and} \quad \frac{4}{5}X \text{ grams of } B.$$

The amounts of A and B remaining at any time are then

$$50 - \frac{X}{5} \quad \text{and} \quad 32 - \frac{4}{5}X,$$

respectively.

Now we know that the rate at which compound C is formed satisfies

$$\frac{dX}{dt} \propto \left(50 - \frac{X}{5}\right)\left(32 - \frac{4}{5}X\right).$$

To simplify the subsequent algebra, we factor $\frac{1}{5}$ from the first term and $\frac{4}{5}$ from the second and then introduce the constant of proportionality:

$$\frac{dX}{dt} = k(250 - X)(40 - X).$$

By separation of variables and partial fractions we can write

$$-\frac{1/210}{250 - X} dX + \frac{1/210}{40 - X} dX = k\, dt.$$

Integrating gives

$$\ln\left|\frac{250 - X}{40 - X}\right| = 210kt + c_1 \quad \text{or} \quad \frac{250 - X}{40 - X} = c_2 e^{210kt}. \tag{10}$$

When $t = 0$, $X = 0$, so it follows at this point that $c_2 = \frac{25}{4}$. Using $X = 30$ g at $t = 10$, we find $210k = \frac{1}{10}\ln\frac{88}{25} = 0.1258$. With this information we solve the last equation in (10) for X:

$$X(t) = 1000\,\frac{1 - e^{-0.1258t}}{25 - 4e^{-0.1258t}}. \tag{11}$$

The behavior of X as a function of time is displayed in Figure 3.15. It is clear from the accompanying table and (11) that $X \to 40$ as $t \to \infty$. This means that 40 grams of compound C is formed, leaving

$$50 - \frac{1}{5}(40) = 42 \text{ g of } A \quad \text{and} \quad 32 - \frac{4}{5}(40) = 0 \text{ g of } B. \quad \blacksquare$$

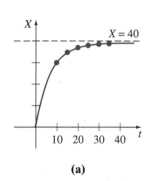

(a)

t (min)	X (g)
10	30 (measured)
15	34.78
20	37.25
25	38.54
30	39.22
35	39.59

(b)

Figure 3.15

Remarks   The indefinite integral $\int du/(a^2 - u^2)$ can be evaluated in terms of logarithms, the inverse hyperbolic tangent, or the inverse hyperbolic cotangent. For example, of the two results

$$\int \frac{du}{a^2 - u^2} = \frac{1}{a} \tanh^{-1} \frac{u}{a} + c, \quad |u| < a \qquad (12)$$

$$\int \frac{du}{a^2 - u^2} = \frac{1}{2a} \ln \left| \frac{a + u}{a - u} \right| + c, \quad |u| \neq a, \qquad (13)$$

(12) may be convenient in Problems 13 and 22 in Exercises 3.2, whereas (13) may be preferable in Problem 23.

EXERCISES 3.2

Answers to odd-numbered problems begin on page AN-2.

Logistic Equation

1. The number $N(t)$ of supermarkets throughout the country that use a computerized checkout system is described by the initial-value problem

$$\frac{dN}{dt} = N(1 - 0.0005N), \quad N(0) = 1.$$

   (a) Use the phase portrait concept of Section 2.1 to predict how many supermarkets are expected to adopt the new procedure over a long period of time. By hand, sketch a solution curve of the given initial-value problem.

   (b) Solve the initial-value problem and then use a graphing utility to verify the solution curve in part (a). How many supermarkets are expected to adopt the new technology when $t = 10$?

2. The number $N(t)$ of people in a community who are exposed to a particular advertisement is governed by the logistic equation. Initially $N(0) = 500$, and it is observed that $N(1) = 1000$. Solve for $N(t)$ if it is predicted that the limiting number of people in the community who will see the advertisement is 50,000.

3. A model for the population $P(t)$ in a suburb of a large city is given by the initial-value problem

$$\frac{dP}{dt} = P(10^{-1} - 10^{-7}P), \quad P(0) = 5000,$$

   where t is measured in months. What is the limiting value of the population? At what time will the population be equal to one-half of this limiting value?

4. **(a)** Census data for the United States between 1790 and 1950 are given in Table 3.1. Construct a logistic population model using the data from 1790, 1850, and 1910.

TABLE 3.1

Year	Population (in millions)
1790	3.929
1800	5.308
1810	7.240
1820	9.638
1830	12.866
1840	17.069
1850	23.192
1860	31.433
1870	38.558
1880	50.156
1890	62.948
1900	75.996
1910	91.972
1920	105.711
1930	122.775
1940	131.669
1950	150.697

(b) Construct a table comparing the actual census population with the population predicted by the model in part (a). Compute the error and the percentage error for each entry pair.

Modifications of the Logistic Equation

5. **(a)** Suppose $a = b = 1$ in the Gompertz differential equation (7). Since the DE is autonomous, use the phase portrait concept of Section 2.1 to sketch representative solution curves corresponding to the cases $P_0 > e$ and $0 < P_0 < e$.

   (b) Suppose $a = 1$, $b = -1$ in (7). Use a new phase portrait to sketch representative solution curves corresponding to the cases $P_0 > e^{-1}$ and $0 < P_0 < e^{-1}$.

6. Find an explicit solution of equation (7), subject to $P(0) = P_0$.

Chemical Reactions

7. Two chemicals A and B are combined to form a chemical C. The rate, or velocity, of the reaction is proportional to the product of the instantaneous amounts of A and B not converted to chemical C. Initially there are 40 grams of A and 50 grams of B, and for each gram of B, 2 grams of A is used. It is observed that 10 grams of C is formed in 5 minutes. How much is formed in 20 minutes? What is the limiting amount of C after a long time? How much of chemicals A and B remains after a long time?

8. Solve Problem 7 if 100 grams of chemical A is present initially. At what time is chemical C half-formed?

Miscellaneous Nonlinear Models

9. A tank in the form of a right-circular cylinder standing on end is leaking water through a circular hole in its bottom. As we saw in (10) of Section 1.3, when friction and contraction of water at the hole are ignored, the height h of water in the tank is described by

$$\frac{dh}{dt} = -\frac{A_h}{A_w}\sqrt{2gh},$$

   where A_w and A_h are the cross-sectional areas of the water and the hole, respectively.

   (a) Solve for $h(t)$ if the initial height of the water is H. By hand, draw the solution curve and give the interval I of its definition in terms of the symbols A_w, A_h, and H. Use $g = 32$ ft/s^2.

   (b) Suppose the tank is 10 feet high and has radius 2 feet and the circular hole has radius $\frac{1}{2}$ inch. If the tank is initially full, how long will it take to empty?

10. When friction and contraction of the water at the hole are taken into account, the model in Problem 9 becomes

$$\frac{dh}{dt} = -c\frac{A_h}{A_w}\sqrt{2gh},$$

   where $0 < c < 1$. How long will it take the tank in Problem 9(b) to empty if $c = 0.6$? See Problem 11 in Exercises 1.3.

11. A tank in the form of a right-circular cone standing on end, vertex down, is leaking water through a circular hole in its bottom.
    (a) Suppose the tank is 20 feet high and has radius 8 feet and the circular hole has radius 2 inches. In Problem 12 in Exercises 1.3 you were asked to show that the differential equation governing the height h of water leaking from a tank is

$$\frac{dh}{dt} = -\frac{5}{6h^{3/2}}.$$

In this model, friction and contraction of the water at the hole were taken into account with $c = 0.6$ and g was taken to be 32 ft/s². See Figure 1.26. If the tank is initially full, how long will it take the tank to empty?
    (b) Suppose the tank has a vertex angle of 60° and the circular hole has radius 2 inches. Determine the differential equation governing the height h of water. Use $c = 0.6$ and $g = 32$ ft/s². If the height of the water is initially 9 feet, how long will it take the tank to empty?

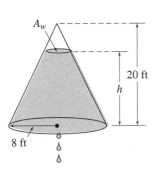

A_w

20 ft

h

8 ft

Figure 3.16

12. Suppose that the conical tank in Problem 11(a) is inverted, as shown in Figure 3.16, and that water leaks out a circular hole of radius 2 inches in the center of the circular base. Is the time it will take to empty a full tank the same as for the tank with vertex down in Problem 11? Take the friction/contraction coefficient to be $c = 0.6$ and $g = 32$ ft/s².

13. A differential equation governing the velocity v of a falling mass m subjected to air resistance proportional to the square of the instantaneous velocity is

$$m\frac{dv}{dt} = mg - kv^2,$$

where $k > 0$ is a constant of proportionality. The positive direction is downward.
    (a) Solve the equation subject to the initial condition $v(0) = v_0$.
    (b) Use the solution in part (a) to determine the limiting, or terminal, velocity of the mass. We saw how to determine the terminal velocity without solving the DE in Problem 25 in Exercises 2.1.
    (c) If the distance s, measured from the point where the mass was released above ground, is related to velocity by $ds/dt = v$, find an explicit expression for s if $s(0) = 0$.

14. Consider the 16-pound cannonball in Problem 30 in Exercises 3.1, shot vertically upward with an initial velocity of 300 ft/s. Determine the maximum height attained by the cannonball if air resistance is assumed to be proportional to the square of the instantaneous velocity. Assume the positive direction is upward and take $k = 0.0003$. [*Hint:* You will have to slightly modify the DE in Problem 13.]

15. (a) Determine a differential equation for the velocity $v(t)$ of a mass m sinking in water that imparts a resistance proportional to the square of the instantaneous velocity and also exerts an upward buoyant force whose magnitude is given by Archimedes' principle. See Problem 16 in Exercises 1.3. Assume the positive direction is downward.

(b) Solve the differential equation in part (a).

(c) Determine the limiting, or terminal, velocity of the sinking mass.

16. The differential equation

$$\frac{dy}{dx} = \frac{-x + \sqrt{x^2 + y^2}}{y}$$

describes the shape of a plane curve C that will reflect all incoming light beams to the same point. See Problem 27 in Exercises 1.3. There are several ways of solving this equation.

(a) Verify that the differential equation is homogeneous (see Section 2.5). Show that the substitution $y = ux$ yields

$$\frac{u\,du}{\sqrt{1 + u^2}\,(1 - \sqrt{1 + u^2})} = \frac{dx}{x}.$$

Use a CAS (or another judicious substitution) to integrate the left-hand side of the equation. Show that the curve C must be a parabola with focus at the origin and is symmetric with respect to the x-axis.

(b) Show that the differential equation in (a) can also be solved by means of the substitution $u = x^2 + y^2$.

17. (a) A simple model for the shape of a tsunami, or tidal wave, is given by

$$\frac{dW}{dx} = W\sqrt{4 - 2W},$$

where $W(x) > 0$ is the height of the wave expressed as a function of its position relative to a point offshore. By inspection, find all constant solutions of the DE.

(b) Solve the differential equation in part (a). A CAS may be useful for integration.

(c) Use a graphing utility to obtain the graphs of all solutions that satisfy the initial condition $W(0) = 2$.

18. An outdoor decorative pond in Palms Springs, California is in the shape of hemispherical tank and is to be filled with water pumped in through an inlet in the bottom of the tank. Suppose that the radius of the tank is $R = 10$ ft, that water is pumped in at a rate of π ft^3/min, and that the tank is initially empty. See Figure 3.17. As the tank fills, it loses water through evaporation. Assume that the rate of evaporation is proportional to the area A of the surface of the water and that the constant of proportionality is $k = 0.01$.

(a) The rate of change dV/dt of the volume of the water at time t is a net rate. Use this net rate to determine a differential equation for the height h of the water at time t. The volume of the water shown in the figure is $V = \pi R h^2 - \frac{1}{3}\pi h^3$, where $R = 10$. Express the area of the surface of the water $A = \pi r^2$ in terms of h.

(b) Solve the differential equation in part (a). Graph the solution.

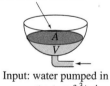

Output: water evaporates at rate proportional to area A of surface

Input: water pumped in at rate π ft^3/min

(a) hemispherical tank

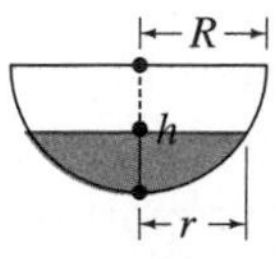

(b) cross-section of tank

Figure 3.17

(c) If there were no evaporation, how long would it take the tank to fill?

(d) With evaporation, what will be the depth of the water at the time found in part (c)? Will the tank ever be full? Prove your assertion.

Computer Lab Assignments

19. Read the documentation for your CAS on *scatter plots* (or *scatter diagrams*) and *least-squares linear fit*. The straight line that best fits a set of data points is called a **regression line** or a **least-squares line.** Your task is to construct a logistic model for the population of the United States, defining $f(P)$ in (2) as an equation of a regression line based on the population data in the table in Problem 4. One way of doing this is to approximate the left-hand side $\frac{1}{P}\frac{dP}{dt}$ of the first equation in (2), using the forward difference quotient in place of dP/dt:

$$Q(t) = \frac{1}{P(t)}\frac{P(t+h) - P(t)}{h}.$$

(a) Make a table of the values t, $P(t)$, and $Q(t)$, using $t = 0, 10, 20,$ $\dots, 160$ and $h = 10$. For example, the first line of the table should contain $t = 0$, $P(0)$, and $Q(0)$. With $P(0) = 3.929$ and $P(10) = 5.308$,

$$Q(0) = \frac{1}{P(0)}\frac{P(10) - P(0)}{10} = 0.035.$$

Note that $Q(160)$ depends on the 1960 census population $P(170)$. Look up this value.

(b) Use a CAS to obtain a scatter plot of the data $(P(t), Q(t))$ computed in part (a). Also use a CAS to find an equation of the regression line and to superimpose its graph on the scatter plot.

(c) Construct a logistic model $dP/dt = Pf(P)$, where $f(P)$ is the equation of the regression line found in part (b).

(d) Solve the model in part (c) using the initial condition $P(0) = 3.929$.

(e) Use a CAS to obtain another scatter plot, this time of the ordered pairs $(t, P(t))$ from your table in part (a). Use your CAS to superimpose the graph of the solution in part (d) on the scatter plot.

(f) Look up the U.S. census data for 1970, 1980, and 1990. What population does the logistic model in part (c) predict for these years? What does the model predict for the U.S. population $P(t)$ as $t \to \infty$?

20. (a) In Examples 3 and 4 of Section 2.1, we saw that any solution $P(t)$ of (4) possesses the asymptotic behavior $P(t) \to a/b$ as $t \to \infty$ for $P_0 > a/b$ and for $0 < P_0 < a/b$; as a consequence, the equilibrium solution $P = a/b$ is called an attractor. Use a root-finding application of a CAS (or a graphic calculator) to approximate the equilibrium solution of the immigration model $P' = P(1 - P) + 0.3e^{-P}$.

(b) Use a graphing utility to graph the function $F(P) = P(1 - P) + 0.3e^{-P}$. Explain how this graph can be used to determine whether the number found in part (a) is an attractor.

(c) Use a numerical solver to compare the solution curves for the IVPs

$$\frac{dP}{dt} = P(1 - P), \quad P(0) = P_0$$

for $P_0 = 0.2$ and $P_0 = 1.2$ with the solution curves for the IVPs

$$\frac{dP}{dt} = P(1 - P) + 0.3e^{-P}, \quad P(0) = P_0$$

for $P_0 = 0.2$ and $P_0 = 1.2$. Superimpose all curves on the same coordinate axes but, if possible, use a different color for the curves of the second initial-value problem. Over a long period of time, what percentage increase does the immigration model predict in the population, compared to the logistic model?

21. In Problem 14, let t_a be the time it takes the cannonball to attain its maximum height and let t_d be the time it takes the cannonball to fall from its maximum height to the ground. Compare the value of t_a with t_d and compare the magnitude of the impact velocity v_i with the initial velocity v_0. See Problem 42 in Exercises 3.1. A root-finding application of a CAS (or a graphic calculator) may be useful here. [*Hint:* Use the model of the cannonball falling in Problem 13.]

22. A skydiver is equipped with a stop watch and an altimeter. As shown in Figure 3.18, he opens his parachute 25 seconds after exiting a plane flying at an altitude of 20,000 feet and observes that his altitude is 14,800 feet. Assume that air resistance is proportional to the square of the instantaneous velocity, his initial velocity upon leaving the plane is zero, and $g = 32$ ft/s².

(a) Find the distance $s(t)$, measured from the plane, that the skydiver has traveled during freefall in time t. [*Hint:* The constant of proportionality k in the model given in Problem 13 is not specified. Use the expression for terminal velocity v_t obtained in part (b) of Problem 13 to eliminate k from the IVP. Then eventually solve for v_t.]

(b) How far does the skydiver fall and what is his velocity at $t = 15$ s?

14,800 ft

$s(t)$

25 s

Figure 3.18

23. A helicopter hovers 500 feet above a large open tank full of liquid (not water). A dense compact object weighing 160 pounds is dropped (released from rest) from the helicopter into the liquid. Assume that air resistance is proportional to v while the object is in the air and that viscous damping is proportional to v^2 after the object has entered the liquid. For air take $k = \frac{1}{4}$ and for the liquid $k = 0.1$. Assume the positive direction is downward. If the tank is 75 feet high, determine the time and the impact velocity when the object hits the bottom of the tank. [*Hint:* Think in terms of two distinct IVPs. Also, do not be discouraged by messy algebra. If you use (13), be careful in removing the absolute value sign. You might compare the velocity when the object hits the liquid (the initial velocity for the second problem) with the terminal velocity v_t of the object falling through the liquid.]

3.3 SYSTEMS OF LINEAR AND NONLINEAR DIFFERENTIAL EQUATIONS

• Systems of DEs as mathematical models   • Linear and nonlinear systems   • Radioactive decay series   • Mixtures   • Lotka-Volterra predator-prey model   • Competition models   • Electrical networks

In this section we are going to discuss mathematical models based on some of the topics explored in the preceding two sections. This section is similar to Section 1.3 in that we are just going to discuss certain mathematical models that are systems of first-order differential equations; we are not going to develop any methods for solving these systems. There are reasons for not solving systems at this point: First, we do not yet possess the necessary mathematical tools for solving systems. Second, some of the systems that we discuss simply cannot be solved in terms of elementary functions. We shall examine solution methods for systems of linear first-order equations in Chapter 8 and for systems of linear higher-order differential equations in Chapters 4 and 7.

Linear/Nonlinear Systems   Up to now all the mathematical models that we have considered have been single differential equations. A single differential equation can describe a single population in an environment; but if there are, say, two interacting and perhaps competing species living in the same environment (for example, rabbits and foxes), then a model for their populations $x(t)$ and $y(t)$ *might* be a system of two first-order differential equations such as

$$\frac{dx}{dt} = g_1(t, x, y)$$
$$\frac{dy}{dt} = g_2(t, x, y). \tag{1}$$

When g_1 and g_2 are linear in the variables x and y—that is, g_1 and g_2 have the forms

$$g_1(x, y) = c_1 x + c_2 y + f_1(t) \quad \text{and} \quad g_2(x, y) = c_3 x + c_4 y + f_2(t),$$

where the coefficients c_i could depend on t—then (1) is said to be a **linear system.** A system of differential equations that is not linear is said to be **nonlinear.**

Radioactive Series   In the discussion of radioactive decay in Sections 1.3 and 3.1, we assumed that the rate of decay was proportional to the number $A(t)$ of nuclei of the substance present at time t. When a substance decays by radioactivity it usually doesn't just transmute in one step into a stable substance; rather, the first substance decays into another radioactive substance, which in turn decays into a third substance, and so on. This process, called a **radioactive decay series,** continues until a stable element is reached. For example, the uranium decay series is U-238 $\rightarrow$ Th-234 $\rightarrow$ $\cdots$ $\rightarrow$ Pb-206, where Pb-206 is a stable isotope of lead. The half-lives of the various elements in a radioactive series can range from billions of years (4.5×10^9 years for U-238) to a fraction of a second. Suppose a radioactive series is described schematically by $X \xrightarrow{-\lambda_1} Y \xrightarrow{-\lambda_2} Z$, where $k_1 = -\lambda_1 < 0$ and $k_2 = -\lambda_2 < 0$ are the decay constants for substances X and Y, respectively, and Z is a stable element. Suppose, too, that $x(t)$, $y(t)$, and $z(t)$ denote amounts of substances X, Y, and Z, respectively, remaining at time t. The decay of element X is described by

$$\frac{dx}{dt} = -\lambda_1 x,$$

whereas the rate at which the second element Y decays is the net rate

$$\frac{dy}{dt} = \lambda_1 x - \lambda_2 y,$$

since Y is *gaining* atoms from the decay of X and at the same time *losing* atoms because of its own decay. Since Z is a stable element, it is simply gaining atoms from the decay of element Y:

$$\frac{dz}{dt} = \lambda_2 y.$$

In other words, a model of the radioactive decay series for three elements is the linear system of three first-order differential equations

$$\frac{dx}{dt} = -\lambda_1 x$$

$$\frac{dy}{dt} = \lambda_1 x - \lambda_2 y \qquad \qquad \textbf{(2)}$$

$$\frac{dz}{dt} = \lambda_2 y.$$

Mixtures   Consider the two tanks shown in Figure 3.19. Let us suppose for the sake of discussion that tank A contains 50 gallons of water in which 25 pounds of salt is dissolved. Suppose tank B contains 50 gallons of pure water. Liquid is pumped in and out of the tanks as indicated in the figure; the mixture exchanged between the two tanks and the liquid pumped out of tank B are assumed to be well stirred. We wish to construct

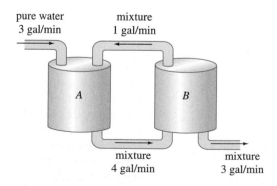

Figure 3.19

a mathematical model that describes the number of pounds $x_1(t)$ and $x_2(t)$ of salt in tanks A and B, respectively, at time t.

By an analysis similar to that on page 26 in Section 1.3 and Example 5 of Section 3.1, we see that the net rate of change of $x_1(t)$ for tank A is

$$\frac{dx_1}{dt} = \overbrace{(3 \text{ gal/min}) \cdot (0 \text{ lb/gal}) + (1 \text{ gal/min}) \cdot \left(\frac{x_2}{50} \text{ lb/gal}\right)}^{\substack{\text{input rate} \\ \text{of salt}}} - \overbrace{(4 \text{ gal/min}) \cdot \left(\frac{x_1}{50} \text{ lb/gal}\right)}^{\substack{\text{output rate} \\ \text{of salt}}}$$

$$= -\frac{2}{25} x_1 + \frac{1}{50} x_2.$$

Similarly, for tank B the net rate of change of $x_2(t)$ is

$$\frac{dx_2}{dt} = 4 \cdot \frac{x_1}{50} - 3 \cdot \frac{x_2}{50} - 1 \cdot \frac{x_2}{50}$$

$$= \frac{2}{25} x_1 - \frac{2}{25} x_2.$$

Thus we obtain the linear system

$$\frac{dx_1}{dt} = -\frac{2}{25} x_1 + \frac{1}{50} x_2$$

$$\frac{dx_2}{dt} = \frac{2}{25} x_1 - \frac{2}{25} x_2. \tag{3}$$

Observe that the foregoing system is accompanied by the initial conditions $x_1(0) = 25$, $x_2(0) = 0$.

A Predator-Prey Model   Suppose that two different species of animals interact within the same environment or ecosystem, and suppose further that the first species eats only vegetation and the second eats only the first species. In other words, one species is a predator and the other is a prey. For example, wolves hunt grass-eating caribou, sharks devour little fish, and the snowy owl pursues an arctic rodent called the lemming. For the sake of discussion, let us imagine that the predators are foxes and the prey are rabbits.

Let $x(t)$ and $y(t)$ denote, respectively, the fox and rabbit populations at any time t. If there were no rabbits, then one might expect that the foxes, lacking an adequate food supply, would decline in number according to

$$\frac{dx}{dt} = -ax, \quad a > 0. \tag{4}$$

When rabbits are present in the environment, however, it seems reasonable that the number of encounters or interactions between these two species per unit time is jointly proportional to their populations x and y—that is, proportional to the product xy. Thus when rabbits are present there is a supply of food, and so foxes are added to the system at a rate $bxy, b > 0$. Adding this last rate to (4) gives a model for the fox population:

$$\frac{dx}{dt} = -ax + bxy. \tag{5}$$

On the other hand, were there no foxes, then the rabbits would, with an added assumption of unlimited food supply, grow at a rate that is proportional to the number of rabbits present at time t:

$$\frac{dy}{dt} = dy, \quad d > 0. \tag{6}$$

But when foxes are present, a model for the rabbit population is (6) decreased by cxy, $c > 0$—that is, decreased by the rate at which the rabbits are eaten during their encounters with the foxes:

$$\frac{dy}{dt} = dy - cxy. \tag{7}$$

Equations (5) and (7) constitute a system of nonlinear differential equations

$$\frac{dx}{dt} = -ax + bxy = x(-a + by)$$
$$\frac{dy}{dt} = dy - cxy = y(d - cx), \tag{8}$$

where $a, b, c,$ and d are positive constants. This famous system of equations is known as the **Lotka-Volterra predator-prey model.**

Except for two constant solutions, $x(t) = 0$, $y(t) = 0$ and $x(t) = d/c$, $y(t) = a/b$, the nonlinear system (8) cannot be solved in terms of elementary functions. However, we can analyze such systems quantitatively and qualitatively. See Chapter 9, Numerical Solutions of Differential Equations, and Chapter 10, Plane Autonomous Systems and Stability.*

EXAMPLE 1   Predator-Prey Model

Suppose

$$\frac{dx}{dt} = -0.16x + 0.08xy$$

$$\frac{dy}{dt} = 4.5y - 0.9xy$$

represents a predator-prey model. Since we are dealing with populations, we have $x(t) \geq 0, y(t) \geq 0$. Figure 3.20, obtained with the aid of a numerical solver, shows typical population curves of the predators and prey for this model superimposed on the same coordinate axes. The initial conditions used were $x(0) = 4, y(0) = 4$. The curve in black represents the population $x(t)$ of the predator (foxes), and the colored curve is the population $y(t)$ of the prey (rabbits). Observe that the model seems to predict that both populations $x(t)$ and $y(t)$ are periodic in time. This makes intuitive sense since, as the number of prey decreases, the predator population eventually decreases because of a diminished food supply; but attendant to a decrease

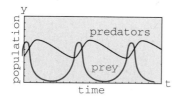

Figure 3.20

---

*Chapters 10–15 are in the expanded version of this text, *Differential Equations with Boundary-Value Problems.*

in the number of predators is an increase in the number of prey; this in turn gives rise to an increased number of predators, which ultimately brings about another decrease in the number of prey. ■

Competition Models Now suppose two different species of animals occupy the same ecosystem, not as predator and prey but rather as competitors for the same resources (such as food and living space) in the system. In the absence of the other, let us assume that the rate at which each population grows is given by

$$\frac{dx}{dt} = ax \quad \text{and} \quad \frac{dy}{dt} = cy, \tag{9}$$

respectively.

Since the two species compete, another assumption might be that each of these rates is diminished simply by the influence, or existence, of the other population. Thus a model for the two populations is given by the linear system

$$\frac{dx}{dt} = ax - by$$

$$\frac{dy}{dt} = cy - dx, \tag{10}$$

where a, b, c, and d are positive constants.

On the other hand, we might assume, as we did in (5), that each growth rate in (9) should be reduced by a rate proportional to the number of interactions between the two species:

$$\frac{dx}{dt} = ax - bxy$$

$$\frac{dy}{dt} = cy - dxy. \tag{11}$$

Inspection shows that this nonlinear system is similar to the Lotka-Volterra predator-prey model. Lastly, it might be more realistic to replace the rates in (9), which indicate that the population of each species in isolation grows exponentially, with rates indicating that each population grows logistically (that is, over a long time the population is bounded):

$$\frac{dx}{dt} = a_1 x - b_1 x^2 \quad \text{and} \quad \frac{dy}{dt} = a_2 y - b_2 y^2. \tag{12}$$

When these new rates are decreased by rates proportional to the number of interactions, we obtain another nonlinear model

$$\frac{dx}{dt} = a_1 x - b_1 x^2 - c_1 xy = x(a_1 - b_1 x - c_1 y)$$

$$\frac{dy}{dt} = a_2 y - b_2 y^2 - c_2 xy = y(a_2 - b_2 y - c_2 x), \tag{13}$$

where all coefficients are positive. The linear system (10) and the nonlinear systems (11) and (13) are, of course, called **competition models.**

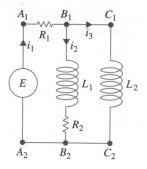

Figure 3.21

Networks   An electrical network having more than one loop also gives rise to simultaneous differential equations. As shown in Figure 3.21, the current $i_1(t)$ splits in the directions shown at point B_1, called a *branch point* of the network. By Kirchhoff's first law we can write

$$i_1(t) = i_2(t) + i_3(t). \tag{14}$$

In addition, we can also apply **Kirchhoff's second law** to each loop. For loop $A_1B_1B_2A_2A_1$, summing the voltage drops across each part of the loop gives

$$E(t) = i_1R_1 + L_1\frac{di_2}{dt} + i_2R_2. \tag{15}$$

Similarly, for loop $A_1B_1C_1C_2B_2A_2A_1$ we find

$$E(t) = i_1R_1 + L_2\frac{di_3}{dt}. \tag{16}$$

Using (14) to eliminate i_1 in (15) and (16) yields two linear first-order equations for the currents $i_2(t)$ and $i_3(t)$:

$$L_1\frac{di_2}{dt} + (R_1 + R_2)i_2 + R_1i_3 = E(t)$$

$$L_2\frac{di_3}{dt} + \qquad R_1i_2 + R_1i_3 = E(t). \tag{17}$$

We leave it as an exercise (see Problem 14) to show that the system of differential equations describing the currents $i_1(t)$ and $i_2(t)$ in the network containing a resistor, an inductor, and a capacitor shown in Figure 3.22 is

$$L\frac{di_1}{dt} + Ri_2 \quad = E(t)$$

$$RC\frac{di_2}{dt} + i_2 - i_1 = 0. \tag{18}$$

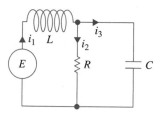

Figure 3.22

EXERCISES 3.3

Answers to odd-numbered problems begin on page AN-3.

Radioactive Series

1. We have not discussed methods by which systems of first-order differential equations can be solved. Nevertheless, systems such as (2) can be solved with no knowledge other than how to solve a single linear first-order equation. Find a solution of (2) subject to the initial conditions $x(0) = x_0$, $y(0) = 0$, $z(0) = 0$.

2. In Problem 1, suppose that time is measured in days, that the decay constants are $k_1 = -0.138629$ and $k_2 = -0.004951$, and that $x_0 = 20$. Use a graphing utility to obtain the graphs of the solutions $x(t)$, $y(t)$, and $z(t)$ on the same set of coordinate axes. Use the graphs to approximate the half-lives of substances X and Y.

3. Use the graphs in Problem 2 to approximate the times when the amounts $x(t)$ and $y(t)$ are the same, the times when the amounts $x(t)$ and $z(t)$ are the same, and the times when the amounts $y(t)$ and $z(t)$ are the same. Why does the time determined when the amounts $y(t)$ and $z(t)$ are the same make intuitive sense?

4. Construct a mathematical model for a radioactive series of four elements W, X, Y, and Z, where Z is a stable element.

Mixtures

5. Consider two tanks A and B, with liquid being pumped in and out at the same rates, as described by the system of equations (3). What is the system of differential equations if, instead of pure water, a brine solution containing 2 pounds of salt per gallon is pumped into tank A?

6. Use the information given in Figure 3.23 to construct a mathematical model for the number of pounds of salt $x_1(t)$, $x_2(t)$, and $x_3(t)$ at time t in tanks A, B, and C, respectively.

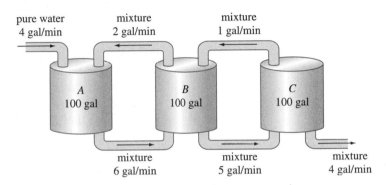

Figure 3.23

7. Two large tanks A and B each hold 100 gallons of brine. Initially, 100 pounds of salt is dissolved in the solution in tank A and 50 pounds of salt is dissolved in the solution in tank B. The system is closed in that well-stirred liquid is pumped only between the tanks, as shown in Figure 3.24.

   (a) Use the information given in the figure to construct a mathematical model for the number of pounds of salt $x_1(t)$ and $x_2(t)$ at time t in tanks A and B, respectively.

   (b) Find a relationship between the variables $x_1(t)$ and $x_2(t)$ that holds at time t. Explain why this relationship makes intuitive sense. Use this relationship to help find the amount of salt in tank B at $t = 30$ min.

8. Three large tanks contain brine, as shown in Figure 3.25. Use the information in the figure to construct a mathematical model for the number of pounds of salt $x_1(t)$, $x_2(t)$, and $x_3(t)$ at time t in tanks A, B, and C, respectively. Without solving the system, predict limiting values of $x_1(t)$, $x_2(t)$, and $x_3(t)$ as $t \to \infty$.

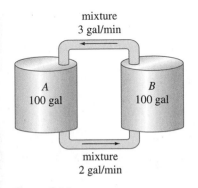

Figure 3.24

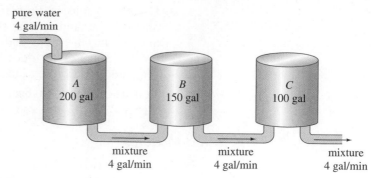

Figure 3.25

Predator-Prey Models

9. Consider the Lotka-Volterra predator-prey model defined by

$$\frac{dx}{dt} = -0.1x + 0.02xy$$

$$\frac{dy}{dt} = 0.2y - 0.025xy,$$

where the populations $x(t)$ (predators) and $y(t)$ (prey) are measured in thousands. Suppose $x(0) = 6$ and $y(0) = 6$. Use a numerical solver to graph $x(t)$ and $y(t)$. Use the graphs to approximate the time $t > 0$ when the two populations are first equal. Use the graphs to approximate the period of each population.

Competition Models

10. Consider the competition model defined by

$$\frac{dx}{dt} = x(2 - 0.4x - 0.3y)$$

$$\frac{dy}{dt} = y(1 - 0.1y - 0.3x),$$

where the populations $x(t)$ and $y(t)$ are measured in thousands and t in years. Use a numerical solver to analyze the populations over a long period of time for each of the following cases:
(a) $x(0) = 1.5, \quad y(0) = 3.5$ **(b)** $x(0) = 1, \quad y(0) = 1$
(c) $x(0) = 2, \quad y(0) = 7$ **(d)** $x(0) = 4.5, \quad y(0) = 0.5$

11. Consider the competition model defined by

$$\frac{dx}{dt} = x(1 - 0.1x - 0.05y)$$

$$\frac{dy}{dt} = y(1.7 - 0.1y - 0.15x),$$

where the populations $x(t)$ and $y(t)$ are measured in thousands and t in years. Use a numerical solver to analyze the populations over a long period of time for each of the following cases:
(a) $x(0) = 1, \quad y(0) = 1$ **(b)** $x(0) = 4, \quad y(0) = 10$
(c) $x(0) = 9, \quad y(0) = 4$ **(d)** $x(0) = 5.5, \quad y(0) = 3.5$

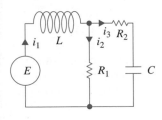

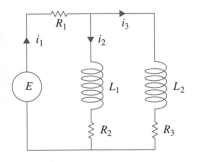

Figure 3.26

Figure 3.27

Networks

12. Show that a system of differential equations that describes the currents $i_2(t)$ and $i_3(t)$ in the electrical network shown in Figure 3.26 is

$$L\frac{di_2}{dt} + L\frac{di_3}{dt} + R_1 i_2 = E(t)$$

$$-R_1\frac{di_2}{dt} + R_2\frac{di_3}{dt} + \frac{1}{C}i_3 = 0.$$

13. Determine a system of first-order differential equations that describes the currents $i_2(t)$ and $i_3(t)$ in the electrical network shown in Figure 3.27.

14. Show that the linear system given in (18) describes the currents $i_1(t)$ and $i_2(t)$ in the network shown in Figure 3.22. [*Hint: dq/dt = i_3.*]

Miscellaneous Models

15. A communicable disease is spread throughout a small community, with a fixed population of n people, by contact between infected persons and persons who are susceptible to the disease. Suppose that everyone is susceptible to the disease initially and that no one leaves the community while the epidemic is spreading. At time t, let $s(t)$, $i(t)$, and $r(t)$ denote, in turn, the number of people in the community (measured in hundreds) who are *susceptible* to the disease but not yet infected, the number of people who are *infected* with the disease, and the number of people who have *recovered* from the disease. Explain why the system of differential equations

$$\frac{ds}{dt} = -k_1 si$$

$$\frac{di}{dt} = -k_2 i + k_1 si$$

$$\frac{dr}{dt} = k_2 i,$$

where k_1 (called the *infection rate*) and k_2 (called the *removal rate*) are positive constants, is a reasonable mathematical model, commonly called a **SIR model,** for the spread of the epidemic throughout the community. Give plausible initial conditions associated with this system of equations.

16. (a) Explain why, in Problem 15, it is sufficient to analyze only

$$\frac{ds}{dt} = -k_1 si$$

$$\frac{di}{dt} = -k_2 i + k_1 si.$$

(b) Suppose $k_1 = 0.2$, $k_2 = 0.7$, and $n = 10$. Choose various values of $i(0) = i_0$, $0 < i_0 < 10$. Use a numerical solver to determine what the model predicts about the epidemic in the two cases $s_0 > k_2/k_1$ and $s_0 \le k_2/k_1$. In the case of an epidemic, estimate the number of people who are eventually infected.

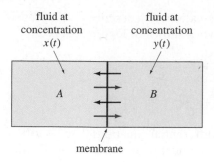

fluid at concentration $x(t)$

fluid at concentration $y(t)$

A   B

membrane

Figure 3.28

Discussion Problems

17. Suppose compartments A and B shown in Figure 3.28 are filled with fluids and are separated by a permeable membrane. The figure is a compartmental representation of the exterior and interior of a cell. Suppose, too, that a nutrient necessary for cell growth passes through the membrane. A model for the concentrations $x(t)$ and $y(t)$ of the nutrient in compartments A and B, respectively, at time t is given by the linear system of differential equations

$$\frac{dx}{dt} = \frac{\kappa}{V_A}(y - x)$$

$$\frac{dy}{dt} = \frac{\kappa}{V_B}(x - y),$$

where V_A and V_B are the volumes of the compartments and $\kappa > 0$ is a permeability factor. Let $x(0) = x_0$ and $y(0) = y_0$ denote the initial concentrations of the nutrient. Based solely on the equations in the system and the assumption $x_0 > y_0 > 0$, sketch, on the same set of coordinate axes, possible solution curves of the system. Explain your reasoning. Discuss the behavior of the solutions over a long period of time.

18. The system in Problem 17, like the system in (2), can be solved with no advanced knowledge. Solve for $x(t)$ and $y(t)$ and compare their graphs with your conjecture in Problem 18. [*Hint:* Subtract the two equations and let $z(t) = x(t) - y(t)$.] Determine the limiting values of $x(t)$ and $y(t)$ as $t \to \infty$. Explain why the answer to the last question makes intuitive sense.

19. Based solely on the physical description of the mixture problem on pages 122–123 and in Figure 3.19, discuss the nature of the functions $x_1(t)$ and $x_2(t)$. What is the behavior of each function over a long period of time? Sketch possible graphs of $x_1(t)$ and $x_2(t)$. Check your conjectures by using a numerical solver to obtain the solution curves of (3) subject to $x_1(0) = 25$, $x_2(0) = 0$.

CHAPTER 3 IN REVIEW

Answers to odd-numbered problems begin on page AN-3.

1. In March 1976 the world population reached 4 billion. A popular news magazine predicted that with an average yearly growth rate of 1.8%, the world population would be 8 billion in 45 years. How does this value compare with that predicted by the model that says the rate of increase is proportional to the population at any time?

2. Air containing 0.06% carbon dioxide is pumped into a room whose volume is 8000 ft^3. The air is pumped in at a rate of 2000 ft^3/min, and the circulated air is then pumped out at the same rate. If there is an initial concentration of 0.2% carbon dioxide, determine the subsequent amount in the room at any time. What is the concentration at 10 minutes? What is the steady-state, or equilibrium, concentration of carbon dioxide?

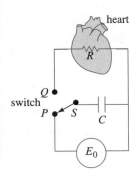

heart

R

Q

switch

P   S

C

E_0

Figure 3.29

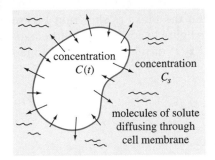

concentration
$C(t)$

concentration
C_s

molecules of solute
diffusing through
cell membrane

Figure 3.30

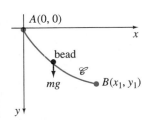

A(0, 0)

x

bead

$\mathscr{C}$

mg $B(x_1, y_1)$

y

Figure 3.31

3. A heart pacemaker, as shown in Figure 3.29, consists of a battery of constant voltage E_0, a capacitor, and the heart as a resistor. Over a time interval of length t_1, $0 \le t < t_1$, switch S is at position P and the capacitor is charging. At time t_1 switch S is moved to position Q; the capacitor then discharges, sending an electrical impulse to the heart over a time interval of length t_2, $t_1 \le t < t_1 + t_2$. The voltage applied to the heart over these two intervals is described by the piecewise-defined differential equation

$$\frac{dE}{dt} = \begin{cases} 0, & 0 \le t < t_1 \\ -\dfrac{1}{RC}E, & t_1 \le t < t_1 + t_2. \end{cases}$$

The charging and discharging over time intervals of the same length is repeated indefinitely.
   (a) Suppose $t_1 = 4$ s, $t_2 = 2$ s, and $E_0 = 12$ V. Solve for $E(t)$ over the interval $0 \le t \le 18$.
   (b) For the sake of illustration, suppose $R = C = 1$. Graph your solution in part (a) on the interval $0 \le t \le 18$.

4. Suppose a cell is suspended in a solution containing a solute of constant concentration C_s. Suppose further that the cell has constant volume V and that the area of its permeable membrane is the constant A. By Fick's law the rate of change of its mass m is directly proportional to the area A and the difference $C_s - C(t)$, where $C(t)$ is the concentration of the solute inside the cell at any time t. Find $C(t)$ if $m = VC(t)$ and $C(0) = C_0$. See Figure 3.30.

5. Consider Newton's law of cooling $dT/dt = k(T - T_m)$, $k < 0$, where the temperature of the surrounding medium T_m changes with time. Suppose that the initial temperature of a body is T_1, that the initial temperature of the surrounding medium is T_2, and that $T_m = T_2 + B(T_1 - T)$, where $B > 0$ is a constant.
   (a) Find the temperature of the body at any time t.
   (b) What is the limiting value of the temperature as $t \to \infty$?
   (c) What is the limiting value of T_m as $t \to \infty$?

6. An LR series circuit has a variable inductor with the inductance defined by

$$L = \begin{cases} 1 - \dfrac{t}{10}, & 0 \le t < 10 \\ 0, & t \ge 10. \end{cases}$$

Find the current $i(t)$ if the resistance is 0.2 ohm, the impressed voltage is $E(t) = 4$, and $i(0) = 0$. Graph $i(t)$.

7. A classical problem in the calculus of variations is to find the shape of a curve $\mathscr{C}$ such that a bead, under the influence of gravity, will slide from point $A(0, 0)$ to point $B(x_1, y_1)$ in the least time. See Figure 3.31. It can be shown that a nonlinear differential equation for the shape $y(x)$ of the path is $y[1 + (y')^2] = k$, where k is a constant. First solve for dx in terms of y and dy, and then use the substitution $y = k \sin^2 \theta$ to obtain the parametric form of the solution. The curve $\mathscr{C}$ turns out to be a cycloid.

8. When all the curves in a family $G(x, y, c_1) = 0$ intersect orthogonally all the curves in another family $H(x, y, c_2) = 0$, the families are said

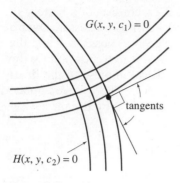

$G(x, y, c_1) = 0$

tangents

$H(x, y, c_2) = 0$

Figure 3.32

to be **orthogonal trajectories** of each other. See Figure 3.32. If $dy/dx = f(x, y)$ is the differential equation of one family, then the differential equation for the orthogonal trajectories of this family is $dy/dx = -1/f(x, y)$.

(a) Find a differential equation for the family $y = -x - 1 + c_1e^x$. Find the orthogonal trajectories of this family.

(b) Find the orthogonal trajectories for the family $y = 1/(x + c_1)$. Use a graphing utility to graph both families on the same set of coordinate axes.

9. The populations of two species of animals are described by the nonlinear system of first-order differential equations

$$\frac{dx}{dt} = k_1x(\alpha - x)$$

$$\frac{dy}{dt} = k_2xy.$$

Solve for x and y in terms of t.

10. Initially, two large tanks A and B each hold 100 gallons of brine. The well-stirred liquid is pumped between the tanks as shown in Figure 3.33. Use the information given in the figure to construct a mathematical model for the numbers $x_1(t)$ and $x_2(t)$ of pounds of salt at time t in tanks A and B, respectively.

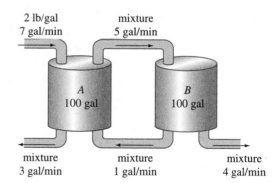

2 lb/gal
7 gal/min

mixture
5 gal/min

A
100 gal

B
100 gal

mixture
3 gal/min

mixture
1 gal/min

mixture
4 gal/min

Figure 3.33

Project Module

by
Gilbert N. Lewis

HARVESTING OF RENEWABLE NATURAL RESOURCES

There are many renewable natural resources that humans desire to use. Examples are salmon and halibut from the Pacific Ocean, lake trout from the Great Lakes, trees from our forests, and the deer and waterfowl that are hunted annually. Nonrenewable resources, such as coal, oil, gas, or minerals, are not considered here. It is desirable that a policy be developed that will allow a maximal harvest of a renewable natural resource and yet not deplete that resource below a sustainable level. The simple mathematical model developed here provides some insights into the planning process. We follow the development given by Clark.*

Without human intervention, we assume that the population would behave logistically—that is, follow the solution given by equation (3) and Figure 3.13 in Section 3.2 of this text. Recall that if $P(t)$ represents the population (expressed in terms of either biomass or number of individuals) at time t (measured in years), then the logistic equation is

$$\frac{dP}{dt} = P\left(r - \frac{r}{K}P\right) = F(P), \tag{1}$$

where $r > 0$ is the **intrinsic growth rate** and K is the **environmental carrying capacity** (also known as the **saturation level** or the **limiting population,** since the population approaches this value over time). The values of these constants are determined experimentally.

In this module we will further assume that humans will be harvesting individuals from an animal population.

Constant Harvesting Rate As a first example, we assume that the **harvest rate** is a constant h. The modification of the differential equation (1) is then

$$\frac{dP}{dt} = P\left(r - \frac{r}{K}P\right) - h = F(P) - h = G(P). \tag{2}$$

Note that the function $G(P) = (-r/K)P^2 + rP - h$ is a quadratic polynomial in P whose graph is concave down. In the normal situation where

---

*Colin W. Clark, *Mathematical Bioeconomics: The Optimal Management of Renewable Resources,* 2nd ed. (New York: John Wiley & Sons, 1990).

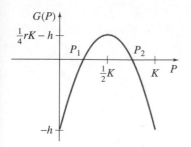

Figure 1

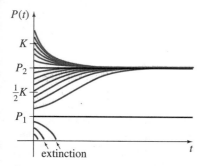

Figure 2

the harvest rate is not too high (that is, $h < \max F(P) = F(\frac{1}{2}K) = \frac{1}{4}rK$), the function G has two real zeros on the interval $[0, K]$. The values of the two zeros P_1 and P_2, shown in Figure 1, are found from the quadratic formula:

$$P_{1,2} = \frac{K \pm \sqrt{K^2 - 4Kh/r}}{2}.$$

We observe that $P(t) = P_1$ and $P(t) = P_2$ are constant solutions, called equilibrium solutions, of (2). Now from Figure 1 we see that the derivative dP/dt is positive on the interval $P_1 < P < P_2$, and so P will increase on that interval, while dP/dt is negative for all other values of P. In addition, by differentiating (2) we get

$$\frac{d^2P}{dt^2} = r\left(1 - \frac{2}{K}P\right)\frac{dP}{dt} = r\left(1 - \frac{2}{K}P\right)G(P),$$

which implies that the graph of $P(t)$ is concave downward for $P < P_1$ and for $\frac{1}{2}K < P < P_2$ and concave upward otherwise. Without solving the differential equation, therefore, we can still deduce the qualitative behavior of the solutions. See Figure 2. Note that if the initial population P_0 is less than P_1, the population $P(t)$ decreases to zero (extinction) in finite time, while otherwise the population $P(t)$ approaches P_2, a value less than K (the limiting population without harvesting). Mathematicians refer to the number P_2 as an **asymptotically stable equilibrium** or an **attractor,** since other solutions that start close to P_2 approach the horizontal line $P = P_2$; the number P_1 is called an **unstable equilibrium** or a **repeller.** Thus we can conclude that the harvest cannot be too large without depleting the resource.

We now ask the next obvious question: How large can the harvest be and still allow a sustainable (that is, long-term) harvest? We observed earlier that there are two real solutions to $G(P) = 0$ if $h < \frac{1}{4}rK$. On the other hand, if $h > \frac{1}{4}rK$, or $\frac{1}{4}rK - h < 0$, it can be seen from Figure 1 that $dP/dt = G(P) < 0$ and so $P(t)$ will decrease to zero. Finally, for $h = \frac{1}{4}rK$ the equation $G(P) = 0$ has the single root $P_1 = \frac{1}{2}K$. This value of P is also a constant solution of the differential equation. The value $h = \frac{1}{4}rK$ is called the **maximum sustainable yield** (MSY). It allows for a constant population of $P_1 = \frac{1}{2}K$ and a constant harvest equal to the MSY. The MSY, in other words, is equal to the population added annually due to reproduction minus death.

A word of caution is in order for anyone using this model to calculate the MSY in actual management practice. The values of r and K may be known only to within an accuracy of 10%. The value $h = \frac{1}{4}rK$ calculated for the MSY might, in fact, be too large for the given population, resulting in a decline to extinction.

So far, we have deduced much qualitative information about solutions of (2) without actually solving the differential equation. A solution of (2) subject to the initial condition $P(0) = P_0$ is easy to find by separation of variables. Since P_1 and P_2 are zeros, $P - P_1$ and $P - P_2$ must be factors of $G(P)$. Writing

$$\frac{dP}{dt} = -\frac{r}{K}(P - P_1)(P - P_2) \quad \text{as} \quad \frac{dP}{(P - P_1)(P - P_2)} = -\frac{r}{K}dt,$$

using partial fractions, and then integrating yields

$$\frac{1}{P_2 - P_1} \ln\left|\frac{P - P_2}{P - P_1}\right| = -\frac{r}{K}t + c. \tag{3}$$

By applying $P(0) = P_0$ to (3) and solving for $P(t)$, we arrive at

$$P(t) = \frac{P_2(P_0 - P_1) - P_1(P_0 - P_2)e^{-\alpha t}}{P_0 - P_1 - (P_0 - P_2)e^{-\alpha t}}, \tag{4}$$

where $\alpha = r(P_2 - P_1)/K = r\sqrt{1 - 4h/Kr}$. It should be clear from the explicit solution (4) that $P(t)$ approaches P_2 as time t increases.

Harvesting Proportional to Population In our next model, we assume that the harvest is proportional to the size of the population. The modification of (1) is then

$$\frac{dP}{dt} = F(P) - EP = P\left(r - \frac{r}{K}P\right) - EP = G(P), \tag{5}$$

where $E > 0$ is a constant referred to as the **effort,** since it is a measure of the effort that goes into harvesting the resource. As before, we consider the equilibrium solutions obtained from the equation $G(P) = P(r - E - rP/K) = 0$. We find one positive solution $P_1 = K(1 - E/r)$, as long as the effort E does not exceed the growth rate r. Since (5) can be written $dP/dt = -(r/K) P(P - P_1)$, we see that if $0 < P < P_1$ then $dP/dt > 0$, while if $P > P_1$ then $dP/dt < 0$. This indicates that a solution will always approach the equilibrium solution $P(t) = P_1$, making P_1 an asymptotically stable equilibrium, or an attractor. The equilibrium harvest, or **sustainable yield,** in this case is $EP_1 = KE(1 - E/r)$. Note that the left-hand side of this last expression is quadratically dependent upon E (the right-hand side) and has a maximum when $E = \frac{1}{2}r$ and $P_1 = \frac{1}{2}K$. For these latter values the number EP_1 is the maximum sustainable yield, or MSY.

To conclude this model, we note that we can solve (5) analytically. In fact, equation (5) in the form

$$\frac{dP}{dt} = P\left(r - E - \frac{r}{K}P\right) \tag{6}$$

is recognized as the logistic equation (4) of Section 3.2 with $a = r - E$ and $b = r/K$. Thus from (5) of Section 3.2 we obtain the solution

$$P(t) = \frac{(r - E)P_0}{rP_0/K + (r - E - rP_0/K)e^{-(r-E)t}}. \tag{7}$$

It is seen from (7) that the limiting population as $t \to \infty$ is $K(1 - E/r)$.

Antarctic fin whale

EXAMPLE 1 **Antarctic Fin Whale**

As an example, Clark estimates the values $r = 0.08$ and $K = 400,000$ for the Antarctic fin whale, with 1976 corresponding to $t = 0$ and $P_0 = P(0) = 70,000$.

(a) In the *constant harvesting model,* we find that the MSY is given by $h = \frac{1}{4}rK = 8000$, with a fixed population of $P_1 = \frac{1}{2}K = 200,000$. However, since the initial population is $P_0 = 70,000 < P_1$, the population will decline to zero, since reproduction the first year (population times growth rate $= 70{,}000 \times 0.08 = 5600$) is less than the harvest. In order to guarantee that $P_0 > P_1$, we need to choose h less than 4620 (see part (b) of Problem 3 in the Related Exercises). In this case, the population $P(t)$ would slowly approach $P_2 = 330{,}000$ over time. One management technique for increasing the number of whales faster is to severely limit the harvest initially, with the possibility of a larger annual harvest later.

(b) For the *constant effort model*, let us assume that the effort is one-half the growth rate in part (a)—that is, $E = \frac{1}{2}r = 0.04$. Then $P_1 = \frac{1}{2}K = 200{,}000$, and the first-year harvest is $EP(0) = 0.04 \times 70{,}000 = 2800$. The MSY is $EP_1 = 0.04 \times 200{,}000 = 8000$. ∎

RELATED EXERCISES

1. **(a)** Consider the constant harvesting model

$$\frac{dP}{dt} = P(5 - P) - 4, \quad P(0) = P_0.$$

Since the differential equation is autonomous, use the phase portrait concept discussed in Section 2.1 to sketch representative solution curves corresponding to the cases $P_0 > 4$, $1 < P_0 < 4$, and $0 < P_0 < 1$. Determine the long-term behavior of the population in each case.

   (b) Solve the IVP in part (a). Use a graphing utility to verify your results in part (a).

   (c) Use the information in parts (a) and (b) to determine whether the population becomes extinct in finite time. If so, find that time.

2. As in Problem 1, investigate the given constant harvesting models both qualitatively and analytically. Determine whether the population becomes extinct in finite time. If so, find that time.

   (a) $\dfrac{dP}{dt} = P(5 - P) - \frac{25}{4}$ **(b)** $\dfrac{dP}{dt} = P(5 - P) - 7$

3. This problem concerns the Antarctic fin whale discussed in Example 1.
   (a) If no harvesting were allowed, how long would it take for the population to pass $\frac{1}{2}K = 200{,}000$? Plot the solution for $0 < t < 100$, using $P_0 = 70{,}000$.

   (b) In the constant harvesting case, calculate the value of $h = h_0$ that allows $P_0 = P_1$. Plot the solution for $0 < t < 100$, using $P_0 = 70{,}000$ and $h = \frac{1}{2}h_0$.

   (c) In the constant effort case, what effort E_0 yields the MSY? What is the yield? What is the limiting population? Plot the solution for $0 < t < 100$, using $P_0 = 70{,}000$ and $E = \frac{1}{2}E_0$.

4. **(a)** The data in Table 3.2 (taken from the 1976 edition of the book by Clark) give a portion of the statistical history of the Peruvian anchovy (*Engraulis ringens*) fishery. We can consider the product of the number of boats B times the number of fishing days D to be a measure of the fishing effort. We let $E = cBD$, where c is a constant of proportionality. If the catch C (harvest) is proportional to both the effort and the size of the population, then

TABLE 3.2

Year	Number of boats	Number of fishing days	Catch (millions of tons)
1959	414	294	1.91
1960	667	279	2.93
1961	756	298	4.58
1962	1069	294	6.27
1963	1655	269	6.42
1964	1744	297	8.86
1965	1623	265	7.23
1966	1650	190	8.53
1967	1569	170	9.82
1968	1490	167	10.26
1969	1455	162	8.96
1970	1499	180	12.27
1971	1473	89	10.28
1972	1399	89	4.45
1973	1256	27	1.78

$C = EP(t) = cBDP(t)$, where $P(t)$ is the population (measured in millions of tons) at time t, and $P(t) = C/(cBD)$. Use a CAS to plot the data (P vs. years) for 1959–1973 with $c = 2 \times 10^{-7}$.

(b) Taking into account the natural scatter in the data and ignoring anomalous points, we could consider the graph produced in (a) as representing a logistic function of the form

$$P(t) = \frac{P_0 K}{P_0 + (K - P_0)e^{-rt}},$$

where P_0, K, and r are the initial population, the limiting population, and the growth rate, respectively. Find values of P_0, K, and r for which the graph of $P(t)$ approximates your plot of the data points in part (a) reasonably well. Superimpose the graph of $P(t)$ on the graph of the data.

(c) For the values of P_0, K, and r found in part (b) and for a harvest $h = \frac{1}{8}rK$, plot the solution (4) of the differential equation in the constant harvest case on the interval $0 \le t \le 100$.

(d) For the values of P_0, K, and r found in part (b) and for an effort $E = \frac{1}{2}r$, plot the solution (7) of the differential equation in the constant effort case on the interval $0 \le t \le 100$.

Solution of a Cauchy-Euler DE; see page 197.

4 HIGHER-ORDER DIFFERENTIAL EQUATIONS

INTRODUCTION We turn now to the solution of differential equations of order two or higher. In the first seven sections of this chapter we will examine the underlying theory and some of the methods for solving certain kinds of *linear* equations. The elimination method for solving systems of linear ordinary differential equations is introduced in Section 4.8 because this basic method simply uncouples a system into individual linear higher-order equations in each dependent variable. The chapter concludes with a brief examination of *nonlinear* higher-order equations.

4.1 PRELIMINARY THEORY: LINEAR EQUATIONS

• Linear higher-order DEs   • Initial-value problem   • Existence and uniqueness   • Homogeneous and nonhomogeneous DEs   • Linear differential operator   • Linear dependence/independence   • Wronskian • Fundamental set   • Superposition principles   • General solution • Complementary function   • Particular solution

In Chapter 2 we saw that we could solve a few first-order differential equations by recognizing them as separable, linear, exact, homogeneous, or perhaps Bernoulli equations. Even though the solutions of these equations were in the form of a one-parameter family, this family, with one exception, did not represent the general solution of the differential equation. Only in the case of linear first-order equations were we able to obtain general solutions, by paying attention to certain continuity conditions. Recall that a general solution is a family of solutions defined on some interval that contains all the solutions of the DE that are defined on that interval. Because our primary goal in the present chapter is to find general solutions of linear higher-order DEs, we first need to examine some theory.

4.1.1 INITIAL-VALUE AND BOUNDARY-VALUE PROBLEMS

Initial-Value Problem   In Section 1.2 we defined an initial-value problem for a general nth-order differential equation. For a linear differential equation, an **nth-order initial-value problem** is

Solve:
$$a_n(x)\frac{d^n y}{dx^n} + a_{n-1}(x)\frac{d^{n-1}y}{dx^{n-1}} + \cdots + a_1(x)\frac{dy}{dx} + a_0(x)y = g(x) \tag{1}$$

Subject to: $y(x_0) = y_0, \quad y'(x_0) = y_1, \quad \ldots, \quad y^{(n-1)}(x_0) = y_{n-1}.$

Recall that for a problem such as this one we seek a function defined on some interval I, containing x_0, that satisfies the differential equation and the n initial conditions specified at x_0: $y(x_0) = y_0, y'(x_0) = y_1, \ldots,$ $y^{(n-1)}(x_0) = y_{n-1}$. We have already seen that in the case of a second-order initial-value problem, a solution curve must pass through the point (x_0, y_0) and have slope y_1 at this point.

Existence and Uniqueness   In Section 1.2 we stated a theorem that gave conditions under which the existence and uniqueness of a solution of a first-order initial-value problem were guaranteed. The theorem that follows gives sufficient conditions for the existence of a unique solution of the problem in (1).

THEOREM 4.1	**Existence of a Unique Solution**

Let $a_n(x), a_{n-1}(x), \ldots, a_1(x), a_0(x)$ and $g(x)$ be continuous on an interval I, and let $a_n(x) \neq 0$ for every x in this interval. If $x = x_0$ is any point in this interval, then a solution $y(x)$ of the initial-value problem (1) exists on the interval and is unique.

EXAMPLE 1   Unique Solution of an IVP

The initial-value problem

$$3y''' + 5y'' - y' + 7y = 0, \quad y(1) = 0, \quad y'(1) = 0, \quad y''(1) = 0$$

possesses the trivial solution $y = 0$. Since the third-order equation is linear with constant coefficients, it follows that all the conditions of Theorem 4.1 are fulfilled. Hence $y = 0$ is the *only* solution on any interval containing $x = 1$. ∎

EXAMPLE 2   Unique Solution of an IVP

You should verify that the function $y = 3e^{2x} + e^{-2x} - 3x$ is a solution of the initial-value problem

$$y'' - 4y = 12x, \quad y(0) = 4, \quad y'(0) = 1.$$

Now the differential equation is linear, the coefficients as well as $g(x) = 12x$ are continuous, and $a_2(x) = 1 \neq 0$ on any interval I containing $x = 0$. We conclude from Theorem 4.1 that the given function is the unique solution on I. ∎

The requirements in Theorem 4.1 that $a_i(x)$, $i = 0, 1, 2, \ldots, n$ be continuous and $a_n(x) \neq 0$ for every x in I are both important. Specifically, if $a_n(x) = 0$ for some x in the interval, then the solution of a linear initial-value problem may not be unique or even exist. For example, you should verify that the function $y = cx^2 + x + 3$ is a solution of the initial-value problem

$$x^2 y'' - 2xy' + 2y = 6, \quad y(0) = 3, \quad y'(0) = 1$$

on the interval $(-\infty, \infty)$ for any choice of the parameter c. In other words, there is no unique solution of the problem. Although most of the conditions of Theorem 4.1 are satisfied, the obvious difficulties are that $a_2(x) = x^2$ is zero at $x = 0$ and that the initial conditions are also imposed at $x = 0$.

Boundary-Value Problem   Another type of problem consists of solving a linear differential equation of order two or greater in which the dependent variable y or its derivatives are specified at *different points*. A problem such as

$$\text{Solve:} \quad a_2(x)\frac{d^2 y}{dx^2} + a_1(x)\frac{dy}{dx} + a_0(x)y = g(x)$$

$$\text{Subject to:} \quad y(a) = y_0, \quad y(b) = y_1$$

is called a **boundary-value problem (BVP).** The prescribed values $y(a) = y_0$ and $y(b) = y_1$ are called **boundary conditions.** A solution of the foregoing problem is a function satisfying the differential equation on some interval I, containing a and b, whose graph passes through the two points (a, y_0) and (b, y_1). See Figure 4.1.

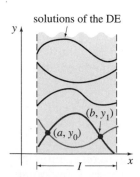

solutions of the DE

Figure 4.1

For a second-order differential equation, other pairs of boundary conditions could be

$$y'(a) = y_0, \quad y(b) = y_1$$
$$y(a) = y_0, \quad y'(b) = y_1$$
$$y'(a) = y_0, \quad y'(b) = y_1,$$

where y_0 and y_1 denote arbitrary constants. These three pairs of conditions are just special cases of the general boundary conditions

$$\alpha_1 y(a) + \beta_1 y'(a) = \gamma_1$$
$$\alpha_2 y(b) + \beta_2 y'(b) = \gamma_2.$$

The next example shows that even when the conditions of Theorem 4.1 are fulfilled, a boundary-value problem may have several solutions (as suggested in Figure 4.1), a unique solution, or no solution at all.

EXAMPLE 3 **A BVP Can Have Many, One, or No Solutions**

In Example 4 of Section 1.1 we saw that the two-parameter family of solutions of the differential equation $x'' + 16x = 0$ is

$$x = c_1 \cos 4t + c_2 \sin 4t. \tag{2}$$

(a) Suppose we now wish to determine that solution of the equation that further satisfies the boundary conditions $x(0) = 0$, $x(\pi/2) = 0$. Observe that the first condition $0 = c_1 \cos 0 + c_2 \sin 0$ implies $c_1 = 0$, so $x = c_2 \sin 4t$. But when $t = \pi/2$, $0 = c_2 \sin 2\pi$ is satisfied for any choice of c_2 since $\sin 2\pi = 0$. Hence the boundary-value problem

$$x'' + 16x = 0, \quad x(0) = 0, \quad x\left(\frac{\pi}{2}\right) = 0 \tag{3}$$

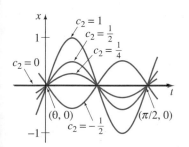

Figure 4.2

has infinitely many solutions. Figure 4.2 shows the graphs of some of the members of the one-parameter family $x = c_2 \sin 4t$ that pass through the two points $(0, 0)$ and $(\pi/2, 0)$.

(b) If the boundary-value problem in (3) is changed to

$$x'' + 16x = 0, \quad x(0) = 0, \quad x\left(\frac{\pi}{8}\right) = 0 \tag{4}$$

then $x(0) = 0$ still requires $c_1 = 0$ in the solution (2). But applying $x(\pi/8) = 0$ to $x = c_2 \sin 4t$ demands that $0 = c_2 \sin(\pi/2) = c_2 \cdot 1$. Hence $x = 0$ is a solution of this new boundary-value problem. Indeed, it can be proved that $x = 0$ is the *only* solution of (4).

(c) Finally, if we change the problem to

$$x'' + 16x = 0, \quad x(0) = 0, \quad x\left(\frac{\pi}{2}\right) = 1 \tag{5}$$

we find again from $x(0) = 0$ that $c_1 = 0$, but applying $x(\pi/2) = 1$ to $x = c_2 \sin 4t$ leads to the contradiction $1 = c_2 \sin 2\pi = c_2 \cdot 0 = 0$. Hence the boundary-value problem (5) has no solution. ■

4.1.2   HOMOGENEOUS EQUATIONS

A linear nth-order differential equation of the form

$$a_n(x)\frac{d^n y}{dx^n} + a_{n-1}(x)\frac{d^{n-1} y}{dx^{n-1}} + \cdots + a_1(x)\frac{dy}{dx} + a_0(x)y = 0 \tag{6}$$

is said to be **homogeneous,** whereas an equation

$$a_n(x)\frac{d^n y}{dx^n} + a_{n-1}(x)\frac{d^{n-1} y}{dx^{n-1}} + \cdots + a_1(x)\frac{dy}{dx} + a_0(x)y = g(x), \tag{7}$$

with $g(x)$ not identically zero, is said to be **nonhomogeneous.** For example, $2y'' + 3y' - 5y = 0$ is a homogeneous linear second-order differential equation, whereas $x^3y''' + 6y' + 10y = e^x$ is a nonhomogeneous linear third-order differential equation. The word *homogeneous* in this context does not refer to coefficients that are homogeneous functions, as in Section 2.5.

We shall see that in order to solve a nonhomogeneous linear equation (7), we must first be able to solve the **associated homogeneous equation** (6).

Note   To avoid needless repetition throughout the remainder of this text, we shall, as a matter of course, make the following important assumptions when stating definitions and theorems about the linear equations (6) and (7): On some common interval I,

- the coefficients $a_i(x)$, $i = 0, 1, 2, \ldots, n$ are continuous;
- the right-hand member $g(x)$ is continuous; and
- $a_n(x) \neq 0$ for every x in the interval.

Differential Operators   In calculus, differentiation is often denoted by the capital letter D—that is, $dy/dx = Dy$. The symbol D is called a **differential operator** since it transforms a differentiable function into another function. For example, $D(\cos 4x) = -4 \sin 4x$ and $D(5x^3 - 6x^2) = 15x^2 - 12x$. Higher-order derivatives can be expressed in terms of D in a natural manner:

$$\frac{d}{dx}\left(\frac{dy}{dx}\right) = \frac{d^2 y}{dx^2} = D(Dy) = D^2 y \quad \text{and, in general,} \quad \frac{d^n y}{dx^n} = D^n y,$$

where y represents a sufficiently differentiable function. Polynomial expressions involving D, such as $D + 3$, $D^2 + 3D - 4$, and $5x^3D^3 - 6x^2D^2 + 4xD + 9$, are also differential operators. In general, we define an ***n*th-order differential operator** to be

$$L = a_n(x)D^n + a_{n-1}(x)D^{n-1} + \cdots + a_1(x)D + a_0(x). \tag{8}$$

As a consequence of two basic properties of differentiation, $D(cf(x)) = cDf(x)$, c is a constant, and $D\{f(x) + g(x)\} = Df(x) + Dg(x)$, the differential operator L possesses a linearity property; that is, L operating on a linear combination of two differentiable functions is the same as the linear combination of L operating on the individual functions. In symbols this means that

$$L\{\alpha f(x) + \beta g(x)\} = \alpha L(f(x)) + \beta L(g(x)), \tag{9}$$

where α and β are constants. Because of (9) we say that the nth-order differential operator L is a **linear operator.**

Differential Equations Any linear differential equation can be expressed in terms of the D notation. For example, the differential equation $y'' + 5y' + 6y = 5x - 3$ can be written as $D^2y + 5Dy + 6y = 5x - 3$ or $(D^2 + 5D + 6)y = 5x - 3$. Using (8), we can write the linear nth-order differential equations (6) and (7) compactly as

$$L(y) = 0 \quad \text{and} \quad L(y) = g(x),$$

respectively.

Superposition Principle In the next theorem we see that the sum, or **superposition,** of two or more solutions of a homogeneous linear differential equation is also a solution.

THEOREM 4.2   **Superposition Principle—Homogeneous Equations**

Let $y_1, y_2, \ldots, y_k$ be solutions of the homogeneous nth-order differential equation (6) on an interval I. Then the linear combination

$$y = c_1 y_1(x) + c_2 y_2(x) + \cdots + c_k y_k(x),$$

where the c_i, $i = 1, 2, \ldots, k$ are arbitrary constants, is also a solution on the interval.

Proof We prove the case $k = 2$. Let L be the differential operator defined in (8), and let $y_1(x)$ and $y_2(x)$ be solutions of the homogeneous equation $L(y) = 0$. If we define $y = c_1 y_1(x) + c_2 y_2(x)$, then by linearity of L we have

$$L(y) = L\{c_1 y_1(x) + c_2 y_2(x)\} = c_1 L(y_1) + c_2 L(y_2) = c_1 \cdot 0 + c_2 \cdot 0 = 0. \quad \blacksquare$$

Corollaries to Theorem 4.2

(A) A constant multiple $y = c_1 y_1(x)$ of a solution $y_1(x)$ of a homogeneous linear differential equation is also a solution.

(B) A homogeneous linear differential equation always possesses the trivial solution $y = 0$.

EXAMPLE 4   **Superposition—Homogeneous DE**

The functions $y_1 = x^2$ and $y_2 = x^2 \ln x$ are both solutions of the homogeneous linear equation $x^3 y''' - 2xy' + 4y = 0$ on the interval $(0, \infty)$. By the superposition principle, the linear combination

$$y = c_1 x^2 + c_2 x^2 \ln x$$

is also a solution of the equation on the interval. ∎

The function $y = e^{7x}$ is a solution of $y'' - 9y' + 14y = 0$. Since the differential equation is linear and homogeneous, the constant multiple $y = ce^{7x}$ is also a solution. For various values of c we see that $y = 9e^{7x}$, $y = 0$, $y = -\sqrt{5}e^{7x}, \ldots$ are all solutions of the equation.

Linear Dependence and Linear Independence The next two concepts are basic to the study of linear differential equations.

DEFINITION 4.1 **Linear Dependence/Independence**

A set of functions $f_1(x), f_2(x), \ldots, f_n(x)$ is said to be **linearly dependent** on an interval I if there exist constants $c_1, c_2, \ldots, c_n$, not all zero, such that

$$c_1 f_1(x) + c_2 f_2(x) + \cdots + c_n f_n(x) = 0$$

for every x in the interval. If the set of functions is not linearly dependent on the interval, it is said to be **linearly independent.**

In other words, a set of functions is linearly independent on an interval I if the only constants for which

$$c_1 f_1(x) + c_2 f_2(x) + \cdots + c_n f_n(x) = 0$$

for every x in the interval are $c_1 = c_2 = \cdots = c_n = 0$.

It is easy to understand these definitions for a set consisting of two functions $f_1(x)$ and $f_2(x)$. If the set of functions is linearly dependent on an interval, then there exist constants c_1 and c_2 that are not both zero such that, for every x in the interval, $c_1 f_1(x) + c_2 f_2(x) = 0$. Therefore, if we assume that $c_1 \neq 0$, it follows that $f_1(x) = (-c_2/c_1)f_2(x)$; that is, *if a set of two functions is linearly dependent, then one function is simply a constant multiple of the other.* Conversely, if $f_1(x) = c_2 f_2(x)$ for some constant c_2, then $(-1) \cdot f_1(x) + c_2 f_2(x) = 0$ for every x in the interval. Hence the set of functions is linearly dependent because at least one of the constants (namely, $c_1 = -1$) is not zero. We conclude that *a set of two functions $f_1(x)$ and $f_2(x)$ is linearly independent when neither function is a constant multiple of the other* on the interval. For example, the set of functions $f_1(x) = \sin 2x$, $f_2(x) = \sin x \cos x$ is linearly dependent on $(-\infty, \infty)$ since $f_1(x)$ is a constant multiple of $f_2(x)$. Recall from the double angle formula for the sine that $\sin 2x = 2 \sin x \cos x$. On the other hand, the set of functions $f_1(x), f_2(x) = |x|$ is linearly independent on $(-\infty, \infty)$. Inspection of Figure 4.3 should convince you that neither function is a constant multiple of the other on the interval.

It follows from the preceding discussion that the quotient $f_2(x)/f_1(x)$

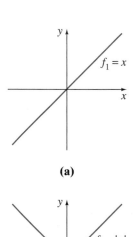

(a)

(b)

Figure 4.3

is not a constant on an interval on which the set $f_1(x)$, $f_2(x)$ is linearly independent. This little fact will be used in the next section.

EXAMPLE 5   Linearly Dependent Set of Functions

The set of functions $f_1(x) = \cos^2 x$, $f_2(x) = \sin^2 x$, $f_3(x) = \sec^2 x$, $f_4(x) = \tan^2 x$ is linearly dependent on the interval $(-\pi/2, \pi/2)$ since

$$c_1 \cos^2 x + c_2 \sin^2 x + c_3 \sec^2 x + c_4 \tan^2 x = 0$$

when $c_1 = c_2 = 1$, $c_3 = -1$, $c_4 = 1$. We used here $\cos^2 x + \sin^2 x = 1$ and $1 + \tan^2 x = \sec^2 x$. ∎

A set of functions $f_1(x)$, $f_2(x)$, ..., $f_n(x)$ is linearly dependent on an interval if at least one function can be expressed as a linear combination of the remaining functions.

EXAMPLE 6   Linearly Dependent Set of Functions

The set of functions $f_1(x) = \sqrt{x} + 5$, $f_2(x) = \sqrt{x} + 5x$, $f_3(x) = x - 1$, $f_4(x) = x^2$ is linearly dependent on the interval $(0, \infty)$ since f_2 can be written as a linear combination of f_1, f_3, and f_4. Observe that

$$f_2(x) = 1 \cdot f_1(x) + 5 \cdot f_3(x) + 0 \cdot f_4(x)$$

for every x in the interval $(0, \infty)$. ∎

Solutions of Differential Equations   We are primarily interested in linearly independent functions or, more to the point, linearly independent solutions of a linear differential equation. Although we could always appeal directly to Definition 4.1, it turns out that the question of whether the set of n solutions y_1, y_2, ..., y_n of a homogeneous linear nth-order differential equation (6) is linearly independent can be settled somewhat mechanically using a determinant.

DEFINITION 4.2   Wronskian

Suppose each of the functions $f_1(x)$, $f_2(x)$, ..., $f_n(x)$ possesses at least $n - 1$ derivatives. The determinant

$$W(f_1, f_2, \ldots, f_n) = \begin{vmatrix} f_1 & f_2 & \cdots & f_n \\ f_1' & f_2' & \cdots & f_n' \\ \vdots & \vdots & & \vdots \\ f_1^{(n-1)} & f_2^{(n-1)} & \cdots & f_n^{(n-1)} \end{vmatrix},$$

where the primes denote derivatives, is called the **Wronskian** of the functions.

> **THEOREM 4.3**   **Criterion for Linearly Independent Solutions**
>
> Let $y_1, y_2, \ldots, y_n$ be n solutions of the homogeneous linear nth-order differential equation (6) on an interval I. Then the set of solutions is **linearly independent** on I if and only if $W(y_1, y_2, \ldots, y_n) \neq 0$ for every x in the interval.

It follows from Theorem 4.3 that when $y_1, y_2, \ldots, y_n$ are n solutions of (6) on an interval I, the Wronskian $W(y_1, y_2, \ldots, y_n)$ is either identically zero or never zero on the interval.

A set of n linearly independent solutions of a homogeneous linear nth-order differential equation is given a special name.

> **DEFINITION 4.3**   **Fundamental Set of Solutions**
>
> Any set $y_1, y_2, \ldots, y_n$ of n linearly independent solutions of the homogeneous linear nth-order differential equation (6) on an interval I is said to be a **fundamental set of solutions** on the interval.

The basic question of whether a fundamental set of solutions exists for a linear equation is answered in the next theorem.

> **THEOREM 4.4**   **Existence of a Fundamental Set**
>
> There exists a fundamental set of solutions for the homogeneous linear nth-order differential equation (6) on an interval I.

Analogous to the fact that any vector in three dimensions can be expressed as a linear combination of the *linearly independent* vectors **i**, **j**, **k**, any solution of an nth-order homogeneous linear differential equation on an interval I can be expressed as a linear combination of n linearly independent solutions on I. In other words, n linearly independent solutions $y_1, y_2, \ldots, y_n$ are the basic building blocks for the general solution of the equation.

> **THEOREM 4.5**   **General Solution—Homogeneous Equations**
>
> Let $y_1, y_2, \ldots, y_n$ be a fundamental set of solutions of the homogeneous linear nth-order differential equation (6) on an interval I. Then the **general solution** of the equation on the interval is
>
> $$y = c_1 y_1(x) + c_2 y_2(x) + \cdots + c_n y_n(x),$$
>
> where c_i, $i = 1, 2, \ldots, n$ are arbitrary constants.

Theorem 4.5 states that if $Y(x)$ is any solution of (6) on the interval, then constants $C_1, C_2, \ldots, C_n$ can always be found so that

$$Y(x) = C_1 y_1(x) + C_2 y_2(x) + \cdots + C_n y_n(x).$$

We will prove the case when $n = 2$.

Proof   Let Y be a solution and y_1 and y_2 be linearly independent solutions of $a_2 y'' + a_1 y' + a_0 y = 0$ on an interval I. Suppose $x = t$ is a point in I for which $W(y_1(t), y_2(t)) \neq 0$. Suppose also that $Y(t) = k_1$ and $Y'(t) = k_2$. If we now examine the equations

$$C_1 y_1(t) + C_2 y_2(t) = k_1$$

$$C_1 y_1'(t) + C_2 y_2'(t) = k_2,$$

it follows that we can determine C_1 and C_2 uniquely, provided that the determinant of the coefficients satisfies

$$\begin{vmatrix} y_1(t) & y_2(t) \\ y_1'(t) & y_2'(t) \end{vmatrix} \neq 0.$$

But this determinant is simply the Wronskian evaluated at $x = t$, and, by assumption, $W \neq 0$. If we define $G(x) = C_1 y_1(x) + C_2 y_2(x)$, we observe that $G(x)$ satisfies the differential equation since it is a superposition of two known solutions; $G(x)$ satisfies the initial conditions

$$G(t) = C_1 y_1(t) + C_2 y_2(t) = k_1 \quad \text{and} \quad G'(t) = C_1 y_1'(t) + C_2 y_2'(t) = k_2;$$

and $Y(x)$ satisfies the *same* linear equation and the *same* initial conditions. Since the solution of this linear initial-value problem is unique (Theorem 4.1), we have $Y(x) = G(x)$ or $Y(x) = C_1 y_1(x) + C_2 y_2(x)$.   ∎

EXAMPLE 7   **General Solution of a Homogeneous DE**

The functions $y_1 = e^{3x}$ and $y_2 = e^{-3x}$ are both solutions of the homogeneous linear equation $y'' - 9y = 0$ on the interval $(-\infty, \infty)$. By inspection, the solutions are linearly independent on the x-axis. This fact can be corroborated by observing that the Wronskian

$$W(e^{3x}, e^{-3x}) = \begin{vmatrix} e^{3x} & e^{-3x} \\ 3e^{3x} & -3e^{-3x} \end{vmatrix} = -6 \neq 0$$

for every x. We conclude that y_1 and y_2 form a fundamental set of solutions and, consequently $y = c_1 e^{3x} + c_2 e^{-3x}$ is the general solution of the equation on the interval.   ■

EXAMPLE 8   **A Solution Obtained from a General Solution**

The function $y = 4 \sinh 3x - 5e^{3x}$ is a solution of the differential equation in Example 7. (Verify this.) In view of Theorem 4.5, we must be able to obtain this solution from the general solution $y = c_1 e^{3x} + c_2 e^{-3x}$. Observe that if we choose $c_1 = 2$ and $c_2 = -7$, then $y = 2e^{3x} - 7e^{-3x}$ can be rewritten as

$$y = 2e^{3x} - 2e^{-3x} - 5e^{-3x} = 4\left(\frac{e^{3x} - e^{-3x}}{2}\right) - 5e^{-3x}.$$

The last expression is recognized as $y = 4 \sinh 3x - 5e^{-3x}$. ■

EXAMPLE 9　**General Solution of a Homogeneous DE**

The functions $y_1 = e^x$, $y_2 = e^{2x}$, and $y_3 = e^{3x}$ satisfy the third-order equation $y''' - 6y'' + 11y' - 6y = 0$. Since

$$W(e^x, e^{2x}, e^{3x}) = \begin{vmatrix} e^x & e^{2x} & e^{3x} \\ e^x & 2e^{2x} & 3e^{3x} \\ e^x & 4e^{2x} & 9e^{3x} \end{vmatrix} = 2e^{6x} \neq 0$$

for every real value of x, the functions y_1, y_2, and y_3 form a fundamental set of solutions on $(-\infty, \infty)$. We conclude that $y = c_1 e^x + c_2 e^{2x} + c_3 e^{3x}$ is the general solution of the differential equation on the interval. ■

4.1.3　Nonhomogeneous Equations

Any function y_p, free of arbitrary parameters, that satisfies (7) is said to be a **particular solution** or **particular integral** of the equation. For example, it is a straightforward task to show that the constant function $y_p = 3$ is a particular solution of the nonhomogeneous equation $y'' + 9y = 27$.

　　Now if $y_1, y_2, \ldots, y_k$ are solutions of (6) on an interval I and y_p is any particular solution of (7) on I, then the linear combination

$$y = c_1 y_1(x) + c_2 y_2(x) + \cdots + c_k y_k(x) + y_p \tag{10}$$

is also a solution of the nonhomogeneous equation (7). If you think about it, this makes sense, because the linear combination $c_1 y_1(x) + c_2 y_2(x) + \cdots + c_k y_k(x)$ is transformed into 0 by the operator $L = a_n D^n + a_{n-1} D^{n-1} + \cdots + a_1 D + a_0$, whereas y_p is transformed into $g(x)$. If we use $k = n$ linearly independent solutions of the nth-order equation (6), then the expression in (10) becomes the general solution of (7).

THEOREM 4.6　**General Solution—Nonhomogeneous Equations**

Let y_p be any particular solution of the nonhomogeneous linear nth-order differential equation (7) on an interval I, and let $y_1, y_2, \ldots, y_n$ be a fundamental set of solutions of the associated homogeneous differential equation (6) on I. Then the **general solution** of the equation on the interval is

$$y = c_1 y_1(x) + c_2 y_2(x) + \cdots + c_n y_n(x) + y_p,$$

where the c_i, $i = 1, 2, \ldots, n$ are arbitrary constants.

Proof   Let L be the differential operator defined in (8), and let $Y(x)$ and $y_p(x)$ be particular solutions of the nonhomogeneous equation $L(y) = g(x)$. If we define $u(x) = Y(x) - y_p(x)$, then by linearity of L we have

$$L(u) = L\{Y(x) - y_p(x)\} = L(Y(x)) - L(y_p(x)) = g(x) - g(x) = 0.$$

This shows that $u(x)$ is a solution of the homogeneous equation $L(y) = 0$. Hence, by Theorem 4.5, $u(x) = c_1 y_1(x) + c_2 y_2(x) + \cdots + c_n y_n(x)$, and so

$$Y(x) - y_p(x) = c_1 y_1(x) + c_2 y_2(x) + \cdots + c_n y_n(x)$$

or
$$Y(x) = c_1 y_1(x) + c_2 y_2(x) + \cdots + c_n y_n(x) + y_p(x). \qquad \blacksquare$$

Complementary Function   We see in Theorem 4.6 that the general solution of a nonhomogeneous linear equation consists of the sum of two functions:

$$y = c_1 y_1(x) + c_2 y_2(x) + \cdots + c_n y_n(x) + y_p(x) = y_c(x) + y_p(x).$$

The linear combination $y_c(x) = c_1 y_1(x) + c_2 y_2(x) + \cdots + c_n y_n(x)$, which is the general solution of (6), is called the **complementary function** for equation (7). In other words, to solve a nonhomogeneous linear differential equation we first solve the associated homogeneous equation and then find any particular solution of the nonhomogeneous equation. The general solution of the nonhomogeneous equation is then

$$y = complementary\ function + any\ particular\ solution$$
$$= y_c + y_p.$$

EXAMPLE 10   **General Solution of a Nonhomogeneous DE**

By substitution, the function $y_p = -\frac{11}{12} - \frac{1}{2}x$ is readily shown to be a particular solution of the nonhomogeneous equation

$$y''' - 6y'' + 11y' - 6y = 3x. \qquad \textbf{(11)}$$

In order to write the general solution of (11), we must also be able to solve the associated homogeneous equation

$$y''' - 6y'' + 11y' - 6y = 0.$$

But in Example 9 we saw that the general solution of this latter equation on the interval $(-\infty, \infty)$ was $y_c = c_1 e^x + c_2 e^{2x} + c_3 e^{3x}$. Hence the general solution of (11) on the interval is

$$y = y_c + y_p = c_1 e^x + c_2 e^{2x} + c_3 e^{3x} - \frac{11}{12} - \frac{1}{2}x. \qquad \blacksquare$$

Another Superposition Principle   The last theorem of this discussion will be useful in Section 4.4 when we consider a method for finding particular solutions of nonhomogeneous equations.

| THEOREM 4.7 | Superposition Principle—Nonhomogeneous Equations |

Let $y_{p_1}, y_{p_2}, \ldots, y_{p_k}$ be k particular solutions of the nonhomogeneous linear nth-order differential equation (7) on an interval I corresponding, in turn, to k distinct functions $g_1, g_2, \ldots, g_k$. That is, suppose y_{p_i} denotes a particular solution of the corresponding differential equation

$$a_n(x)y^{(n)} + a_{n-1}(x)y^{(n-1)} + \cdots + a_1(x)y' + a_0(x)y = g_i(x), \qquad (12)$$

where $i = 1, 2, \ldots, k$. Then

$$y_p = y_{p_1}(x) + y_{p_2}(x) + \cdots + y_{p_k}(x) \qquad (13)$$

is a particular solution of

$$a_n(x)y^{(n)} + a_{n-1}(x)y^{(n-1)} + \cdots + a_1(x)y' + a_0(x)y$$
$$= g_1(x) + g_2(x) + \cdots + g_k(x). \qquad (14)$$

Proof   We prove the case $k = 2$. Let L be the differential operator defined in (8), and let $y_{p_1}(x)$ and $y_{p_2}(x)$ be particular solutions of the nonhomogeneous equations $L(y) = g_1(x)$ and $L(y) = g_2(x)$, respectively. If we define $y_p = y_{p_1}(x) + y_{p_2}(x)$, we want to show that y_p is a particular solution of $L(y) = g_1(x) + g_2(x)$. The result follows again by the linearity of the operator L:

$$L(y_p) = L\{y_{p_1}(x) + y_{p_2}(x)\} = L(y_{p_1}(x)) + L(y_{p_2}(x)) = g_1(x) + g_2(x).$$

∎

| EXAMPLE 11 | Superposition—Nonhomogeneous DE |

You should verify that

$y_{p_1} = -4x^2$ is a particular solution of $y'' - 3y' + 4y = -16x^2 + 24x - 8$,

$y_{p_2} = e^{2x}$ is a particular solution of $y'' - 3y' + 4y = 2e^{2x}$,

$y_{p_3} = xe^x$ is a particular solution of $y'' - 3y' + 4y = 2xe^x - e^x$.

It follows from (13) of Theorem 4.7 that the superposition of y_{p_1}, y_{p_2}, and y_{p_3},

$$y = y_{p_1} + y_{p_2} + y_{p_3} = -4x^2 + e^{2x} + xe^x,$$

is a solution of

$$y'' - 3y' + 4y = \underbrace{-16x^2 + 24x - 8}_{g_1(x)} + \underbrace{2e^{2x}}_{g_2(x)} + \underbrace{2xe^x - e^x}_{g_3(x)}. \qquad ∎$$

Note   If the y_{p_i} are particular solutions of (12) for $i = 1, 2, \ldots, k$, then the linear combination

$$y_p = c_1 y_{p_1} + c_2 y_{p_2} + \cdots + c_k y_{p_k},$$

where the c_i are constants, is also a particular solution of (14) when the right-hand member of the equation is the linear combination

$$c_1 g_1(x) + c_2 g_2(x) + \cdots + c_k g_k(x).$$

Before we actually start solving homogeneous and nonhomogeneous linear differential equations, we need one additional bit of theory presented in the next section.

Remarks   This remark is a continuation of the brief discussion of dynamical systems given at the end of Section 1.3.

A dynamical system whose rule or mathematical model is a linear nth-order differential equation

$$a_n(t) y^{(n)} + a_{n-1}(t) y^{(n-1)} + \cdots + a_1(t) y' + a_0(t) y = g(t)$$

is said to be an nth-order **linear system.** The n time-dependent functions $y(t), y'(t), \ldots, y^{(n-1)}(t)$ are the **state variables** of the system. Recall that their values at some time t give the **state of the system.** The function g is variously called the **input function, forcing function, or excitation function.** A solution $y(t)$ of the differential equation is said to be the **output** or **response of the system.** Under the conditions stated in Theorem 4.1, the output or response $y(t)$ is uniquely determined by the input and the state of the system prescribed at a time t_0—that is, by the initial conditions $y(t_0), y'(t_0), \ldots, y^{(n-1)}(t_0)$.

In order for a dynamical system to be a linear system, it is necessary that the superposition principle (Theorem 4.7) holds in the system; that is, the response of the system to a superposition of inputs is a superposition of outputs. We have already examined some simple linear systems in Section 3.1 (linear first-order equations); in Section 5.1 we examine linear systems in which the mathematical models are second-order differential equations.

EXERCISES 4.1

Answers to odd-numbered problems begin on page AN-3.

4.1.1   Initial-Value and Boundary-Value Problems

In Problems 1–4 the given family of functions is the general solution of the differential equation on the indicated interval. Find a member of the family that is a solution of the initial-value problem.

1. $y = c_1 e^x + c_2 e^{-x}$, $(-\infty, \infty)$; $y'' - y = 0$, $y(0) = 0$, $y'(0) = 1$

2. $y = c_1 e^{4x} + c_2 e^{-x}$, $(-\infty, \infty)$; $y'' - 3y' - 4y = 0$, $y(0) = 1$, $y'(0) = 2$

3. $y = c_1 x + c_2 x \ln x$, $(0, \infty)$; $x^2 y'' - xy' + y = 0$, $y(1) = 3$, $y'(1) = -1$

4. $y = c_1 + c_2 \cos x + c_3 \sin x$, $(-\infty, \infty)$; $y''' + y' = 0$, $y(\pi) = 0$, $y'(\pi) = 2$, $y''(\pi) = -1$

5. Given that $y = c_1 + c_2 x^2$ is a two-parameter family of solutions of $xy'' - y' = 0$ on the interval $(-\infty, \infty)$, show that constants c_1 and c_2

cannot be found so that a member of the family satisfies the initial conditions $y(0) = 0$, $y'(0) = 1$. Explain why this does not violate Theorem 4.1.

6. Find two members of the family of solutions in Problem 5 that satisfy the initial conditions $y(0) = 0$, $y'(0) = 0$.

7. Given that $x(t) = c_1 \cos \omega t + c_2 \sin \omega t$ is the general solution of $x'' + \omega^2 x = 0$ on the interval $(-\infty, \infty)$, show that a solution satisfying the initial conditions $x(0) = x_0$, $x'(0) = x_1$ is given by

$$x(t) = x_0 \cos \omega t + \frac{x_1}{\omega} \sin \omega t.$$

8. Use the general solution of $x'' + \omega^2 x = 0$ given in Problem 7 to show that a solution satisfying the initial conditions $x(t_0) = x_0$, $x'(t_0) = x_1$ is the solution given in Problem 7 shifted by an amount t_0:

$$x(t) = x_0 \cos \omega(t - t_0) + \frac{x_1}{\omega} \sin \omega(t - t_0).$$

In Problems 9 and 10 find an interval centered about $x = 0$ for which the given initial-value problem has a unique solution.

9. $(x - 2)y'' + 3y = x$, $y(0) = 0$, $y'(0) = 1$

10. $y'' + (\tan x)y = e^x$, $y(0) = 1$, $y'(0) = 0$

11. Use the family in Problem 1 to find a solution of $y'' - y = 0$ that satisfies the boundary conditions $y(0) = 0$, $y(1) = 1$.

12. Use the family in Problem 5 to find a solution of $xy'' - y' = 0$ that satisfies the boundary conditions $y(0) = 1$, $y'(1) = 6$.

In Problems 13 and 14 the given two-parameter family is a solution of the indicated differential equation on the interval $(-\infty, \infty)$. Determine whether a member of the family can be found that satisfies the boundary conditions.

13. $y = c_1 e^x \cos x + c_2 e^x \sin x$; $y'' - 2y' + 2y = 0$
   (a) $y(0) = 1$, $y'(\pi) = 0$ **(b)** $y(0) = 1$, $y(\pi) = -1$
   (c) $y(0) = 1$, $y\left(\dfrac{\pi}{2}\right) = 1$ **(d)** $y(0) = 0$, $y(\pi) = 0$.

14. $y = c_1 x^2 + c_2 x^4 + 3$; $x^2 y'' - 5xy' + 8y = 24$
   (a) $y(-1) = 0$, $y(1) = 4$ **(b)** $y(0) = 1$, $y(1) = 2$
   (c) $y(0) = 3$, $y(1) = 0$ **(d)** $y(1) = 3$, $y(2) = 15$

4.1.2 **Homogeneous Equations**

In Problems 15–22 determine whether the given set of functions is linearly independent on the interval $(-\infty, \infty)$.

15. $f_1(x) = x$, $f_2(x) = x^2$, $f_3(x) = 4x - 3x^2$

16. $f_1(x) = 0$, $f_2(x) = x$, $f_3(x) = e^x$

17. $f_1(x) = 5$, $f_2(x) = \cos^2 x$, $f_3(x) = \sin^2 x$

18. $f_1(x) = \cos 2x$, $f_2(x) = 1$, $f_3(x) = \cos^2 x$

19. $f_1(x) = x$, $f_2(x) = x - 1$, $f_3(x) = x + 3$

20. $f_1(x) = 2 + x$, $f_2(x) = 2 + |x|$

21. $f_1(x) = 1 + x$, $f_2(x) = x$, $f_3(x) = x^2$

22. $f_1(x) = e^x$, $f_2(x) = e^{-x}$, $f_3(x) = \sinh x$

In Problems 23–30 verify that the given functions form a fundamental set of solutions of the differential equation on the indicated interval. Form the general solution.

23. $y'' - y' - 12y = 0$; $e^{-3x}, e^{4x}, (-\infty, \infty)$

24. $y'' - 4y = 0$; $\cosh 2x, \sinh 2x, (-\infty, \infty)$

25. $y'' - 2y' + 5y = 0$; $e^x \cos 2x, e^x \sin 2x, (-\infty, \infty)$

26. $4y'' - 4y' + y = 0$; $e^{x/2}, xe^{x/2}, (-\infty, \infty)$

27. $x^2 y'' - 6xy' + 12y = 0$; $x^3, x^4, (0, \infty)$

28. $x^2 y'' + xy' + y = 0$; $\cos(\ln x), \sin(\ln x), (0, \infty)$

29. $x^3 y''' + 6x^2 y'' + 4xy' - 4y = 0$; $x, x^{-2}, x^{-2} \ln x, (0, \infty)$

30. $y^{(4)} + y'' = 0$; $1, x, \cos x, \sin x, (-\infty, \infty)$

4.1.3   Nonhomogeneous Equations

In Problems 31–34 verify that the given two-parameter family of functions is the general solution of the nonhomogeneous differential equation on the indicated interval.

31. $y'' - 7y' + 10y = 24e^x$;
$y = c_1 e^{2x} + c_2 e^{5x} + 6e^x, (-\infty, \infty)$

32. $y'' + y = \sec x$;
$y = c_1 \cos x + c_2 \sin x + x \sin x + (\cos x) \ln(\cos x), (-\pi/2, \pi/2)$

33. $y'' - 4y' + 4y = 2e^{2x} + 4x - 12$;
$y = c_1 e^{2x} + c_2 xe^{2x} + x^2 e^{2x} + x - 2, (-\infty, \infty)$

34. $2x^2 y'' + 5xy' + y = x^2 - x$;
$y = c_1 x^{-1/2} + c_2 x^{-1} + \frac{1}{15} x^2 - \frac{1}{6} x, (0, \infty)$

35. (a) Verify that $y_{p_1} = 3e^{2x}$ and $y_{p_2} = x^2 + 3x$ are, respectively, particular solutions of

$$y'' - 6y' + 5y = -9e^{2x}$$
and
$$y'' - 6y' + 5y = 5x^2 + 3x - 16.$$

(b) Use part (a) to find particular solutions of

$$y'' - 6y' + 5y = 5x^2 + 3x - 16 - 9e^{2x}$$
and
$$y'' - 6y' + 5y = -10x^2 - 6x + 32 + e^{2x}.$$

36. (a) By inspection find a particular solution of $y'' + 2y = 10$.
(b) By inspection find a particular solution of $y'' + 2y = -4x$.
(c) Find a particular solution of $y'' + 2y = -4x + 10$.
(d) Find a particular solution of $y'' + 2y = 8x + 5$.

Discussion Problems

37. Let $n = 1, 2, 3, \ldots$. Discuss how the observations $D^n x^{n-1} = 0$ and $D^n x^n = n!$ can be used to find the general solutions of the given differential equations.

(a) $y'' = 0$ **(b)** $y''' = 0$
(c) $y^{(4)} = 0$ **(d)** $y'' = 2$
(e) $y''' = 6$ **(f)** $y^{(4)} = 24$

38. Suppose that $y_1 = e^x$ and $y_2 = e^{-x}$ are two solutions of a homogeneous linear differential equation. Explain why $y_3 = \cosh x$ and $y_4 = \sinh x$ are also solutions of the equation.

39. (a) Verify that $y_1 = x^3$ and $y_2 = |x|^3$ are linearly independent solutions of the differential equation $x^2 y'' - 4xy' + 6y = 0$ on the interval $(-\infty, \infty)$.

   (b) Show that $W(y_1, y_2) = 0$ for every real number x. Does this result violate Theorem 4.3? Explain.

   (c) Verify that $Y_1 = x^3$ and $Y_2 = x^2$ are also linearly independent solutions of the differential equation in part (a) on the interval $(-\infty, \infty)$.

   (d) Find a solution of the differential equation satisfying $y(0) = 0$, $y'(0) = 0$.

   (e) By the superposition principle, Theorem 4.2, both linear combinations $y = c_1 y_1 + c_2 y_2$ and $Y = c_1 Y_1 + c_2 Y_2$ are solutions of the differential equation. Discuss whether one, both, or neither of the linear combinations is a general solution of the differential equation on the interval $(-\infty, \infty)$.

40. Is the set of functions $f_1(x) = e^{x+2}$, $f_2(x) = e^{x-3}$ linearly dependent or linearly independent on $(-\infty, \infty)$? Discuss.

41. Suppose $y_1, y_2, \ldots, y_k$ are k linearly independent solutions on $(-\infty, \infty)$ of a homogeneous linear nth-order differential equation with constant coefficients. By Theorem 4.2 it follows that $y_{k+1} = 0$ is also a solution of the differential equation. Is the set of solutions $y_1, y_2, \ldots, y_k, y_{k+1}$ linearly dependent or linearly independent on $(-\infty, \infty)$? Discuss.

42. Suppose that $y_1, y_2, \ldots, y_k$ are k nontrivial solutions of a homogeneous linear nth-order differential equation with constant coefficients and that $k = n + 1$. Is the set of solutions $y_1, y_2, \ldots, y_k$ linearly dependent or linearly independent on $(-\infty, \infty)$? Discuss.

4.2 REDUCTION OF ORDER

- *Reducing a second-order DE to a first-order DE*
- *Standard form of a homogeneous linear second-order DE*

It is one of the more interesting, and important, facts of mathematical life in the study of linear *second-order* differential equations that we can construct a second solution y_2 of the homogeneous equation

$$a_2(x)y'' + a_1(x)y' + a_0(x)y = 0 \tag{1}$$

on an interval I whenever we know a nontrivial solution y_1 defined on I. The basic idea in the discussion that follows is that equation (1) can be reduced to a linear *first-order* DE by means of a substitution involving y_1. A second solution y_2 of (1) is then apparent after this first-order DE is solved.

Reduction of Order   Suppose that y_1 denotes a nontrivial solution of equation (1) and that y_1 is defined on an interval I. We seek a second solution, y_2, so that y_1, y_2 is a linearly independent set on I. Recall from Section 4.1 that if y_1 and y_2 are linearly independent, then their quotient y_2/y_1 is nonconstant on I—that is, $y_2(x)/y_1(x) = u(x)$ or $y_2(x) = u(x)y_1(x)$. The function $u(x)$ can be found by substituting $y_2(x) = u(x)y_1(x)$ into the given differential equation. This method is called **reduction of order** since we must solve a linear first-order differential equation to find u.

EXAMPLE 1   **A Second Solution by Reduction of Order**

Given that $y_1 = e^x$ is a solution of $y'' - y = 0$ on the interval $(-\infty, \infty)$, use reduction of order to find a second solution y_2.

Solution   If $y = u(x)y_1(x) = u(x)e^x$, then the product rule gives

$$y' = ue^x + e^x u', \quad y'' = ue^x + 2e^x u' + e^x u'',$$

and so $$y'' - y = e^x(u'' + 2u') = 0.$$

Since $e^x \neq 0$, the last equation requires $u'' + 2u' = 0$. If we make the substitution $w = u'$, this linear second-order equation in u becomes $w' + 2w = 0$, which is a linear first-order equation in w. Using the integrating factor e^{2x}, we can write $\dfrac{d}{dx}[e^{2x}w] = 0$. After integrating, we get $w = c_1 e^{-2x}$ or $u' = c_1 e^{-2x}$. Integrating again then yields $u = -\frac{1}{2}c_1 e^{-2x} + c_2$. Thus

$$y = u(x)e^x = -\frac{c_1}{2}e^{-x} + c_2 e^x. \tag{2}$$

By picking $c_2 = 0$ and $c_1 = -2$ we obtain the desired second solution, $y_2 = e^{-x}$. Because $W(e^x, e^{-x}) \neq 0$ for every x, the solutions are linearly independent on $(-\infty, \infty)$.   ∎

Since we have shown that $y_1 = e^x$ and $y_2 = e^{-x}$ are linearly independent solutions of a linear second-order equation, the expression in (2) is actually the general solution of $y'' - y = 0$ on $(-\infty, \infty)$.

General Case   Suppose we divide by $a_2(x)$ in order to put equation (1) in the **standard form**

$$y'' + P(x)y' + Q(x)y = 0, \tag{3}$$

where $P(x)$ and $Q(x)$ are continuous on some interval I. Let us suppose further that $y_1(x)$ is a known solution of (3) on I and that $y_1(x) \neq 0$ for every x in the interval. If we define $y = u(x)y_1(x)$, it follows that

$$y' = uy_1' + y_1 u', \quad y'' = uy_1'' + 2y_1' u' + y_1 u''$$

$$y'' + Py' + Qy = u[\underbrace{y_1'' + Py_1' + Qy_1}_{\text{zero}}] + y_1 u'' + (2y_1' + Py_1)u' = 0.$$

This implies that we must have

$$y_1 u'' + (2y_1' + Py_1)u' = 0 \quad \text{or} \quad y_1 w' + (2y_1' + Py_1)w = 0, \tag{4}$$

where we have let $w = u'$. Observe that the last equation in (4) is both linear and separable. Separating variables and integrating, we obtain

$$\frac{dw}{w} + 2\frac{y_1'}{y_1}\,dx + P\,dx = 0$$

$$\ln|wy_1^2| = -\int P\,dx + c \quad \text{or} \quad wy_1^2 = c_1 e^{-\int P\,dx}.$$

We solve the last equation for w, use $w = u'$, and integrate again:

$$u = c_1 \int \frac{e^{-\int P\,dx}}{y_1^2}\,dx + c_2.$$

By choosing $c_1 = 1$ and $c_2 = 0$ we find from $y = u(x)y_1(x)$ that a second solution of equation (3) is

$$y_2 = y_1(x) \int \frac{e^{-\int P(x)\,dx}}{y_1^2(x)}\,dx. \tag{5}$$

It makes a good review of differentiation to verify that the function $y_2(x)$ defined in (5) satisfies equation (3) and that y_1 and y_2 are linearly independent on any interval on which $y_1(x)$ is not zero.

EXAMPLE 2   **A Second Solution by Formula (5)**

The function $y_1 = x^2$ is a solution of $x^2 y'' - 3xy' + 4y = 0$. Find the general solution of the differential equation on the interval $(0, \infty)$.

Solution   From the standard form of the equation,

$$y'' - \frac{3}{x}y' + \frac{4}{x^2}y = 0,$$

$$y_2 = x^2 \int \frac{e^{3\int dx/x}}{x^4}\,dx \qquad \leftarrow e^{3\int dx/x} = e^{\ln x^3} = x^3$$

we find from (5)

$$= x^2 \int \frac{dx}{x} = x^2 \ln x.$$

The general solution on $(0, \infty)$ is given by $y = c_1 y_1 + c_2 y_2$; that is, $y = c_1 x^2 + c_2 x^2 \ln x.$ ■

Remarks   (*i*) The derivation and use of formula (5) have been illustrated here because this formula appears again in the next section and in Sections 4.7 and 6.2. We use (5) simply to save time in obtaining a desired result. Your instructor will tell you whether you should memorize (5) or whether you should know the first principles of reduction of order.

(*ii*) Reduction of order can be used to find the general solution of a nonhomogeneous equation $a_2(x)y'' + a_1(x)y' + a_0(x)y = g(x)$ whenever a solution y_1 of the associated homogeneous equation is known. See Problems 17–20 in Exercises 4.2.

EXERCISES 4.2

Answers to odd-numbered problems begin on page AN-3.

In Problems 1–16 the indicated function $y_1(x)$ is a solution of the given differential equation. Use reduction of order or formula (5), as instructed, to find a second solution $y_2(x)$.

1. $y'' - 4y' + 4y = 0$; $y_1 = e^{2x}$
2. $y'' + 2y' + y = 0$; $y_1 = xe^{-x}$

3. $y'' + 16y = 0$; $y_1 = \cos 4x$
4. $y'' + 9y = 0$; $y_1 = \sin 3x$

5. $y'' - y = 0$; $y_1 = \cosh x$
6. $y'' - 25y = 0$; $y_1 = e^{5x}$

7. $9y'' - 12y' + 4y = 0$; $y_1 = e^{2x/3}$
8. $6y'' + y' - y = 0$; $y_1 = e^{x/3}$

9. $x^2y'' - 7xy' + 16y = 0$; $y_1 = x^4$
10. $x^2y'' + 2xy' - 6y = 0$; $y_1 = x^2$

11. $xy'' + y' = 0$; $y_1 = \ln x$
12. $4x^2y'' + y = 0$; $y_1 = x^{1/2} \ln x$

13. $x^2y'' - xy' + 2y = 0$; $y_1 = x \sin(\ln x)$

14. $x^2y'' - 3xy' + 5y = 0$; $y_1 = x^2 \cos(\ln x)$

15. $(1 - 2x - x^2)y'' + 2(1 + x)y' - 2y = 0$; $y_1 = x + 1$

16. $(1 - x^2)y'' + 2xy' = 0$; $y_1 = 1$

In Problems 17–20 the indicated function $y_1(x)$ is a solution of the associated homogeneous equation. Use the method of reduction of order to find a second solution $y_2(x)$ of the homogeneous equation and a particular solution of the given nonhomogeneous equation.

17. $y'' - 4y = 2$; $y_1 = e^{-2x}$
18. $y'' + y' = 1$; $y_1 = 1$

19. $y'' - 3y' + 2y = 5e^{3x}$; $y_1 = e^x$
20. $y'' - 4y' + 3y = x$; $y_1 = e^x$

Discussion Problems

21. (a) Give a convincing demonstration that the second-order equation $ay'' + by' + cy = 0$, a, b, and c constants, always possesses at least one solution of the form $y_1 = e^{m_1 x}$, m_1 a constant.

(b) Explain why the differential equation in part (a) must then have a second solution either of the form $y_2 = e^{m_2 x}$ or of the form $y_2 = xe^{m_1 x}$, m_1 and m_2 constants.

(c) Reexamine Problems 1–8. Can you explain why the statements in parts (a) and (b) above are not contradicted by the answers to Problems 3–5?

22. Verify that $y_1(x) = x$ is a solution of $xy'' - xy' + y = 0$. Use reduction of order to find a second solution $y_2(x)$ in the form of an infinite series. Conjecture an interval of definition for $y_2(x)$.

Computer Lab Assignments

23. (a) Verify that $y_1(x) = e^x$ is a solution of $xy'' - (x + 10)y' + 10y = 0$.

(b) Use (5) to find a second solution $y_2(x)$. Use a CAS to carry out the required integration.

(c) Explain, using Corollary (A) of Theorem 4.2, why the second solution can be written compactly as $y_2(x) = \sum_{n=0}^{10} \frac{1}{n!} x^n$.

4.3 HOMOGENEOUS LINEAR EQUATIONS WITH CONSTANT COEFFICIENTS

• Exponential solutions • Auxiliary equation • Roots of a quadratic auxiliary equation • Euler's formula • General solutions for second-order DEs • Higher-order DEs • Roots of auxiliary equations of degree greater than two • General solutions for higher-order DEs

We have seen that the linear first-order DE $y' + ay = 0$, where a is a constant, possesses the exponential solution $y = c_1 e^{-ax}$ on the interval $(-\infty, \infty)$. Therefore, it is natural to ask whether exponential solutions exist on $(-\infty, \infty)$ for homogeneous linear higher-order DEs

$$a_n y^{(n)} + a_{n-1} y^{(n-1)} + \cdots + a_2 y'' + a_1 y' + a_0 y = 0, \tag{1}$$

where the coefficients a_i, $i = 0, 1, \ldots, n$ are real constants and $a_n \neq 0$. The surprising fact is that *all* solutions of (1) are exponential functions or are constructed out of exponential functions.

Auxiliary Equation   We begin by considering the special case of the second-order equation

$$ay'' + by' + cy = 0. \tag{2}$$

If we try a solution of the form $y = e^{mx}$, then after substituting $y' = me^{mx}$ and $y'' = m^2 e^{mx}$ equation (2) becomes

$$am^2 e^{mx} + bme^{mx} + ce^{mx} = 0 \quad \text{or} \quad e^{mx}(am^2 + bm + c) = 0.$$

Since e^{mx} is never zero for real values of x, it is apparent that the only way this exponential function can satisfy the differential equation (2) is if m is chosen as a root of the quadratic equation

$$am^2 + bm + c = 0. \tag{3}$$

This last equation is called the **auxiliary equation** of the differential equation (2). Since the two roots of (3) are $m_1 = (-b + \sqrt{b^2 - 4ac})/2a$ and $m_2 = (-b - \sqrt{b^2 - 4ac})/2a$, there will be three forms of the general solution of (2) corresponding to the three cases:

- m_1 and m_2 real and distinct ($b^2 - 4ac > 0$),
- m_1 and m_2 real and equal ($b^2 - 4ac = 0$), and
- m_1 and m_2 conjugate complex numbers ($b^2 - 4ac < 0$).

We discuss each of these cases in turn.

Case I: Distinct Real Roots   Under the assumption that the auxiliary equation (3) has two unequal real roots m_1 and m_2, we find two solutions, $y_1 = e^{m_1 x}$ and $y_2 = e^{m_2 x}$. We see that these functions are linearly independent on $(-\infty, \infty)$ and hence form a fundamental set. It follows that the general solution of (2) on this interval is

$$y = c_1 e^{m_1 x} + c_2 e^{m_2 x}. \tag{4}$$

Case II: Repeated Real Roots   When $m_1 = m_2$ we necessarily obtain only one exponential solution, $y_1 = e^{m_1 x}$. From the quadratic formula we find that $m_1 = -b/2a$ since the only way to have $m_1 = m_2$ is to have $b^2 - 4ac = 0$. It follows from (5) in Section 4.2 that a second solution of the equation is

$$y_2 = e^{m_1 x} \int \frac{e^{2m_1 x}}{e^{2m_1 x}} \, dx = e^{m_1 x} \int dx = x e^{m_1 x}. \tag{5}$$

In (5) we have used the fact that $-b/a = 2m_1$. The general solution is then

$$y = c_1 e^{m_1 x} + c_2 x e^{m_1 x}. \tag{6}$$

Case III: Conjugate Complex Roots   If m_1 and m_2 are complex, then we can write $m_1 = \alpha + i\beta$ and $m_2 = \alpha - i\beta$, where α and $\beta > 0$ are real and $i^2 = -1$. Formally, there is no difference between this case and Case I, and hence

$$y = C_1 e^{(\alpha + i\beta)x} + C_2 e^{(\alpha - i\beta)x}.$$

However, in practice we prefer to work with real functions instead of complex exponentials. To this end we use Euler's formula:

$$e^{i\theta} = \cos \theta + i \sin \theta,$$

where θ is any real number.* It follows from this formula that

$$e^{i\beta x} = \cos \beta x + i \sin \beta x \quad \text{and} \quad e^{-i\beta x} = \cos \beta x - i \sin \beta x, \tag{7}$$

where we have used $\cos(-\beta x) = \cos \beta x$ and $\sin(-\beta x) = -\sin \beta x$. Note that by first adding and then subtracting the two equations in (7), we obtain, respectively,

$$e^{i\beta x} + e^{-i\beta x} = 2 \cos \beta x \quad \text{and} \quad e^{i\beta x} - e^{-i\beta x} = 2i \sin \beta x.$$

Since $y = C_1 e^{(\alpha + i\beta)x} + C_2 e^{(\alpha - i\beta)x}$ is a solution of (2) for any choice of the constants C_1 and C_2, the choices $C_1 = C_2 = 1$ and $C_1 = 1$, $C_2 = -1$ give, in turn, two solutions:

$$y_1 = e^{(\alpha + i\beta)x} + e^{(\alpha - i\beta)x} \quad \text{and} \quad y_2 = e^{(\alpha + i\beta)x} - e^{(\alpha - i\beta)x}.$$

But $$y_1 = e^{\alpha x}(e^{i\beta x} + e^{-i\beta x}) = 2e^{\alpha x} \cos \beta x$$

and $$y_2 = e^{\alpha x}(e^{i\beta x} - e^{-i\beta x}) = 2i e^{\alpha x} \sin \beta x.$$

Hence from Corollary (A) of Theorem 4.2 the last two results show that $e^{\alpha x} \cos \beta x$ and $e^{\alpha x} \sin \beta x$ are *real* solutions of (2). Moreover, these solutions form a fundamental set on $(-\infty, \infty)$. Consequently, the general solution is

$$y = c_1 e^{\alpha x} \cos \beta x + c_2 e^{\alpha x} \sin \beta x = e^{\alpha x}(c_1 \cos \beta x + c_2 \sin \beta x). \tag{8}$$

---

*A formal derivation of Euler's formula can be obtained from the Maclaurin series $e^x = \sum_{n=0}^{\infty} \frac{x^n}{n!}$ by substituting $x = i\theta$, using $i^2 = -1$, $i^3 = -i$, ..., and then separating the series into real and imaginary parts. The plausibility thus established, we can adopt $\cos \theta + i \sin \theta$ as the *definition* of $e^{i\theta}$.

EXAMPLE 1   Second-Order DEs

Solve the following differential equations.

(a) $2y'' - 5y' - 3y = 0$ (b) $y'' - 10y' + 25y = 0$ (c) $y'' + 4y' + 7y = 0$

Solution   We give the auxiliary equations, the roots, and the corresponding general solutions.

(a) $2m^2 - 5m - 3 = (2m + 1)(m - 3)$, $m_1 = -\frac{1}{2}$, $m_2 = 3$

From (4), $y = c_1 e^{-x/2} + c_2 e^{3x}$.

(b) $m^2 - 10m + 25 = (m - 5)^2$, $m_1 = m_2 = 5$

From (6), $y = c_1 e^{5x} + c_2 x e^{5x}$.

(c) $m^2 + 4m + 7 = 0$, $m_1 = -2 + \sqrt{3}\,i$, $m_2 = -2 - \sqrt{3}\,i$

From (8) with $\alpha = -2$, $\beta = \sqrt{3}$, $y = e^{-2x}(c_1 \cos \sqrt{3}\,x + c_2 \sin \sqrt{3}\,x)$.   ■

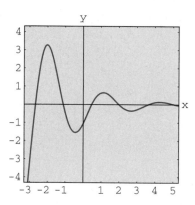

Figure 4.4

EXAMPLE 2   An Initial-Value Problem

Solve $4y'' + 4y' + 17y = 0$, $y(0) = -1$, $y'(0) = 2$.

Solution   By the quadratic formula we find that the roots of the auxiliary equation $4m^2 + 4m + 17 = 0$ are $m_1 = -\frac{1}{2} + 2i$ and $m_2 = -\frac{1}{2} - 2i$. Thus from (8) we have $y = e^{-x/2}(c_1 \cos 2x + c_2 \sin 2x)$. Applying the condition $y(0) = -1$, we see from $e^0(c_1 \cos 0 + c_2 \sin 0) = -1$ that $c_1 = -1$. Differentiating $y = e^{-x/2}(-\cos 2x + c_2 \sin 2x)$ and then using $y'(0) = 2$ gives $2c_2 + \frac{1}{2} = 2$ or $c_2 = \frac{3}{4}$. Hence the solution of the IVP is $y = e^{-x/2}(-\cos 2x + \frac{3}{4} \sin 2x)$. In Figure 4.4 we see that the solution is oscillatory but $y \to 0$ as $x \to \infty$ and $|y| \to \infty$ as $x \to -\infty$.   ■

Two Equations Worth Knowing   The two differential equations

$$y'' + k^2 y = 0 \quad \text{and} \quad y'' - k^2 y = 0,$$

k real, are important in applied mathematics. For $y'' + k^2 y = 0$, the auxiliary equation $m^2 + k^2 = 0$ has imaginary roots $m_1 = ki$ and $m_2 = -ki$. With $\alpha = 0$ and $\beta = k$ in (8), the general solution of the DE is seen to be

$$y = c_1 \cos kx + c_2 \sin kx. \tag{9}$$

On the other hand, the auxiliary equation $m^2 - k^2 = 0$ for $y'' - k^2 y = 0$ has distinct real roots $m_1 = k$ and $m_2 = -k$, and so by (4) the general solution of the DE is

$$y = c_1 e^{kx} + c_2 e^{-kx}. \tag{10}$$

Notice that if we choose $c_1 = c_2 = \frac{1}{2}$ and $c_1 = \frac{1}{2}$, $c_2 = -\frac{1}{2}$ in (10) we get the particular solutions $y = \frac{1}{2}(e^{kx} + e^{-kx}) = \cosh kx$ and $y = \frac{1}{2}(e^{kx} - e^{-kx}) = \sinh kx$. Since $\cosh kx$ and $\sinh kx$ are linearly independent on any interval of the x-axis, an alternative form for the general solution of

$y'' - k^2 y = 0$ is

$$y = c_1 \cosh kx + c_2 \sinh kx. \tag{11}$$

See Problems 41, 42, and 49 in Exercises 4.3.

Higher-Order Equations   In general, to solve an nth-order differential equation (1) where the a_i, $i = 0, 1, \ldots, n$ are real constants, we must solve an nth-degree polynomial equation

$$a_n m^n + a_{n-1} m^{n-1} + \cdots + a_2 m^2 + a_1 m + a_0 = 0. \tag{12}$$

If all the roots of (12) are real and distinct, then the general solution of (1) is

$$y = c_1 e^{m_1 x} + c_2 e^{m_2 x} + \cdots + c_n e^{m_n x}.$$

It is somewhat harder to summarize the analogues of Cases II and III because the roots of an auxiliary equation of degree greater than two can occur in many combinations. For example, a fifth-degree equation could have five distinct real roots, or three distinct real and two complex roots, or one real and four complex roots, or five real but equal roots, or five real roots but two of them equal, and so on. When m_1 is a root of multiplicity k of an nth-degree auxiliary equation (that is, k roots are equal to m_1), it can be shown that the linearly independent solutions are

$$e^{m_1 x}, x e^{m_1 x}, x^2 e^{m_1 x}, \ldots, x^{k-1} e^{m_1 x}$$

and the general solution must contain the linear combination

$$c_1 e^{m_1 x} + c_2 x e^{m_1 x} + c_3 x^2 e^{m_1 x} + \cdots + c_k x^{k-1} e^{m_1 x}.$$

Lastly, it should be remembered that when the coefficients are real, complex roots of an auxiliary equation always appear in conjugate pairs. Thus, for example, a cubic polynomial equation can have at most two complex roots.

EXAMPLE 3   **Third-Order DE**

Solve $y''' + 3y'' - 4y = 0$.

Solution   It should be apparent from inspection of $m^3 + 3m^2 - 4 = 0$ that one root is $m_1 = 1$ and so $m - 1$ is a factor of $m^3 + 3m^2 - 4$. By division we find

$$m^3 + 3m^2 - 4 = (m - 1)(m^2 + 4m + 4) = (m - 1)(m + 2)^2,$$

and so the other roots are $m_2 = m_3 = -2$. Thus the general solution is $y = c_1 e^x + c_2 e^{-2x} + c_3 x e^{-2x}$. ∎

EXAMPLE 4   **Fourth-Order DE**

Solve $\dfrac{d^4 y}{dx^4} + 2\dfrac{d^2 y}{dx^2} + y = 0$.

Solution   The auxiliary equation $m^4 + 2m^2 + 1 = (m^2 + 1)^2 = 0$ has roots $m_1 = m_3 = i$ and $m_2 = m_4 = -i$. Thus from Case II the solution is

$$y = C_1 e^{ix} + C_2 e^{-ix} + C_3 x e^{ix} + C_4 x e^{-ix}.$$

By Euler's formula the grouping $C_1 e^{ix} + C_2 e^{-ix}$ can be rewritten as

$$c_1 \cos x + c_2 \sin x$$

after a relabeling of constants. Similarly, $x(C_3 e^{ix} + C_4 e^{-ix})$ can be expressed as $x(c_3 \cos x + c_4 \sin x)$. Hence the general solution is

$$y = c_1 \cos x + c_2 \sin x + c_3 x \cos x + c_4 x \sin x. \qquad \blacksquare$$

Example 4 illustrates a special case when the auxiliary equation has repeated complex roots. In general, if $m_1 = \alpha + i\beta$, $\beta > 0$ is a complex root of multiplicity k of an auxiliary equation with real coefficients, then its conjugate $m_2 = \alpha - i\beta$ is also a root of multiplicity k. From the $2k$ complex-valued solutions

$$e^{(\alpha+i\beta)x}, \quad xe^{(\alpha+i\beta)x}, \quad x^2 e^{(\alpha+i\beta)x}, \quad \ldots, \quad x^{k-1} e^{(\alpha+i\beta)x}$$
$$e^{(\alpha-i\beta)x}, \quad xe^{(\alpha-i\beta)x}, \quad x^2 e^{(\alpha-i\beta)x}, \quad \ldots, \quad x^{k-1} e^{(\alpha-i\beta)x}$$

we conclude, with the aid of Euler's formula, that the general solution of the corresponding differential equation must then contain a linear combination of the $2k$ real linearly independent solutions

$$e^{\alpha x} \cos \beta x, \quad xe^{\alpha x} \cos \beta x, \quad x^2 e^{\alpha x} \cos \beta x, \quad \ldots, \quad x^{k-1} e^{\alpha x} \cos \beta x$$
$$e^{\alpha x} \sin \beta x, \quad xe^{\alpha x} \sin \beta x, \quad x^2 e^{\alpha x} \sin \beta x, \quad \ldots, \quad x^{k-1} e^{\alpha x} \sin \beta x.$$

In Example 4 we identify $k = 2$, $\alpha = 0$, and $\beta = 1$.

Of course the most difficult aspect of solving constant-coefficient differential equations is finding roots of auxiliary equations of degree greater than two. For example, to solve $3y''' + 5y'' + 10y' - 4y = 0$ we must solve $3m^3 + 5m^2 + 10m - 4 = 0$. Something we can try is to test the auxiliary equation for rational roots. Recall that if $m_1 = p/q$ is a rational root (expressed in lowest terms) of an auxiliary equation $a_n m^n + \cdots + a_1 m + a_0 = 0$ with integer coefficients, then p is a factor of a_0 and q is a factor of a_n. For our specific cubic auxiliary equation, all the factors of $a_0 = -4$ and $a_n = 3$ are p: $\pm 1, \pm 2, \pm 4$ and q: $\pm 1, \pm 3$, so the possible rational roots are p/q: $\pm 1, \pm 2, \pm 4, \pm\frac{1}{3}, \pm\frac{2}{3}, \pm\frac{4}{3}$. Each of these numbers can then be tested—say, by synthetic division. In this way we discover both the root $m_1 = \frac{1}{3}$ and the factorization

$$3m^3 + 5m^2 + 10m - 4 = (m - \tfrac{1}{3})(3m^2 + 6m + 12).$$

The quadratic formula then yields the remaining roots $m_2 = -1 + \sqrt{3}\, i$ and $m_3 = -1 - \sqrt{3}\, i$. Therefore the general solution of $3y''' + 5y'' + 10y' - 4y = 0$ is $y = c_1 e^{x/3} + e^{-x}(c_2 \cos \sqrt{3}x + c_3 \sin \sqrt{3}x)$.

Use of Computers   Finding roots or approximations of roots of polynomial equations is a routine problem with an appropriate calculator or computer software. The computer algebra systems *Mathematica* and *Maple* can solve polynomial equations (in one variable) of degree less than five in terms of algebraic formulas. For the auxiliary equation in the preceding paragraph, the commands

Solve[3 m^3 + 5 m^2 + 10 m − 4 == 0, m]   (in *Mathematica*)

solve(3*m^3 + 5*m^2 + 10*m − 4, m);   (in *Maple*) yield immediately their representations of the roots $\frac{1}{3}$, $-1 + \sqrt{3}\,i$, $-1 - \sqrt{3}\,i$. For auxiliary equations of higher degree it may be necessary to resort to numerical commands such as NSolve and FindRoot in *Mathematica*. Because of their capability of solving polynomial equations, it is not surprising that some computer algebra systems are also able to give explicit solutions of homogeneous linear constant-coefficient differential equations. For example, the inputs

DSolve [y″[x] + 2 y′[x] + 2 y[x] = = 0, y[x], x]   (in *Mathematica*)

dsolve(diff(y(x),x\$2) + 2*diff(y(x),x) +2*y(x) = 0, y(x));   (in *Maple*)

give, respectively,

$$y[x] \,-\!> \frac{C[2]\ Cos\ [x] - C[1]\ Sin\ [x]}{E^x} \tag{13}$$

and   **y(x) = _C1 exp(−x) sin(x) + _C2 exp(−x) cos(x)**

Translated, this means $y = c_2 e^{-x} \cos x + c_1 e^{-x} \sin x$ is a solution of $y'' + 2y' + 2y = 0$.

In the classic text *Differential Equations,* by Ralph Palmer Agnew* (used by the author as a student), the following statement is made:

> *It is not reasonable to expect students in this course to have computing skill and equipment necessary for efficient solving of equations such as*

$$4.317\frac{d^4y}{dx^4} + 2.179\frac{d^3y}{dx^3} + 1.416\frac{d^2y}{dx^2} + 1.295\frac{dy}{dx} + 3.169y = 0. \tag{14}$$

Although it is debatable whether computing skills have improved in the intervening years, it is a certainty that technology has. If one has access to a computer algebra system, equation (14) could be considered reasonable. After simplification and some relabeling of the output, *Mathematica* yields the (approximate) general solution

$$y = c_1 e^{-0.728852x} \cos(0.618605x) + c_2 e^{-0.728852x} \sin(0.618605x)$$
$$+ c_3 e^{0.476478x} \cos(0.759081x) + c_4 e^{0.476478x} \sin(0.759081x).$$

We note in passing that the DSolve and dsolve commands in *Mathematica* and *Maple,* like most aspects of any CAS, have their limitations.

Finally, if we are faced with an initial-value problem consisting of, say, a fourth-order equation, then to fit the general solution of the DE to the four initial conditions we must solve a system of four linear equations in four unknowns (the c_1, c_2, c_3, c_4 in the general solution). Using a CAS to solve the system can save lots of time. See Problems 63 and 64 in Exercises 4.3 and Problem 35 in Chapter 4 in Review.

EXERCISES 4.3

Answers to odd-numbered problems begin on page AN-3.

In Problems 1–14 find the general solution of the given second-order differential equation.

1. $4y'' + y' = 0$ **2.** $y'' - 36y = 0$

---

*McGraw-Hill, New York, 1960.

3. $y'' - y' - 6y = 0$ 4. $y'' - 3y' + 2y = 0$

5. $y'' + 8y' + 16y = 0$ 6. $y'' - 10y' + 25y = 0$

7. $12y'' - 5y' - 2y = 0$ 8. $y'' + 4y' - y = 0$

9. $y'' + 9y = 0$ 10. $3y'' + y = 0$

11. $y'' - 4y' + 5y = 0$ 12. $2y'' + 2y' + y = 0$

13. $3y'' + 2y' + y = 0$ 14. $2y'' - 3y' + 4y = 0$

In Problems 15–28 find the general solution of the given higher-order differential equation.

15. $y''' - 4y'' - 5y' = 0$ 16. $y''' - y = 0$

17. $y''' - 5y'' + 3y' + 9y = 0$ 18. $y''' + 3y'' - 4y' - 12y = 0$

19. $\dfrac{d^3u}{dt^3} + \dfrac{d^2u}{dt^2} - 2u = 0$ 20. $\dfrac{d^3x}{dt^3} - \dfrac{d^2x}{dt^2} - 4x = 0$

21. $y''' + 3y'' + 3y' + y = 0$ 22. $y''' - 6y'' + 12y' - 8y = 0$

23. $y^{(4)} + y''' + y'' = 0$ 24. $y^{(4)} - 2y'' + y = 0$

25. $16\dfrac{d^4y}{dx^4} + 24\dfrac{d^2y}{dx^2} + 9y = 0$ 26. $\dfrac{d^4y}{dx^4} - 7\dfrac{d^2y}{dx^2} - 18y = 0$

27. $\dfrac{d^5u}{dr^5} + 5\dfrac{d^4u}{dr^4} - 2\dfrac{d^3u}{dr^3} - 10\dfrac{d^2u}{dr^2} + \dfrac{du}{dr} + 5u = 0$

28. $2\dfrac{d^5x}{ds^5} - 7\dfrac{d^4x}{ds^4} + 12\dfrac{d^3x}{ds^3} + 8\dfrac{d^2x}{ds^2} = 0$

In Problems 29–36 solve the given initial-value problem.

29. $y'' + 16y = 0$, $y(0) = 2$, $y'(0) = -2$

30. $\dfrac{d^2y}{d\theta^2} + y = 0$, $y\left(\dfrac{\pi}{3}\right) = 0$, $y'\left(\dfrac{\pi}{3}\right) = 2$

31. $\dfrac{d^2y}{dt^2} - 4\dfrac{dy}{dt} - 5y = 0$, $y(1) = 0$, $y'(1) = 2$

32. $4y'' - 4y' - 3y = 0$, $y(0) = 1$, $y'(0) = 5$

33. $y'' + y' + 2y = 0$, $y(0) = y'(0) = 0$

34. $y'' - 2y' + y = 0$, $y(0) = 5$, $y'(0) = 10$

35. $y''' + 12y'' + 36y' = 0$, $y(0) = 0$, $y'(0) = 1$, $y''(0) = -7$

36. $y''' + 2y'' - 5y' - 6y = 0$, $y(0) = y'(0) = 0$, $y''(0) = 1$

In Problems 37–40 solve the given boundary-value problem.

37. $y'' - 10y' + 25y = 0$, $y(0) = 1$, $y(1) = 0$

38. $y'' + 4y = 0$, $y(0) = 0$, $y(\pi) = 0$

39. $y'' + y = 0$, $y'(0) = 0$, $y'\left(\dfrac{\pi}{2}\right) = 2$

40. $y'' - 2y' + 2y = 0$, $y(0) = 1$, $y(\pi) = 1$

In Problems 41 and 42 solve the given problem first by using the form of the general solution given in (10). Solve again, this time using the form given in (11).

41. $y'' - 3y = 0$, $y(0) = 1$, $y'(0) = 5$

42. $y'' - y = 0$, $y(0) = 1$, $y'(1) = 0$

In Problems 43–48 each figure represents the graph of a particular solution of one of the following differential equations:

(a) $y'' - 3y' - 4y = 0$ **(b)** $y'' + 4y = 0$ **(c)** $y'' + 2y' + y = 0$
(d) $y'' + y = 0$ **(e)** $y'' + 2y' + 2y = 0$ **(f)** $y'' - 3y' + 2y = 0$

Match each solution curve with one of the differential equations. Explain your reasoning.

43.

Figure 4.5

44.

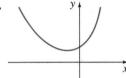

Figure 4.6

45.

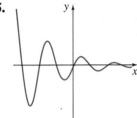

Figure 4.7

46.

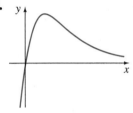

Figure 4.8

47.

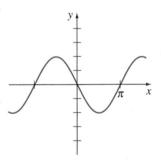

Figure 4.9

48.

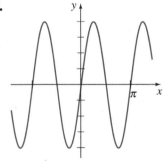

Figure 4.10

49. Reread the discussion on the slipping chain in Section 1.3 and review Figure 1.21 on page 29.

    (a) Use the form of the solution given in (11) of this section to find the general solution of equation (16) of Section 1.3,

$$\frac{d^2x}{dt^2} - \frac{64}{L}x = 0.$$

    (b) Find a particular solution that satisfies the initial conditions stated in the discussion on page 29.

    (c) Suppose that the total length of the chain is $L = 20$ ft and that $x_0 = 1$. Find the velocity at which the slipping chain will leave the supporting peg.

Discussion Problems

50. Reexamine equation (13). Notice that when we rewrote this computer-generated solution, $-C[1]$ was changed to $+c_1$. Why can we do that?

51. The roots of a cubic auxiliary equation are $m_1 = 4$ and $m_2 = m_3 = -5$. What is the corresponding homogeneous linear differential equation? Is your answer unique? Discuss.

52. Two roots of a cubic auxiliary equation with real coefficients are $m_1 = -\frac{1}{2}$ and $m_2 = 3 + i$. What is the corresponding homogeneous linear differential equation?

53. Find the general solution of $y''' + 6y'' + y' - 34y = 0$ if it is known that $y_1 = e^{-4x} \cos x$ is one solution.

54. Consider the differential equation $xy'' + y' + xy = 0$ or $y'' + (1/x)y' + y = 0$ for $x > 0$. Discuss how this last equation enables us to discern the qualitative behavior of the solutions as $x \to \infty$. Verify your conjectures using a numerical solver.

55. In order to solve $y^{(4)} + y = 0$ we must find the roots of $m^4 + 1 = 0$. This is a trivial problem with a CAS, but it can also be done by hand by working with complex numbers. Observe that $m^4 + 1 = (m^2 + 1)^2 - 2m^2$. How does this help? Solve the differential equation.

56. Consider the second-order equation with constant coefficients $y'' + by' + cy = 0$.
   (a) If $y(x)$ is a solution of the equation, discuss what conditions should be put on b and c so that $\lim_{x \to \infty} y(x) = 0$.
   (b) Discuss what conditions should be put on b and c so that the equation possesses a nontrivial solution satisfying the boundary conditions $y(0) = 0$, $y(1) = 0$.

57. Consider the boundary-value problem $y'' + \lambda y = 0$, $y(0) = 0$, $y(\pi/2) = 0$. Discuss whether it is possible to determine values of λ so that the problem possesses **(a)** trivial solutions, **(b)** nontrivial solutions.

58. In the study of techniques of integration in calculus, certain indefinite integrals of the form $\int e^{ax} f(x)\, dx$ can be evaluated by applying integration by parts twice, recovering the original integral on the right-hand side, solving for the original integral, and obtaining a constant multiple $k \int e^{ax} f(x)\, dx$ on the left-hand side. Then the value of the integral is found by dividing by k. Discuss for what kind of functions f the described procedure works. Your solution should lead to a differential equation. Carefully analyze this equation and solve for f.

Computer Lab Assignments

In Problems 59–62 use a computer either as an aid in solving the auxiliary equation or as a means of directly obtaining the general solution of the given differential equation. If you use a CAS to obtain the general solution, simplify the output and, if necessary, write the solution in terms of real functions.

59. $y''' - 6y'' + 2y' + y = 0$

60. $6.11y''' + 8.59y'' + 7.93y' + 0.778y = 0$

61. $3.15y^{(4)} - 5.34y'' + 6.33y' - 2.03y = 0$

62. $y^{(4)} + 2y'' - y' + 2y = 0$

In Problems 63 and 64 use a CAS as an aid in solving the auxiliary equation. Form the general solution of the differential equation. Then use a CAS as an aid in solving the system of equations for the coefficients c_i, $i = 1, 2, 3, 4$ that result when the initial conditions are applied to the general solution.

63. $2y^{(4)} + 3y''' - 16y'' + 15y' - 4y = 0$, $y(0) = -2$, $y'(0) = 6$,
$y''(0) = 3$, $y'''(0) = \frac{1}{2}$

64. $y^{(4)} - 3y''' + 3y'' - y' = 0$, $y(0) = y'(0) = 0$, $y''(0) = y'''(0) = 1$

4.4 UNDETERMINED COEFFICIENTS— SUPERPOSITION APPROACH*

• *General solution of a nonhomogeneous linear DE* • *The form of a particular solution* • *Superposition principle for nonhomogeneous linear DEs* • *Cases for using undetermined coefficients*

To solve a nonhomogeneous linear differential equation

$$a_n y^{(n)} + a_{n-1} y^{(n-1)} + \cdots + a_1 y' + a_0 y = g(x) \qquad \textbf{(1)}$$

we must do two things: (*i*) find the complementary function y_c and (*ii*) find *any* particular solution of (1). Then, as discussed in Section 4.1, the general solution of (1) on an interval is $y = y_c + y_p$. In this section we examine one method for obtaining a particular solution y_p.

Method of Undetermined Coefficients The complementary function y_c of (1) is the general solution of the associated homogeneous equation $a_n y^{(n)} + a_{n-1} y^{(n-1)} + \cdots + a_1 y' + a_0 y = 0$. In the last section we saw how to solve these kinds of equations when the coefficients were constants. The first of two ways we shall consider for finding a particular solution y_p is called the **method of undetermined coefficients.** The underlying idea in this method is a conjecture, or educated guess, about the form of y_p that is motivated by the kinds of functions that comprise the input function $g(x)$. Basically straightforward, the method is regrettably limited to nonhomogeneous linear equations such as (1) where

- the coefficients a_i, $i = 0, 1, \ldots, n$ are constants and
- $g(x)$ is a constant k, a polynomial function, an exponential function $e^{\alpha x}$, a sine or cosine function $\sin \beta x$ or $\cos \beta x$, or finite sums and products of these functions.

Note Strictly speaking, $g(x) = k$ (constant) is a polynomial function. Since a constant function is probably not the first thing that comes to

---

*Note to the Instructor: In this section the method of undetermined coefficients is developed from the viewpoint of the superposition principle for nonhomogeneous equations (Theorem 4.7). In Section 4.5 an entirely different approach will be presented, one utilizing the concept of differential annihilator operators. Take your pick.

mind when you think of polynomial functions, for emphasis we shall continue to use the redundancy "constant functions, polynomials, ..."

The following functions are some examples of the types of inputs $g(x)$ that are appropriate for this discussion:

$$g(x) = 10, \quad g(x) = x^2 - 5x, \quad g(x) = 15x - 6 + 8e^{-x},$$
$$g(x) = \sin 3x - 5x \cos 2x, \quad g(x) = xe^x \sin x + (3x^2 - 1)e^{-4x}.$$

That is, $g(x)$ is a linear combination of functions of the type

$$P(x) = a_n x^n + a_{n-1}x^{n-1} + \cdots + a_1 x + a_0, \quad P(x)e^{\alpha x}, \quad P(x)e^{\alpha x} \sin \beta x, \quad \text{and} \quad P(x)e^{\alpha x} \cos \beta x,$$

where n is a nonnegative integer and α and β are real numbers. The method of undetermined coefficients is not applicable to equations of form (1) when

$$g(x) = \ln x, \quad g(x) = \frac{1}{x}, \quad g(x) = \tan x, \quad g(x) = \sin^{-1}x,$$

and so on. Differential equations in which the input $g(x)$ is a function of this last kind will be considered in Section 4.6.

The set of functions that consists of constants, polynomials, exponentials $e^{\alpha x}$, sines, and cosines has the remarkable property that derivatives of their sums and products are again sums and products of constants, polynomials, exponentials $e^{\alpha x}$, sines, and cosines. Since the linear combination of derivatives $a_n y_p^{(n)} + a_{n-1}y_p^{(n-1)} + \cdots + a_1 y_p' + a_0 y_p$ must be identical to $g(x)$, it seems reasonable to assume that y_p *has the same form as $g(x)$*.

The next two examples illustrate the basic method.

EXAMPLE 1   **General Solution Using Undetermined Coefficients**

Solve $y'' + 4y' - 2y = 2x^2 - 3x + 6$.   **(2)**

Solution   **Step 1.** We first solve the associated homogeneous equation $y'' + 4y' - 2y = 0$. From the quadratic formula we find that the roots of the auxiliary equation $m^2 + 4m - 2 = 0$ are $m_1 = -2 - \sqrt{6}$ and $m_2 = -2 + \sqrt{6}$. Hence the complementary function is

$$y_c = c_1 e^{-(2+\sqrt{6})x} + c_2 e^{(-2+\sqrt{6})x}.$$

Step 2. Now, since the function $g(x)$ is a quadratic polynomial, let us assume a particular solution that is also in the form of a quadratic polynomial:

$$y_p = Ax^2 + Bx + C.$$

We seek to determine *specific* coefficients A, B, and C for which y_p is a solution of (2). Substituting y_p and the derivatives

$$y_p' = 2Ax + B \quad \text{and} \quad y_p'' = 2A$$

into the given differential equation (2), we get

$$y_p'' + 4y_p' - 2y_p = 2A + 8Ax + 4B - 2Ax^2 - 2Bx - 2C$$
$$= 2x^2 - 3x + 6.$$

Since the last equation is supposed to be an identity, the coefficients of like powers of x must be equal:

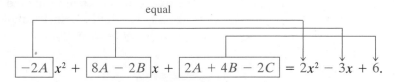

That is,

$$-2A = 2, \quad 8A - 2B = -3, \quad 2A + 4B - 2C = 6.$$

Solving this system of equations leads to the values $A = -1$, $B = -\frac{5}{2}$, and $C = -9$. Thus a particular solution is

$$y_p = -x^2 - \frac{5}{2}x - 9.$$

Step 3. The general solution of the given equation is

$$y = y_c + y_p = c_1 e^{-(2+\sqrt{6})x} + c_2 e^{(-2+\sqrt{6})x} - x^2 - \frac{5}{2}x - 9. \qquad \blacksquare$$

EXAMPLE 2   **Particular Solution Using Undetermined Coefficients**

Find a particular solution of $y'' - y' + y = 2 \sin 3x$.

Solution   A natural first guess for a particular solution would be $A \sin 3x$. But since successive differentiations of $\sin 3x$ produce $\sin 3x$ *and* $\cos 3x$, we are prompted instead to assume a particular solution that includes both of these terms:

$$y_p = A \cos 3x + B \sin 3x.$$

Differentiating y_p and substituting the results into the differential equation gives, after regrouping,

$$y_p'' - y_p' + y_p = (-8A - 3B) \cos 3x + (3A - 8B) \sin 3x = 2 \sin 3x$$

or

$$\boxed{-8A - 3B}\ \cos 3x + \boxed{3A - 8B}\ \sin 3x = 0 \cos 3x + 2 \sin 3x.$$

From the resulting system of equations,

$$-8A - 3B = 0, \quad 3A - 8B = 2,$$

we get $A = \frac{6}{73}$ and $B = -\frac{16}{73}$. A particular solution of the equation is

$$y_p = \frac{6}{73} \cos 3x - \frac{16}{73} \sin 3x. \qquad \blacksquare$$

As we mentioned, the form that we assume for the particular solution y_p is an educated guess; it is not a blind guess. This educated guess must take into consideration not only the types of functions that make up $g(x)$ but also, as we shall see in Example 4, the functions that make up the complementary function y_c.

EXAMPLE 3 **Forming y_p by Superposition**

Solve $y'' - 2y' - 3y = 4x - 5 + 6xe^{2x}$. $\hspace{2cm}$ **(3)**

Solution **Step 1.** First, the solution of the associated homogeneous equation $y'' - 2y' - 3y = 0$ is found to be $y_c = c_1 e^{-x} + c_2 e^{3x}$.

Step 2. Next, the presence of $4x - 5$ in $g(x)$ suggests that the particular solution includes a linear polynomial. Furthermore, since the derivative of the product xe^{2x} produces $2xe^{2x}$ and e^{2x}, we also assume that the particular solution includes both xe^{2x} and e^{2x}. In other words, g is the sum of two basic kinds of functions:

$$g(x) = g_1(x) + g_2(x) = polynomial + exponentials.$$

Correspondingly, the superposition principle for nonhomogeneous equations (Theorem 4.7) suggests that we seek a particular solution

$$y_p = y_{p_1} + y_{p_2},$$

where $y_{p_1} = Ax + B$ and $y_{p_2} = Cxe^{2x} + Ee^{2x}$. Substituting

$$y_p = Ax + B + Cxe^{2x} + Ee^{2x}$$

into the given equation (3) and grouping like terms gives

$$y_p'' - 2y_p' - 3y_p = -3Ax - 2A - 3B - 3Cxe^{2x} + (2C - 3E)e^{2x} = 4x - 5 + 6xe^{2x}. \hspace{1cm} \textbf{(4)}$$

From this identity we obtain the four equations

$$-3A = 4, \quad -2A - 3B = -5, \quad -3C = 6, \quad 2C - 3E = 0.$$

The last equation in this system results from the interpretation that the coefficient of e^{2x} in the right member of (4) is zero. Solving, we find $A = -\frac{4}{3}$, $B = \frac{23}{9}$, $C = -2$, and $E = -\frac{4}{3}$. Consequently,

$$y_p = -\frac{4}{3}x + \frac{23}{9} - 2xe^{2x} - \frac{4}{3}e^{2x}.$$

Step 3. The general solution of the equation is

$$y = c_1 e^{-x} + c_2 e^{3x} - \frac{4}{3}x + \frac{23}{9} - \left(2x + \frac{4}{3}\right)e^{2x}. \hspace{1cm} \blacksquare$$

In light of the superposition principle (Theorem 4.7) we can also approach Example 3 from the viewpoint of solving two simpler problems.

You should verify that substituting

$$y_{p_1} = Ax + B \qquad \text{into} \quad y'' - 2y' - 3y = 4x - 5$$

and $\qquad y_{p_2} = Cxe^{2x} + Ee^{2x} \quad \text{into} \quad y'' - 2y' - 3y = 6xe^{2x}$

yields, in turn, $y_{p_1} = -\frac{4}{3}x + \frac{23}{9}$ and $y_{p_2} = -(2x + \frac{4}{3})e^{2x}$. A particular solution of (3) is then $y_p = y_{p_1} + y_{p_2}$.

The next example illustrates that sometimes the "obvious" assumption for the form of y_p is not a correct assumption.

EXAMPLE 4   A Glitch in the Method

Find a particular solution of $y'' - 5y' + 4y = 8e^x$.

Solution   Differentiation of e^x produces no new functions. Thus, proceeding as we did in the earlier examples, we can reasonably assume a particular solution of the form $y_p = Ae^x$. But substitution of this expression into the differential equation yields the contradictory statement $0 = 8e^x$, and so we have clearly made the wrong guess for y_p.

The difficulty here is apparent upon examining the complementary function $y_c = c_1e^x + c_2e^{4x}$. Observe that our assumption Ae^x is already present in y_c. This means that e^x is a solution of the associated homogeneous differential equation, and a constant multiple Ae^x when substituted into the differential equation necessarily produces zero.

What then should be the form of y_p? Inspired by Case II of Section 4.3, let's see whether we can find a particular solution of the form

$$y_p = Axe^x.$$

Substituting $y_p' = Axe^x + Ae^x$ and $y_p'' = Axe^x + 2Ae^x$ into the differential equation and simplifying gives

$$y_p'' - 5y_p' + 4y_p = -3Ae^x = 8e^x.$$

From the last equality we see that the value of A is now determined as $A = -\frac{8}{3}$. Therefore a particular solution of the given equation is $y_p = -\frac{8}{3}xe^x$.   ■

The difference in the procedures used in Examples 1–3 and in Example 4 suggests that we consider two cases. The first case reflects the situation in Examples 1–3.

Case I   No function in the assumed particular solution is a solution of the associated homogeneous differential equation.

In Table 4.1 we illustrate some specific examples of $g(x)$ in (1) along with the corresponding form of the particular solution. We are, of course, taking for granted that no function in the assumed particular solution y_p is duplicated by a function in the complementary function y_c.

TABLE 4.1   Trial Particular Solutions

$g(x)$	Form of y_p
1. 1 (any constant)	A
2. $5x + 7$	$Ax + B$
3. $3x^2 - 2$	$Ax^2 + Bx + C$
4. $x^3 - x + 1$	$Ax^3 + Bx^2 + Cx + E$
5. $\sin 4x$	$A \cos 4x + B \sin 4x$
6. $\cos 4x$	$A \cos 4x + B \sin 4x$
7. e^{5x}	Ae^{5x}
8. $(9x - 2)e^{5x}$	$(Ax + B)e^{5x}$
9. $x^2 e^{5x}$	$(Ax^2 + Bx + C)e^{5x}$
10. $e^{3x} \sin 4x$	$Ae^{3x} \cos 4x + Be^{3x} \sin 4x$
11. $5x^2 \sin 4x$	$(Ax^2 + Bx + C) \cos 4x + (Ex^2 + Fx + G) \sin 4x$
12. $xe^{3x} \cos 4x$	$(Ax + B)e^{3x} \cos 4x + (Cx + E)e^{3x} \sin 4x$

EXAMPLE 5   **Forms of Particular Solutions—Case I**

Determine the form of a particular solution of

(a) $y'' - 8y' + 25y = 5x^3 e^{-x} - 7e^{-x}$ **(b)** $y'' + 4y = x \cos x$

Solution   **(a)** We can write $g(x) = (5x^3 - 7)e^{-x}$. Using entry 9 in Table 4.1 as a model, we assume a particular solution of the form

$$y_p = (Ax^3 + Bx^2 + Cx + E)e^{-x}.$$

Note that there is no duplication between the terms in y_p and the terms in the complementary function $y_c = e^{4x}(c_1 \cos 3x + c_2 \sin 3x)$.

(b) The function $g(x) = x \cos x$ is similar to entry 11 in Table 4.1 except, of course, that we use a linear rather than a quadratic polynomial and $\cos x$ and $\sin x$ instead of $\cos 4x$ and $\sin 4x$ in the form of y_p:

$$y_p = (Ax + B) \cos x + (Cx + E) \sin x.$$

Again observe that there is no duplication of terms between y_p and $y_c = c_1 \cos 2x + c_2 \sin 2x$.   ∎

If $g(x)$ consists of a sum of, say, m terms of the kind listed in the table, then (as in Example 3) the assumption for a particular solution y_p consists of the sum of the trial forms $y_{p_1}, y_{p_2}, \ldots, y_{p_m}$ corresponding to these terms:

$$y_p = y_{p_1} + y_{p_2} + \cdots + y_{p_m}.$$

The foregoing sentence can be put another way.

Form Rule for Case I   *The form of y_p is a linear combination of all linearly independent functions that are generated by repeated differentiations of $g(x)$.*

EXAMPLE 6   **Forming y_p by Superposition—Case I**

Determine the form of a particular solution of

$$y'' - 9y' + 14y = 3x^2 - 5\sin 2x + 7xe^{6x}.$$

Solution

Corresponding to $3x^2$ we assume $y_{p_1} = Ax^2 + Bx + C.$

Corresponding to $-5\sin 2x$ we assume $y_{p_2} = E\cos 2x + F\sin 2x.$

Corresponding to $7xe^{6x}$ we assume $y_{p_3} = (Gx + H)e^{6x}.$

The assumption for the particular solution is then

$$y_p = y_{p_1} + y_{p_2} + y_{p_3} = Ax^2 + Bx + C + E\cos 2x + F\sin 2x + (Gx + H)e^{6x}.$$

No term in this assumption duplicates a term in $y_c = c_1 e^{2x} + c_2 e^{7x}.$ ▬

Case II   A function in the assumed particular solution is also a solution of the associated homogeneous differential equation.

The next example is similar to Example 4.

EXAMPLE 7   **Particular Solution—Case II**

Find a particular solution of $y'' - 2y' + y = e^x.$

Solution   The complementary function is $y_c = c_1 e^x + c_2 x e^x.$ As in Example 4, the assumption $y_p = Ae^x$ will fail since it is apparent from y_c that e^x is a solution of the associated homogeneous equation $y'' - 2y' + y = 0.$ Moreover, we will not be able to find a particular solution of the form $y_p = Axe^x$ since the term xe^x is also duplicated in $y_c.$ We next try

$$y_p = Ax^2 e^x.$$

Substituting into the given differential equation yields $2Ae^x = e^x$ and so $A = \frac{1}{2}.$ Thus a particular solution is $y_p = \frac{1}{2}x^2 e^x.$ ▬

Suppose again that $g(x)$ consists of m terms of the kind given in Table 4.1, and suppose further that the usual assumption for a particular solution is

$$y_p = y_{p_1} + y_{p_2} + \cdots + y_{p_m},$$

where the y_{p_i}, $i = 1, 2, \ldots, m$ are the trial particular solution forms corresponding to these terms. Under the circumstances described in Case II, we can make up the following general rule.

> ***Multiplication Rule for Case II*** *If any* y_{p_i} *contains terms that duplicate terms in* y_c, *then that* y_{p_i} *must be multiplied by* x^n, *where n is the smallest positive integer that eliminates that duplication.*

EXAMPLE 8   An Initial-Value Problem

Solve $y'' + y = 4x + 10 \sin x$, $y(\pi) = 0$, $y'(\pi) = 2$.

Solution   The solution of the associated homogeneous equation $y'' + y = 0$ is $y_c = c_1 \cos x + c_2 \sin x$. Since $g(x) = 4x + 10 \sin x$ is the sum of a linear polynomial and a sine function, our normal assumption for y_p, from entries 2 and 5 of Table 4.1, would be the sum of $y_{p_1} = Ax + B$ and $y_{p_2} = C \cos x + E \sin x$:

$$y_p = Ax + B + C \cos x + E \sin x. \tag{5}$$

But there is an obvious duplication of the terms $\cos x$ and $\sin x$ in this assumed form and two terms in the complementary function. This duplication can be eliminated by simply multiplying y_{p_2} by x. Instead of (5) we now use

$$y_p = Ax + B + Cx \cos x + Ex \sin x. \tag{6}$$

Differentiating this expression and substituting the results into the differential equation gives

$$y_p'' + y_p = Ax + B - 2C \sin x + 2E \cos x = 4x + 10 \sin x,$$

and so $A = 4$, $B = 0$, $-2C = 10$, and $2E = 0$. The solutions of the system are immediate: $A = 4$, $B = 0$, $C = -5$, and $E = 0$. Therefore from (6) we obtain $y_p = 4x - 5x \cos x$. The general solution of the given equation is

$$y = y_c + y_p = c_1 \cos x + c_2 \sin x + 4x - 5x \cos x.$$

We now apply the prescribed initial conditions to the general solution of the equation. First, $y(\pi) = c_1 \cos \pi + c_2 \sin \pi + 4\pi - 5\pi \cos \pi = 0$ yields $c_1 = 9\pi$ since $\cos \pi = -1$ and $\sin \pi = 0$. Next, from the derivative

$$y' = -9\pi \sin x + c_2 \cos x + 4 + 5x \sin x - 5 \cos x$$

and $y'(\pi) = -9\pi \sin \pi + c_2 \cos \pi + 4 + 5\pi \sin \pi - 5 \cos \pi = 2$

we find $c_2 = 7$. The solution of the initial value is then

$$y = 9\pi \cos x + 7 \sin x + 4x - 5x \cos x. \qquad \blacksquare$$

EXAMPLE 9   Using the Multiplication Rule

Solve $y'' - 6y' + 9y = 6x^2 + 2 - 12e^{3x}$.

Solution   The complementary function is $y_c = c_1 e^{3x} + c_2 x e^{3x}$. And so, based on entries 3 and 7 of Table 4.1, the usual assumption for a particular solution would be

$$y_p = \underbrace{Ax^2 + Bx + C}_{y_{p_1}} + \underbrace{Ee^{3x}}_{y_{p_2}}.$$

Inspection of these functions shows that the one term in y_{p_2} is duplicated in y_c. If we multiply y_{p_2} by x, we note that the term xe^{3x} is still part of y_c. But multiplying y_{p_2} by x^2 eliminates all duplications. Thus the operative form of a particular solution is

$$y_p = Ax^2 + Bx + C + Ex^2 e^{3x}.$$

Differentiating this last form, substituting into the differential equation, and collecting like terms gives

$$y_p'' - 6y_p' + 9y_p = 9Ax^2 + (-12A + 9B)x + 2A - 6B + 9C + 2Ee^{3x} = 6x^2 + 2 - 12e^{3x}.$$

It follows from this identity that $A = \frac{2}{3}$, $B = \frac{8}{9}$, $C = \frac{2}{3}$, and $E = -6$. Hence the general solution $y = y_c + y_p$ is

$$y = c_1 e^{3x} + c_2 x e^{3x} + \frac{2}{3}x^2 + \frac{8}{9}x + \frac{2}{3} - 6x^2 e^{3x}. \qquad \blacksquare$$

EXAMPLE 10   Third-Order DE—Case I

Solve $y''' + y'' = e^x \cos x$.

Solution   From the characteristic equation $m^3 + m^2 = 0$ we find $m_1 = m_2 = 0$ and $m_3 = -1$. Hence the complementary function of the equation is $y_c = c_1 + c_2 x + c_3 e^{-x}$. With $g(x) = e^x \cos x$, we see from entry 10 of Table 4.1 that we should assume

$$y_p = Ae^x \cos x + Be^x \sin x.$$

Since there are no functions in y_p that duplicate functions in the complementary solution, we proceed in the usual manner. From

$$y_p''' + y_p'' = (-2A + 4B)e^x \cos x + (-4A - 2B)e^x \sin x = e^x \cos x$$

we get $-2A + 4B = 1$ and $-4A - 2B = 0$. This system gives $A = -\frac{1}{10}$ and $B = \frac{1}{5}$, so a particular solution is $y_p = -\frac{1}{10}e^x \cos x + \frac{1}{5}e^x \sin x$. The general solution of the equation is

$$y = y_c + y_p = c_1 + c_2 x + c_3 e^{-x} - \frac{1}{10}e^x \cos x + \frac{1}{5}e^x \sin x. \qquad \blacksquare$$

EXAMPLE 11   Fourth-Order DE—Case II

Determine the form of a particular solution of $y^{(4)} + y''' = 1 - x^2 e^{-x}$.

Solution   Comparing $y_c = c_1 + c_2 x + c_3 x^2 + c_4 e^{-x}$ with our normal assumption for a particular solution

$$y_p = \underbrace{A}_{y_{p_1}} + \underbrace{Bx^2 e^{-x} + Cxe^{-x} + Ee^{-x}}_{y_{p_2}},$$

we see that the duplications between y_c and y_p are eliminated when y_{p_1} is multiplied by x^3 and y_{p_2} is multiplied by x. Thus the correct assumption for a particular solution is

$$y_p = Ax^3 + Bx^3e^{-x} + Cx^2e^{-x} + Exe^{-x}.$$ ▄

Remarks In Problems 27–36 of Exercises 4.4 you are asked to solve initial-value problems, and in Problems 37 and 38, boundary-value problems. As illustrated in Example 8, be sure to apply the initial conditions or the boundary conditions to the general solution $y = y_c + y_p$. Students often make the mistake of applying these conditions to only the complementary function y_c since it is that part of the solution that contains the constants.

EXERCISES 4.4

Answers to odd-numbered problems begin on page AN-4.

In Problems 1–26 solve the given differential equation by undetermined coefficients.

1. $y'' + 3y' + 2y = 6$

2. $4y'' + 9y = 15$

3. $y'' - 10y' + 25y = 30x + 3$

4. $y'' + y' - 6y = 2x$

5. $\dfrac{1}{4}y'' + y' + y = x^2 - 2x$

6. $y'' - 8y' + 20y = 100x^2 - 26xe^x$

7. $y'' + 3y = -48x^2e^{3x}$

8. $4y'' - 4y' - 3y = \cos 2x$

9. $y'' - y' = -3$

10. $y'' + 2y' = 2x + 5 - e^{-2x}$

11. $y'' - y' + \dfrac{1}{4}y = 3 + e^{x/2}$

12. $y'' - 16y = 2e^{4x}$

13. $y'' + 4y = 3 \sin 2x$

14. $y'' + 4y = (x^2 - 3) \sin 2x$

15. $y'' + y = 2x \sin x$

16. $y'' - 5y' = 2x^3 - 4x^2 - x + 6$

17. $y'' - 2y' + 5y = e^x \cos 2x$

18. $y'' - 2y' + 2y = e^{2x}(\cos x - 3 \sin x)$

19. $y'' + 2y' + y = \sin x + 3 \cos 2x$

20. $y'' + 2y' - 24y = 16 - (x + 2)e^{4x}$

21. $y''' - 6y'' = 3 - \cos x$

22. $y''' - 2y'' - 4y' + 8y = 6xe^{2x}$

23. $y''' - 3y'' + 3y' - y = x - 4e^x$

24. $y''' - y'' - 4y' + 4y = 5 - e^x + e^{2x}$

25. $y^{(4)} + 2y'' + y = (x - 1)^2$

26. $y^{(4)} - y'' = 4x + 2xe^{-x}$

In Problems 27–36 solve the given initial-value problem.

27. $y'' + 4y = -2, \quad y\left(\dfrac{\pi}{8}\right) = \dfrac{1}{2}, y'\left(\dfrac{\pi}{8}\right) = 2$

28. $2y'' + 3y' - 2y = 14x^2 - 4x - 11, \quad y(0) = 0, y'(0) = 0$

29. $5y'' + y' = -6x, \quad y(0) = 0, y'(0) = -10$

30. $y'' + 4y' + 4y = (3 + x)e^{-2x}, \quad y(0) = 2, y'(0) = 5$

31. $y'' + 4y' + 5y = 35e^{-4x}, \quad y(0) = -3, y'(0) = 1$

32. $y'' - y = \cosh x, \quad y(0) = 2, y'(0) = 12$

33. $\dfrac{d^2x}{dt^2} + \omega^2 x = F_0 \sin \omega t, \quad x(0) = 0, x'(0) = 0$

34. $\dfrac{d^2x}{dt^2} + \omega^2 x = F_0 \cos \gamma t, \quad x(0) = 0, x'(0) = 0$

35. $y''' - 2y'' + y' = 2 - 24e^x + 40e^{5x}, \quad y(0) = \frac{1}{2}, y'(0) = \frac{5}{2}, y''(0) = -\frac{9}{2}$

36. $y''' + 8y = 2x - 5 + 8e^{-2x}, \quad y(0) = -5, y'(0) = 3, y''(0) = -4$

In Problems 37 and 38 solve the given boundary-value problem.

37. $y'' + y = x^2 + 1, \quad y(0) = 5, y(1) = 0$

38. $y'' - 2y' + 2y = 2x - 2, \quad y(0) = 0, y(\pi) = \pi$

In Problems 39 and 40 solve the given initial-value problem in which the input function $g(x)$ is discontinuous. [*Hint:* Solve each problem on two intervals, and then find a solution so that y and y' are continuous at $x = \pi/2$ (Problem 39) and at $x = \pi$ (Problem 40).]

39. $y'' + 4y = g(x), \quad y(0) = 1, y'(0) = 2$, where

$$g(x) = \begin{cases} \sin x, & 0 \le x \le \dfrac{\pi}{2} \\ 0, & x > \dfrac{\pi}{2} \end{cases}$$

40. $y'' - 2y' + 10y = g(x), \quad y(0) = 0, y'(0) = 0$, where

$$g(x) = \begin{cases} 20, & 0 \le x \le \pi \\ 0, & x > \pi \end{cases}$$

Discussion Problems

41. Consider the differential equation $ay'' + by' + cy = e^{kx}$, where a, b, c, and k are constants. The auxiliary equation of the associated homogeneous equation is $am^2 + bm + c = 0$.

   (a) If k is not a root of the auxiliary equation, show that we can find a particular solution of the form $y_p = Ae^{kx}$, where $A = 1/(ak^2 + bk + c)$.

   (b) If k is a root of the auxiliary equation of multiplicity one, show that we can find a particular solution of the form $y_p = Axe^{kx}$, where $A = 1/(2ak + b)$. Explain how we know that $k \neq -b/2a$.

   (c) If k is a root of the auxiliary equation of multiplicity two, show that we can find a particular solution of the form $y = Ax^2e^{kx}$, where $A = 1/(2a)$.

42. Discuss how the method of this section can be used to find a particular solution of $y'' + y = \sin x \cos 2x$. Carry out your idea.

43. Without solving, match a solution curve of $y'' + y = f(x)$ shown in the figure with one of the following functions:

    (i) $f(x) = 1$,   *(ii)* $f(x) = e^{-x}$,   *(iii)* $f(x) = e^x$,

    (iv) $f(x) = \sin 2x$,   *(v)* $f(x) = e^x \sin x$,   *(vi)* $f(x) = \sin x$.

Briefly discuss your reasoning.

(a)

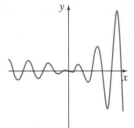

Figure 4.11

(b)

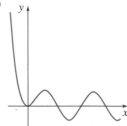

Figure 4.12

(c)

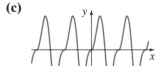

Figure 4.13

(d)

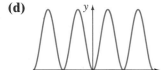

Figure 4.14

Computer Lab Assignments

In Problems 44 and 45 find a particular solution of the given differential equation. Use a CAS as an aid in carrying out differentiations, simplifications, and algebra.

44. $y'' - 4y' + 8y = (2x^2 - 3x)e^{2x} \cos 2x + (10x^2 - x - 1)e^{2x} \sin 2x$

45. $y^{(4)} + 2y'' + y = 2 \cos x - 3x \sin x$

4.5 UNDETERMINED COEFFICIENTS—ANNIHILATOR APPROACH

• Differential operator • Factoring a differential operator • Annihilator operator • Determining the form of a particular solution • Undetermined coefficients

We saw in Section 4.1 that an nth-order differential equation can be written

$$a_n D^n y + a_{n-1} D^{n-1} y + \cdots + a_1 D y + a_0 y = g(x), \tag{1}$$

where $D^k y = d^k y/dx^k$, $k = 0, 1, \ldots, n$. When it suits our purpose, (1) is also written as $L(y) = g(x)$, where L denotes the linear nth-order differential operator

$$L = a_n D^n + a_{n-1} D^{n-1} + \cdots + a_1 D + a_0. \tag{2}$$

Not only is the operator notation a helpful shorthand, but on a very practical level the *application* of differential operators enables us to obtain a particular solution of certain kinds of nonhomogeneous linear differential equations. Before investigating this, we need to examine two concepts.

Factoring Operators   When the a_i, $i = 0, 1, \ldots, n$ are real constants, a linear differential operator (1) can be *factored* whenever the characteristic polynomial $a_n m^n + a_{n-1} m^{n-1} + \cdots + a_1 m + a_0$ factors. In other words, if r_1 is a root of the auxiliary equation

$$a_n m^n + a_{n-1} m^{n-1} + \cdots + a_1 m + a_0 = 0,$$

then $L = (D - r_1)P(D)$, where the polynomial expression $P(D)$ is a linear differential operator of order $n - 1$. For example, if we treat D as an algebraic quantity, then the operator $D^2 + 5D + 6$ can be factored as $(D + 2)(D + 3)$ or as $(D + 3)(D + 2)$. Thus if a function $y = f(x)$ possesses a second derivative,

$$(D^2 + 5D + 6)y = (D + 2)(D + 3)y = (D + 3)(D + 2)y.$$

This illustrates a general property:

Factors of a linear differential operator with constant coefficients commute.

A differential equation such as $y'' + 4y' + 4y = 0$ can be written as

$$(D^2 + 4D + 4)y = 0 \quad \text{or} \quad (D + 2)(D + 2)y = 0 \quad \text{or} \quad (D + 2)^2 y = 0.$$

Annihilator Operator   If L is a linear differential operator with constant coefficients and f is a sufficiently differentiable function such that

$$L(f(x)) = 0,$$

then L is said to be an **annihilator** of the function. For example, a constant function $y = k$ is annihilated by D since $Dk = 0$. The function $y = x$ is annihilated by the differential operator D^2 since the first and second derivatives of x are 1 and 0, respectively. Similarly, $D^3 x^2 = 0$, and so on.

> The differential operator D^n annihilates each of the functions
>
> $$1, \quad x, \quad x^2, \quad \ldots, \quad x^{n-1}.$$

(3)

As an immediate consequence of (3) and the fact that differentiation can be done term by term, a polynomial

$$c_0 + c_1 x + c_2 x^2 + \cdots + c_{n-1} x^{n-1}$$

(4)

can be annihilated by finding an operator that annihilates the highest power of x.

The functions that are annihilated by a linear nth-order differential operator L are simply those functions that can be obtained from the general solution of the homogeneous differential equation $L(y) = 0$.

> The differential operator $(D - \alpha)^n$ annihilates each of the functions
>
> $$e^{\alpha x}, \quad xe^{\alpha x}, \quad x^2e^{\alpha x}, \quad \ldots, \quad x^{n-1}e^{\alpha x}.$$
>
> (5)

To see this, note that the auxiliary equation of the homogeneous equation $(D - \alpha)^n y = 0$ is $(m - \alpha)^n = 0$. Since α is a root of multiplicity n, the general solution is

$$y = c_1e^{\alpha x} + c_2xe^{\alpha x} + \cdots + c_nx^{n-1}e^{\alpha x}. \tag{6}$$

EXAMPLE 1   Annihilator Operators

Find a differential operator that annihilates the given function.

(a) $1 - 5x^2 + 8x^3$ **(b)** e^{-3x} **(c)** $4e^{2x} - 10xe^{2x}$

Solution   **(a)** From (3) we know that $D^4x^3 = 0$, and so it follows from (4) that

$$D^4(1 - 5x^2 + 8x^3) = 0.$$

(b) From (5), with $\alpha = -3$ and $n = 1$, we see that

$$(D + 3)e^{-3x} = 0.$$

(c) From (5) and (6), with $\alpha = 2$ and $n = 2$, we have

$$(D - 2)^2(4e^{2x} - 10xe^{2x}) = 0. \qquad \blacksquare$$

When α and β, $\beta > 0$ are real numbers, the quadratic formula reveals that $[m^2 - 2\alpha m + (\alpha^2 + \beta^2)]^n = 0$ has complex roots $\alpha + i\beta$, $\alpha - i\beta$, both of multiplicity n. From the discussion at the end of Section 4.3 we have the next result.

> The differential operator $[D^2 - 2\alpha D + (\alpha^2 + \beta^2)]^n$ annihilates each of the functions
>
> $$e^{\alpha x} \cos \beta x, \quad xe^{\alpha x} \cos \beta x, \quad x^2e^{\alpha x} \cos \beta x, \quad \ldots, \quad x^{n-1}e^{\alpha x} \cos \beta x,$$
> $$e^{\alpha x} \sin \beta x, \quad xe^{\alpha x} \sin \beta x, \quad x^2e^{\alpha x} \sin \beta x, \quad \ldots, \quad x^{n-1}e^{\alpha x} \sin \beta x.$$
>
> (7)

EXAMPLE 2 **Annihilator Operator**

Find a differential operator that annihilates $5e^{-x} \cos 2x - 9e^{-x} \sin 2x$.

Solution   Inspection of the functions $e^{-x} \cos 2x$ and $e^{-x} \sin 2x$ shows that $\alpha = -1$ and $\beta = 2$. Hence, from (7) we conclude that $D^2 + 2D + 5$ will annihilate each function. Since $D^2 + 2D + 5$ is a linear operator, it will annihilate *any* linear combination of these functions such as $5e^{-x} \cos 2x - 9e^{-x} \sin 2x$.   ■

When $\alpha = 0$ and $n = 1$, a special case of (7) is

$$(D^2 + \beta^2) \begin{cases} \cos \beta x \\ \sin \beta x \end{cases} = 0. \tag{8}$$

For example, $D^2 + 16$ will annihilate any linear combination of $\sin 4x$ and $\cos 4x$.

We are often interested in annihilating the sum of two or more functions. As just seen in Examples 1 and 2, if L is a linear differential operator such that $L(y_1) = 0$ and $L(y_2) = 0$, then L will annihilate the linear combination $c_1 y_1(x) + c_2 y_2(x)$. This is a direct consequence of Theorem 4.2. Let us now suppose that L_1 and L_2 are linear differential operators with constant coefficients such that L_1 annihilates $y_1(x)$ and L_2 annihilates $y_2(x)$, but $L_1(y_2) \neq 0$ and $L_2(y_1) \neq 0$. Then the *product* of differential operators $L_1 L_2$ annihilates the sum $c_1 y_1(x) + c_2 y_2(x)$. We can easily demonstrate this, using linearity and the fact that $L_1 L_2 = L_2 L_1$:

$$\begin{aligned} L_1 L_2(y_1 + y_2) &= L_1 L_2(y_1) + L_1 L_2(y_2) \\ &= L_2 L_1(y_1) + L_1 L_2(y_2) \\ &= L_2 [\underbrace{L_1(y_1)}_{\text{zero}}] + L_1 [\underbrace{L_2(y_2)}_{\text{zero}}] = 0. \end{aligned}$$

For example, we know from (3) that D^2 annihilates $7 - x$ and from (8) that $D^2 + 16$ annihilates $\sin 4x$. Therefore the product of operators $D^2(D^2 + 16)$ will annihilate the linear combination $7 - x + 6 \sin 4x$.

Note   The differential operator that annihilates a function is not unique. We saw in part (b) of Example 1 that $D + 3$ will annihilate e^{-3x}, but so will differential operators of higher order as long as $D + 3$ is one of the factors of the operator. For example, $(D + 3)(D + 1)$, $(D + 3)^2$, and $D^3(D + 3)$ all annihilate e^{-3x}. (Verify this.) As a matter of course, when we seek a differential annihilator for a function $y = f(x)$ we want the operator of *lowest possible order* that does the job.

Undetermined Coefficients   This brings us to the point of the preceding discussion. Suppose that $L(y) = g(x)$ is a linear differential equation with constant coefficients and that the input $g(x)$ consists of finite sums and products of the functions listed in (3), (5), and (7)—that is, $g(x)$ is a linear combination of functions of the form

$$k \text{ (constant)}, \quad x^m, \quad x^m e^{\alpha x}, \quad x^m e^{\alpha x} \cos \beta x, \quad \text{and} \quad x^m e^{\alpha x} \sin \beta x,$$

where m is a nonnegative integer and α and β are real numbers. We now know that such a function $g(x)$ can be annihilated by a differential operator L_1 of lowest order, consisting of a product of the operators D^n, $(D - \alpha)^n$, and $(D^2 - 2\alpha D + \alpha^2 + \beta^2)^n$. Applying L_1 to both sides of the equation $L(y) = g(x)$ yields $L_1 L(y) = L_1(g(x)) = 0$. By solving the *homogeneous higher-order* equation $L_1 L(y) = 0$, we can discover the *form* of a particular solution y_p for the original *nonhomogeneous* equation $L(y) = g(x)$. We then substitute this assumed form into $L(y) = g(x)$ to find an explicit particular solution. This procedure for determining y_p, called the **method of undetermined coefficients,** is illustrated in the next several examples.

Before proceeding, recall that the general solution of a nonhomogeneous linear differential equation $L(y) = g(x)$ is $y = y_c + y_p$, where y_c is the complementary function—that is, the general solution of the associated homogeneous equation $L(y) = 0$. The general solution of each equation $L(y) = g(x)$ is defined on the interval $(-\infty, \infty)$.

EXAMPLE 3   General Solution Using Undetermined Coefficients

Solve $y'' + 3y' + 2y = 4x^2$.   **(9)**

Solution   **Step 1.** First, we solve the homogeneous equation $y'' + 3y' + 2y = 0$. Then, from the auxiliary equation $m^2 + 3m + 2 = (m + 1)(m + 2) = 0$ we find $m_1 = -1$ and $m_2 = -2$, and so the complementary function is

$$y_c = c_1 e^{-x} + c_2 e^{-2x}.$$

Step 2. Now, since $4x^2$ is annihilated by the differential operator D^3, we see that $D^3(D^2 + 3D + 2)y = 4D^3 x^2$ is the same as

$$D^3(D^2 + 3D + 2)y = 0. \qquad \textbf{(10)}$$

The auxiliary equation of the fifth-order equation in (10),

$$m^3(m^2 + 3m + 2) = 0 \quad \text{or} \quad m^3(m + 1)(m + 2) = 0,$$

has roots $m_1 = m_2 = m_3 = 0$, $m_4 = -1$, and $m_5 = -2$. Thus its general solution must be

$$y = c_1 + c_2 x + c_3 x^2 + \boxed{c_4 e^{-x} + c_5 e^{-2x}}. \qquad \textbf{(11)}$$

The terms in the shaded box in (11) constitute the complementary function of the original equation (9). We can then argue that a particular solution y_p of (9) should also satisfy equation (10). This means that the terms remaining in (11) must be the basic form of y_p:

$$y_p = A + Bx + Cx^2, \qquad \textbf{(12)}$$

where, for convenience, we have replaced c_1, c_2, and c_3 by A, B, and C, respectively. For (12) to be a particular solution of (9), it is necessary to find *specific* coefficients A, B, and C. Differentiating (12), we have

$$y_p' = B + 2Cx, \quad y_p'' = 2C,$$

and substitution into (9) then gives

$$y_p'' + 3y_p' + 2y_p = 2C + 3B + 6Cx + 2A + 2Bx + 2Cx^2 = 4x^2.$$

Since the last equation is supposed to be an identity, the coefficients of like powers of x must be equal:

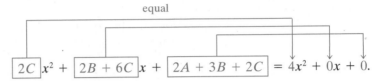

That is, $2C = 4, \quad 2B + 6C = 0, \quad 2A + 3B + 2C = 0.$ **(13)**

Solving the equations in (13) gives $A = 7$, $B = -6$, and $C = 2$. Thus $y_p = 7 - 6x + 2x^2$.

Step 3. The general solution of the equation in (9) is $y = y_c + y_p$ or

$$y = c_1 e^{-x} + c_2 e^{-2x} + 7 - 6x + 2x^2.$$ ■

EXAMPLE 4   General Solution Using Undetermined Coefficients

Solve $y'' - 3y' = 8e^{3x} + 4\sin x.$ **(14)**

Solution   **Step 1.** The auxiliary equation for the associated homogeneous equation $y'' - 3y' = 0$ is $m^2 - 3m = m(m - 3) = 0$, and so $y_c = c_1 + c_2 e^{3x}$.

Step 2. Now, since $(D - 3)e^{3x} = 0$ and $(D^2 + 1)\sin x = 0$, we apply the differential operator $(D - 3)(D^2 + 1)$ to both sides of (14):

$$(D - 3)(D^2 + 1)(D^2 - 3D)y = 0.$$ **(15)**

The auxiliary equation of (15) is

$$(m - 3)(m^2 + 1)(m^2 - 3m) = 0 \quad \text{or} \quad m(m - 3)^2(m^2 + 1) = 0.$$

Thus $y = \boxed{c_1 + c_2 e^{3x}} + c_3 x e^{3x} + c_4 \cos x + c_5 \sin x.$

After excluding the linear combination of terms in the box that corresponds to y_c, we arrive at the form of y_p:

$$y_p = Axe^{3x} + B\cos x + C\sin x.$$

Substituting y_p in (14) and simplifying yields

$$y_p'' - 3y_p' = 3Ae^{3x} + (-B - 3C)\cos x + (3B - C)\sin x = 8e^{3x} + 4\sin x.$$

Equating coefficients gives $3A = 8$, $-B - 3C = 0$, and $3B - C = 4$. We find $A = \frac{8}{3}$, $B = \frac{6}{5}$, and $C = -\frac{2}{5}$, and consequently

$$y_p = \frac{8}{3}xe^{3x} + \frac{6}{5}\cos x - \frac{2}{5}\sin x.$$

Step 3. The general solution of (14) is then

$$y = c_1 + c_2 e^{3x} + \frac{8}{3} x e^{3x} + \frac{6}{5} \cos x - \frac{2}{5} \sin x.$$ ∎

EXAMPLE 5 **General Solution Using Undetermined Coefficients**

Solve $y'' + y = x \cos x - \cos x$.   (16)

Solution   The complementary function is $y_c = c_1 \cos x + c_2 \sin x$. Now by comparing $\cos x$ and $x \cos x$ with the functions in the first row of (7) we see that $\alpha = 0$ and $n = 1$, and so $(D^2 + 1)^2$ is an annihilator for the right-hand member of the equation in (16). Applying this operator to the differential equation gives

$$(D^2 + 1)^2(D^2 + 1)y = 0 \quad \text{or} \quad (D^2 + 1)^3 y = 0.$$

Since i and $-i$ are both complex roots of multiplicity 3 of the auxiliary equation of the last differential equation, we conclude that

$$y = \boxed{c_1 \cos x + c_2 \sin x} + c_3 x \cos x + c_4 x \sin x + c_5 x^2 \cos x + c_6 x^2 \sin x.$$

We substitute

$$y_p = A x \cos x + B x \sin x + C x^2 \cos x + E x^2 \sin x$$

into (16) and simplify:

$$y_p'' + y_p = 4E x \cos x - 4C x \sin x + (2B + 2C) \cos x + (-2A + 2E) \sin x$$
$$= x \cos x - \cos x.$$

Equating coefficients gives the equations $4E = 1, -4C = 0, 2B + 2C = -1$, and $-2A + 2E = 0$, from which we find $A = \frac{1}{4}, B = -\frac{1}{2}, C = 0$, and $E = \frac{1}{4}$. Hence the general solution of (16) is

$$y = c_1 \cos x + c_2 \sin x + \frac{1}{4} x \cos x - \frac{1}{2} x \sin x + \frac{1}{4} x^2 \sin x.$$ ∎

EXAMPLE 6 **Form of a Particular Solution**

Determine the form of a particular solution for

$$y'' - 2y' + y = 10 e^{-2x} \cos x.$$ (17)

Solution   The complementary function for the given equation is $y_c = c_1 e^x + c_2 x e^x$.
Now from (7), with $\alpha = -2, \beta = 1$, and $n = 1$, we know that

$$(D^2 + 4D + 5)e^{-2x} \cos x = 0.$$

Applying the operator $D^2 + 4D + 5$ to (17) gives

$$(D^2 + 4D + 5)(D^2 - 2D + 1)y = 0.$$ (18)

Since the roots of the auxiliary equation of (18) are $-2 - i$, $-2 + i$, 1, and 1, we see from

$$y = \boxed{c_1 e^x + c_2 x e^x} + c_3 e^{-2x} \cos x + c_4 e^{-2x} \sin x$$

that a particular solution of (17) can be found with the form

$$y_p = A e^{-2x} \cos x + B e^{-2x} \sin x.$$ ∎

EXAMPLE 7   Form of a Particular Solution

Determine the form of a particular solution for

$$y''' - 4y'' + 4y' = 5x^2 - 6x + 4x^2 e^{2x} + 3e^{5x}. \tag{19}$$

Solution   Observe that

$$D^3(5x^2 - 6x) = 0, \quad (D - 2)^3 x^2 e^{2x} = 0, \quad \text{and} \quad (D - 5)e^{5x} = 0.$$

Therefore $D^3(D - 2)^3(D - 5)$ applied to (19) gives

$$D^3(D - 2)^3(D - 5)(D^3 - 4D^2 + 4D)y = 0$$

or

$$D^4(D - 2)^5(D - 5)y = 0.$$

The roots of the auxiliary equation for the last differential equation are easily seen to be 0, 0, 0, 0, 2, 2, 2, 2, 2, and 5. Hence

$$y = \boxed{c_1} + c_2 x + c_3 x^2 + c_4 x^3 + \boxed{c_5 e^{2x} + c_6 x e^{2x}} + c_7 x^2 e^{2x} + c_8 x^3 e^{2x} + c_9 x^4 e^{2x} + c_{10} e^{5x}. \tag{20}$$

Since the linear combination $c_1 + c_5 e^{2x} + c_6 x e^{2x}$ corresponds to the complementary function of (19), the remaining terms in (20) give the form of a particular solution of the differential equation:

$$y_p = Ax + Bx^2 + Cx^3 + Ex^2 e^{2x} + Fx^3 e^{2x} + Gx^4 e^{2x} + He^{5x}.$$ ∎

Summary of the Method   For your convenience the method of undetermined coefficients is summarized here.

Undetermined Coefficients–Annihilator Approach

The differential equation $L(y) = g(x)$ has constant coefficients, and the function $g(x)$ consists of finite sums and products of constants, polynomials, exponential functions $e^{\alpha x}$, sines, and cosines.

(*i*) Find the complementary solution y_c for the homogeneous equation $L(y) = 0$.

(*ii*) Operate on both sides of the nonhomogeneous equation $L(y) = g(x)$ with a differential operator L_1 that annihilates the function $g(x)$.

(*iii*) Find the general solution of the higher-order homogeneous differential equation $L_1 L(y) = 0$.

(*iv*) Delete from the solution in step (*iii*) all those terms that are duplicated in the complementary solution y_c found in step (*i*).

Form a linear combination y_p of the terms that remain. This is the form of a particular solution of $L(y) = g(x)$.

(v) Substitute y_p found in step (iv) into $L(y) = g(x)$. Match coefficients of the various functions on each side of the equality, and solve the resulting system of equations for the unknown coefficients in y_p.

(vi) With the particular solution found in step (v), form the general solution $y = y_c + y_p$ of the given differential equation.

Remarks   The method of undetermined coefficients is not applicable to linear differential equations with variable coefficients nor is it applicable to linear equations with constant coefficients when $g(x)$ is a function such as

$$g(x) = \ln x, \quad g(x) = \frac{1}{x}, \quad g(x) = \tan x, \quad g(x) = \sin^{-1}x,$$

and so on. Differential equations in which the input $g(x)$ is a function of this last kind will be considered in the next section.

SECTION 4.5 EXERCISES

Answers to odd-numbered problems begin on page AN-4.

In Problems 1–10 write the given differential equation in the form $L(y) = g(x)$, where L is a linear differential operator with constant coefficients. If possible, factor L.

1. $9y'' - 4y = \sin x$

2. $y'' - 5y = x^2 - 2x$

3. $y'' - 4y' - 12y = x - 6$

4. $2y'' - 3y' - 2y = 1$

5. $y''' + 10y'' + 25y' = e^x$

6. $y''' + 4y' = e^x \cos 2x$

7. $y''' + 2y'' - 13y' + 10y = xe^{-x}$

8. $y''' + 4y'' + 3y' = x^2 \cos x - 3x$

9. $y^{(4)} + 8y' = 4$

10. $y^{(4)} - 8y'' + 16y = (x^3 - 2x)e^{4x}$

In Problems 11–14 verify that the given differential operator annihilates the indicated functions.

11. $D^4; \quad y = 10x^3 - 2x$

12. $2D - 1; \quad y = 4e^{x/2}$

13. $(D - 2)(D + 5); \quad y = e^{2x} + 3e^{-5x}$

14. $D^2 + 64; \quad y = 2 \cos 8x - 5 \sin 8x$

In Problems 15–26 find a linear differential operator that annihilates the given function.

15. $1 + 6x - 2x^3$

16. $x^3(1 - 5x)$

17. $1 + 7e^{2x}$

18. $x + 3xe^{6x}$

19. $\cos 2x$

20. $1 + \sin x$

21. $13x + 9x^2 - \sin 4x$

22. $8x - \sin x + 10 \cos 5x$

23. $e^{-x} + 2xe^x - x^2e^x$

24. $(2 - e^x)^2$

25. $3 + e^x \cos 2x$

26. $e^{-x} \sin x - e^{2x} \cos x$

In Problems 27–34 find linearly independent functions that are annihilated by the given differential operator.

27. D^5

28. $D^2 + 4D$

29. $(D - 6)(2D + 3)$

30. $D^2 - 9D - 36$

31. $D^2 + 5$

32. $D^2 - 6D + 10$

33. $D^3 - 10D^2 + 25D$

34. $D^2(D - 5)(D - 7)$

In Problems 35–64 solve the given differential equation by undetermined coefficients.

35. $y'' - 9y = 54$

36. $2y'' - 7y' + 5y = -29$

37. $y'' + y' = 3$

38. $y''' + 2y'' + y' = 10$

39. $y'' + 4y' + 4y = 2x + 6$

40. $y'' + 3y' = 4x - 5$

41. $y''' + y'' = 8x^2$

42. $y'' - 2y' + y = x^3 + 4x$

43. $y'' - y' - 12y = e^{4x}$

44. $y'' + 2y' + 2y = 5e^{6x}$

45. $y'' - 2y' - 3y = 4e^x - 9$

46. $y'' + 6y' + 8y = 3e^{-2x} + 2x$

47. $y'' + 25y = 6 \sin x$

48. $y'' + 4y = 4 \cos x + 3 \sin x - 8$

49. $y'' + 6y' + 9y = -xe^{4x}$

50. $y'' + 3y' - 10y = x(e^x + 1)$

51. $y'' - y = x^2 e^x + 5$

52. $y'' + 2y' + y = x^2 e^{-x}$

53. $y'' - 2y' + 5y = e^x \sin x$

54. $y'' + y' + \dfrac{1}{4}y = e^x(\sin 3x - \cos 3x)$

55. $y'' + 25y = 20 \sin 5x$

56. $y'' + y = 4 \cos x - \sin x$

57. $y'' + y' + y = x \sin x$

58. $y'' + 4y = \cos^2 x$

59. $y''' + 8y'' = -6x^2 + 9x + 2$

60. $y''' - y'' + y' - y = xe^x - e^{-x} + 7$

61. $y''' - 3y'' + 3y' - y = e^x - x + 16$

62. $2y''' - 3y'' - 3y' + 2y = (e^x + e^{-x})^2$

63. $y^{(4)} - 2y''' + y'' = e^x + 1$

64. $y^{(4)} - 4y'' = 5x^2 - e^{2x}$

In Problems 65–72 solve the given initial-value problem.

65. $y'' - 64y = 16$, $y(0) = 1, y'(0) = 0$

66. $y'' + y' = x$, $y(0) = 1, y'(0) = 0$

67. $y'' - 5y' = x - 2$, $y(0) = 0, y'(0) = 2$

68. $y'' + 5y' - 6y = 10e^{2x}$, $y(0) = 1, y'(0) = 1$

69. $y'' + y = 8 \cos 2x - 4 \sin x$, $y\left(\dfrac{\pi}{2}\right) = -1, y'\left(\dfrac{\pi}{2}\right) = 0$

70. $y''' - 2y'' + y' = xe^x + 5$, $y(0) = 2, y'(0) = 2, y''(0) = -1$

71. $y'' - 4y' + 8y = x^3$, $y(0) = 2, y'(0) = 4$

72. $y^{(4)} - y''' = x + e^x$, $y(0) = 0, y'(0) = 0, y''(0) = 0, y'''(0) = 0$

Discussion Problems

73. Suppose L is a linear differential operator that factors but has variable coefficients. Do the factors of L commute? Defend your answer.

4.6 VARIATION OF PARAMETERS

• Standard form of a nonhomogeneous linear second-order ODE   • The Wronskian returns   • Finding variable parameters   • Particular solutions for second-order equations   • Higher-order equations

The procedure that we used in Section 2.3 to find a particular solution of a linear first-order differential equation on an interval is applicable to linear higher-order equations as well. To adapt the method of **variation of parameters** to a linear second-order differential equation

$$a_2(x)y'' + a_1(x)y' + a_0(x)y = g(x), \tag{1}$$

we begin as we did in Section 4.2—we put the differential equation in the standard form

$$y'' + P(x)y' + Q(x)y = f(x) \tag{2}$$

by dividing through by the lead coefficient $a_2(x)$. Equation (2) is the second-order analogue of the linear first-order equation $dy/dx + P(x)y = f(x)$. In (2) we suppose that $P(x)$, $Q(x)$, and $f(x)$ are continuous on some common interval I. As we have already seen in Section 4.3, there is no difficulty in obtaining the complementary function y_c of (2) when the coefficients are constants.

The Assumptions   Corresponding to the assumption $y_p = u_1(x)y_1(x)$ that we used in Section 2.3 to find a particular solution y_p of $dy/dx + P(x)y = f(x)$, for the linear second-order equation (2) we seek a solution of the form

$$y_p = u_1(x)y_1(x) + u_2(x)y_2(x), \tag{3}$$

where y_1 and y_2 form a fundamental set of solutions on I of the associated homogeneous form of (1). Using the product rule to differentiate y_p twice, we get

$$y_p' = u_1 y_1' + y_1 u_1' + u_2 y_2' + y_2 u_2'$$
$$y_p'' = u_1 y_1'' + y_1' u_1' + y_1 u_1'' + u_1' y_1' + u_2 y_2'' + y_2' u_2' + y_2 u_2'' + u_2' y_2'.$$

Substituting (3) and the foregoing derivatives into (2) and grouping terms yields

$$y_p'' + P(x)y_p' + Q(x)y_p = u_1 \underbrace{[y_1'' + Py_1' + Qy_1]}_{\text{zero}} + u_2 \underbrace{[y_2'' + Py_2' + Qy_2]}_{\text{zero}}$$
$$+ y_1 u_1'' + u_1' y_1' + y_2 u_2'' + u_2' y_2' + P[y_1 u_1' + y_2 u_2'] + y_1' u_1' + y_2' u_2'$$
$$= \frac{d}{dx}[y_1 u_1'] + \frac{d}{dx}[y_2 u_2'] + P[y_1 u_1' + y_2 u_2'] + y_1' u_1' + y_2' u_2'$$
$$= \frac{d}{dx}[y_1 u_1' + y_2 u_2'] + P[y_1 u_1' + y_2 u_2'] + y_1' u_1' + y_2' u_2' = f(x). \tag{4}$$

Because we seek to determine two unknown functions u_1 and u_2, reason dictates that we need two equations. We can obtain these equations by making the further assumption that the functions u_1 and u_2 satisfy $y_1 u_1' + y_2 u_2' = 0$. This assumption does not come out of the blue but is prompted by the first two terms in (4) since, if we demand that $y_1 u_1' + y_2 u_2' = 0$, then (4) reduces to $y_1' u_1' + y_2' u_2' = f(x)$. We now have our desired two equations, albeit two equations for determining the derivatives u_1' and u_2'. By Cramer's rule, the solution of the system

$$y_1 u_1' + y_2 u_2' = 0$$
$$y_1' u_1' + y_2' u_2' = f(x)$$

can be expressed in terms of determinants:

$$u_1' = \frac{W_1}{W} = -\frac{y_2 f(x)}{W} \quad \text{and} \quad u_2' = \frac{W_2}{W} = \frac{y_1 f(x)}{W}, \tag{5}$$

where

$$W = \begin{vmatrix} y_1 & y_2 \\ y_1' & y_2' \end{vmatrix}, \quad W_1 = \begin{vmatrix} 0 & y_2 \\ f(x) & y_2' \end{vmatrix}, \quad W_2 = \begin{vmatrix} y_1 & 0 \\ y_1' & f(x) \end{vmatrix}. \tag{6}$$

The functions u_1 and u_2 are found by integrating the results in (5). The determinant W is recognized as the Wronskian of y_1 and y_2. By linear independence of y_1 and y_2 on I, we know that $W(y_1(x), y_2(x)) \neq 0$ for every x in the interval.

Summary of the Method   Usually it is not a good idea to memorize formulas in lieu of understanding a procedure. However, the foregoing procedure is too long and complicated to use each time we wish to solve a differential equation. In this case it is more efficient to simply use the formulas in (5). Thus to solve $a_2 y'' + a_1 y' + a_0 y = g(x)$, first find the complementary function $y_c = c_1 y_1 + c_2 y_2$ and then compute the Wronskian $W(y_1(x), y_2(x))$. By dividing by a_2, we put the equation into the standard form $y'' + Py' + Qy = f(x)$ to determine $f(x)$. We find u_1 and u_2 by integrating $u_1' = W_1/W$ and $u_2' = W_2/W$, where W_1 and W_2 are defined as in (6). A particular solution is $y_p = u_1 y_1 + u_2 y_2$. The general solution of the equation is then $y = y_c + y_p$.

EXAMPLE 1   General Solution Using Variation of Parameters

Solve $y'' - 4y' + 4y = (x + 1)e^{2x}$.

Solution   From the auxiliary equation $m^2 - 4m + 4 = (m - 2)^2 = 0$ we have $y_c = c_1 e^{2x} + c_2 x e^{2x}$. With the identifications $y_1 = e^{2x}$ and $y_2 = xe^{2x}$, we next compute the Wronskian:

$$W(e^{2x}, xe^{2x}) = \begin{vmatrix} e^{2x} & xe^{2x} \\ 2e^{2x} & 2xe^{2x} + e^{2x} \end{vmatrix} = e^{4x}.$$

Since the given differential equation is already in form (2) (that is, the coefficient of y'' is 1), we identify $f(x) = (x + 1)e^{2x}$. From (6) we obtain

$$W_1 = \begin{vmatrix} 0 & xe^{2x} \\ (x + 1)e^{2x} & 2xe^{2x} + e^{2x} \end{vmatrix} = -(x + 1)xe^{4x}, \quad W_2 = \begin{vmatrix} e^{2x} & 0 \\ 2e^{2x} & (x + 1)e^{2x} \end{vmatrix} = (x + 1)e^{4x},$$

and so from (5)

$$u_1' = -\frac{(x+1)xe^{4x}}{e^{4x}} = -x^2 - x, \quad u_2' = \frac{(x+1)e^{4x}}{e^{4x}} = x + 1.$$

It follows that $u_1 = -\frac{1}{3}x^3 - \frac{1}{2}x^2$ and $u_2 = \frac{1}{2}x^2 + x$. Hence

$$y_p = \left(-\frac{1}{3}x^3 - \frac{1}{2}x^2\right)e^{2x} + \left(\frac{1}{2}x^2 + x\right)xe^{2x} = \frac{1}{6}x^3e^{2x} + \frac{1}{2}x^2e^{2x}$$

and

$$y = y_c + y_p = c_1e^{2x} + c_2xe^{2x} + \frac{1}{6}x^3e^{2x} + \frac{1}{2}x^2e^{2x}.$$ ∎

EXAMPLE 2   General Solution Using Variation of Parameters

Solve $4y'' + 36y = \csc 3x$.

Solution   We first put the equation in the standard form (2) by dividing by 4:

$$y'' + 9y = \frac{1}{4}\csc 3x.$$

Since the roots of the auxiliary equation $m^2 + 9 = 0$ are $m_1 = 3i$ and $m_2 = -3i$, the complementary function is $y_c = c_1 \cos 3x + c_2 \sin 3x$. Using $y_1 = \cos 3x$, $y_2 = \sin 3x$, and $f(x) = \frac{1}{4}\csc 3x$, we obtain

$$W(\cos 3x, \sin 3x) = \begin{vmatrix} \cos 3x & \sin 3x \\ -3\sin 3x & 3\cos 3x \end{vmatrix} = 3,$$

$$W_1 = \begin{vmatrix} 0 & \sin 3x \\ \frac{1}{4}\csc 3x & 3\cos 3x \end{vmatrix} = -\frac{1}{4}, \quad W_2 = \begin{vmatrix} \cos 3x & 0 \\ -3\sin 3x & \frac{1}{4}\csc 3x \end{vmatrix} = \frac{1}{4}\frac{\cos 3x}{\sin 3x}.$$

Integrating

$$u_1' = \frac{W_1}{W} = -\frac{1}{12} \quad \text{and} \quad u_2' = \frac{W_2}{W} = \frac{1}{12}\frac{\cos 3x}{\sin 3x}$$

gives $u_1 = -\frac{1}{12}x$ and $u_2 = \frac{1}{36}\ln|\sin 3x|$. Thus a particular solution is

$$y_p = -\frac{1}{12}x\cos 3x + \frac{1}{36}(\sin 3x)\ln|\sin 3x|.$$

The general solution of the equation is

$$y = y_c + y_p = c_1\cos 3x + c_2\sin 3x - \frac{1}{12}x\cos 3x + \frac{1}{36}(\sin 3x)\ln|\sin 3x|. \quad \textbf{(7)} \ \blacksquare$$

Equation (7) represents the general solution of the differential equation on, say, the interval $(0, \pi/6)$.

Constants of Integration   When computing the indefinite integrals of u_1' and u_2', we need not introduce any constants. This is because

$$y = y_c + y_p = c_1y_1 + c_2y_2 + (u_1 + a_1)y_1 + (u_2 + b_1)y_2$$
$$= (c_1 + a_1)y_1 + (c_2 + b_1)y_2 + u_1y_1 + u_2y_2$$
$$= C_1y_1 + C_2y_2 + u_1y_1 + u_2y_2.$$

EXAMPLE 3 **General Solution Using Variation of Parameters**

Solve $y'' - y = \dfrac{1}{x}$.

Solution The auxiliary equation $m^2 - 1 = 0$ yields $m_1 = -1$ and $m_2 = 1$. Therefore $y_c = c_1 e^x + c_2 e^{-x}$. Now $W(e^x, e^{-x}) = -2$ and

$$u_1' = -\frac{e^{-x}(1/x)}{-2}, \quad u_1 = \frac{1}{2}\int_{x_0}^x \frac{e^{-t}}{t}\,dt,$$

$$u_2' = \frac{e^x(1/x)}{-2}, \quad u_2 = -\frac{1}{2}\int_{x_0}^x \frac{e^t}{t}\,dt.$$

Since the foregoing integrals are nonelementary, we are forced to write

$$y_p = \frac{1}{2}e^x\int_{x_0}^x \frac{e^{-t}}{t}\,dt - \frac{1}{2}e^{-x}\int_{x_0}^x \frac{e^t}{t}\,dt,$$

and so

$$y = y_c + y_p = c_1 e^x + c_2 e^{-x} + \frac{1}{2}e^x\int_{x_0}^x \frac{e^{-t}}{t}\,dt - \frac{1}{2}e^{-x}\int_{x_0}^x \frac{e^t}{t}\,dt. \qquad \blacksquare$$

In Example 3 we can integrate on any interval $x_0 \le t \le x$ not containing the origin.

Higher-Order Equations The method we have just examined for nonhomogeneous second-order differential equations can be generalized to linear nth-order equations that have been put into the standard form

$$y^{(n)} + P_{n-1}(x)y^{(n-1)} + \cdots + P_1(x)y' + P_0(x)y = f(x). \qquad \textbf{(8)}$$

If $y_c = c_1 y_1 + c_2 y_2 + \cdots + c_n y_n$ is the complementary function for (8), then a particular solution is

$$y_p = u_1(x)y_1(x) + u_2(x)y_2(x) + \cdots + u_n(x)y_n(x),$$

where the u_k', $k = 1, 2, \ldots, n$ are determined by the n equations

$$
\begin{aligned}
y_1 u_1' + \ y_2 u_2' + \cdots + \ y_n u_n' &= 0 \\
y_1' u_1' + \ y_2' u_2' + \cdots + \ y_n' u_n' &= 0 \\
&\ \ \vdots \\
y_1^{(n-1)} u_1' + y_2^{(n-1)} u_2' + \cdots + y_n^{(n-1)} u_n' &= f(x).
\end{aligned}
\qquad \textbf{(9)}
$$

The first $n - 1$ equations in this system, like $y_1 u_1' + y_2 u_2' = 0$ in (4), are assumptions made to simplify the resulting equation after $y_p = u_1(x)y_1(x) + \cdots + u_n(x)y_n(x)$ is substituted in (8). In this case Cramer's rule gives

$$u_k' = \frac{W_k}{W}, \quad k = 1, 2, \ldots, n,$$

where W is the Wronskian of $y_1, y_2, \ldots, y_n$ and W_k is the determinant obtained by replacing the kth column of the Wronskian by the column consisting of the right-hand side of (9)—that is, the column consisting of $(0, 0, \ldots, f(x))$. When $n = 2$ we get (5).

Remarks (*i*) Variation of parameters has a distinct advantage over the method of undetermined coefficients in that it will *always* yield a particular solution y_p provided the related homogeneous equation can be solved. The present method is not limited to a function $f(x)$, which is a combination of the four types of functions listed on page 168. Also, variation of parameters, unlike undetermined coefficients, is applicable to differential equations with variable coefficients.

(*ii*) In the problems that follow do not hesitate to simplify the form of y_p. Depending on how the antiderivatives of u_1' and u_2' are found, you may not obtain the same y_p as given in the answer section. For example, in Problem 3 in Exercises 4.6 both $y_p = \frac{1}{2}\sin x - \frac{1}{2}x\cos x$ and $y_p = \frac{1}{4}\sin x - \frac{1}{2}x\cos x$ are valid answers. In either case the general solution $y = y_c + y_p$ simplifies to $y = c_1\cos x + c_2\sin x - \frac{1}{2}x\cos x$. Why?

EXERCISES 4.6

Answers to odd-numbered problems begin on page AN-4.

In Problems 1–18 solve each differential equation by variation of parameters.

1. $y'' + y = \sec x$

2. $y'' + y = \tan x$

3. $y'' + y = \sin x$

4. $y'' + y = \sec\theta\tan\theta$

5. $y'' + y = \cos^2 x$

6. $y'' + y = \sec^2 x$

7. $y'' - y = \cosh x$

8. $y'' - y = \sinh 2x$

9. $y'' - 4y = \dfrac{e^{2x}}{x}$

10. $y'' - 9y = \dfrac{9x}{e^{3x}}$

11. $y'' + 3y' + 2y = \dfrac{1}{1 + e^x}$

12. $y'' - 2y' + y = \dfrac{e^x}{1 + x^2}$

13. $y'' + 3y' + 2y = \sin e^x$

14. $y'' - 2y' + y = e^t \arctan t$

15. $y'' + 2y' + y = e^{-t}\ln t$

16. $2y'' + 2y' + y = 4\sqrt{x}$

17. $3y'' - 6y' + 6y = e^x \sec x$

18. $4y'' - 4y' + y = e^{x/2}\sqrt{1 - x^2}$

In Problems 19–22 solve each differential equation by variation of parameters, subject to the initial conditions $y(0) = 1$, $y'(0) = 0$.

19. $4y'' - y = xe^{x/2}$

20. $2y'' + y' - y = x + 1$

21. $y'' + 2y' - 8y = 2e^{-2x} - e^{-x}$

22. $y'' - 4y' + 4y = (12x^2 - 6x)e^{2x}$

In Problems 23 and 24 the indicated functions are known linearly independent solutions of the associated homogeneous differential equation on $(0, \infty)$. Find the general solution of the given nonhomogeneous equation.

23. $x^2 y'' + xy' + (x^2 - \frac{1}{4})y = x^{3/2}$; $y_1 = x^{-1/2}\cos x$, $y_2 = x^{-1/2}\sin x$

24. $x^2 y'' + xy' + y = \sec(\ln x)$; $y_1 = \cos(\ln x)$, $y_2 = \sin(\ln x)$

In Problems 25 and 26 solve the given third-order differential equation by variation of parameters.

25. $y''' + y' = \tan x$

26. $y''' + 4y' = \sec 2x$

Discussion Problems

In Problems 27 and 28 discuss how the methods of undetermined coefficients and variation of parameters can be combined to solve the given differential equation. Carry out your ideas.

27. $3y'' - 6y' + 30y = 15 \sin x + e^x \tan 3x$

28. $y'' - 2y' + y = 4x^2 - 3 + x^{-1}e^x$

29. What are the intervals of definition of the general solutions in Problems 1, 7, 9, and 18? Discuss why the interval of definition of the general solution in Problem 24 is *not* $(0, \infty)$.

30. Find the general solution of $x^4y'' + x^3y' - 4x^2y = 1$ given that $y_1 = x^2$ is a solution of the associated homogeneous equation.

4.7 CAUCHY-EULER EQUATION

• A linear DE with variable coefficients   • Auxiliary equation   • Roots of a quadratic auxiliary equation   • Forms of the general solution of a homogeneous linear second-order Cauchy-Euler DE   • Using variation of parameters   • Higher-order DEs

The relative ease with which we were able to find explicit solutions of higher-order linear differential equations with constant coefficients in the preceding sections does not, in general, carry over to linear equations with variable coefficients. We shall see in Chapter 6 that when a linear differential equation has variable coefficients, the best that we can *usually* expect is to find a solution in the form of an infinite series. However, the type of differential equation considered in this section is an exception to this rule; it is an equation with variable coefficients whose general solution can always be expressed in terms of powers of x, sines, cosines, and logarithmic functions. Moreover, its method of solution is quite similar to that for constant-coefficient equations in that an auxiliary equation must be solved.

Cauchy-Euler Equation   A linear differential equation of the form

$$a_n x^n \frac{d^n y}{dx^n} + a_{n-1} x^{n-1} \frac{d^{n-1}y}{dx^{n-1}} + \cdots + a_1 x \frac{dy}{dx} + a_0 y = g(x),$$

where the coefficients $a_n, a_{n-1}, \ldots, a_0$ are constants, is known as a **Cauchy-Euler equation**. The observable characteristic of this type of equation is that the degree $k = n, n - 1, \ldots, 1, 0$ of the monomial coefficients x^k matches the order k of differentiation $d^k y/dx^k$:

$$\overset{\substack{\text{same} \\ \downarrow}}{a_n x^n} \frac{d^n y}{dx^n} + \overset{\substack{\text{same} \\ \downarrow}}{a_{n-1} x^{n-1}} \frac{d^{n-1}y}{dx^{n-1}} + \cdots.$$

As in Section 4.3, we start the discussion with a detailed examination of the forms of the general solutions of the homogeneous second-order equation

$$ax^2 \frac{d^2y}{dx^2} + bx \frac{dy}{dx} + cy = 0.$$

The solution of higher-order equations follows analogously. Also, we can solve the nonhomogeneous equation $ax^2y'' + bxy' + cy = g(x)$ by variation of parameters, once we have determined the complementary function y_c.

Note   The coefficient ax^2 of y'' is zero at $x = 0$. Hence in order to guarantee that the fundamental results of Theorem 4.1 are applicable to the Cauchy-Euler equation, we confine our attention to finding the general solutions defined on the interval $(0, \infty)$. Solutions on the interval $(-\infty, 0)$ can be obtained by substituting $t = -x$ into the differential equation. See Problems 35 and 36 in Exercises 4.7.

Method of Solution   We try a solution of the form $y = x^m$, where m is to be determined. Analogous to what happened when we substituted e^{mx} into a linear equation with constant coefficients, when we substitute x^m each term of a Cauchy-Euler equation becomes a polynomial in m times x^m, since

$$a_k x^k \frac{d^k y}{dx^k} = a_k x^k m(m - 1)(m - 2) \cdots (m - k + 1)x^{m-k} = a_k m(m - 1)(m - 2) \cdots (m - k + 1)x^m.$$

For example, when we substitute $y = x^m$ the second-order equation becomes

$$ax^2 \frac{d^2y}{dx^2} + bx \frac{dy}{dx} + cy = am(m - 1)x^m + bmx^m + cx^m = (am(m - 1) + bm + c)x^m.$$

Thus $y = x^m$ is a solution of the differential equation whenever m is a solution of the **auxiliary equation**

$$am(m - 1) + bm + c = 0 \quad \text{or} \quad am^2 + (b - a)m + c = 0. \quad \textbf{(1)}$$

There are three different cases to be considered, depending on whether the roots of this quadratic equation are real and distinct, real and equal, or complex. In the last case the roots appear as a conjugate pair.

Case I: Distinct Real Roots   Let m_1 and m_2 denote the real roots of (1) such that $m_1 \neq m_2$. Then $y_1 = x^{m_1}$ and $y_2 = x^{m_2}$ form a fundamental set of solutions. Hence the general solution is

$$y = c_1 x^{m_1} + c_2 x^{m_2}. \quad \textbf{(2)}$$

EXAMPLE 1   **Distinct Roots**

Solve $x^2 \dfrac{d^2y}{dx^2} - 2x \dfrac{dy}{dx} - 4y = 0$.

Solution   Rather than just memorizing equation (1), it is preferable to assume $y = x^m$ as the solution a few times in order to understand the origin and the difference between this new form of the auxiliary equation and that obtained in Section 4.3. Differentiate twice,

$$\frac{dy}{dx} = mx^{m-1}, \quad \frac{d^2y}{dx^2} = m(m-1)x^{m-2},$$

and substitute back into the differential equation:

$$x^2\frac{d^2y}{dx^2} - 2x\frac{dy}{dx} - 4y = x^2 \cdot m(m-1)x^{m-2} - 2x \cdot mx^{m-1} - 4x^m$$

$$= x^m(m(m-1) - 2m - 4) = x^m(m^2 - 3m - 4) = 0$$

if $m^2 - 3m - 4 = 0$. Now $(m+1)(m-4) = 0$ implies $m_1 = -1$, $m_2 = 4$, so $y = c_1x^{-1} + c_2x^4$.   ■

Case II: Repeated Real Roots   If the roots of (1) are repeated (that is, $m_1 = m_2$), then we obtain only one solution—namely, $y = x^{m_1}$. When the roots of the quadratic equation $am^2 + (b-a)m + c = 0$ are equal, the discriminant of the coefficients is necessarily zero. It follows from the quadratic formula that the root must be $m_1 = -(b-a)/2a$.

Now we can construct a second solution y_2, using (5) of Section 4.2. We first write the Cauchy-Euler equation in the standard form

$$\frac{d^2y}{dx^2} + \frac{b}{ax}\frac{dy}{dx} + \frac{c}{ax^2}y = 0$$

and make the identifications $P(x) = b/ax$ and $\int (b/ax)\,dx = (b/a)\ln x$. Thus

$$y_2 = x^{m_1}\int \frac{e^{-(b/a)\ln x}}{x^{2m_1}}\,dx$$

$$= x^{m_1}\int x^{-b/a} \cdot x^{-2m_1}\,dx \qquad \leftarrow e^{-(b/a)\ln x} = e^{\ln x^{-b/a}} - x^{-b/a}$$

$$= x^{m_1}\int x^{-b/a} \cdot x^{(b-a)/a}\,dx \qquad \leftarrow -2m_1 = (b-a)/a$$

$$= x^{m_1}\int \frac{dx}{x} = x^{m_1}\ln x.$$

The general solution is then

$$y = c_1x^{m_1} + c_2x^{m_1}\ln x. \tag{3}$$

EXAMPLE 2   **Repeated Roots**

Solve $4x^2\frac{d^2y}{dx^2} + 8x\frac{dy}{dx} + y = 0.$

Solution   The substitution $y = x^m$ yields

$$4x^2\frac{d^2y}{dx^2} + 8x\frac{dy}{dx} + y = x^m(4m(m-1) + 8m + 1) = x^m(4m^2 + 4m + 1) = 0$$

when $4m^2 + 4m + 1 = 0$ or $(2m+1)^2 = 0$. Since $m_1 = -\frac{1}{2}$, the general solution is $y = c_1x^{-1/2} + c_2x^{-1/2}\ln x$.   ■

For higher-order equations, if m_1 is a root of multiplicity k, then it can be shown that

$$x^{m_1}, \quad x^{m_1}\ln x, \quad x^{m_1}(\ln x)^2, \quad \ldots, \quad x^{m_1}(\ln x)^{k-1}$$

are k linearly independent solutions. Correspondingly, the general solution of the differential equation must then contain a linear combination of these k solutions.

Case III: Conjugate Complex Roots

If the roots of (1) are the conjugate pair $m_1 = \alpha + i\beta$, $m_2 = \alpha - i\beta$, where α and $\beta > 0$ are real, then a solution is

$$y = C_1 x^{\alpha+i\beta} + C_2 x^{\alpha-i\beta}.$$

But when the roots of the auxiliary equation are complex, as in the case of equations with constant coefficients, we wish to write the solution in terms of real functions only. We note the identity

$$x^{i\beta} = (e^{\ln x})^{i\beta} = e^{i\beta\ln x},$$

which, by Euler's formula, is the same as

$$x^{i\beta} = \cos(\beta \ln x) + i \sin(\beta \ln x).$$

Similarly, $$x^{-i\beta} = \cos(\beta \ln x) - i \sin(\beta \ln x).$$

Adding and subtracting the last two results yields

$$x^{i\beta} + x^{-i\beta} = 2 \cos(\beta \ln x) \quad \text{and} \quad x^{i\beta} - x^{-i\beta} = 2i \sin(\beta \ln x),$$

respectively. From the fact that $y = C_1 x^{\alpha+i\beta} + C_2 x^{\alpha-i\beta}$ is a solution for any values of the constants, we see, in turn, for $C_1 = C_2 = 1$ and $C_1 = 1$, $C_2 = -1$ that

$$y_1 = x^\alpha(x^{i\beta} + x^{-i\beta}) \quad \text{and} \quad y_2 = x^\alpha(x^{i\beta} - x^{-i\beta})$$

or $$y_1 = 2x^\alpha \cos(\beta \ln x) \quad \text{and} \quad y_2 = 2ix^\alpha \sin(\beta \ln x)$$

are also solutions. Since $W(x^\alpha \cos(\beta \ln x), x^\alpha \sin(\beta \ln x)) = \beta x^{2\alpha-1} \neq 0$, $\beta > 0$ on the interval $(0, \infty)$, we conclude that

$$y_1 = x^\alpha \cos(\beta \ln x) \quad \text{and} \quad y_2 = x^\alpha \sin(\beta \ln x)$$

constitute a fundamental set of real solutions of the differential equation. Hence the general solution is

$$y = x^\alpha[c_1 \cos(\beta \ln x) + c_2 \sin(\beta \ln x)]. \tag{4}$$

EXAMPLE 3   An Initial-Value Problem

Solve $4x^2y'' + 17y = 0$, $y(1) = -1$, $y'(1) = 0$.

Solution   The y' term is missing in the given Cauchy-Euler equation; nevertheless, the substitution $y = x^m$ yields

$$4x^2y'' + 17y = x^m(4m(m-1) + 17) = x^m(4m^2 - 4m + 17) = 0$$

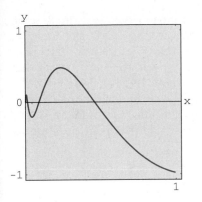

(a) solution on $0 < x \leq 1$

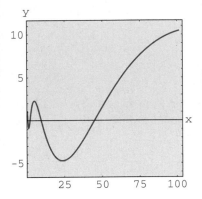

(b) solution on $0 < x \leq 100$

Figure 4.15

when $4m^2 - 4m + 17 = 0$. From the quadratic formula we find that the roots are $m_1 = \frac{1}{2} + 2i$ and $m_2 = \frac{1}{2} - 2i$. With the identifications $\alpha = \frac{1}{2}$ and $\beta = 2$, we see from (4) that the general solution of the differential equation is

$$y = x^{1/2}\left[c_1 \cos(2 \ln x) + c_2 \sin(2 \ln x)\right].$$

By applying the initial conditions $y(1) = -1$, $y'(1) = 0$ to the foregoing solution and using $\ln 1 = 0$, we then find, in turn, that $c_1 = -1$ and $c_2 = 0$. Hence the solution of the initial-value problem is $y = -x^{1/2} \cos(2 \ln x)$. The graph of this function, obtained with the aid of computer software, is given in Figure 4.15. The particular solution is seen to be oscillatory and unbounded as $x \to \infty$. ∎

The next example illustrates the solution of a third-order Cauchy-Euler equation.

EXAMPLE 4   **Third-Order Equation**

Solve $x^3 \dfrac{d^3y}{dx^3} + 5x^2 \dfrac{d^2y}{dx^2} + 7x \dfrac{dy}{dx} + 8y = 0$.

Solution   The first three derivatives of $y = x^m$ are

$$\frac{dy}{dx} = mx^{m-1}, \quad \frac{d^2y}{dx^2} = m(m-1)x^{m-2}, \quad \frac{d^3y}{dx^3} = m(m-1)(m-2)x^{m-3},$$

so the given differential equation becomes

$$x^3 \frac{d^3y}{dx^3} + 5x^2 \frac{d^2y}{dx^2} + 7x \frac{dy}{dx} + 8y = x^3m(m-1)(m-2)x^{m-3} + 5x^2m(m-1)x^{m-2} + 7xmx^{m-1} + 8x^m$$

$$= x^m(m(m-1)(m-2) + 5m(m-1) + 7m + 8)$$
$$= x^m(m^3 + 2m^2 + 4m + 8) = x^m(m+2)(m^2 + 4) = 0.$$

In this case we see that $y = x^m$ will be a solution of the differential equation for $m_1 = -2$, $m_2 = 2i$, and $m_3 = -2i$. Hence the general solution is $y = c_1x^{-2} + c_2 \cos(2 \ln x) + c_3 \sin(2 \ln x)$. ∎

The method of undetermined coefficients described in Sections 4.5 and 4.6 does not carry over, *in general*, to linear differential equations with variable coefficients. Consequently, in our next example the method of variation of parameters is employed.

EXAMPLE 5   **Variation of Parameters**

Solve $x^2y'' - 3xy' + 3y = 2x^4e^x$.

Solution   Since the equation is nonhomogeneous, we first solve the associated homogeneous equation. From the auxiliary equation $(m - 1)(m - 3) = 0$ we find $y_c = c_1x + c_2x^3$. Now before using variation of parameters to find a particular solution $y_p = u_1y_1 + u_2y_2$, recall that the formulas $u_1' = W_1/W$ and $u_2' = W_2/W$, where W_1, W_2, and W are the determinants defined on page 189, were derived under the assumption that the differential equation has been put into the standard form $y'' + P(x)y' + Q(x)y = f(x)$. Therefore we divide the given equation by x^2, and from

$$y'' - \frac{3}{x}y' + \frac{3}{x^2}y = 2x^2e^x$$

we make the identification $f(x) = 2x^2e^x$. Now with $y_1 = x$, $y_2 = x^3$ and

$$W = \begin{vmatrix} x & x^3 \\ 1 & 3x^2 \end{vmatrix} = 2x^3, \quad W_1 = \begin{vmatrix} 0 & x^3 \\ 2x^2e^x & 3x^2 \end{vmatrix} = -2x^5e^x, \quad W_2 = \begin{vmatrix} x & 0 \\ 1 & 2x^2e^x \end{vmatrix} = 2x^3e^x,$$

we find $u_1' = -\dfrac{2x^5e^x}{2x^3} = -x^2e^x$ and $u_2' = \dfrac{2x^3e^x}{2x^3} = e^x.$

The integral of the last function is immediate, but in the case of u_1' we integrate by parts twice. The results are $u_1 = -x^2e^x + 2xe^x - 2e^x$ and $u_2 = e^x$. Hence $y_p = u_1y_1 + u_2y_2$ is

$$y_p = (-x^2e^x + 2xe^x - 2e^x)x + e^xx^3 = 2x^2e^x - 2xe^x.$$

Finally $y = y_c + y_p = c_1x + c_2x^3 + 2x^2e^x - 2xe^x.$ ∎

Reduction to Constant Coefficients   The similarities between the forms of solutions of Cauchy-Euler equations and solutions of linear equations with constant coefficients are not just a coincidence. For example, when the roots of the auxiliary equations for $ay'' + by' + cy = 0$ and $ax^2y'' + bxy' + cy = 0$ are distinct and real, the respective general solutions are

$$y = c_1e^{m_1x} + c_2e^{m_2x} \quad \text{and} \quad y = c_1x^{m_1} + c_2x^{m_2}, \quad x > 0. \tag{5}$$

In view of the identity $e^{\ln x} = x$, $x > 0$, the second solution given in (5) can be expressed in the same form as the first solution:

$$y = c_1e^{m_1\ln x} + c_2e^{m_2\ln x} = c_1e^{m_1t} + c_2e^{m_2t},$$

where $t = \ln x$. This last result illustrates the fact that any Cauchy-Euler equation can *always* be rewritten as a linear differential equation with constant coefficients by means of the substitution $x = e^t$. The idea is to solve the new differential equation in terms of the variable t, using the methods of the previous sections, and once the general solution is obtained, resubstitute $t = \ln x$. This method, illustrated in the last example, requires the use of the Chain Rule of differentiation.

EXAMPLE 6   **Changing to Constant Coefficients**

Solve $x^2y'' - xy' + y = \ln x.$

Solution   With the substitution $x = e^t$ or $t = \ln x$, it follows that

$$\frac{dy}{dx} = \frac{dy}{dt}\frac{dt}{dx} = \frac{1}{x}\frac{dy}{dt} \qquad \leftarrow\text{Chain Rule}$$

$$\frac{d^2y}{dx^2} = \frac{1}{x}\frac{d}{dx}\left(\frac{dy}{dt}\right) + \frac{dy}{dt}\left(-\frac{1}{x^2}\right) \qquad \leftarrow\text{Product Rule and Chain Rule}$$

$$= \frac{1}{x}\left(\frac{d^2y}{dt^2}\frac{1}{x}\right) + \frac{dy}{dt}\left(-\frac{1}{x^2}\right) = \frac{1}{x^2}\left(\frac{d^2y}{dt^2} - \frac{dy}{dt}\right).$$

Substituting in the given differential equation and simplifying yields

$$\frac{d^2y}{dt^2} - 2\frac{dy}{dt} + y = t.$$

Since this last equation has constant coefficients, its auxiliary equation is $m^2 - 2m + 1 = 0$, or $(m - 1)^2 = 0$. Thus we obtain $y_c = c_1e^t + c_2te^t$.

By undetermined coefficients we try a particular solution of the form $y_p = A + Bt$. This assumption leads to $-2B + A + Bt = t$, so $A = 2$ and $B = 1$. Using $y = y_c + y_p$, we get

$$y = c_1e^t + c_2te^t + 2 + t,$$

and so the general solution of the original differential equation on the interval $(0, \infty)$ is $y = c_1x + c_2x \ln x + 2 + \ln x$.   ■

EXERCISES 4.7 ──────────────────────────────

Answers to odd-numbered problems begin on page AN-4.

In Problems 1–18 solve the given differential equation.

1. $x^2y'' - 2y = 0$ **2.** $4x^2y'' + y = 0$

3. $xy'' + y' = 0$ **4.** $xy'' - 3y' = 0$

5. $x^2y'' + xy' + 4y = 0$ **6.** $x^2y'' + 5xy' + 3y = 0$

7. $x^2y'' - 3xy' - 2y = 0$ **8.** $x^2y'' + 3xy' - 4y = 0$

9. $25x^2y'' + 25xy' + y = 0$ **10.** $4x^2y'' + 4xy' - y = 0$

11. $x^2y'' + 5xy' + 4y = 0$ **12.** $x^2y'' + 8xy' + 6y = 0$

13. $3x^2y'' + 6xy' + y = 0$ **14.** $x^2y'' - 7xy' + 41y = 0$

15. $x^3y''' - 6y = 0$ **16.** $x^3y''' + xy' - y = 0$

17. $xy^{(4)} + 6y''' = 0$

18. $x^4y^{(4)} + 6x^3y''' + 9x^2y'' + 3xy' + y = 0$

In Problems 19–22 solve the given differential equation by variation of parameters.

19. $xy'' - 4y' = x^4$ **20.** $2x^2y'' + 5xy' + y = x^2 - x$

21. $x^2y'' - xy' + y = 2x$ **22.** $x^2y'' - 2xy' + 2y = x^4e^x$

In Problems 23–28 solve the given initial-value problem. Use a graphing utility to plot the solution curve.

23. $x^2y'' + 3xy' = 0, \quad y(1) = 0, y'(1) = 4$

24. $x^2y'' - 5xy' + 8y = 0, \quad y(2) = 32, y'(2) = 0$

25. $x^2y'' + xy' + y = 0$, $y(1) = 1$, $y'(1) = 2$

26. $x^2y'' - 3xy' + 4y = 0$, $y(1) = 5$, $y'(1) = 3$

27. $xy'' + y' = x$, $y(1) = 1$, $y'(1) = -\frac{1}{2}$

28. $x^2y'' - 5xy' + 8y = 8x^6$, $y(\frac{1}{2}) = 0$, $y'(\frac{1}{2}) = 0$

In Problems 29–34 use the substitution $x = e^t$ to transform the given Cauchy-Euler equation to a differential equation with constant coefficients. Solve the original equation by using the procedures in Sections 4.3–4.5 to solve the new equation.

29. $x^2y'' + 9xy' - 20y = 0$ **30.** $x^2y'' - 9xy' + 25y = 0$

31. $x^2y'' + 10xy' + 8y = x^2$ **32.** $x^2y'' - 4xy' + 6y = \ln x^2$

33. $x^2y'' - 3xy' + 13y = 4 + 3x$

34. $x^3y''' - 3x^2y'' + 6xy' - 6y = 3 + \ln x^3$

In Problems 35 and 36 solve the given initial-value problem on the interval $(-\infty, 0)$.

35. $4x^2y'' + y = 0$, $y(-1) = 2$, $y'(-1) = 4$

36. $x^2y'' - 4xy' + 6y = 0$, $y(-2) = 8$, $y'(-2) = 0$

Discussion Problems

37. How would you use the method of this section to solve

$$(x + 2)^2y'' + (x + 2)y' + y = 0?$$

Carry out your ideas. State an interval over which the solution is defined.

38. Can a Cauchy-Euler differential equation of lowest order with real coefficients be found if it is known that 2 and $1 - i$ are roots of its auxiliary equation? Carry out your ideas.

39. The initial conditions $y(0) = y_0$, $y'(0) = y_1$ apply to each of the following differential equations:

$$x^2y'' = 0, \quad x^2y'' - 2xy' + 2y = 0, \quad x^2y'' - 4xy' + 6y = 0.$$

For what values of y_0 and y_1 does each initial-value problem have a solution?

40. What are the x-intercepts of the solution curve shown in Figure 4.15? How many x-intercepts are there in the interval $0 < x < 0.5$?

Computer Lab Assignments

In Problems 41–44 solve the given differential equation by using a root-finding application of a CAS to find the (approximate) roots of the auxiliary equation.

41. $2x^3y''' - 10.98x^2y'' + 8.5xy' + 1.3y = 0$

42. $x^3y''' + 4x^2y'' + 5xy' - 9y = 0$

43. $x^4y^{(4)} + 6x^3y''' + 3x^2y'' - 3xy' + 4y = 0$

44. $x^4y^{(4)} - 6x^3y''' + 33x^2y'' - 105xy' + 169y = 0$

4.8 SOLVING SYSTEMS OF LINEAR EQUATIONS BY ELIMINATION

- *Systematic elimination* • *Solution of a system*
- *Linear differential operators*

Simultaneous ordinary differential equations involve two or more equations that contain derivatives of two or more dependent variables—the unknown functions—with respect to a single independent variable. The method of **systematic elimination** for solving systems of linear equations with constant coefficients is based on the algebraic principle of elimination of variables. We shall see that the analogue of *multiplying* an algebraic equation by a constant is *operating* on an ODE with some combination of derivatives. Since this method simply uncouples the system into distinct linear ODEs in each independent variable, this section gives you an opportunity to practice what you learned earlier in the chapter.

Systematic Elimination   The elimination of an unknown in a system of linear differential equations is expedited by rewriting each equation in the system in differential operator notation. Recall from Section 4.1 that a single linear equation

$$a_n y^{(n)} + a_{n-1} y^{(n-1)} + \cdots + a_1 y' + a_0 y = g(t),$$

where the a_i, $i = 0, 1, \ldots, n$ are constants, can be written as

$$(a_n D^n + a_{n-1} D^{(n-1)} + \cdots + a_1 D + a_0) y = g(t).$$

If the nth-order differential operator $a_n D^n + a_{n-1} D^{(n-1)} + \cdots + a_1 D + a_0$ factors into differential operators of lower order, then the factors commute. Now, for example, to rewrite the system

$$x'' + 2x' + y'' = x + 3y + \sin t$$
$$x' + y' = -4x + 2y + e^{-t}$$

in terms of the operator D, we first bring all terms involving the dependent variables to one side and group the same variables:

$$\begin{aligned} x'' + 2x' - x + y'' - 3y &= \sin t \\ x' - 4x + y' - 2y &= e^{-t} \end{aligned} \quad \text{is the same as} \quad \begin{aligned} (D^2 + 2D - 1)x + (D^2 - 3)y &= \sin t \\ (D - 4)x + (D - 2)y &= e^{-t}. \end{aligned}$$

Solution of a System   A **solution** of a system of differential equations is a set of sufficiently differentiable functions $x = \phi_1(t)$, $y = \phi_2(t)$, $z = \phi_3(t)$, and so on, that satisfies each equation in the system on some common interval I.

Method of Solution   Consider the simple system of linear first-order equations

$$\frac{dx}{dt} = 3y \qquad \text{or, equivalently,} \qquad \begin{aligned} Dx - 3y &= 0 \\ 2x - Dy &= 0. \end{aligned} \qquad \textbf{(1)}$$
$$\frac{dy}{dt} = 2x$$

Operating on the first equation in (1) by D while multiplying the second by -3 and then adding eliminates y from the system and gives $D^2x - 6x = 0$. Since the roots of the auxiliary equation of the last DE are $m_1 = \sqrt{6}$ and $m_2 = -\sqrt{6}$, we obtain

$$x(t) = c_1 e^{-\sqrt{6}t} + c_2 e^{\sqrt{6}t}. \tag{2}$$

Multiplying the first equation in (1) by 2 while operating on the second by D and then subtracting gives the differential equation for y, $D^2y - 6y = 0$. It follows immediately that

$$y(t) = c_3 e^{-\sqrt{6}t} + c_4 e^{\sqrt{6}t}. \tag{3}$$

Now (2) and (3) do not satisfy the system (1) for every choice of c_1, c_2, c_3, and c_4 because the system itself puts a constraint on the number of parameters in a solution that can be chosen arbitrarily. To see this, observe that substituting $x(t)$ and $y(t)$ into the first equation of the original system (1) gives, after simplification,

$$(-\sqrt{6}c_1 - 3c_3)e^{-\sqrt{6}t} + (\sqrt{6}c_2 - 3c_4)e^{\sqrt{6}t} = 0.$$

Since the latter expression is to be zero for all values of t, we must have $-\sqrt{6}c_1 - 3c_3 = 0$ and $\sqrt{6}c_2 - 3c_4 = 0$. These two equations enable us to write c_3 as a multiple of c_1 and c_4 as a multiple of c_2:

$$c_3 = -\frac{\sqrt{6}}{3}c_1 \quad \text{and} \quad c_4 = \frac{\sqrt{6}}{3}c_2. \tag{4}$$

Hence we conclude that a solution of the system must be

$$x(t) = c_1 e^{-\sqrt{6}t} + c_2 e^{\sqrt{6}t}, \quad y(t) = -\frac{\sqrt{6}}{3}c_1 e^{-\sqrt{6}t} + \frac{\sqrt{6}}{3}c_2 e^{\sqrt{6}t}.$$

You are urged to substitute (2) and (3) into the second equation of (1) and verify that the same relationship (4) holds between the constants.

EXAMPLE 1   Solution by Elimination

Solve
$$\begin{aligned} Dx + (D + 2)y &= 0 \\ (D - 3)x - \qquad 2y &= 0. \end{aligned} \tag{5}$$

Solution   Operating on the first equation by $D - 3$ and on the second by D and then subtracting eliminates x from the system. It follows that the differential equation for y is

$$[(D - 3)(D + 2) + 2D]y = 0 \quad \text{or} \quad (D^2 + D - 6)y = 0.$$

Since the characteristic equation of this last differential equation is $m^2 + m - 6 = (m - 2)(m + 3) = 0$, we obtain the solution

$$y(t) = c_1 e^{2t} + c_2 e^{-3t}. \tag{6}$$

Eliminating y in a similar manner yields $(D^2 + D - 6)x = 0$, from which we find

$$x(t) = c_3 e^{2t} + c_4 e^{-3t}. \tag{7}$$

As we noted in the foregoing discussion, a solution of (5) does not contain four independent constants. Substituting (6) and (7) into the first equation of (5) gives

$$(4c_1 + 2c_3)e^{2t} + (-c_2 - 3c_4)e^{-3t} = 0.$$

From $4c_1 + 2c_3 = 0$ and $-c_2 - 3c_4 = 0$ we get $c_3 = -2c_1$ and $c_4 = -\frac{1}{3}c_2$. Accordingly, a solution of the system is

$$x(t) = -2c_1 e^{2t} - \frac{1}{3}c_2 e^{-3t}, \quad y(t) = c_1 e^{2t} + c_2 e^{-3t}.$$ ∎

Since we could just as easily solve for c_3 and c_4 in terms of c_1 and c_2, the solution in Example 1 can be written in the alternative form

$$x(t) = c_3 e^{2t} + c_4 e^{-3t}, \quad y(t) = -\frac{1}{2}c_3 e^{2t} - 3c_4 e^{-3t}.$$

Note   It sometimes pays to keep one's eyes open when solving systems. Had we solved for x first, then y could be found, along with the relationship between the constants, by using the last equation in (5). You should verify that substituting $x(t)$ into $y = \frac{1}{2}(Dx - 3x)$ yields $y = -\frac{1}{2}c_3 e^{2t} - 3c_4 e^{-3t}$.

EXAMPLE 2   **Solution by Elimination**

Solve

$$x' - 4x + y'' = t^2$$
$$x' + x + y' = 0. \tag{8}$$

Solution   First we write the system in differential operator notation:

$$(D - 4)x + D^2 y = t^2$$
$$(D + 1)x + Dy = 0. \tag{9}$$

Then, by eliminating x, we obtain

$$[(D + 1)D^2 - (D - 4)D]y = (D + 1)t^2 - (D - 4)0$$

or

$$(D^3 + 4D)y = t^2 + 2t.$$

Since the roots of the auxiliary equation $m(m^2 + 4) = 0$ are $m_1 = 0$, $m_2 = 2i$, and $m_3 = -2i$, the complementary function is

$$y_c = c_1 + c_2 \cos 2t + c_3 \sin 2t.$$

To determine the particular solution y_p we use undetermined coefficients by assuming $y_p = At^3 + Bt^2 + Ct$. Therefore

$$y'_p = 3At^2 + 2Bt + C, \quad y''_p = 6At + 2B, \quad y'''_p = 6A,$$
$$y'''_p + 4y'_p = 12At^2 + 8Bt + 6A + 4C = t^2 + 2t.$$

The last equality implies $12A = 1$, $8B = 2$, and $6A + 4C = 0$; hence $A = \frac{1}{12}$, $B = \frac{1}{4}$, and $C = -\frac{1}{8}$. Thus

$$y = y_c + y_p = c_1 + c_2 \cos 2t + c_3 \sin 2t + \frac{1}{12}t^3 + \frac{1}{4}t^2 - \frac{1}{8}t. \tag{10}$$

Eliminating y from the system (9) leads to

$$[(D - 4) - D(D + 1)]x = t^2 \quad \text{or} \quad (D^2 + 4)x = -t^2.$$

It should be obvious that

$$x_c = c_4 \cos 2t + c_5 \sin 2t$$

and that undetermined coefficients can be applied to obtain a particular solution of the form $x_p = At^2 + Bt + C$. In this case the usual differentiations and algebra yield $x_p = -\frac{1}{4}t^2 + \frac{1}{8}$, and so

$$x = x_c + x_p = c_4 \cos 2t + c_5 \sin 2t - \frac{1}{4}t^2 + \frac{1}{8}. \tag{11}$$

Now c_4 and c_5 can be expressed in terms of c_2 and c_3 by substituting (10) and (11) into either equation of (8). By using the second equation, we find, after combining terms,

$$(c_5 - 2c_4 - 2c_2) \sin 2t + (2c_5 + c_4 + 2c_3) \cos 2t = 0,$$

so $c_5 - 2c_4 - 2c_2 = 0$ and $2c_5 + c_4 + 2c_3 = 0$. Solving for c_4 and c_5 in terms of c_2 and c_3 gives

$$c_4 = -\frac{1}{5}(4c_2 + 2c_3) \quad \text{and} \quad c_5 = \frac{1}{5}(2c_2 - 4c_3).$$

Finally, a solution of (8) is found to be

$$x(t) = -\frac{1}{5}(4c_2 + 2c_3) \cos 2t + \frac{1}{5}(2c_2 - 4c_3) \sin 2t - \frac{1}{4}t^2 + \frac{1}{8}$$

$$y(t) = c_1 + c_2 \cos 2t + c_3 \sin 2t + \frac{1}{12}t^3 + \frac{1}{4}t^2 - \frac{1}{8}t. \qquad \blacksquare$$

EXAMPLE 3 **A Mathematical Model Revisited**

In (3) Section 3.3 we saw that the system of linear first-order differential equations describes the number of pounds of salt $x_1(t)$ and $x_2(t)$ of a brine mixture that flows between two tanks. At that time we were not able to solve the system. But now, in terms of differential operators, the system is

$$\left(D + \frac{2}{25}\right)x_1 - \frac{1}{50}x_2 = 0$$

$$-\frac{2}{25}x_1 + \left(D + \frac{2}{25}\right)x_2 = 0.$$

Operating on the first equation by $D + \frac{2}{25}$, multiplying the second equation by $\frac{1}{50}$, adding, and then simplifying gives

$$(625D^2 + 100D + 3)x_1 = 0.$$

From the auxiliary equation

$$625m^2 + 100m + 3 = (25m + 1)(25m + 3) = 0$$

we see immediately that

$$x_1(t) = c_1 e^{-t/25} + c_2 e^{-3t/25}.$$

In like manner we find $(625D^2 + 100D + 3)x_2 = 0$, and so

$$x_2(t) = c_3 e^{-t/25} + c_4 e^{-3t/25}.$$

Substituting $x_1(t)$ and $x_2(t)$ into, say, the first equation of the system then gives

$$(2c_1 - c_3)e^{-t/25} + (-2c_2 - c_4)e^{-3t/25} = 0.$$

From this last equation we find $c_3 = 2c_1$ and $c_4 = -2c_2$. Thus a solution of the system is

$$x_1(t) = c_1 e^{-t/25} + c_2 e^{-3t/25}, \quad x_2(t) = 2c_1 e^{-t/25} - 2c_2 e^{-3t/25}.$$

In the original discussion we assumed that the initial conditions were $x_1(0) = 25$ and $x_2(0) = 0$. Applying these conditions to the solution yields $c_1 + c_2 = 25$ and $2c_1 - 2c_2 = 0$. Solving these equations simultaneously gives $c_1 = c_2 = \frac{25}{2}$. Finally, a solution of the initial-value problem is

$$x_1(t) = \frac{25}{2} e^{-t/25} + \frac{25}{2} e^{-3t/25}, \quad x_2(t) = 25e^{-t/25} - 25e^{-3t/25}. \qquad \blacksquare$$

EXERCISES 4.8

Answers to odd-numbered problems begin on page AN-4.

In Problems 1–20 solve the given system of differential equations by systematic elimination.

1. $\dfrac{dx}{dt} = 2x - y$

$\dfrac{dy}{dt} = x$

2. $\dfrac{dx}{dt} = 4x + 7y$

$\dfrac{dy}{dt} = x - 2y$

3. $\dfrac{dx}{dt} = -y + t$

$\dfrac{dy}{dt} = x - t$

4. $\dfrac{dx}{dt} - 4y = 1$

$\dfrac{dy}{dt} + x = 2$

5. $(D^2 + 5)x - \qquad 2y = 0$
$\qquad -2x + (D^2 + 2)y = 0$

6. $(D + 1)x + (D - 1)y = 2$
$\qquad 3x + (D + 2)y = -1$

7. $\dfrac{d^2x}{dt^2} = 4y + e^t$

$\dfrac{d^2y}{dt^2} = 4x - e^t$

8. $\dfrac{d^2x}{dt^2} + \dfrac{dy}{dt} = -5x$

$\dfrac{dx}{dt} + \dfrac{dy}{dt} = -x + 4y$

9. $\qquad Dx + \qquad D^2y = e^{3t}$
$\qquad (D + 1)x + (D - 1)y = 4e^{3t}$

10. $\qquad D^2x - \qquad Dy = t$
$\qquad (D + 3)x + (D + 3)y = 2$

11. $(D^2 - 1)x - y = 0$
$(D - 1)x + Dy = 0$

12. $(2D^2 - D - 1)x - (2D + 1)y = 1$
$(D - 1)x + Dy = -1$

13. $2\dfrac{dx}{dt} - 5x + \dfrac{dy}{dt} = e^t$

$\dfrac{dx}{dt} - x + \dfrac{dy}{dt} = 5e^t$

14. $\dfrac{dx}{dt} + \dfrac{dy}{dt} = e^t$

$-\dfrac{d^2x}{dt^2} + \dfrac{dx}{dt} + x + y = 0$

15. $(D - 1)x + (D^2 + 1)y = 1$
$(D^2 - 1)x + (D + 1)y = 2$

16. $D^2x - 2(D^2 + D)y = \sin t$
$x + Dy = 0$

17. $Dx = y$
$Dy = z$
$Dz = x$

18. $Dx + z = e^t$
$(D - 1)x + Dy + Dz = 0$
$x + 2y + Dz = e^t$

19. $\dfrac{dx}{dt} = 6y$

$\dfrac{dy}{dt} = x + z$

$\dfrac{dz}{dt} = x + y$

20. $\dfrac{dx}{dt} = -x + z$

$\dfrac{dy}{dt} = -y + z$

$\dfrac{dz}{dt} = -x + y$

In Problems 21 and 22 solve the given initial-value problem.

21. $\dfrac{dx}{dt} = -5x - y$

$\dfrac{dy}{dt} = 4x - y$

$x(1) = 0, y(1) = 1$

22. $\dfrac{dx}{dt} = y - 1$

$\dfrac{dy}{dt} = -3x + 2y$

$x(0) = 0, y(0) = 0$

23. A projectile shot from a gun has weight $w = mg$ and velocity $\mathbf{v}$ tangent to its path of motion. Ignoring air resistance and all other forces except its weight, determine a system of differential equations that describes the motion. See Figure 4.16. Solve the system. [*Hint:* Use Newton's second law of motion in the x and y directions.]

24. Determine a system of differential equations that describes the motion in Problem 23 if the projectile encounters a retarding force $\mathbf{k}$ (of magnitude k) acting tangent to the path but opposite to the motion. See Figure 4.17. Solve the system. [*Hint:* $\mathbf{k}$ is a multiple of velocity— say, $c\mathbf{v}$.]

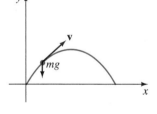

Figure 4.16

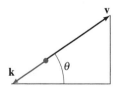

Figure 4.17

Discussion Problems

25. Discuss the following system:

$$Dx - 2Dy = t^2$$
$$(D + 1)x - 2(D + 1)y = 1.$$

4.9 NONLINEAR EQUATIONS

• Differences between linear and nonlinear ODEs • Solution by substitution • Use of Taylor series • Use of numerical solvers • Autonomous equations

There are several significant differences between linear and nonlinear differential equations. We saw in Section 4.1 that homogeneous linear equations of order two or higher have the property that a linear combination of solutions is also a solution (Theorem 4.2). Nonlinear equations do not possess this property of superposability. (See Problems 1 and 20 in Exercises 4.9.) We can find general solutions of linear first-order DEs and higher-order equations with constant coefficients. Even when we can solve a nonlinear first-order differential equation in the form of a one-parameter family, this family does not, as a rule, represent a general solution. Stated another way, nonlinear first-order DEs can possess singular solutions whereas linear equations cannot. But the major difference between linear and nonlinear equations of order two or higher lies in the realm of solvability. Given a linear equation, there is a chance that we can find some form of a solution that we can look at—an explicit solution or perhaps a solution in the form of an infinite series (see Chapter 6). On the other hand, nonlinear higher-order differential equations virtually defy solution by analytical methods. Although this may sound disheartening, there are still things that can be done. As pointed out at the end of Section 1.3, we can always analyze a nonlinear DE qualitatively and numerically.

Let us make it clear at the outset that nonlinear higher-order differential equations are important—dare we say even more important than linear equations?—because as we fine tune the mathematical model of, say, a physical system, we also increase the likelihood that this higher-resolution model will be nonlinear.

We begin by illustrating an analytical method that *occasionally* enables us to find explicit/implicit solutions of special kinds of nonlinear second-order differential equations.

Reduction of Order Nonlinear second-order differential equations $F(x, y', y'') = 0$, where the dependent variable y is missing, and $F(y, y', y'') = 0$, where the independent variable x is missing, can sometimes be solved using first-order methods. Each equation can be reduced to a first-order equation by means of the substitution $u = y'$.

The next example illustrates the substitution technique for an equation of the form $F(x, y', y'') = 0$. If $u = y'$, then the differential equation becomes $F(x, u, u') = 0$. If we can solve this last equation for u, we can find y by integration. Note that since we are solving a second-order equation, its solution will contain two arbitrary constants.

EXAMPLE 1 Dependent Variable y Is Missing

Solve $y'' = 2x(y')^2$.

Solution   If we let $u = y'$, then $du/dx = y''$. After substituting, the second-order equation reduces to a first-order equation with separable variables; the independent variable is x and the dependent variable is u:

$$\frac{du}{dx} = 2xu^2 \quad \text{or} \quad \frac{du}{u^2} = 2x\,dx$$

$$\int u^{-2}\,du = \int 2x\,dx$$

$$-u^{-1} = x^2 + c_1^2.$$

The constant of integration is written as c_1^2 for convenience. The reason should be obvious in the next few steps. Since $u^{-1} = 1/y'$, it follows that

$$\frac{dy}{dx} = -\frac{1}{x^2 + c_1^2},$$

and so $$y = -\int \frac{dx}{x^2 + c_1^2} \quad \text{or} \quad y = -\frac{1}{c_1}\tan^{-1}\frac{x}{c_1} + c_2. \qquad \blacksquare$$

Next we show how to solve an equation that has the form $F(y, y', y'') = 0$. Once more we let $u = y'$, but since the independent variable x is missing we use this substitution to transform the differential equation into one in which the independent variable is y and the dependent variable is u. To this end we use the Chain Rule to compute the second derivative of y:

$$y'' = \frac{du}{dx} = \frac{du}{dy}\frac{dy}{dx} = u\frac{du}{dy}.$$

In this case the first-order equation that we must now solve is

$$F\left(y, u, u\frac{du}{dy}\right) = 0.$$

EXAMPLE 2   **Independent Variable x Is Missing**

Solve $yy'' = (y')^2$.

Solution   With the aid of $u = y'$, the Chain Rule shown above, and separation of variables, the given differential equation becomes

$$y\left(u\frac{du}{dy}\right) = u^2 \quad \text{or} \quad \frac{du}{u} = \frac{dy}{y}.$$

Integrating the last equation then yields $\ln|u| = \ln|y| + c_1$, which, in turn, gives $u = c_2 y$, where the constant $\pm e^{c_1}$ has been relabeled as c_2. We now resubstitute $u = dy/dx$, separate variables once again, integrate, and relabel constants a second time:

$$\int \frac{dy}{y} = c_2 \int dx \quad \text{or} \quad \ln|y| = c_2 x + c_3 \quad \text{or} \quad y = c_4 e^{c_2 x}. \qquad \blacksquare$$

Use of Taylor Series   In some instances a solution of a nonlinear initial-value problem, in which the initial conditions are specified at x_0, can be approximated by a Taylor series centered at x_0.

EXAMPLE 3	**Taylor Series Solution of an IVP**

Let us assume that a solution of the initial-value problem

$$y'' = x + y - y^2, \quad y(0) = -1, \quad y'(0) = 1 \qquad \textbf{(1)}$$

exists. If we further assume that the solution $y(x)$ of the problem is analytic at 0, then $y(x)$ possesses a Taylor series expansion centered at 0:

$$y(x) = y(0) + \frac{y'(0)}{1!}x + \frac{y''(0)}{2!}x^2 + \frac{y'''(0)}{3!}x^3 + \frac{y^{(4)}(0)}{4!}x^4 + \frac{y^{(5)}(0)}{5!}x^5 + \cdots. \qquad \textbf{(2)}$$

Note that the values of the first and second terms in the series (2) are known since those values are the specified initial conditions $y(0) = -1$, $y'(0) = 1$. Moreover, the differential equation itself defines the value of the second derivative at 0: $y''(0) = 0 + y(0) - y(0)^2 = 0 + (-1) - (-1)^2 = -2$. We can then find expressions for the higher derivatives y''', $y^{(4)}$, ... by calculating the successive derivatives of the differential equation:

$$y'''(x) = \frac{d}{dx}(x + y - y^2) = 1 + y' - 2yy', \qquad \textbf{(3)}$$

$$y^{(4)}(x) = \frac{d}{dx}(1 + y' - 2yy') = y'' - 2yy'' - 2(y')^2, \qquad \textbf{(4)}$$

$$y^{(5)}(x) = \frac{d}{dx}(y'' - 2yy'' - 2(y')^2) = y''' - 2yy''' - 6y'y'', \qquad \textbf{(5)}$$

and so on. Now using $y(0) = -1$ and $y'(0) = 1$, we find from (3) that $y'''(0) = 4$. From the values $y(0) = -1$, $y'(0) = 1$, and $y''(0) = -2$ we find $y^{(4)}(0) = -8$ from (4). With the additional information that $y'''(0) = 4$, we then see from (5) that $y^{(5)}(0) = 24$. Hence from (2) the first six terms of a series solution of the initial-value problem (1) are

$$y(x) = -1 + x - x^2 + \frac{2}{3}x^3 - \frac{1}{3}x^4 + \frac{1}{5}x^5 + \cdots. \qquad \blacksquare$$

Use of a Numerical Solver   Numerical methods, such as Euler's method or the Runge-Kutta method, are developed solely for first-order differential equations and then are extended to systems of first-order equations. In order to analyze an nth-order initial-value problem numerically, we express the nth-order ODE as a system of n first-order equations. In brief, here how it is done for a second-order initial-value problem: First, solve for y''—that is, put the DE into normal form $y'' = f(x, y, y')$—and then let $y' = u$. For example, if we substitute $y' = u$ in

$$\frac{d^2y}{dx^2} = f(x, y, y'), \quad y(x_0) = y_0, \quad y'(x_0) = u_0, \qquad \textbf{(6)}$$

then $y'' = u'$ and $y'(x_0) = u(x_0)$, so the initial-value problem (6) becomes

$$\textit{Solve:} \quad \begin{cases} y' = u \\ u' = f(x, y, u) \end{cases}$$

$$\textit{Subject to:} \quad y(x_0) = y_0, u(x_0) = u_0.$$

However, it should be noted that a commercial numerical solver *may not* require* that you supply the system.

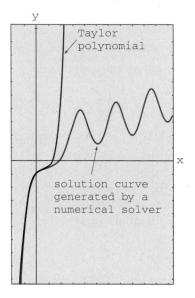

Figure 4.18 Comparison of two approximate solutions

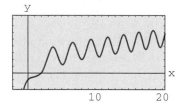

Figure 4.19

EXAMPLE 4   Graphical Analysis of Example 3

Following the foregoing procedure, we find that the second-order initial-value problem in Example 3 is equivalent to

$$\frac{dy}{dx} = u$$

$$\frac{du}{dx} = x + y - y^2$$

with initial conditions $y(0) = -1$, $u(0) = 1$. With the aid of a numerical solver, we get the solution curve shown in color in Figure 4.18. For comparison, the graph of the fifth-degree Taylor polynomial $T_5(x) = -1 + x - x^2 + \frac{2}{3}x^3 - \frac{1}{3}x^4 + \frac{1}{5}x^5$ is shown in black. Although we do not know the interval of convergence of the Taylor series obtained in Example 3, the closeness of the two curves in a neighborhood of the origin suggests that the power series may converge on the interval $(-1, 1)$.

Qualitative Questions The colored graph in Figure 4.18 raises some questions of a qualitative nature: Is the solution of the original initial-value problem oscillatory as $x \to \infty$? The graph generated by a numerical solver on the larger interval shown in Figure 4.19 would seem to *suggest* that the answer is yes. But this single example—or even an assortment of examples—does not answer the basic question as to whether *all* solutions of the differential equation $y'' = x + y - y^2$ are oscillatory in nature. Also, what is happening to the solution curve in Figure 4.19 when x is near -1? What is the behavior of solutions of the differential equation as $x \to -\infty$? Are solutions bounded as $x \to \infty$? Questions such as these are not easily answered, in general, for nonlinear second-order differential equations. But certain kinds of second-order equations lend themselves to a systematic qualitative analysis, and these, like their first-order relatives encountered in Section 2.1, are the kind that have no explicit dependence on the independent variable. Second-order ODEs of the form

$$F(y, y', y'') = 0 \quad \text{or} \quad \frac{d^2y}{dx^2} = f(y, y'),$$

equations free of the independent variable x, are called **autonomous.** The differential equation in Example 2 is autonomous, and because of the

---

*Some numerical solvers only require that a second-order differential equation be expressed in normal form $y'' = f(x, y, y')$. The translation of the single equation into a system of two equations is then built into the computer program, since the first equation of the system is always $y' = u$ and the second equation is $u' = f(x, y, u)$.

presence of the x term on its right-hand side, the equation in Example 3 is nonautonomous. For an in-depth treatment of the topic of stability of autonomous second-order differential equations and autonomous systems of differential equations refer to Chapter 10 in *Differential Equations with Boundary-Value Problems*.

EXERCISES 4.9

Answers to odd-numbered problems begin on page AN-5.

In Problems 1 and 2 verify that y_1 and y_2 are solutions of the given differential equation but that $y = c_1 y_1 + c_2 y_2$ is, in general, not a solution.

1. $(y'')^2 = y^2$; $y_1 = e^x, y_2 = \cos x$ **2.** $yy'' = \dfrac{1}{2}(y')^2$; $y_1 = 1, y_2 = x^2$

In Problems 3–8 solve the given differential equation by using the substitution $u = y'$.

3. $y'' + (y')^2 + 1 = 0$ **4.** $y'' = 1 + (y')^2$

5. $x^2 y'' + (y')^2 = 0$ **6.** $(y + 1)y'' = (y')^2$

7. $y'' + 2y(y')^3 = 0$ **8.** $y^2 y'' = y'$

9. Consider the initial-value problem $y'' + yy' = 0$, $y(0) = 1$, $y'(0) = -1$.
  (a) Use the DE and a numerical solver to graph the solution curve.
  (b) Find an explicit solution of the IVP. Use a graphing utility to graph this solution.
  (c) Find an interval of definition for the solution in part (b).

10. Find two solutions of the initial-value problem
$$(y'')^2 + (y')^2 = 1, \quad y\!\left(\frac{\pi}{2}\right) = \frac{1}{2}, \quad y'\!\left(\frac{\pi}{2}\right) = \frac{\sqrt{3}}{2}.$$
Use a numerical solver to graph the solution curves.

In Problems 11 and 12 show that the substitution $u = y'$ leads to a Bernoulli equation. Solve this equation (see Section 2.5).

11. $xy'' = y' + (y')^3$ **12.** $xy'' = y' + x(y')^2$

In Problems 13–16 proceed as in Example 3 and obtain the first six nonzero terms of a Taylor series solution, centered at 0, of the given initial-value problem. Use a numerical solver and a graphing utility to compare the solution curve with the graph of the Taylor polynomial.

13. $y'' = x + y^2$, $y(0) = 1$, $y'(0) = 1$

14. $y'' + y^2 = 1$, $y(0) = 2$, $y'(0) = 3$

15. $y'' = x^2 + y^2 - 2y'$, $y(0) = 1$, $y'(0) = 1$

16. $y'' = e^y$, $y(0) = 0$, $y'(0) = -1$

17. In calculus the curvature of a curve that is defined by a function $y = f(x)$ is defined as
$$\kappa = \frac{y''}{[1 + (y')^2]^{3/2}}.$$
Find $y = f(x)$ for which $\kappa = 1$. [*Hint:* For simplicity, ignore constants of integration.]

18. A mathematical model for the position $x(t)$ of a body moving rectilinearly on the x-axis in an inverse-square force field is given by

$$\frac{d^2x}{dt^2} = -\frac{k^2}{x^2}.$$

Suppose at $t = 0$ the body starts from rest from the position $x = x_0$, $x_0 > 0$. Show that the velocity of the body at any time is given by $v^2 = 2k^2(1/x - 1/x_0)$. Use this equation and a CAS to carry out the integration to express time t in terms of x.

Discussion Problems

19. A mathematical model for the position $x(t)$ of a moving object is

$$\frac{d^2x}{dt^2} + \sin x = 0.$$

Use a numerical solver to graphically investigate the solutions of the equation subject to $x(0) = 0$, $x'(0) = x_1$, $x_1 \geq 0$. Discuss the motion of the object for $t \geq 0$ and various choices of x_1. Investigate the equation

$$\frac{d^2x}{dt^2} + \frac{dx}{dt} + \sin x = 0$$

in the same manner. Discuss a possible physical interpretation of the dx/dt term.

20. In Problem 1 we saw that $\cos x$ and e^x were solutions of the nonlinear equation $(y'')^2 - y^2 = 0$. Verify that $\sin x$ and e^{-x} are also solutions. Without attempting to solve the differential equation, discuss how these explicit solutions can be found using knowledge about linear equations. Without attempting to verify, discuss why linear combinations such as $y = c_1 e^x + c_2 e^{-x} + c_3 \cos x + c_4 \sin x$ and $y = c_2 e^{-x} + c_4 \sin x$ are not, in general, solutions, but the two special linear combinations $y = c_1 e^x + c_2 e^{-x}$ and $y = c_3 \cos x + c_4 \sin x$ *must* satisfy the differential equation.

21. Discuss how the method of reduction of order considered in this section can be applied to the third-order differential equation $y''' = \sqrt{1 + (y'')^2}$. Carry out your ideas and solve the equation.

CHAPTER 4 IN REVIEW

Answers to odd-numbered problems begin on page AN-5.

Answer Problems 1–4 without referring back to the text. Fill in the blank or answer true or false.

1. The only solution of the initial-value problem $y'' + x^2 y = 0$, $y(0) = 0$, $y'(0) = 0$ is _____.

2. For the method of undetermined coefficients, the assumed form of the particular solution y_p for $y'' - y = 1 + e^x$ is _____.

3. A constant multiple of a solution of a linear differential equation is also a solution.

4. If the set consisting of two functions f_1 and f_2 is linearly independent on an interval I, then the Wronskian $W(f_1, f_2) \neq 0$ for all x in I.

5. Give an interval over which the set of two functions $f_1(x) = x^2$ and $f_2(x) = x|x|$ is linearly independent. Then give an interval over which the set consisting of f_1 and f_2 is linearly dependent.

6. Without the aid of the Wronskian, determine whether the given set of functions is linearly independent or linearly dependent on the indicated interval.
  (a) $f_1(x) = \ln x$, $f_2(x) = \ln x^2$, $(0, \infty)$
  (b) $f_1(x) = x^n$, $f_2(x) = x^{n+1}$, $n = 1, 2, \ldots$, $(-\infty, \infty)$
  (c) $f_1(x) = x$, $f_2(x) = x + 1$, $(-\infty, \infty)$
  (d) $f_1(x) = \cos\left(x + \dfrac{\pi}{2}\right)$, $f_2(x) = \sin x$, $(-\infty, \infty)$
  (e) $f_1(x) = 0$, $f_2(x) = x$, $(-5, 5)$
  (f) $f_1(x) = 2$, $f_2(x) = 2x$, $(-\infty, \infty)$
  (g) $f_1(x) = x^2$, $f_2(x) = 1 - x^2$, $f_3(x) = 2 + x^2$, $(-\infty, \infty)$
  (h) $f_1(x) = xe^{x+1}$, $f_2(x) = (4x - 5)e^x$, $f_3(x) = xe^x$, $(-\infty, \infty)$

7. Suppose $m_1 = 3$, $m_2 = -5$, and $m_3 = 1$ are roots of multiplicity one, two, and three, respectively, of an auxiliary equation. Write down the general solution of the corresponding homogeneous linear DE if it is
  (a) an equation with constant coefficients,
  (b) a Cauchy-Euler equation.

8. Consider the differential equation $ay'' + by' + cy = g(x)$, where a, b, and c are constants. Choose the input functions $g(x)$ for which the method of undetermined coefficients is applicable and the input functions for which the method of variation of parameters is applicable.
  (a) $g(x) = e^x \ln x$ **(b)** $g(x) = x^3 \cos x$
  (c) $g(x) = \dfrac{\sin x}{e^x}$ **(d)** $g(x) = 2x^{-2}e^x$
  (e) $g(x) = \sin^2 x$ **(f)** $g(x) = \dfrac{e^x}{\sin x}$

In Problems 9–24 use the procedures developed in this chapter to find the general solution of each differential equation.

9. $y'' - 2y' - 2y = 0$ **10.** $2y'' + 2y' + 3y = 0$

11. $y''' + 10y'' + 25y' = 0$ **12.** $2y''' + 9y'' + 12y' + 5y = 0$

13. $3y''' + 10y'' + 15y' + 4y = 0$

14. $2y^{(4)} + 3y''' + 2y'' + 6y' - 4y = 0$

15. $y'' - 3y' + 5y = 4x^3 - 2x$ **16.** $y'' - 2y' + y = x^2 e^x$

17. $y''' - 5y'' + 6y' = 8 + 2 \sin x$ **18.** $y''' - y'' = 6$

19. $y'' - 2y' + 2y = e^x \tan x$ **20.** $y'' - y = \dfrac{2e^x}{e^x + e^{-x}}$.

21. $6x^2 y'' + 5xy' - y = 0$

22. $2x^3 y''' + 19x^2 y'' + 39xy' + 9y = 0$

23. $x^2 y'' - 4xy' + 6y = 2x^4 + x^2$ **24.** $x^2 y'' - xy' + y = x^3$

25. Write down the form of the general solution $y = y_c + y_p$ of the given differential equation in the two cases $\omega \neq \alpha$ and $\omega = \alpha$. Do not determine the coefficients in y_p.
  (a) $y'' + \omega^2 y = \sin \alpha x$ **(b)** $y'' - \omega^2 y = e^{\alpha x}$

26. (a) Given that $y = \sin x$ is a solution of $y^{(4)} + 2y''' + 11y'' + 2y' + 10y = 0$, find the general solution of the DE *without the aid of a calculator or a computer.*

  (b) Find a linear second-order differential equation with constant coefficients for which $y_1 = 1$ and $y_2 = e^{-x}$ are solutions of the associated homogeneous equation and $y_p = \frac{1}{2}x^2 - x$ is a particular solution of the nonhomogeneous equation.

27. (a) Write the general solution of the fourth-order DE $y^{(4)} - 2y'' + y = 0$ entirely in terms of hyperbolic functions.

  (b) Write down the form of a particular solution of $y^{(4)} - 2y'' + y = \sinh x$.

28. Consider the differential equation $x^2 y'' - (x^2 + 2x)y' + (x + 2)y = x^3$. Verify that $y_1 = x$ is one solution of the associated homogeneous equation. Then show that the method of reduction of order discussed in Section 4.2 leads to a second solution y_2 of the homogeneous equation as well as a particular solution y_p of the nonhomogeneous equation. Form the general solution of the DE on the interval $(0, \infty)$.

In Problems 29–34 solve the given differential equation subject to the indicated conditions.

29. $y'' - 2y' + 2y = 0, \quad y\left(\dfrac{\pi}{2}\right) = 0, y(\pi) = -1$

30. $y'' + 2y' + y = 0, \quad y(-1) = 0, y'(0) = 0$

31. $y'' - y = x + \sin x, \quad y(0) = 2, y'(0) = 3$

32. $y'' + y = \sec^3 x, \quad y(0) = 1, y'(0) = \dfrac{1}{2}$

33. $y'y'' = 4x, \quad y(1) = 5, y'(1) = 2$

34. $2y'' = 3y^2, \quad y(0) = 1, y'(0) = 1$

35. (a) Use a CAS as an aid in finding the roots of the auxiliary equation for $12y^{(4)} + 64y''' + 59y'' - 23y' - 12y = 0$. Give the general solution of the equation.

  (b) Solve the DE in part (a) subject to the initial conditions $y(0) = -1, y'(0) = 2, y''(0) = 5, y'''(0) = 0$. Use a CAS as an aid in solving the resulting systems of four equations in four unknowns.

36. Find a member of the family of solutions of $xy'' + y' + \sqrt{x} = 0$ whose graph is tangent to the x-axis at $x = 1$. Use a graphing utility to obtain the solution curve.

In Problems 37–40 use systematic elimination to solve the given system.

37. $\dfrac{dx}{dt} + \dfrac{dy}{dt} = 2x + 2y + 1$

$\dfrac{dx}{dt} + 2\dfrac{dy}{dt} = y + 3$

38. $\dfrac{dx}{dt} = 2x + y + t - 2$

$\dfrac{dy}{dt} = 3x + 4y - 4t$

39. $(D - 2)x \qquad\qquad -y = -e^t$

$\qquad -3x + (D - 4)y = -7e^t$

40. $(D + 2)x + (D + 2)y = \sin 2t$

$\qquad 5x + (D + 3)y = \cos 2t$

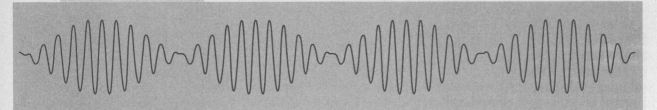

Beats oscillations; see page 235.

5 MODELING WITH HIGHER-ORDER DIFFERENTIAL EQUATIONS

INTRODUCTION We have seen that a single differential equation can serve as a mathematical model for different phenomena. For this reason we examine one application, the motion of a mass attached to a spring, in great detail in Section 5.1. We shall see that, except for terminology and physical interpretations of the four terms in the linear equation $ay'' + by' + cy = g(t)$, the mathematics of, say, an electrical series circuit is identical to that of a vibrating spring/mass system. Forms of this linear second-order differential equation appear in the analysis of problems in many diverse areas of science and engineering. In Section 5.1 we deal exclusively with initial-value problems, whereas in Section 5.2 we examine applications described by boundary-value problems. In Section 5.2 we also see how some boundary-value problems lead to the concepts of eigenvalues and eigenfunctions. Section 5.3 begins with a discussion of the differences between linear and nonlinear springs; we then show how the simple pendulum and a suspended wire lead to nonlinear models.

5.1 LINEAR EQUATIONS: INITIAL-VALUE PROBLEMS

• Linear dynamical system • Hooke's law • Newton's second law of motion • Spring/mass system • Free undamped motion • Simple harmonic motion • Equation of motion • Amplitude • Phase angle • Aging spring • Free damped motion • Driven motion • Transient and steady-state terms • Pure resonance • Series circuits

In this section we consider several linear dynamical systems in which each mathematical model is a second-order differential equation with constant coefficients

$$a_2 \frac{d^2 y}{dt^2} + a_1 \frac{dy}{dt} + a_0 y = g(t).$$

Recall that the function g is the **input** or **forcing function** of the system. A solution $y(t)$ of the differential equation on an interval containing t_0 and satisfies prescribed initial conditions $y(t_0) = y_0$, $y'(t_0) = y_1$ is the **output** or **response** of the system.

5.1.1 SPRING/MASS SYSTEMS: FREE UNDAMPED MOTION

Hooke's Law Suppose that a flexible spring is suspended vertically from a rigid support and then a mass m is attached to its free end. The amount of stretch, or elongation, of the spring will of course depend on the mass; masses with different weights stretch the spring by differing amounts. By Hooke's law, the spring itself exerts a restoring force F opposite to the direction of elongation and proportional to the amount of elongation s. Simply stated, $F = ks$, where k is a constant of proportionality called the **spring constant.** The spring is essentially characterized by the number k. For example, if a mass weighing 10 pounds stretches a spring $\frac{1}{2}$ foot, then $10 = k(\frac{1}{2})$ implies $k = 20$ lb/ft. Necessarily then, a mass weighing, say, 8 pounds stretches the same spring only $\frac{2}{5}$ foot.

Newton's Second Law After a mass m is attached to a spring, it stretches the spring by an amount s and attains a position of equilibrium at which its weight W is balanced by the restoring force ks. Recall that weight is defined by $W = mg$, where mass is measured in slugs, kilograms, or grams and $g = 32$ ft/s², 9.8 m/s², or 980 cm/s², respectively. As indicated in Figure 5.1(b), the condition of equilibrium is $mg = ks$ or $mg - ks = 0$. If the mass is displaced by an amount x from its equilibrium position, the restoring force of the spring is then $k(x + s)$. Assuming that there are no retarding forces acting on the system and assuming that the mass vibrates free of other external forces—**free motion**—we can equate Newton's second law with the net, or resultant, force of the restoring force and the weight:

$$m \frac{d^2 x}{dt^2} = -k(s + x) + mg = -kx + \underbrace{mg - ks}_{\text{zero}} = -kx. \tag{1}$$

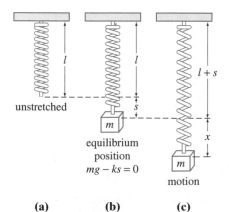

unstretched

equilibrium position
$mg - ks = 0$

$l + s$

x

m

motion

(a)   **(b)**   **(c)**

Figure 5.1

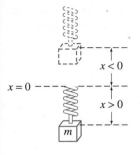

Figure 5.2

The negative sign in (1) indicates that the restoring force of the spring acts opposite to the direction of motion. Furthermore, we can adopt the convention that displacements measured *below* the equilibrium position are positive. See Figure 5.2.

DE of Free Undamped Motion By dividing (1) by the mass m we obtain the second-order differential equation $d^2x/dt^2 + (k/m)x = 0$ or

$$\frac{d^2x}{dt^2} + \omega^2 x = 0, \tag{2}$$

where $\omega^2 = k/m$. Equation (2) is said to describe **simple harmonic motion** or **free undamped motion.** Two obvious initial conditions associated with (2) are $x(0) = x_0$, the amount of initial displacement, and $x'(0) = x_1$, the initial velocity of the mass. For example, if $x_0 > 0$, $x_1 < 0$, the mass starts from a point *below* the equilibrium position with an imparted *upward* velocity. When $x_1 = 0$ the mass is said to be released from *rest*. For example, if $x_0 < 0$, $x_1 = 0$, the mass is released from rest from a point $|x_0|$ units *above* the equilibrium position.

Solution and Equation of Motion To solve equation (2) we note that the solutions of the auxiliary equation $m^2 + \omega^2 = 0$ are the complex numbers $m_1 = \omega i$, $m_2 = -\omega i$. Thus from (8) of Section 4.3 we find the general solution of (2) to be

$$x(t) = c_1 \cos \omega t + c_2 \sin \omega t. \tag{3}$$

The **period** of free vibrations described by (3) is $T = 2\pi/\omega$, and the **frequency** is $f = 1/T = \omega/2\pi$. For example, for $x(t) = 2 \cos 3t - 4 \sin 3t$ the period is $2\pi/3$ and the frequency is $3/2\pi$. The former number means that the graph of $x(t)$ repeats every $2\pi/3$ units; the latter number means that there are 3 cycles of the graph every 2π units or, equivalently, that the mass undergoes $3/2\pi$ complete vibrations per unit time. In addition, it can be shown that the period $2\pi/\omega$ is the time interval between two successive maxima of $x(t)$. Keep in mind that a maximum of $x(t)$ is a positive displacement corresponding to the mass's attaining a maximum distance *below* the equilibrium position, whereas a minimum of $x(t)$ is a negative displacement corresponding to the mass's attaining a maximum height *above* the equilibrium position. We refer to either case as an **extreme displacement** of the mass. Finally, when the initial conditions are used to determine the constants c_1 and c_2 in (3), we say that the resulting particular solution or response is the **equation of motion.**

EXAMPLE 1 **Free Undamped Motion**

A mass weighing 2 pounds stretches a spring 6 inches. At $t = 0$ the mass is released from a point 8 inches below the equilibrium position with an upward velocity of $\frac{4}{3}$ ft/s. Determine the equation of free motion.

Solution Because we are using the engineering system of units, the measurements given in terms of inches must be converted into feet: 6 in. $= \frac{1}{2}$ ft; 8 in. $= \frac{2}{3}$ ft. In addition, we must convert the units of weight given in pounds into units of mass. From $m = W/g$ we have $m = \frac{2}{32} = \frac{1}{16}$ slug. Also, from Hooke's law, $2 = k(\frac{1}{2})$ implies that the spring constant is $k = 4$ lb/ft. Hence (1) gives

$$\frac{1}{16}\frac{d^2x}{dt^2} = -4x \quad \text{or} \quad \frac{d^2x}{dt^2} + 64x = 0.$$

The initial displacement and initial velocity are $x(0) = \frac{2}{3}$, $x'(0) = -\frac{4}{3}$, where the negative sign in the last condition is a consequence of the fact that the mass is given an initial velocity in the negative, or upward, direction.

Now $\omega^2 = 64$ or $\omega = 8$, so the general solution of the differential equation is

$$x(t) = c_1 \cos 8t + c_2 \sin 8t. \tag{4}$$

Applying the initial conditions to $x(t)$ and $x'(t)$ gives $c_1 = \frac{2}{3}$ and $c_2 = -\frac{1}{6}$. Thus the equation of motion is

$$x(t) = \frac{2}{3}\cos 8t - \frac{1}{6}\sin 8t. \tag{5} \quad \blacksquare$$

Alternative Form of $x(t)$ When $c_1 \neq 0$ and $c_2 \neq 0$, the actual **amplitude** A of free vibrations is not obvious from inspection of equation (3). For example, although the mass in Example 1 is initially displaced $\frac{2}{3}$ foot beyond the equilibrium position, the amplitude of vibrations is a number larger than $\frac{2}{3}$. Hence it is often convenient to convert a solution of form (3) to the simpler form

$$x(t) = A \sin(\omega t + \phi), \tag{6}$$

where $A = \sqrt{c_1^2 + c_2^2}$ and ϕ is a **phase angle** defined by

$$\left.\begin{array}{c} \sin \phi = \dfrac{c_1}{A} \\[2mm] \cos \phi = \dfrac{c_2}{A} \end{array}\right\} \quad \tan \phi = \dfrac{c_1}{c_2}. \tag{7}$$

To verify this we expand (6) by the addition formula for the sine function:

$$A \sin \omega t \cos \phi + A \cos \omega t \sin \phi = (A \sin \phi) \cos \omega t + (A \cos \phi) \sin \omega t. \tag{8}$$

It follows from Figure 5.3 that if ϕ is defined by

$$\sin \phi = \frac{c_1}{\sqrt{c_1^2 + c_2^2}} = \frac{c_1}{A}, \quad \cos \phi = \frac{c_2}{\sqrt{c_1^2 + c_2^2}} = \frac{c_2}{A},$$

then (8) becomes

$$A\frac{c_1}{A}\cos \omega t + A\frac{c_2}{A}\sin \omega t = c_1 \cos \omega t + c_2 \sin \omega t = x(t).$$

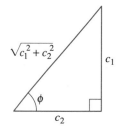

Figure 5.3

EXAMPLE 2   **Alternative Form of Solution (5)**

In view of the foregoing discussion, we can write solution (5), $x(t) = \frac{2}{3}\cos 8t - \frac{1}{6}\sin 8t$, in the alternative form $x(t) = A \sin(8t + \phi)$. Computation of the amplitude is straightforward, $A = \sqrt{(\frac{2}{3})^2 + (-\frac{1}{6})^2} = \sqrt{\frac{17}{36}} \approx 0.69$ ft, but some care should be exercised when computing the phase angle ϕ defined by (7). With $c_1 = \frac{2}{3}$ and $c_2 = -\frac{1}{6}$ we find $\tan \phi = -4$, and a calculator then gives $\tan^{-1}(-4) = -1.326$ rad. This is *not* the phase angle, since $\tan^{-1}(-4)$ is located in the *fourth quadrant* and therefore contradicts the fact that $\sin \phi > 0$ and $\cos \phi < 0$ because $c_1 > 0$ and $c_2 < 0$. Hence we must take ϕ to be the *second-quadrant* angle $\phi = \pi + (-1.326) = 1.816$ rad. Thus we have

$$x(t) = \frac{\sqrt{17}}{6} \sin(8t + 1.816). \tag{9}$$

The period of this function is $T = 2\pi/8 = \pi/4$.   ▬

Figure 5.4(a) illustrates the mass in Example 2 going through approximately two complete cycles of motion. Reading from left to right, the first five positions (marked with black dots) correspond to the initial position of the mass below the equilibrium position ($x = \frac{2}{3}$), the mass passing through the equilibrium position for the first time heading upward

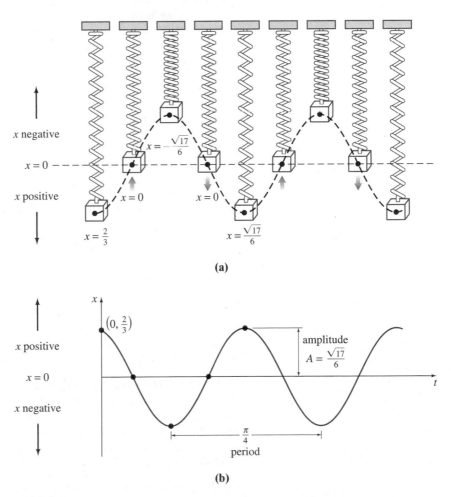

(a)

(b)

Figure 5.4

($x = 0$), the mass at its extreme displacement above the equilibrium position ($x = -\sqrt{17}/6$), the mass at the equilibrium position for the second time heading downward ($x = 0$), and the mass at its extreme displacement below the equilibrium position ($x = \sqrt{17}/6$). The dots on the graph of (9), given in Figure 5.4(b), also agree with the five positions just given. Note, however, that in Figure 5.4(b) the positive direction in the tx-plane is the usual upward direction and so is opposite to the positive direction indicated in Figure 5.4(a). Hence the solid color graph representing the motion of the mass in Figure 5.4(b) is the mirror image through the t-axis of the black dashed curve in Figure 5.4(a).

Form (6) is very useful, since it is easy to find values of time for which the graph of $x(t)$ crosses the positive t-axis (the line $x = 0$). We observe that $\sin(\omega t + \phi) = 0$ when $\omega t + \phi = n\pi$, where n is a nonnegative integer.

Systems with Variable Spring Constants In the model discussed above we assumed an ideal world—a world in which the physical characteristics of the spring do not change over time. In the non-ideal world, however, it seems reasonable to expect that when a spring/mass system is in motion for a long period, the spring will weaken; in other words, the "spring constant" will vary—or, more specifically, decay—with time. In one model for the **aging spring** the spring constant k in (1) is replaced by the decreasing function $K(t) = ke^{-\alpha t}$, $k > 0$, $\alpha > 0$. The linear differential equation $mx'' + ke^{-\alpha t}x = 0$ cannot be solved by the methods considered in Chapter 4. Nevertheless, we can obtain two linearly independent solutions using the methods in Chapter 6. See Problem 15 in Exercises 5.1, Example 3 in Section 6.3, and Problems 25 and 26 in Exercises 6.3.

When a spring/mass system is subjected to an environment in which the temperature is rapidly decreasing, it might make sense to replace the constant k with $K(t) = kt$, $k > 0$, a function that increases with time. The resulting model, $mx'' + ktx = 0$, is a form of **Airy's differential equation.** Like the equation for an aging spring, Airy's equation can be solved by the methods of Chapter 6. See Problem 16 in Exercises 5.1, Example 2 in Section 6.1, and Problems 27–29 in Exercises 6.3.

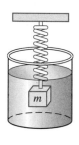

(a)

5.1.2 SPRING/MASS SYSTEMS: FREE DAMPED MOTION

The concept of free harmonic motion is somewhat unrealistic, since the motion described by equation (1) assumes that there are no retarding forces acting on the moving mass. Unless the mass is suspended in a perfect vacuum, there will be at least a resisting force due to the surrounding medium. As Figure 5.5 shows, the mass could be suspended in a viscous medium or connected to a dashpot damping device.

DE of Free Damped Motion In the study of mechanics, damping forces acting on a body are considered to be proportional to a power of the instantaneous velocity. In particular, we shall assume throughout the subsequent discussion that this force is given by a constant multiple of dx/dt. When no other external forces are impressed on the system, it follows from Newton's second law that

(b)

Figure 5.5

$$m\frac{d^2x}{dt^2} = -kx - \beta\frac{dx}{dt}, \tag{10}$$

where β is a positive *damping constant* and the negative sign is a consequence of the fact that the damping force acts in a direction opposite to the motion.

Dividing (10) by the mass m, we find that the differential equation of **free damped motion** is $d^2x/dt^2 + (\beta/m)dx/dt + (k/m)x = 0$ or

$$\frac{d^2x}{dt^2} + 2\lambda \frac{dx}{dt} + \omega^2 x = 0, \tag{11}$$

where

$$2\lambda = \frac{\beta}{m}, \quad \omega^2 = \frac{k}{m}. \tag{12}$$

The symbol 2λ is used only for algebraic convenience since the auxiliary equation is $m^2 + 2\lambda m + \omega^2 = 0$ and the corresponding roots are then

$$m_1 = -\lambda + \sqrt{\lambda^2 - \omega^2}, \quad m_2 = -\lambda - \sqrt{\lambda^2 - \omega^2}.$$

We can now distinguish three possible cases depending on the algebraic sign of $\lambda^2 - \omega^2$. Since each solution contains the *damping factor* $e^{-\lambda t}$, $\lambda > 0$, the displacements of the mass become negligible as time t increases.

Case I: $\lambda^2 - \omega^2 > 0$ In this situation the system is said to be **over-damped** since the damping coefficient β is large when compared to the spring constant k. The corresponding solution of (11) is $x(t) = c_1 e^{m_1 t} + c_2 e^{m_2 t}$ or

$$x(t) = e^{-\lambda t}(c_1 e^{\sqrt{\lambda^2 - \omega^2}\,t} + c_2 e^{-\sqrt{\lambda^2 - \omega^2}\,t}). \tag{13}$$

This equation represents a smooth and nonoscillatory motion. Figure 5.6 shows two possible graphs of $x(t)$.

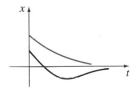

Figure 5.6

Case II: $\lambda^2 - \omega^2 = 0$ The system is said to be **critically damped** since any slight decrease in the damping force would result in oscillatory motion. The general solution of (11) is $x(t) = c_1 e^{m_1 t} + c_2 t e^{m_1 t}$ or

$$x(t) = e^{-\lambda t}(c_1 + c_2 t). \tag{14}$$

Some graphs of typical motion are given in Figure 5.7. Notice that the motion is quite similar to that of an overdamped system. It is also apparent from (14) that the mass can pass through the equilibrium position at most one time.

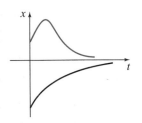

Figure 5.7

Case III: $\lambda^2 - \omega^2 < 0$ In this case the system is said to be **underdamped** since the damping coefficient is small compared to the spring constant. The roots m_1 and m_2 are now complex:

$$m_1 = -\lambda + \sqrt{\omega^2 - \lambda^2}\, i, \quad m_2 = -\lambda - \sqrt{\omega^2 - \lambda^2}\, i.$$

Thus the general solution of equation (11) is

$$x(t) = e^{-\lambda t}(c_1 \cos \sqrt{\omega^2 - \lambda^2}\,t + c_2 \sin \sqrt{\omega^2 - \lambda^2}\,t). \tag{15}$$

As indicated in Figure 5.8, the motion described by (15) is oscillatory; but because of the coefficient $e^{-\lambda t}$, the amplitudes of vibration $\to 0$ as $t \to \infty$.

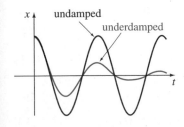

Figure 5.8

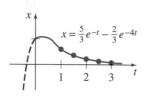

$x = \frac{5}{3}e^{-t} - \frac{2}{3}e^{-4t}$

(a)

t	$x(t)$
1	0.601
1.5	0.370
2	0.225
2.5	0.137
3	0.083

(b)

Figure 5.9

EXAMPLE 3   Overdamped Motion

It is readily verified that the solution of the initial-value problem

$$\frac{d^2x}{dt^2} + 5\frac{dx}{dt} + 4x = 0, \quad x(0) = 1, \quad x'(0) = 1$$

is

$$x(t) = \frac{5}{3}e^{-t} - \frac{2}{3}e^{-4t}. \tag{16}$$

The problem can be interpreted as representing the overdamped motion of a mass on a spring. The mass starts from a position 1 unit *below* the equilibrium position with a *downward* velocity of 1 ft/s.

To graph $x(t)$ we find the value of t for which the function has an extremum—that is, the value of time for which the first derivative (velocity) is zero. Differentiating (16) gives $x'(t) = -\frac{5}{3}e^{-t} + \frac{8}{3}e^{-4t}$, so that $x'(t) = 0$ implies $e^{3t} = \frac{8}{5}$ or $t = \frac{1}{3}\ln\frac{8}{5} = 0.157$. It follows from the first derivative test, as well as our physical intuition, that $x(0.157) = 1.069$ ft is actually a maximum. In other words, the mass attains an extreme displacement of 1.069 feet below the equilibrium position.

We should also check to see whether the graph crosses the t-axis—that is, whether the mass passes through the equilibrium position. This cannot happen in this instance since the equation $x(t) = 0$, or $e^{3t} = \frac{2}{5}$, has the physically irrelevant solution $t = \frac{1}{3}\ln\frac{2}{5} = -0.305$.

The graph of $x(t)$, along with some other pertinent data, is given in Figure 5.9.   ■

EXAMPLE 4   Critically Damped Motion

An 8-pound weight stretches a spring 2 feet. Assuming that a damping force numerically equal to 2 times the instantaneous velocity acts on the system, determine the equation of motion if the weight is released from the equilibrium position with an upward velocity of 3 ft/s.

Solution   From Hooke's law we see that $8 = k(2)$ gives $k = 4$ lb/ft and that $W = mg$ gives $m = \frac{8}{32} = \frac{1}{4}$ slug. The differential equation of motion is then

$$\frac{1}{4}\frac{d^2x}{dt^2} = -4x - 2\frac{dx}{dt} \quad \text{or} \quad \frac{d^2x}{dt^2} + 8\frac{dx}{dt} + 16x = 0. \tag{17}$$

The auxiliary equation for (17) is $m^2 + 8m + 16 = (m + 4)^2 = 0$, so $m_1 = m_2 = -4$. Hence the system is critically damped and

$$x(t) = c_1 e^{-4t} + c_2 t e^{-4t}. \tag{18}$$

Applying the initial conditions $x(0) = 0$ and $x'(0) = -3$, we find, in turn, that $c_1 = 0$ and $c_2 = -3$. Thus the equation of motion is

$$x(t) = -3te^{-4t}. \tag{19}$$

To graph $x(t)$ we proceed as in Example 3. From $x'(t) = -3e^{-4t}(1 - 4t)$ we see that $x'(t) = 0$ when $t = \frac{1}{4}$. The corresponding extreme displacement

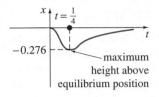

Figure 5.10

is $x(\tfrac{1}{4}) = -3(\tfrac{1}{4})e^{-1} = -0.276$ ft. As shown in Figure 5.10, we interpret this value to mean that the weight reaches a maximum height of 0.276 foot above the equilibrium position. ■

EXAMPLE 5   **Underdamped Motion**

A 16-pound weight is attached to a 5-foot-long spring. At equilibrium the spring measures 8.2 feet. If the weight is pushed up and released from rest at a point 2 feet above the equilibrium position, find the displacements $x(t)$ if it is further known that the surrounding medium offers a resistance numerically equal to the instantaneous velocity.

Solution   The elongation of the spring after the weight is attached is $8.2 - 5 = 3.2$ ft, so it follows from Hooke's law that $16 = k(3.2)$ or $k = 5$ lb/ft. In addition, $m = \tfrac{16}{32} = \tfrac{1}{2}$ slug, so the differential equation is given by

$$\frac{1}{2}\frac{d^2x}{dt^2} = -5x - \frac{dx}{dt} \quad \text{or} \quad \frac{d^2x}{dt^2} + 2\frac{dx}{dt} + 10x = 0. \tag{20}$$

Proceeding, we find that the roots of $m^2 + 2m + 10 = 0$ are $m_1 = -1 + 3i$ and $m_2 = -1 - 3i$, which then implies that the system is underdamped and

$$x(t) = e^{-t}(c_1 \cos 3t + c_2 \sin 3t). \tag{21}$$

Finally, the initial conditions $x(0) = -2$ and $x'(0) = 0$ yield $c_1 = -2$ and $c_2 = -\tfrac{2}{3}$, so the equation of motion is

$$x(t) = e^{-t}\left(-2 \cos 3t - \frac{2}{3}\sin 3t\right). \tag{22} \quad ■$$

Alternative Form of $x(t)$ In a manner identical to the procedure used on page 218, we can write any solution

$$x(t) = e^{-\lambda t}(c_1 \cos \sqrt{\omega^2 - \lambda^2}\,t + c_2 \sin \sqrt{\omega^2 - \lambda^2}\,t)$$

in the alternative form

$$x(t) = Ae^{-\lambda t} \sin(\sqrt{\omega^2 - \lambda^2}\,t + \phi), \tag{23}$$

where $A = \sqrt{c_1^2 + c_2^2}$ and the phase angle ϕ is determined from the equations

$$\sin \phi = \frac{c_1}{A}, \quad \cos \phi = \frac{c_2}{A}, \quad \tan \phi = \frac{c_1}{c_2}.$$

The coefficient $Ae^{-\lambda t}$ is sometimes called the **damped amplitude** of vibrations. Because (23) is not a periodic function, the number $2\pi/\sqrt{\omega^2 - \lambda^2}$ is called the **quasi period** and $\sqrt{\omega^2 - \lambda^2}/2\pi$ is the **quasi frequency.** The quasi period is the time interval between two successive maxima of $x(t)$. You should verify, for the equation of motion in Example 5, that $A = 2\sqrt{10}/3$ and $\phi = 4.391$. Therefore an equivalent form of (22) is

$$x(t) = \frac{2\sqrt{10}}{3}e^{-t} \sin(3t + 4.391).$$

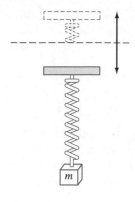

Figure 5.11

5.1.3 SPRING/MASS SYSTEMS: DRIVEN MOTION

DE of Driven Motion with Damping Suppose we now take into consideration an external force $f(t)$ acting on a vibrating mass on a spring. For example, $f(t)$ could represent a driving force causing an oscillatory vertical motion of the support of the spring. See Figure 5.11. The inclusion of $f(t)$ in the formulation of Newton's second law gives the differential equation of **driven** or **forced motion:**

$$m\frac{d^2x}{dt^2} = -kx - \beta\frac{dx}{dt} + f(t). \tag{24}$$

Dividing (24) by m gives

$$\frac{d^2x}{dt^2} + 2\lambda\frac{dx}{dt} + \omega^2 x = F(t), \tag{25}$$

where $F(t) = f(t)/m$ and, as in the preceding section, $2\lambda = \beta/m$, $\omega^2 = k/m$. To solve the latter nonhomogeneous equation we can use either the method of undetermined coefficients or variation of parameters.

EXAMPLE 6 | **Interpretation of an Initial-Value Problem**

Interpret and solve the initial-value problem

$$\frac{1}{5}\frac{d^2x}{dt^2} + 1.2\frac{dx}{dt} + 2x = 5\cos 4t, \quad x(0) = \frac{1}{2}, \quad x'(0) = 0. \tag{26}$$

Solution We can interpret the problem to represent a vibrational system consisting of a mass ($m = \frac{1}{5}$ slug or kilogram) attached to a spring ($k = 2$ lb/ft or N/m). The mass is released from rest $\frac{1}{2}$ unit (foot or meter) below the equilibrium position. The motion is damped ($\beta = 1.2$) and is being driven by an external periodic ($T = \pi/2$ s) force beginning at $t = 0$. Intuitively we would expect that even with damping the system would remain in motion until such time as the forcing function was "turned off," in which case the amplitudes would diminish. However, as the problem is given, $f(t) = 5\cos 4t$ will remain "on" forever.

We first multiply the differential equation in (26) by 5 and solve

$$\frac{dx^2}{dt^2} + 6\frac{dx}{dt} + 10x = 0$$

by the usual methods. Since $m_1 = -3 + i$, $m_2 = -3 - i$, it follows that $x_c(t) = e^{-3t}(c_1\cos t + c_2\sin t)$. Using the method of undetermined coefficients, we assume a particular solution of the form $x_p(t) = A\cos 4t + B\sin 4t$. Differentiating $x_p(t)$ and substituting into the DE gives

$$x_p'' + 6x_p' + 10x_p = (-6A + 24B)\cos 4t + (-24A - 6B)\sin 4t = 25\cos 4t.$$

The resulting system of equations

$$-6A + 24B = 25, \quad -24A - 6B = 0$$

yields $A = -\frac{25}{102}$ and $B = \frac{50}{51}$. It follows that

$$x(t) = e^{-3t}(c_1\cos t + c_2\sin t) - \frac{25}{102}\cos 4t + \frac{50}{51}\sin 4t. \tag{27}$$

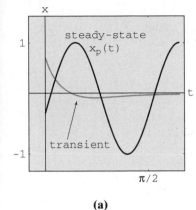

(a)

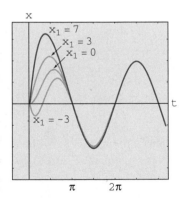

(b)

Figure 5.12

When we set $t = 0$ in the above equation, we obtain $c_1 = \frac{38}{51}$. By differentiating the expression and then setting $t = 0$, we also find that $c_2 = -\frac{86}{51}$. Therefore the equation of motion is

$$x(t) = e^{-3t}\left(\frac{38}{51}\cos t - \frac{86}{51}\sin t\right) - \frac{25}{102}\cos 4t + \frac{50}{51}\sin 4t. \quad \textbf{(28)} \quad \blacksquare$$

Transient and Steady-State Terms When F is a periodic function, such as $F(t) = F_0 \sin \gamma t$ or $F(t) = F_0 \cos \gamma t$, the general solution of (25) for $\lambda > 0$ is the sum of a nonperiodic function $x_c(t)$ and a periodic function $x_p(t)$. Moreover, $x_c(t)$ dies off as time increases—that is, $\lim_{t\to\infty} x_c(t) = 0$. Thus for large values of time, the displacements of the mass are closely approximated by the particular solution $x_p(t)$. The complementary function $x_c(t)$ is said to be a **transient term** or **transient solution,** and the function $x_p(t)$, the part of the solution that remains after an interval of time, is called a **steady-state term** or **steady-state solution.** Note therefore that the effect of the initial conditions on a spring/mass system driven by F is transient. In the particular solution (28), $e^{-3t}(\frac{38}{51}\cos t - \frac{86}{51}\sin t)$ is a transient term and $x_p(t) = -\frac{25}{102}\cos 4t + \frac{50}{51}\sin 4t$ is a steady-state term. The graphs of these two terms and the solution (28) are given in Figures 5.12(a) and (b), respectively.

EXAMPLE 7   Transient/Steady-State Solutions

The solution of the initial-value problem

$$\frac{d^2x}{dt^2} + 2\frac{dx}{dt} + 2x = 4\cos t + 2\sin t, \quad x(0) = 0, \quad x'(0) = x_1,$$

where x_1 is constant, is given by

$$x(t) = \underbrace{(x_1 - 2)e^{-t}\sin t}_{\text{transient}} + \underbrace{2\sin t}_{\text{steady-state}}.$$

Solution curves for selected values of the initial velocity x_1 are shown in Figure 5.13. The graphs show that the influence of the transient term is negligible for about $t > 3\pi/2$. $\blacksquare$

DE of Driven Motion Without Damping With a periodic impressed force and no damping force, there is no transient term in the solution of a problem. Also, we shall see that a periodic impressed force with a frequency near or the same as the frequency of free undamped vibrations can cause a severe problem in any oscillatory mechanical system.

EXAMPLE 8   Undamped Forced Motion

Solve the initial-value problem

$$\frac{d^2x}{dt^2} + \omega^2 x = F_0 \sin \gamma t, \quad x(0) = 0, \quad x'(0) = 0, \quad \textbf{(29)}$$

where F_0 is a constant and $\gamma \neq \omega$.

Figure 5.13

Solution The complementary function is $x_c(t) = c_1 \cos \omega t + c_2 \sin \omega t$. To obtain a particular solution we assume $x_p(t) = A \cos \gamma t + B \sin \gamma t$ so that

$$x_p'' + \omega^2 x_p = A(\omega^2 - \gamma^2) \cos \gamma t + B(\omega^2 - \gamma^2) \sin \gamma t = F_0 \sin \gamma t.$$

Equating coefficients immediately gives $A = 0$ and $B = F_0/(\omega^2 - \gamma^2)$. Therefore

$$x_p(t) = \frac{F_0}{\omega^2 - \gamma^2} \sin \gamma t.$$

Applying the given initial conditions to the general solution

$$x(t) = c_1 \cos \omega t + c_2 \sin \omega t + \frac{F_0}{\omega^2 - \gamma^2} \sin \gamma t$$

yields $c_1 = 0$ and $c_2 = -\gamma F_0/\omega(\omega^2 - \gamma^2)$. Thus the solution is

$$x(t) = \frac{F_0}{\omega(\omega^2 - \gamma^2)} (-\gamma \sin \omega t + \omega \sin \gamma t), \quad \gamma \neq \omega. \quad \textbf{(30)} \quad \blacksquare$$

Pure Resonance Although equation (30) is not defined for $\gamma = \omega$, it is interesting to observe that its limiting value as $\gamma \to \omega$ can be obtained by applying L'Hôpital's rule. This limiting process is analogous to "tuning in" the frequency of the driving force $(\gamma/2\pi)$ to the frequency of free vibrations $(\omega/2\pi)$. Intuitively we expect that over a length of time we should be able to substantially increase the amplitudes of vibration. For $\gamma = \omega$ we define the solution to be

$$x(t) = \lim_{\gamma \to \omega} F_0 \frac{-\gamma \sin \omega t + \omega \sin \gamma t}{\omega(\omega^2 - \gamma^2)} = F_0 \lim_{\gamma \to \omega} \frac{\dfrac{d}{d\gamma}(-\gamma \sin \omega t + \omega \sin \gamma t)}{\dfrac{d}{d\gamma}(\omega^3 - \omega\gamma^2)}$$

$$= F_0 \lim_{\gamma \to \omega} \frac{-\sin \omega t + \omega t \cos \gamma t}{-2\omega\gamma}$$

$$= F_0 \frac{-\sin \omega t + \omega t \cos \omega t}{-2\omega^2}$$

$$= \frac{F_0}{2\omega^2} \sin \omega t - \frac{F_0}{2\omega} t \cos \omega t. \quad \textbf{(31)}$$

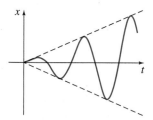

Figure 5.14

As suspected, when $t \to \infty$ the displacements become large; in fact, $|x(t_n)| \to \infty$ when $t_n = n\pi/\omega$, $n = 1, 2, \ldots$. The phenomenon we have just described is known as **pure resonance**. The graph given in Figure 5.14 shows typical motion in this case.

In conclusion it should be noted that there is no actual need to use a limiting process on (30) to obtain the solution for $\gamma = \omega$. Alternatively, equation (31) follows by solving the initial-value problem

$$\frac{d^2x}{dt^2} + \omega^2 x = F_0 \sin \omega t, \quad x(0) = 0, \quad x'(0) = 0$$

directly by conventional methods.

If the displacements of a spring/mass system were actually described by a function such as (31), the system would necessarily fail. Large oscillations of the mass would eventually force the spring beyond its elastic limit. One might argue too that the resonating model presented in Figure 5.14

is completely unrealistic, because it ignores the retarding effects of ever-present damping forces. Although it is true that pure resonance cannot occur when the smallest amount of damping is taken into consideration, large and equally destructive amplitudes of vibration (although bounded as $t \to \infty$) can occur. See Problem 43 in Exercises 5.1.

5.1.4 SERIES CIRCUIT ANALOGUE

***LRC* Series Circuits**   As mentioned in the introduction to this chapter, many different physical systems can be described by a linear second-order differential equation similar to the differential equation of forced motion with damping:

$$m\frac{d^2x}{dt^2} + \beta\frac{dx}{dt} + kx = f(t). \tag{32}$$

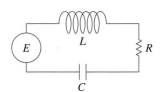

Figure 5.15

If $i(t)$ denotes current in the ***LRC* series electrical circuit** shown in Figure 5.15, then the voltage drops across the inductor, resistor, and capacitor are as shown in Figure 1.18. By Kirchhoff's second law, the sum of these voltages equals the voltage $E(t)$ impressed on the circuit; that is,

$$L\frac{di}{dt} + Ri + \frac{1}{C}q = E(t). \tag{33}$$

But the charge $q(t)$ on the capacitor is related to the current $i(t)$ by $i = dq/dt$, and so (33) becomes the linear second-order differential equation

$$L\frac{d^2q}{dt^2} + R\frac{dq}{dt} + \frac{1}{C}q = E(t). \tag{34}$$

The nomenclature used in the analysis of circuits is similar to that used to describe spring-mass systems.

If $E(t) = 0$, the **electrical vibrations** of the circuit are said to be **free.** Since the auxiliary equation for (34) is $Lm^2 + Rm + 1/C = 0$, there will be three forms of the solution with $R \neq 0$, depending on the value of the discriminant $R^2 - 4L/C$. We say that the circuit is

overdamped if	$R^2 - 4L/C > 0$,
critically damped if	$R^2 - 4L/C = 0$,

and   **underdamped** if $R^2 - 4L/C < 0$.

In each of these three cases the general solution of (34) contains the factor $e^{-Rt/2L}$, and so $q(t) \to 0$ as $t \to \infty$. In the underdamped case when $q(0) = q_0$, the charge on the capacitor oscillates as it decays; in other words, the capacitor is charging and discharging as $t \to \infty$. When $E(t) = 0$ and $R = 0$, the circuit is said to be undamped and the electrical vibrations do not approach zero as t increases without bound; the response of the circuit is **simple harmonic.**

EXAMPLE 9   Underdamped Series Circuit

Find the charge $q(t)$ on the capacitor in an *LRC* series circuit when $L = 0.25$ henry (h), $R = 10$ ohms (Ω), $C = 0.001$ farad (f), $E(t) = 0$, $q(0) = q_0$ coulombs (C), and $i(0) = 0$.

Solution   Since $1/C = 1000$, equation (34) becomes

$$\frac{1}{4}q'' + 10q' + 1000q = 0 \quad \text{or} \quad q'' + 40q' + 4000q = 0.$$

Solving this homogeneous equation in the usual manner, we find that the circuit is underdamped and $q(t) = e^{-20t}(c_1 \cos 60t + c_2 \sin 60t)$. Applying the initial conditions, we find $c_1 = q_0$ and $c_2 = \frac{1}{3}q_0$. Thus

$$q(t) = q_0 e^{-20t}\left(\cos 60t + \frac{1}{3}\sin 60t\right).$$

Using (23), we can write the foregoing solution as

$$q(t) = \frac{q_0\sqrt{10}}{3}e^{-20t}\sin(60t + 1.249). \qquad \blacksquare$$

When there is an impressed voltage $E(t)$ on the circuit, the electrical vibrations are said to be **forced.** In the case when $R \neq 0$, the complementary function $q_c(t)$ of (34) is called a **transient solution.** If $E(t)$ is periodic or a constant, then the particular solution $q_p(t)$ of (34) is a **steady-state solution.**

EXAMPLE 10   Steady-State Current

Find the steady-state solution $q_p(t)$ and the **steady-state current** in an *LRC* series circuit when the impressed voltage is $E(t) = E_0 \sin \gamma t$.

Solution   The steady-state solution $q_p(t)$ is a particular solution of the differential equation

$$L\frac{d^2q}{dt^2} + R\frac{dq}{dt} + \frac{1}{C}q = E_0 \sin \gamma t.$$

Using the method of undetermined coefficients, we assume a particular solution of the form $q_p(t) = A \sin \gamma t + B \cos \gamma t$. Substituting this expression into the differential equation, simplifying, and equating coefficients gives

$$A = \frac{E_0\left(L\gamma - \dfrac{1}{C\gamma}\right)}{-\gamma\left(L^2\gamma^2 - \dfrac{2L}{C} + \dfrac{1}{C^2\gamma^2} + R^2\right)}, \quad B = \frac{E_0 R}{-\gamma\left(L^2\gamma^2 - \dfrac{2L}{C} + \dfrac{1}{C^2\gamma^2} + R^2\right)}.$$

It is convenient to express A and B in terms of some new symbols.

If $\qquad X = L\gamma - \dfrac{1}{C\gamma}$, $\quad$ then $\quad X^2 = L^2\gamma^2 - \dfrac{2L}{C} + \dfrac{1}{C^2\gamma^2}$.

If $\qquad Z = \sqrt{X^2 + R^2}$, $\quad$ then $\quad Z^2 = L^2\gamma^2 - \dfrac{2L}{C} + \dfrac{1}{C^2\gamma^2} + R^2$.

Therefore $A = E_0 X/(-\gamma Z^2)$ and $B = E_0 R/(-\gamma Z^2)$, so the steady-state charge is

$$q_p(t) = -\frac{E_0 X}{\gamma Z^2}\sin \gamma t - \frac{E_0 R}{\gamma Z^2}\cos \gamma t.$$

Now the steady-state current is given by $i_p(t) = q_p'(t)$:

$$i_p(t) = \frac{E_0}{Z}\left(\frac{R}{Z}\sin \gamma t - \frac{X}{Z}\cos \gamma t\right). \qquad \textbf{(35)} \quad \blacksquare$$

The quantities $X = L\gamma - 1/C\gamma$ and $Z = \sqrt{X^2 + R^2}$ defined in Example 11 are called, respectively, the **reactance** and **impedance** of the circuit. Both the reactance and the impedance are measured in ohms.

EXERCISES 5.1

Answers to odd-numbered problems begin on page AN-5.

5.1.1   Spring/Mass Systems: Free Undamped Motion

1. A 4-pound weight is attached to a spring whose spring constant is 16 lb/ft. What is the period of simple harmonic motion?

2. A 20-kilogram mass is attached to a spring. If the frequency of simple harmonic motion is $2/\pi$ vibrations/second, what is the spring constant k? What is the frequency of simple harmonic motion if the original mass is replaced with an 80-kilogram mass?

3. A 24-pound weight, attached to the end of a spring, stretches it 4 inches. Find the equation of motion if the weight is released from rest from a point 3 inches above the equilibrium position.

4. Determine the equation of motion if the weight in Problem 3 is released from the equilibrium position with an initial downward velocity of 2 ft/s.

5. A 20-pound weight stretches a spring 6 inches. The weight is released from rest 6 inches below the equilibrium position.
   (a) Find the position of the weight at $t = \pi/12$, $\pi/8$, $\pi/6$, $\pi/4$, and $9\pi/32$ s.
   (b) What is the velocity of the weight when $t = 3\pi/16$ s? In which direction is the weight heading at this instant?
   (c) At what times does the weight pass through the equilibrium position?

6. A force of 400 newtons stretches a spring 2 meters. A mass of 50 kilograms is attached to the end of the spring and released from the equilibrium position with an upward velocity of 10 m/s. Find the equation of motion.

7. Another spring whose constant is 20 N/m is suspended from the same rigid support but parallel to the spring/mass system in Problem 6. A mass of 20 kilograms is attached to the second spring, and both masses are released from the equilibrium position with an upward velocity of 10 m/s.
   (a) Which mass exhibits the greater amplitude of motion?

(b) Which mass is moving faster at $t = \pi/4$ s? at $\pi/2$ s?

(c) At what times are the two masses in the same position? Where are the masses at these times? In which directions are they moving?

8. A 32-pound weight stretches a spring 2 feet. Determine the amplitude and period of motion if the weight is released 1 foot above the equilibrium position with an initial upward velocity of 2 ft/s. How many complete vibrations will the weight have completed at the end of 4π seconds?

9. An 8-pound weight attached to a spring exhibits simple harmonic motion. Determine the equation of motion if the spring constant is 1 lb/ft and if the weight is released 6 inches below the equilibrium position with a downward velocity of $\frac{3}{2}$ ft/s. Express the solution in form (6).

10. A mass weighing 10 pounds stretches a spring $\frac{1}{4}$ foot. This mass is removed and replaced with a mass of 1.6 slugs, which is released $\frac{1}{3}$ foot above the equilibrium position with a downward velocity of $\frac{5}{4}$ ft/s. Express the solution in form (6). At what times does the mass attain a displacement below the equilibrium position numerically equal to $\frac{1}{2}$ the amplitude?

11. A 64-pound weight attached to the end of a spring stretches it 0.32 foot. From a position 8 inches above the equilibrium position the weight is given a downward velocity of 5 ft/s.

(a) Find the equation of motion.

(b) What are the amplitude and period of motion?

(c) How many complete vibrations will the weight have completed at the end of 3π seconds?

(d) At what time does the weight pass through the equilibrium position heading downward for the second time?

(e) At what time does the weight attain its extreme displacement on either side of the equilibrium position?

(f) What is the position of the weight at $t = 3$ s?

(g) What is the instantaneous velocity at $t = 3$ s?

(h) What is the acceleration at $t = 3$ s?

(i) What is the instantaneous velocity at the times when the weight passes through the equilibrium position?

(j) At what times is the weight 5 inches below the equilibrium position?

(k) At what times is the weight 5 inches below the equilibrium position heading in the upward direction?

12. A mass of 1 slug is suspended from a spring whose characteristic spring constant is 9 lb/ft. Initially the mass starts from a point 1 foot above the equilibrium position with an upward velocity of $\sqrt{3}$ ft/s. Find the times for which the mass is heading downward at a velocity of 3 ft/s.

13. Under some circumstances when two parallel springs, with constants k_1 and k_2, support a single weight W, the **effective spring constant** of the system is given by $k = 4k_1k_2/(k_1 + k_2)$. A 20-pound weight stretches one spring 6 inches and another spring 2 inches. The springs are attached to a common rigid support and then to a metal plate. As

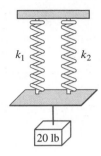

Figure 5.16

shown in Figure 5.16, the 20-pound weight is attached to the center of the plate in the double-spring arrangement. Determine the effective spring constant of this system. Find the equation of motion if the weight is released from the equilibrium position with a downward velocity of 2 ft/s.

14. A certain weight stretches one spring $\frac{1}{3}$ foot and another spring $\frac{1}{2}$ foot. The two springs are attached to a common rigid support in the manner indicated in Problem 13 and Figure 5.16. The first weight is set aside, an 8-pound weight is attached to the double-spring arrangement, and the system is set in motion. If the period of motion is $\pi/15$ second, determine the numerical value of the first weight.

15. By inspection of the differential equation only, discuss the behavior of a spring/mass system described by $4x'' + e^{-0.1t}x = 0$ over a long period of time.

16. By inspection of the differential equation only, discuss the behavior of a spring/mass system described by $4x'' + tx = 0$ over a long period of time.

5.1.2 Spring/Mass Systems: Free Damped Motion

In Problems 17–20 the given figure represents the graph of an equation of motion for a mass on a spring. The spring/mass system is damped. Use the graph to determine

(a) whether the initial displacement of the mass is above or below the equilibrium position and

(b) whether the mass is initially released from rest, heading downward, or heading upward.

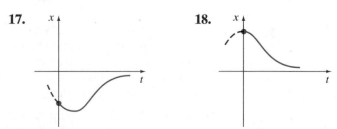

17. **18.**

Figure 5.17 **Figure 5.18**

19. **20.**

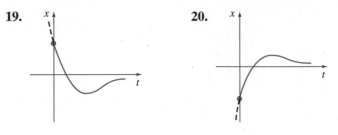

Figure 5.19   Figure 5.20

21. A 4-pound weight is attached to a spring whose constant is 2 lb/ft. The medium offers a resistance to the motion of the weight numerically equal to the instantaneous velocity. If the weight is released from a point 1 foot above the equilibrium position with a downward velocity of 8 ft/s, determine the time at which the weight passes through the equilibrium position. Find the time at which the weight attains its extreme displacement from the equilibrium position. What is the position of the weight at this instant?

22. A 4-foot spring measures 8 feet long after an 8-pound weight is attached to it. The medium through which the weight moves offers a resistance numerically equal to $\sqrt{2}$ times the instantaneous velocity. Find the equation of motion if the weight is released from the equilibrium position with a downward velocity of 5 ft/s. Find the time at which the weight attains its extreme displacement from the equilibrium position. What is the position of the weight at this instant?

23. A 1-kilogram mass is attached to a spring whose constant is 16 N/m, and the entire system is then submerged in a liquid that imparts a damping force numerically equal to 10 times the instantaneous velocity. Determine the equations of motion if
  (a) the weight is released from rest 1 meter below the equilibrium position and
  (b) the weight is released 1 meter below the equilibrium position with an upward velocity of 12 m/s.

24. In parts (a) and (b) of Problem 23 determine whether the weight passes through the equilibrium position. In each case find the time at which the weight attains its extreme displacement from the equilibrium position. What is the position of the weight at this instant?

25. A force of 2 pounds stretches a spring 1 foot. A 3.2-pound weight is attached to the spring, and the system is then immersed in a medium that imparts a damping force numerically equal to 0.4 times the instantaneous velocity.
  (a) Find the equation of motion if the weight is released from rest 1 foot above the equilibrium position.
  (b) Express the equation of motion in the form given in (23).
  (c) Find the first time at which the weight passes through the equilibrium position heading upward.

26. After a 10-pound weight is attached to a 5-foot spring, the spring measures 7 feet long. The 10-pound weight is removed and replaced with an 8-pound weight, and the entire system is placed in a medium offering a resistance numerically equal to the instantaneous velocity.

(a) Find the equation of motion if the weight is released $\frac{1}{2}$ foot below the equilibrium position with a downward velocity of 1 ft/s.

(b) Express the equation of motion in the form given in (23).

(c) Find the times at which the weight passes through the equilibrium position heading downward.

(d) Graph the equation of motion.

27. A 10-pound weight attached to a spring stretches it 2 feet. The weight is attached to a dashpot damping device that offers a resistance numerically equal to β ($\beta > 0$) times the instantaneous velocity. Determine the values of the damping constant β so that the subsequent motion is (a) overdamped, (b) critically damped, and (c) underdamped.

28. A 24-pound weight stretches a spring 4 feet. The subsequent motion takes place in a medium offering a resistance numerically equal to β ($\beta > 0$) times the instantaneous velocity. If the weight starts from the equilibrium position with an upward velocity of 2 ft/s, show that if $\beta > 3\sqrt{2}$ the equation of motion is

$$x(t) = \frac{-3}{\sqrt{\beta^2 - 18}} e^{-2\beta t/3} \sinh \frac{2}{3} \sqrt{\beta^2 - 18} t.$$

5.1.3   Spring/Mass Systems: Driven Motion

29. A 16-pound weight stretches a spring $\frac{8}{3}$ feet. Initially the weight starts from rest 2 feet below the equilibrium position, and the subsequent motion takes place in a medium that offers a damping force numerically equal to $\frac{1}{2}$ the instantaneous velocity. Find the equation of motion if the weight is driven by an external force equal to $f(t) = 10 \cos 3t$.

30. A mass of 1 slug is attached to a spring whose constant is 5 lb/ft. Initially the mass is released 1 foot below the equilibrium position with a downward velocity of 5 ft/s, and the subsequent motion takes place in a medium that offers a damping force numerically equal to 2 times the instantaneous velocity.

(a) Find the equation of motion if the mass is driven by an external force equal to $f(t) = 12 \cos 2t + 3 \sin 2t$.

(b) Graph the transient and steady-state solutions on the same coordinate axes.

(c) Graph the equation of motion.

31. A mass of 1 slug, when attached to a spring, stretches it 2 feet and then comes to rest in the equilibrium position. Starting at $t = 0$, an external force equal to $f(t) = 8 \sin 4t$ is applied to the system. Find the equation of motion if the surrounding medium offers a damping force numerically equal to 8 times the instantaneous velocity.

32. In Problem 31 determine the equation of motion if the external force is $f(t) = e^{-t} \sin 4t$. Analyze the displacements for $t \to \infty$.

33. When a mass of 2 kilograms is attached to a spring whose constant is 32 N/m, it comes to rest in the equilibrium position. Starting at $t = 0$, a force equal to $f(t) = 68e^{-2t} \cos 4t$ is applied to the system. Find the equation of motion in the absence of damping.

34. In Problem 33 write the equation of motion in the form $x(t) = A \sin(\omega t + \phi) + Be^{-2t} \sin(4t + \theta)$. What is the amplitude of vibrations after a very long time?

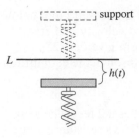

Figure 5.21

35. A mass m is attached to the end of a spring whose constant is k. After the mass reaches equilibrium, its support begins to oscillate vertically about a horizontal line L according to a formula $h(t)$. The value of h represents the distance in feet measured from L. See Figure 5.21.
   (a) Determine the differential equation of motion if the entire system moves through a medium offering a damping force numerically equal to $\beta(dx/dt)$.
   (b) Solve the differential equation in part (a) if the spring is stretched 4 feet by a weight of 16 pounds and $\beta = 2$, $h(t) = 5 \cos t$, $x(0) = x'(0) = 0$.

36. A mass of 100 grams is attached to a spring whose constant is 1600 dynes/cm. After the mass reaches equilibrium, its support oscillates according to the formula $h(t) = \sin 8t$, where h represents displacement from its original position. See Problem 35 and Figure 5.21.
   (a) In the absence of damping, determine the equation of motion if the mass starts from rest from the equilibrium position.
   (b) At what times does the mass pass through the equilibrium position?
   (c) At what times does the mass attain its extreme displacements?
   (d) What are the maximum and minimum displacements?
   (e) Graph the equation of motion.

In Problems 37 and 38 solve the given initial-value problem.

37. $\dfrac{d^2x}{dt^2} + 4x = -5 \sin 2t + 3 \cos 2t$, $x(0) = -1$, $x'(0) = 1$

38. $\dfrac{d^2x}{dt^2} + 9x = 5 \sin 3t$, $x(0) = 2$, $x'(0) = 0$

39. (a) Show that the solution of the initial-value problem

$$\frac{d^2x}{dt^2} + \omega^2 x = F_0 \cos \gamma t, \quad x(0) = 0, \quad x'(0) = 0$$

is $x(t) = \dfrac{F_0}{\omega^2 - \gamma^2}(\cos \gamma t - \cos \omega t)$.

   (b) Evaluate $\lim\limits_{\gamma \to \omega} \dfrac{F_0}{\omega^2 - \gamma^2}(\cos \gamma t - \cos \omega t)$.

40. Compare the result obtained in part (b) of Problem 39 with the solution obtained using variation of parameters when the external force is $F_0 \cos \omega t$.

41. (a) Show that $x(t)$ given in part (a) of Problem 39 can be written in the form

$$x(t) = \frac{-2F_0}{\omega^2 - \gamma^2} \sin \frac{1}{2}(\gamma - \omega)t \sin \frac{1}{2}(\gamma + \omega)t.$$

   (b) If we define $\varepsilon = \frac{1}{2}(\gamma - \omega)$, show that when ε is small an approximate solution is

$$x(t) = \frac{F_0}{2\varepsilon\gamma} \sin \varepsilon t \sin \gamma t.$$

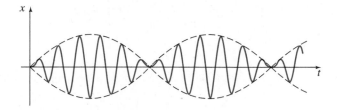

Figure 5.22

When ε is small, the frequency $\gamma/2\pi$ of the impressed force is close to the frequency $\omega/2\pi$ of free vibrations. When this occurs, the motion is as indicated in Figure 5.22. Oscillations of this kind are called **beats** and are due to the fact that the frequency of $\sin \varepsilon t$ is quite small in comparison to the frequency of $\sin \gamma t$. The dashed curves, or envelope of the graph of $x(t)$, are obtained from the graphs of $\pm(F_0/2\varepsilon\gamma) \sin \varepsilon t$. Use a graphing utility with various values of F_0, ε, and γ to verify the graph in Figure 5.22.

Computer Lab Assignments

42. Can there be beats when a damping force is added to the model in part (a) of Problem 39? Defend your position with graphs obtained either from the explicit solution of the problem

$$\frac{d^2x}{dt^2} + 2\lambda \frac{dx}{dt} + \omega^2 x = F_0 \cos \gamma t, \quad x(0) = 0, \quad x'(0) = 0$$

or from solution curves obtained using a numerical solver.

43. (a) Show that the general solution of

$$\frac{d^2x}{dt^2} + 2\lambda \frac{dx}{dt} + \omega^2 x = F_0 \sin \gamma t$$

is

$$x(t) = Ae^{-\lambda t} \sin(\sqrt{\omega^2 - \lambda^2}\,t + \phi) + \frac{F_0}{\sqrt{(\omega^2 - \gamma^2)^2 + 4\lambda^2\gamma^2}} \sin(\gamma t + \theta),$$

where $A = \sqrt{c_1^2 + c_2^2}$ and the phase angles ϕ and θ are, respectively, defined by $\sin \phi = c_1/A$, $\cos \phi = c_2/A$ and

$$\sin \theta = \frac{-2\lambda\gamma}{\sqrt{(\omega^2 - \gamma^2)^2 + 4\lambda^2\gamma^2}}, \quad \cos \theta = \frac{\omega^2 - \gamma^2}{\sqrt{(\omega^2 - \gamma^2)^2 + 4\lambda^2\gamma^2}}.$$

(b) The solution in part (a) has the form $x(t) = x_c(t) + x_p(t)$. Inspection shows that $x_c(t)$ is transient, and hence, for large values of time, the solution is approximated by $x_p(t) = g(\gamma) \sin(\gamma t + \theta)$, where

$$g(\gamma) = \frac{F_0}{\sqrt{(\omega^2 - \gamma^2)^2 + 4\lambda^2\gamma^2}}.$$

Although the amplitude $g(\gamma)$ of $x_p(t)$ is bounded as $t \to \infty$, show that the maximum oscillations will occur at the value $\gamma_1 = \sqrt{\omega^2 - 2\lambda^2}$. What is the maximum value of g? The number $\sqrt{\omega^2 - 2\lambda^2}/2\pi$ is said to be the **resonance frequency** of the system.

(c) When $F_0 = 2$, $m = 1$, and $k = 4$, g becomes

$$g(\gamma) = \frac{2}{\sqrt{(4 - \gamma^2)^2 + \beta^2\gamma^2}}.$$

Construct a table of the values of γ_1 and $g(\gamma_1)$ corresponding to the damping coefficients $\beta = 2$, $\beta = 1$, $\beta = \frac{3}{4}$, $\beta = \frac{1}{2}$, and $\beta = \frac{1}{4}$. Use a graphing utility to obtain the graphs of g corresponding to these damping coefficients. Use the same coordinate axes. This family of graphs is called the **resonance curve** or **frequency response curve** of the system. What is γ_1 approaching as $\beta \to 0$? What is happening to the resonance curve as $\beta \to 0$?

44. Consider a driven undamped spring/mass system described by the initial-value problem

$$\frac{d^2x}{dt^2} + \omega^2 x = F_0 \sin^n \gamma t, \quad x(0) = 0, \quad x'(0) = 0.$$

(a) For $n = 2$, discuss why there is a single frequency $\gamma_1/2\pi$ at which the system is in pure resonance.

(b) For $n = 3$, discuss why there are two frequencies $\gamma_1/2\pi$ and $\gamma_2/2\pi$ at which the system is in pure resonance.

(c) Suppose $\omega = 1$ and $F_0 = 1$. Use a numerical solver to obtain the graph of the solution of the initial-value problem for $n = 2$ and $\gamma = \gamma_1$ in part (a). Obtain the graph of the solution of the initial-value problem for $n = 3$ corresponding, in turn, to $\gamma = \gamma_1$ and $\gamma = \gamma_2$ in part (b).

5.1.4   Series Circuit Analogue

45. Find the charge on the capacitor in an LRC series circuit at $t = 0.01$ s when $L = 0.05$ h, $R = 2$ Ω, $C = 0.01$ f, $E(t) = 0$ V, $q(0) = 5$ C, and $i(0) = 0$ A. Determine the first time at which the charge on the capacitor is equal to zero.

46. Find the charge on the capacitor in an LRC series circuit when $L = \frac{1}{4}$ h, $R = 20$ Ω, $C = \frac{1}{300}$ f, $E(t) = 0$ V, $q(0) = 4$ C, and $i(0) = 0$ A. Is the charge on the capacitor ever equal to zero?

In Problems 47 and 48 find the charge on the capacitor and the current in the given LRC series circuit. Find the maximum charge on the capacitor.

47. $L = \frac{5}{3}$ h, $R = 10$ Ω, $C = \frac{1}{30}$ f, $E(t) = 300$ V, $q(0) = 0$ C, $i(0) = 0$ A

48. $L = 1$ h, $R = 100$ Ω, $C = 0.0004$ f, $E(t) = 30$ V, $q(0) = 0$ C, $i(0) = 2$ A

49. Find the steady-state charge and the steady-state current in an LRC series circuit when $L = 1$ h, $R = 2$ Ω, $C = 0.25$ f, and $E(t) = 50 \cos t$ V.

50. Show that the amplitude of the steady-state current in the LRC series circuit in Example 10 is given by E_0/Z, where Z is the impedance of the circuit.

51. Show that the steady-state current in an LRC series circuit when $L = \frac{1}{2}$ h, $R = 20$ Ω, $C = 0.001$ f, and $E(t) = 100 \sin 60t$ V is given by $i_p(t) = 4.160 \sin(60t - 0.588)$. [*Hint:* Use Problem 50.]

52. Find the steady-state current in an *LRC* series circuit when $L = \frac{1}{2}$ h, $R = 20\ \Omega$, $C = 0.001$ f, and $E(t) = 100 \sin 60t + 200 \cos 40t$ V.

53. Find the charge on the capacitor in an *LRC* series circuit when $L = \frac{1}{2}$ h, $R = 10\ \Omega$, $C = 0.01$ f, $E(t) = 150$ V, $q(0) = 1$ C, and $i(0) = 0$ A. What is the charge on the capacitor after a long time?

54. Show that if L, R, C, and E_0 are constant, then the amplitude of the steady-state current in Example 10 is a maximum when $\gamma = 1/\sqrt{LC}$. What is the maximum amplitude?

55. Show that if L, R, E_0, and γ are constant, then the amplitude of the steady-state current in Example 10 is a maximum when the capacitance is $C = 1/L\gamma^2$.

56. Find the charge on the capacitor and the current in an *LC* circuit when $L = 0.1$ h, $C = 0.1$ f, $E(t) = 100 \sin \gamma t$ V, $q(0) = 0$ C, and $i(0) = 0$ A.

57. Find the charge on the capacitor and the current in an *LC* circuit when $E(t) = E_0 \cos \gamma t$ V, $q(0) = q_0$ C, and $i(0) = i_0$ A.

58. In Problem 57 find the current when the circuit is in resonance.

5.2 LINEAR EQUATIONS: BOUNDARY-VALUE PROBLEMS

- *DE for the deflection of a beam* • *Boundary conditions*
- *Eigenvalues and eigenfunctions* • *Nontrivial solutions* • *Buckling of a thin column* • *Euler load* • *DE of a rotating string*

The preceding section was devoted to systems in which a second-order mathematical model was accompanied by prescribed initial conditions—that is, side conditions specified on the unknown function and its first derivative at a single point. But often the mathematical description of a physical system demands that we solve a differential equation subject to boundary conditions—that is, conditions specified on the unknown function, or on one of its derivatives, or even on a linear combination of the unknown function and one of its derivatives at two (or more) different points.

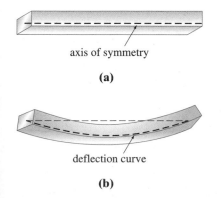

axis of symmetry

(a)

deflection curve

(b)

Figure 5.23

Deflection of a Beam Many structures are constructed using girders, or beams, and these beams deflect or distort under their own weight or under the influence of some external force. As we shall now see, this deflection $y(x)$ is governed by a relatively simple linear fourth-order differential equation.

To begin, let us assume that a beam of length L is homogeneous and has uniform cross-sections along its length. In the absence of any load on the beam (including its weight), a curve joining the centroids of all its cross-sections is a straight line called the **axis of symmetry.** See Figure 5.23(a). If a load is applied to the beam in a vertical plane containing the axis of symmetry, the beam, as shown in Figure 5.23(b), undergoes a distortion, and the curve connecting the centroids of all cross-sections is called the **deflection curve** or **elastic curve.** The deflection curve approximates the shape of the beam. Now suppose that the x-axis coincides with the axis of symmetry and that the deflection $y(x)$, measured from this axis, is positive if downward. In the theory of elasticity it is shown that the bending moment $M(x)$ at a point x along the beam is related to the load per unit length $w(x)$ by the equation

$$\frac{d^2M}{dx^2} = w(x). \tag{1}$$

In addition, the bending moment $M(x)$ is proportional to the curvature κ of the elastic curve

$$M(x) = EI\kappa, \tag{2}$$

where E and I are constants; E is Young's modulus of elasticity of the material of the beam, and I is the moment of inertia of a cross-section of the beam (about an axis known as the neutral axis). The product EI is called the **flexural rigidity** of the beam.

Now, from calculus, curvature is given by $\kappa = y''/[1 + (y')^2]^{3/2}$. When the deflection $y(x)$ is small, the slope $y' \approx 0$ and so $[1 + (y')^2]^{3/2} \approx 1$. If we let $\kappa = y''$, equation (2) becomes $M = EIy''$. The second derivative of this last expression is

$$\frac{d^2M}{dx^2} = EI\frac{d^2}{dx^2}y'' = EI\frac{d^4y}{dx^4}. \tag{3}$$

Using the given result in (1) to replace d^2M/dx^2 in (3), we see that the deflection $y(x)$ satisfies the first-order differential equation

$$EI\frac{d^4y}{dx^4} = w(x). \tag{4}$$

Boundary conditions associated with equation (4) depend on how the ends of the beam are supported. A cantilever beam is **embedded** or **clamped** at one end and **free** at the other. A diving board, an outstretched arm, an airplane wing, and a balcony are common examples of such beams, but even trees, flagpoles, skyscrapers, and the George Washington monument can act as cantilever beams because they are embedded at one end and are subject to the bending force of the wind. For a cantilever beam the deflection $y(x)$ must satisfy the following two conditions at the embedded end $x = 0$:

- $y(0) = 0$ since there is no deflection, and
- $y'(0) = 0$ since the deflection curve is tangent to the x-axis (in other words, the slope of the deflection curve is zero at this point).

At $x = L$ the free-end conditions are

- $y''(L) = 0$ since the bending moment is zero, and
- $y'''(L) = 0$ since the shear force is zero.

The function $F(x) = dM/dx = EI\, d^3y/dx^3$ is called the shear force. If an end of a beam is **simply supported** (also called **pin supported, fulcrum**

$x = 0$ $x = L$

(a) embedded at both ends

$x = 0$ $x = L$

(b) cantilever beam: embedded at the left end, free at the right end

$x = 0$ $x = L$

(c) simply supported at both ends

Figure 5.24

supported, and **hinged**), then we must have $y = 0$ and $y'' = 0$ at that end. See Figure 5.24. Table 5.1 summarizes the boundary conditions that are associated with (4).

TABLE 5.1

Ends of the Beam	Boundary Conditions
embedded	$y = 0, \ y' = 0$
free	$y'' = 0, \ y''' = 0$
simply supported	$y = 0, \ y'' = 0$

EXAMPLE 1 **An Embedded Beam**

A beam of length L is embedded at both ends. Find the deflection of the beam if a constant load w_0 is uniformly distributed along its length—that is, $w(x) = w_0, \ 0 < x < L$.

Solution From the preceding discussion we see that the deflection $y(x)$ satisfies

$$EI \frac{d^4 y}{dx^4} = w_0.$$

Because the beam is embedded at both its left end ($x = 0$) and its right end ($x = L$), there is no vertical deflection and the line of deflection is horizontal at these points. Thus the boundary conditions are

$$y(0) = 0, \quad y'(0) = 0, \quad \text{and} \quad y(L) = 0, \quad y'(L) = 0.$$

We can solve the nonhomogeneous differential equation in the usual manner (find y_c by observing that $m = 0$ is a root of multiplicity four of the auxiliary equation $m^4 = 0$ and then find a particular solution y_p by undetermined coefficients), or we can simply integrate the equation $d^4 y/dx^4 = w_0/EI$ four times in succession. Either way, we find the general solution of the equation to be

$$y(x) = c_1 + c_2 x + c_3 x^2 + c_4 x^3 + \frac{w_0}{24 EI} x^4.$$

Now the conditions $y(0) = 0$ and $y'(0) = 0$ give, in turn, $c_1 = 0$ and $c_2 = 0$, whereas the remaining conditions $y(L) = 0$ and $y'(L) = 0$ applied to $y(x) = c_3 x^2 + c_4 x^3 + \frac{w_0}{24 EI} x^4$ yield the equations

$$c_3 L^2 + c_4 L^3 + \frac{w_0}{24 EI} L^4 = 0$$

$$2 c_3 L + 3 c_4 L^2 + \frac{w_0}{6 EI} L^3 = 0.$$

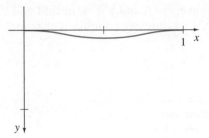

Figure 5.25

Solving this system gives $c_3 = w_0 L^2/24EI$ and $c_4 = -w_0 L/12EI$. Thus the deflection is

$$y(x) = \frac{w_0 L^2}{24EI} x^2 - \frac{w_0 L}{12EI} x^3 + \frac{w_0}{24EI} x^4 = \frac{w_0}{24EI} x^2 (x - L)^2.$$

By choosing $w_0 = 24EI$ and $L = 1$, we obtain the graph of the deflection curve in Figure 5.25. ∎

Eigenvalues and Eigenfunctions Many applied problems demand that we solve a two-point boundary-value problem involving a linear differential equation that contains a parameter λ. We seek the values of λ for which the boundary-value problem has nontrivial solutions.

EXAMPLE 2 **Nontrivial Solutions of BVP**

Solve the boundary-value problem

$$y'' + \lambda y = 0, \quad y(0) = 0, \quad y(L) = 0.$$

Solution We consider three cases: $\lambda = 0$, $\lambda < 0$, and $\lambda > 0$.

Case I. For $\lambda = 0$ the solution of $y'' = 0$ is $y = c_1 x + c_2$. The conditions $y(0) = 0$ and $y(L) = 0$ imply, in turn, $c_2 = 0$ and $c_1 = 0$. Hence for $\lambda = 0$ the only solution of the boundary-value problem is the trivial solution $y = 0$.

Case II. For $\lambda < 0$ we have $y = c_1 \cosh \sqrt{-\lambda} x + c_2 \sinh \sqrt{-\lambda} x.$* Again, $y(0) = 0$ gives $c_1 = 0$, and so $y = c_2 \sinh \sqrt{-\lambda} x$. The second condition $y(L) = 0$ dictates that $c_2 \sinh \sqrt{-\lambda} L = 0$. Since $\sinh \sqrt{-\lambda} L \neq 0$, we must have $c_2 = 0$. Thus $y = 0$.

Case III. For $\lambda > 0$ the general solution of $y'' + \lambda y = 0$ is given by $y = c_1 \cos \sqrt{\lambda} x + c_2 \sin \sqrt{\lambda} x$. As before, $y(0) = 0$ yields $c_1 = 0$, but $y(L) = 0$ implies

$$c_2 \sin \sqrt{\lambda} L = 0.$$

If $c_2 = 0$, then necessarily $y = 0$. However, if $c_2 \neq 0$, then $\sin \sqrt{\lambda} L = 0$. The last condition implies that the argument of the sine function must be an integer multiple of π:

$$\sqrt{\lambda} L = n\pi \quad \text{or} \quad \lambda = \frac{n^2 \pi^2}{L^2}, \quad n = 1, 2, 3, \ldots .$$

Therefore for any real nonzero c_2, $y = c_2 \sin(n\pi x/L)$ is a solution of the problem for each n. Since the differential equation is homogeneous, we may, if desired, not write c_2. In other words, for a given number in the sequence

$$\frac{\pi^2}{L^2}, \frac{4\pi^2}{L^2}, \frac{9\pi^2}{L^2}, \ldots ,$$

---

*$\sqrt{-\lambda}$ looks a little strange, but bear in mind that $\lambda < 0$ is equivalent to $-\lambda > 0$.

the *corresponding* function in the sequence

$$\sin\frac{\pi}{L}x, \sin\frac{2\pi}{L}x, \sin\frac{3\pi}{L}x, \ldots$$

is a nontrivial solution of the original problem.   ■

The numbers $\lambda_n = n^2\pi^2/L^2$, $n = 1, 2, 3, \ldots$ for which the boundary-value problem in Example 2 has a nontrivial solution are known as **characteristic values** or, more commonly, **eigenvalues.** The solutions depending on these values of λ_n, $y_n = c_2 \sin(n\pi x/L)$ or simply $y_n = \sin(n\pi x/L)$, are called **characteristic functions** or **eigenfunctions.**

Buckling of a Thin Vertical Column   In the eighteenth century Leonhard Euler was one of the first mathematicians to study an eigenvalue problem in analyzing how a thin elastic column buckles under a compressive axial force.

Consider a long slender vertical column of uniform cross-section and length L. Let $y(x)$ denote the deflection of the column when a constant vertical compressive force, or load, P is applied to its top, as shown in Figure 5.26. By comparing bending moments at any point along the column we obtain

$$EI\frac{d^2y}{dx^2} = -Py \quad \text{or} \quad EI\frac{d^2y}{dx^2} + Py = 0, \tag{5}$$

where E is Young's modulus of elasticity and I is the moment of inertia of a cross-section about a vertical line through its centroid.

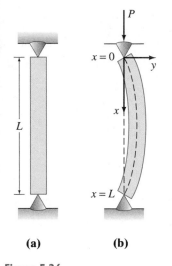

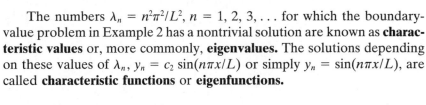

(a)   **(b)**

Figure 5.26

| EXAMPLE 3 | **The Euler Load** |

Find the deflection of a thin vertical homogeneous column of length L subjected to a constant axial load P if the column is hinged at both ends.

Solution   The boundary-value problem to be solved is

$$EI\frac{d^2y}{dx^2} + Py = 0, \quad y(0) = 0, \quad y(L) = 0.$$

First note that $y = 0$ is a perfectly good solution of this problem. This solution has a simple intuitive interpretation: If the load P is not great enough, there is no deflection. The question then is this: For what values of P will the column bend? In mathematical terms: For what values of P does the given boundary-value problem possess nontrivial solutions?

By writing $\lambda = P/EI$ we see that

$$y'' + \lambda y = 0, \quad y(0) = 0, \quad y(L) = 0$$

is identical to the problem in Example 2. From Case III of that discussion we see that the deflection curves are $y_n(x) = c_2 \sin(n\pi x/L)$, corresponding to the eigenvalues $\lambda_n = P_n/EI = n^2\pi^2/L^2$, $n = 1, 2, 3, \ldots$. Physically this means that the column will buckle or deflect only when the compressive

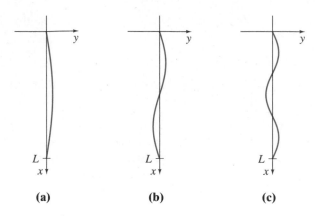

Figure 5.27

force is one of the values $P_n = n^2\pi^2 EI/L^2, n = 1, 2, 3, \ldots$. These different forces are called **critical loads.** The deflection curve corresponding to the smallest critical load $P_1 = \pi^2 EI/L^2$, called the **Euler load,** is $y_1(x) = c_2 \sin(\pi x/L)$ and is known as the **first buckling mode.** ▪

The deflection curves in Example 3 corresponding to $n = 1, n = 2$, and $n = 3$ are shown in Figure 5.27. Note that if the original column has some sort of physical restraint put on it at $x = L/2$, then the smallest critical load will be $P_2 = 4\pi^2 EI/L^2$ and the deflection curve will be as shown in Figure 5.27(b). If restraints are put on the column at $x = L/3$ and at $x = 2L/3$, then the column will not buckle until the critical load $P_3 = 9\pi^2 EI/L^2$ is applied and the deflection curve will be as shown in Figure 5.27(c). See Problem 23 in Exercises 5.2.

Rotating String   The simple linear second-order differential equation

$$y'' + \lambda y = 0 \tag{6}$$

occurs again and again as a mathematical model. In Section 5.1 we saw (6) in the forms $d^2x/dt^2 + (k/m)x = 0$ and $d^2q/dt^2 + (1/LC)q = 0$ as models for, respectively, the simple harmonic motion of a spring/mass system and the simple harmonic response of a series circuit. It is apparent when the model for the deflection of a thin column in (5) is written as $d^2y/dx^2 + (P/EI)y = 0$ that it is the same as (6). We encounter the basic equation (6) one more time in this section: as a model that defines the deflection curve or the shape $y(x)$ assumed by a rotating string. The physical situation is analogous to when two persons hold a jump rope and twirl it in a synchronous manner. See Figure 5.28, parts (a) and (b).

Suppose a string of length L with constant linear density ρ (mass per unit length) is stretched along the x-axis and fixed at $x = 0$ and $x = L$. Suppose the string is then rotated about that axis at a constant angular speed ω. Consider a portion of the string on the interval $[x, x + \Delta x]$, where Δx is small. If the magnitude T of the tension $\mathbf{T}$, acting tangential to the string, is constant along the string, then the desired differential equation can be obtained by equating two different formulations of the net force acting on the string on the interval $[x, x + \Delta x]$. First, we see

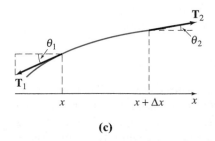

(a)

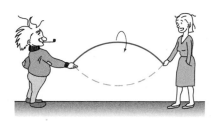

(b)

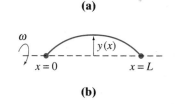

(c)

Figure 5.28

from Figure 5.28(c) that the net vertical force is

$$F = T \sin \theta_2 - T \sin \theta_1. \tag{7}$$

When angles θ_1 and θ_2 (measured in radians) are small, we have $\sin \theta_2 \approx$ $\tan \theta_2$ and $\sin \theta_1 \approx \tan \theta_1$. Moreover, since $\tan \theta_2$ and $\tan \theta_1$ are, in turn, slopes of the lines containing the vectors $\mathbf{T}_2$ and $\mathbf{T}_1$, we can also write

$$\tan \theta_2 = y'(x + \Delta x) \quad \text{and} \quad \tan \theta_1 = y'(x).$$

Thus (7) becomes

$$F \approx T[y'(x + \Delta x) - y'(x)]. \tag{8}$$

Second, we can obtain a different form of this same net force using Newton's second law, $F = ma$. Here the mass of string on the interval is $m = \rho \, \Delta x$; the centripetal acceleration of a body rotating with angular speed ω in a circle of radius r is $a = r\omega^2$. With Δx small we take $r = y$. Thus the net vertical force is also approximated by

$$F \approx -(\rho \, \Delta x) y \omega^2, \tag{9}$$

where the minus sign comes from the fact that the acceleration points in the direction opposite to the positive y-direction. Now by equating (8) and (9) we have

$$T[y'(x + \Delta x) - y'(x)] \approx -(\rho \, \Delta x) y \omega^2 \quad \text{or} \quad T \frac{y'(x + \Delta x) - y'(x)}{\Delta x} \approx -\rho \omega^2 y. \tag{10}$$

For Δx close to zero the difference quotient $[y'(x + \Delta x) - y'(x)]/\Delta x$ in (10) is approximated by the second derivative d^2y/dx^2. Finally we arrive at the model

$$T \frac{d^2y}{dx^2} = -\rho \omega^2 y \quad \text{or} \quad T \frac{d^2y}{dx^2} + \rho \omega^2 y = 0. \tag{11}$$

Since the string is anchored at its ends $x = 0$ and $y = L$, we expect that the solution $y(x)$ of the last equation in (11) should also satisfy the boundary conditions $y(0) = 0$ and $y(L) = 0$.

EXERCISES 5.2

Answers to odd-numbered problems begin on page AN-6.

Deflection of a Beam

In Problems 1–5 solve equation (4) subject to the appropriate boundary conditions. The beam is of length L, and w_0 is a constant.

1. **(a)** The beam is embedded at its left end and free at its right end and $w(x) = w_0$, $0 < x < L$.
   (b) Use a graphing utility to obtain the graph of the deflection curve of the beam when $w_0 = 24EI$ and $L = 1$.
2. **(a)** The beam is simply supported at both ends and $w(x) = w_0$, $0 < x < L$.
   (b) Use a graphing utility to obtain the graph of the deflection curve of the beam when $w_0 = 24EI$ and $L = 1$.

3. **(a)** The beam is embedded at its left end and simply supported at its right end and $w(x) = w_0, 0 < x < L$.
   (b) Use a graphing utility to obtain the graph of the deflection curve of the beam when $w_0 = 48EI$ and $L = 1$.

4. **(a)** The beam is embedded at its left end and simply supported at its right end and $w(x) = w_0 \sin(\pi x/L), 0 < x < L$.
   (b) Use a graphing utility to obtain the graph of the deflection curve of the beam when $w_0 = 2\pi^3 EI$ and $L = 1$.
   (c) Use a root-finding application of a CAS (or a graphic calculator) to approximate the point in the graph in part (b) at which the maximum deflection occurs. What is the maximum deflection?

5. **(a)** The beam is simply supported at both ends and $w(x) = w_0 x$, $0 < x < L$.
   (b) Use a graphing utility to obtain the graph of the deflection curve of the beam where $w_0 = 36EI$ and $L = 1$.
   (c) Use a root-finding application of a CAS (or a graphic calculator) to approximate the point in the graph in part (b) at which the maximum deflection occurs. What is the maximum deflection?

6. **(a)** Find the maximum deflection of the cantilever beam in Problem 1.
   (b) How does the maximum deflection of a beam that is half as long compare with the value in part (a)?
   (c) Find the maximum deflection of the simply supported beam in Problem 2.
   (d) How does the maximum deflection of the simply supported beam in part (c) compare with the value of maximum deflection of the embedded beam in Example 1?

7. A cantilever beam of length L is embedded at its right end, and a horizontal tensile force of P pounds is applied to its free left end. When the origin is taken at its free end, as shown in Figure 5.29, the deflection $y(x)$ of the beam can be shown to satisfy the differential equation

$$EIy'' = Py - w(x)\frac{x}{2}.$$

Find the deflection of the cantilever beam if $w(x) = w_0 x$, $0 < x < L$ and $y(0) = 0, y'(L) = 0$.

8. When a compressive instead of a tensile force is applied at the free end of the beam in Problem 7, the differential equation of the deflection is

$$EIy'' = -Py - w(x)\frac{x}{2}.$$

Solve this equation if $w(x) = w_0 x, 0 < x < L$ and $y(0) = 0, y'(L) = 0$.

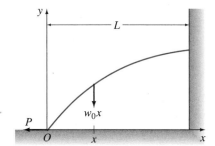

Figure 5.29

Eigenvalues and Eigenfunctions

In Problems 9–22 find the eigenvalues and eigenfunctions for the given boundary-value problem.

9. $y'' + \lambda y = 0, \quad y(0) = 0, \quad y(\pi) = 0$

10. $y'' + \lambda y = 0, \quad y(0) = 0, \quad y\left(\frac{\pi}{4}\right) = 0$

11. $y'' + \lambda y = 0$, $y'(0) = 0$, $y(L) = 0$

12. $y'' + \lambda y = 0$, $y(0) = 0$, $y'\left(\dfrac{\pi}{2}\right) = 0$

13. $y'' + \lambda y = 0$, $y'(0) = 0$, $y'(\pi) = 0$

14. $y'' + \lambda y = 0$, $y(-\pi) = 0$, $y(\pi) = 0$

15. $y'' + 2y' + (\lambda + 1)y = 0$, $y(0) = 0$, $y(5) = 0$

16. $y'' + (\lambda + 1)y = 0$, $y'(0) = 0$, $y'(1) = 0$

17. $y'' + \lambda^2 y = 0$, $y(0) = 0$, $y(L) = 0$

18. $y'' + \lambda^2 y = 0$, $y(0) = 0$, $y'(3\pi) = 0$

19. $x^2 y'' + xy' + \lambda y = 0$, $y(1) = 0$, $y(e^\pi) = 0$

20. $x^2 y'' + xy' + \lambda y = 0$, $y'(e^{-1}) = 0$, $y(1) = 0$

21. $x^2 y'' + xy' + \lambda y = 0$, $y'(1) = 0$, $y'(e^2) = 0$

22. $x^2 y'' + 2xy' + \lambda y = 0$, $y(1) = 0$, $y(e^2) = 0$

Buckling of a Thin Column

23. Consider Figure 5.27. Where should physical restraints be placed on the column if we want the critical load to be P_4? Sketch the deflection curve corresponding to this load.

24. The critical loads of thin columns depend on the end conditions of the column. The value of the Euler load P_1 in Example 3 was derived under the assumption that the column was hinged at both ends. Suppose that a thin vertical homogeneous column is embedded at its base ($x = 0$) and free at its top ($x = L$) and that an axial load P is applied to its free end. This load either causes a small deflection δ as shown in Figure 5.30 or does not cause such a deflection. In either case the differential equation for the deflection $y(x)$ is

$$EI\frac{d^2 y}{dx^2} + Py = P\delta.$$

   (a) What is the predicted deflection when $\delta = 0$?

   (b) When $\delta \neq 0$, show that the Euler load for this column is one-fourth of the Euler load for the hinged column in Example 3.

Rotating String

25. Consider the boundary-value problem introduced in the construction of the mathematical model for the shape of a rotating string:

$$T\frac{d^2 y}{dx^2} + \rho\omega^2 y = 0, \quad y(0) = 0, \quad y(L) = 0.$$

   For constant T and ρ, define the critical speeds of angular rotation ω_n as the values of ω for which the boundary-value problem has nontrivial solutions. Find the critical speeds ω_n and the corresponding deflection curves $y_n(x)$.

26. When the magnitude of tension T is not constant, then a model for the deflection curve or shape $y(x)$ assumed by a rotating string is given by

$$\frac{d}{dx}\left[T(x)\frac{dy}{dx}\right] + \rho\omega^2 y = 0.$$

   Suppose that $1 < x < e$ and $T(x) = x^2$.

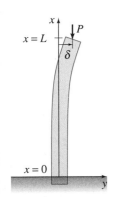

x

$x = L$

P

δ

$x = 0$

y

Figure 5.30

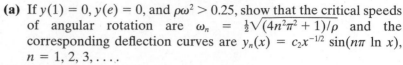

(a) If $y(1) = 0$, $y(e) = 0$, and $\rho\omega^2 > 0.25$, show that the critical speeds of angular rotation are $\omega_n = \frac{1}{2}\sqrt{(4n^2\pi^2 + 1)/\rho}$ and the corresponding deflection curves are $y_n(x) = c_2 x^{-1/2} \sin(n\pi \ln x)$, $n = 1, 2, 3, \ldots$.

(b) Use a graphing utility to graph the deflection curves on the interval $[1, e]$ for $n = 1, 2, 3$. Choose $c_2 = 1$.

Miscellaneous Boundary-Value Problems

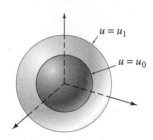

27. Consider two concentric spheres of radius $r = a$ and $r = b$, $a < b$, as shown in Figure 5.31 The temperature $u(r)$ in the region between the spheres is determined from the boundary-value problem

$$r\frac{d^2u}{dr^2} + 2\frac{du}{dr} = 0, \qquad u(a) = u_0, \quad u(b) = u_1,$$

where u_0 and u_1 are constants. Solve for $u(r)$.

28. The temperature $u(r)$ in the circular ring shown in Figure 5.32 is determined from the boundary-value problem

$$r\frac{d^2u}{dr^2} + \frac{du}{dr} = 0, \quad u(a) = u_0, \quad u(b) = u_1,$$

where u_0 and u_1 are constants. Show that

$$u(r) = \frac{u_0 \ln(r/b) - u_1 \ln(r/a)}{\ln(a/b)}.$$

Figure 5.31

Discussion Problems

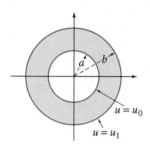

29. Consider the boundary-value problem $y'' + 16y = 0$, $y(0) = y_0$, $y(\pi/2) = y_1$. Determine whether it is possible to find values of y_0 and y_1 so that the problem possesses **(a)** precisely one nontrivial solution, **(b)** more than one solution, **(c)** no solution, **(d)** the trivial solution.

30. Consider the boundary-value problem $y'' + 16y = 0$, $y(0) = 1$, $y(L) = 1$. Determine whether it is possible to find values of $L > 0$ so that the problem possesses **(a)** precisely one nontrivial solution, **(b)** more than one solution, **(c)** no solution, **(d)** the trivial solution.

Figure 5.32

31. Consider the boundary-value problem

$$y'' + \lambda^2 y = 0, \quad y(-\pi) = y(\pi), \quad y'(-\pi) = y'(\pi).$$

(a) Discuss a geometric interpretation of the boundary conditions.

(b) Find the eigenvalues and eigenfunctions of the problem.

(c) Use a graphing utility to graph some of the eigenfunctions. Verify the geometric interpretation of the boundary conditions.

Computer Lab Assignments

32. **(a)** Show that the eigenvalues and eigenfunctions of the boundary-value problem

$$y'' + \lambda y = 0, \quad y(0) = 0, \quad y(1) + y'(1) = 0$$

are, respectively, $\lambda_n = x_n^2$, where x_n, $n = 1, 2, 3, \ldots$ are the consecutive *positive* roots of the equation $\tan\sqrt{\lambda} = -\sqrt{\lambda}$, and $y_n = \sin\sqrt{\lambda_n}\, x$.

(b) Use a graphing utility to convince yourself that the equation $\tan x = -x$ has an infinite number of roots. Explain why the negative roots of the equation can be ignored. Explain why $\lambda = 0$ is not an eigenvalue even though $x = 0$ is an obvious root of the equation.

(c) Use a root-finding application of a CAS (or a graphic calculator) to approximate the first four nonnegative roots of the equation $\tan x = -x$. Use this information to approximate the eigenvalues $\lambda_1, \lambda_2, \lambda_3$, and λ_4 of the boundary-value problem in part (a). Write down the corresponding eigenfunctions.

5.3 NONLINEAR EQUATIONS

• Linear and nonlinear springs • Hard and soft springs • Model of a nonlinear pendulum • Linearization • Model of a suspended wire • Catenary • Rocket motion • Newton's second law when the mass is variable

The mathematical model in (1) of Section 5.1 has the form

$$m\frac{d^2x}{dt^2} + F(x) = 0, \tag{1}$$

where $F(x) = kx$. Since x denotes the displacement of the mass from its equilibrium position, $F(x) = kx$ is Hooke's law—that is, the force exerted by the spring that tends to restore the mass to the equilibrium position. A spring acting under a linear restoring force $F(x) = kx$ is naturally referred to as a **linear spring.** But springs are seldom perfectly linear. Depending on how it is constructed and the material used, a spring can range from "mushy," or soft, to "stiff," or hard, so its restorative force may vary from something below to something above that given by the linear law.

Nonlinear Springs In the case of free motion, if we assume that a non-aging spring possesses some nonlinear characteristics, then it might be reasonable to assume that the restorative force of a spring—that is, $F(x)$ in (1)—is proportional to, say, the cube of the displacement x of the mass beyond its equilibrium position or that $F(x)$ is a linear combination of powers of the displacement such as that given by the nonlinear function $F(x) = kx + k_1x^3$. A spring whose mathematical model incorporates a nonlinear restorative force, such as

$$m\frac{d^2x}{dt^2} + kx^3 = 0 \quad \text{or} \quad m\frac{d^2x}{dt^2} + kx + k_1x^3 = 0, \tag{2}$$

is called a **nonlinear spring.** In addition, we examined mathematical models in which damping imparted to the motion was proportional to the instantaneous velocity dx/dt and the restoring force of a spring was given by the linear function $F(x) = kx$. But these were simply assumptions; in more realistic situations damping could be proportional to some power of the instantaneous velocity dx/dt. The nonlinear differential equation

$$m\frac{d^2x}{dt^2} + \beta\left|\frac{dx}{dt}\right|\frac{dx}{dt} + kx = 0 \tag{3}$$

is one model of a free spring/mass system with damping proportional to the square of the velocity. One can then envision other kinds of models:

linear damping and nonlinear restoring force, nonlinear damping and nonlinear restoring force, and so on. The point is that nonlinear characteristics of a physical system lead to a mathematical model that is nonlinear.

Notice in (2) that both $F(x) = kx^3$ and $F(x) = kx + k_1x^3$ are odd functions of x. To see why a polynomial function containing only odd powers of x provides a reasonable model for the restoring force, let us express F as a power series centered at the equilibrium position $x = 0$:

$$F(x) = c_0 + c_1x + c_2x^2 + c_3x^3 + \cdots.$$

When the displacements x are small, the values of x^n are negligible for n sufficiently large. If we truncate the power series with, say, the fourth term, then

$$F(x) = c_0 + c_1x + c_2x^2 + c_3x^3.$$

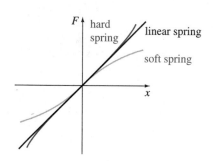

Figure 5.33

In order for the force at $x > 0$ $(F(x) = c_0 + c_1x + c_2x^2 + c_3x^3)$ and the force at $-x < 0$ $(F(-x) = c_0 - c_1x + c_2x^2 - c_3x^3)$ to have the same magnitude but act in opposite directions, we must have $F(-x) = -F(x)$. Since this means F is an odd function, we must have $c_0 = 0$ and $c_2 = 0$ and so $F(x) = c_1x + c_3x^3$. Had we used only the first two terms in the series, the same argument would yield the linear function $F(x) = c_1x$. For discussion purposes we shall write $c_1 = k$ and $c_2 = k_1$. A restoring force with mixed powers, such as $F(x) = kx + k_1x^2$, and the corresponding vibrations are said to be unsymmetrical.

Hard and Soft Springs   Let us take a closer look at the equation in (1) in the case where the restoring force is given by $F(x) = kx + k_1x^3$, $k > 0$. The spring is said to be **hard** if $k_1 > 0$ and **soft** if $k_1 < 0$. Graphs of three types of restoring forces are illustrated in Figure 5.33. The next example illustrates these two special cases of the differential equation $m\, d^2x/dt^2 + kx + k_1x^3 = 0$, $m > 0$, $k > 0$.

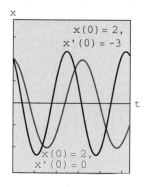

(a) hard spring

EXAMPLE 1   Comparison of Hard and Soft Springs

The differential equations

$$\frac{d^2x}{dt^2} + x + x^3 = 0 \qquad (4)$$

and

$$\frac{d^2x}{dt^2} + x - x^3 = 0 \qquad (5)$$

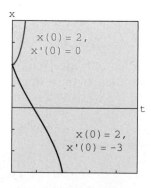

(b) soft spring

Figure 5.34

are special cases of (2) and are models of a hard spring and a soft spring, respectively. Figure 5.34(a) shows two solutions of (4) and Figure 5.34(b) shows two solutions of (5) obtained from a numerical solver. The curves shown in black are solutions satisfying the initial conditions $x(0) = 2$, $x'(0) = -3$; the two curves in color are solutions satisfying $x(0) = 2$, $x'(0) = 0$. These solution curves certainly suggest that the motion of a mass on the hard spring is oscillatory, whereas the motion of a mass on the soft spring is not oscillatory. But we must be careful about drawing conclusions based on a couple of solution curves. A more complete picture of the nature of the solutions of both of these equations can be obtained from the qualitative analysis discussed in Chapter 10 of *Differential Equations with Boundary-Value Problems*.   ■

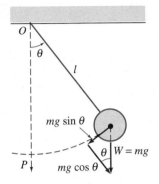

Figure 5.35

Nonlinear Pendulum Any object that swings back and forth is called a **physical pendulum.** The **simple pendulum** is a special case of the physical pendulum and consists of a rod of length l to which a mass m is attached at one end. In describing the motion of a simple pendulum in a vertical plane, we make the simplifying assumptions that the mass of the rod is negligible and that no external damping or driving forces act on the system. The displacement angle θ of the pendulum, measured from the vertical as shown in Figure 5.35, is considered positive when measured to the right of OP and negative when measured to the left of OP. Now recall that the arc s of a circle of radius l is related to the central angle θ by the formula $s = l\theta$. Hence angular acceleration is

$$a = \frac{d^2s}{dt^2} = l\frac{d^2\theta}{dt^2}.$$

From Newton's second law we then have

$$F = ma = ml\frac{d^2\theta}{dt^2}.$$

From Figure 5.35 we see that the magnitude of the tangential component of the force due to the weight W is $mg \sin\theta$. In direction this force is $-mg \sin\theta$, since it points to the left for $\theta > 0$ and to the right for $\theta < 0$. We equate the two different versions of the tangential force to obtain $ml\, d^2\theta/dt^2 = -mg \sin\theta$ or

$$\frac{d^2\theta}{dt^2} + \frac{g}{l}\sin\theta = 0. \tag{6}$$

Linearization Because of the presence of $\sin\theta$, the model in (6) is nonlinear. In an attempt to understand the behavior of the solutions of nonlinear higher-order differential equations, we sometimes try to simplify the problem by replacing nonlinear terms by certain approximations. For example, the Maclaurin series for $\sin\theta$ is given by

$$\sin\theta = \theta - \frac{\theta^3}{3!} + \frac{\theta^5}{5!} - \cdots,$$

and if we use the approximation $\sin\theta \approx \theta - \theta^3/6$, equation (6) becomes $d^2\theta/dt^2 + (g/l)\theta + (g/6l)\theta^3 = 0$. Observe that this last equation is the same as the second nonlinear equation in (2) with $m = 1$, $k = g/l$, and $k_1 = -g/6l$. However, if we assume that the displacements θ are small enough to justify using the replacement $\sin\theta \approx \theta$, then (6) becomes

$$\frac{d^2\theta}{dt^2} + \frac{g}{l}\theta = 0. \tag{7}$$

See Problem 22 in Exercises 5.3. If we set $\omega^2 = g/l$, we recognize (7) as the differential equation (2) of Section 5.1, which is a model for the free undamped vibrations of a linear spring/mass system. In other words, (7) is again the basic linear equation $y'' + \lambda y = 0$ discussed on page 240 of Section 5.2. As a consequence, we say that equation (7) is a **linearization** of equation (6). Since the general solution of (7) is $\theta(t) = c_1 \cos\omega t + c_2 \sin\omega t$, this linearization suggests that for initial conditions amenable to small oscillations the motion of the pendulum described by (6) will be periodic.

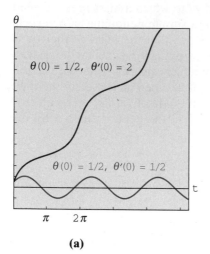

$\theta(0) = 1/2, \ \theta'(0) = 2$

$\theta(0) = 1/2, \ \theta'(0) = 1/2$

π　2π

(a)

(b) $\theta(0) = \frac{1}{2},$
$\theta'(0) = \frac{1}{2}$

(c) $\theta(0) = \frac{1}{2},$
$\theta'(0) = 2$

Figure 5.36

EXAMPLE 2　Two Initial-Value Problems

The graphs in Figure 5.36(a) were obtained with the aid of a numerical solver and represent solution curves of equation (6) when $\omega^2 = 1$. The color curve depicts the solution of (6) that satisfies the initial conditions $\theta(0) = \frac{1}{2}, \ \theta'(0) = \frac{1}{2}$, whereas the black curve is the solution of (6) that satisfies $\theta(0) = \frac{1}{2}, \ \theta'(0) = 2$. The color curve represents a periodic solution—the pendulum oscillating back and forth as shown in Figure 5.36(b) with an apparent amplitude $A \le 1$. The black curve shows that θ increases without bound as time increases—the pendulum, starting from the same initial displacement, is given an initial velocity of magnitude great enough to send it over the top; in other words, the pendulum is whirling about its pivot as shown in Figure 5.36(c). In the absence of damping, the motion in each case is continued indefinitely.　∎

Suspended Wire　Suppose that a suspended wire hangs under its own weight. As Figure 5.37(a) shows, a physical model for this could be a long telephone wire strung between two posts. Our goal is to construct a mathematical model that describes the shape that the hanging wire assumes.

To begin, let us suppose that the y-axis in Figure 5.37(b) is chosen to pass through the lowest point P_1 on the curve and that the x-axis is a units below P_1. Let us also agree to examine only a portion of the wire between the lowest point P_1 and any arbitrary point P_2. Three forces are acting on the wire: the weight of the segment P_1P_2 and the tensions $\mathbf{T}_1$ and $\mathbf{T}_2$ in the wire at P_1 and P_2, respectively. If w is the linear density of the wire (measured, say, in lb/ft) and s is the length of the segment P_1P_2, then its weight is ws. Now the tension $\mathbf{T}_2$ resolves into the horizontal and vertical components (scalar quantities) $T_2 \cos \theta$ and $T_2 \sin \theta$. Because of equilibrium we can write

$$|\mathbf{T}_1| = T_1 = T_2 \cos \theta \quad \text{and} \quad ws = T_2 \sin \theta.$$

Dividing the last two equations, we find $\tan \theta = ws/T_1$. That is,

$$\frac{dy}{dx} = \frac{ws}{T_1}. \tag{8}$$

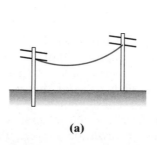

(a)

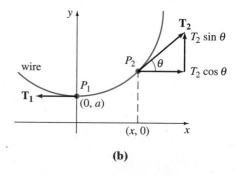

(b)

Figure 5.37

Now since the arc length between points P_1 and P_2 is given by

$$s = \int_0^x \sqrt{1 + \left(\frac{dy}{dx}\right)^2}\, dx, \tag{9}$$

it follows from the fundamental theorem of calculus that the derivative of (9) is

$$\frac{ds}{dx} = \sqrt{1 + \left(\frac{dy}{dx}\right)^2}. \tag{10}$$

Differentiating (8) with respect to x and using (10) leads to

$$\frac{d^2y}{dx^2} = \frac{w}{T_1}\frac{ds}{dx} \quad \text{or} \quad \frac{d^2y}{dx^2} = \frac{w}{T_1}\sqrt{1 + \left(\frac{dy}{dx}\right)^2}. \tag{11}$$

One might conclude from Figure 5.37 that the shape the hanging wire assumes is parabolic. The next example shows that this is not the case—the curve assumed by the suspended cable is called a **catenary.** Before proceeding, observe that the nonlinear second-order differential equation (11) is one of those equations of the form $F(x, y', y'') = 0$ discussed in Section 4.9. Recall that we have a chance of solving this type of equation by reducing the order of the equation by means of the substitution $u = y'$.

EXAMPLE 3   **An Initial-Value Problem**

From the position of the y-axis in Figure 5.37(b) it is apparent that initial conditions associated with the second differential equation in (11) are $y(0) = a$ and $y'(0) = 0$. If we substitute $u = y'$, the last equation in (11) becomes $\dfrac{du}{dx} = \dfrac{w}{T_1}\sqrt{1 + u^2}$. Separating variables, we find that

$$\int \frac{du}{\sqrt{1 + u^2}} = \frac{w}{T_1}\int dx \qquad \text{gives} \qquad \sinh^{-1} u = \frac{w}{T_1}x + c_1.$$

Now $y'(0) = 0$ is equivalent to $u(0) = 0$. Since $\sinh^{-1} 0 = 0$, $c_1 = 0$ and so $u = \sinh(wx/T_1)$. Finally, by integrating both sides of

$$\frac{dy}{dx} = \sinh \frac{w}{T_1}x \qquad \text{we get} \qquad y = \frac{T_1}{w}\cosh \frac{w}{T_1}x + c_2.$$

If we use $y(0) = a$, $\cosh 0 = 1$, the last equation implies that $c_2 = a - T_1/w$. Thus we see that the shape of the hanging wire is defined by

$$y = \frac{T_1}{w}\cosh \frac{w}{T_1}x + a - \frac{T_1}{w}. \qquad \blacksquare$$

In Example 3, had we been clever enough at the start to choose $a = T_1/w$, then the solution of the problem would have been simply the hyperbolic cosine $y = (T_1/w)\cosh(wx/T_1)$.

Rocket Motion   In Section 1.3 we saw that the differential equation of a free-falling body of mass m near the surface of the earth is given by

$$m\frac{d^2s}{dt^2} = -mg \qquad \text{or simply} \qquad \frac{d^2s}{dt^2} = -g,$$

where s represents the distance from the surface of the earth to the object and the positive direction is considered to be upward. In other words, the underlying assumption here is that the distance s to the object is small when compared with the radius R of the earth; put yet another way, the distance y from the center of the earth to the object is approximately the same as R. If, on the other hand, the distance y to an object—such as a rocket or a space probe—is large compared to R, then we combine Newton's second law of motion and his universal law of gravitation to derive a differential equation in the variable y.

Suppose a rocket is launched vertically upward from the ground as shown in Figure 5.38. If the positive direction is upward and air resistance is ignored, then the differential equation of motion after fuel burnout is

$$m\frac{d^2y}{dt^2} = -k\frac{Mm}{y^2} \quad \text{or} \quad \frac{d^2y}{dt^2} = -k\frac{M}{y^2}, \tag{12}$$

where k is a constant of proportionality, y is the distance from the center of the earth to the rocket, M is the mass of the earth, and m is the mass of the rocket. To determine the constant k we use the fact that when $y = R$, $kMm/R^2 = mg$ or $k = gR^2/M$. Thus the last equation in (12) becomes

$$\frac{d^2y}{dt^2} = -g\frac{R^2}{y^2}. \tag{13}$$

See Problem 14 in Exercises 5.3.

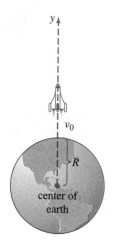

Figure 5.38

Variable Mass   In the preceding discussion we described the motion of the rocket after it has burned all its fuel, when presumably its mass m is constant. Of course during its powered ascent the total mass of the rocket varies as its fuel is being expended. The second law of motion, as originally advanced by Newton, states that when a body of mass m moves through a force field with velocity v, the time rate of change of the momentum mv of the body is equal to the applied or net force F acting on the body:

$$F = \frac{d}{dt}(mv). \tag{14}$$

If m is constant, then (14) yields the more familiar form $F = m\,dv/dt = ma$, where a is acceleration. We use the form of Newton's second law given in (14) in the next example, in which the mass m of the body is variable.

EXAMPLE 4   Chain Pulled Upward by a Constant Force

A uniform 10-foot-long chain is coiled loosely on the ground. One end of the chain is pulled vertically upward by means of a constant force of 5 pounds. The chain weighs 1 pound per foot. Determine the height of the end above ground level at time t. See Figure 1.32 and Problem 19 in Exercises 1.3.

Solution   Let us suppose that $x = x(t)$ denotes the height of the end of the chain in the air at time t, $v = dx/dt$, and the positive direction is upward. For that portion of the chain in the air at time t we have the following variable quantities:

$$\textit{weight:} \quad W = (x \text{ ft}) \cdot (1 \text{ lb/ft}) = x,$$
$$\textit{mass:} \quad m = W/g = x/32,$$
$$\textit{net force:} \quad F = 5 - W = 5 - x.$$

Thus from (14) we have

↓ product rule

$$\frac{d}{dt}\left(\frac{x}{32}v\right) = 5 - x \quad \text{or} \quad x\frac{dv}{dt} + v\frac{dx}{dt} = 160 - 32x. \qquad \textbf{(15)}$$

Since $v = dx/dt$, the last equation becomes

$$x\frac{d^2x}{dt^2} + \left(\frac{dx}{dt}\right)^2 + 32x = 160. \qquad \textbf{(16)}$$

The nonlinear second-order differential equation (16) has the form $F(x, x', x'') = 0$, which is the second of the two forms considered in Section 4.9 that can possibly be solved by reduction of order. In order to solve (16) we revert back to (15) and use $v = x'$ along with the Chain Rule. From $\dfrac{dv}{dt} = \dfrac{dv}{dx}\dfrac{dx}{dt} = v\dfrac{dv}{dx}$ the second equation in (15) can be rewritten as

$$xv\frac{dv}{dx} + v^2 = 160 - 32x. \qquad \textbf{(17)}$$

On inspection, (17) might appear as intractable as (16) since it cannot be characterized as any of the first-order equations that were solved in Chapter 2. However, by rewriting (17) in the differential form $M(x, v)dx + N(x, v)dv = 0$ we observe that, although the equation

$$(v^2 + 32x - 160) \, dx + xv \, dv = 0 \qquad \textbf{(18)}$$

is not exact, it can be transformed into an exact equation by means of an integrating factor. From $(M_v - N_x)/N = 1/x$ we see from (13) of Section 2.4 that an integrating factor is $e^{\int dx/x} = e^{\ln x} = x$. When (18) is multiplied by $\mu(x) = x$, the resulting equation is exact. (Verify.) By identifying $\partial f/\partial x = xv^2 + 32x^2 - 160x$, $\partial f/\partial v = x^2v$ and proceeding as in Section 2.4, we obtain

$$\frac{1}{2}x^2v^2 + \frac{32}{3}x^3 - 80x^2 = c_1. \qquad \textbf{(19)}$$

Since we have assumed that all of the chain is on the floor initially, we have $x(0) = 0$. This last condition applied to (19) yields $c_1 = 0$. By solving the algebraic equation $\frac{1}{2}x^2v^2 + \frac{32}{3}x^3 - 80x^2 = 0$ for $v = dx/dt > 0$, we get another first-order differential equation,

$$\frac{dx}{dt} = \sqrt{160 - \frac{64}{3}x}.$$

The last equation can be solved by separation of variables. You should verify that

$$-\frac{3}{32}\left(160 - \frac{64}{3}x\right)^{1/2} = t + c_2. \qquad \textbf{(20)}$$

This time the initial condition $x(0) = 0$ implies $c_2 = -3\sqrt{10}/8$. Finally, by squaring both sides of (20) and solving for x, we arrive at the desired result

$$x(t) = \frac{15}{2} - \frac{15}{2}\left(1 - \frac{4\sqrt{10}}{15}t\right)^2. \qquad (21) \quad \blacksquare$$

See Problem 15 in Exercises 5.3.

EXERCISES 5.3

Answers to odd-numbered problems begin on page AN-6.

Nonlinear Springs

In Problems 1–4 the given differential equation is a model of an undamped spring/mass system in which the restoring force $F(x)$ in (1) is nonlinear. For each equation use a numerical solver to plot the solution curves satisfying the given initial conditions. If the solutions appear to be periodic, use the solution curve to estimate the period T of oscillations.

1. $\dfrac{d^2x}{dt^2} + x^3 = 0$, $x(0) = 1$, $x'(0) = 1$; $x(0) = \frac{1}{2}$, $x'(0) = -1$

2. $\dfrac{d^2x}{dt^2} + 4x - 16x^3 = 0$, $x(0) = 1$, $x'(0) = 1$; $x(0) = -2$, $x'(0) = 2$

3. $\dfrac{d^2x}{dt^2} + 2x - x^2 = 0$, $x(0) = 1$, $x'(0) = 1$; $x(0) = \frac{3}{2}$, $x'(0) = -1$

4. $\dfrac{d^2x}{dt^2} + xe^{0.01x} = 0$, $x(0) = 1$, $x'(0) = 1$; $x(0) = 3$, $x'(0) = -1$

5. In Problem 3, suppose the mass is released from the initial position $x(0) = 1$ with an initial velocity $x'(0) = x_1$. Use a numerical solver to estimate the smallest value of $|x_1|$ at which the motion of the mass is nonperiodic.

6. In Problem 3, suppose the mass is released from an initial position $x(0) = x_0$ with the initial velocity $x'(0) = 1$. Use a numerical solver to estimate an interval $a \le x_0 \le b$ for which the motion is oscillatory.

7. Find a linearization of the differential equation in Problem 4.

8. Consider the model of an undamped nonlinear spring/mass system given by

$$\frac{d^2x}{dt^2} + 8x - 6x^3 + x^5 = 0.$$

Use a numerical solver to discuss the nature of the oscillations of the system corresponding to the initial conditions:

$x(0) = 1, x'(0) = 1$; $x(0) = -2, x'(0) = 0.5$; $x(0) = \sqrt{2}, x'(0) = 1$;
$x(0) = 2, x'(0) = 0.5$; $x(0) = 2, x'(0) = 0$; $x(0) = -\sqrt{2}, x'(0) = -1$.

In Problems 9 and 10 the given differential equation is a model of a damped nonlinear spring/mass system. Predict the behavior of each system as $t \to \infty$. For each equation use a numerical solver to obtain the solution curves satisfying the given initial conditions.

9. $\dfrac{d^2x}{dt^2} + \dfrac{dx}{dt} + x + x^3 = 0$, $x(0) = -3$, $x'(0) = 4$; $x(0) = 0$, $x'(0) = -8$

10. $\dfrac{d^2x}{dt^2} + \dfrac{dx}{dt} + x - x^3 = 0$, $x(0) = 0$, $x'(0) = \frac{3}{2}$; $x(0) = -1$, $x'(0) = 1$

11. The model $mx'' + kx + k_1x^3 = F_0 \cos \omega t$ of an undamped periodically driven spring/mass system is called **Duffing's differential equation.** Consider the initial-value problem $x'' + x + k_1x^3 = 5 \cos t$, $x(0) = 1$, $x'(0) = 0$. Use a numerical solver to investigate the behavior of the system for values of $k_1 > 0$ ranging from $k_1 = 0.01$ to $k_1 = 100$. State your conclusions.

12. (a) Find values of $k_1 < 0$ for which the system in Problem 11 is oscillatory.
   (b) Consider the initial-value problem $x'' + x + k_1x^3 = \cos \frac{3}{2}t$, $x(0) = 0$, $x'(0) = 0$. Find values for $k_1 < 0$ for which the system is oscillatory.

Nonlinear Pendulum

13. Consider the model of the free damped nonlinear pendulum given by

$$\frac{d^2\theta}{dt^2} + 2\lambda \frac{d\theta}{dt} + \omega^2 \sin \theta = 0.$$

Use a numerical solver to investigate whether the motion in the two cases $\lambda^2 - \omega^2 > 0$ and $\lambda^2 - \omega^2 < 0$ corresponds, respectively, to the overdamped and underdamped cases discussed in Section 5.1 for spring/mass systems. Choose appropriate initial conditions and values of λ and ω.

Rocket Motion

14. (a) Use the substitution $v = dy/dt$ to solve (13) for v in terms of y. Assuming that the velocity of the rocket at burnout is $v = v_0$ and $y \approx R$ at that instant, show that the approximate value of the constant c of integration is $c = -gR + \frac{1}{2}v_0^2$.
   (b) Use the solution for v in part (a) to show that the escape velocity of the rocket is given by $v_0 = \sqrt{2gR}$. [*Hint:* Take $y \to \infty$ and assume $v > 0$ for all time t.]
   (c) The result in part (b) holds for any body in the solar system. Use the values $g = 32$ ft/s^2 and $R = 4000$ mi to show that the escape velocity from the earth is (approximately) $v_0 = 25{,}000$ mi/h.
   (d) Find the escape velocity from the moon if the acceleration of gravity is $0.165g$ and $R = 1080$ mi.

Variable Mass

15. (a) In Example 4, how much of the chain would you intuitively expect the constant 5-pound force to be able to lift?

(b) What is the initial velocity of the chain?

(c) Use a graphing utility to plot the solution $x(t)$ given in (21) of Example 4. What is the interval I of definition of the solution? How much chain is actually lifted? Explain any difference between this answer and your prediction in part (a).

16. A uniform chain of length L, measured in feet, is held vertically so that the lower end just touches the floor. The chain weighs 2 lb/ft. The upper end that is held is released from rest at $t = 0$ and the chain falls straight down. See Figure 1.33. As we saw in Problem 20 in Exercises 1.3, if $x(t)$ denotes the length of the chain on the floor at time t, air resistance is ignored, and the positive direction is taken to be downward, then

$$(L - x)\frac{d^2x}{dt^2} - \left(\frac{dx}{dt}\right)^2 = Lg.$$

(a) Solve for v in terms of x. Solve for x in terms of t. Express v in terms of t.

(b) Determine how long it takes for the chain to fall completely to the ground.

(c) What velocity does the model in part (a) predict for the upper end of the chain as it hits the ground?

17. A portion of a uniform chain of length 8 feet is loosely coiled around a peg at the edge of a high horizontal platform, and the remaining portion of the chain hangs at rest over the edge of the platform. Suppose that the length of the overhang is 3 feet and that the chain weighs 2 lb/ft. Starting at $t = 0$ the weight of the overhanging portion causes the chain on the platform to uncoil smoothly and fall to the floor.

(a) Ignore any resistive forces and assume that the positive direction is downward. If $x(t)$ denotes the length of the chain overhanging the platform at time $t > 0$ and $v = dx/dt$, find a differential equation that relates v to x.

(b) Proceed as in Example 4 and solve for v in terms of x by finding an appropriate integrating factor.

(c) Express time t in terms of x. Use a CAS as an aid in determining the time it takes for a 7-foot segment of chain to uncoil completely—that is, fall from the platform.

18. A portion of a uniform chain of length 8 feet lies stretched out on a high horizontal platform, and the remaining portion of the chain hangs over the edge of the platform as shown in Figure 5.39. Suppose that the length of the overhang is 3 feet and that the chain weighs 2 lb/ft. The end of the chain on the platform is held until it is released from rest at $t = 0$ and the chain begins to slide off the platform because of the weight of the overhanging portion.

(a) Ignore any resistive forces and assume that the positive direction is downward. If $x(t)$ denotes the length of the chain overhanging the platform at time $t > 0$ and $v = dx/dt$, show that v is related to x by the differential equation $v\frac{dv}{dx} = 4x.$

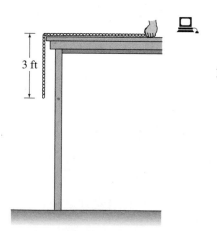

3 ft

Figure 5.39

(b) Solve for v in terms of x. Solve for x in terms of t. Express v in terms of t.

(c) Approximate the time it takes for the rest of the chain to slide off the platform. Find the velocity at which the end of the chain leaves the edge of the platform.

(d) Suppose the chain is L feet long and weighs a total of W pounds. If the overhang at $t = 0$ is x_0 feet, show that the velocity at which the end of the chain leaves the edge of the platform is $v(L) =$

$$\sqrt{\frac{g}{L}(L^2 - x_0^2)}.$$

Miscellaneous Nonlinear Models

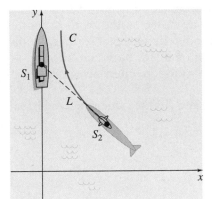

Figure 5.40

19. In a naval exercise a ship S_1 is pursued by a submarine S_2 as shown in Figure 5.40. Ship S_1 departs point $(0, 0)$ at $t = 0$ and proceeds along a straight-line course (the y-axis) at a constant speed v_1. Submarine S_2 keeps ship S_1 in visual contact, indicated by the straight dashed line L in the figure, while traveling at a constant speed v_2 along a curve C. Assume that S_2 starts at the point $(a, 0)$, $a > 0$ at $t = 0$ and that L is tangent to C. Determine a mathematical model that describes the curve C. Find an explicit solution of the differential equation. For convenience define $r = v_1/v_2$. Determine whether the paths of S_1 and S_2 will ever intersect by considering the cases $r > 1$, $r < 1$, and $r = 1$. [*Hint:* $\dfrac{dt}{dx} = \dfrac{dt}{ds}\dfrac{ds}{dx}$, where s is the arc length measured along C.]

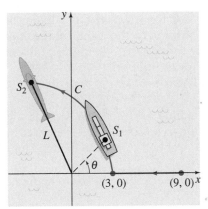

Figure 5.41

20. In another naval exercise a destroyer S_1 pursues a submerged submarine S_2. Suppose that S_1 at $(9, 0)$ on the x-axis detects S_2 at $(0, 0)$ and that S_2 simultaneously detects S_1. The captain of destroyer S_1 assumes that the submarine will take immediate evasive action and conjectures that its likely new course is the straight line L indicated in Figure 5.41. When S_1 is at $(3, 0)$, it changes from its straight-line course toward the origin to pursuit curve C. Assume that the destroyer's speed is a constant 30 mi/h at all times, and the submarine's speed is a constant 15 mi/h.

(a) Explain why the captain waits until S_1 reaches $(3, 0)$ before ordering a course change to C.

(b) Using polar coordinates, find an equation $r = f(\theta)$ for the curve C.

(c) Explain why the time, measured from the initial detection, at which the destroyer intercepts the submarine must be less than $\frac{1}{5}(1 + e^{2\pi/\sqrt{3}})$.

Discussion Problems

21. Discuss why the damping term in equation (3) is written as

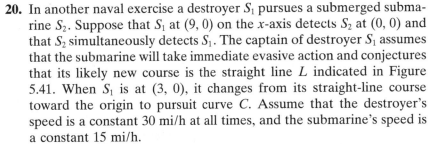

$$\beta \left| \frac{dx}{dt} \right| \frac{dx}{dt} \quad \text{instead of} \quad \beta \left(\frac{dx}{dt} \right)^2.$$

22. (a) Experiment with a calculator to find an interval $0 \le \theta < \theta_1$, where θ is measured in radians, for which you think $\sin \theta \approx \theta$ is a fairly good estimate. Then use a graphing utility to plot the graphs of

$y = x$ and $y = \sin x$ on the same coordinate axes for $0 \le x \le \pi/2$. Do the graphs confirm your observations with the calculator?

(b) Use a numerical solver to plot the solutions curves of the initial-value problems

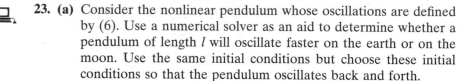

$$\frac{d^2\theta}{dt^2} + \sin\theta = 0, \ \theta(0) = \theta_0, \ \theta'(0) = 0 \quad \text{and} \quad \frac{d^2\theta}{dt^2} + \theta = 0, \ \theta(0) = \theta_0, \ \theta'(0) = 0$$

for several values of θ_0 in the interval $0 \le \theta < \theta_1$ found in part (a). Then plot solutions curves of the initial-value problems for several values of θ_0 for which $\theta_0 > \theta_1$.

23. (a) Consider the nonlinear pendulum whose oscillations are defined by (6). Use a numerical solver as an aid to determine whether a pendulum of length l will oscillate faster on the earth or on the moon. Use the same initial conditions but choose these initial conditions so that the pendulum oscillates back and forth.

(b) For which location in part (a) does the pendulum have greater amplitude?

(c) Are the conclusions in parts (a) and (b) the same when the linear model (7) is used?

Computer Lab Assignments

24. Consider the initial-value problem

$$\frac{d^2\theta}{dt^2} + \sin\theta = 0, \quad \theta(0) = \frac{\pi}{12}, \quad \theta'(0) = -\frac{1}{3}$$

for a nonlinear pendulum. Since we cannot solve the differential equation, we can find no explicit solution of this problem. But suppose we wish to determine the first time $t_1 > 0$ for which the pendulum in Figure 5.35, starting from its initial position to the right, reaches the position OP—that is, the first positive root of $\theta(t) = 0$. In this problem and the next we examine several ways to proceed.

(a) Approximate t_1 by solving the linear problem $d^2\theta/dt^2 + \theta = 0$, $\theta(0) = \pi/12$, $\theta'(0) = -\frac{1}{3}$.

(b) Use the method illustrated in Example 3 of Section 4.9 to find the first four nonzero terms of a Taylor series solution $\theta(t)$ centered at 0 for the nonlinear initial-value problem. Give the exact values of all coefficients.

(c) Use the first two terms of the Taylor series in part (b) to approximate t_1.

(d) Use the first three terms of the Taylor series in part (b) to approximate t_1.

(e) Use a root-finding application of a CAS (or a graphic calculator) and the first four terms of the Taylor series in part (b) to approximate t_1.

(f) In this part of the problem you are lead through the commands in *Mathematica* that enable you to approximate the root t_1. The procedure is easily modified so that any root of $\theta(t) = 0$ can be approximated. (*If you do not have Mathematica, adapt the given procedure by finding the corresponding syntax for the CAS you have on hand.*) Precisely reproduce and then, in turn, execute each line in the given sequence of commands.

```
sol = NDSolve[{y''[t] + Sin[y[t]] == 0, y[0] == Pi/12, y'[0] == -1/3},
            y, {t, 0, 5}]//Flatten
solution = y[t] /. sol
Clear[y]
y[t_] : = Evaluate[solution]
y[t]
gr1 = Plot[y[t], {t, 0, 5}]
root = FindRoot[y[t] == 0, {t, 1}]
```

(g) Appropriately modify the syntax in part (f) and find the next two positive roots of $\theta(t) = 0$.

25. Consider a pendulum that is released from rest from an initial displacement of θ_0 radians. Solving the linear model (7) subject to the initial conditions $\theta(0) = \theta_0$, $\theta'(0) = 0$ gives $\theta(t) = \theta_0 \cos \sqrt{g/l}\,t$. The period of oscillations predicted by this model is given by the familiar formula $T = 2\pi/\sqrt{g/l} = 2\pi\sqrt{l/g}$. The interesting thing about this formula for T is that it does not depend on the magnitude of the initial displacement θ_0. In other words, the linear model predicts that the time it would take the pendulum to swing from an initial displacement of, say, $\theta_0 = \pi/2$ (= 90°) to $-\pi/2$ and back again would be exactly the same as the time it would take to cycle from, say, $\theta_0 = \pi/360$ (= 0.5°) to $-\pi/360$. This is intuitively unreasonable; the actual period must depend on θ_0.

If we assume that $g = 32$ ft/s^2 and $l = 32$ ft, then the period of oscillation of the linear model is $T = 2\pi$ s. Let us compare this last number with the period predicted by the nonlinear model when $\theta_0 = \pi/4$. Using a numerical solver that is capable of generating hard data, approximate the solution of

$$\frac{d^2\theta}{dt^2} + \sin\theta = 0, \quad \theta(0) = \frac{\pi}{4}, \quad \theta'(0) = 0$$

on the interval $0 \le t \le 2$. As in Problem 24, if t_1 denotes the first time the pendulum reaches the position OP in Figure 5.35, then the period of the nonlinear pendulum is $4t_1$. Here is another way of solving the equation $\theta(t) = 0$. Experiment with small step sizes and advance the time, starting at $t = 0$ and ending at $t = 2$. From your hard data observe the time t_1 when $\theta(t)$ changes, for the first time, from positive to negative. Use the value t_1 to determine the true value of the period of the nonlinear pendulum. Compute the percentage relative error in the period estimated by $T = 2\pi$.

CHAPTER 5 IN REVIEW

Answers to odd-numbered problems begin on page AN-6.

Answer Problems 1–8 without referring back to the text. Fill in the blank or answer true or false.

1. If a 10-pound weight stretches a spring 2.5 feet, a 32-pound weight will stretch it _____ feet.

2. The period of simple harmonic motion of an 8-pound weight attached to a spring whose constant is 6.25 lb/ft is _____ seconds.

3. The differential equation of a weight on a spring is $x'' + 16x = 0$. If the weight is released at $t = 0$ from 1 meter above the equilibrium position with a downward velocity of 3 m/s, the amplitude of vibrations is _____ meter.

4. Pure resonance cannot take place in the presence of a damping force. _____

5. In the presence of damping, the displacements of a weight on a spring will always approach zero as $t \to \infty$. _____

6. A weight on a spring whose motion is critically damped can possibly pass through the equilibrium position twice. _____

7. At critical damping any increase in damping will result in an _____ system.

8. If simple harmonic motion is described by $x = (\sqrt{2}/2) \sin(2t + \phi)$, the phase angle ϕ is _____ when $x(0) = -\frac{1}{2}$ and $x'(0) = 1$.

9. A free undamped spring/mass system oscillates with a period of 3 seconds. When 8 pounds are removed from the spring, the system then has a period of 2 seconds. What was the weight of the original mass on the spring?

10. A 12-pound weight stretches a spring 2 feet. The weight is released from a point 1 foot below the equilibrium position with an upward velocity of 4 ft/s.

(a) Find the equation describing the resulting simple harmonic motion.

(b) What are the amplitude, period, and frequency of motion?

(c) At what times does the weight return to the point 1 foot below the equilibrium position?

(d) At what times does the weight pass through the equilibrium position moving upward? moving downward?

(e) What is the velocity of the weight at $t = 3\pi/16$ s?

(f) At what times is the velocity zero?

11. A force of 2 pounds stretches a spring 1 foot. With one end held fixed, an 8-pound weight is attached to the other end. The system lies on a table that imparts a frictional force numerically equal to $\frac{3}{2}$ times the instantaneous velocity. Initially the weight is displaced 4 inches above the equilibrium position and released from rest. Find the equation of motion if the motion takes place along a horizontal straight line that is taken as the x-axis.

12. A 32-pound weight stretches a spring 6 inches. The weight moves through a medium offering a damping force numerically equal to β times the instantaneous velocity. Determine the values of β for which the system will exhibit oscillatory motion.

13. A spring with constant $k = 2$ is suspended in a liquid that offers a damping force numerically equal to 4 times the instantaneous velocity. If a mass m is suspended from the spring, determine the values of m for which the subsequent free motion is nonoscillatory.

14. The vertical motion of a weight attached to a spring is described by the initial-value problem

$$\frac{1}{4}\frac{d^2x}{dt^2} + \frac{dx}{dt} + x = 0, \quad x(0) = 4, \quad x'(0) = 2.$$

Determine the maximum vertical displacement.

15. A 4-pound weight stretches a spring 18 inches. A periodic force equal to $f(t) = \cos \gamma t + \sin \gamma t$ is impressed on the system starting at $t = 0$. In the absence of a damping force, for what value of γ will the system be in a state of pure resonance?

16. Find a particular solution for $\dfrac{d^2x}{dt^2} + 2\lambda \dfrac{dx}{dt} + \omega^2 x = A$, where A is a constant force.

17. A 4-pound weight is suspended from a spring whose constant is 3 lb/ft. The entire system is immersed in a fluid offering a damping force numerically equal to the instantaneous velocity. Beginning at $t = 0$, an external force equal to $f(t) = e^{-t}$ is impressed on the system. Determine the equation of motion if the weight is released from rest at a point 2 feet below the equilibrium position.

18. (a) Two springs are attached in series as shown in Figure 5.42. If the mass of each spring is ignored, show that the effective spring constant k is given by $1/k = 1/k_1 + 1/k_2$.
    (b) A weight of W pounds stretches one spring $\frac{1}{2}$ foot and stretches a different spring $\frac{1}{4}$ foot. The two springs are attached as in the figure, and the weight W is then attached to the double spring. Assume that the motion is free and that there is no damping force present. Determine the equation of motion if the weight is released at a point 1 foot below the equilibrium position with a downward velocity of $\frac{2}{3}$ ft/s.
    (c) Show that the maximum speed of the weight is $\frac{2}{3}\sqrt{3g + 1}$.

19. A series circuit contains an inductance of $L = 1$ h, a capacitance of $C = 10^{-4}$ f, and an electromotive force of $E(t) = 100 \sin 50t$ V. Initially the charge q and current i are zero.

    (a) Find the equation for the charge at any time.
    (b) Find the equation for the current at any time.
    (c) Find the times for which the charge on the capacitor is zero.

20. (a) Show that the current $i(t)$ in an LRC series circuit satisfies the differential equation $L\dfrac{d^2i}{dt^2} + R\dfrac{di}{dt} + \dfrac{1}{C}i = E'(t)$, where $E'(t)$ denotes the derivative of $E(t)$.
    (b) Two initial conditions $i(0)$ and $i'(0)$ can be specified for the DE in part (a). If $i(0) = i_0$ and $q(0) = q_0$, what is $i'(0)$?

21. Consider the boundary-value problem $y'' + \lambda y = 0$, $y(0) = y(2\pi)$, $y'(0) = y'(2\pi)$. Show that except for the case $\lambda = 0$, there are two independent eigenfunctions corresponding to each eigenvalue.

22. A bead is constrained to slide along a frictionless rod of length L. The rod is rotating in a vertical plane with a constant angular velocity ω about a pivot P fixed at the midpoint of the rod, but the design of

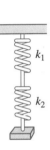

Figure 5.42

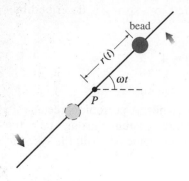

Figure 5.43

the pivot allows the bead to move along the entire length of the rod. Let $r(t)$ denote the position of the bead relative to this rotating coordinate system, as shown in Figure 5.43. In order to apply Newton's second law of motion to this rotating frame of reference, it is necessary to use the fact that the net force acting on the bead is the sum of the real forces (in this case, the force due to gravity) and the inertial forces (coriolis, transverse, and centrifugal). The mathematics is a little complicated, so we just give the resulting differential equation for r:

$$m\frac{d^2r}{dt^2} = m\omega^2 r - mg\,\sin(\omega t).$$

(a) Solve the foregoing DE subject to the initial conditions $r(0) = r_0$, $r'(0) = v_0$.

(b) Determine the initial conditions for which the bead exhibits simple harmonic motion. What is the minimum length L of the rod for which it can accommodate simple harmonic motion of the bead?

(c) For initial conditions other than those obtained in part (b), the bead must eventually fly off the rod. Explain, using the solution $r(t)$ in part (a).

(d) Suppose $\omega = 1$ rad/s. Use a graphing utility to plot the graph of the solution $r(t)$ for the initial conditions $r(0) = 0$, $r'(0) = v_0$, where v_0 is 0, 10, 15, 16, 16.1, and 17.

(e) Suppose the length of the rod is $L = 40$ ft. For each pair of initial conditions in part (d), use a root-finding application to find the total time that the bead stays on the rod.

THE COLLAPSE OF THE TACOMA NARROWS SUSPENSION BRIDGE

Tacoma Narrows suspension bridge collapse, November, 1940

In the summer of 1940, the Tacoma Narrows suspension bridge was completed. Almost immediately, observers noted that sometimes the wind appeared to set up large vertical oscillations of the roadbed. The bridge became a tourist attraction as people came to watch, and perhaps ride, the undulating bridge. Finally, on November 7, 1940, during a powerful storm, the oscillations increased beyond any previously observed. Soon the vertical oscillations became rotational, as observed by looking down the roadway. The entire span was eventually shaken apart by the large oscillations, and the bridge collapsed.

A New Model For almost fifty years the generally accepted hypothesis was that a resonance episode was the probable cause of the collapse. See Figure 1. But as can be seen from equation (31) of Section 5.1, resonance is a *linear* phenomenon. In addition, for resonance to occur, there must be an exact match between the frequency of the forcing function and the natural frequency of the bridge. Furthermore, there must be absolutely no damping in the system. It should not be surprising, then, that resonance was *not* the culprit in the collapse. If not resonance, what did cause the crash? In recent research, the mathematicians Lazer and McKenna contend that nonlinear effects, and not linear resonance, were the main factors leading to the large oscillations of the bridge.* Although the theory involves partial differential equations, a simplified model leading to a nonlinear ordinary differential equation can be constructed. This model is not exactly the same as that of Lazer and McKenna, but it results in a similar differential equation. The example is a new one, and it shows another way that amplitudes of oscillation can increase.

Consider a single vertical cable of a suspension bridge. We assume that it acts like a spring, but with different characteristics in tension and compression. When it is stretched, the cable acts like a spring with Hooke's constant b, but when it is compressed, Hooke's constant is a. We also assume that, when compressed, the cable exerts a smaller force on the roadway than when stretched the same distance, so $0 < a < b$. Let the vertical deflection (positive direction downward) of the slice of the road-

---

*A. C. Lazer and P. J. McKenna, Large amplitude periodic oscillations in suspension bridges: Some new connections with nonlinear analysis, *SIAM Review* 32 (December 1990): 537–578.

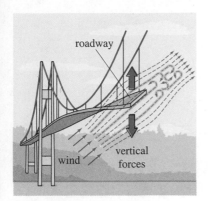

Figure 1 The collapse of the Tacoma Narrows bridge was attributed to a resonance condition induced when wind blowing across the roadway separated into vortices (called von Karman vortices), thereby setting up a periodic vertical force acting in the same direction and with the same frequency as the natural vibrations of the bridge.

way attached to this cable be $x(t)$, where t represents time and $x = 0$ represents the equilibrium position. As the roadway oscillates under the influence of an applied force (the von Karman vortices), the cable provides an upward restoring force equal to bx when $x \geq 0$, and a downward restoring force equal to ax when $x < 0$. In the absence of damping, a nonlinear model for the driven motion (see (1) in Section 5.3) is

$$mx'' + F(x) = g(t), \tag{1}$$

where $F(x)$ is the piecewise-defined function

$$F(x) = \begin{cases} bx, & x \geq 0 \\ ax, & x < 0 \end{cases}$$

$g(t)$ is the applied force, and m is the mass of the section of the roadway. Note that the differential equation (1) is linear on any interval on which x does not change sign.

EXAMPLE 1   A Solution of (1)

Now, let us see what a typical solution of this problem would look like. We will assume that $m = 1$, $b = 4$, $a = 1$, $g(t) = \sin 4t$, and the roadway initially starts at the equilibrium position with a downward velocity $\alpha > 0$. The initial-value problem is

$$x'' + F(x) = \sin 4t, \quad x(0) = 0, \quad x'(0) = \alpha, \quad \text{where} \quad F(x) = \begin{cases} 4x, & x \geq 0 \\ x, & x < 0 \end{cases} \tag{2}$$

Note that the frequency of the forcing function g is larger than the natural frequencies of the cable in both tension and compression, so we do not expect resonance to occur.

It is left as an exercise (see Problem 1) to verify each of the following steps in the solution of (2) on the interval $[0, 3\pi]$.

Because of the positive downward initial velocity and the positive applied force, we solve (2) first for $x \geq 0$. The solution of $x'' + 4x = \sin 4t$, $x(0) = 0$, $x'(0) = \alpha$ is

$$x(t) = \sin 2t[\tfrac{1}{2}(\alpha + \tfrac{1}{3}) - \tfrac{1}{6}\cos 2t]. \tag{3}$$

Note in (3) that the first positive value of t for which $x(t)$ is again zero is $t = \pi/2$. At that time, $x'(\pi/2) = -(\alpha + \tfrac{2}{3}) < 0$. In other words, the roadway has passed through the equilibrium heading upward. Thus for $x < 0$ we now solve the new initial-value problem $x'' + x = \sin 4t$, $x(\pi/2) = 0$, $x'(\pi/2) = -(\alpha + \tfrac{2}{3})$. The solution is

$$x(t) = \cos t[(\alpha + \tfrac{2}{5}) - \tfrac{4}{15}\sin t \cos 2t]. \tag{4}$$

We then see from (4) that the next positive value of t for which $x(t)$ is zero is $t = 3\pi/2$. At that time, $x'(3\pi/2) = \alpha + \tfrac{2}{15} > 0$.

In this manner we conclude that the roadway has gone through one complete cycle in the time interval $[0, 3\pi/2]$: The section of roadway started from the equilibrium position with positive velocity heading downward, eventually turned around and came back to and through the equilibrium position with negative velocity, and then turned around and came back to the equilibrium position with positive velocity. This pattern continues indefinitely, with each cycle of length $3\pi/2$ time units. We summarize the solution of (2) over *two* cycles:

$$\text{first cycle}\quad x(t) = \begin{cases} \sin 2t[\tfrac{1}{2}(\alpha + \tfrac{1}{3}) - \tfrac{1}{6}\cos 2t], & 0 \le t \le \dfrac{\pi}{2} \\[2mm] \cos t[(\alpha + \tfrac{2}{5}) - \tfrac{4}{15}\sin t \cos 2t], & \dfrac{\pi}{2} \le t \le \dfrac{3\pi}{2} \end{cases} \tag{5}$$

$$\text{second cycle}\quad x(t) = \begin{cases} \sin 2t[-\tfrac{1}{2}(\alpha + \tfrac{7}{15}) - \tfrac{1}{6}\cos 2t], & \dfrac{3\pi}{2} \le t \le 2\pi \\[2mm] \sin t[-(\alpha + \tfrac{8}{15}) - \tfrac{4}{15}\cos t \cos 2t], & 2\pi \le t \le 3\pi \end{cases} \tag{6}$$

It is instructive to note that the velocity at the beginning of the second cycle is $\alpha + \tfrac{2}{15}$, while at the beginning of the third cycle it is $\alpha + \tfrac{4}{15}$. In fact, at the beginning of each cycle the velocity is $\tfrac{2}{15}$ greater than at the beginning of the previous cycle. Hence *the amplitudes of oscillations will increase over time,* since the amplitude of one term in the solution during any one cycle is directly proportional to the velocity at the beginning of the cycle. ▄

It must be remembered that the model presented here is a very simplified one-dimensional model that cannot take into account all of the intricate interactions of real bridges. More recently, McKenna has refined that model to provide a different viewpoint of the torsional oscillations in the Tacoma Narrows bridge.*

Research on the behavior of bridges under external forces continues. It is likely that the models will be refined over time, and new insights will be gained from this research. However, it should be clear at this point that the large oscillations causing the destruction of the Tacoma Narrows suspension bridge were not the result of resonance.

RELATED EXERCISES

1. **(a)** Use the methods of Sections 4.3 and 4.4 to obtain the solutions (5) that describe the first cycle of the roadway. [*Hint:* Use the double angle formula from trigonometry.]
   (b) Show that the velocity of the portion of the roadway at the start of the second cycle is $\alpha + \tfrac{2}{15}$.
   (c) Using the information in part (b), obtain the solutions (6) that describe the second cycle of the roadway.
   (d) Use a graphing utility to plot $x(t)$ over two cycles for $\alpha = 1$ and $\alpha = 2$.

2. Solve the following initial-value problems and plot the solutions for $0 \le t \le 6\pi$. Note that resonance occurs in the first problem but not in the second.
   (a) $x'' + x = \cos t,\ x(0) = 0,\ x'(0) = 0$
   (b) $x'' + x = \cos 2t,\ x(0) = 0,\ x'(0) = 0$

---

*P. J. McKenna, Large torsional oscillations in suspension bridges revisited: Fixing an old approximation, *American Mathematical Monthly* 106: (1999): 1–18.

3. (a) Solve the initial-value problem

$$x'' + F(x) = \sin 4t, \quad x(0) = 0, \quad x'(0) = 1, \quad \text{where} \quad F(x) = \begin{cases} bx, & x \geq 0 \\ ax, & x < 0 \end{cases}$$

in the three cases $b = 1$, $a = 4$; $b = 64$, $a = 4$; and $b = 36$, $a = 25$. Note that in the first case the condition $0 < a < b$ assumed in the discussion is not satisfied.

(b) Plot the solutions in part (a). What happens to $x(t)$ in each case as t increases?

(c) What would happen in each case if the initial velocity were $x'(0) = \alpha$, $\alpha > 0$? Can you make any conclusions similar to those in the discussion regarding the solution over the long term?

4. (a) The term $\beta x'$, $\beta > 0$ in the initial-value problem

$$x'' + \beta x' + F(x) = \sin 4t, \quad x(0) = 0, \quad x'(0) = 1, \quad \text{where} \quad F(x) = \begin{cases} 4x, & x \geq 0 \\ x, & x < 0 \end{cases}$$

represents damping. Solve the initial-value problem in the three cases $\beta = 0.01$, $\beta = 0.1$, and $\beta = 0.5$.

(b) Plot the solutions in part (a). What happens to $x(t)$ in each case as t increases?

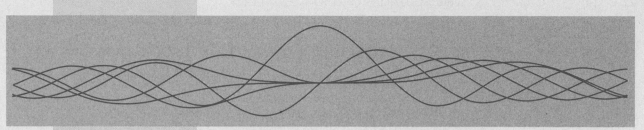

Graphs of Bessel functions; see page 294.

6 SERIES SOLUTIONS OF LINEAR EQUATIONS

INTRODUCTION   Up to now we have primarily solved linear differential equations of order two or higher when the equation had constant coefficients. The only exception was the Cauchy-Euler equation. In applications, higher-order linear equations with variable coefficients are just as important as, if not more important than, differential equations with constant coefficients. As mentioned in Section 4.7, even a simple linear second-order equation with variable coefficients such as $y'' + xy = 0$ does not possess elementary solutions. We *can* find two linearly independent solutions of $y'' + xy = 0$, but, as we shall see in Sections 6.1 and 6.3, the solutions of this equation are defined by infinite series.

6.1 SOLUTIONS ABOUT ORDINARY POINTS

• Power series   • Convergence/divergence   • Interval of convergence   • Radius of convergence   • Analyticity at a point   • Ordinary points   • Singular points   • Power series solutions of ODEs   • Recurrence relation

Section 4.7 notwithstanding, most linear higher-order ordinary differential equations with variable coefficients cannot be solved in terms of elementary functions. The usual strategy for solving differential equations of this sort is to assume a solution in the form of an infinite series and proceed in a manner similar to the method of undetermined coefficients (Section 4.4). Because these series solutions often turn out to be power series, it is appropriate to list some of the more important facts about power series. For a more in-depth review of this theory, you are encouraged to consult a calculus text.

6.1.1   REVIEW OF POWER SERIES

- **Definition**   A power series in $x - a$ is an infinite series of the form

$$c_0 + c_1(x - a) + c_2(x - a)^2 + \cdots = \sum_{n=0}^{\infty} c_n(x - a)^n.$$

Such a series is also said to be a power series centered at a. For example, the power series $\sum_{n=0}^{\infty} (x + 1)^n$ is centered at $a = -1$. In this section we are concerned principally with power series in x—that is, power series centered at $a = 0$. For example, $\sum_{n=1}^{\infty} 2^n x^n$ is a power series in x.

- **Convergence**   A power series $\sum_{n=0}^{\infty} c_n(x - a)^n$ is convergent at a specified value of x if its sequence of partial sums $\{S_N(x)\}$ converges— that is, $\lim_{N \to \infty} S_N(x) = \lim_{N \to \infty} \sum_{n=0}^{N} c_n(x - a)^n$ exists. If the limit does not exist at x, the series is said to be divergent.

- **Interval of Convergence**   Every power series has an interval of convergence. The interval of convergence is the set of all real numbers x for which the series converges.

- **Radius of Convergence**   Every power series has a radius of convergence R. If $R > 0$, then a power series $\sum_{n=0}^{\infty} c_n(x - a)^n$ converges for $|x - a| < R$ and diverges for $|x - a| > R$. If the series converges only at its center a, then $R = 0$. If the series converges for all x, then we write $R = \infty$. Recall that $|x - a| < R$ is equivalent to $a - R < x < a + R$. A power series may or may not converge at the endpoints $a - R$ and $a + R$ of this interval.

- **Absolute Convergence**   Within its interval of convergence a power series converges absolutely. In other words, if x is in the interval of convergence and is not an endpoint of the interval, then the series of absolute values $\sum_{n=0}^{\infty} |c_n(x - a)^n|$ converges.

- **Ratio Test**   Convergence of a power series $\sum_{n=0}^{\infty} c_n(x - a)^n$ can often be determined by the ratio test. Suppose that $c_n \neq 0$ for all *n* and that

$$\lim_{n \to \infty} \left| \frac{c_{n+1}(x-a)^{n+1}}{c_n(x-a)^n} \right| = |x-a| \lim_{n \to \infty} \left| \frac{c_{n+1}}{c_n} \right| = L.$$

If $L < 1$ the series converges absolutely, if $L > 1$ the series diverges, and if $L = 1$ the test is inconclusive. For example, for the power series $\sum_{n=1}^{\infty}(x-3)^n/2^n n$, the ratio test gives

$$\lim_{n \to \infty} \left| \frac{\dfrac{(x-3)^{n+1}}{2^{n+1}(n+1)}}{\dfrac{(x-3)^n}{2^n n}} \right| = |x-3| \lim_{n \to \infty} \frac{n+1}{2n} = \frac{1}{2}|x-3|;$$

the series converges absolutely for $\frac{1}{2}|x-3| < 1$ or $|x-3| < 2$ or $1 < x < 5$. This last interval is referred to as the *open* interval of convergence. The series diverges for $|x-3| > 2$—that is, for $x > 5$ or $x < 1$. At the left endpoint $x = 1$ of the open interval of convergence, the series of constants $\sum_{n=1}^{\infty}((-1)^n/n)$ is convergent by the alternating series test. At the right endpoint $x = 5$, the series $\sum_{n=1}^{\infty}(1/n)$ is the divergent harmonic series. The interval of convergence of the series is $[1, 5)$, and the radius of convergence is $R = 2$.

- **A Power Series Defines a Function** A power series defines a function $f(x) = \sum_{n=0}^{\infty} c_n(x-a)^n$ whose domain is the interval of convergence of the series. If the radius of convergence is $R > 0$, then f is continuous, differentiable, and integrable on the interval $(a - R, a + R)$. Moreover, $f'(x)$ and $\int f(x)dx$ can be found by term-by-term differentiation and integration. Convergence at an endpoint may be either lost by differentiation or gained through integration. If $y = \sum_{n=0}^{\infty} c_n x^n$ is a power series in x, then the first two derivatives are $y' = \sum_{n=0}^{\infty} nx^{n-1}$ and $y'' = \sum_{n=0}^{\infty} n(n-1)x^{n-2}$. Notice that the first term in the first derivative and the first two terms in the second derivative are zero. We omit these zero terms and write

$$y' = \sum_{n=1}^{\infty} c_n n x^{n-1} \quad \text{and} \quad y'' = \sum_{n=2}^{\infty} c_n n(n-1)x^{n-2}. \tag{1}$$

These results are important and will be used shortly.

- **Identity Property** If $\sum_{n=0}^{\infty} c_n(x-a)^n = 0$, $R > 0$ for all numbers x in the interval of convergence, then $c_n = 0$ for all n.

- **Analytic at a Point** A function f is analytic at a point a if it can be represented by a power series in $x - a$ with a positive or infinite radius of convergence. In calculus it is seen that functions such as e^x, $\cos x$, $\sin x$, $\ln(x-1)$, and so on can be represented by Taylor series. Recall, for example, that

$$e^x = 1 + \frac{x}{1!} + \frac{x^2}{2!} + \cdots, \quad \sin x = x - \frac{x^3}{3!} + \frac{x^5}{5!} - \cdots, \quad \cos x = 1 - \frac{x^2}{2!} + \frac{x^4}{4!} - \frac{x^6}{6!} + \cdots \tag{2}$$

for $|x| < \infty$. These Taylor series centered at 0, called Maclaurin series, show that e^x, $\sin x$, and $\cos x$ are analytic at $x = 0$.

- **Arithmetic of Power Series** Power series can be combined through the operations of addition, multiplication, and division. The procedures for powers series are similar to those by which two polynomials are added, multiplied, and divided—that is, we add coefficients of like powers of x, use the distributive law and collect like terms, and perform long division. For example, using the series in (2) we have

$$e^x \sin x = \left(1 + x + \frac{x^2}{2} + \frac{x^3}{6} + \frac{x^4}{24} + \cdots\right)\left(x - \frac{x^3}{6} + \frac{x^5}{120} - \frac{x^7}{5040} + \cdots\right)$$

$$= (1)x + (1)x^2 + \left(-\frac{1}{6} + \frac{1}{2}\right)x^3 + \left(-\frac{1}{6} + \frac{1}{6}\right)x^4 + \left(\frac{1}{120} - \frac{1}{12} + \frac{1}{24}\right)x^5 + \cdots$$

$$= x + x^2 + \frac{x^3}{3} - \frac{x^5}{30} - \cdots.$$

Since the power series for e^x and $\sin x$ converge for $|x| < \infty$, the product series converges on the same interval. Problems involving multiplication or division of power series can be done with minimal fuss by using a computer algebra system.

Shifting the Summation Index   For the remainder of this section, as well as this chapter, it is important that you become adept at simplifying the sum of two or more power series, each expressed in summation (sigma) notation, to an expression with a single Σ. As the next example illustrates, combining two or more summations as a single summation often requires a reindexing—that is, a shift in the index of summation.

EXAMPLE 1   **Adding Two Power Series**

Write $\sum_{n=2}^{\infty} n(n-1)c_n x^{n-2} + \sum_{n=0}^{\infty} c_n x^{n+1}$ as one power series.

Solution   In order to add the two series, it is necessary that both summation indices start with the same number and the powers of x in each series be "in phase"; that is, if one series starts with a multiple of, say, x to the first power, then we want the other series to start with the same power. Note that in the given problem the first series starts with x^0 whereas the second series starts with x^1. By writing the first term of the first series outside the summation notation,

$$\underset{n=2}{\overset{\infty}{\sum}} n(n-1)c_n x^{n-2} + \underset{n=0}{\overset{\infty}{\sum}} c_n x^{n+1} = 2 \cdot 1 c_2 x^0 + \overset{\substack{\text{series starts} \\ \text{with } x \\ \text{for } n=3 \\ \downarrow}}{\underset{n=3}{\overset{\infty}{\sum}} n(n-1)c_n x^{n-2}} + \overset{\substack{\text{series starts} \\ \text{with } x \\ \text{for } n=0 \\ \downarrow}}{\underset{n=0}{\overset{\infty}{\sum}} c_n x^{n+1}},$$

we see that both series on the right-hand side start with the same power of x—namely, x^1. Now to get the same summation index, we are inspired by the exponents of x; we let $k = n - 2$ in the first series and at the same time let $k = n + 1$ in the second series. The right-hand side becomes

$$2c_2 + \overset{\text{same}}{\overbrace{\underset{k=1}{\overset{\infty}{\sum}} (k+2)(k+1)c_{k+2} x^k + \underset{k=1}{\overset{\infty}{\sum}} c_{k-1} x^k}_{\text{same}}}. \tag{3}$$

Remember that the summation index is a "dummy" variable; the fact that $k = n - 1$ in one case and $k = n + 1$ in the other should cause no confusion if you keep in mind that it is the *value* of the summation index that is important. In both cases k takes on the same successive values $k = 1, 2, 3, \ldots$ when n takes on the values $n = 2, 3, 4, \ldots$ for $k = n - 1$ and $n = 0, 1, 2, \ldots$ for $k = n + 1$. We are now in a position to add the series in (3) term by term:

$$\sum_{n=2}^{\infty} n(n-1)c_n x^{n-2} + \sum_{n=0}^{\infty} c_n x^{n+1} = 2c_2 + \sum_{k=1}^{\infty} [(k+2)(k+1)c_{k+2} + c_{k-1}]x^k. \quad \textbf{(4)} \quad \blacksquare$$

If you are not convinced of the result in (4), then write out a few terms on both sides of the equality.

6.1.2 POWER SERIES SOLUTIONS

Suppose the linear second-order differential equation

$$a_2(x)y'' + a_1(x)y' + a_0(x)y = 0 \quad \textbf{(5)}$$

is put into standard form

$$y'' + P(x)y' + Q(x)y = 0 \quad \textbf{(6)}$$

by dividing by the leading coefficient $a_2(x)$. We have the following definition.

DEFINITION 6.1 **Ordinary and Singular Points**

A point x_0 is said to be an **ordinary point** of the differential equation (5) if both $P(x)$ and $Q(x)$ in the standard form (6) are analytic at x_0. A point that is not an ordinary point is said to be a **singular point** of the equation.

Every finite value of x is an ordinary point of the differential equation $y'' + (e^x)y' + (\sin x)y = 0$. In particular, $x = 0$ is an ordinary point since, as we have already seen in (2), both e^x and $\sin x$ are analytic at this point. The negation in the second sentence of Definition 6.1 stipulates that if at least one of the functions $P(x)$ and $Q(x)$ in (6) fails to be analytic at x_0, then x_0 is a singular point. Note that $x = 0$ is a singular point of the differential equation $y'' + (e^x)y' + (\ln x)y = 0$ since $Q(x) = \ln x$ is discontinuous at $x = 0$ and so cannot be represented by a power series in x.

Polynomial Coefficients We shall be interested primarily in the case when (5) has polynomial coefficients. A polynomial is analytic at any value x, and a rational function is analytic *except* at points where its denominator is zero. Thus if $a_2(x)$, $a_1(x)$, and $a_0(x)$ are polynomials with no common factors, then both rational functions $P(x) = a_1(x)/a_2(x)$ and $Q(x) = a_0(x)/a_2(x)$ are analytic except where $a_2(x) = 0$. It follows, then, that $x = x_0$ is an ordinary point of (5) if $a_2(x_0) \neq 0$ whereas $x = x_0$ is a singular point of (5) if $a_2(x_0) = 0$. For example, the only singular points of the equation $(x^2 - 1)y'' + 2xy' + 6y = 0$ are solutions of $x^2 - 1 = 0$ or $x = \pm 1$. All other finite values* of x are ordinary points. Inspection of the Cauchy-Euler equation $ax^2y'' + bxy' + cy = 0$ shows that it has a singular point at $x = 0$. Singular points need not be real numbers. The equation $(x^2 + 1)y'' + xy' - y = 0$ has singular points at the solutions of $x^2 + 1 = 0$—namely, $x = \pm i$. All other values of x, real or complex, are ordinary points.

We state the following theorem about the existence of power series solutions without proof.

| **THEOREM 6.1** | **Existence of Power Series Solutions** |

If $x = x_0$ is an ordinary point of the differential equation (5), we can always find two linearly independent solutions in the form of a power series centered at x_0—that is, $y = \sum_{n=0}^{\infty} c_n(x - x_0)^n$. A series solution converges at least on some interval defined by $|x - x_0| < R$, where R is the distance from x_0 to the closest singular point.

A solution of the form $y = \sum_{n=0}^{\infty} c_n(x - x_0)^n$ is said to be a **solution about the ordinary point x_0**. The distance R in Theorem 6.1 is the *minimum* value for the radius of convergence. For example, the complex numbers $1 \pm 2i$ are singular points of $(x^2 - 2x + 5)y'' + xy' - y = 0$, but since $x = 0$ is an ordinary point of the equation Theorem 6.1 guarantees that we can find two power series solutions centered at 0. That is, the solutions look like $y = \sum_{n=0}^{\infty} c_n x^n$, and moreover, we know without actually finding these solutions that each series must converge *at least* for $|x| < \sqrt{5}$, where $R = \sqrt{5}$ is the distance in the complex plane from 0 to either $1 + 2i$ or $1 - 2i$. However, the differential equation has a solution that is valid for much larger values of x; indeed, this solution is valid on $(-\infty, \infty)$ because it can be shown that one of the two solutions is a polynomial.

Note   In the examples that follow and in Exercises 6.1 we shall, for the sake of simplicity, find power series solutions only about the ordinary point $x = 0$. If it is necessary to find a power series solution of an ODE about an ordinary point $x_0 \neq 0$, we can simply make the change of variable $t = x - x_0$ in the equation (this translates $x = x_0$ to $t = 0$), find solutions of the new equation of the form $y = \sum_{n=0}^{\infty} c_n t^n$, and then resubstitute $t = x - x_0$.

Finding a power series solution of a homogeneous linear second-order ODE has been accurately described as "the method of undetermined *series* coefficients," since the procedure is quite analogous to what we did in Section 4.4. In brief, here is the idea: We substitute $y = \sum_{n=0}^{\infty} c_n x^n$ into the differential equation, combine series as we did in Example 1, and then equate all coefficients to the right-hand side of the equation to determine the coefficients c_n. But since the right-hand side is zero, the last step requires, by the identity property in the previous bulleted list, that all coefficients of x must be equated to zero. No, this does *not* mean

---

*For our purposes, ordinary points and singular points will always be finite points. It is possible for an ODE to have, say, a singular point at infinity.

that all coefficients *are* zero; this would not make sense—after all, Theorem 6.1 guarantees that we can find two solutions. Example 2 illustrates how the single assumption that $y = \sum_{n=0}^{\infty} c_n x^n = c_0 + c_1 x + c_2 x^2 + \cdots$ leads to two sets of coefficients, so we have two distinct power series $y_1(x)$ and $y_2(x)$, both expanded about the ordinary point $x = 0$. The general solution of the differential equation is $y = C_1 y_1(x) + C_2 y_2(x)$; indeed, it can be shown that $C_1 = c_0$ and $C_2 = c_1$.

| **EXAMPLE 2** | **Power Series Solutions** |

Solve $y'' + xy = 0$.

Solution   Since there are no finite singular points, Theorem 6.1 guarantees two power series solutions centered at 0, convergent for $|x| < \infty$. Substituting $y = \sum_{n=0}^{\infty} c_n x^n$ and the second derivative $y'' = \sum_{n=2}^{\infty} n(n-1)c_n x^{n-2}$ (see (1)) into the differential equation gives

$$y'' + xy = \sum_{n=2}^{\infty} c_n n(n-1)x^{n-2} + x \sum_{n=0}^{\infty} c_n x^n = \sum_{n=2}^{\infty} c_n n(n-1)x^{n-2} + \sum_{n=0}^{\infty} c_n x^{n+1}. \quad \textbf{(7)}$$

In Example 1 we already added the last two series on the right-hand side of the equality in (7) by shifting the summation index. From the result given in (4),

$$y'' + xy = 2c_2 + \sum_{k=1}^{\infty} [(k+1)(k+2)c_{k+2} + c_{k-1}]x^k = 0. \quad \textbf{(8)}$$

At this point we invoke the identity property. Since (8) is identically zero, it is necessary that the coefficient of each power of x be set equal to zero—that is, $2c_2 = 0$ (it is the coefficient of x^0), and

$$(k+1)(k+2)c_{k+2} + c_{k-1} = 0, \quad k = 1, 2, 3, \ldots. \quad \textbf{(9)}$$

Now $2c_2 = 0$ obviously dictates that $c_2 = 0$. But the expression in (9), called a **recurrence relation,** determines the c_k in such a manner that we can choose a certain subset of the set of coefficients to be *nonzero*. Since $(k+1)(k+2) \neq 0$ for all values of k, we can solve (9) for c_{k+2} in terms of c_{k-1}:

$$c_{k+2} = -\frac{c_{k-1}}{(k+1)(k+2)}, \quad k = 1, 2, 3, \ldots. \quad \textbf{(10)}$$

This relation generates consecutive coefficients of the assumed solution one at a time as we let k take on the successive integers indicated in (10):

$$k = 1, \quad c_3 = -\frac{c_0}{2 \cdot 3}$$

$$k = 2, \quad c_4 = -\frac{c_1}{3 \cdot 4}$$

$$k = 3, \quad c_5 = -\frac{c_2}{4 \cdot 5} = 0 \quad \leftarrow c_2 \text{ is zero}$$

$$k = 4, \quad c_6 = -\frac{c_3}{5 \cdot 6} = \frac{1}{2 \cdot 3 \cdot 5 \cdot 6}c_0$$

$$k = 5, \quad c_7 = -\frac{c_4}{6 \cdot 7} = \frac{1}{3 \cdot 4 \cdot 6 \cdot 7}c_1$$

$$k = 6, \quad c_8 = -\frac{c_5}{7 \cdot 8} \quad = 0 \quad \leftarrow c_5 \text{ is zero}$$

$$k = 7, \quad c_9 = -\frac{c_6}{8 \cdot 9} \quad = -\frac{1}{2 \cdot 3 \cdot 5 \cdot 6 \cdot 8 \cdot 9} c_0$$

$$k = 8, \quad c_{10} = -\frac{c_7}{9 \cdot 10} \quad = -\frac{1}{3 \cdot 4 \cdot 6 \cdot 7 \cdot 9 \cdot 10} c_1$$

$$k = 9, \quad c_{11} = -\frac{c_8}{10 \cdot 11} \quad = 0 \quad \leftarrow c_8 \text{ is zero}$$

and so on. Now substituting the coefficients just obtained into the original assumption

$$y = c_0 + c_1 x + c_2 x^2 + c_3 x^3 + c_4 x^4 + c_5 x^5 + c_6 x^6 + c_7 x^7 + c_8 x^8 + c_9 x^9 + c_{10} x^{10} + c_{11} x^{11} + \cdots,$$

we get

$$y = c_0 + c_1 x + 0 - \frac{c_0}{2 \cdot 3} x^3 - \frac{c_1}{3 \cdot 4} x^4 + 0 + \frac{c_0}{2 \cdot 3 \cdot 5 \cdot 6} x^6$$

$$+ \frac{c_1}{3 \cdot 4 \cdot 6 \cdot 7} x^7 + 0 - \frac{c_0}{2 \cdot 3 \cdot 5 \cdot 6 \cdot 8 \cdot 9} x^9 - \frac{c_1}{3 \cdot 4 \cdot 6 \cdot 7 \cdot 9 \cdot 10} x^{10} + 0 + \cdots.$$

After grouping the terms containing c_0 and the terms containing c_1, we obtain $y = c_0 y_1(x) + c_1 y_2(x)$, where

$$y_1(x) = 1 - \frac{1}{2 \cdot 3} x^3 + \frac{1}{2 \cdot 3 \cdot 5 \cdot 6} x^6 - \frac{1}{2 \cdot 3 \cdot 5 \cdot 6 \cdot 8 \cdot 9} x^9 + \cdots = 1 + \sum_{k=1}^{\infty} \frac{(-1)^k}{2 \cdot 3 \cdots (3n-1)(3n)} x^{3k}$$

$$y_2(x) = x - \frac{1}{3 \cdot 4} x^4 + \frac{1}{3 \cdot 4 \cdot 6 \cdot 7} x^7 - \frac{1}{3 \cdot 4 \cdot 6 \cdot 7 \cdot 9 \cdot 10} x^{10} + \cdots = x + \sum_{k=1}^{\infty} \frac{(-1)^k}{3 \cdot 4 \cdots (3n)(3n+1)} x^{3k+1}.$$

Since the recursive use of (10) leaves c_0 and c_1 completely undetermined, they can be chosen arbitrarily. As mentioned prior to this example, the linear combination $y = c_0 y_1(x) + c_1 y_2(x)$ actually represents the general solution of the differential equation. Although we know from Theorem 6.1 that each series solution converges for $|x| < \infty$, this fact can also be verified by the ratio test. ▬

The differential equation in Example 2 is called **Airy's equation** and is encountered in the study of diffraction of light, diffraction of radio waves around the surface of the earth, aerodynamics, and the deflection of a uniform thin vertical column that bends under its own weight. Other common forms of Airy's equation are $y'' - xy = 0$ and $y'' + \alpha^2 xy = 0$. See Problem 36 in Exercises 6.3 for an application of the last equation.

EXAMPLE 3 **Power Series Solution**

Solve $(x^2 + 1)y'' + xy' - y = 0$.

Solution As we have already seen on page 272, the given differential equation has singular points at $x = \pm i$, and so a power series solution centered at 0 will converge at least for $|x| < 1$, where 1 is the distance in the complex plane from 0 to either i or $-i$. The assumption $y = \sum_{n=0}^{\infty} c_n x^n$ and its first two derivatives (see (1)) lead to

$$(x^2 + 1) \sum_{n=2}^{\infty} n(n-1)c_n x^{n-2} + x \sum_{n=1}^{\infty} n c_n x^{n-1} - \sum_{n=0}^{\infty} c_n x^n$$

$$= \sum_{n=2}^{\infty} n(n-1)c_n x^n + \sum_{n=2}^{\infty} n(n-1)c_n x^{n-2} + \sum_{n=1}^{\infty} n c_n x^n - \sum_{n=0}^{\infty} c_n x^n$$

$$= 2c_0 x^0 - c_0 x^0 + 6c_3 x + c_1 x - c_1 x + \underbrace{\sum_{n=2}^{\infty} n(n-1)c_n x^n}_{k=n}$$

$$+ \underbrace{\sum_{n=4}^{\infty} n(n-1)c_n x^{n-2}}_{k=n-2} + \underbrace{\sum_{n=2}^{\infty} n c_n x^n}_{k=n} - \underbrace{\sum_{n=2}^{\infty} c_n x^n}_{k=n}$$

$$= 2c_2 - c_0 + 6c_3 x + \sum_{n=2}^{\infty} [k(k-1)c_k + (k+2)(k+1)c_{k+2} + kc_k - c_k]x^k$$

$$= 2c_2 - c_0 + 6c_3 x + \sum_{n=2}^{\infty} [(k+1)(k-1)c_k + (k+2)(k+1)c_{k+2}]x^k = 0.$$

From this identity we conclude that $2c_2 - c_0 = 0$, $6c_3 = 0$, and

$$(k+1)(k-1)c_k + (k+2)(k+1)c_{k+2} = 0.$$

Thus
$$c_2 = \frac{1}{2}c_0$$

$$c_3 = 0,$$

$$c_{k+2} = \frac{1-k}{k+2}c_k, \quad k = 2, 3, 4, \ldots.$$

Substituting $k = 2, 3, 4, \ldots$ into the last formula gives

$$c_4 = -\frac{1}{4}c_2 = -\frac{1}{2 \cdot 4}c_0 = -\frac{1}{2^2 2!}c_0$$

$$c_5 = -\frac{2}{5}c_3 = 0 \qquad \leftarrow c_3 \text{ is zero}$$

$$c_6 = -\frac{3}{6}c_4 = \frac{3}{2 \cdot 4 \cdot 6}c_0 = \frac{1 \cdot 3}{2^3 3!}c_0$$

$$c_7 = -\frac{4}{7}c_5 = 0 \qquad \leftarrow c_5 \text{ is zero}$$

$$c_8 = -\frac{5}{8}c_6 = -\frac{3 \cdot 5}{2 \cdot 4 \cdot 6 \cdot 8}c_0 = -\frac{1 \cdot 3 \cdot 5}{2^4 4!}c_0$$

$$c_9 = -\frac{6}{9}c_7 = 0 \qquad \leftarrow c_7 \text{ is zero}$$

$$c_{10} = -\frac{7}{10}c_8 = \frac{3 \cdot 5 \cdot 7}{2 \cdot 4 \cdot 6 \cdot 8 \cdot 10}c_0 = -\frac{1 \cdot 3 \cdot 5 \cdot 7}{2^5 5!}c_0$$

and so on. Therefore

$$y = c_0 + c_1x + c_2x^2 + c_3x^3 + c_4x^4 + c_5x^5 + c_6x^6 + c_7x^7 + c_8x^8 + c_9x^9 + c_{10}x^{10} + \cdots$$

$$= c_0 \left[1 + \frac{1}{2}x^2 - \frac{1}{2^2 2!}x^4 + \frac{1 \cdot 3}{2^3 3!}x^6 - \frac{1 \cdot 3 \cdot 5}{2^4 4!}x^2 + \frac{1 \cdot 3 \cdot 5 \cdot 7}{2^5 5!}x^{10} - \cdots \right] + c_1x$$

$$= c_0 y_1(x) + c_1 y_2(x).$$

The solutions are the polynomial $y_2(x) = x$ and the power series

$$y_1(x) = 1 + \frac{1}{2}x^2 + \sum_{n=2}^{\infty} (-1)^{n-1} \frac{1 \cdot 3 \cdot 5 \cdots (2n - 3)}{2^n n!} x^{2n}, \quad |x| < 1. \quad \blacksquare$$

EXAMPLE 4   Three-Term Recurrence Relation

If we seek a power series solution $y = \sum_{n=0}^{\infty} c_n x^n$ for the differential equation

$$y'' - (1 + x)y = 0,$$

we obtain $c_2 = \frac{1}{2}c_0$ and the three-term recurrence relation

$$c_{k+2} = \frac{c_k + c_{k-1}}{(k + 1)(k + 2)}, \quad k = 1, 2, 3, \ldots.$$

Examination of the formulas shows that the coefficients $c_3, c_4, c_5, \ldots$ are expressed in terms of both c_1 and c_0, and moreover, the algebra becomes a little messy. To simplify life we can first choose $c_0 \neq 0$, $c_1 = 0$; this yields the coefficients for one solution expressed entirely in terms of c_0:

$$c_2 = \frac{1}{2}c_0$$

$$c_3 = \frac{c_1 + c_0}{2 \cdot 3} = \frac{c_0}{2 \cdot 3} = \frac{1}{6}c_0$$

$$c_4 = \frac{c_2 + c_1}{3 \cdot 4} = \frac{c_0}{2 \cdot 3 \cdot 4} = \frac{1}{24}c_0$$

$$c_5 = \frac{c_3 + c_2}{4 \cdot 5} = \frac{c_0}{4 \cdot 5}\left[\frac{1}{6} + \frac{1}{2}\right] = \frac{1}{30}c_0$$

and so on. Next, if we choose $c_0 = 0$, $c_1 \neq 0$, the coefficients for the other solution are expressed in terms of c_1:

$$c_2 = \frac{1}{2}c_0 = 0$$

$$c_3 = \frac{c_1 + c_0}{2 \cdot 3} = \frac{c_1}{2 \cdot 3} = \frac{1}{6}c_1$$

$$c_4 = \frac{c_2 + c_1}{3 \cdot 4} = \frac{c_1}{3 \cdot 4} = \frac{1}{12}c_1$$

$$c_5 = \frac{c_3 + c_2}{4 \cdot 5} = \frac{c_1}{4 \cdot 5 \cdot 6} = \frac{1}{120}c_1$$

and so on. Finally, we see that the general solution of the equation is $y = c_0 y_1(x) + c_1 y_2(x)$, where

$$y_1(x) = 1 + \frac{1}{2}x^2 + \frac{1}{6}x^3 + \frac{1}{24}x^4 + \frac{1}{30}x^5 + \cdots$$

and $$y_2(x) = x + \frac{1}{6}x^3 + \frac{1}{12}x^4 + \frac{1}{120}x^5 + \cdots.$$

Each series converges for all finite values of x. ▄

Nonpolynomial Coefficients The next example illustrates how to find a power series solution about the ordinary point $x_0 = 0$ of a differential equation when its coefficients are not polynomials. In this example we see an application of the multiplication of two power series.

EXAMPLE 5 **ODE with Nonpolynomial Coefficients**

Solve $y'' + (\cos x)y = 0$.

Solution We see that $x = 0$ is an ordinary point of the equation because, as we have already seen, $\cos x$ is analytic at that point. Using the Maclaurin series for $\cos x$ given in (2), along with the usual assumption $y = \sum_{n=0}^{\infty} c_n x^n$ and the results in (1), we find

$$y'' + (\cos x)y = \sum_{n=2}^{\infty} n(n-1)c_n x^{n-2} + \left(1 - \frac{x^2}{2!} + \frac{x^4}{4!} - \frac{x^6}{6!} + \cdots\right)\sum_{n=0}^{\infty} c_n x^n$$

$$= 2c_2 + 6c_3 x + 12c_4 x^2 + 20c_5 x^3 + \cdots + \left(1 - \frac{x^2}{2!} + \frac{x^4}{4!} + \cdots\right)(c_0 + c_1 x + c_2 x^2 + c_3 x^3 + \cdots)$$

$$= 2c_2 + c_0 + (6c_3 + c_1)x + \left(12c_4 + c_2 - \frac{1}{2}c_0\right)x^2 + \left(20c_5 + c_3 - \frac{1}{2}c_1\right)x^3 + \cdots = 0.$$

It follows that

$$2c_2 + c_0 = 0, \quad 6c_3 + c_1 = 0, \quad 12c_4 + c_2 - \frac{1}{2}c_0 = 0, \quad 20c_5 + c_3 - \frac{1}{2}c_1 = 0,$$

and so on. This gives $c_2 = -\frac{1}{2}c_0$, $c_3 = -\frac{1}{6}c_1$, $c_4 = \frac{1}{12}c_0$, $c_5 = \frac{1}{30}c_1, \ldots$ By grouping terms we arrive at the general solution $y = c_0 y_1(x) + c_1 y_2(x)$, where

$$y_1(x) = 1 - \frac{1}{2}x^2 + \frac{1}{12}x^4 - \cdots \quad \text{and} \quad y_2(x) = x - \frac{1}{6}x^3 + \frac{1}{30}x^5 - \cdots.$$

Since the differential equation has no finite singular points, both power series converge for $|x| < \infty$. ▄

Solution Curves The approximate graph of a power series solution $y(x) = \sum_{n=0}^{\infty} c_n x^n$ can be obtained in several ways. We can always resort to graphing the terms in the sequence of partial sums of the series—in other words, the graphs of the polynomials $S_N(x) = \sum_{n=0}^{N} c_n x^n$. For large values of N,

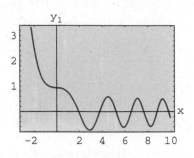

(a) plot of $y_1(x)$ vs. x

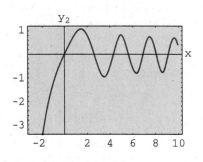

(b) plot of $y_2(x)$ vs. x

Figure 6.1 Solutions of Airy's equation

$S_N(x)$ should give us an indication of the behavior of $y(x)$ near the ordinary point $x = 0$. We can also obtain an approximate or numerical solution curve by using a solver as we did in Section 4.9. For example, if you carefully scrutinize the series solutions of Airy's equation in Example 2, you should see that $y_1(x)$ and $y_2(x)$ are, in turn, the solutions of the initial-value problems

$$y'' + xy = 0, \quad y(0) = 1, \quad y'(0) = 0,$$
$$y'' + xy = 0, \quad y(0) = 0, \quad y'(0) = 1. \tag{11}$$

The specified initial conditions "pick out" the solutions $y_1(x)$ and $y_2(x)$ from $y = c_0 y_1(x) + c_1 y_2(x)$, since it should be apparent from our basic series assumption $y = \sum_{n=0}^{\infty} c_n x^n$ that $y(0) = c_0$ and $y'(0) = c_1$. Now if your numerical solver requires a system of equations, the substitution $y' = u$ in $y'' + xy = 0$ gives $y'' = u' = -xy$, and so a system of two first-order equations equivalent to Airy's equation is

$$y' = u$$
$$u' = -xy. \tag{12}$$

Initial conditions for the system in (12) are the two sets of initial conditions in (11) rewritten as $y(0) = 1$, $u(0) = 0$ and $y(0) = 0$, $u(0) = 1$. The graphs of $y_1(x)$ and $y_2(x)$ shown in Figure 6.1 were obtained with the aid of a numerical solver. The fact that the numerical solution curves appear to be oscillatory is consistent with the fact that Airy's equation appeared in Section 5.1 (page 220) in the form $mx'' + ktx = 0$ as a model of a spring whose "spring constant" $K(t) = kt$ increases with time.

Remarks In the problems that follow, do not expect to be able to write a solution in terms of summation notation in each case. Even though we can generate as many terms as desired in a series solution $y = \sum_{n=0}^{\infty} c_n x^n$ either through the use of a recurrence relation or, as in Example 4, by multiplication, it may not be possible to deduce any general term for the coefficients c_n. We may have to settle, as we did in Examples 4 and 5, for just writing out the first few terms of the series.

EXERCISES 6.1

Answers to odd-numbered problems begin on page AN-7.

6.1.1 **Review of Power Series**

In Problems 1–4 find the radius of convergence and interval of convergence for the given power series.

1. $\displaystyle\sum_{n=1}^{\infty} \frac{2^n}{n} x^n$

2. $\displaystyle\sum_{n=0}^{\infty} \frac{(100)^n}{n!} (x + 7)^n$

3. $\displaystyle\sum_{k=1}^{\infty} \frac{(-1)^k}{10^k} (x - 5)^k$

4. $\displaystyle\sum_{k=0}^{\infty} k!(x - 1)^k$

In Problems 5 and 6 the given function is analytic at $x = 0$. Find the first four terms of a power series in x. Perform the multiplication by hand or use a CAS, as instructed.

5. $\sin x \cos x$ **6.** $e^{-x} \cos x$

In Problems 7 and 8 the given function is analytic at $x = 0$. Find the first four terms of a power series in x. Perform the long division by hand or use a CAS, as instructed. Give the open interval of convergence.

7. $\dfrac{1}{\cos x}$ **8.** $\dfrac{1 - x}{2 + x}$

In Problems 9 and 10 rewrite the given expression as a single power series.

9. $\displaystyle\sum_{n=1}^{\infty} 2nc_n x^{n-1} + \sum_{n=0}^{\infty} 6c_n x^{n+1}$

10. $\displaystyle\sum_{n=2}^{\infty} n(n-1)c_n x^n + 2\sum_{n=2}^{\infty} n(n-1)c_n x^{n-2} + 3\sum_{n=1}^{\infty} nc_n x^n$

In Problems 11 and 12 verify by direct substitution that the given power series is a particular solution of the indicated differential equation.

11. $y = \displaystyle\sum_{n=1}^{\infty} \frac{(-1)^{n+1}}{n} x^n, \quad (x+1)y'' + y' = 0$

12. $y = \displaystyle\sum_{n=0}^{\infty} \frac{(-1)^n}{2^{2n}(n!)^2} x^{2n}, \quad xy'' + y' + xy = 0$

6.1.2   **Power Series Solutions**

In Problems 13–24 find two power series solutions of the given differential equation about the ordinary point $x = 0$.

13. $y'' - xy = 0$ **14.** $y'' + x^2 y = 0$

15. $y'' - 2xy' + y = 0$ **16.** $y'' - xy' + 2y = 0$

17. $y'' + x^2 y' + xy = 0$ **18.** $y'' + 2xy' + 2y = 0$

19. $(x - 1)y'' + y' = 0$ **20.** $(x + 2)y'' + xy' - y = 0$

21. $y'' - (x + 1)y' - y = 0$ **22.** $(x^2 + 1)y'' - 6y = 0$

23. $(x^2 + 2)y'' + 3xy' - y = 0$ **24.** $(x^2 - 1)y'' + xy' - y = 0$

In Problems 25–28 use the power series method to solve the given initial-value problem.

25. $(x - 1)y'' - xy' + y = 0, \quad y(0) = -2, \, y'(0) = 6$

26. $(x + 1)y'' - (2 - x)y' + y = 0, \quad y(0) = 2, \, y'(0) = -1$

27. $y'' - 2xy' + 8y = 0, \quad y(0) = 3, \, y'(0) = 0$

28. $(x^2 + 1)y'' + 2xy' = 0, \quad y(0) = 0, \, y'(0) = 1$

In Problems 29 and 30 use the procedure in Example 5 to find two power series solutions of the given differential equation about the ordinary point $x = 0$.

29. $y'' + (\sin x)y = 0$ **30.** $y'' + e^x y' - y = 0$

Discussion Problems

31. Discuss how the power series method can be used to solve nonhomogeneous equations such as $y'' - xy = 1$ and $y'' - 4xy' - 4y = e^x$. Carry out your ideas by solving both equations.

32. Is $x = 0$ an ordinary or a singular point of the differential equation $xy'' + (\sin x)y = 0$? Defend your answer with sound mathematics.

33. The particular solution given in Problem 12 is analytic at $x = 0$ even though $x = 0$ is a singular point of the equation $xy'' + y' + xy = 0$. It is known, however, that this equation possesses a second solution that is *not* analytic at that point. Make up a linear second-order differential equation with a singular point at $x = 0$ that has *two* solutions $y_1(x)$ and $y_2(x)$ that are analytic at $x = 0$. Do not think profound thoughts.

34. For purposes of this problem ignore the graphs given in Figure 6.1. If Airy's equation is written as $y'' = -xy$, what can we say about the shape of a solution curve if $x > 0$ and $y > 0$? if $x > 0$ and $y < 0$?

Computer Lab Assignments

35. (a) Find two power series solutions for $y'' + xy' + y = 0$ and express the solutions $y_1(x)$ and $y_2(x)$ in terms of summation notation.

(b) Use a CAS to plot the graphs of the partial sums $S_N(x)$ for $y_1(x)$. Use $N = 2, 3, 5, 6, 8, 10$. Repeat using the partial sums $S_N(x)$ for $y_2(x)$.

(c) Compare the graphs obtained in part (b) with the curve obtained using a numerical solver. Use the initial conditions $y_1(0) = 1$, $y_1'(0) = 0$ and $y_2(0) = 0$, $y_2'(0) = 1$.

(d) Reexamine the solution $y_1(x)$ in part (a). Express this series as an elementary function. Then use (5) of Section 4.2 to find a second solution of the equation. Verify that this second solution is the same as the power series solution $y_2(x)$.

36. (a) Find one more nonzero term for each of the solutions $y_1(x)$ and $y_2(x)$ in Example 5.

(b) Find a series solution $y(x)$ of the initial-value problem $y'' + (\cos x)y = 0$, $y(0) = 1$, $y'(0) = 1$.

(c) Use a CAS to plot the graphs of the partial sums $S_N(x)$ for the solution $y(x)$ in part (b). Use $N = 2, 3, 4, 5, 6, 7$.

(d) Compare the graphs obtained in part (c) with the curve obtained using a numerical solver for the initial-value problem in part (b).

6.2 SOLUTIONS ABOUT SINGULAR POINTS

- *Regular singular points* • *Irregular singular points* • *Frobenius' Theorem* • *Series solutions of ODEs* • *Indicial equation* • *Indicial roots*

The two differential equations $y'' + xy = 0$ and $xy'' + y = 0$ are similar only in that they are both examples of simple linear second-order equations with variable coefficients. That is all they have in common. We saw in the preceding section that since $x = 0$ is an *ordinary point* of the first equation, there is no problem in finding two linearly independent power series solutions centered at that point. In contrast, because $x = 0$ is a *singular point* of the second ODE, finding two infinite series solutions of the equation about that point becomes a more difficult task. See Examples 4 and 5.

Regular and Irregular Singular Points   A singular point $x = x_0$ of a linear differential equation

$$a_2(x)y'' + a_1(x)y' + a_0(x)y = 0 \qquad (1)$$

is further classified as either regular or irregular. The classification again depends on the functions P and Q in the standard form

$$y'' + P(x)y' + Q(x)y = 0. \qquad (2)$$

DEFINITION 6.2   **Regular and Irregular Singular Points**

A singular point x_0 is said to be a **regular singular point** of the differential equation (1) if the functions $p(x) = (x - x_0)P(x)$ and $q(x) = (x - x_0)^2Q(x)$ are both analytic at x_0. A singular point that is not regular is said to be an **irregular singular point** of the equation.

The second sentence in Definition 6.2 indicates that if one or both of the functions $p(x) = (x - x_0)P(x)$ and $q(x) = (x - x_0)^2Q(x)$ fail to be analytic at x_0, then x_0 is an irregular singular point.

Polynomial Coefficients   As in Section 6.1, we are mainly interested in linear equations (1) where the coefficients $a_2(x)$, $a_1(x)$, and $a_0(x)$ are polynomials with no common factors. We have already seen that if $a_2(x_0) = 0$ then $x = x_0$ is a singular point of (1), since at least one of the rational functions $P(x) = a_1(x)/a_2(x)$ and $Q(x) = a_0(x)/a_2(x)$ in the standard form (2) fails to be analytic at that point. But since $a_2(x)$ is a polynomial and x_0 is one of its zeros, it follows from the Factor Theorem of algebra that $x - x_0$ is a factor of $a_2(x)$. This means that after $a_1(x)/a_2(x)$ and $a_0(x)/a_2(x)$ are reduced to lowest terms, the factor $x - x_0$ must remain, to some positive integer power, in one or both denominators. Now suppose that $x = x_0$ is a singular point of (1) but both the functions defined by the products $p(x) = (x - x_0)P(x)$ and $q(x) = (x - x_0)^2Q(x)$ are analytic at x_0. We are lead to the conclusion that multiplying $P(x)$ by $x - x_0$ and $Q(x)$ by $(x - x_0)^2$ has the effect (through cancellation) that $x - x_0$ no longer appears in either denominator. We can now determine whether x_0 is regular by a quick visual check of denominators: If $x - x_0$ appears *at most* to the first power in the denominator of $P(x)$ and *at most* to the second power in the denominator of $Q(x)$, then $x = x_0$ is a regular singular point. Moreover, observe that if $x = x_0$ is a regular singular point and we multiply (2) by $(x - x_0)^2$, then the original DE can be put into the form

$$(x - x_0)^2 y'' + (x - x_0)p(x)y' + q(x)y = 0, \tag{3}$$

where p and q are analytic at $x = x_0$.

EXAMPLE 1   Classification of Singular Points

It should be clear that $x = 2$ and $x = -2$ are singular points of

$$(x^2 - 4)^2 y'' + 3(x - 2)y' + 5y = 0.$$

After dividing the equation by $(x^2 - 4)^2 = (x - 2)^2(x + 2)^2$ and reducing the coefficients to lowest terms, we find that

$$P(x) = \frac{3}{(x - 2)(x + 2)^2} \quad \text{and} \quad Q(x) = \frac{5}{(x - 2)^2(x + 2)^2}.$$

We now test $P(x)$ and $Q(x)$ at each singular point.

In order for $x = 2$ to be a regular singular point, the factor $x - 2$ can appear at most to the first power in the denominator of $P(x)$ and at most to the second power in the denominator of $Q(x)$. A check of the denominators of $P(x)$ and $Q(x)$ shows that both these conditions are satisfied, so $x = 2$ is a regular singular point. Alternatively, we are lead to the same conclusion by noting that both rational functions

$$p(x) = (x - 2)P(x) = \frac{3}{(x + 2)^2} \quad \text{and} \quad q(x) = (x - 2)^2 Q(x) = \frac{5}{(x + 2)^2}$$

are analytic at $x = 2$.

Now since the factor $x - (-2) = x + 2$ appears to the second power in the denominator of $P(x)$, we can conclude immediately that $x = -2$ is an irregular singular point of the equation. This also follows from the fact that

$$p(x) = (x + 2)P(x) = \frac{3}{(x - 2)(x + 2)}$$

is not analytic at $x = -2$.   ∎

In Example 1, notice that since $x = 2$ is a regular singular point, the original equation can be written as

$$\overset{\displaystyle \downarrow\, p(x)\text{ analytic}}{\underset{\text{at } x=2}{}} \qquad \overset{\displaystyle \downarrow\, q(x)\text{ analytic}}{\underset{\text{at } x=2}{}}$$

$$(x - 2)^2 y'' + (x - 2)\frac{3}{(x + 2)^2}y' + \frac{5}{(x + 2)^2}y = 0.$$

As another example, we can see that $x = 0$ is an irregular singular point of $x^3 y'' - 2xy' + 8y = 0$ by inspection of the denominators of $P(x) = -2/x^2$ and $Q(x) = 8/x^3$. On the other hand, $x = 0$ is a regular singular point of $xy'' - 2xy' + 8y = 0$, since $x - 0$ and $(x - 0)^2$ do not even appear in the respective denominators of $P(x) = -2$ and $Q(x) = 8/x$. For a singular point $x = x_0$, any nonnegative power of $x - x_0$ less than one (namely, zero) and any nonnegative power less than two (namely, zero and one) in the denominators of $P(x)$ and $Q(x)$, respectively, imply that x_0 is a regular singular point. A singular point can be a complex number. You should verify that $x = 3i$ and $x = -3i$ are two regular singular points of $(x^2 + 9)y'' - 3xy' + (1 - x)y = 0$.

Any second-order Cauchy-Euler equation $ax^2y'' + bxy' + cy = 0$, a, b, and c real constants, has a regular singular point at $x = 0$. You should verify that two solutions of the Cauchy-Euler equation $x^2y'' - 3xy' + 4y = 0$ on the interval $(0, \infty)$ are $y_1 = x^2$ and $y_2 = x^2 \ln x$. If we attempted to find a power series solution about the regular singular point $x = 0$ (namely, $y = \sum_{n=0}^{\infty} c_n x^n$), we would succeed in obtaining only the polynomial solution $y_1 = x^2$. The fact that we would not obtain the second solution is not surprising, since $\ln x$ (and consequently $y_2 = x^2 \ln x$) is not analytic at $x = 0$—that is, y_2 does not possess a Taylor series expansion centered at $x = 0$.

Method of Frobenius To solve a differential equation (1) about a regular singular point we employ the following theorem due to Frobenius.

THEOREM 6.2   Frobenius' Theorem

If $x = x_0$ is a regular singular point of the differential equation (1), then there exists at least one solution of the form

$$y = (x - x_0)^r \sum_{n=0}^{\infty} c_n(x - x_0)^n = \sum_{n=0}^{\infty} c_n(x - x_0)^{n+r}, \qquad \textbf{(4)}$$

where the number r is a constant to be determined. The series will converge at least on some interval $0 < x - x_0 < R$.

Notice the words *at least* in the first sentence of Theorem 6.2. This means that, in contrast to Theorem 6.1, Theorem 6.2 gives us no assurance that two series solutions of the type indicated in (4) can be found. The **method of Frobenius,** finding series solutions about a regular singular point x_0, is similar to the "method of undetermined series coefficients" of the preceding section in that we substitute $y = \sum_{n=0}^{\infty} c_n(x - x_0)^{n+r}$ into the given differential equation and determine the unknown coefficients c_n by a recurrence relation. However, we have an additional task in this procedure; before determining the coefficients, we must find the unknown exponent r. If r is found to be a number that is not a nonnegative integer, then the corresponding solution $y = \sum_{n=0}^{\infty} c_n(x - x_0)^{n+r}$ is not a power series.

As we did in the discussion of solutions about ordinary points, we shall always assume, for the sake of simplicity in solving differential equations, that the regular singular point is $x = 0$.

EXAMPLE 2   Two Series Solutions

Because $x = 0$ is a regular singular point of the differential equation

$$3xy'' + y' - y = 0, \qquad \textbf{(5)}$$

we try to find a solution of the form $y = \sum_{n=0}^{\infty} c_n x^{n+r}$. Now

$$y' = \sum_{n=0}^{\infty} (n + r)c_n x^{n+r-1} \quad \text{and} \quad y'' = \sum_{n=0}^{\infty} (n + r)(n + r - 1)c_n x^{n+r-2},$$

so

$$3xy'' + y' - y = 3 \sum_{n=0}^{\infty} (n+r)(n+r-1)c_n x^{n+r-1} + \sum_{n=0}^{\infty} (n+r)c_n x^{n+r-1} - \sum_{n=0}^{\infty} c_n x^{n+r}$$

$$= \sum_{n=0}^{\infty} (n+r)(3n+3r-2)c_n x^{n+r-1} - \sum_{n=0}^{\infty} c_n x^{n+r}$$

$$= x^r \left[r(3r-2)c_0 x^{-1} + \underbrace{\sum_{n=1}^{\infty} (n+r)(3n+3r-2)c_n x^{n-1}}_{k=n-1} - \underbrace{\sum_{n=0}^{\infty} c_n x^n}_{k=n} \right]$$

$$= x^r \left[r(3r-2)c_0 x^{-1} + \sum_{k=0}^{\infty} [(k+r+1)(3k+3r+1)c_{k+1} - c_k]x^k \right] = 0,$$

which implies $$r(3r-2)c_0 = 0$$

and $$(k+r+1)(3k+3r+1)c_{k+1} - c_k = 0, \quad k = 0, 1, 2, \ldots.$$

Since nothing is gained by taking $c_0 = 0$, we must then have

$$r(3r-2) = 0 \tag{6}$$

and $$c_{k+1} = \frac{c_k}{(k+r+1)(3k+3r+1)}, \quad k = 0, 1, 2, \ldots. \tag{7}$$

When substituted in (7), the two values of r that satisfy the quadratic equation (6), $r_1 = \frac{2}{3}$ and $r_2 = 0$, give two different recurrence relations:

$$r_1 = \tfrac{2}{3}, \quad c_{k+1} = \frac{c_k}{(3k+5)(k+1)}, \quad k = 0, 1, 2, \ldots \tag{8}$$

$$r_2 = 0, \quad c_{k+1} = \frac{c_k}{(k+1)(3k+1)}, \quad k = 0, 1, 2, \ldots. \tag{9}$$

From (8) we find

$$c_1 = \frac{c_0}{5 \cdot 1}$$

$$c_2 = \frac{c_1}{8 \cdot 2} = \frac{c_0}{2!5 \cdot 8}$$

$$c_3 = \frac{c_2}{11 \cdot 3} = \frac{c_0}{3!5 \cdot 8 \cdot 11}$$

$$c_4 = \frac{c_3}{14 \cdot 4} = \frac{c_0}{4!5 \cdot 8 \cdot 11 \cdot 14}$$

.
.
.

$$c_n = \frac{c_0}{n!5 \cdot 8 \cdot 11 \cdots (3n+2)}.$$

From (9) we find

$$c_1 = \frac{c_0}{1 \cdot 1}$$

$$c_2 = \frac{c_1}{2 \cdot 4} = \frac{c_0}{2!1 \cdot 4}$$

$$c_3 = \frac{c_2}{3 \cdot 7} = \frac{c_0}{3!1 \cdot 4 \cdot 7}$$

$$c_4 = \frac{c_3}{4 \cdot 10} = \frac{c_0}{4!1 \cdot 4 \cdot 7 \cdot 10}$$

.
.
.

$$c_n = \frac{(-1)^n c_0}{n!1 \cdot 4 \cdot 7 \cdots (3n-2)}.$$

Here we encounter something that did not happen when we obtained solutions about an ordinary point; we have what looks to be two different sets of coefficients, but each set contains the *same* multiple c_0. If we omit this term, the series solutions are

$$y_1(x) = x^{2/3}\left[1 + \sum_{n=1}^{\infty} \frac{1}{n!5 \cdot 8 \cdot 11 \cdots (3n+2)} x^n\right] \qquad (10)$$

$$y_2(x) = x^0\left[1 + \sum_{n=1}^{\infty} \frac{1}{n!1 \cdot 4 \cdot 7 \cdots (3n-2)} x^n\right]. \qquad (11)$$

By the ratio test it can be demonstrated that both (10) and (11) converge for all finite values of x—that is, $|x| < \infty$. Also, it should be apparent from the form of these solutions that neither series is a constant multiple of the other and, therefore, $y_1(x)$ and $y_2(x)$ are linearly independent on the entire x-axis. Hence, by the superposition principle, $y = C_1 y_1(x) + C_2 y_2(x)$ is another solution of (5). On any interval not containing the origin, such as $(0, \infty)$, this linear combination represents the general solution of the differential equation. ∎

Indicial Equation   Equation (6) is called the **indicial equation** of the problem, and the values $r_1 = \frac{2}{3}$ and $r_2 = 0$ are called the **indicial roots,** or **exponents,** of the singularity $x = 0$. In general, after substituting $y = \sum_{n=0}^{\infty} c_n x^{n+r}$ into the given differential equation and simplifying, the indicial equation is a quadratic equation in r that results from equating the *total coefficient of the lowest power of x to zero*. We solve for the two values of r and substitute these values into a recurrence relation such as (7). Theorem 6.2 guarantees that at least one solution of the assumed series form can be found.

It is possible to obtain the indicial equation in advance of substituting $y = \sum_{n=0}^{\infty} c_n x^{n+r}$ into the differential equation. If $x = 0$ is a regular singular point of (1), then by Definition 6.2 both functions $p(x) = xP(x)$ and $q(x) = x^2 Q(x)$, where P and Q are defined by the standard form (2), are analytic at $x = 0$; that is, the power series expansions

$$p(x) = xP(x) = a_0 + a_1 x + a_2 x^2 + \cdots \quad \text{and} \quad q(x) = x^2 Q(x) = b_0 + b_1 x + b_2 x^2 + \cdots \qquad (12)$$

are valid on intervals that have a positive radius of convergence. By multiplying (2) by x^2, we get the form given in (3):

$$x^2 y'' + x[xP(x)]y' + [x^2 Q(x)]y = 0. \qquad (13)$$

After substituting $y = \sum_{n=0}^{\infty} c_n x^{n+r}$ and the two series in (12) into (13) and carrying out the multiplication of series, we find the general indicial equation to be

$$r(r-1) + a_0 r + b_0 = 0, \qquad (14)$$

where a_0 and b_0 are as defined in (12). See Problems 13 and 14 in Exercises 6.2.

EXAMPLE 3   **Two Series Solutions**

Solve $2xy'' + (1 + x)y' + y = 0$.

Solution   Substituting $y = \sum_{n=0}^{\infty} c_n x^{n+r}$ gives

$$2xy'' + (1 + x)y' + y = 2\sum_{n=0}^{\infty} (n + r)(n + r - 1)c_n x^{n+r-1} + \sum_{n=0}^{\infty} (n + r)c_n x^{n+r-1}$$

$$+ \sum_{n=0}^{\infty} (n + r)c_n x^{n+r} + \sum_{n=0}^{\infty} c_n x^{n+r}$$

$$= \sum_{n=0}^{\infty} (n + r)(2n + 2r - 1)c_n x^{n+r-1} + \sum_{n=0}^{\infty} (n + r + 1)c_n x^{n+r}$$

$$= x^r\left[r(2r - 1)c_0 x^{-1} + \underbrace{\sum_{n=1}^{\infty} (n + r)(2n + 2r - 1)c_n x^{n-1}}_{k=n-1} + \underbrace{\sum_{n=0}^{\infty} (n + r + 1)c_n x^n}_{k=n} \right]$$

$$= x^r\left[r(2r - 1)c_0 x^{-1} + \sum_{k=0}^{\infty} [(k + r + 1)(2k + 2r + 1)c_{k+1} + (k + r + 1)c_k]x^k \right],$$

which implies $$r(2r - 1) = 0 \qquad \textbf{(15)}$$

and $$(k + r + 1)(2k + 2r + 1)c_{k+1} + (k + r + 1)c_k = 0, \qquad \textbf{(16)}$$

$k = 0, 1, 2, \ldots$. From (15) we see that the indicial roots are $r_1 = \frac{1}{2}$ and $r_2 = 0$.

For $r_1 = \frac{1}{2}$, we can divide by $k + \frac{3}{2}$ in (16) to obtain

$$c_{k+1} = \frac{-c_k}{2(k + 1)}, \quad k = 0, 1, 2, \ldots, \qquad \textbf{(17)}$$

whereas for $r_2 = 0$, (16) becomes

$$c_{k+1} = \frac{-c_k}{2k + 1}, \quad k = 0, 1, 2, \ldots. \qquad \textbf{(18)}$$

From (17) we find

$$c_1 = \frac{-c_0}{2 \cdot 1}$$

$$c_2 = \frac{-c_1}{2 \cdot 2} = \frac{c_0}{2^2 \cdot 2!}$$

$$c_3 = \frac{-c_2}{2 \cdot 3} = \frac{-c_0}{2^3 \cdot 3!}$$

$$c_4 = \frac{-c_3}{2 \cdot 4} = \frac{c_0}{2^4 \cdot 4!}$$

$$\vdots$$

$$c_n = \frac{(-1)^n c_0}{2^n n!}.$$

From (18) we find

$$c_1 = \frac{-c_0}{1}$$

$$c_2 = \frac{-c_1}{3} = \frac{c_0}{1 \cdot 3}$$

$$c_3 = \frac{-c_2}{5} = \frac{-c_0}{1 \cdot 3 \cdot 5}$$

$$c_4 = \frac{-c_3}{7} = \frac{c_0}{1 \cdot 3 \cdot 5 \cdot 7}$$

$$\vdots$$

$$c_n = \frac{(-1)^n c_0}{1 \cdot 3 \cdot 5 \cdot 7 \cdots (2n - 1)}.$$

Thus for the indicial root $r_1 = \frac{1}{2}$ we obtain the solution

$$y_1(x) = x^{1/2}\left[1 + \sum_{n=1}^{\infty} \frac{(-1)^n}{2^n n!} x^n\right] = \sum_{n=0}^{\infty} \frac{(-1)^n}{2^n n!} x^{n+1/2},$$

where we have again omitted c_0. The series converges for $x \geq 0$; as given, the series is not defined for negative values of x because of the presence of $x^{1/2}$. For $r_2 = 0$, a second solution is

$$y_2(x) = 1 + \sum_{n=1}^{\infty} \frac{(-1)^n}{1 \cdot 3 \cdot 5 \cdot 7 \cdots (2n-1)} x^n, \quad |x| < \infty.$$

On the interval $(0, \infty)$, the general solution is $y = C_1 y_1(x) + C_2 y_2(x)$. ∎

EXAMPLE 4 **Only One Series Solution**

Solve $xy'' + y = 0$.

Solution From $xP(x) = 0$, $x^2 Q(x) = x$, and the fact that 0 and x are their own power series centered at 0 we conclude that $a_0 = 0$ and $b_0 = 0$, and so from (14) the indicial equation is $r(r - 1) = 0$. You should verify that the two recurrence relations corresponding to the indicial roots $r_1 = 1$ and $r_2 = 0$ yield exactly the same set of coefficients. In other words, in this case the method of Frobenius produces only a single series solution

$$y_1(x) = \sum_{n=0}^{\infty} \frac{(-1)^n}{n!(n+1)!} x^{n+1} = x + \frac{1}{2}x^2 + \frac{1}{12}x^3 + \frac{1}{144}x^4 + \cdots. \quad ∎$$

Three Cases For the sake of discussion let us suppose again that $x = 0$ is a regular singular point of (1) and that indicial roots r_1 and r_2 of the singularity are real, with r_1 denoting the largest root. When using the method of Frobenius we usually distinguish three cases corresponding to the nature of the indicial roots r_1 and r_2.

Case I If r_1 and r_2 are distinct and do not differ by an integer, then there exist two linearly independent solutions $y_1(x)$ and $y_2(x)$ of equation (1) of the form $y = \sum_{n=0}^{\infty} c_n x^{n+r}$. This is the case illustrated in Examples 2 and 3.

In the next case, we see that when the difference of indicial roots $r_1 - r_2$ is a positive integer, the second solution *may* contain a logarithm.

Case II If $r_1 - r_2 = N$, where N is a positive integer, then there exist two linearly independent solutions of equation (1) of the form

$$y_1(x) = \sum_{n=0}^{\infty} c_n x^{n+r_1}, \quad c_0 \neq 0 \tag{19}$$

$$y_2(x) = Cy_1(x) \ln x + \sum_{n=0}^{\infty} b_n x^{n+r_2}, \quad b_0 \neq 0, \tag{20}$$

where C is a constant that could be zero.

When the difference $r_1 - r_2$ is a positive integer, we *may* or *may not* be able to find *two* solutions having the form $y = \sum_{n=0}^{\infty} c_n x^{n+r}$. This is something we do not know in advance but can determine after we have found the indicial roots and have carefully examined the recurrence relation that defines the coefficients c_n. We just may be lucky enough to find two solutions that involve only powers of x—that is, $y_1 = \sum_{n=0}^{\infty} c_n x^{n+r_1}$ (equation (19)) and $y_2 = \sum_{n=0}^{\infty} b_n x^{n+r_2}$ (equation (20) with $C = 0$). See Problem 31 in Exercises 6.2. On the other hand, in Example 4 we had $r_1 - r_2 = 1$, an integer, and the method of Frobenius failed to give a second series solution. In this situation equation (20), with $C \neq 0$, indicates what the second solution looks like. One way to obtain this second solution with the logarithmic term is to use the fact that

$$y_2(x) = y_1(x) \int \frac{e^{-\int P(x)\,dx}}{y_1^2(x)}\,dx \tag{21}$$

is also a solution of $y'' + P(x)y' + Q(x)y = 0$ whenever $y_1(x)$ is the known solution (see (5) in Section 4.2). We will illustrate how to use this formula in Example 5.

In the last case, when the indicial roots are equal, a second solution will *always* contain a logarithm. This situation is analogous to the solution of a Cauchy-Euler equation when the roots of the auxiliary equation are equal.

Case III   If $r_1 = r_2$ then there always exist two linearly independent solutions of equation (1) of the form

$$y_1(x) = \sum_{n=0}^{\infty} c_n x^{n+r_1}, \quad c_0 \neq 0 \tag{22}$$

$$y_2(x) = y_1(x) \ln x + \sum_{n=1}^{\infty} b_n x^{n+r_2}. \tag{23}$$

The solution $y_2(x)$ in (23) can also be obtained using (21).

| EXAMPLE 5 | **Example 4 Revisited Using a CAS** |

Find the general solution of $xy'' + y = 0$.

Solution   From the known solution given in Example 4,

$$y_1(x) = x + \frac{1}{2}x^2 + \frac{1}{12}x^3 + \frac{1}{144}x^4 + \cdots,$$

we can construct a second solution $y_2(x)$ using formula (21). Those with the time, energy, and patience can carry out the drudgery of squaring a series, long division, and integration of the quotient by hand. But all these operations can be done with relative ease with the help of a CAS. We give the results:

$$y_2(x) = y_1(x) \int \frac{e^{-\int 0\,dx}}{[y_1(x)]^2}\,dx = y_1(x) \int \frac{dx}{\left[x + \dfrac{1}{2}x^2 + \dfrac{1}{12}x^3 + \dfrac{1}{144}x^4 + \cdots\right]^2}$$

$$= y_1(x) \int \frac{dx}{\left[x^2 - x^3 + \dfrac{5}{12}x^4 - \dfrac{7}{12}x^5 + \cdots\right]} \qquad \leftarrow \text{after squaring}$$

$$= y_1(x) \int \left[\frac{1}{x^2} + \frac{1}{x} + \frac{7}{12} + \frac{19}{72}x + \cdots\right] dx \qquad \leftarrow \text{after long division}$$

$$= y_1(x) \left[-\frac{1}{x} + \ln x + \frac{7}{12}x + \frac{19}{144}x^2 + \cdots\right] \qquad \leftarrow \text{after integrating}$$

or $\quad y_2(x) = y_1(x) \ln x + y_1(x) \left[-\dfrac{1}{x} + \dfrac{7}{12}x + \dfrac{19}{144}x^2 + \cdots\right].$

On the interval $(0, \infty)$, the general solution is $y = C_1 y_1(x) + C_2 y_2(x)$.

Remarks (*i*) The three different forms of a linear second-order differential equation ((1), (2), and (3)) were used to discuss various theoretical concepts. But on a practical level, when it comes to actually solving a differential equation using the method of Frobenius, it is advisable to work with the form of the DE given in (1).
(*ii*) When the difference of indicial roots $r_1 - r_2$ is a positive integer ($r_1 > r_2$), it sometimes pays to iterate the recurrence relation using the smaller root r_2 first. See Problems 31 and 32 in Exercise 6.2.
(*iii*) Since r is a root of a quadratic equation, it could be complex. We shall not, however, investigate this case.
(*iv*) If $x = 0$ is an irregular singular point, we may not be able to find *any* solution of the form $y = \sum_{n=0}^{\infty} c_n x^{n+r}$.

EXERCISES 6.2

Answers to odd-numbered problems begin on page AN-7.

In Problems 1–10 determine the singular points of the given differential equation. Classify each singular point as regular or irregular.

1. $x^3 y'' + 4x^2 y' + 3y = 0$ **2.** $x(x + 3)^2 y'' - y = 0$

3. $(x^2 - 9)^2 y'' + (x + 3)y' + 2y = 0$

4. $y'' - \dfrac{1}{x}y' + \dfrac{1}{(x - 1)^3}y = 0$

5. $(x^3 + 4x)y'' - 2xy' + 6y = 0$

6. $x^2(x - 5)^2y'' + 4xy' + (x^2 - 25)y = 0$

7. $(x^2 + x - 6)y'' + (x + 3)y' + (x - 2)y = 0$

8. $x(x^2 + 1)^2y'' + y = 0$

9. $x^3(x^2 - 25)(x - 2)^2y'' + 3x(x - 2)y' + 7(x + 5)y = 0$

10. $(x^3 - 2x^2 + 3x)^2y'' + x(x - 3)^2y' - (x + 1)y = 0$

In Problems 11 and 12 put the given differential equation into form (3) for each regular singular point of the equation. Identify the functions $p(x)$ and $q(x)$.

11. $(x^2 - 1)y'' + 5(x + 1)y' + (x^2 - x)y = 0$

12. $xy'' + (x + 3)y' + 7x^2y = 0$

In Problems 13 and 14, $x = 0$ is a regular singular point of the given differential equation. Use the general form of the indicial equation in (14) to find the indicial roots of the singularity. Without solving, discuss the number of series solutions you would expect to find using the method of Frobenius.

13. $x^2y'' + (\tfrac{5}{3}x + x^2)y' - \tfrac{1}{3}y = 0$ **14.** $xy'' + y' + 10y = 0$

In Problems 15–24, $x = 0$ is a regular singular point of the given differential equation. Show that the indicial roots of the singularity do not differ by an integer. Use the method of Frobenius to obtain two linearly independent series solutions about $x = 0$. Form the general solution on $(0, \infty)$.

15. $2xy'' - y' + 2y = 0$

16. $2xy'' + 5y' + xy = 0$

17. $4xy'' + \tfrac{1}{2}y' + y = 0$

18. $2x^2y'' - xy' + (x^2 + 1)y = 0$

19. $3xy'' + (2 - x)y' - y = 0$

20. $x^2y'' - (x - \tfrac{2}{9})y = 0$

21. $2xy'' - (3 + 2x)y' + y = 0$

22. $x^2y'' + xy' + (x^2 - \tfrac{4}{9})y = 0$

23. $9x^2y'' + 9x^2y' + 2y = 0$

24. $2x^2y'' + 3xy' + (2x - 1)y = 0$

 In Problems 25–30, $x = 0$ is a regular singular point of the given differential equation. Show that the indicial roots of the singularity differ by an integer. Use the method of Frobenius to obtain at least one series solution about $x = 0$. Use (21) where necessary and a CAS, if instructed, to find a second solution. Form the general solution on $(0, \infty)$.

25. $xy'' + 2y' - xy = 0$

26. $x^2y'' + xy' + (x^2 - \tfrac{1}{4})y = 0$

27. $xy'' - xy' + y = 0$

28. $y'' + \dfrac{3}{x}y' - 2y = 0$

29. $xy'' + (1 - x)y' - y = 0$

30. $xy'' + y' + y = 0$

In Problems 31 and 32, $x = 0$ is a regular singular point of the given differential equation. Show that the indicial roots of the singularity differ by an integer. Use the recurrence relation found by the method of Frobenius first with the larger root r_1. How many solutions did you find? Next use the recurrence relation with the smaller root r_2. How many solutions did you find?

31. $xy'' + (x - 6)y' - 3y = 0$ **32.** $x(x - 1)y'' + 3y' - 2y = 0$

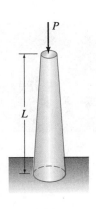

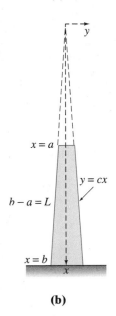

(a)

(b)

Figure 6.2

33. (a) The differential equation $x^4 y'' + \lambda y = 0$ has an irregular singular point at $x = 0$. Show that the substitution $t = 1/x$ yields the differential equation

$$\frac{d^2 y}{dt^2} + \frac{2}{t}\frac{dy}{dt} + \lambda y = 0,$$

which now has a regular singular point at $t = 0$.
(b) Use the method of this section to find two series solutions of the second equation in part (a) about the singular point $t = 0$.
(c) Express each series solution of the original equation in terms of elementary functions.

34. In Example 3 of Section 5.2 we saw that when a constant vertical compressive force, or load, P was applied to a thin column of uniform cross-section, the deflection $y(x)$ satisfied the boundary-value problem

$$EI\frac{d^2 y}{dx^2} + Py = 0, \quad y(0) = 0, \quad y(L) = 0. \qquad \textbf{(24)}$$

The column will buckle or deflect only when the compressive force is a critical load P_n.
(a) Now suppose the column is of length L and has circular cross-sections, but is tapered as shown in Figure 6.2(a). If the column, a truncated cone, has a linear taper $y = cx$ as shown in the cross-section in Figure 6.2(b), the moment of inertia of a cross-section with respect to an axis perpendicular to the xy-plane is $I = \frac{1}{4}\pi r^4$, where $r = y$ and $y = cx$. Hence we can write $I(x) = I_0(x/b)^4$, where $I_0 = I(b) = \frac{1}{4}\pi(cb)^4$. Substituting $I(x)$ into the differential equation (24), we see that the deflection in this case is determined from the boundary-value problem

$$x^4 \frac{d^2 y}{dx^2} + \lambda y = 0, \quad y(a) = 0, \quad y(b) = 0,$$

where $\lambda = Pb^4/EI_0$. Use the results of Problem 33 to find the critical loads P_n for the tapered column. Use an appropriate identity to express the buckling modes $y_n(x)$ as a single function.
(b) Use a graphing utility to plot the graph of the first buckling mode $y_1(x)$ corresponding to the Euler load P_1 when $b = 11$ and $a = 1$.

Discussion Problems

35. Discuss how you would define a regular singular point for the linear third-order differential equation

$$a_3(x)y''' + a_2(x)y'' + a_1(x)y' + a_0(x)y = 0.$$

36. Each of the differential equations $x^3 y'' + y = 0$ and $x^2 y'' + (3x - 1)y' + y = 0$ has an irregular singular point at $x = 0$. Determine whether the method of Frobenius yields a series solution of each differential equation about $x = 0$. Discuss and explain your findings.

37. We have seen that $x = 0$ is a regular singular point of any Cauchy-Euler equation $ax^2 y'' + bxy' + cy = 0$. Are the indicial equation (14) for a Cauchy-Euler equation and its auxiliary equation related? Discuss.

6.3 TWO SPECIAL EQUATIONS

• Bessel's equation   • Legendre's equation   • Bessel functions of the first kind   • Bessel functions of the second kind   • Parametric Bessel ODE   • Differential recurrence relations   • Spherical Bessel functions   • Legendre polynomials   • Recurrence relation

The two differential equations

$$x^2 y'' + xy' + (x^2 - v^2)y = 0 \tag{1}$$

and

$$(1 - x^2)y'' - 2xy' + n(n + 1)y = 0 \tag{2}$$

occur frequently in advanced studies in applied mathematics, physics, and engineering. They are called **Bessel's equation** and **Legendre's equation,** respectively. In solving (1) we shall assume $v \geq 0$, whereas in (2) we shall consider only the case when n is a nonnegative integer. Since we seek series solutions of each equation about $x = 0$, we observe that the origin is a regular singular point of Bessel's equation, but is an ordinary point of Legendre's equation.

Solution of Bessel's Equation   Because $x = 0$ is a regular singular point of Bessel's equation we know that there exists at least one solution of the form $y = \sum_{n=0}^{\infty} c_n x^{n+r}$. Substituting the last expression into (1) gives

$$x^2 y'' + xy' + (x^2 - v^2)y = \sum_{n=0}^{\infty} c_n(n + r)(n + r - 1)x^{n+r} + \sum_{n=0}^{\infty} c_n(n + r)x^{n+r} + \sum_{n=0}^{\infty} c_n x^{n+r+2} - v^2 \sum_{n=0}^{\infty} c_n x^{n+r}$$

$$= c_0(r^2 - r + r - v^2)x^r$$

$$+ x^r \sum_{n=1}^{\infty} c_n[(n + r)(n + r - 1) + (n + r) - v^2]x^n + x^r \sum_{n=0}^{\infty} c_n x^{n+2}$$

$$= c_0(r^2 - v^2)x^r + x^r \sum_{n=1}^{\infty} c_n[(n + r)^2 - v^2]x^n + x^r \sum_{n=0}^{\infty} c_n x^{n+2}. \tag{3}$$

From (3) we see that the indicial equation is $r^2 - v^2 = 0$, so the indicial roots are $r_1 = v$ and $r_2 = -v$. When $r_1 = v$, (3) becomes

$$x^v \sum_{n=1}^{\infty} c_n n(n + 2v)x^n + x^v \sum_{n=0}^{\infty} c_n x^{n+2}$$

$$= x^v \left[(1 + 2v)c_1 x + \underbrace{\sum_{n=2}^{\infty} c_n n(n + 2v)x^n}_{k = n - 2} + \underbrace{\sum_{n=0}^{\infty} c_n x^{n+2}}_{k = n} \right]$$

$$= x^v \left[(1 + 2v)c_1 x + \sum_{k=0}^{\infty} [(k + 2)(k + 2 + 2v)c_{k+2} + c_k]x^{k+2} \right] = 0.$$

Therefore by the usual argument we can write $(1 + 2v)c_1 = 0$ and

$$(k + 2)(k + 2 + 2v)c_{k+2} + c_k = 0$$

or
$$c_{k+2} = \frac{-c_k}{(k+2)(k+2+2\nu)}, \quad k = 0, 1, 2, \ldots. \tag{4}$$

The choice $c_1 = 0$ in (4) implies $c_3 = c_5 = c_7 = \cdots = 0$, so for $k = 0, 2, 4, \ldots$ we find, after letting $k + 2 = 2n$, $n = 1, 2, 3, \ldots$, that

$$c_{2n} = -\frac{c_{2n-2}}{2^2 n(n+\nu)}. \tag{5}$$

Thus
$$c_2 = -\frac{c_0}{2^2 \cdot 1 \cdot (1+\nu)}$$

$$c_4 = -\frac{c_2}{2^2 \cdot 2(2+\nu)} = \frac{c_0}{2^4 \cdot 1 \cdot 2(1+\nu)(2+\nu)}$$

$$c_6 = -\frac{c_4}{2^2 \cdot 3(3+\nu)} = -\frac{c_0}{2^6 \cdot 1 \cdot 2 \cdot 3(1+\nu)(2+\nu)(3+\nu)}$$

$$\vdots$$

$$c_{2n} = \frac{(-1)^n c_0}{2^{2n} n!(1+\nu)(2+\nu)\cdots(n+\nu)}, \quad n = 1, 2, 3, \ldots. \tag{6}$$

It is standard practice to choose c_0 to be a specific value—namely,

$$c_0 = \frac{1}{2^\nu \Gamma(1+\nu)},$$

where $\Gamma(1+\nu)$ is the gamma function. See Appendix I. Since this latter function possesses the convenient property $\Gamma(1+\alpha) = \alpha\Gamma(\alpha)$, we can reduce the indicated product in the denominator of (6) to one term. For example,

$$\Gamma(1+\nu+1) = (1+\nu)\Gamma(1+\nu)$$
$$\Gamma(1+\nu+2) = (2+\nu)\Gamma(2+\nu) = (2+\nu)(1+\nu)\Gamma(1+\nu).$$

Hence we can write (6) as

$$c_{2n} = \frac{(-1)^n}{2^{2n+\nu} n!(1+\nu)(2+\nu)\cdots(n+\nu)\Gamma(1+\nu)} = \frac{(-1)^n}{2^{2n+\nu} n!\Gamma(1+\nu+n)}$$

for $n = 0, 1, 2, \ldots.$

Bessel Functions of the First Kind   Using the coefficients c_{2n} just obtained and $r = \nu$, a series solution of (1) is $y = \sum_{n=0}^{\infty} c_{2n} x^{2n+\nu}$. This solution is usually denoted by $J_\nu(x)$:

$$J_\nu(x) = \sum_{n=0}^{\infty} \frac{(-1)^n}{n!\Gamma(1+\nu+n)} \left(\frac{x}{2}\right)^{2n+\nu}. \tag{7}$$

If $\nu \geq 0$, the series converges at least on the interval $[0, \infty)$. Also, for the second exponent $r_2 = -\nu$ we obtain, in exactly the same manner,

$$J_{-\nu}(x) = \sum_{n=0}^{\infty} \frac{(-1)^n}{n!\Gamma(1-\nu+n)} \left(\frac{x}{2}\right)^{2n-\nu}. \tag{8}$$

The functions $J_\nu(x)$ and $J_{-\nu}(x)$ are called **Bessel functions of the first kind** of order ν and $-\nu$, respectively. Depending on the value of ν, (8) may contain negative powers of x and hence converge on $(0, \infty)$.*

---

*When we replace x by $|x|$, the series given in (7) and (8) converge for $0 < |x| < \infty$.

Now some care must be taken in writing the general solution of (1). When $\nu = 0$, it is apparent that (7) and (8) are the same. If $\nu > 0$ and $r_1 - r_2 = \nu - (-\nu) = 2\nu$ is not a positive integer, it follows from Case I of Section 6.3 that $J_\nu(x)$ and $J_{-\nu}(x)$ are linearly independent solutions of (1) on $(0, \infty)$, and so the general solution of the interval is $y = c_1 J_\nu(x) + c_2 J_{-\nu}(x)$. But we also know from Case II of Section 6.2 that when $r_1 - r_2 = 2\nu$ is a positive integer, a second series solution of (1) *may* exist. In this second case we distinguish two possibilities. When $\nu = m = $ positive integer, $J_{-m}(x)$ defined by (8) and $J_m(x)$ are not linearly independent solutions. It can be shown that J_{-m} is a constant multiple of J_m (see Property (*i*) on page 295). In addition, $r_1 - r_2 = 2\nu$ can be a positive integer when ν is half an odd positive integer. It can be shown in this latter event that $J_\nu(x)$ and $J_{-\nu}(x)$ are linearly independent. In other words, the general solution of (1) on $(0, \infty)$ is

$$y = c_1 J_\nu(x) + c_2 J_{-\nu}(x), \quad \nu \neq \text{integer}. \tag{9}$$

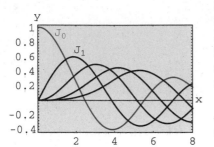

Figure 6.3 Bessel functions of the first kind for $n = 0, 1, 2, 3, 4$

The graphs of $y = J_0(x)$ and $y = J_1(x)$ are given in Figure 6.3.

EXAMPLE 1   **General Solution: ν Not an Integer**

By identifying $\nu^2 = \frac{1}{4}$ and $\nu = \frac{1}{2}$ we can see from (9) that the general solution of the equation $x^2 y'' + xy' + (x^2 - \frac{1}{4})y = 0$ on $(0, \infty)$ is $y = c_1 J_{1/2}(x) + c_2 J_{-1/2}(x)$.   ■

Bessel Functions of the Second Kind   If $\nu \neq $ integer, the function defined by the linear combination

$$Y_\nu(x) = \frac{\cos \nu\pi J_\nu(x) - J_{-\nu}(x)}{\sin \nu\pi} \tag{10}$$

and the function $J_\nu(x)$ are linearly independent solutions of (1). Thus another form of the general solution of (1) is $y = c_1 J_\nu(x) + c_2 Y_\nu(x)$, provided $\nu \neq $ integer. As $\nu \to m$, m an integer, (10) has the indeterminate form 0/0. However, it can be shown by L'Hôpital's rule that $\lim_{\nu \to m} Y_\nu(x)$ exists. Moreover, the function

$$Y_m(x) = \lim_{\nu \to m} Y_\nu(x)$$

and $J_m(x)$ are linearly independent solutions of $x^2 y'' + xy' + (x^2 - m^2)y = 0$. Hence for *any* value of ν the general solution of (1) on $(0, \infty)$ can be written as

$$y = c_1 J_\nu(x) + c_2 Y_\nu(x). \tag{11}$$

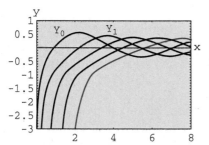

Figure 6.4 Bessel functions of the second kind for $n = 0, 1, 2, 3, 4$

$Y_\nu(x)$ is called the **Bessel function of the second kind** of order ν. Figure 6.4 shows the graphs of $Y_0(x)$ and $Y_1(x)$.

EXAMPLE 2   **General Solution: ν an Integer**

By identifying $\nu^2 = 9$ and $\nu = 3$ we see from (11) that the general solution of the equation $x^2 y'' + xy' + (x^2 - 9)y = 0$ on $(0, \infty)$ is $y = c_1 J_3(x) + c_2 Y_3(x)$.   ■

Sometimes it is possible to transform a given differential equation into equation (1) by means of a change of variable. We can then express the general solution of the original equation in terms of Bessel functions. Example 3 illustrates this technique.

EXAMPLE 3 **The Aging Spring Revisited**

Recall that in Section 5.1 we saw that one mathematical model for the free undamped motion of a mass on an aging spring is given by $mx'' + ke^{-\alpha t}x = 0, \alpha > 0$. We are now in a position to find the general solution of the equation. It is left as a problem to show that the change of variables $s = \dfrac{2}{\alpha}\sqrt{\dfrac{k}{m}}\, e^{-\alpha t/2}$ transforms the differential equation of the aging spring into

$$s^2\frac{d^2x}{ds^2} + s\frac{dx}{ds} + s^2x = 0.$$

The last equation is recognized as (1) with $\nu = 0$ and with the symbols x and s playing the roles of y and x, respectively. The general solution of the new equation is $x = c_1J_0(s) + c_2Y_0(s)$. If we resubstitute s, the general solution of $mx'' + ke^{-\alpha t}x = 0$ is then seen to be

$$x(t) = c_1J_0\left(\frac{2}{\alpha}\sqrt{\frac{k}{m}}\, e^{-\alpha t/2}\right) + c_2Y_0\left(\frac{2}{\alpha}\sqrt{\frac{k}{m}}\, e^{-\alpha t/2}\right).$$

See Problems 25 and 35 in Exercises 6.3. ■

The other model discussed in Section 5.1 of a spring whose characteristics change with time was $mx'' + ktx = 0$. By dividing through by m we see that this equation is Airy's equation, $y'' + \alpha^2xy = 0$. See Example 2 in Section 6.1. The general solution of Airy's differential equation can also be written in terms of Bessel functions. See Problems 26, 27, and 36 in Exercises 6.3.

Parametric Bessel Equation   By replacing x by λx in (1) and using the Chain Rule we obtain an alternative form of Bessel's equation known as the **parametric Bessel equation:**

$$x^2y'' + xy' + (\lambda^2x^2 - \nu^2)y = 0. \tag{12}$$

The general solution of (12) is

$$y = c_1J_\nu(\lambda x) + c_2Y_\nu(\lambda x). \tag{13}$$

Properties   We list below a few of the more useful properties of Bessel functions of order m, $m = 0, 1, 2, \ldots$:

(*i*) $J_{-m}(x) = (-1)^mJ_m(x)$ (*ii*) $J_m(-x) = (-1)^mJ_m(x)$

(*iii*) $J_m(0) = \begin{cases} 0, & m > 0 \\ 1, & m = 0 \end{cases}$ (*iv*) $\lim_{x \to 0^+}Y_m(x) = -\infty$

Note that Property (*ii*) indicates that $J_m(x)$ is an even function if m is an even integer and an odd function if m is an odd integer. The graphs of $Y_0(x)$ and $Y_1(x)$ in Figure 6.4 illustrate Property (*iv*): $Y_m(x)$ is unbounded at the origin. This last fact is not obvious from (10). It can be shown either from (10) or by the methods of Section 6.2 that for $x > 0$,

$$Y_0(x) = \frac{2}{\pi} J_0(x) \left[\gamma + \ln \frac{x}{2} \right] - \frac{2}{\pi} \sum_{k=1}^{\infty} \frac{(-1)^k}{(k!)^2} \left(1 + \frac{1}{2} + \cdots + \frac{1}{k} \right) \left(\frac{x}{2} \right)^{2k},$$

where $\gamma = 0.57721566\ldots$ is **Euler's constant.** Because of the presence of the logarithmic term, $Y_0(x)$ is discontinuous at $x = 0$.

Numerical Values   Some functional values of $J_0(x), J_1(x), Y_0(x)$, and $Y_1(x)$ for selected values of x are given in Table 6.1. The first five nonnegative zeros of $J_0(x), J_1(x), Y_0(x)$, and $Y_1(x)$ are given in Table 6.2.

TABLE 6.1   Numerical Values of J_0, J_1, Y_0, and Y_1

x	$J_0(x)$	$J_1(x)$	$Y_0(x)$	$Y_1(x)$
0	1.0000	0.0000	—	—
1	0.7652	0.4401	0.0883	−0.7812
2	0.2239	0.5767	0.5104	−0.1070
3	−0.2601	0.3391	0.3769	0.3247
4	−0.3971	−0.0660	−0.0169	0.3979
5	−0.1776	−0.3276	−0.3085	0.1479
6	0.1506	−0.2767	−0.2882	−0.1750
7	0.3001	−0.0047	−0.0259	−0.3027
8	0.1717	0.2346	0.2235	−0.1581
9	−0.0903	0.2453	0.2499	0.1043
10	−0.2459	0.0435	0.0557	0.2490
11	−0.1712	−0.1768	−0.1688	0.1637
12	0.0477	−0.2234	−0.2252	−0.0571
13	0.2069	−0.0703	−0.0782	−0.2101
14	0.1711	0.1334	0.1272	−0.1666
15	−0.0142	0.2051	0.2055	0.0211

TABLE 6.2   Zeros of J_0, J_1, Y_0, and Y_1

$J_0(x)$	$J_1(x)$	$Y_0(x)$	$Y_1(x)$
2.4048	0.0000	0.8936	2.1971
5.5201	3.8317	3.9577	5.4297
8.6537	7.0156	7.0861	8.5960
11.7915	10.1735	10.2223	11.7492
14.9309	13.3237	13.3611	14.8974

Differential Recurrence Relation   Recurrence formulas that relate Bessel functions of different orders are important in theory and in applications. In the next example we derive a **differential recurrence relation.**

EXAMPLE 4   **Derivation Using the Series Definition**

Derive the formula $xJ'_\nu(x) = \nu J_\nu(x) - xJ_{\nu+1}(x)$.

Solution   It follows from (7) that

$$xJ'_\nu(x) = \sum_{n=0}^{\infty} \frac{(-1)^n(2n+\nu)}{n!\Gamma(1+\nu+n)} \left(\frac{x}{2}\right)^{2n+\nu}$$

$$= \nu \sum_{n=0}^{\infty} \frac{(-1)^n}{n!\Gamma(1+\nu+n)} \left(\frac{x}{2}\right)^{2n+\nu} + 2 \sum_{n=0}^{\infty} \frac{(-1)^n n}{n!\Gamma(1+\nu+n)} \left(\frac{x}{2}\right)^{2n+\nu}$$

$$= \nu J_\nu(x) + x \underbrace{\sum_{n=1}^{\infty} \frac{(-1)^n}{(n-1)!\Gamma(1+\nu+n)} \left(\frac{x}{2}\right)^{2n+\nu-1}}_{k=n-1}$$

$$= \nu J_\nu(x) - x \sum_{k=0}^{\infty} \frac{(-1)^k}{k!\Gamma(2+\nu+k)} \left(\frac{x}{2}\right)^{2k+\nu+1} = \nu J_\nu(x) - xJ_{\nu+1}(x). \quad \blacksquare$$

The result in Example 4 can be written in an alternative form. Dividing $xJ'_\nu(x) - \nu J_\nu(x) - -xJ_{\nu+1}(x)$ by x gives

$$J'_\nu(x) - \frac{\nu}{x}J_\nu(x) = -J_{\nu+1}(x).$$

This last expression is recognized as a linear first-order differential equation in $J_\nu(x)$. Multiplying both sides of the equality by the integrating factor $x^{-\nu}$ then yields

$$\frac{d}{dx}\left[x^{-\nu}J_\nu(x)\right] = -x^{-\nu}J_{\nu+1}(x). \tag{14}$$

It can be shown in a similar manner that

$$\frac{d}{dx}\left[x^{\nu}J_\nu(x)\right] = x^{\nu}J_{\nu-1}(x). \tag{15}$$

See Problem 19 in Exercises 6.3. The differential recurrence relations (14) and (15) are also valid for the Bessel function of the second kind $Y_\nu(x)$. Observe that when $\nu = 0$ it follows from (14) that

$$J'_0(x) = -J_1(x) \qquad \text{and} \qquad Y'_0(x) = -Y_1(x). \tag{16}$$

An application of these results is given in Problem 35 of Exercises 6.3.

Spherical Bessel Functions   When the order ν is half an odd integer, that is, $\pm\frac{1}{2}$, $\pm\frac{3}{2}$, $\pm\frac{5}{2}$, ..., the Bessel function of the first kind $J_\nu(x)$ can be expressed in terms of the elementary functions $\sin x$, $\cos x$, and powers of x. Such Bessel functions are called **spherical Bessel functions.**

EXAMPLE 5 **Spherical Bessel Function with $\nu = \frac{1}{2}$**

Find an alternative expression for $J_{1/2}(x)$.

Solution With $\nu = \frac{1}{2}$ we have from (7)

$$J_{1/2}(x) = \sum_{n=0}^{\infty} \frac{(-1)^n}{n!\,\Gamma(1 + \frac{1}{2} + n)} \left(\frac{x}{2}\right)^{2n+1/2}.$$

Now in view of the property $\Gamma(1 + \alpha) = \alpha \Gamma(\alpha)$ and the fact that $\Gamma(\frac{1}{2}) = \sqrt{\pi}$ we obtain

$$n = 0: \quad \Gamma(1 + \tfrac{1}{2}) = \tfrac{1}{2}\Gamma(\tfrac{1}{2}) = \tfrac{1}{2}\sqrt{\pi}$$

$$n = 1: \quad \Gamma(1 + \tfrac{3}{2}) = \tfrac{3}{2}\Gamma(\tfrac{3}{2}) = \frac{3}{2^2}\sqrt{\pi} = \frac{3!}{2^3}\sqrt{\pi}$$

$$n = 2: \quad \Gamma(1 + \tfrac{5}{2}) = \tfrac{5}{2}\Gamma(\tfrac{5}{2}) = \frac{5 \cdot 3}{2^3}\sqrt{\pi} = \frac{5 \cdot 4 \cdot 3 \cdot 2 \cdot 1}{2^3 \cdot 4 \cdot 2}\sqrt{\pi} = \frac{5!}{2^5 2!}\sqrt{\pi}$$

$$n = 3: \quad \Gamma(1 + \tfrac{7}{2}) = \tfrac{7}{2}\Gamma(\tfrac{7}{2}) = \frac{7 \cdot 5!}{2^6 2!}\sqrt{\pi} = \frac{7 \cdot 6 \cdot 5!}{2^6 \cdot 6 \cdot 2!}\sqrt{\pi} = \frac{7!}{2^7 3!}\sqrt{\pi}.$$

In general, $$\Gamma(1 + \tfrac{1}{2} + n) = \frac{(2n + 1)!}{2^{2n+1}n!}\sqrt{\pi}.$$

Hence

$$J_{1/2}(x) = \sum_{n=0}^{\infty} \frac{(-1)^n}{n! \dfrac{(2n + 1)!\sqrt{\pi}}{2^{2n+1}n!}} \left(\frac{x}{2}\right)^{2n+1/2} = \sqrt{\frac{2}{\pi x}} \sum_{n=0}^{\infty} \frac{(-1)^n}{(2n + 1)!} x^{2n+1}.$$

Since the series in the last line is the Maclaurin series for $\sin x$, we have shown that

$$J_{1/2}(x) = \sqrt{\frac{2}{\pi x}} \sin x.$$

∎

Solution of Legendre's Equation Since $x = 0$ is an ordinary point of the equation, we substitute the power series $y = \sum_{n=0}^{\infty} c_n x^n$, shift summation indices, and combine series to get

$$(1 - x^2)y'' - 2xy' + n(n + 1)y = [n(n + 1)c_0 + 2c_2] + [(n - 1)(n + 2)c_1 + 6c_3]x$$

$$+ \sum_{j=2}^{\infty} [(j + 2)(j + 1)c_{j+2} + (n - j)(n + j + 1)c_j]x^j = 0,$$

which implies that

$$n(n + 1)c_0 + 2c_2 = 0$$

$$(n - 1)(n + 2)c_1 + 6c_3 = 0$$

$$(j + 2)(j + 1)c_{j+2} + (n - j)(n + j + 1)c_j = 0$$

or

$$c_2 = -\frac{n(n + 1)}{2!} c_0$$

$$c_3 = -\frac{(n - 1)(n + 2)}{3!} c_1$$

$$c_{j+2} = -\frac{(n - j)(n + j + 1)}{(j + 2)(j + 1)} c_j, \quad j = 2, 3, 4, \dots. \tag{17}$$

If we let j take on the values $2, 3, 4, \ldots$, the recurrence relation (17) yields

$$c_4 = -\frac{(n-2)(n+3)}{4 \cdot 3} c_2 = \frac{(n-2)n(n+1)(n+3)}{4!} c_0$$

$$c_5 = -\frac{(n-3)(n+4)}{5 \cdot 4} c_3 = \frac{(n-3)(n-1)(n+2)(n+4)}{5!} c_1$$

$$c_6 = -\frac{(n-4)(n+5)}{6 \cdot 5} c_4 = -\frac{(n-4)(n-2)n(n+1)(n+3)(n+5)}{6!} c_0$$

$$c_7 = -\frac{(n-5)(n+6)}{7 \cdot 6} c_5$$

$$= -\frac{(n-5)(n-3)(n-1)(n+2)(n+4)(n+6)}{7!} c_1$$

and so on. Thus for at least $|x| < 1$ we obtain two linearly independent power series solutions:

$$y_1(x) = c_0 \left[1 - \frac{n(n+1)}{2!} x^2 + \frac{(n-2)n(n+1)(n+3)}{4!} x^4 \right.$$

$$\left. - \frac{(n-4)(n-2)n(n+1)(n+3)(n+5)}{6!} x^6 + \cdots \right]$$

(18)

$$y_2(x) = c_1 \left[x - \frac{(n-1)(n+2)}{3!} x^3 + \frac{(n-3)(n-1)(n+2)(n+4)}{5!} x^5 \right.$$

$$\left. - \frac{(n-5)(n-3)(n-1)(n+2)(n+4)(n+6)}{7!} x^7 + \cdots \right].$$

Notice that if n is an even integer, the first series terminates, whereas $y_2(x)$ is an infinite series. For example, if $n = 4$, then

$$y_1(x) = c_0 \left[1 - \frac{4 \cdot 5}{2!} x^2 + \frac{2 \cdot 4 \cdot 5 \cdot 7}{4!} x^4 \right] = c_0 \left[1 - 10x^2 + \frac{35}{3} x^4 \right].$$

Similarly, when n is an odd integer, the series for $y_2(x)$ terminates with x^n; that is, *when n is a nonnegative integer, we obtain an nth-degree polynomial solution* of Legendre's equation.

Since we know that a constant multiple of a solution of Legendre's equation is also a solution, it is traditional to choose specific values for c_0 or c_1, depending on whether n is an even or odd positive integer, respectively. For $n = 0$ we choose $c_0 = 1$, and for $n = 2, 4, 6, \ldots,$

$$c_0 = (-1)^{n/2} \frac{1 \cdot 3 \cdots (n-1)}{2 \cdot 4 \cdots n};$$

whereas for $n = 1$ we choose $c_1 = 1$, and for $n = 3, 5, 7, \ldots,$

$$c_1 = (-1)^{(n-1)/2} \frac{1 \cdot 3 \cdots n}{2 \cdot 4 \cdots (n-1)}.$$

For example, when $n = 4$ we have

$$y_1(x) = (-1)^{4/2} \frac{1 \cdot 3}{2 \cdot 4} \left[1 - 10x^2 + \frac{35}{3} x^4 \right] = \frac{1}{8} (35x^4 - 30x^2 + 3).$$

Legendre Polynomials   These specific nth-degree polynomial solutions are called **Legendre polynomials** and are denoted by $P_n(x)$. From the series for $y_1(x)$ and $y_2(x)$ and from the above choices of c_0 and c_1 we find that the first several Legendre polynomials are

$$P_0(x) = 1 \qquad\qquad P_1(x) = x$$

$$P_2(x) = \frac{1}{2}(3x^2 - 1) \qquad\qquad P_3(x) = \frac{1}{2}(5x^3 - 3x) \qquad\qquad \textbf{(19)}$$

$$P_4(x) = \frac{1}{8}(35x^4 - 30x^2 + 3) \qquad P_5(x) = \frac{1}{8}(63x^5 - 70x^3 + 15x).$$

Remember, $P_0(x)$, $P_1(x)$, $P_2(x)$, $P_3(x)$, ... are, in turn, particular solutions of the differential equations

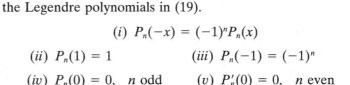

$$
\begin{aligned}
n = 0: & \quad (1 - x^2)y'' - 2xy' = 0 \\
n = 1: & \quad (1 - x^2)y'' - 2xy' + 2y = 0 \\
n = 2: & \quad (1 - x^2)y'' - 2xy' + 6y = 0 \\
n = 3: & \quad (1 - x^2)y'' - 2xy' + 12y = 0 \\
& \quad\quad \vdots \qquad\qquad \vdots
\end{aligned}
\qquad\qquad \textbf{(20)}
$$

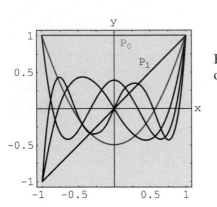

Figure 6.5 Legendre polynomials for $n = 0, 1, 2, 3, 4, 5, 6$

The graphs, on the interval $-1 \le x \le 1$, of the six Legendre polynomials in (19) are given in Figure 6.5.

Properties   You are encouraged to verify the following properties using the Legendre polynomials in (19).

$$(i)\;\; P_n(-x) = (-1)^n P_n(x)$$

$(ii)\;\; P_n(1) = 1 \qquad\qquad (iii)\;\; P_n(-1) = (-1)^n$

$(iv)\;\; P_n(0) = 0, \quad n \text{ odd} \qquad (v)\;\; P_n'(0) = 0, \quad n \text{ even}$

Property (i) indicates, as is apparent in Figure 6.5, that $P_n(x)$ is an even or odd function according to whether n is even or odd.

Recurrence Relation   Recurrence relations that relate Legendre polynomials of different degrees are also important in some aspects of their applications. We state, without proof, the three-term recurrence relation

$$(k + 1)P_{k+1}(x) - (2k + 1)xP_k(x) + kP_{k-1}(x) = 0, \qquad \textbf{(21)}$$

which is valid for $k = 1, 2, 3, \ldots$. In (19) we listed the first six Legendre polynomials. If, say, we wish to find $P_6(x)$, we can use (21) with $k = 5$. This relation expresses $P_6(x)$ in terms of the known $P_4(x)$ and $P_5(x)$. See Problem 30 in Exercises 6.3.

Another formula, although not a recurrence relation, can generate the Legendre polynomials by differentiation. **Rodrigues' formula** for these polynomials is

$$P_n(x) = \frac{1}{2^n n!}\frac{d^n}{dx^n}(x^2 - 1)^n, \quad n = 0, 1, 2, \ldots. \qquad \textbf{(22)}$$

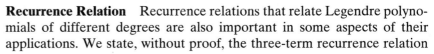

See Problem 40 in Exercises 6.3.

EXERCISES 6.3

Answers to odd-numbered problems begin on page AN-8.

In Problems 1–8 find the general solution of the given differential equation on $(0, \infty)$.

1. $x^2y'' + xy' + (x^2 - \frac{1}{9})y = 0$ **2.** $x^2y'' + xy' + (x^2 - 1)y = 0$

3. $4x^2y'' + 4xy' + (4x^2 - 25)y = 0$

4. $16x^2y'' + 16xy' + (16x^2 - 1)y = 0$

5. $xy'' + y' + xy = 0$ **6.** $\dfrac{d}{dx}[xy'] + \left(x - \dfrac{4}{x}\right)y = 0$

7. $x^2y'' + xy' + (9x^2 - 4)y = 0$ **8.** $x^2y'' + xy' + (36x^2 - \frac{1}{4})y = 0$

9. Use the change of variables $y = x^{-1/2}v(x)$ to find the general solution of

$$x^2y'' + 2xy' + \lambda^2 x^2 y = 0, \quad x > 0.$$

10. Verify that $y = x^n J_n(x)$ is a particular solution of

$$xy'' + (1 - 2n)y' + xy = 0, \quad x > 0.$$

11. Verify that $y = x^{-n} J_n(x)$ is a particular solution of

$$xy'' + (1 + 2n)y' + xy = 0, \quad x > 0.$$

12. Verify that $y = \sqrt{x} J_n(\lambda x)$, $\lambda > 0$ is a particular solution of

$$x^2y'' + (\lambda^2 x^2 - \nu^2 + \tfrac{1}{4})y = 0, \quad x > 0.$$

In Problems 13–18 use the results of Problems 10, 11, and 12 to find a particular solution of the given differential equation on $(0, \infty)$.

13. $y'' + y = 0$ **14.** $xy'' - y' + xy = 0$

15. $xy'' + 3y' + xy = 0$ **16.** $4x^2y'' + (16x^2 + 1)y = 0$

17. $x^2y'' + (x^2 - 2)y = 0$ **18.** $xy'' - 5y' + xy = 0$

19. **(a)** Proceed as in Example 4 to show that $xJ'_\nu(x) = -\nu J_\nu(x) + xJ_{\nu-1}(x)$.

[*Hint:* Write $2n + \nu = 2(n + \nu) - \nu$.]

(b) Use the result in part (a) to derive (15).

20. Use the formula obtained in Example 4 along with part (a) of Problem 19 to derive the recurrence relation $2\nu J_\nu(x) = xJ_{\nu+1}(x) + xJ_{\nu-1}(x)$.

In Problems 21 and 22 use (14) or (15) to obtain the given result.

21. $\displaystyle\int_0^x rJ_0(r)dr = xJ_1(x)$ **22.** $J'_0(x) = J_{-1}(x) = -J_1(x)$

23. Proceed as in Example 5 to express $J_{-1/2}(x)$ in terms of $\cos x$ and a power of x.

24. **(a)** Use Problem 23, the recurrence relation in Problem 20, and the result in Example 5 to express $J_{3/2}(x)$, $J_{-3/2}(x)$, and $J_{5/2}(x)$ in terms of $\sin x$, $\cos x$, and powers of x.

(b) Use a graphing utility to plot the graphs of $J_{1/2}(x)$, $J_{-1/2}(x)$, $J_{3/2}(x)$, $J_{-3/2}(x)$, and $J_{5/2}(x)$.

25. Use the change of variables $s = \dfrac{2}{\alpha}\sqrt{\dfrac{k}{m}}\,e^{-\alpha t/2}$ to show that the differential equation of an aging spring, $mx'' + ke^{-\alpha t}x = 0$, $\alpha > 0$, becomes

$$s^2\frac{d^2x}{ds^2} + s\frac{dx}{ds} + s^2x = 0.$$

26. Show that $y = x^{1/2}w(\tfrac{2}{3}\alpha x^{3/2})$ is a solution of Airy's differential equation $y'' + \alpha^2 xy = 0$, $x > 0$ whenever w is a solution of Bessel's equation $t^2w'' + tw' + (t^2 - \tfrac{1}{9})w = 0$, $t > 0$. [*Hint:* After differentiating, substituting, and simplifying, let $t = \tfrac{2}{3}\alpha x^{3/2}$.]

27. Use the result of Problem 26 to express the general solution of Airy's equation for $x > 0$ in terms of Bessel functions.

28. It can be shown by a change of variables that the general solution of the differential equation $xy'' + \lambda y = 0$ on the interval $(0, \infty)$ is

$$y = c_1\sqrt{x}\,J_1(2\sqrt{\lambda x}) + c_2\sqrt{x}\,Y_1(2\sqrt{\lambda x}).$$

Verify by direct substitution that $y = \sqrt{x}\,J_1(2\sqrt{x})$ is a particular solution of the equation in the case $\lambda = 1$.

29. (a) Use the explicit solutions $y_1(x)$ and $y_2(x)$ of Legendre's equation and the appropriate choice of c_0 and c_1 to find the Legendre polynomials $P_6(x)$ and $P_7(x)$.

　(b) Write the differential equations for which $P_6(x)$ and $P_7(x)$ are particular solutions.

30. Use the recurrence relation (21) and $P_0(x) = 1$, $P_1(x) = x$ to generate the next six Legendre polynomials.

31. Show that the differential equation

$$\sin\theta\frac{d^2y}{d\theta^2} + \cos\theta\frac{dy}{d\theta} + n(n + 1)(\sin\theta)y = 0$$

can be transformed into Legendre's equation by means of the substitution $x = \cos\theta$.

Discussion Problems

32. For the purposes of this problem ignore the graphs given in Figure 6.3. Use the substitution $y = u/\sqrt{x}$ to show that Bessel's equation (1) has the alternative form

$$\frac{d^2u}{dx^2} + \left(1 - \frac{\nu^2 - \tfrac{1}{4}}{x^2}\right)u = 0.$$

This is a form of the differential equation in Problem 12. For a fixed value of ν, discuss how this last equation enables us to discern the qualitative behavior of a solution of (1) as $x \to \infty$.

33. Discuss how $J_0'(x) = -J_1(x)$ (Problem 22) and Rolle's Theorem from calculus can be used to demonstrate that the zeros of $J_0(x)$ and $J_1(x)$ interlace—that is, between any two consecutive zeros of $J_0(x)$ there exists a zero of $J_1(x)$. See Figure 6.3 and Table 6.2.

34. When $\nu = n$, n a nonnegative integer, the Bessel function of the first kind $J_n(x)$ can be represented by the definite integral

$J_n(x) = \dfrac{1}{\pi} \displaystyle\int_0^\pi \cos(x \sin t - nt)\, dt$. How does it follow from this integral that, when n is a nonnegative integer, $|J_n(x)| \leq 1$ for all x? Discuss.

Computer Lab Assignments

35. (a) Use the general solution given in Example 3 to solve the initial-value problem

$$4x'' + e^{-0.1t}x = 0, \quad x(0) = 1, \quad x'(0) = -\tfrac{1}{2}.$$

Also use $J_0'(x) = -J_1(x)$ and $Y_0'(x) = -Y_1(x)$, along with Table 6.1 or a CAS, to evaluate coefficients.

(b) Use a CAS to plot the graph of the solution obtained in part (a) over the interval $0 \leq t \leq 200$.

36. (a) Use the general solution obtained in Problem 27 to solve the initial-value problem

$$4x'' + tx = 0, \quad x(0.1) = 1, \quad x'(0.1) = -\tfrac{1}{2}.$$

Use a CAS to evaluate coefficients.

(b) Use a CAS to plot the graph of the solution obtained in part (a) over the interval $0 \leq t \leq 200$.

37. A uniform thin column, positioned vertically with one end embedded in the ground, will deflect, or bend away, from the vertical under the influence of its own weight when its length is greater than a certain critical height. It can be shown that the angular deflection $\theta(x)$ of the column from the vertical at a point $P(x)$ is a solution of the boundary-value problem

$$EI\frac{d^2\theta}{dx^2} + \delta g(L - x)\theta = 0, \quad \theta(0) = 0, \quad \theta'(0) = 0,$$

where E is Young's modulus, I is the cross-sectional moment of inertia, δ is the constant linear density, and x is distance along the column measured from its base. See Figure 6.6. The column will bend only if this boundary-value problem has a nontrivial solution.

(a) First make the change of variables $t = L - x$ and state the resulting boundary-value problem. Then use the result of Problem 27 to express the general solution of the differential equation in terms of Bessel functions.

(b) With the aid of a CAS, find the critical length L of a solid steel rod of radius $r = 0.05$ in. if $\delta g = 0.28A$ lb/in., $E = 2.6 \times 10^7$ lb/in.^{2}, $A = \pi r^2$, and $I = \tfrac{1}{4}\pi r^4$.

38. In Example 3 of Section 5.2 we saw that when a constant vertical compressive force, or load, P was applied to a thin column of uniform cross-section, the deflection $y(x)$ satisfied the boundary-value problem

$$EI\frac{d^2y}{dx^2} + Py = 0, \quad y(0) = 0, \quad y(L) = 0.$$

(a) If the bending stiffness factor EI is proportional to x, then $EI(x) = kx$, where k is a constant of proportionality. If $EI(L) = kL = M$ is the maximum stiffness factor, then $k = M/L$ and so

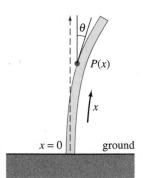

Figure 6.6

(left margin figure labels: θ, $P(x)$, x, $x=0$, ground)

$EI(x) = Mx/L$. Use the information in Problem 28 to find the solution of

$$M\frac{x}{L}\frac{d^2y}{dx^2} + Py = 0, \quad y(0) = 0, \quad y(L) = 0$$

if it is known that $\sqrt{x}\, Y_1(2\sqrt{\lambda x})$ is *not* zero at $x = 0$.

(b) Use Table 6.2 to find the Euler load P_1 for the column.

(c) Use a CAS to plot the graph of the first buckling mode $y_1(x)$ corresponding to the Euler load P_1. For simplicity assume that $c_1 = 1$ and $L = 1$.

39. Using Table 6.2 or a CAS, take the difference between successive zeros of $J_0(x)$. Do the same for $J_1(x)$. Using tables or a CAS, repeat for $J_2(x)$. Try to formulate a conjecture. Discuss how the data tie in with Problem 32.

40. For purposes of this problem ignore the list of Legendre polynomials given on page 300 and the graphs given in Figure 6.5.

(a) Use Rodrigues' formula (22) to generate the Legendre polynomials $P_1(x), P_2(x), \ldots, P_7(x)$. Use a CAS to carry out the differentiation and simplification.

(b) Use a CAS to plot the graphs of $P_1(x), P_2(x), \ldots, P_7(x)$ on the interval $[-1, 1]$.

(c) Use a root-finding application of your CAS to find the zeros of $P_1(x), P_2(x), \ldots, P_7(x)$. If the Legendre polynomials are built-in functions of your CAS, find zeros of Legendre polynomials of higher degree. Form a conjecture about the location of the zeros of *any* Legendre polynomial $P_n(x)$, and then investigate to see whether it is true.

CHAPTER 6 IN REVIEW

Answers to odd-numbered problems begin on page AN-8.

1. Suppose the power series $\sum_{k=0}^{\infty} c_k(x - 4)^k$ is known to converge at -2 and diverge at 13. Discuss whether the series converges at $-7, 0, 7, 10,$ and 11. Possible answers are *does, does not, might.*

2. Use the Maclaurin series for $\sin x$ and $\cos x$ along with long division to find the first three nonzero terms of a power series in x for the function $f(x) = \dfrac{\sin x}{\cos x}$.

In Problems 3 and 4 give the minimum value of the radius R of convergence of a power series solution centered at 0. Do not solve.

3. $(x^2 - 2x + 10)y'' + y = 0$ **4.** $(1 - \sin x)y'' + xy = 0$

In Problems 5 and 6 construct a linear second-order differential equation that has the given properties.

5. A regular singular point at $x = 1$ and an irregular singular point at $x = 0$

6. Regular singular points at $x = 1$ and at $x = -3$

In Problems 7–12 use an appropriate infinite series method about $x = 0$ to find two solutions of the given differential equation.

7. $2xy'' + y' + y = 0$ **8.** $y'' - xy' - y = 0$

9. $(x - 1)y'' + 3y = 0$ **10.** $y'' - x^2y' + xy = 0$

11. $xy'' - (x + 2)y' + 2y = 0$ **12.** $(\cos x)y'' + y = 0$

In Problems 13 and 14 solve the given initial-value problem.

13. $y'' + xy' + 2y = 0$, $y(0) = 3$, $y'(0) = -2$

14. $(x + 2)y'' + 3y = 0$, $y(0) = 0$, $y'(0) = 1$

In Problems 15 and 16 investigate whether $x = 0$ is an ordinary point, singular point, or irregular singular point of the given differential equation. [*Hint:* Recall the Maclaurin series for $\cos x$ and e^x.]

15. $xy'' + (1 - \cos x)y' + x^2y = 0$ **16.** $(e^x - 1 - x)y'' + xy = 0$

17. Note that $x = 0$ is an ordinary point of the differential equation $y'' + x^2y' + 2xy = 5 - 2x + 10x^3$. Use the assumption $y = \sum_{n=0}^{\infty} c_n x^n$ to find the general solution $y = y_c + y_p$ that consists of three power series centered at $x = 0$.

18. The first-order differential equation $dy/dx = x^2 + y^2$ cannot be solved in terms of elementary functions. However, a solution can be expressed in terms of Bessel functions.

   (a) Show that the substitution $y = -\dfrac{1}{u}\dfrac{du}{dx}$ leads to the equation $u'' + x^2u = 0$.

   (b) Show that $u = x^{1/2}w(\frac{1}{2}x^2)$ is a solution of $u'' + x^2u = 0$ whenever w is a solution of Bessel's equation $x^2w'' + xw' + (x^2 - \frac{1}{16})w = 0$.

   (c) Use (14) and (15) in Section 6.3 in the forms

$$J_\nu'(x) = \frac{\nu}{x}J_\nu(x) - J_{\nu+1}(x) \quad \text{and} \quad J_\nu'(x) = -\frac{\nu}{x}J_\nu(x) + J_{\nu-1}(x)$$

   as an aid to show that a one-parameter family of solutions of $dy/dx = x^2 + y^2$ is given by

$$y = x\frac{J_{3/4}(\frac{1}{2}x^2) - cJ_{-3/4}(\frac{1}{2}x^2)}{cJ_{1/4}(\frac{1}{2}x^2) + J_{-1/4}(\frac{1}{2}x^2)}.$$

19. We know that when $n = 1$, $y_1(x) = x$ is a solution to Legendre's differential equation $(1 - x^2)y'' - 2xy' + 2y = 0$. Show that a second linearly independent solution on the interval $-1 < x < 1$ is

$$y_2(x) = \frac{x}{2}\ln\left(\frac{1 + x}{1 - x}\right) - 1.$$

20. (a) Use binomial series to formally show that

$$(1 - 2xt + t^2)^{-1/2} = \sum_{n=0}^{\infty} P_n(x)t^n.$$

   (b) Use the results in part (b) to show that $P_n(1) = 1$ and $P_n(-1) = (-1)^n$.

Meander function; see page 347.

7

THE LAPLACE TRANSFORM

INTRODUCTION

In the linear mathematical model for a physical system such as a spring/mass system or a series electrical circuit, the right-hand member, or input, of the differential equation

$$m\frac{d^2x}{dt^2} + \beta\frac{dx}{dt} + kx = f(t) \quad \text{or} \quad L\frac{d^2q}{dt^2} + R\frac{dq}{dt} + \frac{1}{C}q = E(t)$$

is a driving function and represents either an external force $f(t)$ or an impressed voltage $E(t)$. In Section 5.1 we considered problems in which the functions f and E were continuous. However, discontinuous driving functions are not uncommon. For example, the impressed voltage on a circuit could be piecewise continuous and periodic such as the function shown in the figure above. Solving the differential equation of the circuit in this case is difficult using the techniques of Chapter 4. The Laplace transform studied in this chapter is an invaluable tool that simplifies solving problems such as these (Section 7.4).

7.1 DEFINITION OF THE LAPLACE TRANSFORM

- *Linearity property* • *Integral transform* • *Definition of the Laplace transform* • *Piecewise continuous function* • *Exponential order* • *Existence of the Laplace transform* • *Transforms of some basic functions*

In elementary calculus you learned that differentiation and integration are *transforms*—that is, these operations transform a function into another function. For example, the function $f(x) = x^2$ is transformed, in turn, into a linear function and a family of cubic polynomial functions by the operations of differentiation and integration: $\frac{d}{dx} x^2 = 2x$ and $\int x^2\, dx = \frac{1}{3}x^3 + c$. Moreover, these two transforms possess the **linearity property** that the transform of a linear combination of functions is a linear combination of the transforms. For α and β constants,

$$\frac{d}{dx}[\alpha f(x) + \beta g(x)] = \alpha f'(x) + \beta g'(x)$$

and   **(1)**

$$\int [\alpha f(x) + \beta g(x)]\, dx = \alpha \int f(x)\, dx + \beta \int g(x)\, dx$$

provided each derivative and integral exists. In this section we will examine a special type of integral transform called the **Laplace transform.** In addition to possessing the linearity property, the Laplace transform has many other interesting properties that make it very useful in solving linear initial-value problems.

Integral Transform   If $f(x, y)$ is a function of two variables, then a definite integral of f with respect to one of the variables leads to a function of the other variable. For example, by holding y constant, we see that $\int_1^2 2xy^2\, dx = 3y^2$. Similarly, a definite integral such as $\int_a^b K(s, t)f(t)\, dt$ transforms a function f of the variable t into a function F of the variable s. We are particularly interested in an **integral transform,** where the interval of integration is the unbounded interval $[0, \infty)$. If $f(t)$ is defined for $t \geq 0$, then the improper integral $\int_0^\infty K(s, t)f(t)\, dt$ is defined as a limit:

$$\int_0^\infty K(s, t)f(t)\, dt = \lim_{b \to \infty} \int_0^b K(s, t)f(t)\, dt.$$

If the limit exists, the integral is said to be convergent; if the limit does not exist, the integral is divergent. The foregoing limit will, in general, exist for only certain values of the variable s. The choice $K(s, t) = e^{-st}$ gives us an especially important integral transform.

DEFINITION 7.1   Laplace Transform

Let f be a function defined for $t \geq 0$. Then the integral

$$\mathcal{L}\{f(t)\} = \int_0^\infty e^{-st}f(t)\, dt \qquad\qquad \textbf{(2)}$$

is said to be the **Laplace transform** of f, provided the integral converges.

When the defining integral (2) converges, the result is a function of s. In general discussion we shall use a lowercase letter to denote the function being transformed and the corresponding capital letter to denote its Laplace transform—for example,

$$\mathscr{L}\{f(t)\} = F(s), \quad \mathscr{L}\{g(t)\} = G(s), \quad \mathscr{L}\{y(t)\} = Y(s).$$

EXAMPLE 1   Applying Definition 7.1

Evaluate $\mathscr{L}\{1\}$.

Solution   From (2),

$$\mathscr{L}\{1\} = \int_0^\infty e^{-st}(1)\, dt = \lim_{b \to \infty} \int_0^b e^{-st}\, dt$$

$$= \lim_{b \to \infty} \frac{-e^{-st}}{s} \Big|_0^b = \lim_{b \to \infty} \frac{-e^{-sb} + 1}{s} = \frac{1}{s}$$

provided $s > 0$. In other words, when $s > 0$, the exponent $-sb$ is negative and $e^{-sb} \to 0$ as $b \to \infty$. The integral diverges for $s < 0$. ∎

The use of the limit sign becomes somewhat tedious, so we shall adopt the notation $|_0^\infty$ as a shorthand for writing $\lim_{b \to \infty}(\ )|_0^b$. For example,

$$\mathscr{L}\{1\} = \int_0^\infty e^{-st}(1)\, dt = \frac{-e^{-st}}{s} \Big|_0^\infty = \frac{1}{s}, \quad s > 0.$$

At the upper limit, it is understood that we mean $e^{-st} \to 0$ as $t \to \infty$ for $s > 0$.

EXAMPLE 2   Applying Definition 7.1

Evaluate $\mathscr{L}\{t\}$.

Solution   From Definition 7.1 we have $\mathscr{L}\{t\} = \int_0^\infty e^{-st} t\, dt$. Integrating by parts and using $\lim_{t \to \infty} te^{-st} = 0, s > 0$, along with the result from Example 1, we obtain

$$\mathscr{L}\{t\} = \frac{-te^{-st}}{s} \Big|_0^\infty + \frac{1}{s}\int_0^\infty e^{-st}\, dt = \frac{1}{s}\mathscr{L}\{1\} = \frac{1}{s}\left(\frac{1}{s}\right) = \frac{1}{s^2}. \quad ∎$$

EXAMPLE 3   Applying Definition 7.1

Evaluate $\mathscr{L}\{e^{-3t}\}$.

Solution   From Definition 7.1 we have

$$\mathscr{L}\{e^{-3t}\} = \int_0^\infty e^{-st}e^{-3t}\,dt = \int_0^\infty e^{-(s+3)t}\,dt$$

$$= \left.\frac{-e^{-(s+3)t}}{s+3}\right|_0^\infty$$

$$= \frac{1}{s+3}, \quad s > -3.$$

The result follows from the fact that $\lim_{t\to\infty} e^{-(s+3)t} = 0$ for $s + 3 > 0$ or $s > -3$.   ■

EXAMPLE 4   **Applying Definition 7.1**

Evaluate $\mathscr{L}\{\sin 2t\}$.

Solution   From Definition 7.1 and integration by parts we have

$$\mathscr{L}\{\sin 2t\} = \int_0^\infty e^{-st}\sin 2t\,dt = \left.\frac{-e^{-st}\sin 2t}{s}\right|_0^\infty + \frac{2}{s}\int_0^\infty e^{-st}\cos 2t\,dt$$

$$= \frac{2}{s}\int_0^\infty e^{-st}\cos 2t\,dt, \quad s > 0$$

$$\underset{\underset{\displaystyle\downarrow}{\lim_{t\to\infty} e^{-st}\cos 2t = 0, s > 0}}{} \qquad \underset{\underset{\displaystyle\downarrow}{\text{Laplace transform of }\sin 2t}}{}$$

$$= \frac{2}{s}\left[\left.\frac{-e^{-st}\cos 2t}{s}\right|_0^\infty - \frac{2}{s}\int_0^\infty e^{-st}\sin 2t\,dt\right]$$

$$= \frac{2}{s^2} - \frac{4}{s^2}\,\mathscr{L}\{\sin 2t\}.$$

At this point we have an equation with $\mathscr{L}\{\sin 2t\}$ on both sides of the equality. Solving for that quantity yields the result

$$\mathscr{L}\{\sin 2t\} = \frac{2}{s^2 + 4}, \quad s > 0.$$ ■

$\mathscr{L}$ Is a Linear Transform   For a linear combination of functions, we can write

$$\int_0^\infty e^{-st}[\alpha f(t) + \beta g(t)]\,dt = \alpha \int_0^\infty e^{-st}f(t)\,dt + \beta \int_0^\infty e^{-st}g(t)\,dt$$

whenever both integrals converge for $s > c$. Hence it follows that

$$\mathscr{L}\{\alpha f(t) + \beta g(t)\} = \alpha\mathscr{L}\{f(t)\} + \beta\mathscr{L}\{g(t)\} = \alpha F(s) + \beta G(s). \qquad \textbf{(3)}$$

Because of the property given in (3), $\mathscr{L}$ is said to be a **linear transform**. For example, from Examples 1 and 2,

$$\mathscr{L}\{1 + 5t\} = \mathscr{L}\{1\} + 5\mathscr{L}\{t\} = \frac{1}{s} + \frac{1}{s^2},$$

and from Examples 3 and 4,

$$\mathcal{L}\{4e^{-3t} - 10 \sin 2t\} = 4\mathcal{L}\{e^{-3t}\} - 10\mathcal{L}\{\sin 2t\} = \frac{4}{s+3} - \frac{20}{s^2+4}.$$

We state the generalization of some of the preceding examples by means of the next theorem. From this point on we shall also refrain from stating any restrictions on s; it is understood that s is sufficiently restricted to guarantee the convergence of the appropriate Laplace transform.

THEOREM 7.1   **Transforms of Some Basic Functions**

$$\textbf{(a)} \ \mathcal{L}\{1\} = \frac{1}{s}$$

$$\textbf{(b)} \ \mathcal{L}\{t^n\} = \frac{n!}{s^{n+1}}, \quad n = 1, 2, 3, \ldots \qquad \textbf{(c)} \ \mathcal{L}\{e^{at}\} = \frac{1}{s-a}$$

$$\textbf{(d)} \ \mathcal{L}\{\sin kt\} = \frac{k}{s^2+k^2} \qquad \textbf{(e)} \ \mathcal{L}\{\cos kt\} = \frac{s}{s^2+k^2}$$

$$\textbf{(f)} \ \mathcal{L}\{\sinh kt\} = \frac{k}{s^2-k^2} \qquad \textbf{(g)} \ \mathcal{L}\{\cosh kt\} = \frac{s}{s^2-k^2}$$

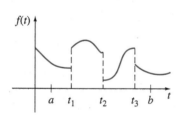

$f(t)$

$a \quad t_1 \quad t_2 \quad t_3 \quad b \quad t$

Figure 7.1

Sufficient Conditions for Existence of $\mathcal{L}\{f(t)\}$ The integral that defines the Laplace transform does not have to converge. For example, neither $\mathcal{L}\{1/t\}$ nor $\mathcal{L}\{e^{t^2}\}$ exists. Sufficient conditions guaranteeing the existence of $\mathcal{L}\{f(t)\}$ are that f be piecewise continuous on $[0, \infty)$ and that f be of exponential order for $t > T$. Recall that a function f is **piecewise continuous** on $[0, \infty)$ if, in any interval $0 \leq a \leq t \leq b$, there are at most a finite number of points t_k, $k = 1, 2, \ldots, n$ $(t_{k-1} < t_k)$ at which f has finite discontinuities and is continuous on each open interval $t_{k-1} < t < t_k$. See Figure 7.1. The concept of **exponential order** is defined in the following manner.

DEFINITION 7.2   **Exponential Order**

A function f is said to be of **exponential order** c if there exist constants c, $M > 0$, and $T > 0$ such that $|f(t)| \leq Me^{ct}$ for all $t > T$.

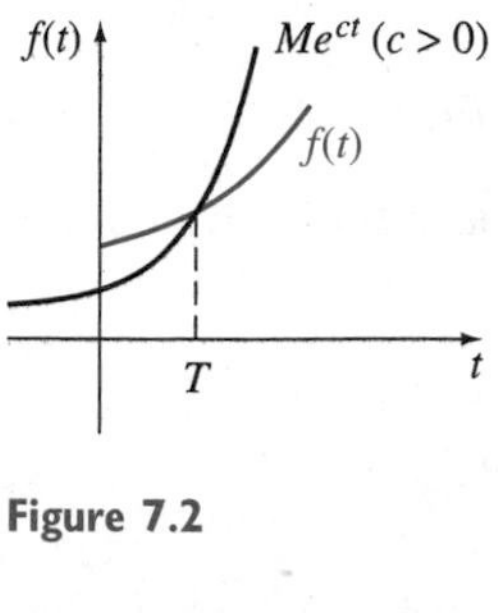

$f(t)$

$Me^{ct} \ (c > 0)$

$f(t)$

T

t

Figure 7.2

If f is an *increasing* function, then the condition $|f(t)| \leq Me^{ct}$, $t > T$ simply states that the graph of f on the interval (T, ∞) does not grow faster than the graph of the exponential function Me^{ct}, where c is a positive constant. See Figure 7.2. The functions $f(t) = t$, $f(t) = e^{-t}$, and $f(t) = 2 \cos t$ are all of exponential order $c = 1$ for $t > 0$ since we have, respectively,

$$|t| \leq e^t, \quad |e^{-t}| \leq e^t, \quad \text{and} \quad |2 \cos t| \leq 2e^t.$$

A comparison of the graphs on the interval $[0, \infty)$ is given in Figure 7.3.

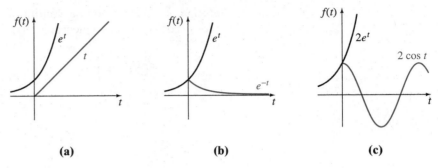

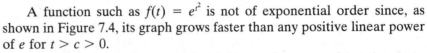

Figure 7.3

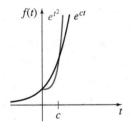

Figure 7.4

A function such as $f(t) = e^{t^2}$ is not of exponential order since, as shown in Figure 7.4, its graph grows faster than any positive linear power of e for $t > c > 0$.

A positive integral power of t is always of exponential order since, for $c > 0$,

$$|t^n| \le Me^{ct} \quad \text{or} \quad \left|\frac{t^n}{e^{ct}}\right| \le M \quad \text{for } t > T$$

is equivalent to showing that $\lim_{t \to \infty} t^n/e^{ct}$ is finite for $n = 1, 2, 3, \ldots$. The result follows by n applications of L'Hôpital's rule.

THEOREM 7.2   Sufficient Conditions for Existence

If $f(t)$ is piecewise continuous on the interval $[0, \infty)$ and of exponential order c for $t > T$, then $\mathcal{L}\{f(t)\}$ exists for $s > c$.

Proof
$$\mathcal{L}\{f(t)\} = \int_0^T e^{-st}f(t)\,dt + \int_T^\infty e^{-st}f(t)\,dt = I_1 + I_2.$$

The integral I_1 exists because it can be written as a sum of integrals over intervals on which $e^{-st}f(t)$ is continuous. Now

$$|I_2| \le \int_T^\infty |e^{-st}f(t)|\,dt \le M\int_T^\infty e^{-st}e^{ct}\,dt$$

$$= M\int_T^\infty e^{-(s-c)t}\,dt = -M\frac{e^{-(s-c)t}}{s-c}\Big|_T^\infty = M\frac{e^{-(s-c)T}}{s-c}$$

for $s > c$. Since $\int_T^\infty Me^{-(s-c)t}\,dt$ converges, the integral $\int_T^\infty |e^{-st}f(t)|\,dt$ converges by the comparison test for improper integrals. This, in turn, implies that I_2 exists for $s > c$. The existence of I_1 and I_2 implies that $\mathcal{L}\{f(t)\} = \int_0^\infty e^{-st}f(t)\,dt$ exists for $s > c$. ∎

EXAMPLE 5   Transform of a Piecewise-Defined Function

Evaluate $\mathcal{L}\{f(t)\}$ for $f(t) = \begin{cases} 0, & 0 \le t < 3 \\ 2, & t \ge 3. \end{cases}$

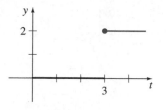

Figure 7.5

Solution　This piecewise continuous function appears in Figure 7.5. Since f is defined in two pieces, $\mathcal{L}\{f(t)\}$ is expressed as the sum of two integrals:

$$\mathcal{L}\{f(t)\} = \int_0^\infty e^{-st} f(t)\, dt = \int_0^3 e^{-st}(0)\, dt + \int_3^\infty e^{-st}(2)\, dt$$

$$= -\left.\frac{2e^{-st}}{s}\right|_3^\infty$$

$$= \frac{2e^{-3s}}{s}, \quad s > 0. \qquad \blacksquare$$

Remarks　Throughout this entire chapter we shall be concerned primarily with functions that are both piecewise continuous and of exponential order. We note, however, that these conditions are sufficient but not necessary for the existence of a Laplace transform. The function $f(t) = t^{-1/2}$ is not piecewise continuous on the interval $[0, \infty)$, but its Laplace transform exists. See Problem 40 in Exercises 7.1.

EXERCISES 7.1

Answers to odd-numbered problems begin on page AN-8.

In Problems 1–18 use Definition 7.1 to find $\mathcal{L}\{f(t)\}$.

1. $f(t) = \begin{cases} -1, & 0 \le t < 1 \\ 1, & t \ge 1 \end{cases}$

2. $f(t) = \begin{cases} 4, & 0 \le t < 2 \\ 0, & t \ge 2 \end{cases}$

3. $f(t) = \begin{cases} t, & 0 \le t < 1 \\ 1, & t \ge 1 \end{cases}$

4. $f(t) = \begin{cases} 2t + 1, & 0 \le t < 1 \\ 0, & t \ge 1 \end{cases}$

5. $f(t) = \begin{cases} \sin t, & 0 \le t < \pi \\ 0, & t \ge \pi \end{cases}$

6. $f(t) = \begin{cases} 0, & 0 \le t < \pi/2 \\ \cos t, & t \ge \pi/2 \end{cases}$

7.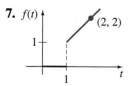

Figure 7.6

8.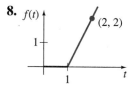

Figure 7.7

9.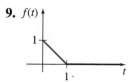

Figure 7.8

10.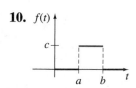

Figure 7.9

11. $f(t) = e^{t+7}$

12. $f(t) = e^{-2t-5}$

13. $f(t) = te^{4t}$

14. $f(t) = t^2 e^{-2t}$

15. $f(t) = e^{-t} \sin t$

16. $f(t) = e^t \cos t$

17. $f(t) = t \cos t$

18. $f(t) = t \sin t$

In Problems 19–36 use Theorem 7.1 to find $\mathcal{L}\{f(t)\}$.

19. $f(t) = 2t^4$

20. $f(t) = t^5$

21. $f(t) = 4t - 10$

22. $f(t) = 7t + 3$

23. $f(t) = t^2 + 6t - 3$

24. $f(t) = -4t^2 + 16t + 9$

25. $f(t) = (t + 1)^3$

26. $f(t) = (2t - 1)^3$

27. $f(t) = 1 + e^{4t}$

28. $f(t) = t^2 - e^{-9t} + 5$

29. $f(t) = (1 + e^{2t})^2$

30. $f(t) = (e^t - e^{-t})^2$

31. $f(t) = 4t^2 - 5 \sin 3t$

32. $f(t) = \cos 5t + \sin 2t$

33. $f(t) = \sinh kt$

34. $f(t) = \cosh kt$

35. $f(t) = e^t \sinh t$

36. $f(t) = e^{-t} \cosh t$

In Problems 37 and 38 find $\mathcal{L}\{f(t)\}$ by using an appropriate trigonometric identity.

37. $f(t) = \sin 2t \cos 2t$

38. $f(t) = \cos^2 t$

39. We used the **gamma function** in the solution of Bessel's equation in Section 6.3. One definition of this function is given by the improper integral $\Gamma(\alpha) = \int_0^\infty t^{\alpha-1}e^{-t}\, dt$, $\alpha > 0$.

   (a) Show that $\Gamma(\alpha + 1) = \alpha\Gamma(\alpha)$.

   (b) Show that $\mathcal{L}\{t^\alpha\} = \dfrac{\Gamma(\alpha + 1)}{s^{\alpha+1}}$, $\alpha > -1$.

40. Use the fact that $\Gamma(\tfrac{1}{2}) = \sqrt{\pi}$ and Problem 39 to find the Laplace transform of

   (a) $f(t) = t^{-1/2}$ **(b)** $f(t) = t^{1/2}$ **(c)** $f(t) = t^{3/2}$.

Discussion Problems

41. Although the proof of part (b) of Theorem 7.1 requires the use of mathematical induction, the result can be justified informally. Show that $\mathcal{L}\{t^n\} = \dfrac{n}{s} \mathcal{L}\{t^{n-1}\}$. See what happens when you let n take on the successive values 1, 2, 3, ...

42. Make up a function $F(t)$ that is of exponential order but where $f(t) = F'(t)$ is not of exponential order. Make up a function f that is not of exponential order but whose Laplace transform exists.

43. Suppose that $\mathcal{L}\{f_1(t)\} = F_1(s)$ for $s > c_1$ and that $\mathcal{L}\{f_2(t)\} = F_2(s)$ for $s > c_2$. When does $\mathcal{L}\{f_1(t) + f_2(t)\} = F_1(s) + F_2(s)$?

44. Figure 7.4 suggests, but does not prove, that the function $f(t) = e^{t^2}$ is not of exponential order. How does the observation that $t^2 > \ln M + ct$, for $M > 0$ and t sufficiently large, show that $e^{t^2} > Me^{ct}$ for any c? Discuss.

45. Use part (c) of Theorem 7.1 to show that $\mathscr{L}\{e^{(a+ib)t}\} = \dfrac{s-a+ib}{(s-a)^2+b^2}$, where a and b are real and $i^2 = -1$. Discuss how Euler's formula (page 159) can then be used to deduce the results

$$\mathscr{L}\{e^{at}\cos bt\} = \frac{s-a}{(s-a)^2+b^2} \quad \text{and} \quad \mathscr{L}\{e^{at}\sin bt\} = \frac{b}{(s-a)^2+b^2}.$$

46. Under what conditions is a linear function $f(x) = mx + b$, $m \neq 0$ a linear transform?

7.2 INVERSE TRANSFORM AND TRANSFORMS OF DERIVATIVES

• Inverse Laplace transform • Linearity property • Use of partial fractions • Transforms of derivatives • Solving linear IVPs

In this section we shall illustrate how to use the Laplace transform to solve differential equations. In order to do this, we first need some important preliminary background on the Laplace transforms of derivatives dy/dt, $d^2y/dt^2, \ldots$ and on the concept of the inverse Laplace transform. We begin with the inverse Laplace transform or, more precisely, the inverse of a Laplace transform $F(s)$. The idea is simply this: Suppose, for example, that $F(s) = \dfrac{-2s+6}{s^2+4}$ is a Laplace transform; find a function $f(t)$ such that $\mathscr{L}\{f(t)\} = F(s)$. See Example 2.

Inverse Transform If $F(s)$ represents the Laplace transform of a function $f(t)$—that is, $\mathscr{L}\{f(t)\} = F(s)$—we then say that $f(t)$ is the **inverse Laplace transform** of $F(s)$ and write $f(t) = \mathscr{L}^{-1}\{F(s)\}$. For example, from Examples 1, 2, and 3 in Section 7.1 we have, respectively,

$$1 = \mathscr{L}^{-1}\left\{\frac{1}{s}\right\}, \quad t = \mathscr{L}^{-1}\left\{\frac{1}{s^2}\right\}, \quad \text{and} \quad e^{-3t} = \mathscr{L}^{-1}\left\{\frac{1}{s+3}\right\}.$$

The analogue of Theorem 7.1 for the inverse transform is presented next.

THEOREM 7.3 **Some Inverse Transforms**

(a) $1 = \mathscr{L}^{-1}\left\{\dfrac{1}{s}\right\}$

(b) $t^n = \mathscr{L}^{-1}\left\{\dfrac{n!}{s^{n+1}}\right\}$, $n = 1, 2, 3, \ldots$ **(c)** $e^{at} = \mathscr{L}^{-1}\left\{\dfrac{1}{s-a}\right\}$

(d) $\sin kt = \mathscr{L}^{-1}\left\{\dfrac{k}{s^2+k^2}\right\}$ **(e)** $\cos kt = \mathscr{L}^{-1}\left\{\dfrac{s}{s^2+k^2}\right\}$

(f) $\sinh kt = \mathscr{L}^{-1}\left\{\dfrac{k}{s^2-k^2}\right\}$ **(g)** $\cosh kt = \mathscr{L}^{-1}\left\{\dfrac{s}{s^2-k^2}\right\}$

When evaluating inverse transforms, it often happens that a function of s under consideration does not match *exactly* the form of a Laplace transform $F(s)$ given in a table. It may be necessary to "fix up" the function of s by multiplying and dividing by an appropriate constant.

EXAMPLE 1   Applying Theorem 7.3

Evaluate   **(a)** $\mathscr{L}^{-1}\left\{\dfrac{1}{s^5}\right\}$ **(b)** $\mathscr{L}^{-1}\left\{\dfrac{1}{s^2+7}\right\}$.

Solution   **(a)** To match the form given in part (b) of Theorem 7.3, we identify $n + 1 = 5$ or $n = 4$ and then multiply and divide by 4!:

$$\mathscr{L}^{-1}\left\{\frac{1}{s^5}\right\} = \frac{1}{4!}\mathscr{L}^{-1}\left\{\frac{4!}{s^5}\right\} = \frac{1}{24}t^4.$$

(b) To match the form given in part (d) of Theorem 7.3, we identify $k^2 = 7$ and so $k = \sqrt{7}$. We fix up the expression by multiplying and dividing by $\sqrt{7}$:

$$\mathscr{L}^{-1}\left\{\frac{1}{s^2+7}\right\} = \frac{1}{\sqrt{7}}\mathscr{L}^{-1}\left\{\frac{\sqrt{7}}{s^2+7}\right\} = \frac{1}{\sqrt{7}}\sin\sqrt{7}t.$$ ■

$\mathscr{L}^{-1}$ **Is a Linear Transform**   The inverse Laplace transform is also a linear transform; that is, for constants α and β,

$$\mathscr{L}^{-1}\{\alpha F(s) + \beta G(s)\} = \alpha\mathscr{L}^{-1}\{F(s)\} + \beta\mathscr{L}^{-1}\{G(s)\},\qquad \textbf{(1)}$$

where F and G are the transforms of some functions f and g. Like (2) of Section 7.1, (1) extends to any finite linear combination of Laplace transforms.

EXAMPLE 2   Termwise Division and Linearity

Evaluate $\mathscr{L}^{-1}\left\{\dfrac{-2s+6}{s^2+4}\right\}$.

Solution   We first rewrite the given function of s as two expressions by means of termwise division and then use (1):

$$\mathscr{L}^{-1}\left\{\frac{-2s+6}{s^2+4}\right\} \underset{\substack{\text{termwise}\\\text{division}\ \downarrow}}{=} \mathscr{L}^{-1}\left\{\frac{-2s}{s^2+4} + \frac{6}{s^2+4}\right\} \underset{\substack{\text{linearity and fixing}\\\text{up constants}\ \downarrow}}{=} -2\mathscr{L}^{-1}\left\{\frac{s}{s^2+4}\right\} + \frac{6}{2}\mathscr{L}^{-1}\left\{\frac{2}{s^2+4}\right\} \qquad \textbf{(2)}$$

$$= -2\cos 2t + 3\sin 2t. \qquad \leftarrow \text{parts (e) and (d)}$$
$$\text{of Theorem 7.3 with } k = 2$$ ■

Partial Fractions   Partial fractions play an important role in finding inverse Laplace transforms. As mentioned in Section 2.2, the decomposition of a rational expression into component fractions can be done quickly by means of a single command on most computer algebra systems. Indeed, some CASs have packages that implement Laplace transform and inverse Laplace transform commands. But for those of you without access to such software, we will review in this and subsequent sections some of the basic algebra in the important cases where the denominator of a Laplace transform $F(s)$ contains distinct linear factors, repeated linear factors, and quadratic polynomials with no real factors. Although we shall examine each of these cases as this chapter develops, it still might be a good idea for you to consult either a calculus text or a current precalculus text for a more comprehensive review of this theory.

The following example illustrates partial fraction decomposition in the case when the denominator of $F(s)$ is factorable into *distinct linear factors*.

EXAMPLE 3   **Partial Fractions: Distinct Linear Factors**

Evaluate $\mathscr{L}^{-1}\left\{\dfrac{s^2 + 6s + 9}{(s-1)(s-2)(s+4)}\right\}$.

Solution   There exist unique real constants A, B, and C so that

$$\frac{s^2 + 6s + 9}{(s-1)(s-2)(s+4)} = \frac{A}{s-1} + \frac{B}{s-2} + \frac{C}{s+4}$$

$$= \frac{A(s-2)(s+4) + B(s-1)(s+4) + C(s-1)(s-2)}{(s-1)(s-2)(s+4)}.$$

Since the denominators are identical, the numerators are identical:

$$s^2 + 6s + 9 = A(s-2)(s+4) + B(s-1)(s+4) + C(s-1)(s-2). \quad \textbf{(3)}$$

By comparing coefficients of powers of s on both sides of the equality, we know that (3) is equivalent to a system of three equations in the three unknowns A, B, and C. However, recall that there is a shortcut for determining these unknowns. If we set $s = 1$, $s = 2$, and $s = -4$ in (3), we obtain, respectively,*

$$16 = A(-1)(5), \quad 25 = B(1)(6), \quad \text{and} \quad 1 = C(-5)(-6),$$

and so $A = -\frac{16}{5}$, $B = \frac{25}{6}$, and $C = \frac{1}{30}$. Hence the partial fraction decomposition is

$$\frac{s^2 + 6s + 9}{(s-1)(s-2)(s+4)} = -\frac{16/5}{s-1} + \frac{25/6}{s-2} + \frac{1/30}{s+4}, \quad \textbf{(4)}$$

and thus, from the linearity of $\mathscr{L}^{-1}$ and part (c) of Theorem 7.3,

$$\mathscr{L}^{-1}\left\{\frac{s^2 + 6s + 9}{(s-1)(s-2)(s+4)}\right\} = -\frac{16}{5}\mathscr{L}^{-1}\left\{\frac{1}{s-1}\right\} + \frac{25}{6}\mathscr{L}^{-1}\left\{\frac{1}{s-2}\right\} + \frac{1}{30}\mathscr{L}^{-1}\left\{\frac{1}{s+4}\right\}$$

$$= -\frac{16}{5}e^t + \frac{25}{6}e^{2t} + \frac{1}{30}e^{-4t}. \quad \textbf{(5)} \quad \blacksquare$$

---

*The numbers 1, 2, and -4 are the zeros of the common denominator $(s-1)(s-2)(s+4)$.

Transforming a Derivative As pointed out in the introduction to this chapter, our immediate goal is to use the Laplace transform to solve differential equations. To that end we need to evaluate quantities such as $\mathcal{L}\{dy/dt\}$ and $\mathcal{L}\{d^2y/dt^2\}$. For example, if f' is continuous for $t \geq 0$, then integration by parts gives

$$\mathcal{L}\{f'(t)\} = \int_0^\infty e^{-st}f'(t)\,dt = e^{-st}f(t)\Big|_0^\infty + s\int_0^\infty e^{-st}f(t)\,dt$$

$$= -f(0) + s\,\mathcal{L}\{f(t)\}$$

or $\qquad \mathcal{L}\{f'(t)\} = sF(s) - f(0).$ $\qquad\qquad$ **(6)**

Here we have assumed that $e^{-st}f(t) \to 0$ as $t \to \infty$. Similarly, with the aid of (6),

$$\mathcal{L}\{f''(t)\} = \int_0^\infty e^{-st}f''(t)\,dt = e^{-st}f'(t)\Big|_0^\infty + s\int_0^\infty e^{-st}f'(t)\,dt$$

$$= -f'(0) + s\mathcal{L}\{f'(t)\}$$

$$= s[sF(s) - f(0)] - f'(0) \qquad \leftarrow \text{from (6)}$$

or $\qquad \mathcal{L}\{f''(t)\} = s^2F(s) - sf(0) - f'(0).$ $\qquad\qquad$ **(7)**

In like manner it can be shown that

$$\mathcal{L}\{f'''(t)\} = s^3F(s) - s^2f(0) - sf'(0) - f''(0).$$ $\qquad\qquad$ **(8)**

The recursive nature of the Laplace transform of the derivatives of a function f should be apparent from the results in (6), (7), and (8). The next theorem gives the Laplace transform of the nth derivative of f. The proof is omitted.

THEOREM 7.4 **Transform of a Derivative**

If $f, f', \ldots, f^{(n-1)}$ are continuous on $[0, \infty)$ and are of exponential order and if $f^{(n)}(t)$ is piecewise continuous on $[0, \infty)$, then

$$\mathcal{L}\{f^{(n)}(t)\} = s^nF(s) - s^{(n-1)}f(0) - s^{(n-2)}f'(0) - \cdots - f^{(n-1)}(0),$$

where $F(s) = \mathcal{L}\{f(t)\}$.

Solving Linear ODEs It is apparent from the general result given in Theorem 7.4 that $\mathcal{L}\{d^ny/dt^n\}$ depends on $Y(s) = \mathcal{L}\{y(t)\}$ and the $n-1$ derivatives of $y(t)$ evaluated at $t = 0$. This property makes the Laplace transform ideally suited for solving linear initial-value problems in which the differential equation has *constant coefficients*. Such a differential equation is simply a linear combination of terms $y, y', y'', \ldots, y^{(n)}$:

$$a_n\frac{d^ny}{dt^n} + a_{n-1}\frac{d^{n-1}y}{dt^{n-1}} + \cdots + a_0y = g(t),$$

$$y(0) = y_0, y'(0) = y_1, \ldots, y^{(n-1)}(0) = y_{n-1},$$

where the a_i, $i = 0, 1, \ldots, n$ and $y_0, y_1, \ldots, y_{n-1}$ are constants. By the linearity property, the Laplace transform of this linear combination is a linear combination of Laplace transforms:

$$a_n \mathcal{L}\left\{\frac{d^n y}{dt^n}\right\} + a_{n-1} \mathcal{L}\left\{\frac{d^{n-1} y}{dt^{n-1}}\right\} + \cdots + a_0 \mathcal{L}\{y\} = \mathcal{L}\{g(t)\}. \qquad (9)$$

From Theorem 7.4, (9) becomes

$$a_n[s^n Y(s) - s^{n-1} y(0) - \cdots - y^{(n-1)}(0)] \qquad (10)$$
$$+ \, a_{n-1}[s^{n-1} Y(s) - s^{n-2} y(0) - \cdots - y^{(n-2)}(0)] + \cdots + a_0 Y(s) = G(s),$$

where $\mathcal{L}\{y(t)\} = Y(s)$ and $\mathcal{L}\{g(t)\} = G(s)$. In other words, *the Laplace transform of a linear differential equation with constant coefficients becomes an algebraic equation in $Y(s)$.* If we solve the general transformed equation (10) for the symbol $Y(s)$, we first obtain $P(s)Y(s) = Q(s) + G(s)$, and then write

$$Y(s) = \frac{Q(s)}{P(s)} + \frac{G(s)}{P(s)}, \qquad (11)$$

where $P(s) = a_n s^n + a_{n-1} s^{n-1} + \cdots + a_0$, $Q(s)$ is a polynomial in s of degree less than or equal to $n - 1$ consisting of the various products of the coefficients a_i, $i = 1, \ldots, n$ and the prescribed initial conditions y_0, $y_1, \ldots, y_{n-1}$, and $G(s)$ is the Laplace transform of $g(t)$.* Typically we put the two terms in (11) over the least common denominator and then decompose the expression into two or more partial fractions. Finally, the solution $y(t)$ of the original initial-value problem is $y(t) = \mathcal{L}^{-1}\{Y(s)\}$, where the inverse transform is done term by term.

The procedure is summarized in the diagram.

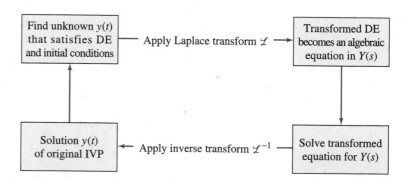

The next example illustrates the foregoing method of solving DEs, as well as partial fraction decomposition in the case when the denominator of $Y(s)$ contains a *quadratic polynomial with no real factors.*

---

*The polynomial $P(s)$ is the same as the nth-degree auxiliary polynomial in (12) in Section 4.3 with the usual symbol m replaced by s.

| EXAMPLE 4 | **Partial Fractions: Quadratic Terms with No Real Factors** |

Use the Laplace transform to solve the initial-value problem

$$\frac{dy}{dt} + 3y = 13 \sin 2t, \quad y(0) = 6.$$

Solution We first take the transform of each member of the differential equation:

$$\mathscr{L}\left\{\frac{dy}{dt}\right\} + 3\mathscr{L}\{y\} = 13\mathscr{L}\{\sin 2t\}. \qquad (12)$$

From (6), $\mathscr{L}\{dy/dt\} = sY(s) - y(0) = sY(s) - 6$, and from part (d) of Theorem 7.1, $\mathscr{L}\{\sin 2t\} = 2/(s^2 + 4)$, and so (12) is the same as

$$sY(s) - 6 + 3Y(s) = \frac{26}{s^2 + 4} \quad \text{or} \quad (s + 3)Y(s) = 6 + \frac{26}{s^2 + 4}.$$

Solving the last equation for $Y(s)$, we get

$$Y(s) = \frac{6}{s + 3} + \frac{26}{(s + 3)(s^2 + 4)} = \frac{6s^2 + 50}{(s + 3)(s^2 + 4)}. \qquad (13)$$

Since the quadratic polynomial $s^2 + 4$ does not factor using real numbers, its assumed numerator in the partial fraction decomposition is a linear polynomial in s:

$$\frac{6s^2 + 50}{(s + 3)(s^2 + 4)} = \frac{A}{s + 3} + \frac{Bs + C}{s^2 + 4}.$$

Putting the right-hand side of the equality over a common denominator and equating numerators gives $6s^2 + 50 = A(s^2 + 4) + (Bs + C)(s + 3)$. Setting $s = -3$ then immediately yields $A = 8$. Since the denominator has no more real zeros, we equate the coefficients of s^2 and s: $6 = A + B$ and $0 = 3B + C$. Using the value of A in the first equation gives $B = -2$, and then using this last value in the second equation gives $C = 6$. Thus

$$Y(s) = \frac{6s^2 + 50}{(s + 3)(s^2 + 4)} = \frac{8}{s + 3} + \frac{-2s + 6}{s^2 + 4}.$$

We are not quite finished because the last rational expression still has to be written as two fractions. This was done by termwise division in Example 2. From (2) of that example,

$$y(t) = 8\mathscr{L}^{-1}\left\{\frac{1}{s + 3}\right\} - 2\mathscr{L}^{-1}\left\{\frac{s}{s^2 + 4}\right\} + 3\mathscr{L}^{-1}\left\{\frac{2}{s^2 + 4}\right\}.$$

It follows from parts (c), (d), and (e) of Theorem 7.3 that the solution of the initial-value problem is $y(t) = 8e^{-3t} - 2\cos 2t + 3\sin 2t.$ ▬

| EXAMPLE 5 | **Solving a Second-Order IVP** |

Solve $y'' - 3y' + 2y = e^{-4t}$, $y(0) = 1$, $y'(0) = 5$.

Solution Proceeding as in Example 4, we transform the DE. We take the sum of the transforms of each term, use (6) and (7), use the given initial conditions, use (c) of Theorem 7.3, and then solve for $Y(s)$:

$$\mathscr{L}\left\{\frac{d^2y}{dt^2}\right\} - 3\mathscr{L}\left\{\frac{dy}{dt}\right\} + 2\mathscr{L}\{y\} = \mathscr{L}\{e^{-4t}\}$$

$$s^2Y(s) - sy(0) - y'(0) - 3[sY(s) - y(0)] + 2Y(s) = \frac{1}{s+4}$$

$$(s^2 - 3s + 2)Y(s) = s + 2 + \frac{1}{s+4}$$

$$Y(s) = \frac{s+2}{s^2-3s+2} + \frac{1}{(s^2-3s+2)(s+4)} = \frac{s^2+6s+9}{(s-1)(s-2)(s+4)}. \quad \textbf{(14)}$$

The details of the partial fraction decomposition of $Y(s)$ have already been carried out in Example 3. In view of the results in (4) and (5), we have the solution of the initial-value problem:

$$y(t) = \mathscr{L}^{-1}\{Y(s)\} = -\frac{16}{5}e^t + \frac{25}{6}e^{2t} + \frac{1}{30}e^{-4t}. \quad \blacksquare$$

Examples 4 and 5 illustrate the basic procedure for using the Laplace transform to solve a linear initial-value problem, but these examples may appear to demonstrate a method that is not much better than the approach to such problems outlined in Sections 2.3 and 4.3–4.6. Don't draw any negative conclusions from only two examples. Yes, there is a lot of algebra inherent in the use of the Laplace transform, *but* observe that we do not have to use variation of parameters or worry about the cases and algebra in the method of undetermined coefficients. Moreover, since the method incorporates the prescribed initial conditions directly into the solution, there is no need for the separate operation of applying the initial conditions to the general solution $y = c_1y_1 + c_2y_2 + \cdots + c_ny_n + y_p$ of the DE to find specific constants in a particular solution of the IVP.

The Laplace transform has many operational properties. In the sections that follow we will examine some of these properties and see how they enable us to solve problems of greater complexity.

We conclude this section with a bit of additional theory related to the types of functions of s that we will, generally, be working with. The next theorem indicates that not every arbitrary function of s is a Laplace transform of a piecewise continuous function of exponential order.

THEOREM 7.5 **Behavior of $F(s)$ as $s \to \infty$**

If f is piecewise continuous on $[0, \infty)$ and of exponential order for $t > T$, then $\lim_{s\to\infty} \mathscr{L}\{f(t)\} = 0$.

Proof Since $f(t)$ is piecewise continuous on $0 \le t \le T$, it is necessarily bounded on the interval; that is, $|f(t)| \le M_1 = M_1e^{0t}$. Also, $|f(t)| \le M_2e^{\gamma t}$ for $t > T$. If M denotes the maximum of $\{M_1, M_2\}$ and c denotes the maximum of $\{0, \gamma\}$, then

$$|\mathscr{L}\{f(t)\}| \leq \int_0^\infty e^{-st} |f(t)|\, dt \leq M \int_0^\infty e^{-st} \cdot e^{ct}\, dt = -M \frac{e^{-(s-c)t}}{s-c} \Big|_0^\infty = \frac{M}{s-c}$$

for $s > c$. As $s \to \infty$, we have $|\mathscr{L}\{f(t)\}| \to 0$ and so $\to \mathscr{L}\{f(t)\} \to 0$. ∎

As a consequence of Theorem 7.5 we can say that functions of s such as $F_1(s) = 1$ and $F_2(s) = s/(s+1)$ are not the Laplace transforms of piecewise continuous functions of exponential order, since $F_1(s) \nrightarrow 0$ and $F_2(s) \nrightarrow 0$ as $s \to \infty$. But you should not conclude from this that $F_1(s)$ and $F_2(s)$ are *not* Laplace transforms. There are other kinds of functions.

Remarks (*i*) The inverse Laplace transform of a function $F(s)$ may not be unique; in other words, it is possible that $\mathscr{L}\{f_1(t)\} = \mathscr{L}\{f_2(t)\}$ and yet $f_1 \neq f_2$. For our purposes this is not anything to be concerned about. If f_1 and f_2 are piecewise continuous on $[0, \infty)$ and of exponential order, then f_1 and f_2 are *essentially* the same. See Problem 41 in Exercises 7.2. However, if f_1 and f_2 are continuous on $[0, \infty)$ and $\mathscr{L}\{f_1(t)\} = \mathscr{L}\{f_2(t)\}$, then $f_1 = f_2$ on the interval.

(*ii*) This remark is for those of you who will be required to do partial fraction decompositions by hand. There is another way of determining the coefficients in a partial fraction decomposition in the special case when $\mathscr{L}\{f(t)\} = F(s)$ is a rational function of s and the denominator of F is a product of *distinct* linear factors. Let us illustrate by reexamining Example 3. Suppose we multiply both sides of the assumed decomposition

$$\frac{s^2 + 6s + 9}{(s-1)(s-2)(s+4)} = \frac{A}{s-1} + \frac{B}{s-2} + \frac{C}{s+4} \qquad \textbf{(15)}$$

by, say, $s - 1$, simplify, and then set $s = 1$. Since the coefficients of B and C on the right-hand side of the equality are zero, we get

$$\frac{s^2 + 6s + 9}{(s-2)(s+4)} \Big|_{s=1} = A \quad \text{or} \quad A = -\frac{16}{5}.$$

Written another way,

$$\frac{s^2 + 6s + 9}{\boxed{(s-1)}(s-2)(s+4)} \Big|_{s=1} = -\frac{16}{5} = A,$$

where we have shaded, or *covered up,* the factor that canceled when the left-hand side was multiplied by $s - 1$. Now to obtain B and C we simply evaluate the left-hand side of (15) while covering up, in turn, $s - 2$ and $s + 4$:

$$\frac{s^2 + 6s + 9}{(s-1)\boxed{(s-2)}(s+4)} \Big|_{s=2} = \frac{25}{6} = B \quad \text{and} \quad \frac{s^2 + 6s + 9}{(s-1)(s-2)\boxed{(s+4)}} \Big|_{s=-4} = \frac{1}{30} = C.$$

The desired decomposition (15) is given in (4). This special technique for determining coefficients is naturally known as the **cover-up method.**

(*iii*) In this remark we continue our introduction to the terminology of dynamical systems. Because of (9) and (10) the Laplace transform is well adapted to *linear* dynamical systems. The polynomial $P(s) = a_n s^n + a_{n-1} s^{n-1} + \cdots + a_0$ in (11) is the total coefficient of $Y(s)$ in (10) and is simply the left-hand side of the DE with the derivatives $d^k y / dt^k$ replaced by powers s^k, $k = 0, 1, \ldots, n$. It is usual practice to call the reciprocal of $P(s)$—namely, $W(s) = 1/P(s)$—the **transfer function** of the system and write (11) as

$$Y(s) = W(s)Q(s) + W(s)G(s). \tag{16}$$

In this manner we have separated, in an additive sense, the effects on the response that are due to the initial conditions (that is, $W(s)Q(s)$) from those due to the input function g (that is, $W(s)G(s)$). See (13) and (14). Hence the response $y(t)$ of the system is a superposition of two responses:

$$y(t) = \mathcal{L}^{-1}\{W(s)Q(s)\} + \mathcal{L}^{-1}\{W(s)G(s)\} = y_0(t) + y_1(t).$$

If the input is $g(t) = 0$, then the solution of the problem is $y_0(t) = \mathcal{L}^{-1}\{W(s)Q(s)\}$. This solution is called the **zero-input response** of the system. On the other hand, the function $y_1(t) = \mathcal{L}^{-1}\{W(s)G(s)\}$ is the output due to the input $g(t)$. Now if the initial state of the system is the zero state (all the initial conditions are zero), then $Q(s) = 0$, and so the only solution of the initial-value problem is $y_1(t)$. The latter solution is called the **zero-state response** of the system. Both $y_0(t)$ and $y_1(t)$ are particular solutions: $y_0(t)$ is a solution of the IVP consisting of the associated homogeneous equation with the given initial conditions, and $y_1(t)$ is a solution of the IVP consisting of the nonhomogeneous equation with zero initial conditions. In Example 5, we see from (14) that the transfer function is $W(s) = 1/(s^2 - 3s + 2)$, the zero-input response is

$$y_0(t) = \mathcal{L}^{-1}\left\{\frac{s+2}{(s-1)(s-2)}\right\} = -3e^t + 4e^{2t},$$

and the zero-state response is

$$y_1(t) = \mathcal{L}^{-1}\left\{\frac{1}{(s-1)(s-2)(s+4)}\right\} = -\frac{1}{5}e^t + \frac{1}{6}e^{2t} + \frac{1}{30}e^{-4t}.$$

Verify that the sum of $y_0(t)$ and $y_1(t)$ is the solution $y(t)$ in Example 5 and that $y_0(0) = 1$, $y_0'(0) = 5$ whereas $y_1(0) = 0$, $y_1'(0) = 0$.

EXERCISES 7.2

Answers to odd-numbered problems begin on page AN-8.

In Problems 1–30 use Theorem 7.3 to find the given inverse transform.

1. $\mathcal{L}^{-1}\left\{\dfrac{1}{s^3}\right\}$

2. $\mathcal{L}^{-1}\left\{\dfrac{1}{s^4}\right\}$

3. $\mathscr{L}^{-1}\left\{\dfrac{1}{s^2} - \dfrac{48}{s^5}\right\}$

4. $\mathscr{L}^{-1}\left\{\left(\dfrac{2}{s} - \dfrac{1}{s^3}\right)^2\right\}$

5. $\mathscr{L}^{-1}\left\{\dfrac{(s+1)^3}{s^4}\right\}$

6. $\mathscr{L}^{-1}\left\{\dfrac{(s+2)^2}{s^3}\right\}$

7. $\mathscr{L}^{-1}\left\{\dfrac{1}{s^2} - \dfrac{1}{s} + \dfrac{1}{s-2}\right\}$

8. $\mathscr{L}^{-1}\left\{\dfrac{4}{s} + \dfrac{6}{s^5} - \dfrac{1}{s+8}\right\}$

9. $\mathscr{L}^{-1}\left\{\dfrac{1}{4s+1}\right\}$

10. $\mathscr{L}^{-1}\left\{\dfrac{1}{5s-2}\right\}$

11. $\mathscr{L}^{-1}\left\{\dfrac{5}{s^2+49}\right\}$

12. $\mathscr{L}^{-1}\left\{\dfrac{10s}{s^2+16}\right\}$

13. $\mathscr{L}^{-1}\left\{\dfrac{4s}{4s^2+1}\right\}$

14. $\mathscr{L}^{-1}\left\{\dfrac{1}{4s^2+1}\right\}$

15. $\mathscr{L}^{-1}\left\{\dfrac{2s-6}{s^2+9}\right\}$

16. $\mathscr{L}^{-1}\left\{\dfrac{s+1}{s^2+2}\right\}$

17. $\mathscr{L}^{-1}\left\{\dfrac{1}{s^2+3s}\right\}$

18. $\mathscr{L}^{-1}\left\{\dfrac{s+1}{s^2-4s}\right\}$

19. $\mathscr{L}^{-1}\left\{\dfrac{s}{s^2+2s-3}\right\}$

20. $\mathscr{L}^{-1}\left\{\dfrac{1}{s^2+s-20}\right\}$

21. $\mathscr{L}^{-1}\left\{\dfrac{0.9s}{(s-0.1)(s+0.2)}\right\}$

22. $\mathscr{L}^{-1}\left\{\dfrac{s-3}{(s-\sqrt{3})(s+\sqrt{3})}\right\}$

23. $\mathscr{L}^{-1}\left\{\dfrac{s}{(s-2)(s-3)(s-6)}\right\}$

24. $\mathscr{L}^{-1}\left\{\dfrac{s^2+1}{s(s-1)(s+1)(s-2)}\right\}$

25. $\mathscr{L}^{-1}\left\{\dfrac{1}{s^3+5s}\right\}$

26. $\mathscr{L}^{-1}\left\{\dfrac{s}{(s+2)(s^2+4)}\right\}$

27. $\mathscr{L}^{-1}\left\{\dfrac{2s-4}{(s^2+s)(s^2+1)}\right\}$

28. $\mathscr{L}^{-1}\left\{\dfrac{1}{s^4-9}\right\}$

29. $\mathscr{L}^{-1}\left\{\dfrac{1}{(s^2+1)(s^2+4)}\right\}$

30. $\mathscr{L}^{-1}\left\{\dfrac{6s+3}{s^4+5s^2+4}\right\}$

In Problems 31–40 use the Laplace transform to solve the given initial-value problem.

31. $\dfrac{dy}{dt} - y = 1, \quad y(0) = 0$

32. $2\dfrac{dy}{dt} + y = 0, \quad y(0) = -3$

33. $y' + 6y = e^{4t}, \quad y(0) = 2$

34. $y' - y = 2\cos 5t, \quad y(0) = 0$

35. $y'' + 5y' + 4y = 0, \quad y(0) = 1, y'(0) = 0$

36. $y'' - 4y' = 6e^{3t} - 3e^{-t}, \quad y(0) = 1, y'(0) = -1$

37. $y'' + y = \sqrt{2}\sin\sqrt{2}t, \quad y(0) = 10, y'(0) = 0$

38. $y'' + 9y = e^t, \quad y(0) = 0, y'(0) = 0$

39. $2y''' + 3y'' - 3y' - 2y = e^{-t}, \quad y(0) = 0, y'(0) = 0, y''(0) = 1$

40. $y''' + 2y'' - y' - 2y = \sin 3t, \quad y(0) = 0, y'(0) = 0, y''(0) = 1$

Discussion Problems

41. Make up two functions f and g that have the same Laplace transform. Do not think profound thoughts.

42. The inverse forms of the results in Problem 45 in Exercises 7.1 are

$$\mathscr{L}^{-1}\left\{\frac{s-a}{(s-a)^2+b^2}\right\} = e^{at}\cos bt \quad \text{and} \quad \mathscr{L}^{-1}\left\{\frac{b}{(s-a)^2+b^2}\right\} = e^{at}\sin bt.$$

Use the Laplace transform and these inverses to solve $y' + y = e^{-3t}\cos 2t$, $y(0) = 0$.

43. Reread Remark (*iii*) on page 322. Find the zero-input and the zero-state responses for the IVP in Problem 36.

7.3 TRANSLATION THEOREMS

• First translation theorem   • Inverse form of first translation theorem   • Unit step function   • Second translation theorem   • Inverse form of second translation theorem   • Alternative form of second translation theorem   • Solving linear IVPs

It is not convenient to use Definition 7.1 each time we wish to find the Laplace transform of a function $f(t)$. For example, the integration by parts involved in evaluating, say, $\mathscr{L}\{e^t t^2 \sin 3t\}$ is formidable to say the least. In this section and the next we present several labor-saving theorems which enable us to build up a more extensive list of transforms (see the table in Appendix III) without resorting to the definition of the Laplace transform.

7.3.1   TRANSLATION ON THE s-AXIS

Evaluating transforms such as $\mathscr{L}\{e^{5t}t^3\}$ and $\mathscr{L}\{e^{-2t}\cos 4t\}$ is straightforward provided we know (and we do) $\mathscr{L}\{t^3\}$ and $\mathscr{L}\{\cos 4t\}$. In general, if we know the Laplace transform of a function f, $\mathscr{L}\{f(t)\} = F(s)$, it is possible to compute the Laplace transform of an exponential multiple of f—that is, $\mathscr{L}\{e^{at}f(t)\}$—with no additional effort other than *translating*, or *shifting*, the transform $F(s)$ to $F(s-a)$. This result is known as the **first translation theorem** or **first shifting theorem.**

THEOREM 7.6   **First Translation Theorem**

If $\mathscr{L}\{f(t)\} = F(s)$ and a is any real number, then

$$\mathscr{L}\{e^{at}f(t)\} = F(s-a).$$

Proof   The proof is immediate, since by Definition 7.1

$$\mathscr{L}\{e^{at}f(t)\} = \int_0^\infty e^{-st}e^{at}f(t)\,dt = \int_0^\infty e^{-(s-a)t}f(t)\,dt = F(s-a). \quad \blacksquare$$

If we consider s a real variable, then the graph of $F(s-a)$ is the graph of $F(s)$ shifted on the s-axis by the amount $|a|$. If $a > 0$, the graph

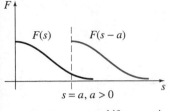

$F(s)$ $F(s - a)$

$s = a, a > 0$

shift on s-axis

Figure 7.10

of $F(s)$ is shifted a units to the right, whereas if $a < 0$, the graph is shifted $|a|$ units to the left. See Figure 7.10.

For emphasis it is sometimes useful to use the symbolism

$$\mathscr{L}\{e^{at}f(t)\} = \mathscr{L}\{f(t)\}|_{s \to s - a},$$

where $s \to s - a$ means that in the Laplace transform $F(s)$ of $f(t)$ we replace the symbol s wherever it appears by $s - a$.

EXAMPLE 1 **Using the First Translation Theorem**

Evaluate   **(a)** $\mathscr{L}\{e^{5t}t^3\}$ **(b)** $\mathscr{L}\{e^{-2t} \cos 4t\}$.

Solution   The results follow from Theorems 7.1 and 7.6.

(a) $\mathscr{L}\{e^{5t}t^3\} = \mathscr{L}\{t^3\}|_{s \to s - 5} = \dfrac{3!}{s^4}\Big|_{s \to s - 5} = \dfrac{6}{(s - 5)^4}$

(b) $\mathscr{L}\{e^{-2t} \cos 4t\} = \mathscr{L}\{\cos 4t\}|_{s \to s - (-2)} = \dfrac{s}{s^2 + 16}\Big|_{s \to s + 2} = \dfrac{s + 2}{(s + 2)^2 + 16}$ ∎

Inverse Form of Theorem 7.6   To compute the inverse of $F(s - a)$ we must recognize $F(s)$, find $f(t)$ by taking the inverse Laplace transform of $F(s)$, and then multiply $f(t)$ by the exponential function e^{at}. This procedure can be summarized symbolically in the following manner:

$$\mathscr{L}^{-1}\{F(s - a)\} = \mathscr{L}^{-1}\{F(s)|_{s \to s - a}\} = e^{at}f(t), \qquad \text{(1)}$$

where $f(t) = \mathscr{L}^{-1}\{F(s)\}$.

The first part of the next example illustrates partial fraction decomposition in the case when the denominator of $Y(s)$ contains *repeated linear factors*.

EXAMPLE 2 **Partial Fractions: Repeated Linear Factors**

Evaluate   **(a)** $\mathscr{L}^{-1}\left\{\dfrac{2s + 5}{(s - 3)^2}\right\}$ **(b)** $\mathscr{L}^{-1}\left\{\dfrac{s/2 + 5/3}{s^2 + 4s + 6}\right\}$.

Solution   **(a)** A repeated linear factor is a term $(s - a)^n$, where a is a real number and n is a positive integer ≥ 2. Recall that if $(s - a)^n$ appears in the denominator of a rational expression, then the assumed decomposition contains n partial fractions with constant numerators and denominators $s - a, (s - a)^2, \ldots, (s - a)^n$. Hence with $a = 3$ and $n = 2$ we write

$$\frac{2s + 5}{(s - 3)^2} = \frac{A}{s - 3} + \frac{B}{(s - 3)^2}.$$

By putting the two terms on the right-hand side over a common denominator, we obtain the numerator $2s + 6 = A(s - 3) + B$, and this identity yields $A = 2$ and $B = 11$. Therefore

$$\frac{2s + 5}{(s - 3)^2} = \frac{2}{s - 3} + \frac{11}{(s - 3)^2} \tag{2}$$

and

$$\mathcal{L}^{-1}\left\{\frac{2s + 5}{(s - 3)^2}\right\} = 2\mathcal{L}^{-1}\left\{\frac{1}{s - 3}\right\} + 11\mathcal{L}^{-1}\left\{\frac{1}{(s - 3)^2}\right\}. \tag{3}$$

Now $1/(s - 3)^2$ is $F(s) = 1/s^2$ shifted 3 units to the right. Since $\mathcal{L}^{-1}\{1/s^2\} = t$, it follows from (1) that

$$\mathcal{L}^{-1}\left\{\frac{1}{(s - 3)^2}\right\} = \mathcal{L}^{-1}\left\{\frac{1}{s^2}\Big|_{s \to s-3}\right\} = e^{-3t}t.$$

Finally, (3) is

$$\mathcal{L}^{-1}\left\{\frac{2s + 5}{(s - 3)^2}\right\} = 2e^{-3t} + 11e^{-3t}t. \tag{4}$$

(b) To start, observe that the quadratic polynomial $s^2 + 4s + 6$ has no real zeros and so has no real linear factors. In this situation we *complete the square*:

$$\frac{s/2 + 5/3}{s^2 + 4s + 6} = \frac{s/2 + 5/3}{(s + 2)^2 + 2}. \tag{5}$$

Our goal here is to recognize the expression on the right-hand side as some Laplace transform $F(s)$ in which s has been replaced throughout by $s + 2$. What we are trying to do is analogous to working part (b) of Example 1 backwards. The denominator in (5) is already in the correct form—that is, $s^2 + 2$ with s replaced by $s + 2$. However, we must fix up the numerator by manipulating the constants: $\frac{1}{2}s + \frac{5}{3} = \frac{1}{2}(s + 2) + \frac{5}{3} - \frac{2}{2} = \frac{1}{2}(s + 2) + \frac{2}{3}$.

Now by termwise division, the linearity of $\mathcal{L}^{-1}$, parts (e) and (d) of Theorem 7.3, and finally (1),

$$\frac{s/2 + 5/3}{(s + 2)^2 + 2} = \frac{\frac{1}{2}(s + 2) + \frac{2}{3}}{(s + 2)^2 + 2} = \frac{1}{2}\frac{s + 2}{(s + 2)^2 + 2} + \frac{2}{3}\frac{1}{(s + 2)^2 + 2}$$

$$\mathcal{L}^{-1}\left\{\frac{s/2 + 5/3}{s^2 + 4s + 6}\right\} = \frac{1}{2}\mathcal{L}^{-1}\left\{\frac{s + 2}{(s + 2)^2 + 2}\right\} + \frac{2}{3}\mathcal{L}^{-1}\left\{\frac{1}{(s + 2)^2 + 2}\right\}$$

$$= \frac{1}{2}\mathcal{L}^{-1}\left\{\frac{s}{s^2 + 2}\Big|_{s \to s+2}\right\} + \frac{2}{3\sqrt{2}}\mathcal{L}^{-1}\left\{\frac{\sqrt{2}}{s^2 + 2}\Big|_{s \to s+2}\right\} \tag{6}$$

$$= \frac{1}{2}e^{-2t}\cos\sqrt{2}t + \frac{3}{\sqrt{2}}e^{-2t}\sin\sqrt{2}t. \tag{7} \blacksquare$$

EXAMPLE 3   An Initial-Value Problem

Solve $y'' - 6y' + 9y = t^2e^{3t}$, $y(0) = 2$, $y'(0) = 6$.

Solution   Before transforming the DE, note that its right-hand side is similar to the function in part (a) of Example 1. After using linearity, Theorem 7.5, and the initial conditions, we simplify and then solve for

$Y(s) = \mathcal{L}\{f(t)\}$:

$$\mathcal{L}\{y''\} - 6\mathcal{L}\{y'\} + 9\mathcal{L}\{y\} = \mathcal{L}\{t^2 e^{3t}\}$$

$$s^2 Y(s) - sy(0) - y'(0) - 6[sY(s) - y(0)] + 9Y(s) = \frac{2}{(s-3)^3}$$

$$(s^2 - 6s + 9)Y(s) = 2s + 5 + \frac{2}{(s-3)^3}$$

$$(s - 3)^2 Y(s) = 2s + 5 + \frac{2}{(s-3)^3}$$

$$Y(s) = \frac{2s + 5}{(s-3)^2} + \frac{2}{(s-3)^5}.$$

The first term on the right-hand side was already decomposed into individual partial fractions in (2) in part (a) of Example 2:

$$Y(s) = \frac{2}{s-3} + \frac{11}{(s-3)^2} + \frac{2}{(s-3)^5}.$$

Thus $\quad y(t) = 2\mathcal{L}^{-1}\left\{\frac{1}{s-3}\right\} + 11\mathcal{L}^{-1}\left\{\frac{1}{(s-3)^2}\right\} + \frac{2}{4!}\mathcal{L}^{-1}\left\{\frac{4!}{(s-3)^5}\right\}.$ **(8)**

From the inverse form (1) of Theorem 7.6, the last two terms in (8) are

$$\mathcal{L}^{-1}\left\{\frac{1}{s^2}\bigg|_{s \to s-3}\right\} = te^{3t} \quad \text{and} \quad \mathcal{L}^{-1}\left\{\frac{4!}{s^5}\bigg|_{s \to s-3}\right\} = t^4 e^{3t}.$$

Thus (8) is $y(t) = 2e^{3t} + 11te^{3t} + \frac{1}{12}t^4 e^{3t}.$ ∎

EXAMPLE 4 **An Initial-Value Problem**

Solve $y'' + 4y' + 6y = 1 + e^{-t}, \quad y(0) = 0, \quad y'(0) = 0.$

Solution $\qquad\qquad \mathcal{L}\{y''\} + 4\mathcal{L}\{y'\} + 6\mathcal{L}\{y\} = \mathcal{L}\{1\} + \mathcal{L}\{e^{-t}\}$

$$s^2 Y(s) - sy(0) - y'(0) + 4[sY(s) - y(0)] + 6Y(s) = \frac{1}{s} + \frac{1}{s+1}$$

$$(s^2 + 4s + 6)Y(s) = \frac{2s+1}{s(s+1)}$$

$$Y(s) = \frac{2s+1}{s(s+1)(s^2+4s+6)}$$

Since the quadratic term in the denominator does not factor into real linear factors, the partial fraction decomposition for $Y(s)$ is found to be

$$Y(s) = \frac{1/6}{s} + \frac{1/3}{s+1} - \frac{s/2 + 5/3}{s^2 + 4s + 6}.$$

Moreover, in preparation for taking the inverse transform, we already manipulated the last term into the necessary form in part (b) of Example 2. So in view of the results in (6) and (7), we have the solution

$$y(t) = \frac{1}{6}\mathcal{L}^{-1}\left\{\frac{1}{s}\right\} + \frac{1}{3}\mathcal{L}^{-1}\left\{\frac{1}{s+1}\right\} - \frac{1}{2}\mathcal{L}^{-1}\left\{\frac{s+2}{(s+2)^2+2}\right\} - \frac{2}{3\sqrt{2}}\mathcal{L}^{-1}\left\{\frac{\sqrt{2}}{(s+2)^2+2}\right\}$$

$$= \frac{1}{6} + \frac{1}{3}e^{-t} - \frac{1}{2}e^{-2t}\cos\sqrt{2}t - \frac{\sqrt{2}}{3}e^{-2t}\sin\sqrt{2}t. \qquad \blacksquare$$

7.3.2   TRANSLATION ON THE t-AXIS

Unit Step Function   In engineering one frequently encounters functions that are either "off" or "on." For example, an external force acting on a mechanical system or a voltage impressed on a circuit can be turned off after a period of time. It is convenient, then, to define a special function that is the number 0 (off) up to a certain time $t = a$ and then the number 1 (on) after that time. This function is called the **unit step function** or the **Heaviside function.**

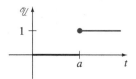

Figure 7.11

DEFINITION 7.3   Unit Step Function

The **unit step function** $\mathcal{U}(t - a)$ is defined to be

$$\mathcal{U}(t - a) = \begin{cases} 0, & 0 \le t < a \\ 1, & t \ge a. \end{cases}$$

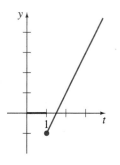

Figure 7.12

Notice that we define $\mathcal{U}(t - a)$ only on the nonnegative t-axis, since this is all that we are concerned with in the study of the Laplace transform. In a broader sense $\mathcal{U}(t - a) = 0$ for $t < a$. The graph of $\mathcal{U}(t - a)$ is given in Figure 7.11.

When a function f defined for $t \ge 0$ is multiplied by $\mathcal{U}(t - a)$, the unit step function "turns off" a portion of the graph of that function. For example, consider the function $f(t) = 2t - 3$. To "turn off" the portion of the graph of f on, say, the interval $0 \le t < 1$, we simply form the product $(2t - 3)\mathcal{U}(t - 1)$. See Figure 7.12. In general, the graph of $f(t)\mathcal{U}(t - a)$ is 0 (off) for $0 \le t < a$ and is the portion of the graph of f (on) for $t \ge a$.

The unit step function can also be used to write piecewise-defined functions in a compact form. For example, if we consider the intervals $0 \le t < 2, 2 \le t < 3$, and $t \ge 3$ and the corresponding values of $\mathcal{U}(t - 2)$ and $\mathcal{U}(t - 3)$, it should be apparent that the piecewise-defined function shown in Figure 7.13 is the same as $f(t) = 2 - 3\mathcal{U}(t - 2) + \mathcal{U}(t - 3)$. Also, a general piecewise-defined function of the type

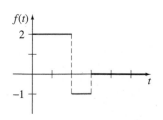

Figure 7.13

$$f(t) = \begin{cases} g(t), & 0 \le t < a \\ h(t), & t \ge a \end{cases} \qquad (9)$$

is the same as

$$f(t) = g(t) - g(t)\,\mathcal{U}(t - a) + h(t)\,\mathcal{U}(t - a). \qquad \textbf{(10)}$$

Similarly, a function of the type

$$f(t) = \begin{cases} 0, & 0 \le t < a \\ g(t), & a \le t < b \\ 0, & t \ge b \end{cases} \qquad \textbf{(11)}$$

can be written

$$f(t) = g(t)[\mathcal{U}(t - a) - \mathcal{U}(t - b)]. \qquad \textbf{(12)}$$

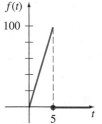

$f(t)$

Figure 7.14

EXAMPLE 5   A Piecewise-Defined Function

Express $f(t) = \begin{cases} 20t, & 0 \le t < 5 \\ 0, & t \ge 5 \end{cases}$ in terms of unit step functions. Graph.

Solution   The graph of f is given in Figure 7.14. Now from (9) and (10) with $a = 5$, $g(t) = 20t$, and $h(t) = 0$ we get $f(t) = 20t - 20t\,\mathcal{U}(t - 5)$.

Consider a general function $y = f(t)$ defined for $t \ge 0$. The piecewise-defined function

$$f(t - a)\,\mathcal{U}(t - a) = \begin{cases} 0, & 0 \le t < a \\ f(t - a), & t \ge a \end{cases} \qquad \textbf{(13)}$$

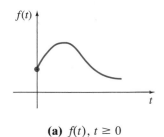

(a) $f(t), t \ge 0$

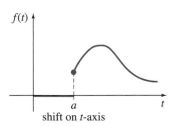

shift on t-axis

(b) $f(t - a)\,\mathcal{U}(t - a)$

Figure 7.15

plays a significant role in the discussion that follows. As shown in Figure 7.15, for $a > 0$ the graph of the function $y = f(t - a)\,\mathcal{U}(t - a)$ coincides with the graph of $y = f(t - a)$ for $t \ge a$ (which is the *entire* graph of $y = f(t), t \ge 0$ shifted a units to the right on the t-axis), but is identically zero for $0 \le t < a$.

We saw in Theorem 7.6 that an exponential multiple of $f(t)$ results in a translation of the transform $F(s)$ on the s-axis. As a consequence of the next theorem, we see that whenever $F(s)$ is multiplied by an exponential function e^{-as}, $a > 0$, the inverse transform of the product $e^{-as}F(s)$ is the function f shifted along the t-axis in the manner illustrated in Figure 7.15(b). This result, presented next in its direct transform version, is called the **second translation theorem** or **second shifting theorem**.

THEOREM 7.7   Second Translation Theorem

If $F(s) = \mathcal{L}\{f(t)\}$ and $a > 0$, then

$$\mathcal{L}\{f(t - a)\,\mathcal{U}(t - a)\} = e^{-as}F(s).$$

Proof   By the additive interval property of integrals,
$\int_0^\infty e^{-st}f(t-a)\mathcal{U}(t-a)\,dt$ can be written as two integrals:

$$\mathcal{L}\{f(t-a)\mathcal{U}(t-a)\} = \int_0^a e^{-st}f(t-a)\underbrace{\mathcal{U}(t-a)}_{\substack{\text{zero for}\\ 0\le t<a}}\,dt + \int_a^\infty e^{-st}f(t-a)\underbrace{\mathcal{U}(t-a)}_{\substack{\text{one for}\\ t\ge a}}\,dt$$

$$= \int_a^\infty e^{-st}f(t-a)\,dt.$$

Now if we let $v = t - a$, $dv = dt$ in the last integral, then

$$\mathcal{L}\{f(t-a)\mathcal{U}(t-a)\} = \int_0^\infty e^{-s(v+a)}f(v)\,dv = e^{-as}\int_0^\infty e^{-sv}f(v)\,dv = e^{-as}\mathcal{L}\{f(t)\}. \quad\blacksquare$$

We often wish to find the Laplace transform of just a unit step function. This can be from either Definition 7.1 or Theorem 7.7. If we identify $f(t) = 1$ in Theorem 7.7, then $f(t - a) = 1$, $F(s) = \mathcal{L}\{1\} = 1/s$, and so

$$\mathcal{L}\{\mathcal{U}(t-a)\} = \frac{e^{-as}}{s}. \tag{14}$$

For example, using (14), the Laplace transform of the function in Figure 7.13 is

$$\mathcal{L}\{f(t)\} = 2\mathcal{L}\{1\} - 3\mathcal{L}\{\mathcal{U}(t-2)\} + \mathcal{L}\{\mathcal{U}(t-3)\}$$

$$= 2\frac{1}{s} - 3\frac{e^{-2s}}{s} + \frac{e^{-3s}}{s}.$$

Inverse Form of Theorem 7.7   If $f(t) = \mathcal{L}^{-1}\{F(s)\}$, the inverse form of Theorem 7.7, $a > 0$, is

$$\mathcal{L}^{-1}\{e^{-as}F(s)\} = f(t-a)\mathcal{U}(t-a). \tag{15}$$

EXAMPLE 6   **Using Formula (15)**

Evaluate   **(a)** $\mathcal{L}^{-1}\left\{\dfrac{1}{s-4}e^{-2s}\right\}$ **(b)** $\mathcal{L}^{-1}\left\{\dfrac{s}{s^2+9}e^{-\pi s/2}\right\}$.

Solution   **(a)** With the identifications $a = 2$, $F(s) = 1/(s - 4)$, and $\mathcal{L}^{-1}\{F(s)\} = e^{4t}$, we have from (15)

$$\mathcal{L}^{-1}\left\{\frac{1}{s-4}e^{-2s}\right\} = e^{4(t-2)}\mathcal{U}(t-2).$$

(b) With $a = \pi/2$, $F(s) = s/(s^2 + 9)$, and $\mathcal{L}^{-1}\{F(s)\} = \cos 3t$, (15) yields

$$\mathcal{L}^{-1}\left\{\frac{s}{s^2+9}e^{-\pi s/2}\right\} = \cos 3\left(t - \frac{\pi}{2}\right)\mathcal{U}\left(t - \frac{\pi}{2}\right).$$

The last expression can be simplified somewhat using the addition formula for the cosine. Verify that the result is the same as $-\sin 3t\,\mathcal{U}\left(t - \dfrac{\pi}{2}\right)$.

Alternative Form of Theorem 7.7 We are frequently confronted with the problem of finding the Laplace transform of a product of a function g and a unit step function $\mathcal{U}(t - a)$ where the function g lacks the precise shifted form $f(t - a)$ in Theorem 7.7. To find the Laplace transform of $g(t)\mathcal{U}(t - a)$, it is possible to fix up $g(t)$ into the required form $f(t - a)$ by algebraic manipulations. For example, if we wanted to use Theorem 7.7 to find the Laplace transform of $t^2\mathcal{U}(t - 2)$, we would have to force $g(t) = t^2$ into the form $f(t - 2)$. You should work through the details and verify that $t^2 = (t - 2)^2 + 4(t - 2) + 4$ is an identity. Therefore

$$\mathcal{L}\{t^2\mathcal{U}(t - 2)\} = \mathcal{L}\{(t - 2)^2\mathcal{U}(t - 2) + 4(t - 2)\mathcal{U}(t - 2) + 4\mathcal{U}(t - 2)\},$$

where each term on the right-hand side can now be evaluated by Theorem 7.7. But since these manipulations are time consuming and often not obvious, it is simpler to devise an alternative version of Theorem 7.7. Using Definition 7.1, the definition of $\mathcal{U}(t - a)$, and the substitution $u = t - a$, we obtain

$$\mathcal{L}\{g(t)\mathcal{U}(t - a)\} = \int_a^\infty e^{-st}g(t)\,dt = \int_0^\infty e^{-s(u+a)}g(u + a)\,du.$$

That is,
$$\mathcal{L}\{g(t)\mathcal{U}(t - a)\} = e^{-as}\mathcal{L}\{g(t + a)\}. \tag{16}$$

EXAMPLE 7 **Second Translation Theorem—Alternative Form**

Evaluate $\mathcal{L}\{\sin t\,\mathcal{U}(t - \pi)\}$.

Solution With $g(t) = \sin t$ and $a = \pi$, then $g(t + \pi) = \sin(t + \pi) = -\cos t$ by the addition formula for the sine function. Hence by (16),

$$\mathcal{L}\{\sin t\,\mathcal{U}(t - \pi)\} = -e^{-\pi s}\mathcal{L}\{\cos t\} = -\frac{s}{s^2 + 1}e^{-\pi s}. \qquad ■$$

EXAMPLE 8 **An Initial-Value Problem**

Solve $y' + y = f(t), \quad y(0) = 5, \quad$ where $f(t) = \begin{cases} 0, & 0 \le t < \pi \\ 3\sin t, & t \ge \pi. \end{cases}$

Solution The function f can be written as $f(t) = 3\sin t\,\mathcal{U}(t - \pi)$, and so by linearity, the results of Example 7, and the usual partial fractions, we have

$$\mathcal{L}\{y'\} + \mathcal{L}\{y\} = 3\mathcal{L}\{\sin t\,\mathcal{U}(t - \pi)\}$$

$$sY(s) - y(0) + Y(s) = -3\frac{s}{s^2 + 1}e^{-\pi s}$$

$$(s + 1)Y(s) = 5 - \frac{3s}{s^2 + 1}e^{-\pi s}$$

$$Y(s) = \frac{5}{s + 1} - \frac{3}{2}\left[-\frac{1}{s + 1}e^{-\pi s} + \frac{1}{s^2 + 1}e^{-\pi s} + \frac{s}{s^2 + 1}e^{-\pi s}\right]. \tag{17}$$

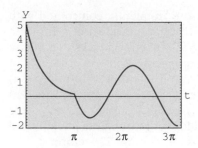

Figure 7.16

Now proceeding as we did in Example 6, it follows from (15) with $a = \pi$ that the inverses of the terms inside the brackets are

$$\mathcal{L}^{-1}\left\{\frac{1}{s+1}e^{-\pi s}\right\} = e^{-(t-\pi)}\mathcal{U}(t-\pi), \quad \mathcal{L}^{-1}\left\{\frac{1}{s^2+1}e^{-\pi s}\right\} = \sin(t-\pi)\mathcal{U}(t-\pi),$$

and

$$\mathcal{L}^{-1}\left\{\frac{s}{s^2+1}e^{-\pi s}\right\} = \cos(t-\pi)\mathcal{U}(t-\pi).$$

Thus the inverse of (17) is

$$y(t) = 5e^{-t} + \frac{3}{2}e^{-(t-\pi)}\mathcal{U}(t-\pi) - \frac{3}{2}\sin(t-\pi)\mathcal{U}(t-\pi) - \frac{3}{2}\cos(t-\pi)\mathcal{U}(t-\pi)$$

$$= 5e^{-t} + \frac{3}{2}[e^{-(t-\pi)} + \sin t + \cos t]\mathcal{U}(t-\pi) \quad \leftarrow \text{trigonometric identities}$$

$$= \begin{cases} 5e^{-t}, & 0 \le t < \pi \\ 5e^{-t} + \frac{3}{2}e^{-(t-\pi)} + \frac{3}{2}\sin t + \frac{3}{2}\cos t, & t \ge \pi. \end{cases} \tag{18}$$

We obtained the graph of (18) shown in Figure 7.16 by using a graphing utility.   ∎

Beams   In Section 5.2 we saw that the static deflection $y(x)$ of a uniform beam of length L carrying load $w(x)$ per unit length is found from the linear fourth-order differential equation

$$EI\frac{d^4y}{dx^4} = w(x), \tag{19}$$

where E is Young's modulus of elasticity and I is a moment of inertia of a cross-section of the beam. The Laplace transform is particularly useful in solving (19) when $w(x)$ is piecewise-defined. However, in order to use the Laplace transform, we must tacitly assume that $y(x)$ and $w(x)$ are defined on $(0, \infty)$ rather than on $(0, L)$. Note, too, that the next example is a boundary-value problem rather than an initial-value problem.

EXAMPLE 9   **A Boundary-Value Problem**

A beam of length L is embedded at both ends, as shown in Figure 7.17. Find the deflection of the beam when the load is given by

$$w(x) = \begin{cases} w_0\left(1 - \dfrac{2}{L}x\right), & 0 < x < L/2 \\ 0, & L/2 < x < L. \end{cases}$$

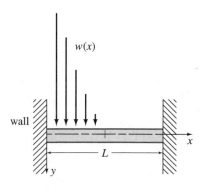

Figure 7.17

Solution   Recall that since the beam is embedded at both ends the boundary conditions are $y(0) = 0$, $y'(0) = 0$, $y(L) = 0$, $y'(L) = 0$. Now by (10) we can express $w(x)$ in terms of the unit step function:

$$w(x) = w_0\left(1 - \frac{2}{L}x\right) - w_0\left(1 - \frac{2}{L}x\right)\mathcal{U}\left(x - \frac{L}{2}\right)$$

$$= \frac{2w_0}{L}\left[\frac{L}{2} - x + \left(x - \frac{L}{2}\right)\mathcal{U}\left(x - \frac{L}{2}\right)\right].$$

Transforming (19) with respect to the variable x gives

$$EI\left(s^4Y(s) - s^3y(0) - s^2y'(0) - sy''(0) - y'''(0)\right) = \frac{2w_0}{L}\left[\frac{L/2}{s} - \frac{1}{s^2} + \frac{1}{s^2}e^{-Ls/2}\right]$$

or

$$s^4Y(s) - sy''(0) - y'''(0) = \frac{2w_0}{EIL}\left[\frac{L/2}{s} - \frac{1}{s^2} + \frac{1}{s^2}e^{-Ls/2}\right].$$

If we let $c_1 = y''(0)$ and $c_2 = y'''(0)$, then

$$Y(s) = \frac{c_1}{s^3} + \frac{c_2}{s^4} + \frac{2w_0}{EIL}\left[\frac{L/2}{s^5} - \frac{1}{s^6} + \frac{1}{s^6}e^{-Ls/2}\right],$$

and consequently

$$y(x) = \frac{c_1}{2!}\mathcal{L}^{-1}\left\{\frac{2!}{s^3}\right\} + \frac{c_2}{3!}\mathcal{L}^{-1}\left\{\frac{3!}{s^4}\right\}$$

$$+ \frac{2w_0}{EIL}\left[\frac{L/2}{4!}\mathcal{L}^{-1}\left\{\frac{4!}{s^5}\right\} - \frac{1}{5!}\mathcal{L}^{-1}\left\{\frac{5!}{s^6}\right\} + \frac{1}{5!}\mathcal{L}^{-1}\left\{\frac{5!}{s^6}e^{-Ls/2}\right\}\right]$$

$$= \frac{c_1}{2}x^2 + \frac{c_2}{6}x^3 + \frac{w_0}{60EIL}\left[\frac{5L}{2}x^4 - x^5 + \left(x - \frac{L}{2}\right)^5\mathcal{U}\left(x - \frac{L}{2}\right)\right].$$

Applying the conditions $y(L) = 0$ and $y'(L) = 0$ to the last result yields a system of equations for c_1 and c_2:

$$c_1\frac{L^2}{2} + c_2\frac{L^3}{6} + \frac{49w_0L^4}{1920EI} = 0$$

$$c_1L + c_2\frac{L^2}{2} + \frac{85w_0L^3}{960EI} = 0.$$

Solving, we find $c_1 = 23w_0L^2/960EI$ and $c_2 = -9w_0L/40EI$. Thus the deflection is given by

$$y(x) = \frac{23w_0L^2}{1920EI}x^2 - \frac{3w_0L}{80EI}x^3 + \frac{w_0}{60EIL}\left[\frac{5L}{2}x^4 - x^5 + \left(x - \frac{L}{2}\right)^5\mathcal{U}\left(x - \frac{L}{2}\right)\right].$$

■

EXERCISES 7.3

Answers to odd-numbered problems begin on page AN-8.

7.3.1 Translation on the s-axis

In Problems 1–20 find either $F(s)$ or $f(t)$, as indicated.

1. $\mathcal{L}\{te^{10t}\}$
2. $\mathcal{L}\{te^{-6t}\}$
3. $\mathcal{L}\{t^3e^{-2t}\}$
4. $\mathcal{L}\{t^{10}e^{-7t}\}$
5. $\mathcal{L}\{t(e^t + e^{2t})^2\}$
6. $\mathcal{L}\{e^{2t}(t - 1)^2\}$
7. $\mathcal{L}\{e^t \sin 3t\}$
8. $\mathcal{L}\{e^{-2t} \cos 4t\}$

9. $\mathcal{L}\{(1 - e^t + 3e^{-4t}) \cos 5t\}$

10. $\mathcal{L}\left\{e^{3t}\left(9 - 4t + 10 \sin\dfrac{t}{2}\right)\right\}$

11. $\mathcal{L}^{-1}\left\{\dfrac{1}{(s + 2)^3}\right\}$

12. $\mathcal{L}^{-1}\left\{\dfrac{1}{(s - 1)^4}\right\}$

13. $\mathcal{L}^{-1}\left\{\dfrac{1}{s^2 - 6s + 10}\right\}$

14. $\mathcal{L}^{-1}\left\{\dfrac{1}{s^2 + 2s + 5}\right\}$

15. $\mathcal{L}^{-1}\left\{\dfrac{s}{s^2 + 4s + 5}\right\}$

16. $\mathcal{L}^{-1}\left\{\dfrac{2s + 5}{s^2 + 6s + 34}\right\}$

17. $\mathcal{L}^{-1}\left\{\dfrac{s}{(s + 1)^2}\right\}$

18. $\mathcal{L}^{-1}\left\{\dfrac{5s}{(s - 2)^2}\right\}$

19. $\mathcal{L}^{-1}\left\{\dfrac{2s - 1}{s^2(s + 1)^3}\right\}$

20. $\mathcal{L}^{-1}\left\{\dfrac{(s + 1)^2}{(s + 2)^4}\right\}$

In Problems 21–30 use the Laplace transform to solve the given initial-value problem.

21. $y' + 4y = e^{-4t}, \quad y(0) = 2$

22. $y' - y = 1 + te^t, \quad y(0) = 0$

23. $y'' + 2y' + y = 0, \quad y(0) = 1, y'(0) = 1$

24. $y'' - 4y' + 4y = t^3 e^{2t}, \quad y(0) = 0, y'(0) = 0$

25. $y'' - 6y' + 9y = t, \quad y(0) = 0, y'(0) = 1$

26. $y'' - 4y' + 4y = t^3, \quad y(0) = 1, y'(0) = 0$

27. $y'' - 6y' + 13y = 0, \quad y(0) = 0, y'(0) = -3$

28. $2y'' + 20y' + 51y = 0, \quad y(0) = 2, y'(0) = 0$

29. $y'' - y' = e^t \cos t, \quad y(0) = 0, y'(0) = 0$

30. $y'' - 2y' + 5y = 1 + t, \quad y(0) = 0, y'(0) = 4$

In Problems 31 and 32 use the Laplace transform and the procedure outlined in Example 9 to solve the given boundary-value problem.

31. $y'' + 2y' + y = 0, \quad y'(0) = 2, y(1) = 2$

32. $y'' + 8y' + 20y = 0, \quad y(0) = 0, y'(\pi) = 0$

33. A 4-pound weight stretches a spring 2 feet. The weight is released from rest 18 inches above the equilibrium position, and the resulting motion takes place in a medium offering a damping force numerically equal to $\frac{7}{8}$ times the instantaneous velocity. Use the Laplace transform to find the equation of motion $x(t)$.

34. Recall that the differential equation for the instantaneous charge $q(t)$ on the capacitor in an *LRC* series circuit is given by

$$L\dfrac{d^2q}{dt^2} + R\dfrac{dq}{dt} + \dfrac{1}{C}q = E(t). \qquad (20)$$

See Section 5.1. Use the Laplace transform to find $q(t)$ when $L = 1$ h, $R = 20\ \Omega$, $C = 0.005$ f, $E(t) = 150$ V, $t > 0$, $q(0) = 0$, and $i(0) = 0$. What is the current $i(t)$?

35. Consider a battery of constant voltage E_0 that charges the capacitor shown in Figure 7.18. Divide equation (20) by L and define $2\lambda = R/L$

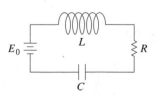

Figure 7.18

and $\omega^2 = 1/LC$. Use the Laplace transform to show that the solution $q(t)$ of $q'' + 2\lambda q' + \omega^2 q = E_0/L$ subject to $q(0) = 0$, $i(0) = 0$ is

$$q(t) = \begin{cases} E_0 C\left[1 - e^{-\lambda t}\left(\cosh\sqrt{\lambda^2 - \omega^2}\,t + \dfrac{\lambda}{\sqrt{\lambda^2 - \omega^2}}\sinh\sqrt{\lambda^2 - \omega^2}\,t\right)\right], & \lambda > \omega \\ E_0 C[1 - e^{-\lambda t}(1 + \lambda t)], & \lambda = \omega \\ E_0 C\left[1 - e^{-\lambda t}\left(\cos\sqrt{\omega^2 - \lambda^2}\,t + \dfrac{\lambda}{\sqrt{\omega^2 - \lambda^2}}\sin\sqrt{\omega^2 - \lambda^2}\,t\right)\right], & \lambda < \omega \end{cases}$$

36. Use the Laplace transform to find the charge $q(t)$ in an RC series circuit when $q(0) = 0$ and $E(t) = E_0 e^{-kt}$, $k > 0$. Consider two cases: $k \neq 1/RC$ and $k = 1/RC$.

7.3.2 Translation on the *t*-axis

In Problems 37–48 find either $F(s)$ or $f(t)$, as indicated.

37. $\mathcal{L}\{(t - 1)\mathcal{U}(t - 1)\}$ **38.** $\mathcal{L}\{e^{2-t}\mathcal{U}(t - 2)\}$

39. $\mathcal{L}\{t\mathcal{U}(t - 2)\}$ **40.** $\mathcal{L}\{(3t + 1)\mathcal{U}(t - 1)\}$

41. $\mathcal{L}\{\cos 2t\,\mathcal{U}(t - \pi)\}$ **42.** $\mathcal{L}\left\{\sin t\,\mathcal{U}\left(t - \dfrac{\pi}{2}\right)\right\}$

43. $\mathcal{L}^{-1}\left\{\dfrac{e^{-2s}}{s^3}\right\}$ **44.** $\mathcal{L}^{-1}\left\{\dfrac{(1 + e^{-2s})^2}{s + 2}\right\}$

45. $\mathcal{L}^{-1}\left\{\dfrac{e^{-\pi s}}{s^2 + 1}\right\}$ **46.** $\mathcal{L}^{-1}\left\{\dfrac{se^{-\pi s/2}}{s^2 + 4}\right\}$

47. $\mathcal{L}^{-1}\left\{\dfrac{e^{-s}}{s(s + 1)}\right\}$ **48.** $\mathcal{L}^{-1}\left\{\dfrac{e^{-2s}}{s^2(s - 1)}\right\}$

In Problems 49–54 match the given graph with one of the functions in (a)–(f). The graph of $f(t)$ is given in Figure 7.19.

(a) $f(t) - f(t)\mathcal{U}(t - a)$

(b) $f(t - b)\mathcal{U}(t - b)$

(c) $f(t)\mathcal{U}(t - a)$

(d) $f(t) - f(t)\mathcal{U}(t - b)$

(e) $f(t)\mathcal{U}(t - a) - f(t)\mathcal{U}(t - b)$

(f) $f(t - a)\mathcal{U}(t - a) - f(t - a)\mathcal{U}(t - b)$

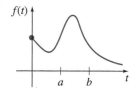

Figure 7.19

49.
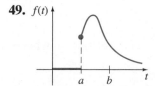
Figure 7.20

50.
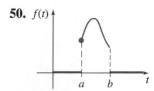
Figure 7.21

51. $f(t)$

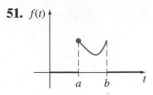

Figure 7.22

52. $f(t)$

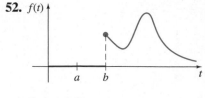

Figure 7.23

53. $f(t)$

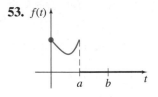

Figure 7.24

54. $f(t)$

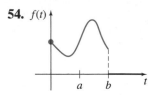

Figure 7.25

In Problems 55–62 write each function in terms of unit step functions. Find the Laplace transform of the given function.

55. $f(t) = \begin{cases} 2, & 0 \le t < 3 \\ -2, & t \ge 3 \end{cases}$

56. $f(t) = \begin{cases} 1, & 0 \le t < 4 \\ 0, & 4 \le t < 5 \\ 1, & t \ge 5 \end{cases}$

57. $f(t) = \begin{cases} 0, & 0 \le t < 1 \\ t^2, & t \ge 1 \end{cases}$

58. $f(t) = \begin{cases} 0, & 0 \le t < 3\pi/2 \\ \sin t, & t \ge 3\pi/2 \end{cases}$

59. $f(t) = \begin{cases} t, & 0 \le t < 2 \\ 0, & t \ge 2 \end{cases}$

60. $f(t) = \begin{cases} \sin t, & 0 \le t < 2\pi \\ 0, & t \ge 2\pi \end{cases}$

61. $f(t)$

1

a　b　t

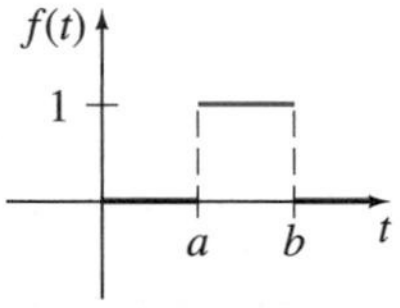

rectangular pulse

Figure 7.26

62. $f(t)$

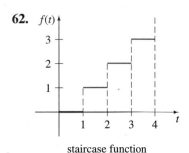

staircase function

Figure 7.27

In Problems 63–70 use the Laplace transform to solve the given initial-value problem.

63. $y' + y = f(t)$, $y(0) = 0$, where $f(t) = \begin{cases} 0, & 0 \le t < 1 \\ 5, & t \ge 1 \end{cases}$

64. $y' + y = f(t)$, $y(0) = 0$, where $f(t) = \begin{cases} 1, & 0 \le t < 1 \\ -1, & t \ge 1 \end{cases}$

65. $y' + 2y = f(t)$, $y(0) = 0$, where $f(t) = \begin{cases} t, & 0 \le t < 1 \\ 0, & t \ge 1 \end{cases}$

66. $y'' + 4y = f(t)$, $y(0) = 0$, $y'(0) = -1$, where $f(t) = \begin{cases} 1, & 0 \le t < 1 \\ 0, & t \ge 1 \end{cases}$

67. $y'' + 4y = \sin t\, \mathcal{U}(t - 2\pi)$, $y(0) = 1$, $y'(0) = 0$

68. $y'' - 5y' + 6y = \mathcal{U}(t - 1)$, $y(0) = 0$, $y'(0) = 1$

69. $y'' + y = f(t)$, $y(0) = 0$, $y'(0) = 1$, where $f(t) = \begin{cases} 0, & 0 \le t < \pi \\ 1, & \pi \le t < 2\pi \\ 0, & t \ge 2\pi \end{cases}$

70. $y'' + 4y' + 3y = 1 - \mathcal{U}(t - 2) - \mathcal{U}(t - 4) + \mathcal{U}(t - 6)$,
$y(0) = 0$, $y'(0) = 0$

71. Suppose a 32-pound weight stretches a spring 2 feet. If the weight is released from rest at the equilibrium position, find the equation of motion $x(t)$ if an impressed force $f(t) = 20t$ acts on the system for $0 \le t < 5$ and is then removed (see Example 5). Ignore any damping forces. Use a graphing utility to obtain the graph $x(t)$ on the interval $[0, 10]$.

72. Solve Problem 71 if the impressed force $f(t) = \sin t$ acts on the system for $0 \le t < 2\pi$ and is then removed.

In Problems 73 and 74 use the Laplace transform to find the charge $q(t)$ on the capacitor in an RC series circuit subject to the given conditions.

73. $q(0) = 0$, $R = 2.5\ \Omega$, $C = 0.08$ f, $E(t)$ given in Figure 7.28

74. $q(0) = q_0$, $R = 10\ \Omega$, $C = 0.1$ f, $E(t)$ given in Figure 7.29

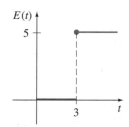

Figure 7.28

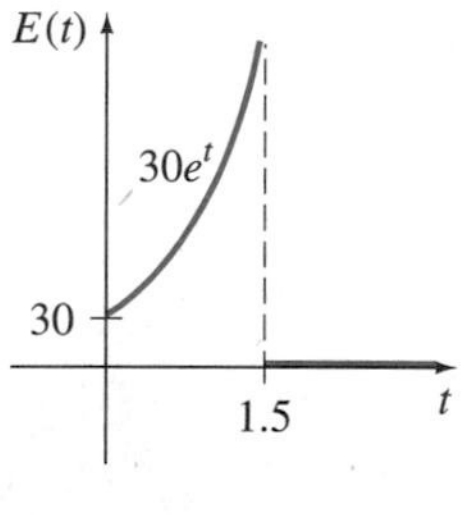

Figure 7.29

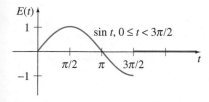

Figure 7.30

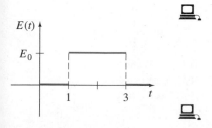

Figure 7.31

75. (a) Use the Laplace transform to find the current $i(t)$ in a single-loop LR series circuit when $i(0) = 0$, $L = 1$ h, $R = 10\ \Omega$, and $E(t)$ is as given in Figure 7.30.

(b) Use a computer graphing program to graph $i(t)$ on the interval $0 \le t \le 6$. Use the graph to estimate i_{max} and i_{min}, the maximum and minimum values of the current.

76. (a) Use the Laplace transform to find the charge $q(t)$ on the capacitor in an RC series circuit when $q(0) = 0$, $R = 50\ \Omega$, $C = 0.01$ f, and $E(t)$ is as given in Figure 7.31.

(b) Assume $E_0 = 100$ V. Use a computer graphing program to graph $q(t)$ on the interval $0 \le t \le 6$. Use the graph to estimate q_{max}, the maximum value of the charge.

77. A cantilever beam is embedded at its left end and free at its right end. Use the Laplace transform to find the deflection $y(x)$ when the load is given by

$$w(x) = \begin{cases} w_0, & 0 < x < L/2 \\ 0, & L/2 \leq x < L \end{cases}$$

78. Solve Problem 77 when the load is given by

$$w(x) = \begin{cases} 0, & 0 < x < L/3 \\ w_0, & L/3 < x < 2L/3 \\ 0, & 2L/3 < x < L \end{cases}$$

79. Find the deflection $y(x)$ of a cantilever beam embedded at its left end and free at its right end when the load is as given in Example 9.

80. A beam is embedded at its left end and simply supported at its right end. Find the deflection $y(x)$ when the load is as given in Problem 77.

Discussion Problems

81. Discuss how you would fix up each of the following functions so that Theorem 7.7 could be used directly to find the given Laplace transform. Check your answers using (16) of this section.
(a) $\mathscr{L}\{(2t + 1)\mathscr{U}(t - 1)\}$ **(b)** $\mathscr{L}\{e^t\mathscr{U}(t - 5)\}$
(c) $\mathscr{L}\{\cos t\,\mathscr{U}(t - \pi)\}$ **(d)** $\mathscr{L}\{(t^2 - 3t)\mathscr{U}(t - 2)\}$

82. (a) Assume that Theorem 7.6 holds when the symbol a is replaced by ki, where k is a real number and $i^2 = -1$. Show that $\mathscr{L}\{te^{kti}\}$ can be used to deduce

$$\mathscr{L}\{t \cos kt\} = \frac{s^2 - k^2}{(s^2 + k^2)^2} \quad \text{and} \quad \mathscr{L}\{t \sin kt\} = \frac{2ks}{(s^2 + k^2)^2}.$$

(b) Now use the Laplace transform to solve the initial-value problem $x'' + \omega^2 x = \cos \omega t$, $x(0) = 0$, $x'(0) = 0$.

7.4 ADDITIONAL OPERATIONAL PROPERTIES

• *Derivatives of transforms* • *Convolution of two functions*
• *Convolution theorem* • *Inverse form of convolution theorem*
• *Transform of an integral* • *Integral equation* • *Integrodifferential equation* • *Transform of a periodic function*

In this section we examine some additional labor-saving operational properties of the Laplace transform. We find how to obtain the Laplace transform of a function that is multiplied by a positive integer power of t, a certain kind of integral, and a periodic function. We also use the Laplace transform to solve some new types of equations—Volterra integral equations, integrodifferential equations, and ordinary differential equations in which the driving function is a periodic piecewise-defined function.

Multiplying a Function by t^n The Laplace transform of the product of a function $f(t)$ with t can be found by differentiating the Laplace transform of $f(t)$. If $F(s) = \mathcal{L}\{f(t)\}$ and if we assume that the interchanging of differentiation and integration is possible, then

$$\frac{d}{ds} F(s) = \frac{d}{ds} \int_0^\infty e^{-st} f(t)\, dt = \int_0^\infty \frac{\partial}{\partial s} [e^{-st} f(t)]\, dt = -\int_0^\infty e^{-st} tf(t)\, dt = -\mathcal{L}\{tf(t)\};$$

that is, $$\mathcal{L}\{tf(t)\} = -\frac{d}{ds} \mathcal{L}\{f(t)\}.$$

We can use the last result to find the Laplace transform of $t^2 f(t)$:

$$\mathcal{L}\{t^2 f(t)\} = \mathcal{L}\{t \cdot t f(t)\} = -\frac{d}{ds} \mathcal{L}\{tf(t)\} = -\frac{d}{ds} \left(-\frac{d}{ds} \mathcal{L}\{f(t)\} \right) = \frac{d^2}{ds^2} \mathcal{L}\{f(t)\}.$$

The preceding two cases suggest the general result for $\mathcal{L}\{t^n f(t)\}$.

THEOREM 7.8 **Derivatives of Transforms**

If $F(s) = \mathcal{L}\{f(t)\}$ and $n = 1, 2, 3, \ldots$, then

$$\mathcal{L}\{t^n f(t)\} = (-1)^n \frac{d^n}{ds^n} F(s).$$

EXAMPLE 1 **Using Theorem 7.8**

Evaluate $\mathcal{L}\{t \sin kt\}$.

Solution With $f(t) = \sin kt$, $F(s) = k/(s^2 + k^2)$, and $n = 1$, Theorem 7.8 gives

$$\mathcal{L}\{t \sin kt\} = -\frac{d}{ds} \mathcal{L}\{\sin kt\} = -\frac{d}{ds} \left(\frac{k}{s^2 + k^2} \right) = \frac{2ks}{(s^2 + k^2)^2}. \qquad \blacksquare$$

If we want to evaluate $\mathcal{L}\{t^2 \sin kt\}$ and $\mathcal{L}\{t^3 \sin kt\}$, all we need do, in turn, is take the negative of the derivative with respect to s of the result in Example 1 and then take the negative of the derivative with respect to s of $\mathcal{L}\{t^2 \sin kt\}$.

Note To find transforms of functions $t^n e^{at}$ we can use either Theorem 7.6 or Theorem 7.8. For example,

Theorem 7.6: $\quad \mathcal{L}\{te^{3t}\} = \mathcal{L}\{t\}_{s \to s-3} = \left. \frac{1}{s^2} \right|_{s \to s-3} = \frac{1}{(s-3)^2}$

Theorem 7.8: $\quad \mathcal{L}\{te^{3t}\} = -\frac{d}{ds} \mathcal{L}\{e^{3t}\} = -\frac{d}{ds} \frac{1}{s-3} = (s-3)^{-2} = \frac{1}{(s-3)^2}.$

EXAMPLE 2   **An Initial-Value Problem**

Solve $x'' + 16x = \cos 4t$, $x(0) = 0$, $x'(0) = 1$.

Solution   The initial-value problem could describe the forced, un-damped, and resonant motion of a mass on a spring. The mass starts with an initial velocity of 1 ft/s in the downward direction from the equilibrium position.

Transforming the differential equation gives

$$(s^2 + 16)X(s) = 1 + \frac{s}{s^2 + 16} \quad \text{or} \quad X(s) = \frac{1}{s^2 + 16} + \frac{s}{(s^2 + 16)^2}.$$

Now we just saw in Example 1 that

$$\mathcal{L}^{-1}\left\{\frac{2ks}{(s^2 + k^2)^2}\right\} = t \sin kt, \tag{1}$$

and so with the identification $k = 4$ in (1) and in part (d) of Theorem 7.3, we obtain

$$x(t) = \frac{1}{4}\mathcal{L}^{-1}\left\{\frac{4}{s^2 + 16}\right\} + \frac{1}{8}\mathcal{L}^{-1}\left\{\frac{8s}{(s^2 + 16)^2}\right\}$$

$$= \frac{1}{4}\sin 4t + \frac{1}{8}t \sin 4t. \qquad \blacksquare$$

Convolution   If functions f and g are piecewise continuous on $[0, \infty)$, then a special product, denoted by $f * g$, is defined by the integral

$$f * g = \int_0^t f(\tau)g(t - \tau)\,d\tau \tag{2}$$

and is called the **convolution** of f and g. The convolution $f * g$ is a function of t. For example,

$$e^t * \sin t = \int_0^t e^\tau \sin(t - \tau)\,d\tau = \frac{1}{2}(-\sin t - \cos t + e^t). \tag{3}$$

It is left as an exercise to show that

$$\int_0^t f(\tau)g(t - \tau)\,d\tau = \int_0^t f(t - \tau)g(\tau)\,d\tau;$$

that is, $f * g = g * f$. This means that the convolution of two functions is commutative. See Problem 54 in Exercises 7.4.

It is *not* true that the integral of a product of functions is the product of the integrals. However, it *is* true that the Laplace transform of the special product (2) is the product of the Laplace transforms of f and g. This means that it is possible to find the Laplace transform of the convolution of two functions without actually evaluating the integral as we did in (3). The result that follows is known as the **convolution theorem**.

THEOREM 7.9   **Convolution Theorem**

If $f(t)$ and $g(t)$ are piecewise continuous on $[0, \infty)$ and of exponential order, then

$$\mathcal{L}\{f * g\} = \mathcal{L}\{f(t)\}\, \mathcal{L}\{g(t)\} = F(s)G(s).$$

Proof   Let $F(s) = \mathcal{L}\{f(t)\} = \displaystyle\int_0^\infty e^{-s\tau} f(\tau)\, d\tau$

and $G(s) = \mathcal{L}\{g(t)\} = \displaystyle\int_0^\infty e^{-s\beta} g(\beta)\, d\beta.$

Proceeding formally, we have

$$F(s)G(s) = \left(\int_0^\infty e^{-s\tau} f(\tau)\, d\tau\right)\left(\int_0^\infty e^{-s\beta} g(\beta)\, d\beta\right)$$

$$= \int_0^\infty \int_0^\infty e^{-s(\tau+\beta)} f(\tau)g(\beta)\, d\tau\, d\beta$$

$$= \int_0^\infty f(\tau)\, d\tau \int_0^\infty e^{-s(\tau+\beta)} g(\beta)\, d\beta.$$

Holding τ fixed, we let $t = \tau + \beta$, $dt = d\beta$ so that

$$F(s)G(s) = \int_0^\infty f(\tau)\, d\tau \int_\tau^\infty e^{-st} g(t - \tau)\, dt.$$

In the $t\tau$-plane we are integrating over the shaded region in Figure 7.32. Since f and g are piecewise continuous on $[0, \infty)$ and of exponential order, it is possible to interchange the order of integration:

$$F(s)G(s) = \int_0^\infty e^{-st}\, dt \int_0^t f(\tau)g(t - \tau)\, d\tau = \int_0^\infty e^{-st}\left\{\int_0^t f(\tau)g(t - \tau)\, d\tau\right\} dt = \mathcal{L}\{f * g\}. \quad\blacksquare$$

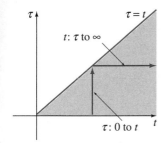

$\tau = t$

$t : \tau$ to ∞

$\tau : 0$ to t

Figure 7.32

EXAMPLE 3   **Transform of a Convolution**

Evaluate $\mathcal{L}\left\{\displaystyle\int_0^t e^\tau \sin(t - \tau)\, d\tau\right\}$.

Solution   With $f(t) = e^t$ and $g(t) = \sin t$, the convolution theorem states that the Laplace transform of the convolution of f and g is the product of their Laplace transforms:

$$\mathcal{L}\left\{\int_0^t e^\tau \sin(t - \tau)\, d\tau\right\} = \mathcal{L}\{e^t\} \cdot \mathcal{L}\{\sin t\} = \frac{1}{s - 1} \cdot \frac{1}{s^2 + 1} = \frac{1}{(s - 1)(s^2 + 1)}. \quad\blacksquare$$

Inverse Form of Theorem 7.9   The convolution theorem is sometimes useful in finding the inverse Laplace transform of the product of two Laplace transforms. From Theorem 7.9 we have

$$\mathcal{L}^{-1}\{F(s)G(s)\} = f * g. \tag{4}$$

Many of the results in the table of Laplace transforms in Appendix III can be derived using (4). For example, in the next example we obtain entry 25 of the table:

$$\mathcal{L}\{\sin kt - kt \cos kt\} = \frac{2k^3}{(s^2 + k^2)^2}. \tag{5}$$

EXAMPLE 4   **Inverse Transform as a Convolution**

Evaluate $\mathcal{L}^{-1}\left\{\dfrac{1}{(s^2 + k^2)^2}\right\}$.

Solution   Let $F(s) = G(s) = \dfrac{1}{s^2 + k^2}$

so that $f(t) = g(t) = \dfrac{1}{k}\mathcal{L}^{-1}\left\{\dfrac{k}{s^2 + k^2}\right\} = \dfrac{1}{k}\sin kt.$

In this case (4) gives

$$\mathcal{L}^{-1}\left\{\frac{1}{(s^2 + k^2)^2}\right\} = \frac{1}{k^2}\int_0^t \sin k\tau \sin k(t - \tau)\, d\tau. \tag{6}$$

Now recall from trigonometry that

$$\cos(A + B) = \cos A \cos B - \sin A \sin B$$

and $\cos(A - B) = \cos A \cos B + \sin A \sin B.$

Subtracting the first from the second gives the identity

$$\sin A \sin B = \frac{1}{2}[\cos(A - B) - \cos(A + B)].$$

If we set $A = k\tau$ and $B = k(t - \tau)$, we can carry out the integration in (6):

$$\mathcal{L}^{-1}\left\{\frac{1}{(s^2 + k^2)^2}\right\} = \frac{1}{2k^2}\int_0^t [\cos k(2\tau - t) - \cos kt]\, d\tau$$

$$= \frac{1}{2k^2}\left[\frac{1}{2k}\sin k(2\tau - t) - \tau \cos kt\right]_0^t$$

$$= \frac{\sin kt - kt \cos kt}{2k^3}.$$

Multiplying both sides by $2k^3$ gives the inverse form of (5).   ∎

Transform of an Integral   When $g(t) = 1$ and $\mathcal{L}\{g(t)\} = G(s) = 1/s$, the convolution theorem implies that the Laplace transform of the integral of f is

$$\mathcal{L}\left\{\int_0^t f(\tau)\, d\tau\right\} = \frac{F(s)}{s}. \tag{7}$$

The inverse form of (7),

$$\int_0^t f(\tau)\, d\tau = \mathcal{L}^{-1}\left\{\frac{F(s)}{s}\right\},\tag{8}$$

can be used in lieu of partial fractions when s^n is a factor of the denominator and $f(t) = \mathcal{L}^{-1}\{F(s)\}$ is easy to integrate. For example, we know for $f(t) = \sin t$ that $F(s) = 1/(s^2 + 1)$, and so by (8)

$$\mathcal{L}^{-1}\left\{\frac{1}{s(s^2+1)}\right\} = \int_0^t \sin \tau\, d\tau = 1 - \cos t$$

$$\mathcal{L}^{-1}\left\{\frac{1}{s^2(s^2+1)}\right\} = \int_0^t (1 - \cos \tau)\, d\tau = t - \sin t$$

$$\mathcal{L}^{-1}\left\{\frac{1}{s^3(s^2+1)}\right\} = \int_0^t (\tau - \sin \tau)\, d\tau = \frac{1}{2}t^2 - 1 + \cos t$$

and so on.

Volterra Integral Equation   The convolution theorem and the result in (7) are useful in solving other types of equations in which an unknown function appears under an integral sign. In the next example we solve a **Volterra integral equation** for $f(t)$,

$$f(t) = g(t) + \int_0^t f(\tau)h(t - \tau)\, d\tau.\tag{9}$$

The functions $g(t)$ and $h(t)$ are known. Notice that the integral in (9) has the convolution form (2) with the symbol h playing the part of g.

EXAMPLE 5   **An Integral Equation**

Solve $\;f(t) = 3t^2 - e^{-t} - \displaystyle\int_0^t f(\tau)e^{t-\tau}\, d\tau\;$ for $f(t)$.

Solution   In the integral we identify $h(t - \tau) = e^{t-\tau}$ so that $h(t) = e^t$. We take the Laplace transform of each term; in particular, by Theorem 7.9 the transform of the integral is the product of $\mathcal{L}\{f(t)\} = F(s)$ and $\mathcal{L}\{e^t\} = 1/(s - 1)$:

$$F(s) = 3 \cdot \frac{2}{s^3} - \frac{1}{s+1} - F(s) \cdot \frac{1}{s-1}.$$

After solving the last equation for $F(s)$ and carrying out the partial fraction decomposition, we find

$$F(s) = \frac{6}{s^3} - \frac{6}{s^4} + \frac{1}{s} - \frac{2}{s+1}.$$

The inverse transform then gives

$$f(t) = 3\mathcal{L}^{-1}\left\{\frac{2!}{s^3}\right\} - \mathcal{L}^{-1}\left\{\frac{3!}{s^4}\right\} + \mathcal{L}^{-1}\left\{\frac{1}{s}\right\} - 2\mathcal{L}^{-1}\left\{\frac{1}{s+1}\right\}$$

$$= 3t^2 - t^3 + 1 - 2e^{-t}.$$

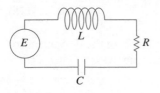

Figure 7.33

Series Circuits In a single-loop or series circuit, Kirchhoff's second law states that the sum of the voltage drops across an inductor, resistor, and capacitor is equal to the impressed voltage $E(t)$. Now it is known that the voltage drops across an inductor, resistor, and capacitor are, respectively,

$$L\frac{di}{dt}, \quad Ri(t), \quad \text{and} \quad \frac{1}{C}\int_0^t i(\tau)\,d\tau,$$

where $i(t)$ is the current and L, R, and C are constants. It follows that the current in a circuit, such as that shown in Figure 7.33, is governed by the **integrodifferential equation**

$$L\frac{di}{dt} + Ri(t) + \frac{1}{C}\int_0^t i(\tau)\,d\tau = E(t). \tag{10}$$

EXAMPLE 6   **An Integrodifferential Equation**

Determine the current $i(t)$ in a single-loop LRC circuit when $L = 0.1$ h, $R = 2\ \Omega$, $C = 0.1$ f, $i(0) = 0$, and the impressed voltage is

$$E(t) = 120t - 120t\ \mathcal{U}(t-1).$$

Solution   With the given data, equation (10) becomes

$$0.1\frac{di}{dt} + 2i + 10\int_0^t i(\tau)\,d\tau = 120t - 120t\ \mathcal{U}(t-1).$$

Now by (7), $\mathcal{L}\left\{\int_0^t i(\tau)\,d\tau\right\} = I(s)/s$, where $I(s) = \mathcal{L}\{i(t)\}$. Thus the Laplace transform of the integrodifferential equation is

$$0.1sI(s) + 2I(s) + 10\frac{I(s)}{s} = 120\left[\frac{1}{s^2} - \frac{1}{s^2}e^{-s} - \frac{1}{s}e^{-s}\right]. \qquad \leftarrow \text{by (16) of Section 7.3}$$

Multiplying this equation by $10s$, using $s^2 + 20s + 100 = (s + 10)^2$, and then solving for $I(s)$ gives

$$I(s) = 1200\left[\frac{1}{s(s+10)^2} - \frac{1}{s(s+10)^2}e^{-s} - \frac{1}{(s+10)^2}e^{-s}\right].$$

By partial fractions,

$$I(s) = 1200\left[\frac{1/100}{s} - \frac{1/100}{s+10} - \frac{1/10}{(s+10)^2} - \frac{1/100}{s}e^{-s}\right.$$

$$\left. + \frac{1/100}{s+10}e^{-s} + \frac{1/10}{(s+10)^2}e^{-s} - \frac{1}{(s+10)^2}e^{-s}\right].$$

From the inverse form of the second translation theorem, (15) of Section 7.3, we finally obtain

$$i(t) = 12[1 - \mathcal{U}(t-1)] - 12[e^{-10t} - e^{-10(t-1)}\ \mathcal{U}(t-1)]$$

$$-120te^{-10t} - 1080(t-1)e^{-10(t-1)}\ \mathcal{U}(t-1).$$

Written as a piecewise-defined function, the foregoing solution is

$$i(t) = \begin{cases} 12 - 12e^{-10t} - 120te^{-10t}, & 0 \leq t < 1 \\ -12e^{-10t} + 12e^{-10(t-1)} - 120te^{-10t} - 1080(t-1)e^{-10(t-1)}, & t \geq 1 \end{cases}$$

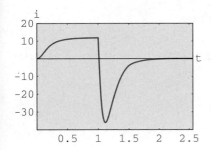

Figure 7.34

Using the last form of the solution and a CAS, we graph $i(t)$ on each of the two intervals and then combine the graphs. Note in Figure 7.34 that even though the input $E(t)$ is discontinuous, the output or response $i(t)$ is a continuous function. ■

Transform of a Periodic Function If a periodic function f has period T, $T > 0$, then $f(t + T) = f(t)$. The Laplace transform of a periodic function can be obtained by integration over one period.

THEOREM 7.10	**Transform of a Periodic Function**

If $f(t)$ is piecewise continuous on $[0, \infty)$, of exponential order, and periodic with period T, then

$$\mathscr{L}\{f(t)\} = \frac{1}{1 - e^{sT}} \int_0^T e^{-st} f(t)\, dt.$$

Proof Write the Laplace transform of f as two integrals:

$$\mathscr{L}\{f(t)\} = \int_0^T e^{-st} f(t)\, dt + \int_T^\infty e^{-st} f(t)\, dt.$$

When we let $t = u + T$, the last integral becomes

$$\int_T^\infty e^{-st} f(t)\, dt = \int_0^\infty e^{-s(u+T)} f(u + T)\, du = e^{-sT} \int_0^\infty e^{-su} f(u)\, du = e^{-sT} \mathscr{L}\{f(t)\}.$$

Therefore $\qquad \mathscr{L}\{f(t)\} = \int_0^T e^{-st} f(t)\, dt + e^{-sT} \mathscr{L}\{f(t)\}.$

Solving the equation in the last line for $\mathscr{L}\{f(t)\}$ proves the theorem. ■

EXAMPLE 7	**Transform of a Periodic Function**

Find the Laplace transform of the periodic function shown in Figure 7.35.

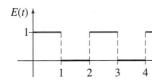

Figure 7.35

Solution The function $E(t)$ is called a square wave and has period $T = 2$. On the interval $0 \le t < 2$, $E(t)$ can be defined by

$$E(t) = \begin{cases} 1, & 0 \le t < 1 \\ 0, & 1 \le t < 2 \end{cases}$$

and outside the interval by $f(t + 2) = f(t)$. Now from Theorem 7.10,

$$\mathscr{L}\{E(t)\} = \frac{1}{1 - e^{-2s}} \int_0^2 e^{-st} E(t)\, dt = \frac{1}{1 - e^{-2s}} \left[\int_0^1 e^{-st} \cdot 1\, dt + \int_1^2 e^{-st} \cdot 0\, dt \right]$$

$$= \frac{1}{1 - e^{-2s}} \frac{1 - e^{-s}}{s} \qquad \leftarrow 1 - e^{-2s} = (1 + e^{-s})(1 - e^{-s})$$

$$= \frac{1}{s(1 + e^{-s})}. \tag{11} ■$$

| EXAMPLE 8 | A Periodic Impressed Voltage |

The differential equation for the current $i(t)$ in a single-loop LR series circuit is

$$L\frac{di}{dt} + Ri = E(t). \tag{12}$$

Determine the current $i(t)$ when $i(0) = 0$ and $E(t)$ is the square wave function shown in Figure 7.35.

Solution Using the result in (11) of the preceding example, the Laplace transform of the DE is

$$LsI(s) + RI(s) = \frac{1}{s(1 + e^{-s})} \quad \text{or} \quad I(s) = \frac{1/L}{s(s + R/L)} \cdot \frac{1}{1 + e^{-s}}. \tag{13}$$

To find the inverse Laplace transform of the last function, we first make use of geometric series. With the identification $x = e^{-s}$, $s > 0$, the geometric series

$$\frac{1}{1 + x} = 1 - x + x^2 - x^3 + \cdots \quad \text{becomes} \quad \frac{1}{1 + e^{-s}} = 1 - e^{-s} + e^{-2s} - e^{-3s} + \cdots.$$

From

$$\frac{1}{s(s + R/L)} = \frac{L/R}{s} - \frac{L/R}{s + R/L}$$

we can then rewrite (13) as

$$I(s) = \frac{1}{R}\left(\frac{1}{s} - \frac{1}{s + R/L}\right)(1 - e^{-s} + e^{-2s} - e^{-3s} + \cdots)$$

$$= \frac{1}{R}\left(\frac{1}{s} - \frac{e^{-s}}{s} + \frac{e^{-2s}}{s} - \frac{e^{-3s}}{s} + \cdots\right) - \frac{1}{R}\left(\frac{1}{s + R/L} - \frac{1}{s + R/L}e^{-s} + \frac{e^{-2s}}{s + R/L} - \frac{e^{-3s}}{s + R/L} + \cdots\right).$$

By applying the form of the second translation theorem to each term of both series, we obtain

$$i(t) = \frac{1}{R}(1 - \mathcal{U}(t - 1) + \mathcal{U}(t - 2) - \mathcal{U}(t - 3) + \cdots)$$

$$- \frac{1}{R}(e^{-Rt/L} - e^{-R(t-1)/L}\mathcal{U}(t - 1) + e^{-R(t-2)/L}\mathcal{U}(t - 2) - e^{-R(t-3)/L}\mathcal{U}(t - 3) + \cdots)$$

or, equivalently,

$$i(t) = \frac{1}{R}(1 - e^{-Rt/L}) + \frac{1}{R}\sum_{n=1}^{\infty}(-1)^n(1 - e^{-R(t-n)/L})\mathcal{U}(t - n).$$

To interpret the solution, let us assume for the sake of illustration that $R = 1$, $L = 1$, and $0 \le t < 4$. In this case

$$i(t) = 1 - e^{-t} - (1 - e^{t-1})\mathcal{U}(t - 1) + (1 - e^{-(t-2)})\mathcal{U}(t - 2) - (1 - e^{-(t-3)})\mathcal{U}(t - 3);$$

in other words,

$$i(t) = \begin{cases} 1 - e^{-t}, & 0 \le t < 1 \\ -e^{-t} + e^{-(t-1)}, & 1 \le t < 2 \\ 1 - e^{-t} + e^{-(t-1)} - e^{-(t-2)}, & 2 \le t < 3 \\ -e^{-t} + e^{-(t-1)} - e^{-(t-2)} + e^{-(t-3)}, & 3 \le t < 4. \end{cases}$$

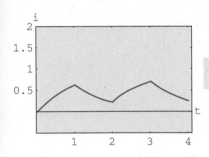

Figure 7.36

The graph of $i(t)$ on the interval $0 \le t < 4$, given in Figure 7.36, was obtained with the help of a CAS. ∎

EXERCISES 7.4

Answers to odd-numbered problems begin on page AN-9.

In Problems 1–10 find $F(s)$.

1. $\mathscr{L}\{t \cos 2t\}$ **2.** $\mathscr{L}\{t \sinh 3t\}$

3. $\mathscr{L}\{t^2 \sinh t\}$ **4.** $\mathscr{L}\{t^2 \cos t\}$

5. $\mathscr{L}\{te^{2t} \sin 6t\}$ **6.** $\mathscr{L}\{te^{-3t} \cos 3t\}$

7. $\mathscr{L}\{1 * t^3\}$ **8.** $\mathscr{L}\{t^2 * te^t\}$

9. $\mathscr{L}\{e^{-t} * e^t \cos t\}$ **10.** $\mathscr{L}\{e^{2t} * \sin t\}$

In Problems 11–18 evaluate the given Laplace transform without evaluating the integral.

11. $\mathscr{L}\left\{\int_0^t e^\tau \, d\tau\right\}$ **12.** $\mathscr{L}\left\{\int_0^t \cos \tau \, d\tau\right\}$

13. $\mathscr{L}\left\{\int_0^t e^{-\tau} \cos \tau \, d\tau\right\}$ **14.** $\mathscr{L}\left\{\int_0^t \tau \sin \tau \, d\tau\right\}$

15. $\mathscr{L}\left\{\int_0^t \tau e^{t-\tau} \, d\tau\right\}$ **16.** $\mathscr{L}\left\{\int_0^t \sin \tau \cos(t - \tau) \, d\tau\right\}$

17. $\mathscr{L}\left\{t \int_0^t \sin \tau \, d\tau\right\}$ **18.** $\mathscr{L}\left\{t \int_0^t \tau e^{-\tau} \, d\tau\right\}$

19. Use (8) to evaluate the given inverse transform.

     (a) $\mathscr{L}^{-1}\left\{\dfrac{1}{s(s - 1)}\right\}$ **(b)** $\mathscr{L}^{-1}\left\{\dfrac{1}{s^2(s - 1)}\right\}$ **(c)** $\mathscr{L}^{-1}\left\{\dfrac{1}{s^3(s - 1)}\right\}$

20. The table in Appendix III does not contain an entry for

$$\mathscr{L}^{-1}\left\{\frac{8k^3s}{(s^2 + k^2)^3}\right\}.$$

     (a) Use (4) along with the results in (5) to evaluate this inverse transform. Use a CAS as an aid in evaluating the convolution integral.

     (b) Reexamine your answer in part (a). Could you have obtained the result in a different manner?

In Problems 21–26 use Theorem 7.10 to find the Laplace transform of the given periodic function.

21.

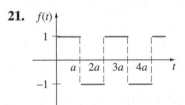

meander function

Figure 7.37

22.

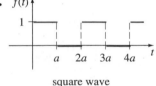

square wave

Figure 7.38

23.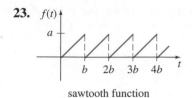

sawtooth function

Figure 7.39

24.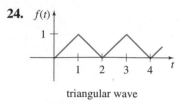

triangular wave

Figure 7.40

25.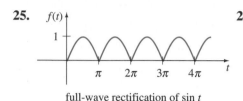

full-wave rectification of sin t

Figure 7.41

26.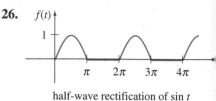

half-wave rectification of sin t

Figure 7.42

For the remaining problems in this exercise set use the table of Laplace transforms in Appendix III as needed.

In Problems 27–32 use the Laplace transform to solve the given initial-value problem.

27. $y' + y = t \sin t$, $y(0) = 0$ **28.** $y' - y = te^t \sin t$, $y(0) = 0$

29. $y'' + 9y = \cos 3t$, $y(0) = 2, y'(0) = 5$

30. $y'' + y = \sin t$, $y(0) = 1, y'(0) = -1$

31. $y'' + 16y = f(t)$, $y(0) = 0, y'(0) = 1$, where $f(t) = \begin{cases} \cos 4t, & 0 \le t < \pi \\ 0, & t \ge \pi \end{cases}$

32. $y'' + y = f(t)$, $y(0) = 1, y'(0) = 0$, where $f(t) = \begin{cases} 1, & 0 \le t < \pi/2 \\ \sin t, & t \ge \pi/2 \end{cases}$

33. Use the Laplace transform and the results obtained in Problem 20 to solve the initial-value problem $y'' + y = \sin t + t \sin t$, $y(0) = 0$, $y'(0) = 0$. Use a graphing utility to obtain the graph of the solution.

34. Use a graphing utility to obtain the graph of the indicated solution.
  (a) $y(t)$ of Problem 31 on the interval $0 \le t < 2\pi$
  (b) $y(t)$ of Problem 32 on the interval $0 \le t < 3\pi$

In Problems 35–44 use the Laplace transform to solve the given integral equation or integrodifferential equation.

35. $f(t) + \int_0^t (t - \tau)f(\tau)\, d\tau = t$ **36.** $f(t) = 2t - 4\int_0^t \sin \tau f(t - \tau)\, d\tau$

37. $f(t) = te^t + \int_0^t \tau f(t - \tau)\, d\tau$

38. $f(t) + 2\int_0^t f(\tau) \cos(t - \tau)\, d\tau = 4e^{-t} + \sin t$

39. $f(t) + \int_0^t f(\tau)\, d\tau = 1$ **40.** $f(t) = \cos t + \int_0^t e^{-\tau}f(t - \tau)\, d\tau$

41. $f(t) = 1 + t - \dfrac{8}{3}\int_0^t (\tau - t)^3 f(\tau)\, d\tau$

42. $t - 2f(t) = \int_0^t (e^\tau - e^{-\tau}) f(t - \tau) \, d\tau$

43. $y'(t) = 1 - \sin t - \int_0^t y(\tau) \, d\tau, \quad y(0) = 0$

44. $\dfrac{dy}{dt} + 6y(t) + 9 \int_0^t y(\tau) \, d\tau = 1, \quad y(0) = 0$

In Problems 45 and 46 solve equation (10) subject to $i(0) = 0$ with L, R, C, and $E(t)$ as given. Use a graphing utility to obtain the graph of the solution on the interval $0 \le t \le 3$.

45. $L = 0.1$ h, $R = 3 \ \Omega$, $C = 0.05$ f, $E(t) = 100[\mathcal{U}(t - 1) - \mathcal{U}(t - 2)]$

46. $L = 0.005$ h, $R = 1 \ \Omega$, $C = 0.02$ f, $E(t) = 100[t - (t - 1)\mathcal{U}(t - 1)]$

In Problems 47 and 48 solve equation (12) subject to $i(0) = 0$ with $E(t)$ as given. Use a graphing utility to obtain the graph of the solution on the interval $0 \le t < 4$ in the case when $L - 1$ and $R = 1$.

47. $E(t)$ is the meander function in Problem 21 with amplitude 1 and $a = 1$.

48. $E(t)$ is the sawtooth function in Problem 23 with amplitude 1 and $b = 1$.

In Problems 49 and 50 solve the model for a driven spring/mass system with damping

$$m \frac{d^2x}{dt^2} + \beta \frac{dx}{dt} + kx = f(t), \quad x(0) = 0, x'(0) = 0,$$

where the driving function f is as specified. Use a graphing utility to obtain the graph of $x(t)$ on the indicated interval.

49. $m = \frac{1}{2}$, $\beta = 1$, $k - 5$, f is the meander function in Problem 21 with amplitude 10, and $a = \pi$, $0 \le t < 2\pi$.

50. $m = 1$, $\beta = 2$, $k = 1$, f is the square wave in Problem 22 with amplitude 5, and $a = \pi$, $0 \le t < 4\pi$.

Discussion Problems

51. Discuss how Theorem 7.8 can be used to find

$$f(t) = \mathcal{L}^{-1} \left\{ \ln \frac{s - 3}{s + 1} \right\}.$$

52. Explain: $t * \mathcal{U}(t - a) = \dfrac{1}{2}(t - a)^2 \mathcal{U}(t - a)$.

53. In (6) we saw that the result $\mathcal{L} \left\{ \int_0^t f(\tau) \, d\tau \right\} = F(s)/s$, where $F(s) = \mathcal{L}\{f(t)\}$, followed from the convolution theorem when $g(t) = 1$. Using the definitions and theorems of this chapter, find two more ways of obtaining the same result.

54. Discuss how you would go about proving that $f * g = g * f$. Carry out your ideas.

55. The Laplace transform is not well suited to solving differential equations with variable coefficients; nevertheless, it can be used in some circumstances.

(a) Laguerre's differential equation $ty'' + (1 - t)y' + ny = 0$ is known to possess polynomial solutions when n is a nonnegative integer. These solutions are naturally called **Laguerre polynomials** and are denoted by $L_n(t)$. Use an appropriate theorem from this section to find $L_n(t)$ for $n = 0, 1, 2, 3, 4$ if it is known that $y(0) = 1$.

(b) Show that $\mathcal{L}\left\{\dfrac{e^t}{n!}\dfrac{d^n}{dt^n}t^n e^{-t}\right\} = Y(s)$, where $Y(s) = \mathcal{L}\{y\}$ and $y = L_n(t)$ is a polynomial solution of the DE in part (a). Conclude that

$$L_n(t) = \frac{e^t}{n!}\frac{d^n}{dt^n}t^n e^{-t}, \quad n = 0, 1, 2, \ldots .$$

This last relation for generating the Laguerre polynomials is the analogue of Rodrigues' formula for the Legendre polynomials. See (22) in Section 6.3.

Computer Lab Assignments

56. In part (a) of this problem you are led through the commands in *Mathematica* that enable you to obtain the symbolic Laplace transform of a differential equation and the solution of the initial-value problem by finding the inverse transform. In *Mathematica* the Laplace transform of a function $y(t)$ is obtained using **LaplaceTransform [y[t], t, s]**. In line two of the syntax we replace **LaplaceTransform [y[t], t, s]** by the symbol **Y**. (*If you do not have Mathematica, then adapt the given procedure by finding the corresponding syntax for the CAS you have.*)

(a) Consider the initial-value problem

$$y'' + 6y' + 9y = t \sin t, \quad y(0) = 2, \quad y'(0) = -1.$$

Load the Laplace transform package. Precisely reproduce and then, in turn, execute each line in the given sequence of commands. Either copy the output by hand or print out the results.

diffequat = y″[t] + 6y′[t] + 9y[t] == t Sin[t]

transformdeq = LaplaceTransform [diffequat, t, s] /. {y[0] − > 2,
         y′[0] − > −1, LaplaceTransform [y[t], t, s] − > Y}

soln = Solve[transformdeq, Y] // Flatten

Y = Y/.soln

InverseLaplaceTransform[Y, s, t]

(b) Appropriately modify the procedure of part (a) to find a solution of

$$y''' + 3y' - 4y = 0, \quad y(0) = 0, \quad y'(0) = 0, \quad y''(0) = 1.$$

(c) The charge $q(t)$ on a capacitor in an *LC* series circuit is given by

$$\frac{d^2q}{dt^2} + q = 1 - 4\,\mathcal{U}(t - \pi) + 6\,\mathcal{U}(t - 3\pi), \quad q(0) = 0, \quad q'(0) = 0.$$

Appropriately modify the procedure of part (a) to find $q(t)$. (In *Mathematica* the unit step function $\mathcal{U}(t - a)$ is written **UnitStep[t − a]**.) Graph your solution.

7.5 DIRAC DELTA FUNCTION

• *Unit impulse* • *Dirac delta function* • *Transform of the Dirac delta function* • *Sifting property*

Mechanical systems are often acted upon by an external force (or emf in an electrical circuit) of large magnitude that acts only for a short period of time. For example, a vibrating airplane wing could be struck by a bolt of lightning, a mass on a spring could be given a sharp blow by a ball peen hammer, a ball (baseball, golf ball, tennis ball) could be sent soaring when struck violently by some kind of club (baseball bat, golf club, tennis racket). See Figure 7.43. We start by considering a function that could serve as a mathematical model for such a force.

Figure 7.43 A tennis racket applies a force of large magnitude that acts on the ball for a very short period of time

Unit Impulse The graph of the piecewise-defined function

$$\delta_a(t - t_0) = \begin{cases} 0, & 0 \le t < t_0 - a \\ \dfrac{1}{2a}, & t_0 - a \le t < t_0 + a \\ 0, & t \ge t_0 + a \end{cases} \tag{1}$$

$a > 0$, $t_0 > 0$ is shown in Figure 7.44(a). For a small value of a, $\delta_a(t - t_0)$ is essentially a constant function of large magnitude that is "on" for just a very short period of time, around t_0. The behavior of $\delta_a(t - t_0)$ as $a \to 0$ is illustrated in Figure 7.44(b). The function $\delta_a(t - t_0)$ is called a **unit impulse** since it possesses the integration property $\int_0^\infty \delta_a(t - t_0)\, dt = 1$.

Dirac Delta Function In practice it is convenient to work with another type of unit impulse, a "function" that approximates $\delta_a(t - t_0)$ and is defined by the limit

$$\delta(t - t_0) = \lim_{a \to 0} \delta_a(t - t_0). \tag{2}$$

The latter expression, which is not a function at all, can be characterized by the two properties

$$(i)\ \ \delta(t - t_0) = \begin{cases} \infty, & t = t_0 \\ 0, & t \ne t_0 \end{cases} \quad \text{and} \quad (ii)\ \int_0^\infty \delta(t - t_0)\, dt = 1.$$

The unit impulse $\delta(t - t_0)$ is called the **Dirac delta function.**

It is possible to obtain the Laplace transform of the Dirac delta function by the formal assumption that $\mathscr{L}\{\delta(t - t_0)\} = \lim_{a \to 0} \mathscr{L}\{\delta_a(t - t_0)\}$.

THEOREM 7.11	**Transform of the Dirac Delta Function**

For $t_0 > 0$, $$\mathscr{L}\{\delta(t - t_0)\} = e^{-st_0}. \tag{3}$$

Proof To begin we can write $\delta_a(t - t_0)$ in terms of the unit step function by virtue of (11) and (12) of Section 7.3:

$$\delta_a(t - t_0) = \frac{1}{2a}[\,\mathscr{U}(t - (t_0 - a)) - \mathscr{U}(t - (t_0 + a))].$$

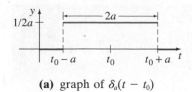

(a) graph of $\delta_a(t - t_0)$

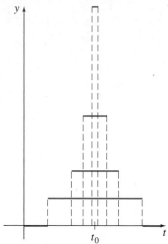

(b) behavior of δ_a as $a \to 0$

Figure 7.44

By linearity and (14) of Section 7.3 the Laplace transform of this last expression is

$$\mathscr{L}\{\delta_a(t - t_0)\} = \frac{1}{2a}\left[\frac{e^{-s(t_0-a)}}{s} - \frac{e^{-s(t_0+a)}}{s}\right] = e^{-st_0}\left(\frac{e^{sa} - e^{-sa}}{2sa}\right). \qquad (4)$$

Since (4) has the indeterminate form 0/0 as $a \to 0$, we apply L'Hôpital's rule:

$$\mathscr{L}\{\delta(t - t_0)\} = \lim_{a \to 0}\mathscr{L}\{\delta_a(t - t_0)\} = e^{-st_0}\lim_{a \to 0}\left(\frac{e^{sa} - e^{-sa}}{2sa}\right) = e^{-st_0}. \qquad \blacksquare$$

Now when $t_0 = 0$, it seems plausible to conclude from (3) that

$$\mathscr{L}\{\delta(t)\} = 1.$$

The last result emphasizes the fact that $\delta(t)$ is not the usual type of function that we have been considering, since we expect from Theorem 7.4 that $\mathscr{L}\{f(t)\} \to 0$ as $s \to \infty$.

EXAMPLE 1   **Two Initial-Value Problems**

Solve $y'' + y = 4\delta(t - 2\pi)$ subject to

(a) $y(0) = 1, \quad y'(0) = 0$ **(b)** $y(0) = 0, \quad y'(0) = 0$.

The two initial-value problems could serve as models for describing the motion of a mass on a spring moving in a medium in which damping is negligible. At $t = 2\pi$ the mass is given a sharp blow. In (a) the mass is released from rest 1 unit below the equilibrium position. In (b) the mass is at rest in the equilibrium position.

Solution   **(a)** From (3) the Laplace transform of the differential equation is

$$s^2 Y(s) - s + Y(s) = 4e^{-2\pi s} \quad \text{or} \quad Y(s) = \frac{s}{s^2 + 1} + \frac{4e^{-2\pi s}}{s^2 + 1}.$$

Using the inverse form of the second translation theorem, we find

$$y(t) = \cos t + 4\sin(t - 2\pi)\,\mathscr{U}(t - 2\pi).$$

Since $\sin(t - 2\pi) = \sin t$, the foregoing solution can be written as

$$y(t) = \begin{cases} \cos t, & 0 \le t < 2\pi \\ \cos t + 4\sin t, & t \ge 2\pi. \end{cases} \qquad (5)$$

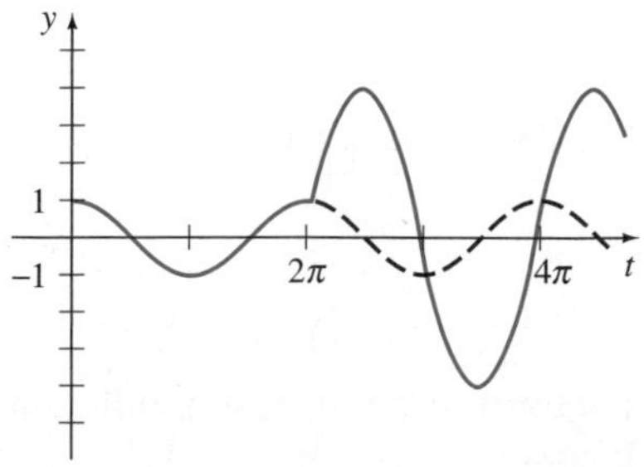

In Figure 7.45 we see from the graph of (5) that the mass is exhibiting simple harmonic motion until it is struck at $t = 2\pi$. The influence of the unit impulse is to increase the amplitude of vibration to $\sqrt{17}$ for $t > 2\pi$.

(b) In this case the transform of the equation is simply

$$Y(s) = \frac{4e^{-2\pi s}}{s^2 + 1},$$

Figure 7.45

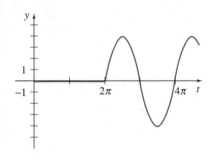

Figure 7.46

and so
$$y(t) = 4 \sin(t - 2\pi)\, \mathcal{U}(t - 2\pi)$$

$$= \begin{cases} 0, & 0 \le t < 2\pi \\ 4 \sin t, & t \ge 2\pi. \end{cases} \qquad (6)$$

The graph of (6) in Figure 7.46 shows, as we would expect from the initial conditions, that the mass exhibits no motion until it is struck at $t = 2\pi$. ∎

Remarks (*i*) If $\delta(t - t_0)$ were a function in the usual sense, then property (*i*) on page 351 would imply $\int_0^\infty \delta(t - t_0)\, dt = 0$ rather than $\int_0^\infty \delta(t - t_0)\, dt = 1$. Since the Dirac delta function did not "behave" like an ordinary function, even though its users produced correct results, it was met initially with great scorn by mathematicians. However, in the 1940s Dirac's controversial function was put on a rigorous footing by the French mathematician Laurent Schwartz in his book *La Théorie de distribution,* and this, in turn, led to an entirely new branch of mathematics known as the **theory of distributions** or **generalized functions**. In this theory, (2) is not an accepted definition of $\delta(t - t_0)$, nor does one speak of a function whose values are either ∞ or 0. Although we shall not pursue this topic any further, suffice it to say that the Dirac delta function is best characterized by its effect on other functions. If f is a continuous function, then

$$\int_0^\infty f(t)\, \delta(t - t_0)\, dt = f(t_0) \qquad (7)$$

can be taken as the *definition* of $\delta(t - t_0)$. This result is known as the **sifting property** since $\delta(t - t_0)$ has the effect of sifting the value $f(t_0)$ out of the set of values of f on $[0, \infty)$. Note that property (*ii*) (with $f(t) = 1$) and (3) (with $f(t) = e^{-st}$) are consistent with (7).

(*ii*) The Remarks in Section 7.2 indicated that the transfer function of a general linear nth-order differential equation with constant coefficients is $W(s) = 1/P(s)$, where $P(s) = a_n s^n + a_{n-1} s^{n-1} + \cdots + a_0$. The transfer function is the Laplace transform of function $w(t)$, called the **weight function** of a linear system. But $w(t)$ can also be characterized in terms of the discussion at hand. For simplicity let us consider a second-order linear system in which the input is a unit impulse at $t = 0$:

$$a_2 y'' + a_1 y' + a_0 y = \delta(t), \quad y(0) = 0, \quad y'(0) = 0.$$

Applying the Laplace transform and using $\mathscr{L}\{\delta(t)\} = 1$ shows that the transform of the response y in this case is the transfer function

$$Y(s) = \frac{1}{a_2 s^2 + a_1 s + a_0} = \frac{1}{P(s)} = W(s) \quad \text{and so} \quad y = \mathscr{L}^{-1}\left\{\frac{1}{P(s)}\right\} = w(t).$$

From this we can see, in general, that the weight function $y = w(t)$ of an nth-order linear system is the zero-state response of the system to a unit impulse. For this reason $w(t)$ is also called the **impulse response** of the system.

EXERCISES 7.5

Answers to odd-numbered problems begin on page AN-9.

In Problems 1–12 use the Laplace transform to solve the given differential equation subject to the indicated initial conditions.

1. $y' - 3y = \delta(t - 2), \quad y(0) = 0$

2. $y' + y = \delta(t - 1), \quad y(0) = 2$

3. $y'' + y = \delta(t - 2\pi), \quad y(0) = 0, y'(0) = 1$

4. $y'' + 16y = \delta(t - 2\pi), \quad y(0) = 0, y'(0) = 0$

5. $y'' + y = \delta(t - \frac{1}{2}\pi) + \delta(t - \frac{3}{2}\pi), \quad y(0) = 0, y'(0) = 0$

6. $y'' + y = \delta(t - 2\pi) + \delta(t - 4\pi), \quad y(0) = 1, y'(0) = 0$

7. $y'' + 2y' = \delta(t - 1), \quad y(0) = 0, y'(0) = 1$

8. $y'' - 2y' = 1 + \delta(t - 2), \quad y(0) = 0, y'(0) = 1$

9. $y'' + 4y' + 5y = \delta(t - 2\pi), \quad y(0) = 0, y'(0) = 0$

10. $y'' + 2y' + y = \delta(t - 1), \quad y(0) = 0, y'(0) = 0$

11. $y'' + 4y' + 13y = \delta(t - \pi) + \delta(t - 3\pi), \quad y(0) = 1, y'(0) = 0$

12. $y'' - 7y' + 6y = e^t + \delta(t - 2) + \delta(t - 4), \quad y(0) = 0, y'(0) = 0$

13. A uniform beam of length L carries a concentrated load w_0 at $x = \frac{1}{2}L$. The beam is embedded at its left end and is free at its right end. Use the Laplace transform to determine the deflection $y(x)$ from

$$EI\frac{d^4y}{dx^4} = w_0\delta(x - \tfrac{1}{2}L),$$

where $y(0) = 0, y'(0) = 0, y''(L) = 0$, and $y'''(L) = 0$.

14. Solve the differential equation in Problem 13 subject to $y(0) = 0$, $y'(0) = 0, y(L) = 0, y'(L) = 0$. In this case the beam is embedded at both ends. See Figure 7.47.

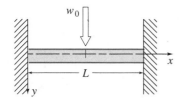

w_0

L

x

y

Figure 7.47

Discussion Problems

15. Use the Laplace transform to solve the initial-value problem $y'' + \omega^2 y = \delta(t), y(0) = 0, y'(0) = 0$. Do you see anything unusual about the solution?

7.6 SYSTEMS OF LINEAR EQUATIONS

• Using the Laplace transform to solve a system of ODEs   • Coupled springs   • Networks   • Double pendulum

When initial conditions are specified, the Laplace transform of each equation in a system of linear differential equations with constant coefficients reduces the system of DEs to a set of simultaneous algebraic equations in the transformed functions. We solve the system of algebraic equations for each of the transformed functions and then find the inverse Laplace transform in the usual manner.

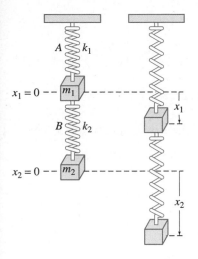

Figure 7.48

Coupled Springs Two masses m_1 and m_2 are connected to two springs A and B of negligible mass having spring constants k_1 and k_2, respectively. In turn the two springs are attached as shown in Figure 7.48. Let $x_1(t)$ and $x_2(t)$ denote the vertical displacements of the masses from their equilibrium positions. When the system is in motion, spring B is subject to both an elongation and a compression; hence its net elongation is $x_2 - x_1$. Therefore it follows from Hooke's law that springs A and B exert forces $-k_1x_1$ and $k_2(x_2 - x_1)$, respectively, on m_1. If no external force is impressed on the system and if no damping force is present, then the net force on m_1 is $-k_1x_1 + k_2(x_2 - x_1)$. By Newton's second law we can write

$$m_1 \frac{d^2x_1}{dt^2} = -k_1x_1 + k_2(x_2 - x_1).$$

Similarly, the net force exerted on mass m_2 is due solely to the net elongation of B; that is, $-k_2(x_2 - x_1)$. Hence we have

$$m_2 \frac{d^2x_2}{dt^2} = -k_2(x_2 - x_1).$$

In other words, the motion of the coupled system is represented by the system of simultaneous second-order differential equations

$$\begin{aligned} m_1 x_1'' &= -k_1x_1 + k_2(x_2 - x_1) \\ m_2 x_2'' &= -k_2(x_2 - x_1). \end{aligned} \tag{1}$$

In the next example we solve (1) under the assumptions that $k_1 = 6$, $k_2 = 4$, $m_1 = 1$, $m_2 = 1$, and that the masses start from their equilibrium positions with opposite unit velocities.

EXAMPLE 1   Coupled Springs

Solve

$$\begin{aligned} x_1'' + 10x_1 \quad\;\; - 4x_2 &= 0 \\ - 4x_1 + x_2'' + 4x_2 &= 0 \end{aligned} \tag{2}$$

subject to $x_1(0) = 0$, $x_1'(0) = 1$, $x_2(0) = 0$, $x_2'(0) = -1$.

Solution   The Laplace transform of each equation is

$$s^2X_1(s) - sx_1(0) - x_1'(0) + 10X_1(s) - 4X_2(s) = 0$$

$$-4X_1(s) + s^2X_2(s) - sx_2(0) - x_2'(0) + 4X_2(s) = 0,$$

where $X_1(s) = \mathscr{L}\{x_1(t)\}$ and $X_2(s) = \mathscr{L}\{x_2(t)\}$. The preceding system is the same as

$$\begin{aligned} (s^2 + 10)X_1(s) - \quad\quad 4X_2(s) &= 1 \\ -4X_1(s) + (s^2 + 4)X_2(s) &= -1. \end{aligned} \tag{3}$$

Solving (3) for $X_1(s)$ and using partial fractions on the result yields

$$X_1(s) = \frac{s^2}{(s^2 + 2)(s^2 + 12)} = -\frac{1/5}{s^2 + 2} + \frac{6/5}{s^2 + 12},$$

and therefore

$$x_1(t) = -\frac{1}{5\sqrt{2}}\mathscr{L}^{-1}\left\{\frac{\sqrt{2}}{s^2+2}\right\} + \frac{6}{5\sqrt{12}}\mathscr{L}^{-1}\left\{\frac{\sqrt{12}}{s^2+12}\right\}$$

$$= -\frac{\sqrt{2}}{10}\sin\sqrt{2}t + \frac{\sqrt{3}}{5}\sin 2\sqrt{3}t.$$

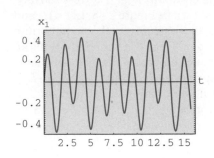

(a) plot of $x_1(t)$ vs. t

Substituting the expression for $X_1(s)$ into the first equation of (3) gives

$$X_2(s) = -\frac{s^2+6}{(s^2+2)(s^2+12)} = -\frac{2/5}{s^2+2} - \frac{3/5}{s^2+12}$$

and

$$x_2(t) = -\frac{2}{5\sqrt{2}}\mathscr{L}^{-1}\left\{\frac{\sqrt{2}}{s^2+2}\right\} - \frac{3}{5\sqrt{12}}\mathscr{L}^{-1}\left\{\frac{\sqrt{12}}{s^2+12}\right\}$$

$$= -\frac{\sqrt{2}}{5}\sin\sqrt{2}t - \frac{\sqrt{3}}{10}\sin 2\sqrt{3}t.$$

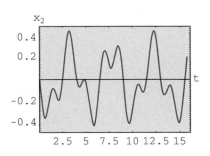

(b) plot of $x_2(t)$ vs. t

Figure 7.49

Finally, the solution to the given system (2) is

$$x_1(t) = -\frac{\sqrt{2}}{10}\sin\sqrt{2}t + \frac{\sqrt{3}}{5}\sin 2\sqrt{3}t$$

$$x_2(t) = -\frac{\sqrt{2}}{5}\sin\sqrt{2}t - \frac{\sqrt{3}}{10}\sin 2\sqrt{3}t. \tag{4}$$

The graphs of x_1 and x_2 in Figure 7.49 reveal the complicated oscillatory motion of each mass. ∎

Networks In (18) of Section 3.3 we saw the currents $i_1(t)$ and $i_2(t)$ in the network shown in Figure 7.50, containing an inductor, a resistor, and a capacitor, were governed by the system of first-order differential equations

$$L\frac{di_1}{dt} + Ri_2 = E(t)$$

$$RC\frac{di_2}{dt} + i_2 - i_1 = 0. \tag{5}$$

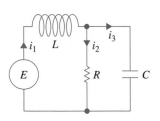

Figure 7.50

We solve this system by the Laplace transform in the next example.

EXAMPLE 2 **An Electrical Network**

Solve the system in (5) under the conditions $E(t) = 60$ V, $L = 1$ h, $R = 50\ \Omega$, $C = 10^{-4}$ f, and the currents i_1 and i_2 are initially zero.

Solution We must solve

$$\frac{di_1}{dt} + 50i_2 = 60$$

$$50(10^{-4})\frac{di_2}{dt} + i_2 - i_1 = 0$$

subject to $i_1(0) = 0$, $i_2(0) = 0$.

Applying the Laplace transform to each equation of the system and simplifying gives

$$sI_1(s) + 50I_2(s) = \frac{60}{s}$$

$$-200I_1(s) + (s + 200)I_2(s) = 0,$$

where $I_1(s) = \mathcal{L}\{i_1(t)\}$ and $I_2(s) = \mathcal{L}\{i_2(t)\}$. Solving the system for I_1 and I_2 and decomposing the results into partial fractions gives

$$I_1(s) = \frac{60s + 12{,}000}{s(s + 100)^2} = \frac{6/5}{s} - \frac{6/5}{s + 100} - \frac{60}{(s + 100)^2}$$

$$I_2(s) = \frac{12{,}000}{s(s + 100)^2} = \frac{6/5}{s} - \frac{6/5}{s + 100} - \frac{120}{(s + 100)^2}.$$

Taking the inverse Laplace transform, we find the currents to be

$$i_1(t) = \frac{6}{5} - \frac{6}{5}e^{-100t} - 60te^{-100t}$$

$$i_2(t) = \frac{6}{5} - \frac{6}{5}e^{-100t} - 120te^{-100t}. \qquad \blacksquare$$

Note that both $i_1(t)$ and $i_2(t)$ in Example 2 tend toward the value $E/R = \frac{6}{5}$ as $t \to \infty$. Furthermore, since the current through the capacitor is $i_3(t) = i_1(t) - i_2(t) = 60te^{-100t}$, we observe that $i_3(t) \to 0$ as $t \to \infty$.

Double Pendulum As shown in Figure 7.51, a double pendulum oscillates in a vertical plane under the influence of gravity. For small displacements $\theta_1(t)$ and $\theta_2(t)$, it can be shown that the system of differential equations describing the motion is

$$(m_1 + m_2)l_1^2\theta_1'' + m_2l_1l_2\theta_2'' + (m_1 + m_2)l_1g\theta_1 = 0$$

$$m_2l_2^2\theta_2'' + m_2l_1l_2\theta_1'' + m_2l_2g\theta_2 = 0. \tag{6}$$

As indicated in Figure 7.51, θ_1 is measured (in radians) from a vertical line extending downward from the pivot of the system and θ_2 is measured from a vertical line extending downward from the center of mass m_1. The positive direction is to the right; the negative direction is to the left.

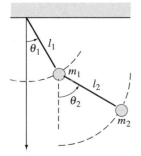

Figure 7.51

EXAMPLE 3   **Double Pendulum**

It is left as an exercise to use the Laplace transform to solve system (6) when $m_1 = 3$, $m_2 = 1$, $l_1 = l_2 = 16$, $\theta_1(0) = 1$, $\theta_2(0) = -1$, $\theta_1'(0) = 0$, and $\theta_2'(0) = 0$. You should find that

$$\theta_1(t) = \frac{1}{4}\cos\frac{2}{\sqrt{3}}t + \frac{3}{4}\cos 2t$$

$$\theta_2(t) = \frac{1}{2}\cos\frac{2}{\sqrt{3}}t - \frac{3}{2}\cos 2t. \tag{7}$$

The positions of the two masses at $t = 0$ and at subsequent times, drawn with the aid of a CAS, are shown in Figure 7.52. See Problem 21 in Exercises 7.6.

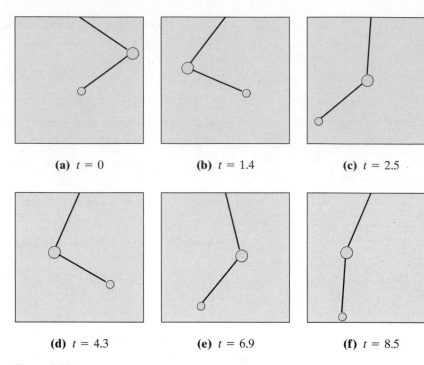

(a) $t = 0$ **(b)** $t = 1.4$ **(c)** $t = 2.5$

(d) $t = 4.3$ **(e)** $t = 6.9$ **(f)** $t = 8.5$

Figure 7.52

EXERCISES 7.6

Answers to odd-numbered problems begin on page AN-9.

In Problems 1–12 use the Laplace transform to solve the given system of differential equations.

1. $\dfrac{dx}{dt} = -x + y$

$\dfrac{dy}{dt} = 2x$

$x(0) = 0,\ y(0) = 1$

2. $\dfrac{dx}{dt} = 2y + e^t$

$\dfrac{dy}{dt} = 8x - t$

$x(0) = 1,\ y(0) = 1$

3. $\dfrac{dx}{dt} = x - 2y$

$\dfrac{dy}{dt} = 5x - y$

$x(0) = -1,\ y(0) = 2$

4. $\dfrac{dx}{dt} + 3x + \dfrac{dy}{dt} = 1$

$\dfrac{dx}{dt} - x + \dfrac{dy}{dt} - y = e^t$

$x(0) = 0,\ y(0) = 0$

5. $2\dfrac{dx}{dt} + \dfrac{dy}{dt} - 2x = 1$

$\dfrac{dx}{dt} + \dfrac{dy}{dt} - 3x - 3y = 2$

$x(0) = 0,\ y(0) = 0$

6. $\dfrac{dx}{dt} + x - \dfrac{dy}{dt} + y = 0$

$\dfrac{dx}{dt} + \dfrac{dy}{dt} + 2y = 0$

$x(0) = 0,\ y(0) = 1$

7. $\dfrac{d^2x}{dt^2} + x - y = 0$

$\dfrac{d^2y}{dt^2} + y - x = 0$

$x(0) = 0, x'(0) = -2,$

$y(0) = 0, y'(0) = 1$

8. $\dfrac{d^2x}{dt^2} + \dfrac{dx}{dt} + \dfrac{dy}{dt} = 0$

$\dfrac{d^2y}{dt^2} + \dfrac{dy}{dt} - 4\dfrac{dx}{dt} = 0$

$x(0) = 1, x'(0) = 0,$

$y(0) = -1, y'(0) = 5$

9. $\dfrac{d^2x}{dt^2} + \dfrac{d^2y}{dt^2} = t^2$

$\dfrac{d^2x}{dt^2} - \dfrac{d^2y}{dt^2} = 4t$

$x(0) = 8, x'(0) = 0,$

$y(0) = 0, y'(0) = 0$

10. $\dfrac{dx}{dt} - 4x + \dfrac{d^3y}{dt^3} = 6\sin t$

$\dfrac{dx}{dt} + 2x - 2\dfrac{d^3y}{dt^3} = 0$

$x(0) = 0, y(0) = 0,$

$y'(0) = 0, y''(0) = 0$

11. $\dfrac{d^2x}{dt^2} + 3\dfrac{dy}{dt} + 3y = 0$

$\dfrac{d^2x}{dt^2} \qquad\quad + 3y = te^{-t}$

$x(0) = 0, x'(0) = 2, y(0) = 0$

12. $\dfrac{dx}{dt} = 4x - 2y + 2\,\mathcal{U}(t - 1)$

$\dfrac{dy}{dt} = 3x - y + \mathcal{U}(t - 1)$

$x(0) = 0, y(0) = \tfrac{1}{2}$

13. Solve system (1) when $k_1 = 3, k_2 = 2, m_1 = 1, m_2 = 1$ and $x_1(0) = 0,$ $x_1'(0) = 1, x_2(0) = 1, x_2'(0) = 0.$

14. Derive the system of differential equations describing the straight-line vertical motion of the coupled springs shown in Figure 7.53. Use the Laplace transform to solve the system when $k_1 = 1, k_2 = 1, k_3 = 1,$ $m_1 = 1, m_2 = 1$ and $x_1(0) = 0, x_1'(0) = -1, x_2(0) = 0, x_2'(0) = 1.$

15. (a) Show that the system of differential equations for the currents $i_2(t)$ and $i_3(t)$ in the electrical network shown in Figure 7.54 is

$$L_1\dfrac{di_2}{dt} + Ri_2 + Ri_3 = E(t)$$

$$L_2\dfrac{di_3}{dt} + Ri_2 + Ri_3 = E(t).$$

(b) Solve the system in part (a) if $R = 5\,\Omega, L_1 = 0.01$ h, $L_2 = 0.0125$ h, $E = 100$ V, $i_2(0) = 0,$ and $i_3(0) = 0.$

(c) Determine the current $i_1(t).$

16. (a) In Problem 12 in Exercises 3.3 you were asked to show that the currents $i_2(t)$ and $i_3(t)$ in the electrical network shown in Figure 7.55 satisfy

$$L\dfrac{di_2}{dt} + L\dfrac{di_3}{dt} + R_1 i_2 = E(t)$$

$$-R_1\dfrac{di_2}{dt} + R_2\dfrac{di_3}{dt} + \dfrac{1}{C}i_3 = 0.$$

Solve the system if $R_1 = 10\,\Omega, R_2 = 5\,\Omega, L = 1$ h, $C = 0.2$ f,

$$E(t) = \begin{cases} 120, & 0 \le t < 2 \\ 0, & t \ge 2 \end{cases}$$

$i_2(0) = 0,$ and $i_3(0) = 0.$

(b) Determine the current $i_1(t).$

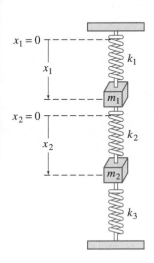

$x_1 = 0$ — k_1 — m_1
$x_2 = 0$ — k_2 — m_2 — k_3

Figure 7.53

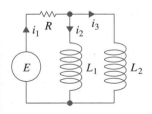

i_1 R i_2 i_3
E L_1 L_2

Figure 7.54

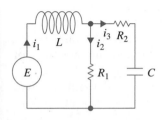

i_1 L i_3 R_2 i_2
E R_1 C

Figure 7.55

17. Solve the system given in (17) of Section 3.3 when $R_1 = 6\ \Omega$, $R_2 = 5\ \Omega$, $L_1 = 1$ h, $L_2 = 1$ h, $E(t) = 50 \sin t$ V, $i_2(0) = 0$, and $i_3(0) = 0$.

18. Solve (5) when $E = 60$ V, $L = \frac{1}{2}$ h, $R = 50\ \Omega$, $C = 10^{-4}$ f, $i_1(0) = 0$, and $i_2(0) = 0$.

19. Solve (5) when $E = 60$ V, $L = 2$ h, $R = 50\ \Omega$, $C = 10^{-4}$ f, $i_1(0) = 0$, and $i_2(0) = 0$.

20. **(a)** Show that the system of differential equations for the charge on the capacitor $q(t)$ and the current $i_3(t)$ in the electrical network shown in Figure 7.56 is

$$R_1 \frac{dq}{dt} + \frac{1}{C}q + R_1 i_3 = E(t)$$

$$L \frac{di_3}{dt} + R_2 i_3 - \frac{1}{C}q = 0.$$

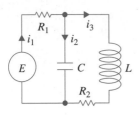

Figure 7.56

(b) Find the charge on the capacitor when $L = 1$ h, $R_1 = 1\ \Omega$, $R_2 = 1\ \Omega$, $C = 1$ f,

$$E(t) = \begin{cases} 0, & 0 < t < 1 \\ 50e^{-t}, & t \geq 1 \end{cases}$$

$i_3(0) = 0$, and $q(0) = 0$.

Computer Lab Assignments

21. **(a)** Use the Laplace transform and the information given in Example 3 to obtain the solution (7) of the system given in (6).
   (b) Use a graphing utility to plot the graphs of $\theta_1(t)$ and $\theta_2(t)$ in the $t\theta$-plane. Which mass has extreme displacements of greater magnitude? Use the graphs to estimate the first time that each mass passes through its equilibrium position. Discuss whether the motion of the pendulums is periodic.
   (c) Graph $\theta_1(t)$ and $\theta_2(t)$ in the $\theta_1\theta_2$-plane as parametric equations. The curve defined by these parametric equations is called a **Lissajous curve**.
   (d) The positions of the masses at $t = 0$ are given in Figure 7.52(a). Note that we have used 1 radian $\approx 57.3°$. Use a calculator or a table application in a CAS to construct a table of values of the angles θ_1 and θ_2 for $t = 1, 2, \ldots, 10$ s. Then plot the positions of the two masses at these times.
   (e) Use a CAS to find the first time that $\theta_1(t) = \theta_2(t)$ and compute the corresponding angular value. Plot the positions of the two masses at these times.
   (f) Utilize the CAS to draw appropriate lines to simulate the pendulum rods, as in Figure 7.52. Use the animation capability of your CAS to make a "movie" of the motion of the double pendulum from $t = 0$ to $t = 10$ using a time increment of 0.1. [*Hint*: Express the coordinates $(x_1(t), y_1(t))$ and $(x_2(t), y_2(t))$ of the masses m_1 and m_2, respectively, in terms of $\theta_1(t)$ and $\theta_2(t)$.]

CHAPTER 7 IN REVIEW

Answers to odd-numbered problems begin on page AN-9.

In Problems 1 and 2 use the definition of the Laplace transform to find $\mathscr{L}\{f(t)\}$.

1. $f(t) = \begin{cases} t, & 0 \le t < 1 \\ 2 - t, & t \ge 1 \end{cases}$

2. $f(t) = \begin{cases} 0, & 0 \le t < 2 \\ 1, & 2 \le t < 4 \\ 0, & t \ge 4 \end{cases}$

In Problems 3–24 fill in the blanks or answer true or false.

3. If f is not piecewise continuous on $[0, \infty)$, then $\mathscr{L}\{f(t)\}$ will not exist. _____

4. The function $f(t) = (e^t)^{10}$ is not of exponential order. _____

5. $F(s) = s^2/(s^2 + 4)$ is not the Laplace transform of a function that is piecewise continuous and of exponential order. _____

6. If $\mathscr{L}\{f(t)\} = F(s)$ and $\mathscr{L}\{g(t)\} = G(s)$, then $\mathscr{L}^{-1}\{F(s)G(s)\} = f(t)g(t)$. _____

7. $\mathscr{L}\{e^{-7t}\} = $ _____

8. $\mathscr{L}\{te^{-7t}\} = $ _____

9. $\mathscr{L}\{\sin 2t\} = $ _____

10. $\mathscr{L}\{e^{-3t} \sin 2t\} = $ _____

11. $\mathscr{L}\{t \sin 2t\} = $ _____

12. $\mathscr{L}\{\sin 2t \, \mathscr{U}(t - \pi)\} = $ _____

13. $\mathscr{L}^{-1}\left\{\dfrac{20}{s^6}\right\} = $ _____

14. $\mathscr{L}^{-1}\left\{\dfrac{1}{3s - 1}\right\} = $ _____

15. $\mathscr{L}^{-1}\left\{\dfrac{1}{(s - 5)^3}\right\} = $ _____

16. $\mathscr{L}^{-1}\left\{\dfrac{1}{s^2 - 5}\right\} = $ _____

17. $\mathscr{L}^{-1}\left\{\dfrac{s}{s^2 - 10s + 29}\right\} = $ _____

18. $\mathscr{L}^{-1}\left\{\dfrac{e^{-5s}}{s^2}\right\} = $ _____

19. $\mathscr{L}^{-1}\left\{\dfrac{s + \pi}{s^2 + \pi^2}e^{-s}\right\} = $ _____

20. $\mathscr{L}^{-1}\left\{\dfrac{1}{L^2s^2 + n^2\pi^2}\right\} = $ _____

21. $\mathscr{L}\{e^{-5t}\}$ exists for $s > $ _____.

22. If $\mathscr{L}\{f(t)\} = F(s)$, then $\mathscr{L}\{te^{8t}f(t)\} = $ _____.

23. If $\mathscr{L}\{f(t)\} = F(s)$ and $k > 0$, then $\mathscr{L}\{e^{at}f(t - k)\mathscr{U}(t - k)\} = $ _____.

24. $\mathscr{L}\left\{\int_0^t e^{a\tau}f(\tau)\,d\tau\right\} = $ _____ whereas $\mathscr{L}\left\{e^{at}\int_0^t f(\tau)\,d\tau\right\} = $ _____.

In Problems 25–28 use the unit step function to find an equation for each graph in terms of the function $y = f(t)$, whose graph is given in Figure 7.57.

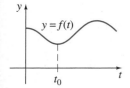

Figure 7.57

25.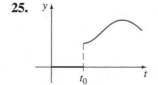

Figure 7.58

26.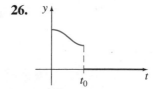

Figure 7.59

27.

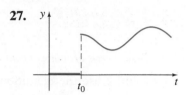

Figure 7.60

28.

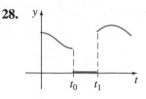

Figure 7.61

In Problems 29–32 express f in terms of unit step functions. Find $\mathscr{L}\{f(t)\}$ and $\mathscr{L}\{e^t f(t)\}$.

29. $f(t)$

Figure 7.62

30. $f(t)$

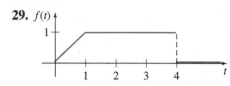

Figure 7.63

31. $f(t)$

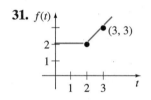

Figure 7.64

32. $f(t)$

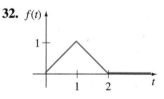

Figure 7.65

In Problems 33–38 use the Laplace transform to solve the given equation.

33. $y'' - 2y' + y = e^t$, $y(0) = 0$, $y'(0) = 5$

34. $y'' - 8y' + 20y = te^t$, $y(0) = 0$, $y'(0) = 0$

35. $y'' + 6y' + 5y = t - t\,\mathscr{U}(t - 2)$, $y(0) = 1$, $y'(0) = 0$

36. $y' - 5y = f(t)$, where $f(t) = \begin{cases} t^2, & 0 \le t < 1 \\ 0, & t \ge 1 \end{cases}$, $y(0) = 1$

37. $y'(t) = \cos t + \int_0^t y(\tau) \cos(t - \tau)\, d\tau$, $y(0) = 1$

38. $\int_0^t f(\tau) f(t - \tau)\, d\tau = 6t^3$

In Problems 39 and 40 use the Laplace transform to solve each system.

39. $x' + y = t$
 $4x + y' = 0$
 $x(0) = 1, \ y(0) = 2$

40. $x'' + y'' = e^{2t}$
 $2x' + y'' = -e^{2t}$
 $x(0) = 0, \ y(0) = 0,$
 $x'(0) = 0, \ y'(0) = 0$

41. The current $i(t)$ in an RC series circuit can be determined from the integral equation

$$Ri + \frac{1}{C}\int_0^t i(\tau)\, d\tau = E(t),$$

where $E(t)$ is the impressed voltage. Determine $i(t)$ when $R = 10\ \Omega$, $C = 0.5$ f, and $E(t) = 2(t^2 + t)$.

42. A series circuit contains an inductor, a resistor, and a capacitor for which $L = \frac{1}{2}$ h, $R = 10$ Ω, and $C = 0.01$ f, respectively. The voltage

$$E(t) = \begin{cases} 10, & 0 \le t < 5 \\ 0, & t \ge 5 \end{cases}$$

is applied to the circuit. Determine the instantaneous charge $q(t)$ on the capacitor for $t > 0$ if $q(0) = 0$ and $q'(0) = 0$.

43. A uniform cantilever beam of length L is embedded at its left end $(x = 0)$ and free at its right end. Find the deflection $y(x)$ if the load per unit length is given by

$$w(x) = \frac{2w_0}{L}\left[\frac{L}{2} - x + \left(x - \frac{L}{2}\right)\mathcal{U}\left(x - \frac{L}{2}\right)\right].$$

44. When a uniform beam is supported by an elastic foundation, the differential equation for its deflection $y(x)$ is

$$EI\frac{d^4y}{dx^4} + ky = w(x),$$

where k is the modulus of the foundation and $-ky$ is the restoring force of the foundation that acts in the direction opposite to that of the load $w(x)$. See Figure 7.66. For algebraic convenience suppose the differential equation is written as

$$\frac{d^4y}{dx^4} + 4a^4y = \frac{w(x)}{EI},$$

where $a = (k/4EI)^{1/4}$. Assume $L = \pi$ and $a = 1$. Find the deflection $y(x)$ of a beam that is supported on an elastic foundation when

(a) the beam is simply supported at both ends and a constant load w_0 is uniformly distributed along its length,

(b) the beam is embedded at both ends and $w(x)$ is a concentrated load w_0 applied at $x = \pi/2$.

[*Hint:* In both parts of this problem use entries 35 and 36 in the table of Laplace transforms in Appendix III.]

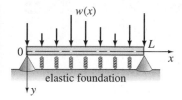

elastic foundation

Figure 7.66

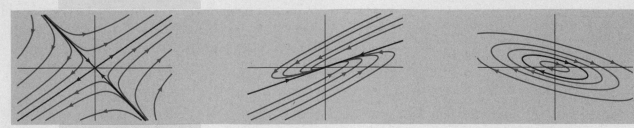

Graphs of a family of solutions of a linear system in the xy-plane are called phase portraits; see page 367.

8 SYSTEMS OF LINEAR FIRST-ORDER DIFFERENTIAL EQUATIONS

INTRODUCTION We saw systems of differential equations in Sections 3.3, 4.8, and 7.6, and we were able to solve some of these systems by means of either systematic elimination or the Laplace transform. In this chapter we are going to concentrate on only systems of linear first-order equations. Although most of the systems considered could be solved using elimination or the Laplace transform, we are going to develop a general theory for these kinds of systems and, in the case of systems with constant coefficients, a method of solution that utilizes some basic concepts from the algebra of matrices. We shall see that this general theory and solution procedure is similar to that of linear higher-order differential equations considered in Sections 4.1, 4.3, and 4.6. This material is fundamental to the analysis of systems of nonlinear first-order equations.

364

8.1 PRELIMINARY THEORY

- *Linear systems* • *Homogeneous and nonhomogeneous systems*
- *Solution vector* • *Initial-value problem* • *Superposition principle* • *Linear dependence* • *Linear independence*
- *Wronskian* • *Fundamental set of solutions* • *General solution*
- *Complementary solution* • *Particular solution*

Recall that in Section 4.8 we illustrated how to use systematic elimination to solve systems of n linear differential equations in n unknowns of the form

$$
\begin{aligned}
P_{11}(D)x_1 + P_{12}(D)x_2 + \cdots + P_{1n}(D)x_n &= b_1(t) \\
P_{21}(D)x_1 + P_{22}(D)x_2 + \cdots + P_{2n}(D)x_n &= b_2(t) \\
&\vdots \\
P_{n1}(D)x_1 + P_{n2}(D)x_2 + \cdots + P_{nn}(D)x_n &= b_n(t),
\end{aligned}
\tag{1}
$$

where the P_{ij} were polynomials of various degrees in the differential operator D. In this chapter we confine our study to systems of first-order differential equations that are special cases of systems that have the **normal form**

$$
\begin{aligned}
\frac{dx_1}{dt} &= g_1(t, x_1, x_2, \ldots, x_n) \\[2mm]
\frac{dx_2}{dt} &= g_2(t, x_1, x_2, \ldots, x_n) \\[2mm]
&\vdots \\[2mm]
\frac{dx_n}{dt} &= g_n(t, x_1, x_2, \ldots, x_n).
\end{aligned}
\tag{2}
$$

A system such as (2) of n first-order equations is called a **first-order system**.

Linear Systems When each of the functions $g_1, g_2, \ldots, g_n$ in (2) is linear in the dependent variables $x_1, x_2, \ldots, x_n$, we get the normal form of a first-order system of linear equations:

$$
\begin{aligned}
\frac{dx_1}{dt} &= a_{11}(t)x_1 + a_{12}(t)x_2 + \cdots + a_{1n}(t)x_n + f_1(t) \\[2mm]
\frac{dx_2}{dt} &= a_{21}(t)x_1 + a_{22}(t)x_2 + \cdots + a_{2n}(t)x_n + f_2(t) \\[2mm]
&\vdots \\[2mm]
\frac{dx_n}{dt} &= a_{n1}(t)x_1 + a_{n2}(t)x_2 + \cdots + a_{nn}(t)x_n + f_n(t).
\end{aligned}
\tag{3}
$$

We refer to a system of form (3) simply as a **linear system**. We assume that the coefficients a_{ij} and the functions f_i are continuous on a common

---

Note to the Student: Matrix notation and properties are used extensively throughout this chapter. You should review Appendix II if you are unfamiliar with these concepts.

interval I. When $f_i(t) = 0$, $i = 1, 2, \ldots, n$, the linear system is said to be **homogeneous**; otherwise it is **nonhomogeneous**.

Matrix Form of a Linear System   If $\mathbf{X}$, $\mathbf{A}(t)$, and $\mathbf{F}(t)$ denote the respective matrices

$$
\mathbf{X} = \begin{pmatrix} x_1(t) \\ x_2(t) \\ \vdots \\ x_n(t) \end{pmatrix}, \quad
\mathbf{A}(t) = \begin{pmatrix} a_{11}(t) & a_{12}(t) & \cdots & a_{1n}(t) \\ a_{21}(t) & a_{22}(t) & \cdots & a_{2n}(t) \\ \vdots & & & \vdots \\ a_{n1}(t) & a_{n2}(t) & \cdots & a_{nn}(t) \end{pmatrix}, \quad
\mathbf{F}(t) = \begin{pmatrix} f_1(t) \\ f_2(t) \\ \vdots \\ f_n(t) \end{pmatrix},
$$

then the system of linear first-order differential equations (3) can be written as

$$
\frac{d}{dt} \begin{pmatrix} x_1 \\ x_2 \\ \vdots \\ x_n \end{pmatrix} = \begin{pmatrix} a_{11}(t) & a_{12}(t) & \cdots & a_{1n}(t) \\ a_{21}(t) & a_{22}(t) & \cdots & a_{2n}(t) \\ \vdots & & & \vdots \\ a_{n1}(t) & a_{n2}(t) & \cdots & a_{nn}(t) \end{pmatrix} \begin{pmatrix} x_1 \\ x_2 \\ \vdots \\ x_n \end{pmatrix} + \begin{pmatrix} f_1(t) \\ f_2(t) \\ \vdots \\ f_n(t) \end{pmatrix}
$$

or simply $$\mathbf{X}' = \mathbf{AX} + \mathbf{F}. \qquad (4)$$

If the system is homogeneous, its matrix form is then

$$\mathbf{X}' = \mathbf{AX}. \qquad (5)$$

EXAMPLE 1   Systems Written in Matrix Notation

(a) If $\mathbf{X} = \begin{pmatrix} x \\ y \end{pmatrix}$, then the matrix form of the homogeneous system

$$
\begin{aligned}
\frac{dx}{dt} &= 3x + 4y \\
\frac{dy}{dt} &= 5x - 7y
\end{aligned}
\quad \text{is} \quad \mathbf{X}' = \begin{pmatrix} 3 & 4 \\ 5 & -7 \end{pmatrix} \mathbf{X}.
$$

(b) If $\mathbf{X} = \begin{pmatrix} x \\ y \\ z \end{pmatrix}$, then the matrix form of the nonhomogeneous system

$$
\begin{aligned}
\frac{dx}{dt} &= 6x + y + z + t \\
\frac{dy}{dt} &= 8x + 7y - z + 10t \\
\frac{dz}{dt} &= 2x + 9y - z + 6t
\end{aligned}
\quad \text{is} \quad \mathbf{X}' = \begin{pmatrix} 6 & 1 & 1 \\ 8 & 7 & -1 \\ 2 & 9 & -1 \end{pmatrix} \mathbf{X} + \begin{pmatrix} t \\ 10t \\ 6t \end{pmatrix}.
$$

■

DEFINITION 8.1 Solution Vector

A solution vector on an interval I is any column matrix

$$\mathbf{X} = \begin{pmatrix} x_1(t) \\ x_2(t) \\ \vdots \\ x_n(t) \end{pmatrix}$$

whose entries are differentiable functions satisfying the system (4) on the interval.

A solution vector of (4) is, of course, equivalent to n scalar equations $x_1 = \phi_1(t), x_2 = \phi_2(t), \ldots, x_n = \phi_n(t)$ and can be interpreted geometrically as a set of parametric equations of a space curve. In the important case $n = 2$, the equations $x_1 = \phi_1(t)$, $x_2 = \phi_2(t)$ represent a curve in the x_1x_2-plane. It is common practice to call a curve in the plane a **trajectory** and to call the x_1x_2-plane the **phase plane**. We will come back to these concepts and illustrate them in the next section.

EXAMPLE 2 Verification of Solutions

Verify that on the interval $(-\infty, \infty)$

$$\mathbf{X}_1 = \begin{pmatrix} 1 \\ -1 \end{pmatrix} e^{-2t} = \begin{pmatrix} e^{-2t} \\ -e^{-2t} \end{pmatrix} \quad \text{and} \quad \mathbf{X}_2 = \begin{pmatrix} 3 \\ 5 \end{pmatrix} e^{6t} = \begin{pmatrix} 3e^{6t} \\ 5e^{6t} \end{pmatrix}$$

are solutions of

$$\mathbf{X}' = \begin{pmatrix} 1 & 3 \\ 5 & 3 \end{pmatrix} \mathbf{X}. \tag{6}$$

Solution From $\mathbf{X}_1' = \begin{pmatrix} -2e^{-2t} \\ 2e^{-2t} \end{pmatrix}$ and $\mathbf{X}_2' = \begin{pmatrix} 18e^{6t} \\ 30e^{6t} \end{pmatrix}$ we see that

$$\mathbf{AX}_1 = \begin{pmatrix} 1 & 3 \\ 5 & 3 \end{pmatrix} \begin{pmatrix} e^{-2t} \\ -e^{-2t} \end{pmatrix} = \begin{pmatrix} e^{-2t} - 3e^{-2t} \\ 5e^{-2t} - 3e^{-2t} \end{pmatrix} = \begin{pmatrix} -2e^{-2t} \\ 2e^{-2t} \end{pmatrix} = \mathbf{X}_1'$$

and

$$\mathbf{AX}_2 = \begin{pmatrix} 1 & 3 \\ 5 & 3 \end{pmatrix} \begin{pmatrix} 3e^{6t} \\ 5e^{6t} \end{pmatrix} = \begin{pmatrix} 3e^{6t} + 15e^{6t} \\ 15e^{6t} + 15e^{6t} \end{pmatrix} = \begin{pmatrix} 18e^{6t} \\ 30e^{6t} \end{pmatrix} = \mathbf{X}_2'. \quad \blacksquare$$

Much of the theory of systems of n linear first-order differential equations is similar to that of linear nth-order differential equations.

Initial-Value Problem Let t_0 denote a point on an interval I and

$$\mathbf{X}(t_0) = \begin{pmatrix} x_1(t_0) \\ x_2(t_0) \\ \vdots \\ x_n(t_0) \end{pmatrix} \quad \text{and} \quad \mathbf{X}_0 = \begin{pmatrix} \gamma_1 \\ \gamma_2 \\ \vdots \\ \gamma_n \end{pmatrix},$$

where the γ_i, $i = 1, 2, \ldots, n$ are given constants. Then the problem

$$\begin{array}{ll} \textit{Solve:} & \mathbf{X}' = \mathbf{A}(t)\mathbf{X} + \mathbf{F}(t) \\ \textit{Subject to:} & \mathbf{X}(t_0) = \mathbf{X}_0 \end{array} \tag{7}$$

is an **initial-value problem** on the interval.

THEOREM 8.1   **Existence of a Unique Solution**

Let the entries of the matrices $\mathbf{A}(t)$ and $\mathbf{F}(t)$ be functions continuous on a common interval I that contains the point t_0. Then there exists a unique solution of the initial-value problem (7) on the interval.

Homogeneous Systems   In the next several definitions and theorems we are concerned only with homogeneous systems. Without stating it, we shall always assume that the a_{ij} and the f_i are continuous functions of t on some common interval I.

Superposition Principle   The following result is a **superposition principle** for solutions of linear systems.

THEOREM 8.2   **Superposition Principle**

Let $\mathbf{X}_1, \mathbf{X}_2, \ldots, \mathbf{X}_k$ be a set of solution vectors of the homogeneous system (5) on an interval I. Then the linear combination

$$\mathbf{X} = c_1\mathbf{X}_1 + c_2\mathbf{X}_2 + \cdots + c_k\mathbf{X}_k,$$

where the c_i, $i = 1, 2, \ldots, k$ are arbitrary constants, is also a solution on the interval.

It follows from Theorem 8.2 that a constant multiple of any solution vector of a homogeneous system of linear first-order differential equations is also a solution.

EXAMPLE 3 **Using the Superposition Principle**

You should practice by verifying that the two vectors

$$\mathbf{X}_1 = \begin{pmatrix} \cos t \\ -\frac{1}{2}\cos t + \frac{1}{2}\sin t \\ -\cos t - \sin t \end{pmatrix} \quad \text{and} \quad \mathbf{X}_2 = \begin{pmatrix} 0 \\ e^t \\ 0 \end{pmatrix}$$

are solutions of the system

$$\mathbf{X}' = \begin{pmatrix} 1 & 0 & 1 \\ 1 & 1 & 0 \\ -2 & 0 & -1 \end{pmatrix} \mathbf{X}. \tag{8}$$

By the superposition principle the linear combination

$$\mathbf{X} = c_1 \mathbf{X}_1 + c_2 \mathbf{X}_2 = c_1 \begin{pmatrix} \cos t \\ -\frac{1}{2}\cos t + \frac{1}{2}\sin t \\ -\cos t - \sin t \end{pmatrix} + c_2 \begin{pmatrix} 0 \\ e^t \\ 0 \end{pmatrix}$$

is yet another solution of the system. ∎

Linear Dependence and Linear Independence We are primarily interested in linearly independent solutions of the homogeneous system (5).

DEFINITION 8.2 **Linear Dependence/Independence**

Let $\mathbf{X}_1, \mathbf{X}_2, \ldots, \mathbf{X}_k$ be a set of solution vectors of the homogeneous system (5) on an interval I. We say that the set is **linearly dependent** on the interval if there exist constants $c_1, c_2, \ldots, c_k$, not all zero, such that

$$c_1 \mathbf{X}_1 + c_2 \mathbf{X}_2 + \cdots + c_k \mathbf{X}_k = \mathbf{0}$$

for every t in the interval. If the set of vectors is not linearly dependent on the interval, it is said to be **linearly independent.**

The case when $k = 2$ should be clear; two solution vectors $\mathbf{X}_1$ and $\mathbf{X}_2$ are linearly dependent if one is a constant multiple of the other, and conversely. For $k > 2$ a set of solution vectors is linearly dependent if we can express at least one solution vector as a linear combination of the remaining vectors.

Wronskian As in our earlier consideration of the theory of a single ordinary differential equation, we can introduce the concept of the **Wronskian** determinant as a test for linear independence. We state the following theorem without proof.

THEOREM 8.3   **Criterion for Linearly Independent Solutions**

Let $\quad \mathbf{X}_1 = \begin{pmatrix} x_{11} \\ x_{21} \\ \vdots \\ x_{n1} \end{pmatrix}, \quad \mathbf{X}_2 = \begin{pmatrix} x_{12} \\ x_{22} \\ \vdots \\ x_{n2} \end{pmatrix}, \quad \dots, \quad \mathbf{X}_n = \begin{pmatrix} x_{1n} \\ x_{2n} \\ \vdots \\ x_{nn} \end{pmatrix}$

be n solution vectors of the homogeneous system (5) on an interval I. Then the set of solution vectors is linearly independent on I if and only if the **Wronskian**

$$W(\mathbf{X}_1, \mathbf{X}_2, \dots, \mathbf{X}_n) = \begin{vmatrix} x_{11} & x_{12} & \cdots & x_{1n} \\ x_{21} & x_{22} & \cdots & x_{2n} \\ \vdots & & & \vdots \\ x_{n1} & x_{n2} & \cdots & x_{nn} \end{vmatrix} \neq 0 \qquad (9)$$

for every t in the interval.

It can be shown that if $\mathbf{X}_1, \mathbf{X}_2, \dots, \mathbf{X}_n$ are solution vectors of (5), then for every t in I either $W(\mathbf{X}_1, \mathbf{X}_2, \dots, \mathbf{X}_n) \neq 0$ or $W(\mathbf{X}_1, \mathbf{X}_2, \dots, \mathbf{X}_n) = 0$. Thus if we can show that $W \neq 0$ for some t_0 in I, then $W \neq 0$ for every t, and hence the solutions are linearly independent on the interval.

Notice that, unlike our definition of the Wronskian in Section 4.1, here the definition of the determinant (9) does not involve differentiation.

EXAMPLE 4   **Linearly Independent Solutions**

In Example 2 we saw that $\mathbf{X}_1 = \begin{pmatrix} 1 \\ -1 \end{pmatrix} e^{-2t}$ and $\mathbf{X}_2 = \begin{pmatrix} 3 \\ 5 \end{pmatrix} e^{6t}$ are solutions of system (6). Clearly $\mathbf{X}_1$ and $\mathbf{X}_2$ are linearly independent on the interval $(-\infty, \infty)$ since neither vector is a constant multiple of the other. In addition, we have

$$W(\mathbf{X}_1, \mathbf{X}_2) = \begin{vmatrix} e^{-2t} & 3e^{6t} \\ -e^{-2t} & 5e^{6t} \end{vmatrix} = 8e^{4t} \neq 0$$

for all real values of t. ∎

DEFINITION 8.3   **Fundamental Set of Solutions**

Any set $\mathbf{X}_1, \mathbf{X}_2, \dots, \mathbf{X}_n$ of n linearly independent solution vectors of the homogeneous system (5) on an interval I is said to be a **fundamental set of solutions** on the interval.

| THEOREM 8.4 | **Existence of a Fundamental Set** |

There exists a fundamental set of solutions for the homogeneous system (5) on an interval I.

    The next two theorems are the linear system equivalents of Theorems 4.5 and 4.6.

| THEOREM 8.5 | **General Solution—Homogeneous Systems** |

Let $\mathbf{X}_1, \mathbf{X}_2, \ldots, \mathbf{X}_n$ be a fundamental set of solutions of the homogeneous system (5) on an interval I. Then the **general solution** of the system on the interval is

$$\mathbf{X} = c_1\mathbf{X}_1 + c_2\mathbf{X}_2 + \cdots + c_n\mathbf{X}_n,$$

where the c_i, $i = 1, 2, \ldots, n$ are arbitrary constants.

| EXAMPLE 5 | **General Solution of System (6)** |

From Example 2 we know that $\mathbf{X}_1 = \begin{pmatrix} 1 \\ -1 \end{pmatrix} e^{-2t}$ and $\mathbf{X}_2 = \begin{pmatrix} 3 \\ 5 \end{pmatrix} e^{6t}$ are linearly independent solutions of (6) on $(-\infty, \infty)$. Hence $\mathbf{X}_1$ and $\mathbf{X}_2$ form a fundamental set of solutions on the interval. The general solution of the system on the interval is then

$$\mathbf{X} = c_1\mathbf{X}_1 + c_2\mathbf{X}_2 = c_1 \begin{pmatrix} 1 \\ -1 \end{pmatrix} e^{-2t} + c_2 \begin{pmatrix} 3 \\ 5 \end{pmatrix} e^{6t}. \qquad \textbf{(10)} \quad \blacksquare$$

| EXAMPLE 6 | **General Solution of System (8)** |

The vectors

$$\mathbf{X}_1 = \begin{pmatrix} \cos t \\ -\frac{1}{2}\cos t + \frac{1}{2}\sin t \\ -\cos t - \sin t \end{pmatrix}, \quad \mathbf{X}_2 = \begin{pmatrix} 0 \\ 1 \\ 0 \end{pmatrix} e^{t}, \quad \mathbf{X}_3 = \begin{pmatrix} \sin t \\ -\frac{1}{2}\sin t - \frac{1}{2}\cos t \\ -\sin t + \cos t \end{pmatrix}$$

are solutions of the system (8) in Example 3 (see Problem 16 in Exercises 8.1). Now

$$W(\mathbf{X}_1, \mathbf{X}_2, \mathbf{X}_3) = \begin{vmatrix} \cos t & 0 & \sin t \\ -\frac{1}{2}\cos t + \frac{1}{2}\sin t & e^{t} & -\frac{1}{2}\sin t - \frac{1}{2}\cos t \\ -\cos t - \sin t & 0 & -\sin t + \cos t \end{vmatrix} = e^{t} \neq 0$$

for all real values of t. We conclude that $\mathbf{X}_1$, $\mathbf{X}_2$, and $\mathbf{X}_3$ form a fundamental set of solutions on $(-\infty, \infty)$. Thus the general solution of the system on the interval is the linear combination $\mathbf{X} = c\mathbf{X}_1 + c_2\mathbf{X}_2 + c_3\mathbf{X}_3$; that is,

$$\mathbf{X} = c_1 \begin{pmatrix} \cos t \\ -\tfrac{1}{2}\cos t + \tfrac{1}{2}\sin t \\ -\cos t - \sin t \end{pmatrix} + c_2 \begin{pmatrix} 0 \\ 1 \\ 0 \end{pmatrix} e^t + c_3 \begin{pmatrix} \sin t \\ -\tfrac{1}{2}\sin t - \tfrac{1}{2}\cos t \\ -\sin t + \cos t \end{pmatrix}. \quad \blacksquare$$

Nonhomogeneous Systems For nonhomogeneous systems a **particular solution** $\mathbf{X}_p$ on an interval I is any vector, free of arbitrary parameters, whose entries are functions that satisfy the system (4).

THEOREM 8.6   **General Solution—Nonhomogeneous Systems**

Let $\mathbf{X}_p$ be a given solution of the nonhomogeneous system (4) on an interval I, and let

$$\mathbf{X}_c = c_1\mathbf{X}_1 + c_2\mathbf{X}_2 + \cdots + c_n\mathbf{X}_n$$

denote the general solution on the same interval of the associated homogeneous system (5). Then the **general solution** of the nonhomogeneous system on the interval is

$$\mathbf{X} = \mathbf{X}_c + \mathbf{X}_p.$$

The general solution $\mathbf{X}_c$ of the homogeneous system (5) is called the **complementary function** of the nonhomogeneous system (4).

EXAMPLE 7   **General Solution—Nonhomogeneous System**

The vector $\mathbf{X}_p = \begin{pmatrix} 3t - 4 \\ -5t + 6 \end{pmatrix}$ is a particular solution of the nonhomogeneous system

$$\mathbf{X}' = \begin{pmatrix} 1 & 3 \\ 5 & 3 \end{pmatrix} \mathbf{X} + \begin{pmatrix} 12t - 11 \\ -3 \end{pmatrix} \quad \textbf{(11)}$$

on the interval $(-\infty, \infty)$. (Verify this.) The complementary function of (11) on the same interval, or the general solution of $\mathbf{X}' = \begin{pmatrix} 1 & 3 \\ 5 & 3 \end{pmatrix}\mathbf{X}$, was seen in (10) of Example 5 to be $\mathbf{X}_c = c_1 \begin{pmatrix} 1 \\ -1 \end{pmatrix} e^{-2t} + c_2 \begin{pmatrix} 3 \\ 5 \end{pmatrix} e^{6t}$. Hence by Theorem 8.6

$$\mathbf{X} = \mathbf{X}_c + \mathbf{X}_p = c_1 \begin{pmatrix} 1 \\ -1 \end{pmatrix} e^{-2t} + c_2 \begin{pmatrix} 3 \\ 5 \end{pmatrix} e^{6t} + \begin{pmatrix} 3t - 4 \\ -5t + 6 \end{pmatrix}$$

is the general solution of (11) on $(-\infty, \infty)$. $\blacksquare$

EXERCISES 8.1

Answers to odd-numbered problems begin on page AN-10.

In Problems 1–6 write the linear system in matrix form.

1. $\dfrac{dx}{dt} = 3x - 5y$

$\dfrac{dy}{dt} = 4x + 8y$

2. $\dfrac{dx}{dt} = 4x - 7y$

$\dfrac{dy}{dt} = 5x$

3. $\dfrac{dx}{dt} = -3x + 4y - 9z$

$\dfrac{dy}{dt} = 6x - y$

$\dfrac{dz}{dt} = 10x + 4y + 3z$

4. $\dfrac{dx}{dt} = x - y$

$\dfrac{dy}{dt} = x + 2z$

$\dfrac{dz}{dt} = -x + z$

5. $\dfrac{dx}{dt} = x - y + z + t - 1$

$\dfrac{dy}{dt} = 2x + y - z - 3t^2$

$\dfrac{dz}{dt} = x + y + z + t^2 - t + 2$

6. $\dfrac{dx}{dt} = -3x + 4y + e^{-t}\sin 2t$

$\dfrac{dy}{dt} = 5x + 9z + 4e^{-t}\cos 2t$

$\dfrac{dz}{dt} = y + 6z - e^{-t}$

In Problems 7–10 write the given system without the use of matrices.

7. $\mathbf{X}' = \begin{pmatrix} 4 & 2 \\ -1 & 3 \end{pmatrix}\mathbf{X} + \begin{pmatrix} 1 \\ -1 \end{pmatrix}e^{t}$

8. $\mathbf{X}' = \begin{pmatrix} 7 & 5 & -9 \\ 4 & 1 & 1 \\ 0 & -2 & 3 \end{pmatrix}\mathbf{X} + \begin{pmatrix} 0 \\ 2 \\ 1 \end{pmatrix}e^{5t} - \begin{pmatrix} 8 \\ 0 \\ 3 \end{pmatrix}e^{-2t}$

9. $\dfrac{d}{dt}\begin{pmatrix} x \\ y \\ z \end{pmatrix} = \begin{pmatrix} 1 & -1 & 2 \\ 3 & -4 & 1 \\ -2 & 5 & 6 \end{pmatrix}\begin{pmatrix} x \\ y \\ z \end{pmatrix} + \begin{pmatrix} 1 \\ 2 \\ 2 \end{pmatrix}e^{-t} - \begin{pmatrix} 3 \\ -1 \\ 1 \end{pmatrix}t$

10. $\dfrac{d}{dt}\begin{pmatrix} x \\ y \end{pmatrix} = \begin{pmatrix} 3 & -7 \\ 1 & 1 \end{pmatrix}\begin{pmatrix} x \\ y \end{pmatrix} + \begin{pmatrix} 4 \\ 8 \end{pmatrix}\sin t + \begin{pmatrix} t - 4 \\ 2t + 1 \end{pmatrix}e^{4t}$

In Problems 11–16 verify that the vector $\mathbf{X}$ is a solution of the given system.

11. $\dfrac{dx}{dt} = 3x - 4y$

$\dfrac{dy}{dt} = 4x - 7y; \quad \mathbf{X} = \begin{pmatrix} 1 \\ 2 \end{pmatrix}e^{-5t}$

12. $\dfrac{dx}{dt} = -2x + 5y$

$\dfrac{dy}{dt} = -2x + 4y; \quad \mathbf{X} = \begin{pmatrix} 5\cos t \\ 3\cos t - \sin t \end{pmatrix}e^{t}$

13. $\mathbf{X}' = \begin{pmatrix} -1 & \frac{1}{4} \\ 1 & -1 \end{pmatrix} \mathbf{X}; \quad \mathbf{X} = \begin{pmatrix} -1 \\ 2 \end{pmatrix} e^{-3t/2}$

14. $\mathbf{X}' = \begin{pmatrix} 2 & 1 \\ -1 & 0 \end{pmatrix} \mathbf{X}; \quad \mathbf{X} = \begin{pmatrix} 1 \\ 3 \end{pmatrix} e^t + \begin{pmatrix} 4 \\ -4 \end{pmatrix} te^t$

15. $\mathbf{X}' = \begin{pmatrix} 1 & 2 & 1 \\ 6 & -1 & 0 \\ -1 & -2 & -1 \end{pmatrix} \mathbf{X}; \quad \mathbf{X} = \begin{pmatrix} 1 \\ 6 \\ -13 \end{pmatrix}$

16. $\mathbf{X}' = \begin{pmatrix} 1 & 0 & 1 \\ 1 & 1 & 0 \\ -2 & 0 & -1 \end{pmatrix} \mathbf{X}; \quad \mathbf{X} = \begin{pmatrix} \sin t \\ -\frac{1}{2}\sin t - \frac{1}{2}\cos t \\ -\sin t + \cos t \end{pmatrix}$

In Problems 17–20 the given vectors are solutions of a system $\mathbf{X}' = \mathbf{AX}$. Determine whether the vectors form a fundamental set on $(-\infty, \infty)$.

17. $\mathbf{X}_1 = \begin{pmatrix} 1 \\ 1 \end{pmatrix} e^{-2t}, \quad \mathbf{X}_2 = \begin{pmatrix} 1 \\ -1 \end{pmatrix} e^{-6t}$

18. $\mathbf{X}_1 = \begin{pmatrix} 1 \\ -1 \end{pmatrix} e^t, \quad \mathbf{X}_2 = \begin{pmatrix} 2 \\ 6 \end{pmatrix} e^t + \begin{pmatrix} 8 \\ -8 \end{pmatrix} te^t$

19. $\mathbf{X}_1 = \begin{pmatrix} 1 \\ -2 \\ 4 \end{pmatrix} + t \begin{pmatrix} 1 \\ 2 \\ 2 \end{pmatrix}, \quad \mathbf{X}_2 = \begin{pmatrix} 1 \\ -2 \\ 4 \end{pmatrix}, \quad \mathbf{X}_3 = \begin{pmatrix} 3 \\ -6 \\ 12 \end{pmatrix} + t \begin{pmatrix} 2 \\ 4 \\ 4 \end{pmatrix}$

20. $\mathbf{X}_1 = \begin{pmatrix} 1 \\ 6 \\ -13 \end{pmatrix}, \quad \mathbf{X}_2 = \begin{pmatrix} 1 \\ -2 \\ -1 \end{pmatrix} e^{-4t}, \quad \mathbf{X}_3 = \begin{pmatrix} 2 \\ 3 \\ -2 \end{pmatrix} e^{3t}$

In Problems 21–24 verify that the vector $\mathbf{X}_p$ is a particular solution of the given system.

21. $\dfrac{dx}{dt} = x + 4y + 2t - 7$

$\dfrac{dy}{dt} = 3x + 2y - 4t - 18; \quad \mathbf{X}_p = \begin{pmatrix} 2 \\ -1 \end{pmatrix} t + \begin{pmatrix} 5 \\ 1 \end{pmatrix}$

22. $\mathbf{X}' = \begin{pmatrix} 2 & 1 \\ 1 & -1 \end{pmatrix} \mathbf{X} + \begin{pmatrix} -5 \\ 2 \end{pmatrix}; \quad \mathbf{X}_p = \begin{pmatrix} 1 \\ 3 \end{pmatrix}$

23. $\mathbf{X}' = \begin{pmatrix} 2 & 1 \\ 3 & 4 \end{pmatrix} \mathbf{X} - \begin{pmatrix} 1 \\ 7 \end{pmatrix} e^t; \quad \mathbf{X}_p = \begin{pmatrix} 1 \\ 1 \end{pmatrix} e^t + \begin{pmatrix} 1 \\ -1 \end{pmatrix} te^t$

24. $\mathbf{X}' = \begin{pmatrix} 1 & 2 & 3 \\ -4 & 2 & 0 \\ -6 & 1 & 0 \end{pmatrix} \mathbf{X} + \begin{pmatrix} -1 \\ 4 \\ 3 \end{pmatrix} \sin 3t; \quad \mathbf{X}_p = \begin{pmatrix} \sin 3t \\ 0 \\ \cos 3t \end{pmatrix}$

25. Prove that the general solution of

$$\mathbf{X}' = \begin{pmatrix} 0 & 6 & 0 \\ 1 & 0 & 1 \\ 1 & 1 & 0 \end{pmatrix} \mathbf{X}$$

on the interval $(-\infty, \infty)$ is

$$\mathbf{X} = c_1 \begin{pmatrix} 6 \\ -1 \\ -5 \end{pmatrix} e^{-t} + c_2 \begin{pmatrix} -3 \\ 1 \\ 1 \end{pmatrix} e^{-2t} + c_3 \begin{pmatrix} 2 \\ 1 \\ 1 \end{pmatrix} e^{3t}.$$

26. Prove that the general solution of

$$\mathbf{X}' = \begin{pmatrix} -1 & -1 \\ -1 & 1 \end{pmatrix} \mathbf{X} + \begin{pmatrix} 1 \\ 1 \end{pmatrix} t^2 + \begin{pmatrix} 4 \\ -6 \end{pmatrix} t + \begin{pmatrix} -1 \\ 5 \end{pmatrix}$$

on the interval $(-\infty, \infty)$ is

$$\mathbf{X} = c_1 \begin{pmatrix} 1 \\ -1 - \sqrt{2} \end{pmatrix} e^{\sqrt{2}t} + c_2 \begin{pmatrix} 1 \\ -1 + \sqrt{2} \end{pmatrix} e^{-\sqrt{2}t} + \begin{pmatrix} 1 \\ 0 \end{pmatrix} t^2 + \begin{pmatrix} -2 \\ 4 \end{pmatrix} t + \begin{pmatrix} 1 \\ 0 \end{pmatrix}.$$

8.2 HOMOGENEOUS LINEAR SYSTEMS WITH CONSTANT COEFFICIENTS

- *Characteristic equation of a square matrix* • *Eigenvalues of a matrix* • *Eigenvectors* • *General solutions of homogeneous linear systems* • *Trajectory* • *Phase plane* • *Phase portrait*

We saw in Example 5 of Section 8.1 that the general solution of the homogeneous system $\mathbf{X}' = \begin{pmatrix} 1 & 3 \\ 5 & 3 \end{pmatrix} \mathbf{X}$ is $\mathbf{X} = c_1\mathbf{X}_1 + c_2\mathbf{X}_2 = c_1 \begin{pmatrix} 1 \\ -1 \end{pmatrix} e^{-2t} + c_2 \begin{pmatrix} 3 \\ 5 \end{pmatrix} e^{6t}$. Since both solution vectors have the form $\mathbf{X}_i = \begin{pmatrix} k_1 \\ k_2 \end{pmatrix} e^{\lambda_i t}$, $i = 1, 2$, where k_1 and k_2 are constants, we are prompted to ask whether we can always find a solution of the form

$$\mathbf{X} = \begin{pmatrix} k_1 \\ k_2 \\ \vdots \\ k_n \end{pmatrix} e^{\lambda t} = \mathbf{K}e^{\lambda t} \tag{1}$$

for the general homogeneous linear first-order system

$$\mathbf{X}' = \mathbf{AX}, \tag{2}$$

where $\mathbf{A}$ is an $n \times n$ matrix of constants.

Eigenvalues and Eigenvectors If (1) is to be a solution vector of the linear system, then $\mathbf{X}' = \mathbf{K}\lambda e^{\lambda t}$ so that (2) becomes $\mathbf{K}\lambda e^{\lambda t} = \mathbf{AK}e^{\lambda t}$. After dividing out $e^{\lambda t}$ and rearranging, we obtain $\mathbf{AK} = \lambda\mathbf{K}$ or $\mathbf{AK} - \lambda\mathbf{K} = \mathbf{0}$. Since $\mathbf{K} = \mathbf{IK}$, the last equation is the same as

$$(\mathbf{A} - \lambda\mathbf{I})\mathbf{K} = \mathbf{0}. \tag{3}$$

The matrix equation (3) is equivalent to the simultaneous algebraic equations

$$
\begin{aligned}
(a_{11} - \lambda)k_1 + & \quad a_{12}k_2 + \cdots + & a_{1n}k_n &= 0 \\
a_{21}k_1 + (a_{22} - \lambda)k_2 + & \cdots + & a_{2n}k_n &= 0 \\
& \quad \vdots & \vdots & \\
a_{n1}k_1 + & \quad a_{n2}k_2 + \cdots + (a_{nn} - \lambda)k_n &= 0.
\end{aligned}
$$

Thus to find a nontrivial solution $\mathbf{X}$ of (2) we must first find a nontrivial solution of the foregoing system; in other words, we must find a nontrivial vector $\mathbf{K}$ that satisfies (3). But in order for (3) to have solutions other than the obvious solution $k_1 = k_2 = \cdots = k_n = 0$, we must have

$$
\det(\mathbf{A} - \lambda\mathbf{I}) = 0.
$$

This polynomial equation in λ is called the **characteristic equation** of the matrix $\mathbf{A}$; its solutions are the **eigenvalues** of $\mathbf{A}$. A solution $\mathbf{K} \neq \mathbf{0}$ of (3) corresponding to an eigenvalue λ is called an **eigenvector** of $\mathbf{A}$. A solution of the homogeneous system (2) is then $\mathbf{X} = \mathbf{K}e^{\lambda t}$.

In the discussion that follows, we examine three cases: real and distinct eigenvalues (that is, no eigenvalues are equal), repeated eigenvalues, and, finally, complex eigenvalues.

8.2.1   DISTINCT REAL EIGENVALUES

When the $n \times n$ matrix $\mathbf{A}$ possesses n distinct real eigenvalues $\lambda_1, \lambda_2, \ldots, \lambda_n$, then a set of n linearly independent eigenvectors $\mathbf{K}_1, \mathbf{K}_2, \ldots, \mathbf{K}_n$ can always be found and

$$
\mathbf{X}_1 = \mathbf{K}_1 e^{\lambda_1 t}, \quad \mathbf{X}_2 = \mathbf{K}_2 e^{\lambda_2 t}, \quad \ldots, \quad \mathbf{X}_n = \mathbf{K}_n e^{\lambda_n t}
$$

is a fundamental set of solutions of (2) on $(-\infty, \infty)$.

THEOREM 8.7   **General Solution—Homogeneous Systems**

Let $\lambda_1, \lambda_2, \ldots, \lambda_n$ be n distinct real eigenvalues of the coefficient matrix $\mathbf{A}$ of the homogeneous system (2), and let $\mathbf{K}_1, \mathbf{K}_2, \ldots, \mathbf{K}_n$ be the corresponding eigenvectors. Then the **general solution** of (2) on the interval $(-\infty, \infty)$ is given by

$$
\mathbf{X} = c_1\mathbf{K}_1 e^{\lambda_1 t} + c_2\mathbf{K}_2 e^{\lambda_2 t} + \cdots + c_n\mathbf{K}_n e^{\lambda_n t}.
$$

EXAMPLE 1   **Distinct Eigenvalues**

Solve

$$
\frac{dx}{dt} = 2x + 3y
$$

$$
\frac{dy}{dt} = 2x + y.
$$

(4)

Solution   We first find the eigenvalues and eigenvectors of the matrix of coefficients.

From the characteristic equation

$$\det(\mathbf{A} - \lambda\mathbf{I}) = \begin{vmatrix} 2 - \lambda & 3 \\ 2 & 1 - \lambda \end{vmatrix} = \lambda^2 - 3\lambda - 4 = (\lambda + 1)(\lambda - 4) = 0$$

we see that the eigenvalues are $\lambda_1 = -1$ and $\lambda_2 = 4$.

Now for $\lambda_1 = -1$, (3) is equivalent to

$$3k_1 + 3k_2 = 0$$
$$2k_1 + 2k_2 = 0.$$

Thus $k_1 = -k_2$. When $k_2 = -1$, the related eigenvector is

$$\mathbf{K}_1 = \begin{pmatrix} 1 \\ -1 \end{pmatrix}.$$

For $\lambda_2 = 4$, we have

$$-2k_1 + 3k_2 = 0$$
$$2k_1 - 3k_2 = 0$$

so that $k_1 = 3k_2/2$, and therefore with $k_2 = 2$ the corresponding eigenvector is

$$\mathbf{K}_2 = \begin{pmatrix} 3 \\ 2 \end{pmatrix}.$$

Since the matrix of coefficients $\mathbf{A}$ is a 2×2 matrix and since we have found two linearly independent solutions of (4),

$$\mathbf{X}_1 = \begin{pmatrix} 1 \\ -1 \end{pmatrix} e^{-t} \quad \text{and} \quad \mathbf{X}_2 = \begin{pmatrix} 3 \\ 2 \end{pmatrix} e^{4t},$$

we conclude that the general solution of the system is

$$\mathbf{X} = c_1\mathbf{X}_1 + c_2\mathbf{X}_2 = c_1 \begin{pmatrix} 1 \\ -1 \end{pmatrix} e^{-t} + c_2 \begin{pmatrix} 3 \\ 2 \end{pmatrix} e^{4t}. \tag{5} \quad \blacksquare$$

For the sake of review, you should keep firmly in mind that writing a solution of a system of linear first-order differential equations in terms of matrices is simply an alternative to the method that we employed in Section 4.8—namely, listing the individual functions and the relationship between the constants. If we add the vectors on the right-hand side of (5) and then equate the entries with the corresponding entries in the vector on the left-hand side, we obtain the more familiar statement

$$x = c_1 e^{-t} + 3c_2 e^{4t}, \quad y = -c_1 e^{-t} + 2c_2 e^{4t}.$$

As pointed out in Section 8.1, we can interpret these equations as parametric equations of a curve in the xy-plane or **phase plane.** The curve is called a **trajectory.** The three graphs shown in Figure 8.1, $x(t)$ in the tx-plane, $y(t)$ in the ty-plane, and the trajectory in the phase plane, correspond to the choice of constants $c_1 = c_2 = 1$ in the solution. A collection of representative trajectories in the phase plane, as shown in Figure 8.2, is said to be a **phase portrait** of the given linear system. What appear to be *two* black lines in Figure 8.2 are actually *four* black half-lines defined parametrically in the first, second, third, and fourth quadrants by the solutions $\mathbf{X}_2$, $-\mathbf{X}_1$,

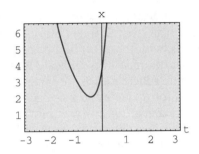

(a) graph of $x = e^{-t} + 3e^{4t}$

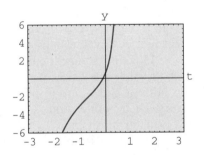

(b) graph of $y = -e^{-t} + 2e^{4t}$

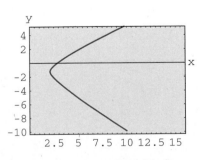

(c) trajectory defined by $x = e^{-t} + 3e^{4t}$, $y = -e^{-t} + 2e^{4t}$ in the phase plane

Figure 8.1

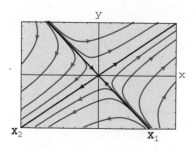

Figure 8.2 A phase portrait of the system (4)

$-\mathbf{X}_2$, and $\mathbf{X}_1$, respectively. For example, the Cartesian equations $y = \frac{2}{3}x$, $x > 0$ and $y = -x$, $x > 0$ of the half-lines in the first and fourth quadrants were obtained by eliminating the parameter t in the solutions $x = 3e^{4t}$, $y = 2e^{4t}$ and $x = e^{-t}$, $y = -e^{-t}$, respectively. Moreover, each eigenvector can be visualized as a two-dimensional vector lying along one of these half-lines. The eigenvector $\mathbf{K}_2 = \begin{pmatrix} 3 \\ 2 \end{pmatrix}$ lies along $y = \frac{2}{3}x$ in the first quadrant, and $\mathbf{K}_1 = \begin{pmatrix} 1 \\ -1 \end{pmatrix}$ lies along $y = -x$ in the fourth quadrant. Each vector starts at the origin; $\mathbf{K}_2$ terminates at the point $(2, 3)$, and $\mathbf{K}_1$ terminates at $(1, -1)$.

The origin is not only a constant solution $x = 0$, $y = 0$ of every 2×2 homogeneous linear system $\mathbf{X}' = \mathbf{AX}$, but also an important point in the qualitative study of such systems. If we think in physical terms, the arrowheads on each trajectory in Figure 8.2 indicate the direction that a particle with coordinates $(x(t), y(t))$ on that trajectory at time t moves as time increases. Observe that the arrowheads, with the exception of only those on the half-lines in the second and fourth quadrants, indicate that a particle moves away from the origin as time t increases. If we imagine time ranging from $-\infty$ to ∞, then inspection of the solution $x = c_1 e^{-t} + 3c_2 e^{4t}$, $y = -c_1 e^{-t} + 2c_2 e^{4t}$, $c_1 \neq 0$, $c_2 \neq 0$ shows that a trajectory, or moving particle, "starts" asymptotic to one of the half-lines defined by $\mathbf{X}_1$ or $-\mathbf{X}_1$ (since e^{4t} is negligible for $t \to -\infty$) and "finishes" asymptotic to one of the half-lines defined by $\mathbf{X}_2$ and $-\mathbf{X}_2$ (since e^{-t} is negligible for $t \to \infty$).

We note in passing that Figure 8.2 represents a phase portrait that is typical of *all* 2×2 homogeneous linear systems $\mathbf{X}' = \mathbf{AX}$ with real eigenvalues of opposite signs. See Problem 17 in Exercises 8.2. Moreover, phase portraits in the two cases when distinct real eigenvalues have the same algebraic sign are typical of all such 2×2 linear systems; the only difference is that the arrowheads indicate that a particle moves away from the origin on any trajectory as $t \to \infty$ when both λ_1 and λ_2 are positive and moves toward the origin on any trajectory when both λ_1 and λ_2 are negative. Consequently, we call the origin a **repeller** in the case $\lambda_1 > 0$, $\lambda_2 > 0$ and an **attractor** in the case $\lambda_1 < 0$, $\lambda_2 < 0$. See Problem 18 in Exercises 8.2. The origin in Figure 8.2 is neither a repeller nor an attractor. Investigation of the remaining case when $\lambda = 0$ is an eigenvalue of a 2×2 homogeneous linear system is left as an exercise. See Problem 49 in Exercises 8.2.

EXAMPLE 2 **Distinct Eigenvalues**

Solve

$$\frac{dx}{dt} = -4x + y + z$$

$$\frac{dy}{dt} = x + 5y - z \qquad \textbf{(6)}$$

$$\frac{dz}{dt} = y - 3z.$$

Solution   Using the cofactors of the third row, we find

$$\det(\mathbf{A} - \lambda\mathbf{I}) = \begin{vmatrix} -4 - \lambda & 1 & 1 \\ 1 & 5 - \lambda & -1 \\ 0 & 1 & -3 - \lambda \end{vmatrix} = -(\lambda + 3)(\lambda + 4)(\lambda - 5) = 0,$$

and so the eigenvalues are $\lambda_1 = -3$, $\lambda_2 = -4$, and $\lambda_3 = 5$.

For $\lambda_1 = -3$, Gauss-Jordan elimination gives

$$(\mathbf{A} + 3\mathbf{I}|\mathbf{0}) = \begin{pmatrix} -1 & 1 & 1 & | & 0 \\ 1 & 8 & -1 & | & 0 \\ 0 & 1 & 0 & | & 0 \end{pmatrix} \xrightarrow[\text{operations}]{\text{row}} \begin{pmatrix} 1 & 0 & -1 & | & 0 \\ 0 & 1 & 0 & | & 0 \\ 0 & 0 & 0 & | & 0 \end{pmatrix}.$$

Therefore $k_1 = k_3$ and $k_2 = 0$. The choice $k_3 = 1$ gives an eigenvector and corresponding solution vector

$$\mathbf{K}_1 = \begin{pmatrix} 1 \\ 0 \\ 1 \end{pmatrix}, \qquad \mathbf{X}_1 = \begin{pmatrix} 1 \\ 0 \\ 1 \end{pmatrix} e^{-3t}. \tag{7}$$

Similarly, for $\lambda_2 = -4$,

$$(\mathbf{A} + 4\mathbf{I}|\mathbf{0}) = \begin{pmatrix} 0 & 1 & 1 & | & 0 \\ 1 & 9 & -1 & | & 0 \\ 0 & 1 & 1 & | & 0 \end{pmatrix} \xrightarrow[\text{operations}]{\text{row}} \begin{pmatrix} 1 & 0 & -10 & | & 0 \\ 0 & 1 & 1 & | & 0 \\ 0 & 0 & 0 & | & 0 \end{pmatrix}$$

implies $k_1 = 10k_3$ and $k_2 = -k_3$. Choosing $k_3 = 1$, we get a second eigenvector and solution vector

$$\mathbf{K}_2 = \begin{pmatrix} 10 \\ -1 \\ 1 \end{pmatrix}, \qquad \mathbf{X}_2 = \begin{pmatrix} 10 \\ -1 \\ 1 \end{pmatrix} e^{-4t}. \tag{8}$$

Finally, when $\lambda_3 = 5$, the augmented matrices

$$(\mathbf{A} - 5\mathbf{I}|\mathbf{0}) = \begin{pmatrix} -9 & 1 & 1 & | & 0 \\ 1 & 0 & -1 & | & 0 \\ 0 & 1 & -8 & | & 0 \end{pmatrix} \xrightarrow[\text{operations}]{\text{row}} \begin{pmatrix} 1 & 0 & -1 & | & 0 \\ 0 & 1 & -8 & | & 0 \\ 0 & 0 & 0 & | & 0 \end{pmatrix}$$

yield

$$\mathbf{K}_3 = \begin{pmatrix} 1 \\ 8 \\ 1 \end{pmatrix}, \qquad \mathbf{X}_3 = \begin{pmatrix} 1 \\ 8 \\ 1 \end{pmatrix} e^{5t}. \tag{9}$$

The general solution of (6) is a linear combination of the solution vectors in (7), (8), and (9):

$$\mathbf{X} = c_1 \begin{pmatrix} 1 \\ 0 \\ 1 \end{pmatrix} e^{-3t} + c_2 \begin{pmatrix} 10 \\ -1 \\ 1 \end{pmatrix} e^{-4t} + c_3 \begin{pmatrix} 1 \\ 8 \\ 1 \end{pmatrix} e^{5t}. \qquad ■$$

Use of Computers   Software packages such as MATLAB, *Mathematica, Maple,* and DERIVE can be real time savers in finding eigenvalues and eigenvectors of a matrix. For example, to find the eigenvalues and eigen-vectors of the matrix of coefficients in (6) using *Mathematica,* we first input the definition of the matrix by rows:

$$\mathbf{m} = \{\{-4, 1, 1\}, \{1, 5, -1\}, \{ob0, 1, -3\}\}.$$

The commands

Eigenvalues[m]   and   **Eigenvectors[m]**

given in sequence yield

$$\{-4, -3, 5\}   \text{and}   \{\{10, -1, 1\}, \{1, 0, 1\}, \{1, 8, 1\}\},$$

respectively. In *Mathematica* eigenvalues and eigenvectors can also be obtained at the same time by using **Eigensystem[m].**

8.2.2   REPEATED EIGENVALUES

Of course, not all of the n eigenvalues $\lambda_1, \lambda_2, \ldots, \lambda_n$ of an $n \times n$ matrix **A** need be distinct—that is, some of the eigenvalues may be repeated. For example, the characteristic equation of the coefficient matrix in the system

$$\mathbf{X}' = \begin{pmatrix} 3 & -18 \\ 2 & -9 \end{pmatrix} \mathbf{X} \tag{10}$$

is readily shown to be $(\lambda + 3)^2 = 0$, and therefore $\lambda_1 = \lambda_2 = -3$ is a root of *multiplicity two.* For this value we find the single eigenvector

$$\mathbf{K}_1 = \begin{pmatrix} 3 \\ 1 \end{pmatrix}, \quad \text{so} \quad \mathbf{X}_1 = \begin{pmatrix} 3 \\ 1 \end{pmatrix} e^{-3t} \tag{11}$$

is one solution of (10). But since we are obviously interested in forming the general solution of the system, we need to pursue the question of finding a second solution.

In general, if m is a positive integer and $(\lambda - \lambda_1)^m$ is a factor of the characteristic equation while $(\lambda - \lambda_1)^{m+1}$ is not a factor, then λ_1 is said to be an **eigenvalue of multiplicity m**. The next three examples illustrate the following cases:

(*i*) For some $n \times n$ matrices **A** it may be possible to find m linearly independent eigenvectors $\mathbf{K}_1, \mathbf{K}_2, \ldots, \mathbf{K}_m$ corresponding to an eigenvalue λ_1 of multiplicity $m \leq n$. In this case the general solution of the system contains the linear combination

$$c_1 \mathbf{K}_1 e^{\lambda_1 t} + c_2 \mathbf{K}_2 e^{\lambda_1 t} + \cdots + c_m \mathbf{K}_m e^{\lambda_1 t}.$$

(*ii*) If there is only one eigenvector corresponding to the eigenvalue λ_1 of multiplicity m, then m linearly independent solutions of the form

$$\mathbf{X}_1 = \mathbf{K}_{11} e^{\lambda_1 t}$$
$$\mathbf{X}_2 = \mathbf{K}_{21} t e^{\lambda_1 t} + \mathbf{K}_{22} e^{\lambda_1 t}$$
$$\vdots$$
$$\mathbf{X}_m = \mathbf{K}_{m1} \frac{t^{m-1}}{(m-1)!} e^{\lambda_1 t} + \mathbf{K}_{m2} \frac{t^{m-2}}{(m-2)!} e^{\lambda_1 t} + \cdots + \mathbf{K}_{mm} e^{\lambda_1 t},$$

where $\mathbf{K}_{ij}$ are column vectors, can always be found.

Eigenvalue of Multiplicity Two   We begin by considering eigenvalues of multiplicity two. In the first example we illustrate a matrix for which we can find two distinct eigenvectors corresponding to a double eigenvalue.

EXAMPLE 3   **Repeated Eigenvalues**

Solve $\mathbf{X}' = \begin{pmatrix} 1 & -2 & 2 \\ -2 & 1 & -2 \\ 2 & -2 & 1 \end{pmatrix} \mathbf{X}$.

Solution   Expanding the determinant in the characteristic equation

$$\det(\mathbf{A} - \lambda\mathbf{I}) = \begin{vmatrix} 1-\lambda & -2 & 2 \\ -2 & 1-\lambda & -2 \\ 2 & -2 & 1-\lambda \end{vmatrix} = 0$$

yields $-(\lambda + 1)^2(\lambda - 5) = 0$. We see that $\lambda_1 = \lambda_2 = -1$ and $\lambda_3 = 5$.

For $\lambda_1 = -1$, Gauss-Jordan elimination immediately gives

$$(\mathbf{A} + \mathbf{I}|\mathbf{0}) = \begin{pmatrix} 2 & -2 & 2 & | & 0 \\ -2 & 2 & -2 & | & 0 \\ 2 & -2 & 2 & | & 0 \end{pmatrix} \xrightarrow[\text{operations}]{\text{row}} \begin{pmatrix} 1 & -1 & 1 & | & 0 \\ 0 & 0 & 0 & | & 0 \\ 0 & 0 & 0 & | & 0 \end{pmatrix}.$$

The first row of the last matrix means $k_1 - k_2 + k_3 = 0$ or $k_1 = k_2 - k_3$. The choices $k_2 = 1$, $k_3 = 0$ and $k_2 = 1$, $k_3 = 1$ yield, in turn, $k_1 = 1$ and $k_1 = 0$. Thus two eigenvectors corresponding to $\lambda_1 = -1$ are

$$\mathbf{K}_1 = \begin{pmatrix} 1 \\ 1 \\ 0 \end{pmatrix} \quad \text{and} \quad \mathbf{K}_2 = \begin{pmatrix} 0 \\ 1 \\ 1 \end{pmatrix}.$$

Since neither eigenvector is a constant multiple of the other, we have found two linearly independent solutions,

$$\mathbf{X}_1 = \begin{pmatrix} 1 \\ 1 \\ 0 \end{pmatrix} e^{-t} \quad \text{and} \quad \mathbf{X}_2 = \begin{pmatrix} 0 \\ 1 \\ 1 \end{pmatrix} e^{-t},$$

corresponding to the same eigenvalue. Last, for $\lambda_3 = 5$, the reduction

$$(\mathbf{A} - 5\mathbf{I}|\mathbf{0}) = \begin{pmatrix} -4 & -2 & 2 & | & 0 \\ -2 & -4 & -2 & | & 0 \\ 2 & -2 & -4 & | & 0 \end{pmatrix} \xrightarrow[\text{operations}]{\text{row}} \begin{pmatrix} 1 & 0 & -1 & | & 0 \\ 0 & 1 & 1 & | & 0 \\ 0 & 0 & 0 & | & 0 \end{pmatrix}$$

implies $k_1 = k_3$ and $k_2 = -k_3$. Picking $k_3 = 1$ gives $k_1 = 1$, $k_2 = -1$, and thus a third eigenvector is

$$\mathbf{K}_3 = \begin{pmatrix} 1 \\ -1 \\ 1 \end{pmatrix}.$$

We conclude that the general solution of the system is

$$\mathbf{X} = c_1 \begin{pmatrix} 1 \\ 1 \\ 0 \end{pmatrix} e^{-t} + c_2 \begin{pmatrix} 0 \\ 1 \\ 1 \end{pmatrix} e^{-t} + c_3 \begin{pmatrix} 1 \\ -1 \\ 1 \end{pmatrix} e^{5t}.$$ ∎

The matrix of coefficients $\mathbf{A}$ in Example 3 is a special kind of matrix known as a symmetric matrix. An $n \times n$ matrix $\mathbf{A}$ is said to be **symmetric** if its transpose $\mathbf{A}^T$ (where the rows and columns are interchanged) is the same as $\mathbf{A}$—that is, if $\mathbf{A}^T = \mathbf{A}$. It can be proved that if the matrix $\mathbf{A}$ in the system $\mathbf{X}' = \mathbf{AX}$ is symmetric and has real entries, then we can always find n linearly independent eigenvectors $\mathbf{K}_1, \mathbf{K}_2, \ldots, \mathbf{K}_n$, and the general solution of such a system is as given in Theorem 8.7. As illustrated in Example 3, this result holds even when some of the eigenvalues are repeated.

Second Solution Now suppose that λ_1 is an eigenvalue of multiplicity two and that there is only one eigenvector associated with this value. A second solution can be found of the form

$$\mathbf{X}_2 = \mathbf{K}te^{\lambda_1 t} + \mathbf{P}e^{\lambda_1 t}, \tag{12}$$

where

$$\mathbf{K} = \begin{pmatrix} k_1 \\ k_2 \\ \vdots \\ k_n \end{pmatrix} \quad \text{and} \quad \mathbf{P} = \begin{pmatrix} p_1 \\ p_2 \\ \vdots \\ p_n \end{pmatrix}.$$

To see this we substitute (12) into the system $\mathbf{X}' = \mathbf{AX}$ and simplify:

$$(\mathbf{AK} - \lambda_1\mathbf{K})te^{\lambda_1 t} + (\mathbf{AP} - \lambda_1\mathbf{P} - \mathbf{K})e^{\lambda_1 t} = \mathbf{0}.$$

Since this last equation is to hold for all values of t, we must have

$$(\mathbf{A} - \lambda_1\mathbf{I})\mathbf{K} = \mathbf{0} \tag{13}$$

and

$$(\mathbf{A} - \lambda_1\mathbf{I})\mathbf{P} = \mathbf{K}. \tag{14}$$

Equation (13) simply states that $\mathbf{K}$ must be an eigenvector of $\mathbf{A}$ associated with λ_1. By solving (13), we find one solution $\mathbf{X}_1 = \mathbf{K}e^{\lambda_1 t}$. To find the second solution $\mathbf{X}_2$ we need only solve the additional system (14) for the vector $\mathbf{P}$.

EXAMPLE 4 **Repeated Eigenvalues**

Find the general solution of the system given in (10).

Solution From (11) we know that $\lambda_1 = -3$ and that one solution is $\mathbf{X}_1 = \begin{pmatrix} 3 \\ 1 \end{pmatrix} e^{-3t}$. Identifying $\mathbf{K} = \begin{pmatrix} 3 \\ 1 \end{pmatrix}$ and $\mathbf{P} = \begin{pmatrix} p_1 \\ p_2 \end{pmatrix}$, we find from (14) that we must now solve

$$(\mathbf{A} + 3\mathbf{I})\mathbf{P} = \mathbf{K} \quad \text{or} \quad \begin{aligned} 6p_1 - 18p_2 &= 3 \\ 2p_1 - 6p_2 &= 1. \end{aligned}$$

Since this system is obviously equivalent to one equation, we have an infinite number of choices for p_1 and p_2. For example, by choosing $p_1 = 1$ we find $p_2 = \frac{1}{6}$. However, for simplicity, we shall choose $p_1 = \frac{1}{2}$ so that $p_2 = 0$. Hence $\mathbf{P} = \begin{pmatrix} \frac{1}{2} \\ 0 \end{pmatrix}$. Thus from (12) we find

$$\mathbf{X}_2 = \begin{pmatrix} 3 \\ 1 \end{pmatrix} te^{-3t} + \begin{pmatrix} \frac{1}{2} \\ 0 \end{pmatrix} e^{-3t}.$$

The general solution of (10) is then

$$\mathbf{X} = c_1 \begin{pmatrix} 3 \\ 1 \end{pmatrix} e^{-3t} + c_2 \left[\begin{pmatrix} 3 \\ 1 \end{pmatrix} te^{-3t} + \begin{pmatrix} \frac{1}{2} \\ 0 \end{pmatrix} e^{-3t} \right]. \qquad \blacksquare$$

By assigning various values to c_1 and c_2 in the solution in Example 4, we can plot trajectories of the system in (10). A phase portrait of (10) is given in Figure 8.3. The solutions $\mathbf{X}_1$ and $-\mathbf{X}_1$ determine two half-lines $y = \frac{1}{3}x, x > 0$ and $y = \frac{1}{3}x, x < 0$, respectively, shown in black in the figure. Because the single eigenvalue is negative and $e^{-3t} \to 0$ as $t \to \infty$ on *every* trajectory, we have $(x(t), y(t)) \to (0, 0)$ as $t \to \infty$. This is why the arrowheads in Figure 8.3 indicate that a particle on any trajectory moves toward the origin as time increases and why the origin is an attractor in this case. Moreover, a moving particle or trajectory $x = 3c_1e^{-3t} + c_2(te^{-3t} + \frac{1}{2}e^{-3t})$, $y = c_1e^{-3t} + c_2te^{-3t}, c_2 \neq 0$ approaches $(0, 0)$ tangentially to one of the half-lines as $t \to \infty$. In contrast, when the repeated eigenvalue is positive, the situation is reversed and the origin is a repeller. See Problem 21 in Exercises 8.2. Analogous to Figure 8.2, Figure 8.3 is typical of *all* 2×2 homogeneous linear systems $\mathbf{X}' = \mathbf{A}\mathbf{X}$ that have two repeated negative eigenvalues. See Problem 32 in Exercises 8.2.

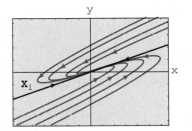

Figure 8.3 A phase portrait of the system (10)

Eigenvalue of Multiplicity Three When the coefficient matrix $\mathbf{A}$ has only one eigenvector associated with an eigenvalue λ_1 of multiplicity three, we can find a second solution of the form (12) and a third solution of the form

$$\mathbf{X}_3 = \mathbf{K}\frac{t^2}{2} e^{\lambda_1 t} + \mathbf{P}te^{\lambda_1 t} + \mathbf{Q}e^{\lambda_1 t}, \qquad (15)$$

where

$$\mathbf{K} = \begin{pmatrix} k_1 \\ k_2 \\ \cdot \\ \cdot \\ \cdot \\ k_n \end{pmatrix}, \quad \mathbf{P} = \begin{pmatrix} p_1 \\ p_2 \\ \cdot \\ \cdot \\ \cdot \\ p_n \end{pmatrix}, \quad \text{and} \quad \mathbf{Q} = \begin{pmatrix} q_1 \\ q_2 \\ \cdot \\ \cdot \\ \cdot \\ q_n \end{pmatrix}.$$

By substituting (15) into the system $\mathbf{X}' = \mathbf{A}\mathbf{X}$ we find that the column vectors $\mathbf{K}$, $\mathbf{P}$, and $\mathbf{Q}$ must satisfy

$$(\mathbf{A} - \lambda_1\mathbf{I})\mathbf{K} = \mathbf{0} \qquad (16)$$

$$(\mathbf{A} - \lambda_1\mathbf{I})\mathbf{P} = \mathbf{K} \qquad (17)$$

and
$$(A - \lambda_1 I)Q = P. \tag{18}$$

Of course, the solutions of (16) and (17) can be used in forming the solutions X_1 and X_2.

EXAMPLE 5   **Repeated Eigenvalues**

Solve $X' = \begin{pmatrix} 2 & 1 & 6 \\ 0 & 2 & 5 \\ 0 & 0 & 2 \end{pmatrix} X$.

Solution   The characteristic equation $(\lambda - 2)^3 = 0$ shows that $\lambda_1 = 2$ is an eigenvalue of multiplicity three. By solving $(A - 2I)K = 0$ we find the single eigenvector

$$K = \begin{pmatrix} 1 \\ 0 \\ 0 \end{pmatrix}.$$

We next solve the systems $(A - 2I)P = K$ and $(A - 2I)Q = P$ in succession and find that

$$P = \begin{pmatrix} 0 \\ 1 \\ 0 \end{pmatrix} \quad \text{and} \quad Q = \begin{pmatrix} 0 \\ -\frac{6}{5} \\ \frac{1}{5} \end{pmatrix}.$$

Using (12) and (15), we see that the general solution of the system is

$$X = c_1 \begin{pmatrix} 1 \\ 0 \\ 0 \end{pmatrix} e^{2t} + c_2 \left[\begin{pmatrix} 1 \\ 0 \\ 0 \end{pmatrix} t e^{2t} + \begin{pmatrix} 0 \\ 1 \\ 0 \end{pmatrix} e^{2t} \right] + c_3 \left[\begin{pmatrix} 1 \\ 0 \\ 0 \end{pmatrix} \frac{t^2}{2} e^{2t} + \begin{pmatrix} 0 \\ 1 \\ 0 \end{pmatrix} t e^{2t} + \begin{pmatrix} 0 \\ -\frac{6}{5} \\ \frac{1}{5} \end{pmatrix} e^{2t} \right]. \quad \blacksquare$$

Remarks   When an eigenvalue λ_1 has multiplicity m, either we can find m linearly independent eigenvectors or the number of corresponding eigenvectors is less than m. Hence the two cases listed on page 380 are not all the possibilities under which a repeated eigenvalue can occur. It can happen, say, that a 5×5 matrix has an eigenvalue of multiplicity five and there exist three corresponding linearly independent eigenvectors. See Problems 31 and 50 in Exercises 8.2.

8.2.3   COMPLEX EIGENVALUES

If $\lambda_1 = \alpha + \beta i$ and $\lambda_2 = \alpha - \beta i$, $\beta > 0$, $i^2 = -1$ are complex eigenvalues of the coefficient matrix A, we can then certainly expect their corresponding eigenvectors to also have complex entries.*

---

*When the characteristic equation has real coefficients, complex eigenvalues always appear in conjugate pairs.

For example, the characteristic equation of the system

$$\frac{dx}{dt} = 6x - y$$

$$\frac{dy}{dt} = 5x + 4y$$

(19)

is $\det(\mathbf{A} - \lambda\mathbf{I}) = \begin{vmatrix} 6 - \lambda & -1 \\ 5 & 4 - \lambda \end{vmatrix} = \lambda^2 - 10\lambda + 29 = 0.$

From the quadratic formula we find $\lambda_1 = 5 + 2i$, $\lambda_2 = 5 - 2i$.

Now for $\lambda_1 = 5 + 2i$ we must solve

$$(1 - 2i)k_1 - \qquad k_2 = 0$$
$$5k_1 - (1 + 2i)k_2 = 0.$$

Since $k_2 = (1 - 2i)k_1$,* the choice $k_1 = 1$ gives the following eigenvector and corresponding solution vector:

$$\mathbf{K}_1 = \begin{pmatrix} 1 \\ 1 - 2i \end{pmatrix}, \quad \mathbf{X}_1 = \begin{pmatrix} 1 \\ 1 - 2i \end{pmatrix} e^{(5+2i)t}.$$

In like manner, for $\lambda_2 = 5 - 2i$ we find

$$\mathbf{K}_2 = \begin{pmatrix} 1 \\ 1 + 2i \end{pmatrix}, \quad \mathbf{X}_2 = \begin{pmatrix} 1 \\ 1 + 2i \end{pmatrix} e^{(5-2i)t}.$$

We can verify by means of the Wronskian that these solution vectors are linearly independent, and so the general solution of (19) is

$$\mathbf{X} = c_1 \begin{pmatrix} 1 \\ 1 - 2i \end{pmatrix} e^{(5+2i)t} + c_2 \begin{pmatrix} 1 \\ 1 + 2i \end{pmatrix} e^{(5-2i)t}.$$

(20)

Note that the entries in $\mathbf{K}_2$ corresponding to λ_2 are the conjugates of the entries in $\mathbf{K}_1$ corresponding to λ_1. The conjugate of λ_1 is, of course, λ_2. We write this as $\lambda_2 = \overline{\lambda}_1$ and $\mathbf{K}_2 = \overline{\mathbf{K}}_1$. We have illustrated the following general result.

THEOREM 8.8   Solutions Corresponding to a Complex Eigenvalue

Let $\mathbf{A}$ be the coefficient matrix having real entries of the homogeneous system (2), and let $\mathbf{K}_1$ be an eigenvector corresponding to the complex eigenvalue $\lambda_1 = \alpha + i\beta$, α and β real. Then

$$\mathbf{K}_1 e^{\lambda_1 t} \quad \text{and} \quad \overline{\mathbf{K}}_1 e^{\overline{\lambda}_1 t}$$

are solutions of (2).

It is desirable and relatively easy to rewrite a solution such as (20) in terms of real functions. To this end we first use Euler's formula to write

$$e^{(5+2i)t} = e^{5t}e^{2ti} = e^{5t}(\cos 2t + i \sin 2t)$$
$$e^{(5-2i)t} = e^{5t}e^{-2ti} = e^{5t}(\cos 2t - i \sin 2t).$$

---

*Note that the second equation is simply $(1 + 2i)$ times the first.

Then, after we multiply complex numbers, collect terms, and replace $c_1 + c_2$ by C_1 and $(c_1 - c_2)i$ by C_2, (20) becomes

$$\mathbf{X} = C_1\mathbf{X}_1 + C_2\mathbf{X}_2, \tag{21}$$

where

$$\mathbf{X}_1 = \left[\begin{pmatrix} 1 \\ 1 \end{pmatrix} \cos 2t - \begin{pmatrix} 0 \\ -2 \end{pmatrix} \sin 2t \right] e^{5t}$$

and

$$\mathbf{X}_2 = \left[\begin{pmatrix} 0 \\ -2 \end{pmatrix} \cos 2t + \begin{pmatrix} 1 \\ 1 \end{pmatrix} \sin 2t \right] e^{5t}.$$

It is now important to realize that the two vectors $\mathbf{X}_1$ and $\mathbf{X}_2$ in (21) are themselves linearly independent *real* solutions of the original system. Consequently we are justified in ignoring the relationship between C_1, C_2 and c_1, c_2, and we can regard C_1 and C_2 as completely arbitrary and real. In other words, the linear combination (21) is an alternative general solution of (19).

The foregoing process can be generalized. Let $\mathbf{K}_1$ be an eigenvector of the coefficient matrix $\mathbf{A}$ (with real entries) corresponding to the complex eigenvalue $\lambda_1 = \alpha + i\beta$. Then the two solution vectors in Theorem 8.8 can be written as

$$\mathbf{K}_1 e^{\lambda_1 t} = \mathbf{K}_1 e^{\alpha t} e^{i\beta t} = \mathbf{K}_1 e^{\alpha t}(\cos \beta t + i \sin \beta t)$$
$$\overline{\mathbf{K}}_1 e^{\overline{\lambda}_1 t} = \overline{\mathbf{K}}_1 e^{\alpha t} e^{-i\beta t} = \overline{\mathbf{K}}_1 e^{\alpha t}(\cos \beta t - i \sin \beta t).$$

By the superposition principle, Theorem 8.2, the following vectors are also solutions:

$$\mathbf{X}_1 = \frac{1}{2}(\mathbf{K}_1 e^{\lambda_1 t} + \overline{\mathbf{K}}_1 e^{\overline{\lambda}_1 t}) = \frac{1}{2}(\mathbf{K}_1 + \overline{\mathbf{K}}_1)e^{\alpha t} \cos \beta t - \frac{i}{2}(-\mathbf{K}_1 + \overline{\mathbf{K}}_1)e^{\alpha t} \sin \beta t$$

$$\mathbf{X}_2 = \frac{i}{2}(-\mathbf{K}_1 e^{\lambda_1 t} + \overline{\mathbf{K}}_1 e^{\overline{\lambda}_1 t}) = \frac{i}{2}(-\mathbf{K}_1 + \overline{\mathbf{K}}_1)e^{\alpha t} \cos \beta t + \frac{1}{2}(\mathbf{K}_1 + \overline{\mathbf{K}}_1)e^{\alpha t} \sin \beta t.$$

Both $\frac{1}{2}(z + \overline{z}) = a$ and $\frac{i}{2}(-z + \overline{z}) = b$ are *real* numbers for *any* complex number $z = a + ib$. Therefore, the entries in the column vectors $\frac{1}{2}(\mathbf{K}_1 + \overline{\mathbf{K}}_1)$ and $\frac{i}{2}(-\mathbf{K}_1 + \overline{\mathbf{K}}_1)$ are real numbers. By defining

$$\mathbf{B}_1 = \frac{1}{2}(\mathbf{K}_1 + \overline{\mathbf{K}}_1) \quad \text{and} \quad \mathbf{B}_2 = \frac{i}{2}(-\mathbf{K}_1 + \overline{\mathbf{K}}_1) \tag{22}$$

we are led to the following theorem.

THEOREM 8.9   **Real Solutions Corresponding to a Complex Eigenvalue**

Let $\lambda_1 = \alpha + i\beta$ be a complex eigenvalue of the coefficient matrix $\mathbf{A}$ in the homogeneous system (2), and let $\mathbf{B}_1$ and $\mathbf{B}_2$ denote the column vectors defined in (22). Then

$$\mathbf{X}_1 = [\mathbf{B}_1 \cos \beta t - \mathbf{B}_2 \sin \beta t]e^{\alpha t}$$
$$\mathbf{X}_2 = [\mathbf{B}_2 \cos \beta t + \mathbf{B}_1 \sin \beta t]e^{\alpha t} \tag{23}$$

are linearly independent solutions of (2) on $(-\infty, \infty)$.

The matrices $\mathbf{B}_1$ and $\mathbf{B}_2$ in (22) are often denoted by

$$\mathbf{B}_1 = \mathrm{Re}(\mathbf{K}_1) \quad \text{and} \quad \mathbf{B}_2 = \mathrm{Im}(\mathbf{K}_1) \tag{24}$$

since these vectors are, respectively, the *real* and *imaginary* parts of the eigenvector $\mathbf{K}_1$. For example, (21) follows from (23) with

$$\mathbf{K}_1 = \begin{pmatrix} 1 \\ 1 - 2i \end{pmatrix} = \begin{pmatrix} 1 \\ 1 \end{pmatrix} + i \begin{pmatrix} 0 \\ -2 \end{pmatrix}$$

$$\mathbf{B}_1 = \mathrm{Re}(\mathbf{K}_1) = \begin{pmatrix} 1 \\ 1 \end{pmatrix} \quad \text{and} \quad \mathbf{B}_2 = \mathrm{Im}(\mathbf{K}_1) = \begin{pmatrix} 0 \\ -2 \end{pmatrix}.$$

EXAMPLE 6   Complex Eigenvalues

Solve the initial-value problem

$$\mathbf{X}' = \begin{pmatrix} 2 & 8 \\ -1 & -2 \end{pmatrix} \mathbf{X}, \quad \mathbf{X}(0) = \begin{pmatrix} 2 \\ -1 \end{pmatrix}. \tag{25}$$

Solution   First we obtain the eigenvalues from

$$\det(\mathbf{A} - \lambda \mathbf{I}) = \begin{vmatrix} 2 - \lambda & 8 \\ -1 & -2 - \lambda \end{vmatrix} = \lambda^2 + 4 = 0.$$

The eigenvalues are $\lambda_1 = 2i$ and $\lambda_2 = \overline{\lambda}_1 = -2i$. For λ_1, the system

$$(2 - 2i)k_1 + \qquad\qquad 8k_2 = 0$$
$$-k_1 + (-2 - 2i)k_2 = 0$$

gives $k_1 = -(2 + 2i)k_2$. By choosing $k_2 = -1$ we get

$$\mathbf{K}_1 = \begin{pmatrix} 2 + 2i \\ -1 \end{pmatrix} = \begin{pmatrix} 2 \\ -1 \end{pmatrix} + i \begin{pmatrix} 2 \\ 0 \end{pmatrix}.$$

Now from (24) we form

$$\mathbf{B}_1 = \mathrm{Re}(\mathbf{K}_1) = \begin{pmatrix} 2 \\ -1 \end{pmatrix} \quad \text{and} \quad \mathbf{B}_2 = \mathrm{Im}(\mathbf{K}_1) = \begin{pmatrix} 2 \\ 0 \end{pmatrix}.$$

Since $\alpha = 0$, it follows from (23) that the general solution of the system is

$$\mathbf{X} = c_1 \left[\begin{pmatrix} 2 \\ -1 \end{pmatrix} \cos 2t - \begin{pmatrix} 2 \\ 0 \end{pmatrix} \sin 2t \right] + c_2 \left[\begin{pmatrix} 2 \\ 0 \end{pmatrix} \cos 2t + \begin{pmatrix} 2 \\ -1 \end{pmatrix} \sin 2t \right]$$

$$= c_1 \begin{pmatrix} 2\cos 2t - 2\sin 2t \\ -\cos 2t \end{pmatrix} + c_2 \begin{pmatrix} 2\cos 2t + 2\sin 2t \\ -\sin 2t \end{pmatrix}. \tag{26}$$

Some graphs of the curves or trajectories defined by solution (26) of the system are illustrated in the phase portrait in Figure 8.4. Now the initial condition $\mathbf{X}(0) = \begin{pmatrix} 2 \\ -1 \end{pmatrix}$ or, equivalently, $x(0) = 2$ and $y(0) = -1$ yields the algebraic system $2c_1 + 2c_2 = 2$, $-c_1 = -1$, whose solution is $c_1 = 1$, $c_2 = 0$. Thus the solution to the problem is $\mathbf{X} = \begin{pmatrix} 2\cos 2t - 2\sin 2t \\ -\cos 2t \end{pmatrix}$. The specific trajectory defined parametrically by the particular solution

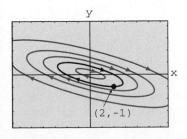

Figure 8.4 A phase portrait of the system (26)

$x = 2\cos 2t - 2\sin 2t$, $y = -\cos 2t$ is the black curve in Figure 8.4. Note that this curve passes through $(2, -1)$. ▬

Remarks In this section we have examined exclusively homogeneous first-order systems of linear equations in normal form $\mathbf{X}' = \mathbf{AX}$. But often the mathematical model of a dynamical physical system is a homogeneous second-order system whose normal form is $\mathbf{X}'' = \mathbf{AX}$. For example, the model for the coupled springs in (1) of Section 7.6,

$$m_1 x_1'' = -k_1 x_1 + k_2(x_2 - x_1)$$

$$m_2 x_2'' = -k_2(x_2 - x_1), \tag{27}$$

can be written as $\qquad \mathbf{MX}'' = \mathbf{KY},$

where

$$\mathbf{M} = \begin{pmatrix} m_1 & 0 \\ 0 & m_2 \end{pmatrix}, \quad \mathbf{K} = \begin{pmatrix} -k_1 - k_2 & k_2 \\ k_2 & -k_2 \end{pmatrix}, \quad \text{and} \quad \mathbf{X} = \begin{pmatrix} x_1(t) \\ x_2(t) \end{pmatrix}.$$

Since $\mathbf{M}$ is nonsingular, we can solve for $\mathbf{X}''$ as $\mathbf{X}'' = \mathbf{AX}$, where $\mathbf{A} = \mathbf{M}^{-1}\mathbf{K}$. Thus (27) is equivalent to

$$\mathbf{X}'' = \begin{pmatrix} -\dfrac{k_1}{m_1} - \dfrac{k_2}{m_1} & \dfrac{k_2}{m_1} \\ \dfrac{k_2}{m_2} & -\dfrac{k_2}{m_2} \end{pmatrix} \mathbf{X}. \tag{28}$$

The methods of this section can be used to solve such a system in two ways:

- First, the original system (27) can be transformed into a first-order system by means of substitutions. If we let $x_1' = x_3$ and $x_2' = x_4$, then $x_3' = x_1''$ and $x_4' = x_2''$ and so (27) is equivalent to a system of *four* linear first-order DEs:

$$x_1' = x_3$$
$$x_2' = x_4$$
$$x_3' = -\left(\frac{k_1}{m_1} + \frac{k_2}{m_1}\right)x_1 + \frac{k_2}{m_1}x_2 \quad \text{or} \quad \mathbf{X}' = \begin{pmatrix} 0 & 0 & 1 & 0 \\ 0 & 0 & 0 & 1 \\ -\dfrac{k_1}{m_1} - \dfrac{k_2}{m_1} & \dfrac{k_2}{m_1} & 0 & 0 \\ \dfrac{k_2}{m_2} & -\dfrac{k_2}{m_2} & 0 & 0 \end{pmatrix} \mathbf{X}.$$
$$x_4' = \frac{k_2}{m_2}x_1 - \frac{k_2}{m_2}x_2$$

$$\tag{29}$$

By finding the eigenvalues and eigenvectors of the coefficient matrix $\mathbf{A}$ in (29), we see that the solution of this first-order system gives the complete state of the physical system—the positions of the masses relative to the equilibrium positions (x_1 and x_2) as well as the velocities of the masses (x_3 and x_4) at time t. See Problem 48(a) in Exercises 8.2.

- Second, because (27) describes free undamped motion, it can be argued that real-valued solutions of the second-order system (28) will have the form

$$\mathbf{X} = \mathbf{V} \cos \omega t \quad \text{and} \quad \mathbf{X} = \mathbf{V} \sin \omega t, \quad \text{(30)}$$

where $\mathbf{V}$ is a column matrix of constants. Substituting either of the functions in (30) into $\mathbf{X}'' = \mathbf{A}\mathbf{X}$ yields $(\mathbf{A} + \omega^2)\mathbf{V} = \mathbf{0}$. (Verify.) By identification with (3) of this section we conclude that $\lambda = -\omega^2$ represents an eigenvalue and $\mathbf{V}$ a corresponding eigenvector of $\mathbf{A}$. It can be shown that the eigenvalues $\lambda_i = -\omega_i^2$, $i = 1, 2$ of $\mathbf{A}$ are negative, and so $\omega_i = \sqrt{-\lambda_i}$ is a real number and represents a (circular) frequency of vibration (see (4) of Section 7.6). By superposition of solutions, the general solution of (28) is then

$$\begin{aligned} \mathbf{X} &= c_1\mathbf{V}_1 \cos \omega_1 t + c_2\mathbf{V}_1 \sin \omega_1 t + c_3\mathbf{V}_2 \cos \omega_2 t + c_4\mathbf{V}_2 \sin \omega_2 t \\ &= (c_1 \cos \omega_1 t + c_2 \sin \omega_1 t)\mathbf{V}_1 + (c_3 \cos \omega_2 t + c_4 \sin \omega_2 t)\mathbf{V}_2, \end{aligned} \quad \text{(31)}$$

where $\mathbf{V}_1$ and $\mathbf{V}_2$ are, in turn, real eigenvectors of $\mathbf{A}$ corresponding to λ_1 and λ_2.

The result given in (31) generalizes. If $-\omega_1^2, -\omega_2^2, \ldots, -\omega_n^2$ are distinct negative eigenvalues and $\mathbf{V}_1, \mathbf{V}_2, \ldots, \mathbf{V}_n$ are corresponding real eigenvectors of the $n \times n$ coefficient matrix $\mathbf{A}$, then the homogeneous second-order system $\mathbf{X}'' = \mathbf{A}\mathbf{X}$ has the general solution

$$\mathbf{X} = \sum_{i=1}^{n} (a_i \cos \omega_i t + b_i \sin \omega_i t)\mathbf{V}_i, \quad \text{(32)}$$

where a_i and b_i represent arbitrary constants. See Problem 48(b) in Exercises 8.2 and the Project Module at the end of this chapter.

EXERCISES 8.2

Answers to odd-numbered problems begin on page AN-10.

8.2.1   Distinct Real Eigenvalues

In Problems 1–12 find the general solution of the given system.

1. $\dfrac{dx}{dt} = x + 2y$

$\dfrac{dy}{dt} = 4x + 3y$

2. $\dfrac{dx}{dt} = 2x + 2y$

$\dfrac{dy}{dt} = x + 3y$

3. $\dfrac{dx}{dt} = -4x + 2y$

$\dfrac{dy}{dt} = -\frac{5}{2}x + 2y$

4. $\dfrac{dx}{dt} = -\frac{5}{2}x + 2y$

$\dfrac{dy}{dt} = \frac{3}{4}x - 2y$

5. $\mathbf{X}' = \begin{pmatrix} 10 & -5 \\ 8 & -12 \end{pmatrix} \mathbf{X}$

6. $\mathbf{X}' = \begin{pmatrix} -6 & 2 \\ -3 & 1 \end{pmatrix} \mathbf{X}$

7. $\dfrac{dx}{dt} = x + y - z$

$\dfrac{dy}{dt} = 2y$

$\dfrac{dz}{dt} = y - z$

8. $\dfrac{dx}{dt} = 2x - 7y$

$\dfrac{dy}{dt} = 5x + 10y + 4z$

$\dfrac{dz}{dt} = 5y + 2z$

9. $\mathbf{X}' = \begin{pmatrix} -1 & 1 & 0 \\ 1 & 2 & 1 \\ 0 & 3 & -1 \end{pmatrix} \mathbf{X}$

10. $\mathbf{X}' = \begin{pmatrix} 1 & 0 & 1 \\ 0 & 1 & 0 \\ 1 & 0 & 1 \end{pmatrix} \mathbf{X}$

11. $\mathbf{X}' = \begin{pmatrix} -1 & -1 & 0 \\ \frac{3}{4} & -\frac{3}{2} & 3 \\ \frac{1}{8} & \frac{1}{4} & -\frac{1}{2} \end{pmatrix} \mathbf{X}$

12. $\mathbf{X}' = \begin{pmatrix} -1 & 4 & 2 \\ 4 & -1 & -2 \\ 0 & 0 & 6 \end{pmatrix} \mathbf{X}$

In Problems 13 and 14 solve the given initial-value problem.

13. $\mathbf{X}' = \begin{pmatrix} \frac{1}{2} & 0 \\ 1 & -\frac{1}{2} \end{pmatrix} \mathbf{X}, \quad \mathbf{X}(0) = \begin{pmatrix} 3 \\ 5 \end{pmatrix}$

14. $\mathbf{X}' = \begin{pmatrix} 1 & 1 & 4 \\ 0 & 2 & 0 \\ 1 & 1 & 1 \end{pmatrix} \mathbf{X}, \quad \mathbf{X}(0) = \begin{pmatrix} 1 \\ 3 \\ 0 \end{pmatrix}$

Computer Lab Assignments

In Problems 15 and 16 use a CAS or linear algebra software as an aid in finding the general solution of the given system.

15. $\mathbf{X}' = \begin{pmatrix} 0.9 & 2.1 & 3.2 \\ 0.7 & 6.5 & 4.2 \\ 1.1 & 1.7 & 3.4 \end{pmatrix} \mathbf{X}$

16. $\mathbf{X}' = \begin{pmatrix} 1 & 0 & 2 & -1.8 & 0 \\ 0 & 5.1 & 0 & -1 & 3 \\ 1 & 2 & -3 & 0 & 0 \\ 0 & 1 & -3.1 & 4 & 0 \\ -2.8 & 0 & 0 & 1.5 & 1 \end{pmatrix} \mathbf{X}$

17. (a) Use computer software to obtain the phase portrait of the system in Problem 5. If possible, include arrowheads as in Figure 8.2. Also include four half-lines in your phase portrait.

    (b) Obtain the Cartesian equations of each of the four half-lines in part (a).

    (c) Draw the eigenvectors on your phase portrait of the system.

18. Find phase portraits for the systems in Problems 2 and 4. For each system find any half-line trajectories and include these lines in your phase portrait.

8.2.2   Repeated Eigenvalues

In Problems 19–28 find the general solution of the given system.

19. $\dfrac{dx}{dt} = 3x - y$

$\dfrac{dy}{dt} = 9x - 3y$

20. $\dfrac{dx}{dt} = -6x + 5y$

$\dfrac{dy}{dt} = -5x + 4y$

21. $\mathbf{X}' = \begin{pmatrix} -1 & 3 \\ -3 & 5 \end{pmatrix} \mathbf{X}$

22. $\mathbf{X}' = \begin{pmatrix} 12 & -9 \\ 4 & 0 \end{pmatrix} \mathbf{X}$

23. $\dfrac{dx}{dt} = 3x - y - z$

$\dfrac{dy}{dt} = x + y - z$

$\dfrac{dz}{dt} = x - y + z$

24. $\dfrac{dx}{dt} = 3x + 2y + 4z$

$\dfrac{dy}{dt} = 2x + 2z$

$\dfrac{dz}{dt} = 4x + 2y + 3z$

25. $\mathbf{X}' = \begin{pmatrix} 5 & -4 & 0 \\ 1 & 0 & 2 \\ 0 & 2 & 5 \end{pmatrix} \mathbf{X}$

26. $\mathbf{X}' = \begin{pmatrix} 1 & 0 & 0 \\ 0 & 3 & 1 \\ 0 & -1 & 1 \end{pmatrix} \mathbf{X}$

27. $\mathbf{X}' = \begin{pmatrix} 1 & 0 & 0 \\ 2 & 2 & -1 \\ 0 & 1 & 0 \end{pmatrix} \mathbf{X}$

28. $\mathbf{X}' = \begin{pmatrix} 4 & 1 & 0 \\ 0 & 4 & 1 \\ 0 & 0 & 4 \end{pmatrix} \mathbf{X}$

In Problems 29 and 30 solve the given initial-value problem.

29. $\mathbf{X}' = \begin{pmatrix} 2 & 4 \\ -1 & 6 \end{pmatrix} \mathbf{X}, \quad \mathbf{X}(0) = \begin{pmatrix} -1 \\ 6 \end{pmatrix}$

30. $\mathbf{X}' = \begin{pmatrix} 0 & 0 & 1 \\ 0 & 1 & 0 \\ 1 & 0 & 0 \end{pmatrix} \mathbf{X}, \quad \mathbf{X}(0) = \begin{pmatrix} 1 \\ 2 \\ 5 \end{pmatrix}$

31. Show that the 5×5 matrix

$$\mathbf{A} = \begin{pmatrix} 2 & 1 & 0 & 0 & 0 \\ 0 & 2 & 0 & 0 & 0 \\ 0 & 0 & 2 & 0 & 0 \\ 0 & 0 & 0 & 2 & 1 \\ 0 & 0 & 0 & 0 & 2 \end{pmatrix}$$

has an eigenvalue λ_1 of multiplicity 5. Show that three linearly independent eigenvectors corresponding to λ_1 can be found.

Computer Lab Assignments

32. Find phase portraits for the systems in Problems 20 and 21. For each system find any half-line trajectories and include these lines in your phase portrait.

8.2.3   Complex Eigenvalues

In Problems 33–44 find the general solution of the given system.

33. $\dfrac{dx}{dt} = 6x - y$

$\dfrac{dy}{dt} = 5x + 2y$

34. $\dfrac{dx}{dt} = x + y$

$\dfrac{dy}{dt} = -2x - y$

35. $\dfrac{dx}{dt} = 5x + y$

$\dfrac{dy}{dt} = -2x + 3y$

36. $\dfrac{dx}{dt} = 4x + 5y$

$\dfrac{dy}{dt} = -2x + 6y$

37. $\mathbf{X'} = \begin{pmatrix} 4 & -5 \\ 5 & -4 \end{pmatrix} \mathbf{X}$

38. $\mathbf{X'} = \begin{pmatrix} 1 & -8 \\ 1 & -3 \end{pmatrix} \mathbf{X}$

39. $\dfrac{dx}{dt} = z$

$\dfrac{dy}{dt} = -z$

$\dfrac{dz}{dt} = y$

40. $\dfrac{dx}{dt} = 2x + y + 2z$

$\dfrac{dy}{dt} = 3x + 6z$

$\dfrac{dz}{dt} = -4x - 3z$

41. $\mathbf{X'} = \begin{pmatrix} 1 & -1 & 2 \\ -1 & 1 & 0 \\ -1 & 0 & 1 \end{pmatrix} \mathbf{X}$

42. $\mathbf{X'} = \begin{pmatrix} 4 & 0 & 1 \\ 0 & 6 & 0 \\ -4 & 0 & 4 \end{pmatrix} \mathbf{X}$

43. $\mathbf{X'} = \begin{pmatrix} 2 & 5 & 1 \\ -5 & -6 & 4 \\ 0 & 0 & 2 \end{pmatrix} \mathbf{X}$

44. $\mathbf{X'} = \begin{pmatrix} 2 & 4 & 4 \\ -1 & -2 & 0 \\ -1 & 0 & -2 \end{pmatrix} \mathbf{X}$

In Problems 45 and 46 solve the given initial-value problem.

45. $\mathbf{X'} = \begin{pmatrix} 1 & -12 & -14 \\ 1 & 2 & -3 \\ 1 & 1 & -2 \end{pmatrix} \mathbf{X}, \quad \mathbf{X}(0) = \begin{pmatrix} 4 \\ 6 \\ -7 \end{pmatrix}$

46. $\mathbf{X'} = \begin{pmatrix} 6 & -1 \\ 5 & 4 \end{pmatrix} \mathbf{X}, \quad \mathbf{X}(0) = \begin{pmatrix} -2 \\ 8 \end{pmatrix}$

Computer Lab Assignments

47. Find phase portraits for the systems in Problems 36, 37, and 38.

48. (a) Solve (2) of Section 7.6 using the first method outlined in the Remarks (page 388)—that is, express (2) of Section 7.6 as a first-order system of four linear equations. Use a CAS or linear algebra software as an aid in finding eigenvalues and eigenvectors of a 4 × 4 matrix. Then apply the initial conditions to your general solution to obtain (4) of Section 7.6.

(b) Solve (2) of Section 7.6 using the second method outlined in the Remarks—that is, express (2) of Section 7.6 as a second-order system of two linear equations. Assume solutions of the form

$\mathbf{X} = \mathbf{V}\sin\omega t$ and $\mathbf{X} = \mathbf{V}\cos\omega t$. Find the eigenvalues and eigenvectors of a 2×2 matrix. As in part (a), obtain (4) of Section 7.6.

Discussion Problems

49. Solve each of the following linear systems.

(a) $\mathbf{X}' = \begin{pmatrix} 1 & 1 \\ 1 & 1 \end{pmatrix}\mathbf{X}$ (b) $\mathbf{X}' = \begin{pmatrix} 1 & 1 \\ -1 & -1 \end{pmatrix}\mathbf{X}$

Find a phase portrait of each system. What is the geometric significance of the line $y = -x$ in each portrait?

50. Consider the 5×5 matrix given in Problem 31. Solve the system $\mathbf{X}' = \mathbf{AX}$ without the aid of matrix methods, but write the general solution using matrix notation. Use the general solution as a basis for a discussion of how the system can be solved using the matrix methods of this section. Carry out your ideas.

51. Obtain a Cartesian equation of the curve defined parametrically by the solution of the linear system in Example 6. Identify the curve passing through $(2, -1)$ in Figure 8.4. [*Hint:* Compute x^2, y^2, and xy.]

52. Examine your phase portraits in Problem 47. Under what conditions will the phase portrait of a 2×2 homogeneous linear system with complex eigenvalues consist of a family of closed curves? consist of a family of spirals? Under what conditions is the origin $(0, 0)$ a repeller? an attractor?

8.3 VARIATION OF PARAMETERS

• Fundamental matrix   *• Finding a particular solution by variation of parameters*

The method of variation of parameters developed in Sections 2.3 and 4.6 to obtain a particular solution of a nonhomogeneous linear differential equation can be extended to linear systems of differential equations. Before developing a matrix version of variation of parameters for nonhomogeneous linear systems $\mathbf{X}' = \mathbf{AX} + \mathbf{F}$, we need to examine a special matrix that is formed out of the solution vectors of the corresponding homogeneous system $\mathbf{X}' = \mathbf{AX}$.

A Fundamental Matrix If $\mathbf{X}_1, \mathbf{X}_2, \dots, \mathbf{X}_n$ is a fundamental set of solutions of the homogeneous system $\mathbf{X}' = \mathbf{AX}$ on an interval I, then its general solution on the interval is $\mathbf{X} = c_1\mathbf{X}_1 + c_2\mathbf{X}_2 + \cdots + c_n\mathbf{X}_n$ or

$$\mathbf{X} = c_1\begin{pmatrix} x_{11} \\ x_{21} \\ \vdots \\ x_{n1} \end{pmatrix} + c_2\begin{pmatrix} x_{12} \\ x_{22} \\ \vdots \\ x_{n2} \end{pmatrix} + \cdots + c_n\begin{pmatrix} x_{1n} \\ x_{2n} \\ \vdots \\ x_{nn} \end{pmatrix} = \begin{pmatrix} c_1x_{11} + c_2x_{12} + \cdots + c_nx_{1n} \\ c_1x_{21} + c_2x_{22} + \cdots + c_nx_{2n} \\ \vdots \\ c_1x_{n1} + c_2x_{n2} + \cdots + c_nx_{nn} \end{pmatrix}. \quad (1)$$

The last matrix in (1) is recognized as the product of an $n \times n$ matrix with an $n \times 1$ matrix. In other words, the general solution (1) can be written as the product

$$\mathbf{X} = \mathbf{\Phi}(t)\mathbf{C}, \tag{2}$$

where $\mathbf{C}$ is an $n \times 1$ column vector of arbitrary constants $c_1, c_2, \ldots, c_n$ and the $n \times n$ matrix, whose columns consist of the entries of the solution vectors of the system $\mathbf{X}' = \mathbf{AX}$,

$$\mathbf{\Phi}(t) = \begin{pmatrix} x_{11} & x_{12} & \cdots & x_{1n} \\ x_{21} & x_{22} & \cdots & x_{2n} \\ \vdots & & & \vdots \\ x_{n1} & x_{n2} & \cdots & x_{nn} \end{pmatrix},$$

is called a **fundamental matrix** of the system on the interval.

In the discussion that follows we need to use two properties of a fundamental matrix:

- A fundamental matrix $\mathbf{\Phi}(t)$ is nonsingular.
- If $\mathbf{\Phi}(t)$ is a fundamental matrix of the system $\mathbf{X}' = \mathbf{AX}$, then

$$\mathbf{\Phi}'(t) = \mathbf{A}\mathbf{\Phi}(t). \tag{3}$$

A reexamination of (9) of Theorem 8.3 shows that det $\mathbf{\Phi}(t)$ is the same as the Wronskian $W(\mathbf{X}_1, \mathbf{X}_2, \ldots, \mathbf{X}_n)$. Hence the linear independence of the columns of $\mathbf{\Phi}(t)$ on the interval I guarantees that det $\mathbf{\Phi}(t) \neq 0$ for every t in the interval. Since $\mathbf{\Phi}(t)$ is nonsingular, the multiplicative inverse $\mathbf{\Phi}^{-1}(t)$ exists for every t in the interval. The result given in (3) follows immediately from the fact that every column of $\mathbf{\Phi}(t)$ is a solution vector of $\mathbf{X}' = \mathbf{AX}$.

Variation of Parameters   Analogous to the procedure in Section 4.6, we ask whether it is possible to replace the matrix of constants $\mathbf{C}$ in (2) by a column matrix of functions

$$\mathbf{U}(t) = \begin{pmatrix} u_1(t) \\ u_2(t) \\ \vdots \\ u_n(t) \end{pmatrix} \quad \text{so that} \quad \mathbf{X}_p = \mathbf{\Phi}(t)\mathbf{U}(t) \tag{4}$$

is a particular solution of the nonhomogeneous system

$$\mathbf{X}' = \mathbf{AX} + \mathbf{F}(t). \tag{5}$$

By the product rule, the derivative of the last expression in (4) is

$$\mathbf{X}_p' = \mathbf{\Phi}(t)\mathbf{U}'(t) + \mathbf{\Phi}'(t)\mathbf{U}(t). \tag{6}$$

Note that the order of the products in (6) is very important. Since $\mathbf{U}(t)$ is a column matrix, the products $\mathbf{U}'(t)\mathbf{\Phi}(t)$ and $\mathbf{U}(t)\mathbf{\Phi}'(t)$ are not defined. Substituting (4) and (6) into (5) gives

$$\mathbf{\Phi}(t)\mathbf{U}'(t) + \mathbf{\Phi}'(t)\mathbf{U}(t) = \mathbf{A}\mathbf{\Phi}(t)\mathbf{U}(t) + \mathbf{F}(t). \tag{7}$$

Now if we use (3) to replace $\mathbf{\Phi}'(t)$, (7) becomes

$$\mathbf{\Phi}(t)\mathbf{U}'(t) + \mathbf{A}\mathbf{\Phi}(t)\mathbf{U}(t) = \mathbf{A}\mathbf{\Phi}(t)\mathbf{U}(t) + \mathbf{F}(t)$$

or
$$\mathbf{\Phi}(t)\mathbf{U}'(t) = \mathbf{F}(t). \qquad (8)$$

Multiplying both sides of equation (8) by $\mathbf{\Phi}^{-1}(t)$ gives

$$\mathbf{U}'(t) = \mathbf{\Phi}^{-1}(t)\,\mathbf{F}(t) \quad\text{and so}\quad \mathbf{U}(t) = \int \mathbf{\Phi}^{-1}(t)\,\mathbf{F}(t)\,dt.$$

Since $\mathbf{X}_p = \mathbf{\Phi}(t)\mathbf{U}(t)$, we conclude that a particular solution of (5) is

$$\mathbf{X}_p = \mathbf{\Phi}(t) \int \mathbf{\Phi}^{-1}(t)\mathbf{F}(t)\,dt. \qquad (9)$$

To calculate the indefinite integral of the column matrix $\mathbf{\Phi}^{-1}(t)\mathbf{F}(t)$ in (9) we integrate each entry. Thus the general solution of the system (5) is $\mathbf{X} = \mathbf{X}_c + \mathbf{X}_p$ or

$$\mathbf{X} = \mathbf{\Phi}(t)\mathbf{C} + \mathbf{\Phi}(t) \int \mathbf{\Phi}^{-1}(t)\mathbf{F}(t)\,dt. \qquad (10)$$

Note that it is not necessary to use a constant of integration in the evaluation of $\int \mathbf{\Phi}^{-1}(t)\mathbf{F}(t)\,dt$ for the same reasons stated in the discussion of variation of parameters in Section 4.6. See page 190.

EXAMPLE 1   Variation of Parameters

Find the general solution of the nonhomogeneous system

$$\mathbf{X}' = \begin{pmatrix} -3 & 1 \\ 2 & -4 \end{pmatrix} \mathbf{X} + \begin{pmatrix} 3t \\ e^{-t} \end{pmatrix} \qquad (11)$$

on the interval $(-\infty, \infty)$.

Solution   We first solve the homogeneous system

$$\mathbf{X}' = \begin{pmatrix} -3 & 1 \\ 2 & -4 \end{pmatrix} \mathbf{X}. \qquad (12)$$

The characteristic equation of the coefficient matrix is

$$\det(\mathbf{A} - \lambda\mathbf{I}) = \begin{vmatrix} -3 - \lambda & 1 \\ 2 & -4 - \lambda \end{vmatrix} = (\lambda + 2)(\lambda + 5) = 0,$$

so the eigenvalues are $\lambda_1 = -2$ and $\lambda_2 = -5$. By the usual method we find that the eigenvectors corresponding to λ_1 and λ_2 are, respectively,

$$\begin{pmatrix} 1 \\ 1 \end{pmatrix} \quad\text{and}\quad \begin{pmatrix} 1 \\ -2 \end{pmatrix}.$$

The solution vectors of the system (11) are then

$$\mathbf{X}_1 = \begin{pmatrix} 1 \\ 1 \end{pmatrix} e^{-2t} = \begin{pmatrix} e^{-2t} \\ e^{-2t} \end{pmatrix} \quad\text{and}\quad \mathbf{X}_2 = \begin{pmatrix} 1 \\ -2 \end{pmatrix} e^{-5t} = \begin{pmatrix} e^{-5t} \\ -2e^{-5t} \end{pmatrix}.$$

The entries in $\mathbf{X}_1$ form the first column of $\mathbf{\Phi}(t)$, and the entries in $\mathbf{X}_2$ form the second column of $\mathbf{\Phi}(t)$. Hence

$$\mathbf{\Phi}(t) = \begin{pmatrix} e^{-2t} & e^{-5t} \\ e^{-2t} & -2e^{-5t} \end{pmatrix} \quad \text{and} \quad \mathbf{\Phi}^{-1}(t) = \begin{pmatrix} \frac{2}{3}e^{2t} & \frac{1}{3}e^{2t} \\ \frac{1}{3}e^{5t} & -\frac{1}{3}e^{5t} \end{pmatrix}.$$

From (9) we obtain

$$\mathbf{X}_p = \mathbf{\Phi}(t) \int \mathbf{\Phi}^{-1}(t)\mathbf{F}(t)\, dt = \begin{pmatrix} e^{-2t} & e^{-5t} \\ e^{-2t} & -2e^{-5t} \end{pmatrix} \int \begin{pmatrix} \frac{2}{3}e^{2t} & \frac{1}{3}e^{2t} \\ \frac{1}{3}e^{5t} & -\frac{1}{3}e^{5t} \end{pmatrix} \begin{pmatrix} 3t \\ e^{-t} \end{pmatrix} dt$$

$$= \begin{pmatrix} e^{-2t} & e^{-5t} \\ e^{-2t} & -2e^{-5t} \end{pmatrix} \int \begin{pmatrix} 2te^{2t} + \frac{1}{3}e^{t} \\ te^{5t} - \frac{1}{3}e^{4t} \end{pmatrix} dt$$

$$= \begin{pmatrix} e^{-2t} & e^{-5t} \\ e^{-2t} & -2e^{-5t} \end{pmatrix} \begin{pmatrix} te^{2t} - \frac{1}{2}e^{2t} + \frac{1}{3}e^{t} \\ \frac{1}{5}te^{5t} - \frac{1}{25}e^{5t} - \frac{1}{12}e^{4t} \end{pmatrix}$$

$$= \begin{pmatrix} \frac{6}{5}t - \frac{27}{50} + \frac{1}{4}e^{-t} \\ \frac{3}{5}t - \frac{21}{50} + \frac{1}{2}e^{-t} \end{pmatrix}.$$

Hence from (10) the general solution of (11) on the interval is

$$\mathbf{X} = \begin{pmatrix} e^{-2t} & e^{-5t} \\ e^{-2t} & -2e^{-5t} \end{pmatrix} \begin{pmatrix} c_1 \\ c_2 \end{pmatrix} + \begin{pmatrix} \frac{6}{5}t - \frac{27}{50} + \frac{1}{4}e^{-t} \\ \frac{3}{5}t - \frac{21}{50} + \frac{1}{2}e^{-t} \end{pmatrix}$$

$$= c_1 \begin{pmatrix} 1 \\ 1 \end{pmatrix} e^{-2t} + c_2 \begin{pmatrix} 1 \\ -2 \end{pmatrix} e^{-5t} + \begin{pmatrix} \frac{6}{5} \\ \frac{3}{5} \end{pmatrix} t - \begin{pmatrix} \frac{27}{50} \\ \frac{21}{50} \end{pmatrix} + \begin{pmatrix} \frac{1}{4} \\ \frac{1}{2} \end{pmatrix} e^{-t}. \qquad \blacksquare$$

Initial-Value Problem   The general solution of (5) on an interval can be written in the alternative manner

$$\mathbf{X} = \mathbf{\Phi}(t)\mathbf{C} + \mathbf{\Phi}(t) \int_{t_0}^{t} \mathbf{\Phi}^{-1}(s)\mathbf{F}(s)\, ds, \qquad \text{(13)}$$

where t and t_0 are points in the interval. This last form is useful in solving (5) subject to an initial condition $\mathbf{X}(t_0) = \mathbf{X}_0$, because the limits of integration are chosen so that the particular solution vanishes at $t = t_0$. Substituting $t = t_0$ into (13) yields $\mathbf{X}_0 = \mathbf{\Phi}(t_0)\mathbf{C}$, from which we get $\mathbf{C} = \mathbf{\Phi}^{-1}(t_0)\mathbf{X}_0$. Substituting this last result into (13) gives the following solution of the initial-value problem:

$$\mathbf{X} = \mathbf{\Phi}(t)\mathbf{\Phi}^{-1}(t_0)\mathbf{X}_0 + \mathbf{\Phi}(t) \int_{t_0}^{t} \mathbf{\Phi}^{-1}(s)\mathbf{F}(s)\, ds. \qquad \text{(14)}$$

Remarks   When the entries in the column matrix $\mathbf{F}(t)$ are constants, polynomials, exponential functions, sines and cosines, or finite sums and products of these functions, we may be able to find a particular solution $\mathbf{X}_p$ of the nonhomogeneous linear system $\mathbf{X}' = \mathbf{AX} + \mathbf{F}$ by **undetermined coefficients**. We have not elaborated on this method because for linear systems it is not simply a straightforward generalization of the method discussed in Section 4.4. See Problems 23–26 and Problem 28 in Exercises 8.3.

EXERCISES 8.3

Answers to odd-numbered problems begin on page AN-11.

In Problems 1–20 use variation of parameters to solve the given system.

1. $\dfrac{dx}{dt} = 3x - 3y + 4$

$\dfrac{dy}{dt} = 2x - 2y - 1$

2. $\dfrac{dx}{dt} = 2x - y$

$\dfrac{dy}{dt} = 3x - 2y + 4t$

3. $\mathbf{X}' = \begin{pmatrix} 3 & -5 \\ \frac{3}{4} & -1 \end{pmatrix} \mathbf{X} + \begin{pmatrix} 1 \\ -1 \end{pmatrix} e^{t/2}$

4. $\mathbf{X}' = \begin{pmatrix} 2 & -1 \\ 4 & 2 \end{pmatrix} \mathbf{X} + \begin{pmatrix} \sin 2t \\ 2\cos 2t \end{pmatrix} e^{2t}$

5. $\mathbf{X}' = \begin{pmatrix} 0 & 2 \\ -1 & 3 \end{pmatrix} \mathbf{X} + \begin{pmatrix} 1 \\ -1 \end{pmatrix} e^{t}$

6. $\mathbf{X}' = \begin{pmatrix} 0 & 2 \\ -1 & 3 \end{pmatrix} \mathbf{X} + \begin{pmatrix} 2 \\ e^{-3t} \end{pmatrix}$

7. $\mathbf{X}' = \begin{pmatrix} 1 & 8 \\ 1 & -1 \end{pmatrix} \mathbf{X} + \begin{pmatrix} 12 \\ 12 \end{pmatrix} t$

8. $\mathbf{X}' = \begin{pmatrix} 1 & 8 \\ 1 & -1 \end{pmatrix} \mathbf{X} + \begin{pmatrix} e^{-t} \\ te^{t} \end{pmatrix}$

9. $\mathbf{X}' = \begin{pmatrix} 3 & 2 \\ -2 & -1 \end{pmatrix} \mathbf{X} + \begin{pmatrix} 2e^{-t} \\ e^{-t} \end{pmatrix}$

10. $\mathbf{X}' = \begin{pmatrix} 3 & 2 \\ -2 & -1 \end{pmatrix} \mathbf{X} + \begin{pmatrix} 1 \\ 1 \end{pmatrix}$

11. $\mathbf{X}' = \begin{pmatrix} 0 & -1 \\ 1 & 0 \end{pmatrix} \mathbf{X} + \begin{pmatrix} \sec t \\ 0 \end{pmatrix}$

12. $\mathbf{X}' = \begin{pmatrix} 1 & -1 \\ 1 & 1 \end{pmatrix} \mathbf{X} + \begin{pmatrix} 3 \\ 3 \end{pmatrix} e^{t}$

13. $\mathbf{X}' = \begin{pmatrix} 1 & -1 \\ 1 & 1 \end{pmatrix} \mathbf{X} + \begin{pmatrix} \cos t \\ \sin t \end{pmatrix} e^{t}$

14. $\mathbf{X}' = \begin{pmatrix} 2 & -2 \\ 8 & -6 \end{pmatrix} \mathbf{X} + \begin{pmatrix} 1 \\ 3 \end{pmatrix} \dfrac{e^{-2t}}{t}$

15. $\mathbf{X}' = \begin{pmatrix} 0 & 1 \\ -1 & 0 \end{pmatrix} \mathbf{X} + \begin{pmatrix} 0 \\ \sec t \tan t \end{pmatrix}$

16. $\mathbf{X}' = \begin{pmatrix} 0 & 1 \\ -1 & 0 \end{pmatrix} \mathbf{X} + \begin{pmatrix} 1 \\ \cot t \end{pmatrix}$

17. $\mathbf{X}' = \begin{pmatrix} 1 & 2 \\ -\frac{1}{2} & 1 \end{pmatrix} \mathbf{X} + \begin{pmatrix} \csc t \\ \sec t \end{pmatrix} e^{t}$

18. $\mathbf{X}' = \begin{pmatrix} 1 & -2 \\ 1 & -1 \end{pmatrix} \mathbf{X} + \begin{pmatrix} \tan t \\ 1 \end{pmatrix}$

19. $\mathbf{X}' = \begin{pmatrix} 1 & 1 & 0 \\ 1 & 1 & 0 \\ 0 & 0 & 3 \end{pmatrix} \mathbf{X} + \begin{pmatrix} e^{t} \\ e^{2t} \\ te^{3t} \end{pmatrix}$

20. $\mathbf{X}' = \begin{pmatrix} 3 & -1 & -1 \\ 1 & 1 & -1 \\ 1 & -1 & 1 \end{pmatrix} \mathbf{X} + \begin{pmatrix} 0 \\ t \\ 2e^{t} \end{pmatrix}$

In Problems 21 and 22 use (14) to solve the given initial-value problem.

21. $\mathbf{X}' = \begin{pmatrix} 3 & -1 \\ -1 & 3 \end{pmatrix} \mathbf{X} + \begin{pmatrix} 4e^{2t} \\ 4e^{4t} \end{pmatrix}, \quad \mathbf{X}(0) = \begin{pmatrix} 1 \\ 1 \end{pmatrix}$

22. $\mathbf{X}' = \begin{pmatrix} 1 & -1 \\ 1 & -1 \end{pmatrix} \mathbf{X} + \begin{pmatrix} 1/t \\ 1/t \end{pmatrix}, \quad \mathbf{X}(1) = \begin{pmatrix} 2 \\ -1 \end{pmatrix}$

In Problems 23–26 determine the coefficients in the particular solution by substituting $\mathbf{X}_p$ into the given system. Find the general solution of the system.

23. $\mathbf{X}' = \begin{pmatrix} 2 & 3 \\ -1 & -2 \end{pmatrix}\mathbf{X} + \begin{pmatrix} -7 \\ 5 \end{pmatrix}; \quad \mathbf{X}_p = \begin{pmatrix} a_1 \\ b_1 \end{pmatrix}$

24. $\mathbf{X}' = \begin{pmatrix} 6 & 1 \\ 4 & 3 \end{pmatrix}\mathbf{X} + \begin{pmatrix} 6t \\ -10t + 4 \end{pmatrix}; \quad \mathbf{X}_p = \begin{pmatrix} a_2 \\ b_2 \end{pmatrix}t + \begin{pmatrix} a_1 \\ b_1 \end{pmatrix}$

25. $\mathbf{X}' = \begin{pmatrix} 4 & \frac{1}{3} \\ 9 & 6 \end{pmatrix}\mathbf{X} + \begin{pmatrix} -3 \\ 10 \end{pmatrix}e^t; \quad \mathbf{X}_p = \begin{pmatrix} a_1 \\ b_1 \end{pmatrix}e^t$

26. $\mathbf{X}' = \begin{pmatrix} -1 & 5 \\ -1 & 1 \end{pmatrix}\mathbf{X} + \begin{pmatrix} \sin t \\ -2\cos t \end{pmatrix}; \quad \mathbf{X}_p = \begin{pmatrix} a_2 \\ b_2 \end{pmatrix}\cos t + \begin{pmatrix} a_1 \\ b_1 \end{pmatrix}\sin t$

27. The system of differential equations for the currents $i_1(t)$ and $i_2(t)$ in the electrical network shown in Figure 8.5 is

$$\frac{d}{dt}\begin{pmatrix} i_1 \\ i_2 \end{pmatrix} = \begin{pmatrix} -(R_1 + R_2)/L_2 & R_2/L_2 \\ R_2/L_1 & -R_2/L_1 \end{pmatrix}\begin{pmatrix} i_1 \\ i_2 \end{pmatrix} + \begin{pmatrix} E/L_2 \\ 0 \end{pmatrix}.$$

Use variation of parameters to solve the system if $R_1 = 8 \ \Omega$, $R_2 = 3 \ \Omega$, $L_1 = 1$ h, $L_2 = 1$ h, $E(t) = 100 \sin t$ V, $i_1(0) = 0$, and $i_2(0) = 0$.

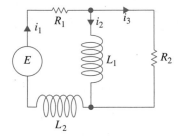

Figure 8.5

Discussion Problems

28. (a) Proceeding analogously to Problem 23, it might be conjectured that the nonhomogeneous system $\mathbf{X}' = \begin{pmatrix} 1 & -1 \\ -1 & 1 \end{pmatrix}\mathbf{X} + \begin{pmatrix} 3 \\ -5 \end{pmatrix}$ possesses a particular solution consisting of a vector with constant entries. However, show that the system possesses no solution of the form $\mathbf{X}_p = \begin{pmatrix} a_1 \\ b_1 \end{pmatrix}$, a_1 and b_1 constants. What is the difference between the system in Problem 23 and the given system?

(b) Based on the procedure of Section 4.4, what would be the logical candidate for the form of a particular solution? Test your conjecture. Find a particular solution $\mathbf{X}_p$ of the system in part (a).

Computer Lab Assignments

29. Solving a nonhomogeneous linear system $\mathbf{X}' = \mathbf{A}\mathbf{X} + \mathbf{F}(t)$ by variation of parameters when $\mathbf{A}$ is a 3×3 (or larger) matrix is almost an impossible task to do by hand. Consider the system

$$\mathbf{X}' = \begin{pmatrix} 2 & -2 & 2 & 1 \\ -1 & 3 & 0 & 3 \\ 0 & 0 & 4 & -2 \\ 0 & 0 & 2 & -1 \end{pmatrix}\mathbf{X} + \begin{pmatrix} te^t \\ e^{-t} \\ e^{2t} \\ 1 \end{pmatrix}.$$

(a) Use a CAS or linear algebra software to find the eigenvalues and eigenvectors of the coefficient matrix.

(b) Form a fundamental matrix $\Phi(t)$ and use the computer to find $\Phi^{-1}(t)$.

(c) Use the computer to compute $\Phi^{-1}(t)\mathbf{F}(t)$, $\int \Phi^{-1}(t)\mathbf{F}(t)\ dt$, $\Phi(t) \int \Phi^{-1}(t)\mathbf{F}(t)\ dt$, $\Phi(t)\mathbf{C}$, and $\Phi(t)\mathbf{C} + \Phi(t) \int \Phi^{-1}(t)\mathbf{F}(t)\ dt$, where $\mathbf{C}$ is a column matrix of constants c_1, c_2, c_3, and c_4.

(d) Rewrite the computer output for the general solution of the system in the form $\mathbf{X} = \mathbf{X}_c + \mathbf{X}_p$, where $\mathbf{X}_c = c_1\mathbf{X}_1 + c_2\mathbf{X}_2 + c_3\mathbf{X}_3 + c_4\mathbf{X}_4$.

8.4 MATRIX EXPONENTIAL

• Homogeneous systems   • Power series for e^{at} • Matrix exponential   • Nonhomogeneous systems

Matrices can be utilized in an entirely different manner to solve a system of linear first-order differential equations. Recall that the simple linear first-order differential equation $x' = ax$, where a is a constant, has the general solution $x = ce^{at}$. It seems natural then to ask whether we can define a matrix exponential $e^{\mathbf{A}t}$ so that $e^{\mathbf{A}t}$ is a solution of the system $\mathbf{X}' = \mathbf{A}\mathbf{X}$.

Homogeneous Systems   We shall now see that it is possible to define a matrix exponential $e^{\mathbf{A}t}$ so that

$$\mathbf{X} = e^{\mathbf{A}t}\mathbf{C} \tag{1}$$

is a solution of the homogeneous system $\mathbf{X}' = \mathbf{A}\mathbf{X}$. Here $\mathbf{A}$ is an $n \times n$ matrix of constants, and $\mathbf{C}$ is an $n \times 1$ column matrix of arbitrary constants. Note in (1) that the matrix $\mathbf{C}$ post multiplies $e^{\mathbf{A}t}$ because we want $e^{\mathbf{A}t}$ to be an $n \times n$ matrix. While the complete development of the meaning and theory of the matrix exponential would require a thorough knowledge of matrix algebra, one way of defining $e^{\mathbf{A}t}$ is inspired by the power series representation of the scalar exponential function e^{at}:

$$e^{at} = 1 + at + a^2\frac{t^2}{2!} + \cdots + a^k\frac{t^k}{k!} + \cdots = \sum_{k=0}^{\infty} a^k\frac{t^k}{k!}. \tag{2}$$

The series in (2) converges for all t. Using this series, with 1 replaced by the identity $\mathbf{I}$ and the constant a replaced by an $n \times n$ matrix $\mathbf{A}$ of constants, we arrive at a definition for the $n \times n$ matrix $e^{\mathbf{A}t}$.

DEFINITION 8.4	Matrix Exponential

For any $n \times n$ matrix $\mathbf{A}$,

$$e^{\mathbf{A}t} = \mathbf{I} + \mathbf{A}t + \mathbf{A}^2\frac{t^2}{2!} + \cdots + \mathbf{A}^k\frac{t^k}{k!} + \cdots = \sum_{k=0}^{\infty} \mathbf{A}^k\frac{t^k}{k!}. \tag{3}$$

It can be shown that the series given in (3) converges to an $n \times n$ matrix for every value of t. Also, $\mathbf{A}^2 = \mathbf{A}\mathbf{A}$, $\mathbf{A}^3 = \mathbf{A}(\mathbf{A}^2)$, and so on.

Derivative of $e^{\mathbf{A}t}$ The derivative of the matrix exponential is analogous to the differentiation property of the scalar exponential $\dfrac{d}{dt}\,e^{at} = ae^{at}$. To justify

$$\frac{d}{dt}\,e^{\mathbf{A}t} = \mathbf{A}e^{\mathbf{A}t}, \tag{4}$$

we differentiate (3) term by term:

$$\frac{d}{dt}\,e^{\mathbf{A}t} = \frac{d}{dt}\left[\mathbf{I} + \mathbf{A}t + \mathbf{A}^2\frac{t^2}{2!} + \cdots + \mathbf{A}^k\frac{t^k}{k!} + \cdots\right] = \mathbf{A} + \mathbf{A}^2 t + \frac{1}{2!}\mathbf{A}^3 t^2 + \cdots$$

$$= \mathbf{A}\left[\mathbf{I} + \mathbf{A}t + \mathbf{A}^2\frac{t^2}{2!} + \cdots\right] = \mathbf{A}e^{\mathbf{A}t}.$$

Because of (4), we can now prove that (1) is a solution of $\mathbf{X}' = \mathbf{A}\mathbf{X}$ for every $n \times 1$ vector $\mathbf{C}$ of constants:

$$\mathbf{X}' = \frac{d}{dt}\,e^{\mathbf{A}t}\mathbf{C} = \mathbf{A}e^{\mathbf{A}t}\mathbf{C} = \mathbf{A}(e^{\mathbf{A}t}\mathbf{C}) = \mathbf{A}\mathbf{X}.$$

$e^{\mathbf{A}t}$ Is a Fundamental Matrix If we denote the matrix exponential $e^{\mathbf{A}t}$ by the symbol $\mathbf{\Psi}(t)$, then (4) is equivalent to the matrix differential equation $\mathbf{\Psi}'(t) = \mathbf{A}\mathbf{\Psi}(t)$ (see (3) of Section 8.3). In addition, it follows immediately from Definition 8.4 that $\mathbf{\Psi}(0) = e^{\mathbf{A}0} = \mathbf{I}$, and so det $\mathbf{\Psi}(0) \neq 0$. It turns out that these two properties are sufficient for us to conclude that $\mathbf{\Psi}(t)$ is a fundamental matrix of the system $\mathbf{X}' = \mathbf{A}\mathbf{X}$.

Nonhomogeneous Systems We saw in (4) of Section 2.4 that the general solution of the single linear first-order differential equation $x' = ax + f(t)$, where a is a constant, can be expressed as

$$x = x_c + x_p = ce^{at} + e^{at}\int_{t_0}^{t} e^{-as}f(s)\,ds.$$

For a nonhomogeneous system of linear first-order differential equations, it can be shown that the general solution of $\mathbf{X}' = \mathbf{A}\mathbf{X} + \mathbf{F}(t)$, where $\mathbf{A}$ is an $n \times n$ matrix of constants, is

$$\mathbf{X} = \mathbf{X}_c + \mathbf{X}_p = e^{\mathbf{A}t}\mathbf{C} + e^{\mathbf{A}t}\int_{t_0}^{t} e^{-\mathbf{A}s}\mathbf{F}(s)\,ds. \tag{5}$$

Since the matrix exponential $e^{\mathbf{A}t}$ is a fundamental matrix, it is always nonsingular and $e^{-\mathbf{A}s} = (e^{\mathbf{A}s})^{-1}$. In practice, $e^{-\mathbf{A}s}$ can be obtained from $e^{\mathbf{A}t}$ by simply replacing t by $-s$.

Computation of $e^{\mathbf{A}t}$ The definition of $e^{\mathbf{A}t}$ given in (3) can, of course, always be used to compute $e^{\mathbf{A}t}$. However, the practical utility of (3) is limited by the fact that the entries in $e^{\mathbf{A}t}$ are power series in t. With a natural desire to work with simple and familiar things, we then try to recognize whether these series define a closed-form function. See Problems 1–4 in Exercises 8.4. Fortunately there are many alternative ways of computing $e^{\mathbf{A}t}$; the following discussion shows how the Laplace transform can be used.

Use of the Laplace Transform   We saw in (5) that $\mathbf{X} = e^{\mathbf{A}t}$ is a solution of $\mathbf{X}' = \mathbf{A}\mathbf{X}$. Indeed, since $e^{\mathbf{A}0} = \mathbf{I}$, $\mathbf{X} = e^{\mathbf{A}t}$ is a solution of the initial-value problem

$$\mathbf{X}' = \mathbf{A}\mathbf{X}, \quad \mathbf{X}(0) = \mathbf{I}. \tag{6}$$

If $\mathbf{x}(s) = \mathcal{L}\{\mathbf{X}(t)\} = \mathcal{L}\{e^{\mathbf{A}t}\}$, then the Laplace transform of (6) is

$$s\mathbf{x}(s) - \mathbf{X}(0) = \mathbf{A}\mathbf{x}(s) \quad \text{or} \quad (s\mathbf{I} - \mathbf{A})\mathbf{x}(s) = \mathbf{I}.$$

Multiplying the last equation by $(s\mathbf{I} - \mathbf{A})^{-1}$ implies $\mathbf{x}(s) = (s\mathbf{I} - \mathbf{A})^{-1}\mathbf{I} = (s\mathbf{I} - \mathbf{A})^{-1}$. In other words,

$$\mathcal{L}\{e^{\mathbf{A}t}\} = (s\mathbf{I} - \mathbf{A})^{-1} \quad \text{or} \quad e^{\mathbf{A}t} = \mathcal{L}^{-1}\{(s\mathbf{I} - \mathbf{A})^{-1}\}. \tag{7}$$

EXAMPLE 1   Matrix Exponential

Use the Laplace transform to compute $e^{\mathbf{A}t}$ for $\mathbf{A} = \begin{pmatrix} 1 & -1 \\ 2 & -2 \end{pmatrix}$.

Solution   First we compute the matrix $s\mathbf{I} - \mathbf{A}$ and find its inverse:

$$s\mathbf{I} - \mathbf{A} = \begin{pmatrix} s - 1 & 1 \\ -2 & s + 2 \end{pmatrix}$$

$$(s\mathbf{I} - \mathbf{A})^{-1} = \begin{pmatrix} s - 1 & 1 \\ -2 & s + 2 \end{pmatrix}^{-1} = \begin{pmatrix} \dfrac{s + 2}{s(s + 1)} & \dfrac{-1}{s(s + 1)} \\ \dfrac{2}{s(s + 1)} & \dfrac{s - 1}{s(s + 1)} \end{pmatrix}.$$

Then we decompose the entries of the last matrix into partial fractions:

$$(s\mathbf{I} - \mathbf{A})^{-1} = \begin{pmatrix} \dfrac{2}{s} - \dfrac{1}{s + 1} & -\dfrac{1}{s} + \dfrac{1}{s + 1} \\ \dfrac{2}{s} - \dfrac{2}{s + 1} & -\dfrac{1}{s} + \dfrac{2}{s + 1} \end{pmatrix}. \tag{8}$$

It follows from (7) that the inverse Laplace transform of (8) gives the desired result,

$$e^{\mathbf{A}t} = \begin{pmatrix} 2 - e^{-t} & -1 + e^{-t} \\ 2 - 2e^{-t} & -1 + 2e^{-t} \end{pmatrix}. \qquad \blacksquare$$

Use of Computers   For those willing to momentarily trade understanding for speed of solution, $e^{\mathbf{A}t}$ can be computed in a mechanical manner with the aid of computer software. For example, to compute the matrix exponential for a square matrix $\mathbf{A}t$, use the function **MatrixExp[A t]** in *Mathematica,* the command **exponential(A, t)** in *Maple,* or the function expm(At) in MATLAB. See Problems 27 and 28 in Exercises 8.4.

EXERCISES 8.4

Answers to odd-numbered problems begin on page AN-11.

In Problems 1 and 2 use (3) to compute $e^{\mathbf{A}t}$ and $e^{-\mathbf{A}t}$.

1. $\mathbf{A} = \begin{pmatrix} 1 & 0 \\ 0 & 2 \end{pmatrix}$ **2.** $\mathbf{A} = \begin{pmatrix} 0 & 1 \\ 1 & 0 \end{pmatrix}$

In Problems 3 and 4 use (3) to compute $e^{\mathbf{A}t}$.

3. $\mathbf{A} = \begin{pmatrix} 1 & 1 & 1 \\ 1 & 1 & 1 \\ -2 & -2 & -2 \end{pmatrix}$ **4.** $\mathbf{A} = \begin{pmatrix} 0 & 0 & 0 \\ 3 & 0 & 0 \\ 5 & 1 & 0 \end{pmatrix}$

In Problems 5–8 use (1) to find the general solution of the given system.

5. $\mathbf{X}' = \begin{pmatrix} 1 & 0 \\ 0 & 2 \end{pmatrix} \mathbf{X}$ **6.** $\mathbf{X}' = \begin{pmatrix} 0 & 1 \\ 1 & 0 \end{pmatrix} \mathbf{X}$

7. $\mathbf{X}' = \begin{pmatrix} 1 & 1 & 1 \\ 1 & 1 & 1 \\ -2 & -2 & -2 \end{pmatrix} \mathbf{X}$ **8.** $\mathbf{X}' = \begin{pmatrix} 0 & 0 & 0 \\ 3 & 0 & 0 \\ 5 & 1 & 0 \end{pmatrix} \mathbf{X}$

In Problems 9–12 use (5) to find the general solution of the given system.

9. $\mathbf{X}' = \begin{pmatrix} 1 & 0 \\ 0 & 2 \end{pmatrix} \mathbf{X} + \begin{pmatrix} 3 \\ -1 \end{pmatrix}$ **10.** $\mathbf{X}' = \begin{pmatrix} 1 & 0 \\ 0 & 2 \end{pmatrix} \mathbf{X} + \begin{pmatrix} t \\ e^{4t} \end{pmatrix}$

11. $\mathbf{X}' = \begin{pmatrix} 0 & 1 \\ 1 & 0 \end{pmatrix} \mathbf{X} + \begin{pmatrix} 1 \\ 1 \end{pmatrix}$ **12.** $\mathbf{X}' = \begin{pmatrix} 0 & 1 \\ 1 & 0 \end{pmatrix} \mathbf{X} + \begin{pmatrix} \cosh t \\ \sinh t \end{pmatrix}$

13. Solve the system in Problem 7 subject to the initial condition
$$\mathbf{X}(0) = \begin{pmatrix} 1 \\ -4 \\ 6 \end{pmatrix}.$$

14. Solve the system in Problem 9 subject to the initial condition
$$\mathbf{X}(0) = \begin{pmatrix} 4 \\ 3 \end{pmatrix}.$$

In Problems 15–18 use the method of Example 1 to compute $e^{\mathbf{A}t}$ for the coefficient matrix. Use (1) to find the general solution of the given system.

15. $\mathbf{X}' = \begin{pmatrix} 4 & 3 \\ -4 & -4 \end{pmatrix} \mathbf{X}$ **16.** $\mathbf{X}' = \begin{pmatrix} 4 & -2 \\ 1 & 1 \end{pmatrix} \mathbf{X}$

17. $\mathbf{X}' = \begin{pmatrix} 5 & -9 \\ 1 & -1 \end{pmatrix} \mathbf{X}$ **18.** $\mathbf{X}' = \begin{pmatrix} 0 & 1 \\ -2 & -2 \end{pmatrix} \mathbf{X}$

Let $\mathbf{P}$ denote a matrix whose columns are eigenvectors $\mathbf{K}_1, \mathbf{K}_2, \ldots, \mathbf{K}_n$ corresponding to distinct eigenvalues $\lambda_1, \lambda_2, \ldots, \lambda_n$ of an $n \times n$ matrix $\mathbf{A}$. Then it can be shown that $\mathbf{A} = \mathbf{PDP}^{-1}$, where $\mathbf{D}$ is defined by

$$\mathbf{D} = \begin{pmatrix} \lambda_1 & 0 & \cdots & 0 \\ 0 & \lambda_2 & \cdots & 0 \\ \vdots & & & \vdots \\ 0 & 0 & \cdots & \lambda_n \end{pmatrix}. \tag{9}$$

In Problems 19 and 20 verify the above result for the given matrix.

19. $\mathbf{A} = \begin{pmatrix} 2 & 1 \\ -3 & 6 \end{pmatrix}$ **20.** $\mathbf{A} = \begin{pmatrix} 2 & 1 \\ 1 & 2 \end{pmatrix}$

21. Suppose $\mathbf{A} = \mathbf{PDP}^{-1}$, where $\mathbf{D}$ is defined as in (9). Use (3) to show that $e^{\mathbf{A}t} = \mathbf{P}e^{\mathbf{D}t}\mathbf{P}^{-1}$.

22. Use (3) to show that

$$e^{\mathbf{D}t} = \begin{pmatrix} e^{\lambda_1 t} & 0 & \cdots & 0 \\ 0 & e^{\lambda_2 t} & \cdots & 0 \\ \vdots & & & \vdots \\ 0 & 0 & \cdots & e^{\lambda_n t} \end{pmatrix},$$

where $\mathbf{D}$ is defined as in (9).

In Problems 23 and 24 use the results of Problems 19–22 to solve the given system.

23. $\mathbf{X}' = \begin{pmatrix} 2 & 1 \\ -3 & 6 \end{pmatrix} \mathbf{X}$ **24.** $\mathbf{X}' = \begin{pmatrix} 2 & 1 \\ 1 & 2 \end{pmatrix} \mathbf{X}$

Discussion Problems

25. Reread the discussion leading to the result given in (7). Does the matrix $s\mathbf{I} - \mathbf{A}$ always have an inverse? Discuss.
[*Hint*: If $\det(s\mathbf{I} - \mathbf{A}) = 0$, then what is s? See (9) of Appendix II.]

26. A matrix $\mathbf{A}$ is said to be **nilpotent** if there exists some integer m such that $\mathbf{A}^m = \mathbf{0}$. Verify that $\mathbf{A} = \begin{pmatrix} -1 & 1 & 1 \\ -1 & 0 & 1 \\ -1 & 1 & 1 \end{pmatrix}$ is nilpotent. Discuss why it is relatively easy to compute $e^{\mathbf{A}t}$ when $\mathbf{A}$ is nilpotent. Compute $e^{\mathbf{A}t}$ and then use (1) to solve the system $\mathbf{X}' = \mathbf{AX}$.

Computer Lab Assignments

27. (a) Use (1) to find the general solution of $\mathbf{X}' = \begin{pmatrix} 4 & 2 \\ 3 & 3 \end{pmatrix} \mathbf{X}$. Use a CAS to find $e^{\mathbf{A}t}$. Then use the computer to find eigenvalues and eigenvectors of the coefficient matrix $\mathbf{A} = \begin{pmatrix} 4 & 2 \\ 3 & 3 \end{pmatrix}$ and form the general solution in the manner of Section 8.2. Finally, reconcile the two forms of the general solution of the system.

(b) Use (1) to find the general solution of $\mathbf{X}' = \begin{pmatrix} -3 & -1 \\ 2 & -1 \end{pmatrix} \mathbf{X}$. Use a CAS to find $e^{\mathbf{A}t}$. In the case of complex output, utilize the software to do the simplification; for example, in *Mathematica*, if **m = MatrixExp[A t]** has complex entries, then try the command **Simplify[ComplexExpand[m]]**.

28. Use (1) to find the general solution of $\mathbf{X}' = \begin{pmatrix} -4 & 0 & 6 & 0 \\ 0 & -5 & 0 & -4 \\ -1 & 0 & 1 & 0 \\ 0 & 3 & 0 & 2 \end{pmatrix} \mathbf{X}$.

Use MATLAB or a CAS to find $e^{\mathbf{A}t}$.

CHAPTER 8 IN REVIEW

Answers to odd-numbered problems begin on page AN-12.

In Problems 1 and 2 fill in the blanks.

1. The vector $\mathbf{X} = k\begin{pmatrix} 4 \\ 5 \end{pmatrix}$ is a solution of $\mathbf{X}' = \begin{pmatrix} 1 & 4 \\ 2 & -1 \end{pmatrix} \mathbf{X} - \begin{pmatrix} 8 \\ 1 \end{pmatrix}$ for $k =$ _____.

2. The vector $\mathbf{X} = c_1\begin{pmatrix} -1 \\ 1 \end{pmatrix} e^{-9t} + c_2\begin{pmatrix} 5 \\ 3 \end{pmatrix} e^{7t}$ is solution of the initial-value problem $\mathbf{X}' = \begin{pmatrix} 1 & 10 \\ 6 & -3 \end{pmatrix} \mathbf{X}, \quad \mathbf{X}(0) = \begin{pmatrix} 2 \\ 0 \end{pmatrix}$ for $c_1 =$ _____ and $c_2 =$ _____.

3. Consider the linear system $\mathbf{X}' = \begin{pmatrix} 4 & 6 & 6 \\ 1 & 3 & 2 \\ -1 & -4 & -3 \end{pmatrix} \mathbf{X}$. Without attempting to solve the system, determine which one of the vectors

$$\mathbf{K}_1 = \begin{pmatrix} 0 \\ 1 \\ 1 \end{pmatrix}, \quad \mathbf{K}_2 = \begin{pmatrix} 1 \\ 1 \\ -1 \end{pmatrix}, \quad \mathbf{K}_3 = \begin{pmatrix} 3 \\ 1 \\ -1 \end{pmatrix}, \quad \text{and} \quad \mathbf{K}_4 = \begin{pmatrix} 6 \\ 2 \\ -5 \end{pmatrix}$$

is an eigenvector of the coefficient matrix. What is the solution of the system corresponding to this eigenvector?

4. Consider the linear system $\mathbf{X}' = \mathbf{AX}$ of two differential equations, where $\mathbf{A}$ is a real coefficient matrix. What is the general solution of the system if it is known that $\lambda_1 = 1 + 2i$ is an eigenvalue and $\mathbf{K}_1 = \begin{pmatrix} 1 \\ i \end{pmatrix}$ is a corresponding eigenvector?

In Problems 5–14 solve the given linear system.

5. $\dfrac{dx}{dt} = 2x + y$

$\dfrac{dy}{dt} = -x$

6. $\dfrac{dx}{dt} = -4x + 2y$

$\dfrac{dy}{dt} = 2x - 4y$

7. $\mathbf{X}' = \begin{pmatrix} 1 & 2 \\ -2 & 1 \end{pmatrix} \mathbf{X}$

8. $\mathbf{X}' = \begin{pmatrix} -2 & 5 \\ -2 & 4 \end{pmatrix} \mathbf{X}$

9. $\mathbf{X}' = \begin{pmatrix} 1 & -1 & 1 \\ 0 & 1 & 3 \\ 4 & 3 & 1 \end{pmatrix} \mathbf{X}$

10. $\mathbf{X}' = \begin{pmatrix} 0 & 2 & 1 \\ 1 & 1 & -2 \\ 2 & 2 & -1 \end{pmatrix} \mathbf{X}$

11. $\mathbf{X}' = \begin{pmatrix} 2 & 8 \\ 0 & 4 \end{pmatrix} \mathbf{X} + \begin{pmatrix} 2 \\ 16t \end{pmatrix}$

12. $\mathbf{X}' = \begin{pmatrix} 1 & 2 \\ -\frac{1}{2} & 1 \end{pmatrix} \mathbf{X} + \begin{pmatrix} 0 \\ e^t \tan t \end{pmatrix}$

13. $\mathbf{X}' = \begin{pmatrix} -1 & 1 \\ -2 & 1 \end{pmatrix} \mathbf{X} + \begin{pmatrix} 1 \\ \cot t \end{pmatrix}$

14. $\mathbf{X}' = \begin{pmatrix} 3 & 1 \\ -1 & 1 \end{pmatrix} \mathbf{X} + \begin{pmatrix} -2 \\ 1 \end{pmatrix} e^{2t}$

15. (a) Consider the linear system $\mathbf{X}' = \mathbf{A}\mathbf{X}$ of three first-order differential equations, where the coefficient matrix is

$$\mathbf{A} = \begin{pmatrix} 5 & 3 & 3 \\ 3 & 5 & 3 \\ -5 & -5 & -3 \end{pmatrix}$$

and $\lambda = 2$ is known to be an eigenvalue of multiplicity two. Find two different solutions of the system corresponding to this eigenvalue without using a special formula (such as (12) of Section 8.2).

(b) Use the procedure of part (a) to solve $\mathbf{X}' = \begin{pmatrix} 1 & 1 & 1 \\ 1 & 1 & 1 \\ 1 & 1 & 1 \end{pmatrix} \mathbf{X}$.

16. Verify that $\mathbf{X} = \begin{pmatrix} c_1 \\ c_2 \end{pmatrix} e^t$ is a solution of the linear system

$$\mathbf{X}' = \begin{pmatrix} 1 & 0 \\ 0 & 1 \end{pmatrix} \mathbf{X}$$

for arbitrary constants c_1 and c_2. By hand, draw a phase portrait of the system.

by
Gilbert N. Lewis

EARTHQUAKE SHAKING OF MULTISTORY BUILDINGS

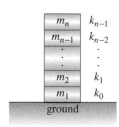

Building collapse in San Francisco, CA during Loma Prieta earthquake

Figure 1

$$k_i(x_{i+1} - x_i)$$

m_i

$$k_{i-1}(x_i - x_{i-1})$$

Figure 2

406

Earthquakes of large magnitude typically have a devastating effect on buildings. For example, the famous 1906 San Francisco earthquake destroyed much of that city. More recently, the same area was hit by the Loma Prieta earthquake, which many people in this country experienced secondhand while watching on television the major league baseball World Series game that was taking place in San Francisco in 1989.

Model for Floor Displacements of a Building   In this project, we attempt to model the effect of an earthquake on a multistory building and then solve and interpret the mathematics. We assume that the ith floor of a building has mass m_i and that successive floors are connected by an elastic connector whose effect resembles that of a spring. Typically, the structural elements in large buildings are made of steel, a highly elastic material. Each connector supplies a restoring force when the floors are displaced relative to each other. We assume that Hooke's law (page 216) holds, with proportionality constant k_i between the ith and the $(i + 1)$st floors. That is, the restoring force between those two floors is

$$F = k_i(x_{i+1} - x_i),$$

where x_i represents the horizontal displacement of the ith floor from equilibrium and $x_{i+1} - x_i$ is the displacement (shift) of the $(i + 1)$st floor relative to the ith floor. We also assume a similar reaction between the first floor and the ground, with proportionality constant k_0. Figure 1 shows a model of a building with n stories, while Figure 2 shows the forces acting on the ith floor. We can apply Newton's second law of motion, $F = ma$, to each section of the building to arrive at the following system of linear differential equations:

$$m_1x_1'' = -k_0x_1 + k_1(x_2 - x_1)$$
$$m_2x_2'' = -k_1(x_2 - x_1) + k_2(x_3 - x_2)$$
$$\vdots \qquad\qquad \vdots$$
$$m_nx_n'' = -k_{n-1}(x_n - x_{n-1}).$$

(1)

EXAMPLE 1 A Two-Story Building

As a simple example, consider a two-story building where each floor has mass $m = 5000$ kg and each restoring force constant has the value $k = 10,000$ kg/s². Then the system of differential equations (1) simplifies to

$$x_1'' = -4x_1 + 2x_2$$
$$x_2'' = \ \ \ 2x_1 - 2x_2. \tag{2}$$

This second-order system can be solved by either the method of Section 4.8 or the method of Section 8.2. It is left to the reader to show that an oscillatory solution of (2) is

$$x_1(t) = c_1 \cos \omega_1 t + c_2 \sin \omega_1 t + c_3 \cos \omega_2 t + c_4 \sin \omega_2 t \tag{3}$$
$$x_2(t) = \tfrac{1}{2}(4 - \omega_1^2)c_1 \cos \omega_1 t + \tfrac{1}{2}(4 - \omega_1^2)c_2 \sin \omega_1 t + \tfrac{1}{2}(4 - \omega_2^2)c_3 \cos \omega_2 t + \tfrac{1}{2}(4 - \omega_2^2)c_2 \sin \omega_2 t.$$

You are asked to find the values of ω_1 and ω_2 in Problem 1 in the Related Exercises. ■

In terms of the matrices

$$\mathbf{X} = \begin{pmatrix} x_1(t) \\ x_2(t) \\ \vdots \\ x_n(t) \end{pmatrix}, \quad \mathbf{M} = \begin{pmatrix} m_1 & 0 & \cdots & 0 \\ 0 & m_2 & \cdots & 0 \\ \vdots & & & \vdots \\ 0 & 0 & \cdots & m_n \end{pmatrix}, \quad \mathbf{K} = \begin{pmatrix} -(k_0 + k_1) & k_1 & 0 & 0 & \cdots & 0 & 0 \\ k_1 & -(k_1 + k_2) & k_2 & 0 & \cdots & 0 & 0 \\ 0 & k_2 & -(k_2 + k_3) & k_3 & \cdots & 0 & 0 \\ \vdots & & & & & & \vdots \\ 0 & 0 & 0 & 0 & \cdots & k_{n-1} & -k_{n-1} \end{pmatrix},$$

the system given in (1) can be written as

$$\mathbf{MX''} = \mathbf{KX}. \tag{4}$$

The $n \times n$ matrices $\mathbf{M}$ and $\mathbf{K}$ are called the **mass matrix** and the **stiffness matrix** of the building, respectively. Note that the matrix $\mathbf{M}$ is a diagonal matrix with the mass of the ith floor of the building as the ith diagonal entry. Since the matrix $\mathbf{M}$ has the inverse*

$$\mathbf{M} = \begin{pmatrix} m_1^{-1} & 0 & \cdots & 0 \\ 0 & m_2^{-1} & & 0 \\ \vdots & & & \vdots \\ 0 & 0 & \cdots & m_n^{-1} \end{pmatrix},$$

we obtain the matrix equation of a homogeneous second-order system in normal form

$$\mathbf{X''} = \mathbf{AX}. \tag{5}$$

The coefficient matrix in (5) is $\mathbf{A} = \mathbf{M}^{-1}\mathbf{K}$.

The eigenvalues of $\mathbf{A}$ reveal the stability of the building during an earthquake. The eigenvalues of $\mathbf{A}$ are negative and distinct. The natural frequencies of the building are the square roots of the negatives of the eigenvalues. If λ_i is the ith eigenvalue of $\mathbf{A}$, then $\omega_i = \sqrt{-\lambda_i}$ is the ith frequency for $i = 1, 2, \ldots, n$. During an earthquake, a large horizontal

*See Problem 56 in Appendix II Exercises.

force is applied to the first floor. If this force is oscillatory in nature, say of the form $\mathbf{F}(t) = \mathbf{G} \cos \gamma t$, where $\mathbf{G}$ is a column matrix of constants, then large displacements may develop in the building, especially if the frequency γ of the forcing term $\mathbf{F}$ is close to one of the natural frequencies ω_i of the building. This is reminiscent of the resonance phenomenon studied in Section 5.1.3.

EXAMPLE 2   A Ten-Story Building

Suppose we have a ten-story building, where each floor has a mass 10,000 kg and each k_i has the value 5000 kg/s². Since the sizes of the matrices $\mathbf{M}$ and $\mathbf{K}$ are both 10×10, the matrix $\mathbf{A}$ is also 10×10. With the help of a CAS we find

$$\mathbf{A} = \mathbf{M}^{-1}\mathbf{K} = \begin{pmatrix} -1 & 0.5 & 0 & 0 & 0 & 0 & 0 & 0 & 0 & 0 \\ 0.5 & -1 & 0.5 & 0 & 0 & 0 & 0 & 0 & 0 & 0 \\ 0 & 0.5 & -1 & 0.5 & 0 & 0 & 0 & 0 & 0 & 0 \\ 0 & 0 & 0.5 & -1 & 0.5 & 0 & 0 & 0 & 0 & 0 \\ 0 & 0 & 0 & 0.5 & -1 & 0.5 & 0 & 0 & 0 & 0 \\ 0 & 0 & 0 & 0 & 0.5 & -1 & 0.5 & 0 & 0 & 0 \\ 0 & 0 & 0 & 0 & 0 & 0.5 & -1 & 0.5 & 0 & 0 \\ 0 & 0 & 0 & 0 & 0 & 0 & 0.5 & -1 & 0.5 & 0 \\ 0 & 0 & 0 & 0 & 0 & 0 & 0 & 0.5 & -1 & 0.5 \\ 0 & 0 & 0 & 0 & 0 & 0 & 0 & 0 & 0.5 & -0.5 \end{pmatrix}.$$

The eigenvalues λ_i of the matrix $\mathbf{A}$ were obtained with the aid of a CAS. The values of λ_i, along with the corresponding frequencies $\omega_i = \sqrt{-\lambda_i}$ and periods $T_i = 2\pi/\omega_i$ (in seconds), are summarized in Table 1.

TABLE 1

λ_i	−1.956	−1.826	−1.623	−1.365	−1.075	−0.777	−0.500	−0.267	−0.099	−0.011
ω_i	1.399	1.351	1.274	1.168	1.037	0.881	0.707	0.517	0.315	0.105
T_i	4.491	4.651	4.932	5.379	6.059	7.132	8.887	12.153	19.947	59.840

From the last row of numbers in the table we see that this building does not seem to be in any danger of developing resonance during a typical earthquake whose period is in the range of 2–3 seconds. See Problem 2 in the Related Exercises. ■

RELATED EXERCISES

1. **(a)** Use the elimination method of Section 4.8 to solve the system (2). Find the exact values of the frequencies ω_1 and ω_2.
   (b) As instructed, use one or both of the methods discussed in the Remarks in Section 8.2 (page 388) to solve the system (2). Use a CAS or linear algebra software to find eigenvalues and eigenvectors.

2. Consider the ten-story building in Example 2. Suppose the stiffness matrix **K** is multiplied by a factor of 10. Show, by constructing a table as in Example 2, that such a building is more likely to suffer damage from a typical earthquake whose period is in the range of 2–3 seconds.

3. Consider the tallest building on your campus. Assume reasonable values for the mass of each floor and for the proportionality constants between the floors. If you have trouble coming up with such values, use the ones in the preceding examples.

   (a) Find the matrices **M**, **K**, and **A**.

   (b) Find the eigenvalues of **A** and the frequencies and periods of oscillation. Is your building safe from a modest-sized period-2 earthquake?

   (c) Is your building safe from a modest-sized period-2 earthquake if the building is made stiffer by a factor of 10? See Problem 2. By what would you have to multiply the stiffness matrix **K** in order to put your building in the danger zone?

4. Solve the initial-value problem $\mathbf{MX}'' = \mathbf{KX} + \mathbf{F}(t)$, $\mathbf{X}(0) = 0$, $\mathbf{X}'(0) = 0$, where $\mathbf{F}(t) = \mathbf{G}\cos\gamma t$, $\mathbf{G} = E\mathbf{B}$, $E = 10{,}000$ lb is the amplitude of the earthquake force acting at ground level,

$$
\mathbf{B} = \begin{pmatrix} 1 \\ 0 \\ \vdots \\ 0 \end{pmatrix},
$$

$\gamma = 3$ is the frequency of the earthquake (a typical earthquake frequency), and **M** and **K** are the matrices from part (a) of Problem 3. See the Remarks in Sections 8.2 and 8.3 and Problems 23–26 in Exercises 8.3. Use a CAS or linear algebra software to carry out all matrix calculations.

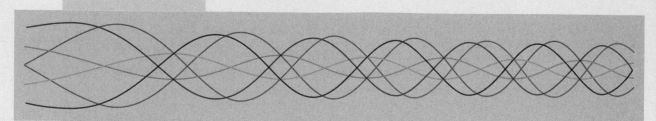

Numerical solution curves for Airy's DE; see pages 278 and 428.

9 NUMERICAL SOLUTIONS OF ORDINARY DIFFERENTIAL EQUATIONS

9.1 Euler Methods and Error Analysis
9.2 Runge-Kutta Methods
9.3 Multistep Methods
9.4 Higher-Order Equations and Systems
9.5 Second-Order Boundary-Value Problems

Chapter 9 in Review

INTRODUCTION A differential equation does not have to have a solution. But even if it can be shown that a solution exists, we may not be able to exhibit it in explicit or implicit form. In many instances we have to be content with an approximation to the solution. As discussed in Section 2.6, if a solution exists, it represents a set of points in the Cartesian plane. In this chapter we continue to explore the basic idea of Section 2.6—that is, utilizing the differential equation to construct algorithms to approximate the coordinates of the points on the actual solution curve.

Our concentration in this chapter is primarily on first-order initial-value problems $dy/dx = f(x, y)$, $y(x_0) = y_0$. We saw in Section 4.9 that the numerical procedures developed for first-order equations extend in a natural way to systems of first-order equations, and so we can approximate solutions of higher-order DEs by expressing the equation as a system of first-order DEs. This chapter concludes with methods for approximating solutions of linear second-order boundary-value problems.

9.1 EULER METHODS AND ERROR ANALYSIS

- *Euler's method* • *Some errors in numerical methods* • *Round-off*
error • *Local truncation error* • *Global truncation*
error • *Improved Euler's method* • *Predictor-corrector method*

In Section 2.6 we examined one of the simplest numerical methods for approximating solutions of first-order initial-value problems $y' = f(x, y)$, $y(x_0) = y_0$. Recall that the backbone of Euler's method is the formula

$$y_{n+1} = y_n + hf(x_n, y_n), \qquad (1)$$

where f is the function obtained from the differential equation $y' = f(x, y)$. The recursive use of (1) for $n = 0, 1, 2, \ldots$ yields the y-coordinates $y_1, y_2, y_3, \ldots$ of points on successive "tangent lines" to the solution curve at $x_1, x_2, x_3, \ldots$ or $x_n = x_0 + nh$, where h is a constant and is the size of the step between x_n and x_{n+1}. The values $y_1, y_2, y_3, \ldots$ approximate the values of a solution $y(x)$ of the IVP at $x_1, x_2, x_3, \ldots$. But whatever advantage (1) has from its simplicity is lost in the crudeness of its approximations.

A Comparison In Problem 4 in Exercises 2.6 you were asked to use Euler's method to obtain the approximate value of $y(1.5)$ for the solution of the initial-value problem $y' = 2xy$, $y(1) = 1$. You should have obtained the analytic solution $y = e^{x^2-1}$ and results similar to those given in Tables 9.1 and 9.2.

TABLE 9.1 Euler's Method with $h = 0.1$

x_n	y_n	True value	Abs. error	% Rel. error
1.00	1.0000	1.0000	0.0000	0.00
1.10	1.2000	1.2337	0.0337	2.73
1.20	1.4640	1.5527	0.0887	5.71
1.30	1.8154	1.9937	0.1784	8.95
1.40	2.2874	2.6117	0.3244	12.42
1.50	2.9278	3.4903	0.5625	16.12

TABLE 9.2 Euler's Method with $h = 0.05$

x_n	y_n	True value	Abs. error	% Rel. error
1.00	1.0000	1.0000	0.0000	0.00
1.05	1.1000	1.1079	0.0079	0.72
1.10	1.2155	1.2337	0.0182	1.47
1.15	1.3492	1.3806	0.0314	2.27
1.20	1.5044	1.5527	0.0483	3.11
1.25	1.6849	1.7551	0.0702	4.00
1.30	1.8955	1.9937	0.0982	4.93
1.35	2.1419	2.2762	0.1343	5.90
1.40	2.4311	2.6117	0.1806	6.92
1.45	2.7714	3.0117	0.2403	7.98
1.50	3.1733	3.4903	0.3171	9.08

In this case, with a step size $h = 0.1$, a 16% relative error in the calculation of the approximation to $y(1.5)$ is totally unacceptable. At the expense of doubling the number of calculations, some improvement in accuracy is obtained by halving the step size to $h = 0.05$.

Errors in Numerical Methods In choosing and using a numerical method for the solution of an initial-value problem, we must be aware of the various sources of errors. For some kinds of computation the accumulation of errors might reduce the accuracy of an approximation to the point of making the computation useless. On the other hand, depending on the use to which a numerical solution may be put, extreme accuracy may not be worth the added expense and complication.

One source of error always present in calculations is **round-off error**. This error results from the fact that any calculator or computer can represent numbers using only a finite number of digits. Suppose, for the sake of illustration, that we have a calculator that uses base 10 arithmetic and carries four digits, so that $\frac{1}{3}$ is represented in the calculator as 0.3333 and $\frac{1}{9}$ is represented as 0.1111. If we use this calculator to compute $(x^2 - \frac{1}{9})/(x - \frac{1}{3})$ for $x = 0.3334$, we obtain

$$\frac{(0.3334)^2 - 0.1111}{0.3334 - 0.3333} = \frac{0.1112 - 0.1111}{0.3334 - 0.3333} = 1.$$

With the help of a little algebra, however, we see that

$$\frac{x^2 - 1/9}{x - 1/3} = \frac{(x - 1/3)(x + 1/3)}{x - 1/3} = x + \frac{1}{3},$$

so when $x = 0.3334$, $(x^2 - \frac{1}{9})/(x - \frac{1}{3}) \approx 0.3334 + 0.3333 = 0.6667$. This example shows that the effects of round-off error can be quite serious unless some care is taken. One way to reduce the effect of round-off error is to minimize the number of calculations. Another technique on a computer is to use double-precision arithmetic to check the results. In general, round-off error is unpredictable and difficult to analyze, and we will neglect it in the error analysis that follows. We will concentrate on investigating the error introduced by using a formula or algorithm to approximate the values of the solution.

Truncation Errors for Euler's Method In the sequence of values y_1, y_2, y_3, ... generated from (1), usually the value of y_1 will not agree with the actual solution evaluated at x_1—namely, $y(x_1)$—because the algorithm gives only a straight-line approximation to the solution. See Figure 2.30. The error is called the **local truncation error**, **formula error**, or **discretization error**. It occurs at each step; that is, if we assume that y_n is accurate, then y_{n+1} will contain local truncation error.

To derive a formula for the local truncation error for Euler's method we use Taylor's formula with remainder. If a function $y(x)$ possesses $k + 1$ derivatives that are continuous on an open interval containing a and x, then

$$y(x) = y(a) + y'(a)\frac{x - a}{1!} + \cdots + y^{(k)}(a)\frac{(x - a)^k}{k!} + y^{(k+1)}(c)\frac{(x - a)^{k+1}}{(k + 1)!},$$

where c is some point between a and x. Setting $k = 1$, $a = x_n$, and $x = x_{n+1} = x_n + h$, we get

$$y(x_{n+1}) = y(x_n) + y'(x_n)\frac{h}{1!} + y''(c)\frac{h^2}{2!}$$

or

$$y(x_{n+1}) = \underbrace{y_n + hf(x_n, y_n)}_{y_{n+1}} + y''(c)\frac{h^2}{2!}.$$

Euler's method (1) is the last formula without the last term; hence the local truncation error in y_{n+1} is

$$y''(c)\frac{h^2}{2!}, \quad \text{where} \quad x_n < c < x_{n+1}.$$

The value of c is usually unknown (it exists theoretically) and so the *exact* error cannot be calculated, but an upper bound on the absolute value of the error is

$$M\frac{h^2}{2!}, \quad \text{where} \quad M = \max_{x_n < x < x_{n+1}} |y''(x)|.$$

In discussing errors arising from the use of numerical methods it is helpful to use the notation $O(h^n)$. To define this concept we let $e(h)$ denote the error in a numerical calculation depending on h. Then $e(h)$ is said to be of order h^n, denoted by $O(h^n)$, if there exist a constant C and a positive integer n such that $|e(h)| \leq Ch^n$ for h sufficiently small. Thus the local truncation error for Euler's method is $O(h^2)$. We note that, in general, if $e(h)$ in a numerical method is of order h^n and h is halved, the new error is approximately $C(h/2)^n = Ch^n/2^n$; that is, the error is reduced by a factor of $1/2^n$.

EXAMPLE 1   **Bound for Local Truncation Errors**

Find a bound for the local truncation errors for Euler's method applied to $y' = 2xy$, $y(1) = 1$.

Solution   From the solution $y = e^{x^2-1}$ we get $y'' = (2 + 4x^2)e^{x^2-1}$, and so the local truncation error is

$$y''(c)\frac{h^2}{2} = (2 + 4c^2)e^{(c^2-1)}\frac{h^2}{2},$$

where c is between x_n and $x_n + h$. In particular, for $h = 0.1$ we can get an upper bound on the local truncation error for y_1 by replacing c by 1.1:

$$[2 + (4)(1.1)^2]e^{((1.1)^2-1)}\frac{(0.1)^2}{2} = 0.0422.$$

From Table 9.1 we see that the error after the first step is 0.0337, less than the value given by the bound.

Similarly, we can get a bound for the local truncation error for any of the five steps given in Table 9.1 by replacing c by 1.5 (this value of c gives the largest value of $y''(c)$ for any of the steps and may be too generous for the first few steps). Doing this gives

$$[2 + (4)(1.5)^2]e^{((1.5)^2-1)}\frac{(0.1)^2}{2} = 0.1920 \tag{2}$$

as an upper bound for the local truncation error in each step.   ∎

Note that if h is halved to 0.05 in Example 1, then the error bound is 0.0480, about one-fourth as much as shown in (2). This is expected because the local truncation error for Euler's method is $O(h^2)$.

In the above analysis we assumed that the value of y_n was exact in the calculation of y_{n+1}, but it is not because it contains local truncation errors from previous steps. The total error in y_{n+1} is an accumulation of the errors in each of the previous steps. This total error is called the **global truncation error**. A complete analysis of the global truncation error is beyond the scope of this text, but it can be shown that the global truncation error for Euler's method is $O(h)$.

We expect that, for Euler's method, if the step size is halved the error will be approximately halved as well. This is borne out in Tables 9.1 and 9.2 where the absolute error at $x = 1.50$ with $h = 0.1$ is 0.5625 and with $h = 0.05$ is 0.3171, approximately half as large.

In general it can be shown that if a method for the numerical solution of a differential equation has local truncation error $O(h^{\alpha+1})$, then the global truncation error is $O(h^\alpha)$.

For the remainder of this section and in the subsequent sections we study methods that give significantly greater accuracy than does Euler's method.

Improved Euler's Method The numerical method defined by the formula

$$y_{n+1} = y_n + h\frac{f(x_n, y_n) + f(x_{n+1}, y_{n+1}^*)}{2}, \tag{3}$$

where

$$y_{n+1}^* = y_n + hf(x_n, y_n), \tag{4}$$

is commonly known as the **improved Euler's method**. In order to compute y_{n+1} for $n = 0, 1, 2, \ldots$ from (3), we must, at each step, first use Euler's method (4) to obtain an initial estimate y_{n+1}^*. For example, with $n = 0$, (4) gives $y_1^* = y_0 + hf(x_0, y_0)$, and then knowing this value we use (3) to get $y_1 = y_0 + h\dfrac{f(x_0, y_0) + f(x_1, y_1^*)}{2}$, where $x_1 = x_0 + h$. These equations can be readily visualized. In Figure 9.1 observe that $m_0 = f(x_0, y_0)$ and $m_1 = f(x_1, y_1^*)$ are slopes of the solid straight lines shown passing through the points (x_0, y_0) and (x_1, y_1^*), respectively. By taking an average of these slopes, that is, $m_{\text{ave}} = \dfrac{f(x_0, y_0) + f(x_1, y_1^*)}{2}$, we obtain the slope of the parallel dashed skew lines. With the first step, rather than advancing along the line through (x_0, y_0) with slope $f(x_0, y_0)$ to the point with y-coordinate y_1^* obtained by Euler's method, we advance instead along the colored dashed line through (x_0, y_0) with slope m_{ave} until we reach x_1. It seems plausible from inspection of the figure that y_1 is an improvement over y_1^*.

In general, the improved Euler's method is an example of a **predictor-corrector method**. The value of y_{n+1}^* given by (4) predicts a value of $y(x_n)$, whereas the value of y_{n+1} defined by formula (3) corrects this estimate.

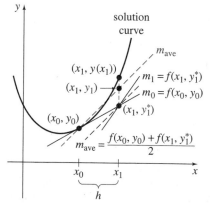

Figure 9.1

EXAMPLE 2 **Improved Euler's Method**

Use the improved Euler's method to obtain the approximate value of $y(1.5)$ for the solution of the initial-value problem $y' = 2xy$, $y(1) = 1$. Compare the results for $h = 0.1$ and $h = 0.05$.

Solution   With $x_0 = 1$, $y_0 = 1$, $f(x_n, y_n) = 2x_n y_n$, $n = 0$, and $h = 0.1$, we first compute (4):

$$y_1^* = y_0 + (0.1)(2x_0 y_0) = 1 + (0.1)2(1)(1) = 1.2.$$

We use this last value in (3) along with $x_1 = 1 + h = 1 + 0.1 = 1.1$:

$$y_1 = y_0 + (0.1)\frac{2x_0 y_0 + 2x_1 y_1^*}{2} = 1 + (0.1)\frac{2(1)(1) + 2(1.1)(1.2)}{2} = 1.232.$$

The comparative values of the calculations for $h = 0.1$ and $h = 0.05$ are given in Tables 9.3 and 9.4, respectively.

TABLE 9.3   Improved Euler's Method with $h = 0.1$

x_n	y_n	True value	Abs. error	% Rel. error
1.00	1.0000	1.0000	0.0000	0.00
1.10	1.2320	1.2337	0.0017	0.14
1.20	1.5479	1.5527	0.0048	0.31
1.30	1.9832	1.9937	0.0106	0.53
1.40	2.5908	2.6117	0.0209	0.80
1.50	3.4509	3.4904	0.0394	1.13

TABLE 9.4   Improved Euler's Method with $h = 0.05$

x_n	y_n	True value	Abs. error	% Rel. error
1.00	1.0000	1.0000	0.0000	0.00
1.05	1.1077	1.1079	0.0002	0.02
1.10	1.2332	1.2337	0.0004	0.04
1.15	1.3798	1.3806	0.0008	0.06
1.20	1.5514	1.5527	0.0013	0.08
1.25	1.7531	1.7551	0.0020	0.11
1.30	1.9909	1.9937	0.0029	0.14
1.35	2.2721	2.2762	0.0041	0.18
1.40	2.6060	2.6117	0.0057	0.22
1.45	3.0038	3.0117	0.0079	0.26
1.50	3.4795	3.4904	0.0108	0.31

A brief word of caution is in order here. We cannot compute all the values of y_n^* first and then substitute these values into formula (3). In other words, we cannot use the data in Table 9.1 to help construct the values in Table 9.3. Why not?

Truncation Errors for the Improved Euler's Method   The local truncation error for the improved Euler's method is $O(h^3)$. The derivation of this result is similar to the derivation of the local truncation error for Euler's method. Since the local truncation error for the improved Euler's method is $O(h^3)$, the global truncation error is $O(h^2)$. This can be seen in Example 2; when the step size is halved from $h = 0.1$ to $h = 0.05$, the absolute error at $x = 1.50$ is reduced from 0.0394 to 0.0108, a reduction of approximately $(\frac{1}{2})^2 = \frac{1}{4}$.

EXERCISES 9.1

Answers to odd-numbered problems begin on page AN-12.

Given the initial-value problems in Problems 1–10, use the improved Euler's method to obtain a four-decimal approximation to the indicated value. First use $h = 0.1$ and then use $h = 0.05$.

1. $y' = 2x - 3y + 1$, $y(1) = 5$; $y(1.5)$

2. $y' = 4x - 2y$, $y(0) = 2$; $y(0.5)$

3. $y' = 1 + y^2$, $y(0) = 0$; $y(0.5)$ **4.** $y' = x^2 + y^2$, $y(0) = 1$; $y(0.5)$

5. $y' = e^{-y}$, $y(0) = 0$; $y(0.5)$ **6.** $y' = x + y^2$, $y(0) = 0$; $y(0.5)$

7. $y' = (x - y)^2$, $y(0) = 0.5$; $y(0.5)$

8. $y' = xy + \sqrt{y}$, $y(0) = 1$; $y(0.5)$

9. $y' = xy^2 - \dfrac{y}{x}$, $y(1) = 1$; $y(1.5)$ **10.** $y' = y - y^2$, $y(0) = 0.5$; $y(0.5)$

11. Consider the initial-value problem $y' = (x + y - 1)^2$, $y(0) = 2$. Use the improved Euler's method with $h = 0.1$ and $h = 0.05$ to obtain approximate values of the solution at $x = 0.5$. At each step compare the approximate value with the exact value of the analytic solution.

12. Although it may not be obvious from the differential equation, its solution could "behave badly" near a point x at which we wish to approximate $y(x)$. Numerical procedures may give widely differing results near this point. Let $y(x)$ be the solution of the initial-value problem $y' = x^2 + y^3$, $y(1) = 1$.

(a) Use a numerical solver to obtain the graph of the solution on the interval $[1, 1.4]$.

(b) Using the step size $h = 0.1$, compare the results obtained from Euler's method with the results from the improved Euler's method in the approximation of $y(1.4)$.

13. Consider the initial-value problem $y' = 2y$, $y(0) = 1$. The analytic solution is $y = e^{2x}$.

(a) Approximate $y(0.1)$ using one step and Euler's method.

(b) Find a bound for the local truncation error in y_1.

(c) Compare the actual error in y_1 with your error bound.

(d) Approximate $y(0.1)$ using two steps and Euler's method.

(e) Verify that the global truncation error for Euler's method is $O(h)$ by comparing the errors in parts (a) and (d).

14. Repeat Problem 13 using the improved Euler's method. Its global truncation error is $O(h^2)$.

15. Repeat Problem 13 using the initial-value problem $y' = x - 2y$, $y(0) = 1$. The analytic solution is $y = \frac{1}{2}x - \frac{1}{4} + \frac{5}{4}e^{-2x}$.

16. Repeat Problem 15 using the improved Euler's method. Its global truncation error is $O(h^2)$.

17. Consider the initial-value problem $y' = 2x - 3y + 1$, $y(1) = 5$. The analytic solution is $y(x) = \frac{1}{9} + \frac{2}{3}x + \frac{38}{9}e^{-3(x-1)}$.

(a) Find a formula involving c and h for the local truncation error in the nth step if Euler's method is used.

(b) Find a bound for the local truncation error in each step if $h = 0.1$ is used to approximate $y(1.5)$.

(c) Approximate $y(1.5)$ using $h = 0.1$ and $h = 0.05$ with Euler's method. See Problem 1 in Exercises 2.6.

(d) Calculate the errors in part (c) and verify that the global truncation error of Euler's method is $O(h)$.

18. Repeat Problem 17 using the improved Euler's method, which has a global truncation error $O(h^2)$. See Problem 1. You may need to keep more than four decimal places to see the effect of reducing the order of the error.

19. Repeat Problem 17 for the initial-value problem $y' = e^{-y}$, $y(0) = 0$. The analytic solution is $y(x) = \ln(x + 1)$. Approximate $y(0.5)$. See Problem 5 in Exercises 2.6.

20. Repeat Problem 19 using the improved Euler's method, which has a global truncation error $O(h^2)$. See Problem 5. You may need to keep more than four decimal places to see the effect of reducing the order of the error.

Discussion Problems

21. Answer the question "Why not?" that follows the three sentences after Example 2 on page 415.

9.2 RUNGE-KUTTA METHODS

• *First-order Runge-Kutta method* • *Second-order Runge-Kutta method* • *Fourth-order Runge-Kutta method* • *Truncation errors* • *Adaptive methods*

Probably one of the more popular as well as most accurate numerical procedures used in obtaining approximate solutions to the initial-value problem $y' = f(x, y)$, $y(x_0) = y_0$ is the **fourth-order Runge-Kutta method.** As the name suggests, there are Runge-Kutta methods of different orders. These methods are derived using the Taylor series expansion with remainder of function $y(x)$.

First-Order Runge-Kutta Method   We saw in the last section that if a function $y(x)$ possesses $k + 1$ derivatives that are continuous on an open interval containing a and x, then we can write

$$y(x) = y(a) + y'(a)\frac{x - a}{1!} + y''(a)\frac{(x - a)^2}{2!} + \cdots + y^{(k+1)}(c)\frac{(x - a)^{k+1}}{(k + 1)!},$$

where c is some number between a and x. If we replace a by x_n and x by $x_{n+1} = x_n + h$, then the foregoing formula becomes

$$y(x_{n+1}) = y(x_n + h) = y(x_n) + hy'(x_n) + \frac{h^2}{2!}y^{(k)}(x_n) + \cdots + \frac{h^{k+1}}{(k + 1)!}y^{(k+1)}(c),$$

where c is now some number between x_n and x_{n+1}. In the case when $k = 1$ and the remainder $\frac{h^2}{2}y''(c)$ is small, we obtain the familiar formula

$$y_{n+1} = y_n + hy_n' = y_n + hf(x_n, y_n).$$

In other words, the basic Euler's method is a **first-order Runge-Kutta procedure**.

Second-Order Runge-Kutta Method   We consider now the **second-order Runge-Kutta procedure**. This consists of finding constants a, b, α, and β so that the formula

$$y_{n+1} = y_n + ak_1 + bk_2, \tag{1}$$

where
$$k_1 = hf(x_n, y_n)$$
$$k_2 = hf(x_n + \alpha h, y_n + \beta k_1),$$

agrees with a Taylor polynomial of degree 2. It can be shown that this can be done whenever the constants satisfy

$$a + b = 1, \quad b\alpha = \frac{1}{2}, \quad \text{and} \quad b\beta = \frac{1}{2}. \tag{2}$$

This is a system of three equations in four unknowns and has infinitely many solutions. Observe that when $a = b = \frac{1}{2}$ and $\alpha = \beta = 1$, (1) reduces to the improved Euler's formula. Since the formula agrees with a Taylor polynomial of degree 2, the local truncation error for this method is $O(h^3)$ and the global truncation error is $O(h^2)$.

Notice that the sum $ak_1 + bk_2$ in (1) is a weighted average of k_1 and k_2 since $a + b = 1$. The numbers k_1 and k_2 are multiples of approximations to the slope of the solution curve $y(x)$ at two different points in the interval from x_n to x_{n+1}.

Fourth-Order Runge-Kutta Method   The **fourth-order Runge-Kutta procedure** consists of finding appropriate constants so that the formula

$$y_{n+1} = y_n + ak_1 + bk_2 + ck_3 + dk_4,$$
where
$$k_1 = hf(x_n, y_n)$$
$$k_2 = hf(x_n + \alpha_1 h, y_n + \beta_1 k_1)$$
$$k_3 = hf(x_n + \alpha_2 h, y_n + \beta_2 k_1 + \beta_3 k_2)$$
$$k_4 = hf(x_n + \alpha_3 h, y_n + \beta_4 k_1 + \beta_5 k_2 + \beta_6 k_3),$$

agrees with a Taylor polynomial of degree 4. This results in 11 equations in 13 unknowns. The most commonly used set of values for the constants yields the following result:

$$y_{n+1} = y_n + \frac{1}{6}(k_1 + 2k_2 + 2k_3 + k_4),$$

$$k_1 = hf(x_n, y_n)$$
$$k_2 = hf(x_n + \tfrac{1}{2}h, y_n + \tfrac{1}{2}k_1)$$
$$k_3 = hf(x_n + \tfrac{1}{2}h, y_n + \tfrac{1}{2}k_2) \tag{3}$$
$$k_4 = hf(x_n + h, y_n + k_3).$$

You are advised to look carefully at the formulas in (3); note that k_2 depends on k_1, k_3 depends on k_2, and k_4 depends on k_3. Also k_2 and k_3 involve approximations to the slope at the midpoint of the interval between x_n and x_{n+1}.

EXAMPLE 1   **Runge-Kutta Method**

Use the Runge-Kutta method with $h = 0.1$ to obtain an approximation to $y(1.5)$ for the solution of $y' = 2xy, y(1) = 1$.

Solution   For the sake of illustration let us compute the case when $n = 0$. From (3) we find

$$k_1 = (0.1)f(x_0, y_0) = (0.1)(2x_0 y_0) = 0.2$$

$$k_2 = (0.1)f(x_0 + \tfrac{1}{2}(0.1), y_0 + \tfrac{1}{2}(0.2))$$
$$= (0.1)2(x_0 + \tfrac{1}{2}(0.1))(y_0 + \tfrac{1}{2}(0.2)) = 0.231$$
$$k_3 = (0.1)f(x_0 + \tfrac{1}{2}(0.1), y_0 + \tfrac{1}{2}(0.231))$$
$$= (0.1)2(x_0 + \tfrac{1}{2}(0.1))(y_0 + \tfrac{1}{2}(0.231)) = 0.234255$$
$$k_4 = (0.1)f(x_0 + 0.1, y_0 + 0.234255)$$
$$= (0.1)2(x_0 + 0.1)(y_0 + 0.234255) = 0.2715361$$

and therefore

$$y_1 = y_0 + \frac{1}{6}(k_1 + 2k_2 + 2k_3 + k_4)$$

$$= 1 + \frac{1}{6}(0.2 + 2(0.231) + 2(0.234255) + 0.2715361) = 1.23367435.$$

The remaining calculations are summarized in Table 9.5, whose entries are rounded to four decimal places.

TABLE 9.5   Runge-Kutta Method with $h = 0.1$

x_n	y_n	True value	Abs. error	% Rel. error
1.00	1.0000	1.0000	0.0000	0.00
1.10	1.2337	1.2337	0.0000	0.00
1.20	1.5527	1.5527	0.0000	0.00
1.30	1.9937	1.9937	0.0000	0.00
1.40	2.6116	2.6117	0.0001	0.00
1.50	3.4902	3.4904	0.0001	0.00

Inspection of Table 9.5 shows why the fourth-order Runge-Kutta method is so popular. If four-decimal-place accuracy is all that we desire, there is no need to use a smaller step size. Table 9.6 compares the results of applying Euler's, the improved Euler's, and the fourth-order Runge-Kutta methods to the initial-value problem $y' = 2xy, y(1) = 1$. (See Tables 9.1 and 9.3.)

TABLE 9.6 $y' = 2xy, y(1) = 1$

	Comparison of numerical methods with $h = 0.1$				Comparison of numerical methods with $h = 0.05$			
x_n	Euler	Improved Euler	Runge-Kutta	True value	x_n	Euler	Improved Euler	Runge-Kutta	True value
1.00	1.0000	1.0000	1.0000	1.0000	1.00	1.0000	1.0000	1.0000	1.0000
1.10	1.2000	1.2320	1.2337	1.2337	1.05	1.1000	1.1077	1.1079	1.1079
1.20	1.4640	1.5479	1.5527	1.5527	1.10	1.2155	1.2332	1.2337	1.2337
1.30	1.8154	1.9832	1.9937	1.9937	1.15	1.3492	1.3798	1.3806	1.3806
1.40	2.2874	2.5908	2.6116	2.6117	1.20	1.5044	1.5514	1.5527	1.5527
1.50	2.9278	3.4509	3.4902	3.4904	1.25	1.6849	1.7531	1.7551	1.7551
					1.30	1.8955	1.9909	1.9937	1.9937
					1.35	2.1419	2.2721	2.2762	2.2762
					1.40	2.4311	2.6060	2.6117	2.6117
					1.45	2.7714	3.0038	3.0117	3.0117
					1.50	3.1733	3.4795	3.4903	3.4904

Truncation Errors for the Runge-Kutta Method   Since the first equation in (3) agrees with a Taylor polynomial of degree 4, the local truncation error for this method is

$$y^{(5)}(c)\frac{h^5}{5!} \quad \text{or} \quad O(h^5),$$

and the global truncation error is thus $O(h^4)$. It is now obvious why this is called the *fourth-order* Runge-Kutta method.

EXAMPLE 2   **Bound for Local/Global Truncation Errors**

Analyze the local and global truncation errors for the fourth-order Runge-Kutta method applied to $y' = 2xy$, $y(1) = 1$.

Solution   By differentiating the known solution $y(x) = e^{x^2-1}$ we get

$$y^{(5)}(c)\frac{h^5}{5!} = (120c + 160c^3 + 32c^5)e^{c^2-1}\frac{h^5}{5!}. \tag{4}$$

Thus with $c = 1.5$, (4) yields a bound of 0.00028 on the local truncation error for each of the five steps when $h = 0.1$. Note that in Table 9.5 the actual error in y_1 is much less than this bound.

Table 9.7 gives the approximations to the solution of the initial-value problem at $x = 1.5$ that are obtained from the fourth-order Runge-Kutta method. By computing the value of the exact solution at $x = 1.5$ we can find the error in these approximations. Because the method is so accurate, many decimal places must be used in the numerical solution to see the effect of halving the step size. Note that when h is halved, from $h = 0.1$ to $h = 0.05$, the error is divided by a factor of about $2^4 = 16$, as expected.

TABLE 9.7   Runge-Kutta Method

h	Approximation	Error
0.1	3.49021064	$1.323210889 \times 10^{-4}$
0.05	3.49033382	$9.137760898 \times 10^{-6}$

Adaptive Methods   We have seen that the accuracy of a numerical method can be improved by decreasing the step size h. Of course, this enhanced accuracy is usually obtained at a cost—namely, increased computation time and greater possibility of round-off error. In general, over the interval of approximation there may be subintervals where a relatively large step size suffices and other subintervals where a smaller step size is necessary in order to keep the truncation error within a desired limit. Numerical methods that use a variable step size are called **adaptive methods.** One of the more popular of these methods for approximating solutions of differential equations is called the **Runge-Kutta-Fehlberg algorithm.**

Answers to odd-numbered problems begin on page AN-13.

1. Use the fourth-order Runge-Kutta method with $h = 0.1$ to obtain a four-decimal-place approximation to the solution of the initial-value problem

$$y' = (x + y - 1)^2, \quad y(0) = 2$$

at $x = 0.5$. Compare the approximate values with the exact values obtained in Problem 11 in Exercises 9.1.

2. Solve the equations in (2) using the assumption $a = \frac{1}{4}$. Use the resulting second-order Runge-Kutta method to obtain a four-decimal-place approximation to the solution of the initial-value problem in Problem 1 at $x = 0.5$. Compare the approximate values with the approximate values obtained in Problem 11 in Exercises 9.1.

Given the initial-value problems in Problems 3–12, use the Runge-Kutta method with $h = 0.1$ to obtain a four-decimal-place approximation to the indicated value.

3. $y' = 2x - 3y + 1, y(1) = 5; \quad y(1.5)$

4. $y' = 4x - 2y, y(0) = 2; \quad y(0.5)$

5. $y' = 1 + y^2, y(0) = 0; \quad y(0.5)$

6. $y' = x^2 + y^2, y(0) = 1; \quad y(0.5)$

7. $y' = e^{-y}, y(0) = 0; \quad y(0.5)$

8. $y' = x + y^2, y(0) = 0; \quad y(0.5)$

9. $y' = (x - y)^2, y(0) = 0.5; \quad y(0.5)$

10. $y' = xy + \sqrt{y}, y(0) = 1; \quad y(0.5)$

11. $y' = xy^2 - \dfrac{y}{x}, y(1) = 1; \quad y(1.5)$

12. $y' = y - y^2, y(0) = 0.5; \quad y(0.5)$

13. If air resistance is proportional to the square of the instantaneous velocity, then the velocity v of a mass m dropped from a height h is determined from

$$m\frac{dv}{dt} = mg - kv^2, \quad k > 0.$$

Let $v(0) = 0$, $k = 0.125$, $m = 5$ slugs, and $g = 32$ ft/s².
   (a) Use the Runge-Kutta method with $h = 1$ to find an approximation to the velocity of the falling mass at $t = 5$ s.
   (b) Use a numerical solver to graph the solution of the initial-value problem.
   (c) Use separation of variables to solve the initial-value problem and find the true value $v(5)$.

14. A mathematical model for the area A (in cm²) that a colony of bacteria (*B. dendroides*) occupies is given by

$$\frac{dA}{dt} = A(2.128 - 0.0432A).*$$

Suppose that the initial area is 0.24 cm².

(a) Use the Runge-Kutta method with $h = 0.5$ to complete the following table.

t (days)	1	2	3	4	5
A (observed)	2.78	13.53	36.30	47.50	49.40
A (approximated)					

(b) Use a numerical solver to graph the solution of the initial-value problem. Estimate the values $A(1)$, $A(2)$, $A(3)$, $A(4)$, and $A(5)$ from the graph.

(c) Use separation of variables to solve the initial-value problem and compute the values $A(1)$, $A(2)$, $A(3)$, $A(4)$, and $A(5)$.

15. Consider the initial-value problem

$$y' = x^2 + y^3, \quad y(1) = 1.$$

(See Problem 12 in Exercises 9.1.)

(a) Compare the results obtained from using the Runge-Kutta method over the interval $[1, 1.4]$ with step sizes $h = 0.1$ and $h = 0.05$.

(b) Use a numerical solver to graph the solution of the initial-value problem on the interval $[1, 1.4]$.

16. Consider the initial-value problem $y' = 2y$, $y(0) = 1$. The analytic solution is $y(x) = e^{2x}$.

(a) Approximate $y(0.1)$ using one step and the fourth-order Runge-Kutta method.

(b) Find a bound for the local truncation error in y_1.

(c) Compare the actual error in y_1 with your error bound.

(d) Approximate $y(0.1)$ using two steps and the fourth-order Runge-Kutta method.

(e) Verify that the global truncation error for the fourth-order Runge-Kutta method is $O(h^4)$ by comparing the errors in parts (a) and (d).

---

*See V. A. Kostitzin, *Mathematical Biology* (London: Harrap, 1939).

17. Repeat Problem 16 using the initial-value problem $y' = -2y + x$, $y(0) = 1$. The analytic solution is $y(x) = \frac{1}{2}x - \frac{1}{4} + \frac{5}{4}e^{-2x}$.

18. Consider the initial-value problem $y' = 2x - 3y + 1$, $y(1) = 5$. The analytic solution is $y(x) = \frac{1}{9} + \frac{2}{3}x + \frac{38}{9}e^{-3(x-1)}$.
  - **(a)** Find a formula involving c and h for the local truncation error in the nth step if the fourth-order Runge-Kutta method is used.
  - **(b)** Find a bound for the local truncation error in each step if $h = 0.1$ is used to approximate $y(1.5)$.
  - **(c)** Approximate $y(1.5)$ using the fourth-order Runge-Kutta method with $h = 0.1$ and $h = 0.05$. See Problem 3. You will need to carry more than six decimal places to see the effect of reducing the step size.

19. Repeat Problem 18 for the initial-value problem $y' = e^{-y}$, $y(0) = 0$. The analytic solution is $y(x) = \ln(x + 1)$. Approximate $y(0.5)$. See Problem 7.

Discussion Problems

20. A count of the number of evaluations of the function f used in solving the initial-value problem $y' = f(x, y)$, $y(x_0) = y_0$ is used as a measure of the computational complexity of a numerical method. Determine the number of evaluations of f required for each step of Euler's, the improved Euler's, and the Runge-Kutta methods. By considering some specific examples, compare the accuracy of these methods when used with comparable computational complexities.

Computer Lab Assignments

21. The Runge-Kutta method for solving an initial-value problem over an interval $[a, b]$ results in a finite set of points that are supposed to approximate points on the graph of the exact solution. In order to expand this set of discrete points to an approximate solution defined at all points on the interval $[a, b]$, we can use an **interpolating function.** This is a function, supported by most computer algebra systems, that agrees with the given data exactly and assumes a smooth transition between data points. These interpolating functions may be polynomials or sets of polynomials joined together smoothly. In *Mathematica* the command **y=Interpolation[data]** can be used to obtain an interpolating function through the points **data** $= \{\{x_0, y_0\}, \{x_1, y_1\}, \ldots, \{x_n, y_n\}\}$. The interpolating function **y[x]** can now be treated like any other function built into the computer algebra system.
  - **(a)** Find the exact solution of the initial-value problem $y' = -y + 10 \sin 3x$, $y(0) = 0$ on the interval $= [0, 2]$. Graph this solution and find its positive roots.
  - **(b)** Use the fourth-order Runge-Kutta method with $h = 0.1$ to approximate a solution of the initial-value problem in part (a). Obtain an interpolating function and graph it. Find the positive roots of the interpolating function of the interval $[0, 2]$.

9.3 MULTISTEP METHODS

• *Single-step methods* • *Multistep methods* • *Predictor-corrector methods* • *Adams-Bashforth/Adams-Moulton method* • *Stability of numerical methods*

Euler's, the improved Euler's, and the Runge-Kutta methods are examples of **single-step** methods. In these methods each successive value y_{n+1} is computed based only on information about the immediately preceding value y_n. A **multistep** or **continuing method**, on the other hand, uses the values from several previously computed steps to obtain the value of y_{n+1}. There are a large number of multistep formulas that can be used to approximate solutions of DEs. Since it is not our intention to survey the vast field of numerical procedures, we will consider only one such method here. Like the improved Euler's method, it is a **predictor-corrector method**; that is, one formula is used to predict a value y_{n+1}^*, which in turn is used to obtain a corrected value y_{n+1}.

Adams-Bashforth/Adams-Moulton Method   One of the more popular multistep methods is the fourth-order **Adams-Bashforth/Adams-Moulton method.** The predictor in this method is the Adams-Bashforth formula

$$y_{n+1}^* = y_n + \frac{h}{24} (55y_n' - 59y_{n-1}' + 37y_{n-2}' - 9y_{n-3}'), \tag{1}$$

$$\begin{aligned} y_n' &= f(x_n, y_n) \\ y_{n-1}' &= f(x_{n-1}, y_{n-1}) \\ y_{n-2}' &= f(x_{n-2}, y_{n-2}) \\ y_{n-3}' &= f(x_{n-3}, y_{n-3}) \end{aligned}$$

for $n \geq 3$. The value of y_{n+1}^* is then substituted into the Adams-Moulton corrector

$$y_{n+1} = y_n + \frac{h}{24} (9y_{n+1}' + 19y_n' - 5y_{n-1}' + y_{n-2}'), \tag{2}$$

$$y_{n+1}' = f(x_{n+1}, y_{n+1}^*).$$

Notice that formula (1) requires that we know the values of y_0, y_1, y_2, and y_3 in order to obtain y_4. The value of y_0 is, of course, the given initial condition. Since the local truncation error of the Adams-Bashforth/Adams-Moulton method is $O(h^5)$, the values of y_1, y_2, and y_3 are generally computed by a method with the same error property, such as the fourth-order Runge-Kutta method.

EXAMPLE 1   **Adams-Bashforth/Adams-Moulton Method**

Use the Adams-Bashforth/Adams-Moulton method with $h = 0.2$ to obtain an approximation to $y(0.8)$ for the solution of

$$y' = x + y - 1, \quad y(0) = 1.$$

Solution   With a step size of $h = 0.2$, $y(0.8)$ will be approximated by y_4. To get started we use the Runge-Kutta method with $x_0 = 0$, $y_0 = 1$, and $h = 0.2$ to obtain

$$y_1 = 1.02140000, \quad y_2 = 1.09181796, \quad y_3 = 1.22210646.$$

Now with the identifications $x_0 = 0$, $x_1 = 0.2$, $x_2 = 0.4$, $x_3 = 0.6$, and $f(x, y) = x + y - 1$, we find

$$\begin{aligned}
y_0' &= f(x_0, y_0) = (0) + (1) - 1 = 0 \\
y_1' &= f(x_1, y_1) = (0.2) + (1.02140000) - 1 = 0.22140000 \\
y_2' &= f(x_2, y_2) = (0.4) + (1.09181796) - 1 = 0.49181796 \\
y_3' &= f(x_3, y_3) = (0.6) + (1.22210646) - 1 = 0.82210646.
\end{aligned}$$

With the foregoing values, the predictor (1) then gives

$$y_4^* = y_3 + \frac{0.2}{24}(55y_3' - 59y_2' + 37y_1' - 9y_0') = 1.42535975.$$

To use the corrector (2) we first need

$$y_4' = f(x_4, y_4^*) = 0.8 + 1.42535975 - 1 = 1.22535975.$$

Finally, (2) yields

$$y_4 = y_3 + \frac{0.2}{24}(9y_4' + 19y_3' - 5y_2' + y_1') = 1.42552788. \qquad \blacksquare$$

You should verify that the exact value of $y(0.8)$ in Example 1 is $y(0.8) = 1.42554093$.

Stability of Numerical Methods   An important consideration in using numerical methods to approximate the solution of an initial-value problem is the stability of the method. Simply stated, a numerical method is **stable** if small changes in the initial condition result in only small changes in the computed solution. A numerical method is said to be **unstable** if it is not stable. The reason that stability considerations are important is that in each step after the first step of a numerical technique we are essentially starting over again with a new initial-value problem, where the initial condition is the approximate solution value computed in the preceding step. Because of the presence of round-off error, this value will almost certainly vary at least slightly from the true value of the solution. Besides round-off error, another common source of error occurs in the initial condition itself; in physical applications the data are often obtained by imprecise measurements.

One possible method for detecting instability in the numerical solution of a specific initial-value problem is to compare the approximate solutions obtained when decreasing step sizes are used. If the numerical method is unstable, the error may actually increase with smaller step sizes. Another way of checking stability is to observe what happens to solutions when the initial condition is slightly perturbed (for example, change $y(0) = 1$ to $y(0) = 0.999$).

For a more detailed and precise discussion of stability, consult a numerical analysis text. In general, all of the methods we have discussed in this chapter have good stability characteristics.

Advantages/Disadvantages of Multistep Methods   Many considerations enter into the choice of a method to solve a differential equation numerically. Single-step methods, particularly the Runge-Kutta method, are often chosen because of their accuracy and the fact that they are easy to program. However, a major drawback is that the right-hand side of the differential equation must be evaluated many times at each step. For instance, the fourth-order Runge-Kutta method requires four function evaluations for each step. On the other hand, if the function evaluations in the previous step have been calculated and stored, a multistep method requires only one new function evaluation for each step. This can lead to great savings in time and expense.

As an example, solving $y' = f(x, y)$, $y(x_0) = y_0$ numerically using n steps by the fourth-order Runge-Kutta method requires $4n$ function evaluations. The Adams-Bashforth multistep method requires 16 function evaluations for the Runge-Kutta fourth-order starter and $n - 4$ for the n Adams-Bashforth steps, giving a total of $n + 12$ function evaluations for this method. In general the Adams-Bashforth multistep method requires slightly more than a quarter of the number of function evaluations required for the fourth-order Runge-Kutta method. If the evaluation of $f(x, y)$ is complicated, the multistep method will be more efficient.

Another issue involved with multistep methods is how many times the Adams-Moulton corrector formula should be repeated in each step. Each time the corrector is used, another function evaluation is done, and so the accuracy is increased at the expense of losing an advantage of the multistep method. In practice, the corrector is calculated once, and if the value of y_{n+1} is changed by a large amount, the entire problem is restarted using a smaller step size. This is often the basis of the variable step size methods, whose discussion is beyond the scope of this text.

EXERCISES 9.3

Answers to odd-numbered problems begin on page AN-14.

1. Find the exact solution of the initial-value problem in Example 1. Compare the exact values of $y(0.2)$, $y(0.4)$, $y(0.6)$, and $y(0.8)$ with the approximations y_1, y_2, y_3, and y_4.

2. Write a computer program to implement the Adams-Bashforth/Adams-Moulton method.

In Problems 3 and 4 use the Adams-Bashforth/Adams-Moulton method to approximate $y(0.8)$, where $y(x)$ is the solution of the given initial-value problem. Use $h = 0.2$ and the Runge-Kutta method to compute y_1, y_2, and y_3.

3. $y' = 2x - 3y + 1$, $y(0) = 1$ 4. $y' = 4x - 2y$, $y(0) = 2$

In Problems 5–8 use the Adams-Bashforth/Adams-Moulton method to approximate $y(1.0)$, where $y(x)$ is the solution of the given initial-value problem. Use $h = 0.2$ and $h = 0.1$ and the Runge-Kutta method to compute y_1, y_2, and y_3.

5. $y' = 1 + y^2, y(0) = 0$ **6.** $y' = y + \cos x, y(0) = 1$

7. $y' = (x - y)^2, y(0) = 0$ **8.** $y' = xy + \sqrt{y}, y(0) = 1$

9.4 HIGHER-ORDER EQUATIONS AND SYSTEMS

- *First-order initial-value problem* • *Second-order initial-value problem* • *A second-order DE as a system of first-order DEs* • *Numerical methods applied to second-order DEs* • *A system of DEs as a system of first-order DEs* • *Numerical methods applied to systems*

In Section 2.6 and the first three sections of this chapter we focused on numerical techniques that can be used to approximate the solution of a first-order initial-value problem $dy/dx = f(x, y)$, $y(x_0) = y_0$. To approximate the solution of a second-order initial-value problem we must, as already discussed in Section 4.9, express a second-order DE as a system of two first-order DEs. To do this we write the second-order DE in normal form (see Section 1.1) by solving for y'' in terms of x, y, and y'.

Second-Order Initial-Value Problems A second-order initial-value problem

$$y'' = f(x, y, y'), \quad y(x_0) = y_0, \quad y'(x_0) = u_0 \tag{1}$$

can be expressed as an initial-value problem for a system of first-order differential equations. If we let $y' = u$, the differential equation in (1) becomes the system

$$\begin{aligned} y' &= u \\ u' &= f(x, y, u). \end{aligned} \tag{2}$$

Since $y'(x_0) = u(x_0)$, the corresponding initial conditions for (2) are then $y(x_0) = y_0, u(x_0) = u_0$. This system of DEs (2) can now be solved numerically by simply applying a particular numerical method to each first-order differential equation in the system. For example, **Euler's method** applied to the system (2) would be

$$\begin{aligned} y_{n+1} &= y_n + hu_n \\ u_{n+1} &= u_n + hf(x_n, y_n, u_n), \end{aligned} \tag{3}$$

whereas the **fourth-order Runge-Kutta method** would be

$$y_{n+1} = y_n + \frac{1}{6}(m_1 + 2m_2 + 2m_3 + m_4)$$

$$u_{n+1} = u_n + \frac{1}{6}(k_1 + 2k_2 + 2k_3 + k_4), \tag{4}$$

where

$$m_1 = hu_n \qquad k_1 = hf(x_n, y_n, u_n)$$
$$m_2 = h(u_n + \tfrac{1}{2}k_1) \quad k_2 = hf(x_n + \tfrac{1}{2}h, y_n + \tfrac{1}{2}m_1, u_n + \tfrac{1}{2}k_1)$$
$$m_3 = h(u_n + \tfrac{1}{2}k_2) \quad k_3 = hf(x_n + \tfrac{1}{2}h, y_n + \tfrac{1}{2}m_2, u_n + \tfrac{1}{2}k_2)$$
$$m_4 = h(u_n + k_3) \quad k_4 = hf(x_n + h, y_n + m_3, u_n + k_3).$$

In general, we can express every nth-order differential equation $y^{(n)} = f(x, y, y', \ldots, y^{(n-1)})$ as a system of n first-order equations using the substitutions $y = u_1, y' = u_2, y'' = u_3, \ldots, y^{(n-1)} = u_n$.

EXAMPLE 1 **Euler's Method**

Use Euler's method to obtain the approximate value of $y(0.2)$, where $y(x)$ is the solution of the initial-value problem

$$y'' + xy' + y = 0, \quad y(0) = 1, \quad y'(0) = 2. \tag{5}$$

Solution In terms of the substitution $y' = u$, the equation is equivalent to the system

$$y' = u$$
$$u' = -xu - y.$$

Thus from (3) we obtain

$$y_{n+1} = y_n + hu_n$$
$$u_{n+1} = u_n + h[-x_nu_n - y_n].$$

Using the step size $h = 0.1$ and $y_0 = 1$, $u_0 = 2$, we find

$$y_1 = y_0 + (0.1)u_0 = 1 + (0.1)2 = 1.2$$
$$u_1 = u_0 + (0.1)[-x_0u_0 - y_0] = 2 + (0.1)[-(0)(2) - 1] = 1.9$$
$$y_2 = y_1 + (0.1)u_1 = 1.2 + (0.1)(1.9) = 1.39$$
$$u_2 = u_1 + (0.1)[-x_1u_1 - y_1] = 1.9 + (0.1)[-(0.1)(1.9) - 1.2] = 1.761.$$

In other words, $y(0.2) \approx 1.39$ and $y'(0.2) \approx 1.761$. ∎

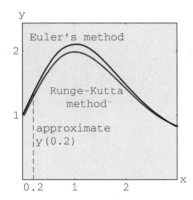

(a) Euler's method (black) and the Runge-Kutta method (color)

With the aid of the graphing feature of a numerical solver, in Figure 9.2(a) we compare the solution curve of (5) generated by Euler's method ($h = 0.1$) on the interval $[0, 3]$ with the solution curve generated by the fourth-order Runge-Kutta method ($h = 0.1$). From Figure 9.2(b), it appears that the solution $y(x)$ of (4) has the property that $y(x) \to 0$ as $x \to \infty$.

If desired we can use the method of Section 6.1 to obtain two power series solutions of the differential equation in (5). But unless this method reveals that the DE possesses an elementary solution, we will still only be able to approximate $y(0.2)$ using a partial sum. Re-inspection of the infinite series solutions of Airy's differential equation $y'' + xy = 0$, given on page 274, does not reveal the oscillatory behavior of the solutions $y_1(x)$ and $y_2(x)$ exhibited in the graphs in Figure 6.1. Those graphs, along with the graphs of six particular solutions of Airy's equation on the opening page of this chapter (page 410), were obtained from a numerical solver using the fourth-order Runge-Kutta method with a step size of $h = 0.1$.

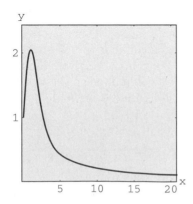

(b) Runge-Kutta method

Figure 9.2

Systems Reduced to First-Order Systems   Using a procedure similar to that just discussed for second-order equations, we can often reduce a system of higher-order differential equations to a system of first-order equations by first solving for the highest-order derivative of each dependent variable and then making appropriate substitutions for the lower-order derivatives.

EXAMPLE 2   **A System Rewritten as a First-Order System**

Write
$$x'' - x' + 5x + 2y'' = e^t$$
$$-2x + y'' + 2y = 3t^2$$

as a system of first-order differential equations.

Solution   Write the system as
$$x'' + 2y'' = e^t - 5x + x'$$
$$y'' = 3t^2 + 2x - 2y$$

and then eliminate y'' by multiplying the second equation by 2 and subtracting. This gives
$$x'' = -9x + 4y + x' + e^t - 6t^2.$$

Since the second equation of the system already expresses the highest-order derivative of y in terms of the remaining functions, we are now in a position to introduce new variables. If we let $x' = u$ and $y' = v$, the expressions for x'' and y'' become, respectively,
$$u' = x'' = -9x + 4y + u + e^t - 6t^2$$
$$v' = y'' = 2x - 2y + 3t^2.$$

The original system can then be written in the form
$$x' = u$$
$$y' = v$$
$$u' = -9x + 4y + u + e^t - 6t^2$$
$$v' = 2x - 2y + 3t^2.$$

It may not always be possible to carry out the reductions illustrated in Example 2.

Numerical Solution of a System   The solution of a system of the form
$$\frac{dx_1}{dt} = f_1(t, x_1, x_2, \ldots, x_n)$$
$$\frac{dx_2}{dt} = f_2(t, x_1, x_2, \ldots, x_n)$$
$$\vdots \qquad \vdots$$
$$\frac{dx_n}{dt} = f_n(t, x_1, x_2, \ldots, x_n)$$

can be approximated by a version of Euler's, the Runge-Kutta, or the Adams-Bashforth/Adams-Moulton method adapted to the system. For example, the **fourth-order Runge-Kutta method** applied to the system

$$x' = f(t, x, y)$$
$$y' = g(t, x, y) \tag{6}$$
$$x(t_0) = x_0, \quad y(t_0) = y_0$$

looks like this:

$$x_{n+1} = x_n + \frac{1}{6}(m_1 + 2m_2 + 2m_3 + m_4)$$

$$y_{n+1} = y_n + \frac{1}{6}(k_1 + 2k_2 + 2k_3 + k_4), \tag{7}$$

where

$$
\begin{aligned}
m_1 &= hf(t_n, x_n, y_n) & k_1 &= hg(t_n, x_n, y_n) \\
m_2 &= hf(t_n + \tfrac{1}{2}h, x_n + \tfrac{1}{2}m_1, y_n + \tfrac{1}{2}k_1) & k_2 &= hg(t_n + \tfrac{1}{2}h, x_n + \tfrac{1}{2}m_1, y_n + \tfrac{1}{2}k_1) \\
m_3 &= hf(t_n + \tfrac{1}{2}h, x_n + \tfrac{1}{2}m_2, y_n + \tfrac{1}{2}k_2) & k_3 &= hg(t_n + \tfrac{1}{2}h, x_n + \tfrac{1}{2}m_2, y_n + \tfrac{1}{2}k_2) \\
m_4 &= hf(t_n + h, x_n + m_3, y_n + k_3) & k_4 &= hg(t_n + h, x_n + m_3, y_n + k_3).
\end{aligned} \tag{8}
$$

EXAMPLE 3 Runge-Kutta Method

Consider the initial-value problem

$$x' = 2x + 4y$$
$$y' = -x + 6y$$
$$x(0) = -1, \quad y(0) = 6.$$

Use the fourth-order Runge-Kutta method to approximate $x(0.6)$ and $y(0.6)$. Compare the results for $h = 0.2$ and $h = 0.1$.

Solution   We illustrate the computations of x_1 and y_1 with the step size $h = 0.2$. With the identifications $f(t, x, y) = 2x + 4y$, $g(t, x, y) = -x + 6y$, $t_0 = 0$, $x_0 = -1$, and $y_0 = 6$, we see from (8) that

$$m_1 = hf(t_0, x_0, y_0) = 0.2f(0, -1, 6) = 0.2[2(-1) + 4(6)] = 4.4000$$

$$k_1 = hg(t_0, x_0, y_0) = 0.2g(0, -1, 6) = 0.2[-1(-1) + 6(6)] = 7.4000$$

$$m_2 = hf(t_0 + \tfrac{1}{2}h, x_0 + \tfrac{1}{2}m_1, y_0 + \tfrac{1}{2}k_1) = 0.2f(0.1, 1.2, 9.7) = 8.2400$$

$$k_2 = hg(t_0 + \tfrac{1}{2}h, x_0 + \tfrac{1}{2}m_1, y_0 + \tfrac{1}{2}k_1) = 0.2g(0.1, 1.2, 9.7) = 11.4000$$

$$m_3 = hf(t_0 + \tfrac{1}{2}h, x_0 + \tfrac{1}{2}m_2, y_0 + \tfrac{1}{2}k_2) = 0.2f(0.1, 3.12, 11.7) = 10.6080$$

$$k_3 = hg(t_0 + \tfrac{1}{2}h, x_0 + \tfrac{1}{2}m_2, y_0 + \tfrac{1}{2}k_2) = 0.2g(0.1, 3.12, 11.7) = 13.4160$$

$$m_4 = hf(t_0 + h, x_0 + m_3, y_0 + k_3) = 0.2f(0.2, 8, 20.216) = 19.3760$$

$$k_4 = hg(t_0 + h, x_0 + m_3, y_0 + k_3) = 0.2g(0.2, 8, 20.216) = 21.3776.$$

Therefore from (7) we get

$$x_1 = x_0 + \frac{1}{6}(m_1 + 2m_2 + 2m_3 + m_4)$$

$$= -1 + \frac{1}{6}(4.4 + 2(8.24) + 2(10.608) + 19.3760) = 9.2453$$

$$y_1 = y_0 + \frac{1}{6}(k_1 + 2k_2 + 2k_3 + k_4)$$

$$= 6 + \frac{1}{6}(7.4 + 2(11.4) + 2(13.416) + 21.3776) = 19.0683,$$

where, as usual, the computed values are rounded to four decimal places. These numbers give us the approximations $x_1 \approx x(0.2)$ and $y_1 \approx y(0.2)$. The subsequent values, obtained with the aid of a computer, are summarized in Tables 9.8 and 9.9.

TABLE 9.8 Runge-Kutta Method with $h = 0.2$

m_1	m_2	m_3	m_4	k_1	k_2	k_3	k_4	t_n	x_n	y_n
								0.00	−1.0000	6.0000
4.4000	8.2400	10.6080	19.3760	7.4000	11.4000	13.4160	21.3776	0.20	9.2453	19.0683
18.9527	31.1564	37.8870	63.6848	21.0329	31.7573	36.9716	57.8214	0.40	46.0327	55.1203
62.5093	97.7863	116.0063	187.3669	56.9378	84.8495	98.0688	151.4191	0.60	158.9430	150.8192

TABLE 9.9 Runge-Kutta Method with $h = 0.1$

m_1	m_2	m_3	m_4	k_1	k_2	k_3	k_4	t_n	x_n	y_n
								0.00	−1.0000	6.0000
2.2000	3.1600	3.4560	4.8720	3.7000	4.7000	4.9520	6.3256	0.10	2.3840	10.8883
4.8321	6.5742	7.0778	9.5870	6.2946	7.9413	8.3482	10.5957	0.20	9.3379	19.1332
9.5208	12.5821	13.4258	17.7609	10.5461	13.2339	13.8872	17.5358	0.30	22.5541	32.8539
17.6524	22.9090	24.3055	31.6554	17.4569	21.8114	22.8549	28.7393	0.40	46.5103	55.4420
31.4788	40.3496	42.6387	54.9202	28.6141	35.6245	37.2840	46.7207	0.50	88.5729	93.3006
54.6348	69.4029	73.1247	93.4107	46.5231	57.7482	60.3774	75.4370	0.60	160.7563	152.0025

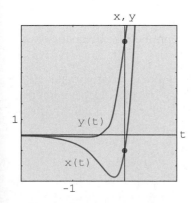

Figure 9.3

You should verify that the solution of the initial-value problem in Example 3 is given by $x(t) = (26t - 1)e^{4t}$, $y(t) = (13t + 6)e^{4t}$. From these equations we see that the exact values are $x(0.6) = 160.9384$ and $y(0.6) = 152.1198$. Shown in Figure 9.3 is the graph of the solution in a neighborhood of $t = 0$, obtained from a numerical solver using the Runge-Kutta method with $h = 0.1$.

In conclusion, we state **Euler's method** for the general system (6):

$$x_{n+1} = x_n + hf(t_n, x_n, y_n)$$
$$y_{n+1} = y_n + hg(t_n, x_n, y_n).$$

EXERCISES 9.4

Answers to odd-numbered problems begin on page AN-14.

1. Use Euler's method to approximate $y(0.2)$, where $y(x)$ is the solution of the initial-value problem

$$y'' - 4y' + 4y = 0, \quad y(0) = -2, \quad y'(0) = 1.$$

Use $h = 0.1$. Find the exact solution of the problem, and compare the exact value of $y(0.2)$ with y_2.

2. Use Euler's method to approximate $y(1.2)$, where $y(x)$ is the solution of the initial-value problem

$$x^2y'' - 2xy' + 2y = 0, \quad y(1) = 4, \quad y'(1) = 9,$$

where $x > 0$. Use $h = 0.1$. Find the exact solution of the problem, and compare the exact value of $y(1.2)$ with y_2.

3. Repeat Problem 1 using the Runge-Kutta method with $h = 0.2$ and $h = 0.1$.

4. Repeat Problem 2 using the Runge-Kutta method with $h = 0.2$ and $h = 0.1$.

5. Use the Runge-Kutta method to obtain the approximate value of $y(0.2)$, where $y(x)$ is a solution of the initial-value problem

$$y'' - 2y' + 2y = e^t \cos t, \quad y(0) = 1, \quad y'(0) = 2.$$

Use $h = 0.2$ and $h = 0.1$.

6. When $E = 100$ V, $R = 10$ Ω, and $L = 1$ h, the system of differential equations for the currents $i_1(t)$ and $i_3(t)$ in the electrical network given in Figure 9.4 is

$$\frac{di_1}{dt} = -20i_1 + 10i_3 + 100$$

$$\frac{di_3}{dt} = 10i_1 - 20i_3,$$

where $i_1(0) = 0$ and $i_3(0) = 0$. Use the Runge-Kutta method to approximate $i_1(t)$ and $i_3(t)$ at $t = 0.1, 0.2, 0.3, 0.4,$ and 0.5. Use $h = 0.1$. Use a numerical solver to obtain the graph of the solution $i_1(t), i_3(t)$ on the interval $0 \le t \le 5$. Use the graph to predict the behavior of $i_1(t)$ and $i_3(t)$ as $t \to \infty$.

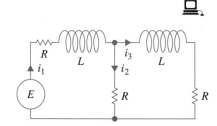

Figure 9.4

In Problems 7–12 use the Runge-Kutta method to approximate $x(0.2)$ and $y(0.2)$. Compare the results for $h = 0.2$ and $h = 0.1$. Use a numerical solver and $h = 0.1$ to obtain the graph of the solution in a neighborhood of $t = 0$.

7. $x' = 2x - y$
   $y' = x$
   $x(0) = 6, y(0) = 2$

8. $x' = x + 2y$
   $y' = 4x + 3y$
   $x(0) = 1, y(0) = 1$

9. $x' = -y + t$
   $y' = x - t$
   $x(0) = -3, y(0) = 5$

10. $x' = 6x + y + 6t$
    $y' = 4x + 3y - 10t + 4$
    $x(0) = 0.5, y(0) = 0.2$

11. $x' + 4x - y' = 7t$
$x' + y' - 2y = 3t$
$x(0) = 1, y(0) = -2$

12. $x' + y' = 4t$
$-x' + y' + y = 6t^2 + 10$
$x(0) = 3, y(0) = -1$

9.5 SECOND-ORDER BOUNDARY-VALUE PROBLEMS

• Difference quotients • Finite differences • Forward difference • Backward difference • Central difference • Interior mesh points • Finite difference method • Shooting method

In Sections 9.1–9.3 we considered techniques that yield an approximation to a solution of a first-order initial-value problem $y' = f(x, y)$, $y(x_0) = y_0$. In Section 9.4 we just saw that we can adapt these approximation techniques to a *second-order initial-value problem* $y'' = f(x, y, y')$, $y(x_0) = y_0$, $y'(x_0) = u_0$ by expressing the second-order DE as a system of first-order DEs. We are now going to examine two methods for approximating a solution to a *second-order boundary-value problem* $y'' = f(x, y, y')$, $y(a) = \alpha$, $y(b) = \beta$. Unlike the procedures used with second-order initial-value problems, the methods for second-order boundary-value problems do not require writing the differential equation as a system.

Finite Difference Approximations The Taylor series expansion of a function $y(x)$ centered at a point a is

$$y(x) = y(a) + y'(a)\frac{x - a}{1!} + y''(a)\frac{(x - a)^2}{2!} + y'''(a)\frac{(x - a)^3}{3!} + \cdots.$$

If we set $h = x - a$, then the preceding line is the same as

$$y(x) = y(a) + y'(a)\frac{h}{1!} + y''(a)\frac{h^2}{2!} + y'''(a)\frac{h^3}{3!} + \cdots.$$

For the subsequent discussion it is convenient then to rewrite this last expression in two alternative forms:

$$y(x + h) = y(x) + y'(x)h + y''(x)\frac{h^2}{2} + y'''(x)\frac{h^3}{6} + \cdots \tag{1}$$

and

$$y(x - h) = y(x) - y'(x)h + y''(x)\frac{h^2}{2} - y'''(x)\frac{h^3}{6} + \cdots. \tag{2}$$

If h is small, we can ignore terms involving h^4, h^5, ... since these values are negligible. Indeed, if we ignore all terms involving h^2 and higher, then (1) and (2) yield, in turn, the following approximations for the first derivative $y'(x)$:

$$y'(x) \approx \frac{1}{h}[y(x + h) - y(x)] \tag{3}$$

$$y'(x) \approx \frac{1}{h}[y(x) - y(x - h)]. \tag{4}$$

Subtracting (1) and (2) also gives

$$y'(x) \approx \frac{1}{2h}[y(x+h) - y(x-h)]. \tag{5}$$

On the other hand, if we ignore terms involving h^3 and higher, then by adding (1) and (2) we obtain an approximation for the second derivative $y''(x)$:

$$y''(x) \approx \frac{1}{h^2}[y(x+h) - 2y(x) + y(x-h)]. \tag{6}$$

The right-hand sides of (3), (4), (5), and (6) are called **difference quotients.** The expressions

$$y(x+h) - y(x), \quad y(x) - y(x-h), \quad y(x+h) - y(x-h)$$

and
$$y(x+h) - 2y(x) + y(x-h)$$

are called **finite differences.** Specifically, $y(x+h) - y(x)$ is called a **forward difference,** $y(x) - y(x-h)$ is a **backward difference,** and both $y(x+h) - y(x-h)$ and $y(x+h) - 2y(x) + y(x-h)$ are called **central differences.** The results given in (5) and (6) are referred to as **central difference approximations** for the derivatives y' and y''.

Finite Difference Method   Consider now a linear second-order boundary-value problem

$$y'' + P(x)y' + Q(x)y = f(x), \quad y(a) = \alpha, \quad y(b) = \beta. \tag{7}$$

Suppose $a = x_0 < x_1 < x_2 < \cdots < x_{n-1} < x_n = b$ represents a regular partition of the interval $[a, b]$—that is, $x_i = a + ih$, where $i = 0, 1, 2, \ldots, n$ and $h = (b - a)/n$. The points

$$x_1 = a + h, \quad x_2 = a + 2h, \quad \ldots, \quad x_{n-1} = a + (n-1)h$$

are called **interior mesh points** of the interval $[a, b]$. If we let

$$y_i = y(x_i), \quad P_i = P(x_i), \quad Q_i = Q(x_i), \quad \text{and} \quad f_i = f(x_i)$$

and if y'' and y' in (7) are replaced by the central difference approximations (5) and (6), we get

$$\frac{y_{i+1} - 2y_i + y_{i-1}}{h^2} + P_i \frac{y_{i+1} - y_{i-1}}{2h} + Q_i y_i = f_i$$

or, after simplifying,

$$\left(1 + \frac{h}{2} P_i\right) y_{i+1} + (-2 + h^2 Q_i) y_i + \left(1 - \frac{h}{2} P_i\right) y_{i-1} = h^2 f_i. \tag{8}$$

The last equation, known as a **finite difference equation,** is an approximation to the differential equation. It enables us to approximate the solution $y(x)$ of (7) at the interior mesh points $x_1, x_2, \ldots, x_{n-1}$ of the interval $[a, b]$. By letting i take on the values $1, 2, \ldots, n - 1$ in (8), we obtain $n - 1$ equations in the $n - 1$ unknowns $y_1, y_2, \ldots, y_{n-1}$. Bear in mind that we know y_0 and y_n since these are the prescribed boundary conditions $y_0 = y(x_0) = y(a) = \alpha$ and $y_n = y(x_n) = y(b) = \beta$.

In Example 1 we consider a boundary-value problem for which we can compare the approximate values found with the exact values of an explicit solution.

EXAMPLE 1 Using the Finite Difference Method

Use the difference equation (8) with $n = 4$ to approximate the solution of the boundary-value problem

$$y'' - 4y = 0, \quad y(0) = 0, \quad y(1) = 5.$$

Solution To use (8) we identify $P(x) = 0$, $Q(x) = -4$, $f(x) = 0$, and $h = (1 - 0)/4 = \frac{1}{4}$. Hence the difference equation is

$$y_{i+1} - 2.25y_i + y_{i-1} = 0. \tag{9}$$

Now the interior points are $x_1 = 0 + \frac{1}{4}$, $x_2 = 0 + \frac{2}{4}$, $x_3 = 0 + \frac{3}{4}$, and so for $i = 1, 2$, and 3, (9) yields the following system for the corresponding y_1, y_2, and y_3:

$$y_2 - 2.25y_1 + y_0 = 0$$
$$y_3 - 2.25y_2 + y_1 = 0$$
$$y_4 - 2.25y_3 + y_2 = 0.$$

With the boundary conditions $y_0 = 0$ and $y_4 = 5$, the foregoing system becomes

$$-2.25y_1 + \quad y_2 \qquad\qquad = 0$$
$$y_1 - 2.25y_2 + \quad y_3 = 0$$
$$y_2 - 2.25y_3 = -5.$$

Solving the system gives $y_1 = 0.7256$, $y_2 = 1.6327$, and $y_3 = 2.9479$.

Now the general solution of the given differential equation is $y = c_1 \cosh 2x + c_2 \sinh 2x$. The condition $y(0) = 0$ implies $c_1 = 0$. The other boundary condition gives c_2. In this way we see that an explicit solution of the boundary-value problem is $y(x) = (5 \sinh 2x)/\sinh 2$. Thus the exact values (rounded to four decimal places) of this solution at the interior points are as follows: $y(0.25) = 0.7184$, $y(0.5) = 1.6201$, and $y(0.75) = 2.9354$. ∎

The accuracy of the approximations in Example 1 can be improved by using a smaller value of h. Of course, the trade-off here is that a smaller value of h necessitates solving a larger system of equations. It is left as an exercise to show that with $h = \frac{1}{8}$, approximations to $y(0.25)$, $y(0.5)$, and $y(0.75)$ are 0.7202, 1.6233, and 2.9386, respectively. See Problem 11 in Exercises 9.5.

EXAMPLE 2 Using the Finite Difference Method

Use the difference equation (8) with $n = 10$ to approximate the solution of

$$y'' + 3y' + 2y = 4x^2, \quad y(1) = 1, \quad y(2) = 6.$$

Solution In this case we identify $P(x) = 3$, $Q(x) = 2$, $f(x) = 4x^2$, and $h = (2 - 1)/10 = 0.1$, and so (8) becomes

$$1.15y_{i+1} - 1.98y_i + 0.85y_{i-1} = 0.04x_i^2. \tag{10}$$

Now the interior points are $x_1 = 1.1$, $x_2 = 1.2$, $x_3 = 1.3$, $x_4 = 1.4$, $x_5 = 1.5$, $x_6 = 1.6$, $x_7 = 1.7$, $x_8 = 1.8$, and $x_9 = 1.9$. For $i = 1, 2, \ldots, 9$ and $y_0 = 1$, $y_{10} = 6$, (10) gives a system of nine equations and nine unknowns:

$$
\begin{aligned}
1.15y_2 - 1.98y_1 &= -0.8016 \\
1.15y_3 - 1.98y_2 + 0.85y_1 &= 0.0576 \\
1.15y_4 - 1.98y_3 + 0.85y_2 &= 0.0676 \\
1.15y_5 - 1.98y_4 + 0.85y_3 &= 0.0784 \\
1.15y_6 - 1.98y_5 + 0.85y_4 &= 0.0900 \\
1.15y_7 - 1.98y_6 + 0.85y_5 &= 0.1024 \\
1.15y_8 - 1.98y_7 + 0.85y_6 &= 0.1156 \\
1.15y_9 - 1.98y_8 + 0.85y_7 &= 0.1296 \\
- 1.98y_9 + 0.85y_8 &= -6.7556.
\end{aligned}
$$

We can solve this large system using Gaussian elimination or, with relative ease, by means of a computer algebra system. The result is found to be $y_1 = 2.4047$, $y_2 = 3.4432$, $y_3 = 4.2010$, $y_4 = 4.7469$, $y_5 = 5.1359$, $y_6 = 5.4124$, $y_7 = 5.6117$, $y_8 = 5.7620$, and $y_9 = 5.8855$. ▬

Shooting Method   Another way of approximating a solution of a boundary-value problem $y'' = f(x, y, y')$, $y(a) = \alpha$, $y(b) = \beta$ is called the **shooting method.** The starting point in this method is the replacement of the boundary-value problem by an initial-value problem

$$y'' = f(x, y, y'), \quad y(a) = \alpha, \quad y'(a) = m_1. \tag{11}$$

The number m_1 in (11) is simply a guess for the unknown slope of the solution curve at the known point $(a, y(a))$. We then apply one of the step-by-step numerical techniques to the second-order equation in (11) to find an approximation β_1 for the value of $y(b)$. If β_1 agrees with the given value $y(b) = \beta$ to some preassigned tolerance, we stop; otherwise the calculations are repeated, starting with a different guess $y'(a) = m_2$ to obtain a second approximation β_2 for $y(b)$. This method can be continued in a trial-and-error manner or the subsequent slopes $m_3, m_4, \ldots$ can be adjusted in some systematic way; linear interpolation is particularly successful when the differential equation in (11) is linear. The procedure is analogous to shooting (the "aim" is the choice of the initial slope) at a target until the bull's-eye $y(b)$ is hit. See Problem 14 in Exercises 9.5.

Of course, underlying the use of these numerical methods is the assumption, which we know is not always warranted, that a solution of the boundary-value problem exists.

Remarks   The approximation method using finite differences can be extended to boundary-value problems in which the first derivative is specified at a boundary—for example, a problem such as $y'' = f(x, y, y')$, $y'(a) = \alpha$, $y(b) = \beta$. See Problem 13 in Exercises 9.5.

Answers to odd-numbered problems begin on page AN-14.

In Problems 1–10 use the finite difference method and the indicated value of n to approximate the solution of the given boundary-value problem.

1. $y'' + 9y = 0$, $y(0) = 4$, $y(2) = 1$; $n = 4$

2. $y'' - y = x^2$, $y(0) = 0$, $y(1) = 0$; $n = 4$

3. $y'' + 2y' + y = 5x$, $y(0) = 0$, $y(1) = 0$; $n = 5$

4. $y'' - 10y' + 25y = 1$, $y(0) = 1$, $y(1) = 0$; $n = 5$

5. $y'' - 4y' + 4y = (x + 1)e^{2x}$, $y(0) = 3$, $y(1) = 0$; $n = 6$

6. $y'' + 5y' = 4\sqrt{x}$, $y(1) = 1$, $y(2) = -1$; $n = 6$

7. $x^2y'' + 3xy' + 3y = 0$, $y(1) = 5$, $y(2) = 0$; $n = 8$

8. $x^2y'' - xy' + y = \ln x$, $y(1) = 0$, $y(2) = -2$; $n = 8$

9. $y'' + (1 - x)y' + xy = x$, $y(0) = 0$, $y(1) = 2$; $n = 10$

10. $y'' + xy' + y = x$, $y(0) = 1$, $y(1) = 0$; $n = 10$

11. Rework Example 1 using $n = 8$.

12. The electrostatic potential u between two concentric spheres of radius $r = 1$ and $r = 4$ is determined from

$$\frac{d^2u}{dr^2} + \frac{2}{r}\frac{du}{dr} = 0, \quad u(1) = 50, \quad u(4) = 100.$$

Use the method of this section with $n = 6$ to approximate the solution of this boundary-value problem.

13. Consider the boundary-value problem $y'' + xy = 0$, $y'(0) = 1$, $y(1) = -1$.
   (a) Find the difference equation corresponding to the differential equation. Show that for $i = 0, 1, 2, \ldots, n - 1$ the difference equation yields n equations in $n + 1$ unknowns $y_{-1}, y_0, y_1, y_2, \ldots, y_{n-1}$. Here y_{-1} and y_0 are unknowns since y_{-1} represents an approximation to y at the exterior point $x = -h$ and y_0 is not specified at $x = 0$.
   (b) Use the central difference approximation (5) to show that $y_1 - y_{-1} = 2h$. Use this equation to eliminate y_{-1} from the system in part (a).
   (c) Use $n = 5$ and the system of equations found in parts (a) and (b) to approximate the solution of the original boundary-value problem.

Computer Lab Assignments

14. Consider the boundary-value problem $y'' = y' - \sin(xy)$, $y(0) = 1$, $y(1) = 1.5$. Use the shooting method to approximate the solution of this problem. (The actual approximation can be obtained using a numerical technique—say, the fourth-order Runge-Kutta method with $h = 0.1$; or, even better, if you have access to a CAS such as *Mathematica* or *Maple,* the **NDSolve** function can be used.)

CHAPTER 9 IN REVIEW

Answers to odd-numbered problems begin on page AN-14.

In Problems 1–4 construct a table comparing the indicated values of $y(x)$ using Euler's method, the improved Euler's method, and the Runge-Kutta method. Compute to four rounded decimal places. Use $h = 0.1$ and $h = 0.05$.

1. $y' = 2 \ln xy, \quad y(1) = 2;$
$y(1.1), y(1.2), y(1.3), y(1.4), y(1.5)$

2. $y' = \sin x^2 + \cos y^2, \quad y(0) = 0;$
$y(0.1), y(0.2), y(0.3), y(0.4), y(0.5)$

3. $y' = \sqrt{x + y}, \quad y(0.5) = 0.5;$
$y(0.6), y(0.7), y(0.8), y(0.9), y(1.0)$

4. $y' = xy + y^2, \quad y(1) = 1;$
$y(1.1), y(1.2), y(1.3), y(1.4), y(1.5)$

5. Use Euler's method to obtain the approximate value of $y(0.2)$, where $y(x)$ is the solution of the initial-value problem $y'' - (2x + 1)y = 1$, $y(0) = 3, y'(0) = 1$. First use one step with $h = 0.2$, and then repeat the calculations using two steps with $h = 0.1$.

6. Use the Adams-Bashforth/Adams-Moulton method to approximate the value of $y(0.4)$, where $y(x)$ is the solution of the initial-value problem $y' = 4x - 2y, y(0) = 2$. Use the Runge-Kutta method and $h = 0.1$ to obtain the values of $y_1, y_2,$ and y_3.

7. Use Euler's method with $h = 0.1$ to approximate values of $x(0.2)$ and $y(0.2)$, where $x(t), y(t)$ is the solution of the initial-value problem

$$x' = x + y$$
$$y' = x - y,$$
$$x(0) = 1, \quad y(0) = 2.$$

8. Use the finite difference method with $n = 10$ to approximate the solution of the boundary-value problem $y'' + 6.55(1 + x)y = 1$, $y(0) = 0, y(1) = 0$.

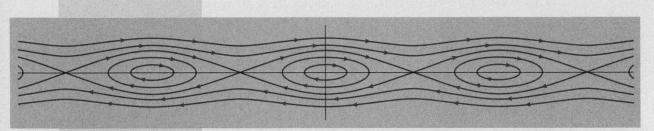

Phase portrait of a model for a nonlinear pendulum; see page 471.

10

PLANE AUTONOMOUS SYSTEMS AND STABILITY

INTRODUCTION In Chapter 8 we used matrix techniques to solve systems of linear first-order differential equations of the form $\mathbf{X}' = \mathbf{AX} + \mathbf{F}(t)$. When a system of differential equations is not linear, it is usually not possible to find solutions in terms of elementary functions. In this chapter we will demonstrate that valuable information on the geometric nature of the solutions of systems can be obtained by first analyzing special constant solutions, called critical points, and then searching for periodic solutions. The important concept of stability will be introduced and illustrated with examples from physics and ecology.

10.1 AUTONOMOUS SYSTEMS, CRITICAL POINTS, AND PERIODIC SOLUTIONS

• Autonomous systems • Second-order DEs as autonomous systems • Dynamical systems • Plane autonomous systems • Vector field interpretation of a plane autonomous system • Critical points • Periodic solutions • Systems in polar coordinates

We introduced the notions of autonomous first-order DEs, critical points of an autonomous DE, and the stability of a critical point in Section 2.1. This earlier consideration of stability was purposely kept at a fairly intuitive level; it is now time to give the precise definition of this concept. To do this we need to examine autonomous systems of first-order DEs. In this section we define critical points of autonomous systems of two first-order DEs; the autonomous systems can be linear or nonlinear.

Autonomous Systems A system of first-order differential equations is said to be **autonomous** when the system can be written in the form

$$\frac{dx_1}{dt} = g_1(x_1, x_2, \ldots, x_n)$$

$$\frac{dx_2}{dt} = g_2(x_1, x_2, \ldots, x_n) \tag{1}$$

$$\vdots \qquad \qquad \vdots$$

$$\frac{dx_n}{dt} = g_n(x_1, x_2, \ldots, x_n).$$

Observe that the independent variable t does not appear explicitly on the right-hand side of each differential equation. Compare (1) with the general system given in (2) of Section 8.1.

<hr>

EXAMPLE 1 **A Nonautonomous System**

The system of differential equations

$$\frac{dx_1}{dt} = x_1 - 3x_2 + t^2$$

$$\frac{dx_2}{dt} = tx_1 \sin x_2$$

is *not* autonomous because of the presence of t^2 and tx_1 on the right-hand sides. ∎

When $n = 1$ in (1), a single first-order differential equation takes on the form $dx/dt = g(x)$. This last equation is equivalent to (1) of Section 2.1

with the symbols x and t playing the parts of y and x, respectively. Explicit solutions can be constructed since the differential equation $dx/dt = g(x)$ is separable, and we will make use of this fact to give illustrations of the concepts in this chapter.

Second-Order DE as a System Any second-order differential equation $x'' = g(x, x')$ can be written as an autonomous system. As we did in Section 4.9, if we let $y = x'$ then $x'' = g(x, x')$ becomes $y' = g(x, y)$. Thus the second-order differential equation becomes the system of two first-order equations

$$x' = y$$
$$y' = g(x, y).$$

| EXAMPLE 2 | **The Pendulum DE as an Autonomous System** |

In (6) of Section 5.3 we showed that the displacement angle θ for a pendulum satisfies the nonlinear second-order differential equation

$$\frac{d^2\theta}{dt^2} + \frac{g}{l}\sin\theta = 0.$$

If we let $x = \theta$ and $y = \theta'$, this second-order differential equation may be rewritten as the autonomous system

$$x' = y$$
$$y' = -\frac{g}{l}\sin x. \qquad \blacksquare$$

Notation If $\mathbf{X}(t)$ and $\mathbf{g}(\mathbf{X})$ denote the respective column vectors

$$\mathbf{X}(t) = \begin{pmatrix} x_1(t) \\ x_2(t) \\ \vdots \\ x_n(t) \end{pmatrix}, \quad \mathbf{g}(\mathbf{X}) = \begin{pmatrix} g_1(x_1, x_2, \ldots, x_n) \\ g_2(x_1, x_2, \ldots, x_n) \\ \vdots \\ g_n(x_1, x_2, \ldots, x_n) \end{pmatrix},$$

then the autonomous system (1) may be written in the compact **column vector form** $\mathbf{X}' = \mathbf{g}(\mathbf{X})$. The homogeneous linear system $\mathbf{X}' = \mathbf{AX}$ studied in Section 8.2 is an important special case.

In this chapter it is also convenient to write (1) using row vectors. If we let

$$\mathbf{X}(t) = (x_1(t), x_2(t), \ldots, x_n(t))$$

and

$$\mathbf{g}(\mathbf{X}) = (g_1(x_1, x_2, \ldots, x_n), g_2(x_1, x_2, \ldots, x_n), \ldots, g_n(x_1, x_2, \ldots, x_n)),$$

then the autonomous system (1) may also be written in the compact **row vector form** $\mathbf{X}' = \mathbf{g}(\mathbf{X})$. *It should be clear from the context whether we are using column or row vector form, and therefore we will not distinguish* *between* **X** *and* $\mathbf{X}^T$, *the transpose of* **X**. In particular, when $n = 2$ it is convenient to use row vector form and write an initial condition as $\mathbf{X}(0) = (x_0, y_0)$.

When the variable t is interpreted as time, we can refer to the system of differential equations in (1) as a **dynamical system** and a solution $\mathbf{X}(t)$ as the **state of the system** or the **response of the system** at time t. With this terminology, a dynamical system is autonomous when the rate $\mathbf{X}'(t)$ at which the system changes depends only on the system's present state $\mathbf{X}(t)$. The linear system $\mathbf{X}' = \mathbf{AX} + \mathbf{F}(t)$ studied in Chapter 8 is then autonomous when $\mathbf{F}(t)$ is constant. In the case $n = 2$ or 3 we can call a solution a **path** or **trajectory** since we may think of $x = x_1(t)$, $y = x_2(t)$, and $z = x_3(t)$ as the parametric equations of a curve.

Vector Field Interpretation   When $n = 2$ the system in (1) is called a **plane autonomous system,** and we write the system as

$$\frac{dx}{dt} = P(x, y)$$

$$\frac{dy}{dt} = Q(x, y). \tag{2}$$

The vector $\mathbf{V}(x, y) = (P(x, y), Q(x, y))$ defines a **vector field** in a region of the plane, and a solution to the system may be interpreted as the resulting path of a particle as it moves through the region. To be more specific, let $\mathbf{V}(x, y) = (P(x, y), Q(x, y))$ denote the velocity of a stream at position (x, y), and suppose that a small particle (such as a cork) is released at a position (x_0, y_0) in the stream. If $\mathbf{X}(t) = (x(t), y(t))$ denotes the position of the particle at time t, then $\mathbf{X}'(t) = (x'(t), y'(t))$ is the velocity vector **v**. When external forces are not present and frictional forces are neglected, the velocity of the particle at time t is the velocity of the stream at position $\mathbf{X}(t)$:

$$\mathbf{X}'(t) = \mathbf{V}(x(t), y(t)) \quad \text{or} \quad \begin{aligned} \frac{dx}{dt} &= P(x(t), y(t)) \\ \frac{dy}{dt} &= Q(x(t), y(t)). \end{aligned}$$

Thus the path of the particle is a solution to the system that satisfies the initial condition $\mathbf{X}(0) = (x_0, y_0)$. We will frequently call on this simple interpretation of a plane autonomous system to illustrate new concepts.

EXAMPLE 3   **Plane Autonomous System of a Vector Field**

A vector field for the steady-state flow of a fluid around a cylinder of radius 1 is given by

$$\mathbf{V}(x, y) = V_0 \left(1 - \frac{x^2 - y^2}{(x^2 + y^2)^2}, \frac{-2xy}{(x^2 + y^2)^2} \right),$$

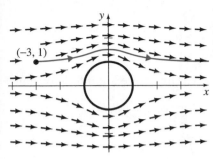

Figure 10.1

where V_0 is the speed of the fluid far from the cylinder. If a small cork is released at $(-3, 1)$, the path $\mathbf{X}(t) = (x(t), y(t))$ of the cork satisfies the plane autonomous system

$$\frac{dx}{dt} = V_0 \left(1 - \frac{x^2 - y^2}{(x^2 + y^2)^2} \right)$$

$$\frac{dy}{dt} = V_0 \left(\frac{-2xy}{(x^2 + y^2)^2} \right)$$

subject to the initial condition $\mathbf{X}(0) = (-3, 1)$. See Figure 10.1. ■

Types of Solutions If $P(x, y)$, $Q(x, y)$, and the first-order partial derivatives $\partial P/\partial x$, $\partial P/\partial y$, $\partial Q/\partial x$, and $\partial Q/\partial y$ are continuous in a region R of the plane, then a solution of the plane autonomous system (2) that satisfies $\mathbf{X}(0) = \mathbf{X}_0$ is unique and of one of three basic types:

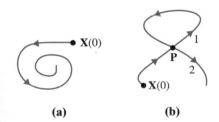

(a)   **(b)**

Figure 10.2

(*i*) A **constant solution** $x(t) = x_0$, $y(t) = y_0$ (or $\mathbf{X}(t) = \mathbf{X}_0$ for all t). A constant solution is called a **critical** or **stationary point.** When the particle is placed at a critical point $\mathbf{X}_0$ (that is, $\mathbf{X}(0) = \mathbf{X}_0$), it remains there indefinitely. For this reason a constant solution is also called an **equilibrium solution.** Note that since $\mathbf{X}'(t) = \mathbf{0}$, a critical point is a solution of the system of algebraic equations

$$P(x, y) = 0$$
$$Q(x, y) = 0.$$

(*ii*) A solution $x = x(t)$, $y = y(t)$ that defines an **arc**—a plane curve that does *not* cross itself. Thus the curve in Figure 10.2(a) can be a solution to a plane autonomous system, whereas the curve in Figure 10.2(b) cannot be a solution. There would be *two solutions* that start from the point $\mathbf{P}$ of intersection.

Figure 10.3

(*iii*) A **periodic solution** $x = x(t)$, $y = y(t)$. A periodic solution is called a **cycle.** If p is the period of the solution, then $\mathbf{X}(t + p) = \mathbf{X}(t)$ and a particle placed on the curve at $\mathbf{X}_0$ will cycle around the curve and return to $\mathbf{X}_0$ in p units of time. See Figure 10.3.

EXAMPLE 4   **Finding Critical Points**

Find all critical points of each of the following plane autonomous systems:

(a) $x' = -x + y$ **(b)** $x' = x^2 + y^2 - 6$ **(c)** $x' = 0.01x(100 - x - y)$
     $y' = x - y$ $y' = x^2 - y$ $y' = 0.05y(60 - y - 0.2x)$

Solution   We find the critical points by setting the right-hand sides of the differential equations equal to zero.

(a) The solution to the system

$$-x + y = 0$$
$$x - y = 0$$

consists of all points on the line $y = x$. Thus there are infinitely many critical points.

(b) To solve the system

$$x^2 + y^2 - 6 = 0$$
$$x^2 - y = 0$$

we substitute the second equation, $x^2 = y$, into the first equation to obtain $y^2 + y - 6 = (y + 3)(y - 2) = 0$. If $y = -3$, then $x^2 = -3$, and so there are no real solutions. If $y = 2$, then $x = \pm\sqrt{2}$, and so the critical points are $(\sqrt{2}, 2)$ and $(-\sqrt{2}, 2)$.

(c) Finding the critical points in part (c) requires a careful consideration of cases. The equation $0.01x(100 - x - y) = 0$ implies

$$x = 0 \quad \text{or} \quad x + y = 100.$$

If $x = 0$, then by substituting in $0.05y(60 - y - 0.2x) = 0$ we have $y(60 - y) = 0$. Thus $y = 0$ or 60, and so $(0, 0)$ and $(0, 60)$ are critical points.

If $x + y = 100$, then $0 = y(60 - y - 0.2(100 - y)) = y(40 - 0.8y)$. It follows that $y = 0$ or 50, and so $(100, 0)$ and $(50, 50)$ are critical points. ∎

When the plane autonomous system is linear, we can use the methods in Chapter 8 to investigate solutions.

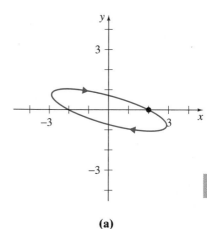

(a)

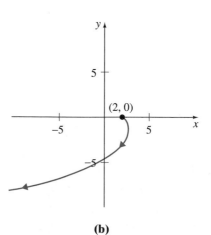

(b)

Figure 10.4

EXAMPLE 5   **Discovering Periodic Solutions**

Determine whether the given linear system possesses a periodic solution:

(a) $x' = 2x + 8y$ **(b)** $x' = x + 2y$

     $y' = -x - 2y$ $y' = -\frac{1}{2}x + y$

In each case sketch the graph of the solution that satisfies $\mathbf{X}(0) = (2, 0)$.

Solution

(a) In Example 6 of Section 8.2 we used the eigenvalue-eigenvector method to show that

$$x = c_1(2\cos 2t - 2\sin 2t) + c_2(2\cos 2t + 2\sin 2t)$$
$$y = c_1(-\cos 2t) - c_2\sin 2t.$$

Thus every solution is periodic with period $p = \pi$. The solution satisfying $\mathbf{X}(0) = (2, 0)$ is

$$x = 2\cos 2t + 2\sin 2t, \quad y = -\sin 2t.$$

This solution generates the ellipse shown in Figure 10.4(a).

(b) Using the eigenvalue-eigenvector method, we can show that

$$x = 2c_1e^t \cos t + 2c_2e^t \sin t, \quad y = -c_1e^t \sin t + c_2e^t \cos t.$$

Because of the presence of e^t in the general solution, there are no periodic solutions (that is, cycles). The solution satisfying $\mathbf{X}(0) = (2, 0)$ is

$$x = 2e^t \cos t, \quad y = -e^t \sin t,$$

and the resulting curve is shown in Figure 10.4(b).   ∎

Changing to Polar Coordinates   Except for the case of constant solutions, it is usually not possible to find explicit expressions for the solutions of a *nonlinear* autonomous system. We can solve some nonlinear systems, however, by changing to polar coordinates. From the formulas $r^2 = x^2 + y^2$ and $\theta = \tan^{-1}(y/x)$ we obtain

$$\frac{dr}{dt} = \frac{1}{r}\left(x\frac{dx}{dt} + y\frac{dy}{dt}\right)$$

$$\frac{d\theta}{dt} = \frac{1}{r^2}\left(-y\frac{dx}{dt} + x\frac{dy}{dt}\right). \tag{3}$$

We can sometimes use (3) to convert a plane autonomous system in rectangular coordinates to a simpler system in polar coordinates.

EXAMPLE 6   **Changing to Polar Coordinates**

Find the solution of the nonlinear plane autonomous system

$$x' = -y - x\sqrt{x^2 + y^2}$$
$$y' = x - y\sqrt{x^2 + y^2}$$

satisfying the initial condition $\mathbf{X}(0) = (3, 3)$.

Solution   Substituting for dx/dt and dy/dt in the expressions for dr/dt and $d\theta/dt$ in (3), we obtain

$$\frac{dr}{dt} = \frac{1}{r}[x(-y - xr) + y(x - yr)] = -r^2$$

$$\frac{d\theta}{dt} = \frac{1}{r^2}[-y(-y - xr) + x(x - yr)] = 1.$$

Since $(3, 3)$ is $(3\sqrt{2}, \pi/4)$ in polar coordinates, the initial condition $\mathbf{X}(0) = (3, 3)$ becomes $r(0) = 3\sqrt{2}$ and $\theta(0) = \pi/4$. Using separation of variables, we see that the solution of the system is

$$r = \frac{1}{t + c_1}, \qquad \theta = t + c_2$$

for $r \neq 0$. (Check this!) Applying the initial condition then gives

$$r = \frac{1}{t + \sqrt{2}/6}, \qquad \theta = t + \frac{\pi}{4}.$$

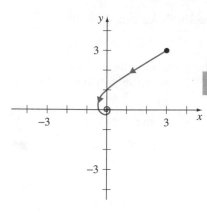

Figure 10.5

The spiral $r = \dfrac{1}{\theta + \sqrt{2}/6 - \pi/4}$ is sketched in Figure 10.5. ■

EXAMPLE 7 Solutions in Polar Coordinates

When expressed in polar coordinates, a plane autonomous system takes the form

$$\frac{dr}{dt} = 0.5(3 - r)$$

$$\frac{d\theta}{dt} = 1.$$

Find and sketch the solutions satisfying $X(0) = (0, 1)$ and $X(0) = (3, 0)$ in rectangular coordinates.

Solution   Applying separation of variables to $dr/dt = 0.5(3 - r)$ and integrating $d\theta/dt$ leads to the solution

$$r = 3 + c_1 e^{-0.5t}, \quad \theta = t + c_2.$$

If $X(0) = (0, 1)$, then $r(0) = 1$ and $\theta(0) = \pi/2$, and so $c_1 = -2$ and $c_2 = \pi/2$. The solution curve is the spiral $r = 3 - 2e^{-0.5(\theta - \pi/2)}$. Note that as $t \to \infty$, θ increases without bound and r approaches 3.

If $X(0) = (3, 0)$, then $r(0) = 3$ and $\theta(0) = 0$. It follows that $c_1 = c_2 = 0$, and so $r = 3$ and $\theta = t$. Hence $x = r \cos \theta = 3 \cos t$ and $y = r \sin \theta = 3 \sin t$, and so the solution is periodic. The solution generates a circle of radius 3 about $(0, 0)$. Both solutions are shown in Figure 10.6. ■

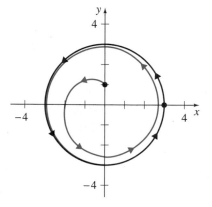

Figure 10.6

EXERCISES 10.1

Answers to odd-numbered problems begin on page AN-15.

In Problems 1–6 write the given nonlinear second-order differential equation as a plane autonomous system. Find all critical points of the resulting system.

1. $x'' + 9 \sin x = 0$

2. $x'' + (x')^2 + 2x = 0$

3. $x'' + x'(1 - x^3) - x^2 = 0$

4. $x'' + 4\dfrac{x}{1 + x^2} + 2x' = 0$

5. $x'' + x = \epsilon x^3$ for $\epsilon > 0$

6. $x'' + x - \epsilon x|x| = 0$ for $\epsilon > 0$

In Problems 7–16 find all critical points of the given plane autonomous system.

7. $x' = x + xy$
$\quad y' = -y - xy$

8. $x' = y^2 - x$
$\quad y' = x^2 - y$

9. $x' = 3x^2 - 4y$
$\quad y' = x - y$

10. $x' = x^3 - y$
$\quad y' = x - y^3$

11. $x' = x(10 - x - \tfrac{1}{2}y)$
$\quad y' = y(16 - y - x)$

12. $x' = -2x + y + 10$
$\quad y' = 2x - y - 15\dfrac{y}{y + 5}$

13. $x' = x^2 e^y$
$y' = y(e^x - 1)$

14. $x' = \sin y$
$y' = e^{x-y} - 1$

15. $x' = x(1 - x^2 - 3y^2)$
$y' = y(3 - x^2 - 3y^2)$

16. $x' = -x(4 - y^2)$
$y' = 4y(1 - x^2)$

In Problems 17–22 the given linear system is taken from Exercises 8.2.

(a) Find the general solution and determine whether there are periodic solutions.

(b) Find the solution satisfying the given initial condition.

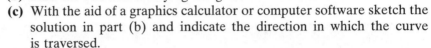

(c) With the aid of a graphics calculator or computer software sketch the solution in part (b) and indicate the direction in which the curve is traversed.

17. $x' = x + 2y$
$y' = 4x + 3y$, $\mathbf{X}(0) = (2, -2)$ (Problem 1, Exercises 8.2)

18. $x' = -6x + 2y$
$y' = -3x + y$, $\mathbf{X}(0) = (3, 4)$ (Problem 6, Exercises 8.2)

19. $x' = 4x - 5y$
$y' = 5x - 4y$, $\mathbf{X}(0) = (4, 5)$ (Problem 37, Exercises 8.2)

20. $x' = x + y$
$y' = -2x - y$, $\mathbf{X}(0) = (-2, 2)$ (Problem 34, Exercises 8.2)

21. $x' = 5x + y$
$y' = -2x + 3y$, $\mathbf{X}(0) = (-1, 2)$ (Problem 35, Exercises 8.2)

22. $x' = x - 8y$
$y' = x - 3y$, $\mathbf{X}(0) = (2, 1)$ (Problem 38, Exercises 8.2)

In Problems 23–26 solve the given nonlinear plane autonomous system by changing to polar coordinates. Describe the geometric behavior of the solution that satisfies the given initial condition(s).

23. $x' = -y - x(x^2 + y^2)^2$
$y' = x - y(x^2 + y^2)^2$, $\mathbf{X}(0) = (4, 0)$

24. $x' = y + x(x^2 + y^2)$
$y' = -x + y(x^2 + y^2)$, $\mathbf{X}(0) = (4, 0)$

25. $x' = -y + x(1 - x^2 - y^2)$
$y' = x + y(1 - x^2 - y^2)$, $\mathbf{X}(0) = (1, 0), \mathbf{X}(0) = (2, 0)$

[*Hint*: The resulting differential equation for r is a Bernoulli differential equation. See Section 2.5.]

26. $x' = y - \dfrac{x}{\sqrt{x^2 + y^2}}(4 - x^2 - y^2)$

$y' = -x - \dfrac{y}{\sqrt{x^2 + y^2}}(4 - x^2 - y^2)$, $\mathbf{X}(0) = (1, 0), \mathbf{X}(0) = (2, 0)$

If a plane autonomous system has a periodic solution, then there must be at least one critical point inside the curve generated by the solution. In Problems 27–30 use this fact together with a numerical solver to investigate the possibility of periodic solutions.

27. $x' = -x + 6y$
$y' = xy + 12$

28. $x' = -x + 6xy$
$y' = -8xy + 2y$

29. $x' = y$
$y' = y(1 - 3x^2 - 2y^2) - x$

30. $x' = xy$
$y' = -1 - x^2 - y^2$

31. If $z = f(x, y)$ is a function with continuous first partial derivatives in a region R, then a flow $\mathbf{V}(x, y) = (P(x, y), Q(x, y))$ in R may be defined by letting $P(x, y) = -\partial f/\partial y$ and $Q(x, y) = \partial f/\partial x$. Show that if $\mathbf{X}(t) = (x(t), y(t))$ is a solution of the plane autonomous system (2) then $f(x(t), y(t)) = c$ for some constant c. Thus a solution curve lies on the level curves of f. [*Hint*: Use the Chain Rule to compute $\frac{d}{dt} f(x(t), y(t))$.]

10.2 STABILITY OF LINEAR SYSTEMS

• Locally stable critical points   • Unstable critical points
• Stability analysis using eigenvalues and eigenvectors   • Saddle
points   • Nodes   • Degenerate nodes   • Centers   • Spiral points

In Section 10.1 we noted that the plane autonomous system

$$\frac{dx}{dt} = P(x, y)$$

$$\frac{dy}{dt} = Q(x, y)$$

gives rise to a vector field $\mathbf{V}(x, y) = (P(x, y), Q(x, y))$, and a solution $\mathbf{X} = \mathbf{X}(t)$ may be interpreted as the resulting path of a particle that is initially placed at position $\mathbf{X}(0) = \mathbf{X}_0$. If $\mathbf{X}_0$ is a critical point, the particle remains stationary. In this section we examine the behavior of solutions when $\mathbf{X}_0$ is chosen close to a critical point of the system.

Some Fundamental Questions   Suppose that $\mathbf{X}_1$ is a critical point of a plane autonomous system and $\mathbf{X} = \mathbf{X}(t)$ is a solution of the system that satisfies $\mathbf{X}(0) = \mathbf{X}_0$. If the solution is interpreted as a path of a moving particle, we are interested in the answers to the following questions when $\mathbf{X}_0$ is placed near $\mathbf{X}_1$:

(*i*) Will the particle return to the critical point? More precisely, does $\lim_{t \to \infty} \mathbf{X}(t) = \mathbf{X}_1$?

(*ii*) If the particle does *not* return to the critical point, does it remain close to the critical point or move away from the critical point? It is conceivable, for example, that the particle may simply circle the critical point, or it may even return to a different critical point or to no critical point at all. See Figure 10.7.

If in some neighborhood of the critical point case (a) or (b) in Figure 10.7 *always* occurs, we call the critical point **locally stable.** If, however, an initial value $\mathbf{X}_0$ that results in behavior similar to (c) can be found in *any* given neighborhood, we call the critical point **unstable.** These concepts will be made more precise in Section 10.3, where questions (*i*) and (*ii*) will be investigated for nonlinear systems.

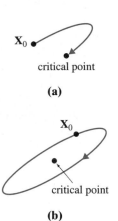

(a)

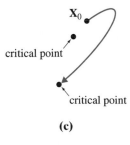

(b)

(c)

Figure 10.7

Stability Analysis   We will first investigate these two stability questions for linear plane autonomous systems and lay the foundation for Section 10.3. The solution methods of Chapter 8 enable us to give a careful geometric analysis of the solutions to

$$x' = ax + by$$
$$y' = cx + dy \tag{1}$$

in terms of the eigenvalues and eigenvectors of the coefficient matrix

$$\mathbf{A} = \begin{pmatrix} a & b \\ c & d \end{pmatrix}.$$

To ensure that $\mathbf{X}_0 = (0, 0)$ is the only critical point, we will assume that the determinant $\Delta = ad - bc \neq 0$. If $\tau = a + d$ is the trace* of matrix $\mathbf{A}$, then the characteristic equation $\det(\mathbf{A} - \lambda\mathbf{I}) = 0$ may be rewritten as

$$\lambda^2 - \tau\lambda + \Delta = 0.$$

Therefore the eigenvalues of $\mathbf{A}$ are $\lambda = (\tau \pm \sqrt{\tau^2 - 4\Delta})/2$, and the usual three cases for these roots occur according to whether $\tau^2 - 4\Delta$ is positive, negative, or zero. In the next example we use a numerical solver to discover the nature of the solutions corresponding to these cases.

EXAMPLE 1   **Eigenvalues and the Shape of Solutions**

Find the eigenvalues of the linear system

$$x' = -x + y$$
$$y' = cx - y$$

in terms of c, and use a numerical solver to discover the shapes of solutions corresponding to the cases $c = \frac{1}{4}, 4, 0$, and -9.

Solution   The coefficient matrix $\begin{pmatrix} -1 & 1 \\ c & -1 \end{pmatrix}$ has trace $\tau = -2$ and determinant $\Delta = 1 - c$, and so the eigenvalues are

$$\lambda = \frac{\tau \pm \sqrt{\tau^2 - 4\Delta}}{2} = \frac{-2 \pm \sqrt{4 - 4(1 - c)}}{2} = -1 \pm \sqrt{c}.$$

The nature of the eigenvalues is therefore determined by the sign of c.

If $c = \frac{1}{4}$, then the eigenvalues are negative and distinct, $\lambda = -\frac{1}{2}$ and $-\frac{3}{2}$. In Figure 10.8(a) we have used a numerical solver to generate solution curves, or trajectories, that correspond to various initial conditions. Note that, except for the trajectories drawn in black in the figure, the trajectories all appear to approach **0** from a fixed direction. Recall from Chapter 8 that a collection of trajectories in the xy-plane, or **phase plane**, is called a **phase portrait** of the system.

When $c = 4$, the eigenvalues have opposite signs, $\lambda = 1$ and -3, and an interesting phenomenon occurs. All trajectories move away from the origin in a fixed direction except for solutions that start along the single

---

*In general, if $\mathbf{A}$ is an $n \times n$ matrix, the **trace** of $\mathbf{A}$ is the sum of the main diagonal entries.

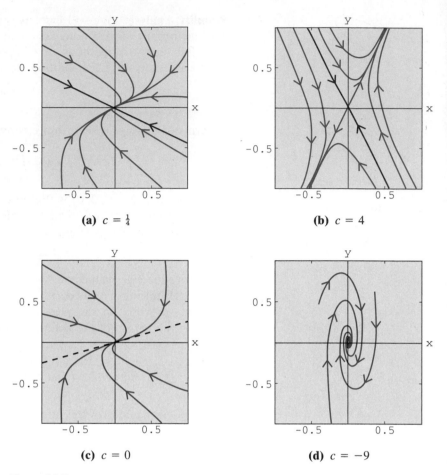

(a) $c = \frac{1}{4}$

(b) $c = 4$

(c) $c = 0$

(d) $c = -9$

Figure 10.8

line drawn in black in Figure 10.8(b). We have already seen behavior like this in the phase portrait given in Figure 8.2. Experiment with your numerical solver and verify these observations.

The selection $c = 0$ leads to a single real eigenvalue $\lambda = -1$. This case is very similar to the case $c = \frac{1}{4}$ with one notable exception. All solution curves in Figure 10.8(c) appear to approach $\mathbf{0}$ from a fixed direction as t increases.

Finally, when $c = -9$, $\lambda = -1 \pm \sqrt{-9} = -1 \pm 3i$. Thus the eigenvalues are conjugate complex numbers with negative real part -1. Figure 10.8(d) shows that solution curves spiral in toward the origin $\mathbf{0}$ as t increases.   ∎

The behaviors of the trajectories observed in the four phase portraits in Figure 10.8 in Example 1 can be explained using the eigenvalue-eigenvector solution results from Chapter 8.

Case I: Real Distinct Eigenvalues ($\tau^2 - 4\Delta > 0$)
According to Theorem 8.7 in Section 8.2, the general solution of (1) is given by

$$\mathbf{X}(t) = c_1 \mathbf{K}_1 e^{\lambda_1 t} + c_2 \mathbf{K}_2 e^{\lambda_2 t}, \tag{2}$$

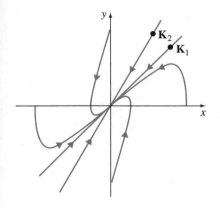

Figure 10.9 Stable node

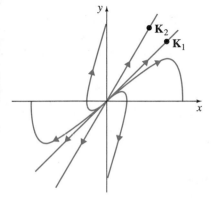

Figure 10.10 Unstable node

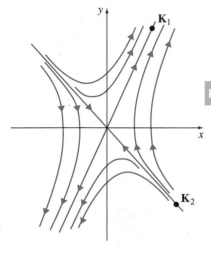

Figure 10.11 Saddle point

where λ_1 and λ_2 are the eigenvalues and $\mathbf{K}_1$ and $\mathbf{K}_2$ are the corresponding eigenvectors. Note that $\mathbf{X}(t)$ can also be written as

$$\mathbf{X}(t) = e^{\lambda_1 t}[c_1\mathbf{K}_1 + c_2\mathbf{K}_2 e^{(\lambda_2 - \lambda_1)t}]. \tag{3}$$

(a) **Both eigenvalues negative** ($\tau^2 - 4\Delta > 0$, $\tau < 0$, and $\Delta > 0$)
Stable Node ($\lambda_2 < \lambda_1 < 0$): Since both eigenvalues are negative, it follows from (2) that $\lim_{t\to\infty} \mathbf{X}(t) = \mathbf{0}$. If we assume that $\lambda_2 < \lambda_1$, then $\lambda_2 - \lambda_1 < 0$ and so $e^{(\lambda_2-\lambda_1)t}$ is an exponential decay function. We may therefore conclude from (3) that $\mathbf{X}(t) \approx c_1\mathbf{K}_1 e^{\lambda_1 t}$ for large values of t. When $c_1 \neq 0$, $\mathbf{X}(t)$ will approach $\mathbf{0}$ from one of the two directions determined by the eigenvector $\mathbf{K}_1$ corresponding to λ_1. If $c_1 = 0$, $\mathbf{X}(t) = c_2\mathbf{K}_2 e^{\lambda_2 t}$ and $\mathbf{X}(t)$ approaches $\mathbf{0}$ along the line determined by the eigenvector $\mathbf{K}_2$. Figure 10.9 shows a collection of solution curves around the origin. A critical point is called a **stable node** when both eigenvalues are negative.

(b) **Both eigenvalues positive** ($\tau^2 - 4\Delta > 0$, $\tau > 0$, and $\Delta > 0$)
Unstable Node ($0 < \lambda_2 < \lambda_1$): The analysis for this case is similar to (a). Again from (2), $\mathbf{X}(t)$ becomes unbounded as t increases. Moreover, again assuming $\lambda_2 < \lambda_1$ and using (3), we see that $\mathbf{X}(t)$ becomes unbounded in one of the directions determined by the eigenvector $\mathbf{K}_1$ (when $c_1 \neq 0$) or along the line determined by the eigenvector $\mathbf{K}_2$ (when $c_1 = 0$). Figure 10.10 shows a typical collection of solution curves. This type of critical point, corresponding to the case when both eigenvalues are positive, is called an **unstable node.**

(c) **Eigenvalues have opposite signs** ($\tau^2 - 4\Delta > 0$ and $\Delta < 0$)
Saddle Point ($\lambda_2 < 0 < \lambda_1$): The analysis of the solutions is identical to (b) with one exception. When $c_1 = 0$, $\mathbf{X}(t) = c_2\mathbf{K}_2 e^{\lambda_2 t}$ and, since $\lambda_2 < 0$, $\mathbf{X}(t)$ will approach $\mathbf{0}$ along the line determined by the eigenvector $\mathbf{K}_2$. If $\mathbf{X}(0)$ does not lie on the line determined by $\mathbf{K}_2$, the line determined by $\mathbf{K}_1$ serves as an asymptote for $\mathbf{X}(t)$. Thus the critical point is unstable even though some solutions approach $\mathbf{0}$ as t increases. This unstable critical point is called a **saddle point.** See Figure 10.11.

EXAMPLE 2 **Real Distinct Eigenvalues**

Classify the critical point $(0, 0)$ of each of the following linear systems $\mathbf{X}' = \mathbf{AX}$ as either a stable node, an unstable node, or a saddle point.

(a) $\mathbf{A} = \begin{pmatrix} 2 & 3 \\ 2 & 1 \end{pmatrix}$ (b) $\mathbf{A} = \begin{pmatrix} -10 & 6 \\ 15 & -19 \end{pmatrix}$

In each case discuss the nature of the solutions in a neighborhood of $(0,0)$.

Solution
(a) Since the trace $\tau = 3$ and the determinant $\Delta = -4$, the eigenvalues are

$$\lambda = \frac{\tau \pm \sqrt{\tau^2 - 4\Delta}}{2} = \frac{3 \pm \sqrt{3^2 - 4(-4)}}{2} = \frac{3 \pm 5}{2} = 4, -1.$$

The eigenvalues have opposite signs, and so $(0, 0)$ is a saddle point. It is not hard to show (see Example 1, Section 8.2) that eigenvectors

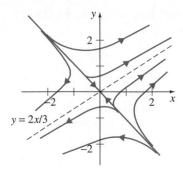

Figure 10.12

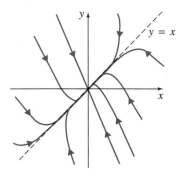

Figure 10.13

corresponding to $\lambda_1 = 4$ and $\lambda_2 = -1$ are

$$\mathbf{K}_1 = \begin{pmatrix} 3 \\ 2 \end{pmatrix} \quad \text{and} \quad \mathbf{K}_2 = \begin{pmatrix} 1 \\ -1 \end{pmatrix},$$

respectively. If $\mathbf{X}(0) = \mathbf{X}_0$ lies on the line $y = -x$, then $\mathbf{X}(t)$ approaches **0**. For any other initial condition, $\mathbf{X}(t)$ will become unbounded in the directions determined by $\mathbf{K}_1$. In other words, the line $y = \frac{2}{3}x$ serves as an asymptote for all these solution curves. See Figure 10.12.

(b) From $\tau = -29$ and $\Delta = 100$ it follows that the eigenvalues of $\mathbf{A}$ are $\lambda_1 = -4$ and $\lambda_2 = -25$. Both eigenvalues are negative, and so $(0, 0)$ is in this case a stable node. Since eigenvectors corresponding to $\lambda_1 = -4$ and $\lambda_2 = -25$ are

$$\mathbf{K}_1 = \begin{pmatrix} 1 \\ 1 \end{pmatrix} \quad \text{and} \quad \mathbf{K}_2 = \begin{pmatrix} 2 \\ -5 \end{pmatrix},$$

respectively, it follows that all solutions approach **0** from the direction defined by $\mathbf{K}_1$ except those solutions for which $\mathbf{X}(0) = \mathbf{X}_0$ lies on the line $y = -\frac{5}{2}x$ determined by $\mathbf{K}_2$. These solutions approach **0** along $y = -\frac{5}{2}x$. See Figure 10.13. ∎

Case II: A Repeated Real Eigenvalue ($\tau^2 - 4\Delta = 0$)

Degenerate Nodes: Recall from Section 8.2 that the general solution takes on one of two different forms depending on whether one or two linearly independent eigenvectors can be found for the repeated eigenvalue λ_1.

(a) Two linearly independent eigenvectors

If $\mathbf{K}_1$ and $\mathbf{K}_2$ are two linearly independent eigenvectors corresponding to λ_1, then the general solution is given by

$$\mathbf{X}(t) = c_1\mathbf{K}_1 e^{\lambda_1 t} + c_2\mathbf{K}_2 e^{\lambda_1 t} = (c_1\mathbf{K}_1 + c_2\mathbf{K}_2)e^{\lambda_1 t}.$$

If $\lambda_1 < 0$, then $\mathbf{X}(t)$ approaches **0** along the line determined by the vector $c_1\mathbf{K}_1 + c_2\mathbf{K}_2$ and the critical point is called a **degenerate stable node** (see Figure 10.14(a)). The arrows in Figure 10.14(a) are reversed when $\lambda_1 > 0$, and we have a **degenerate unstable node.**

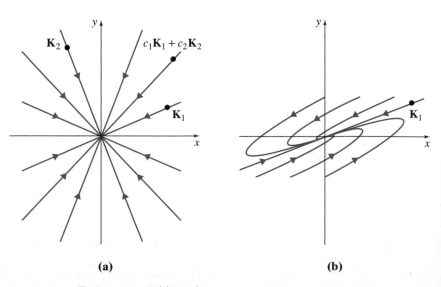

(a)   **(b)**

Figure 10.14 Degenerate stable nodes

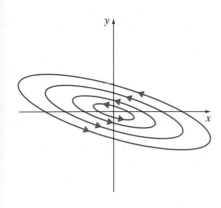

Figure 10.15 Center

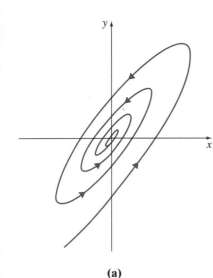

(a)

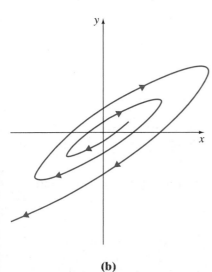

(b)

Figure 10.16 Stable and unstable spiral points

(b) A single linearly independent eigenvector

When only a single linearly independent eigenvector $\mathbf{K}_1$ exists, the general solution is given by

$$\mathbf{X}(t) = c_1\mathbf{K}_1 e^{\lambda_1 t} + c_2(\mathbf{K}_1 t e^{\lambda_1 t} + \mathbf{P}e^{\lambda_1 t}),$$

where $(\mathbf{A} - \lambda_1\mathbf{I})\mathbf{P} = \mathbf{K}_1$ (see Section 8.2, (12)–(14)), and the solution may be rewritten as

$$\mathbf{X}(t) = t e^{\lambda_1 t}\left[c_2\mathbf{K}_1 + \frac{c_1}{t}\mathbf{K}_1 + \frac{c_2}{t}\mathbf{P}\right].$$

If $\lambda_1 < 0$, then $\lim_{t\to\infty} t e^{\lambda_1 t} = 0$ and it follows that $\mathbf{X}(t)$ approaches $\mathbf{0}$ in one of the directions determined by the vector $\mathbf{K}_1$ (see Figure 10.14(b)). The critical point is again called a **degenerate stable node.** When $\lambda_1 > 0$, the solutions look like those in Figure 10.14(b) with the arrows reversed. The line determined by $\mathbf{K}_1$ is an asymptote for *all* solutions. The critical point is again called a **degenerate unstable node.**

Case III: Complex Eigenvalues $(\tau^2 - 4\Delta < 0)$

If $\lambda_1 = \alpha + i\beta$ and $\bar{\lambda}_1 = \alpha - i\beta$ are the complex eigenvalues and $\mathbf{K}_1 = \mathbf{B}_1 + i\mathbf{B}_2$ is a complex eigenvector corresponding to λ_1, the general solution can be written as $\mathbf{X}(t) = c_1\mathbf{X}_1(t) + c_2\mathbf{X}_2(t)$, where

$$\mathbf{X}_1(t) = (\mathbf{B}_1 \cos \beta t - \mathbf{B}_2 \sin \beta t)e^{\alpha t}, \quad \mathbf{X}_2(t) = (\mathbf{B}_2 \cos \beta t + \mathbf{B}_1 \sin \beta t)e^{\alpha t}.$$

See equations (23) and (24) in Section 8.2. A solution can therefore be written in the form

$$x(t) = e^{\alpha t}(c_{11} \cos \beta t + c_{12} \sin \beta t), \quad y(t) = e^{\alpha t}(c_{21} \cos \beta t + c_{22} \sin \beta t), \quad \textbf{(4)}$$

and when $\alpha = 0$ we have

$$x(t) = c_{11} \cos \beta t + c_{12} \sin \beta t, \quad y(t) = c_{21} \cos \beta t + c_{22} \sin \beta t. \quad \textbf{(5)}$$

(a) Pure imaginary roots $(\tau^2 - 4\Delta < 0, \ \tau = 0)$

Center: When $\alpha = 0$, the eigenvalues are pure imaginary and, from (5), all solutions are periodic with period $p = 2\pi/\beta$. Notice that if both c_{12} and c_{21} happened to be 0, then (5) would reduce to

$$x(t) = c_{11} \cos \beta t, \quad y(t) = c_{22} \sin \beta t,$$

which is a standard parametric representation for the ellipse $x^2/c_{11}^2 + y^2/c_{22}^2 = 1$. By solving the system of equations in (4) for $\cos \beta t$ and $\sin \beta t$ and using the identity $\sin^2\beta t + \cos^2\beta t = 1$, it is possible to show that *all solutions are ellipses* with center at the origin. The critical point $(0, 0)$ is called a **center,** and Figure 10.15 shows a typical collection of solution curves. The ellipses are either *all* traversed in the clockwise direction or *all* traversed in the counterclockwise direction.

(b) Nonzero real part $(\tau^2 - 4\Delta < 0, \tau \neq 0)$

Spiral Points: When $\alpha \neq 0$, the effect of the term $e^{\alpha t}$ in (4) is similar to the effect of the exponential term in the analysis of damped motion given in Section 5.1. When $\alpha < 0$, $e^{\alpha t} \to 0$, and the elliptical-like solution spirals closer and closer to the origin. The critical point is called a **stable spiral point.** When $\alpha > 0$, the effect is the opposite. An elliptical-like solution is driven farther and farther from the origin, and the critical point is now called an **unstable spiral point.** See Figure 10.16.

EXAMPLE 3   **Repeated and Complex Eigenvalues**

Classify the critical point (0, 0) of each of the following linear systems $\mathbf{X}' = \mathbf{AX}$:

(a) $\mathbf{A} = \begin{pmatrix} 3 & -18 \\ 2 & -9 \end{pmatrix}$ **(b)** $\mathbf{A} = \begin{pmatrix} -1 & 2 \\ -1 & 1 \end{pmatrix}$

In each case discuss the nature of the solution that satisfies $\mathbf{X}(0) = (1, 0)$. Determine parametric equations for each solution.

Solution
(a) Since $\tau = -6$ and $\Delta = 9$, the characteristic polynomial is $\lambda^2 + 6\lambda + 9 = (\lambda + 3)^2$, and so (0, 0) is a degenerate stable node. For the repeated eigenvalue $\lambda = -3$ we find a single eigenvector $\mathbf{K}_1 = \begin{pmatrix} 3 \\ 1 \end{pmatrix}$, and so the solution $\mathbf{X}(t)$ that satisfies $\mathbf{X}(0) = (1, 0)$ approaches (0, 0) from the direction specified by the line $y = x/3$.

(b) Since $\tau = 0$ and $\Delta = 1$, the eigenvalues are $\lambda = \pm i$, and so (0, 0) is a center. The solution $\mathbf{X}(t)$ that satisfies $\mathbf{X}(0) = (1, 0)$ is an ellipse that circles the origin every 2π units of time.

From Example 4 of Section 8.2 the general solution of the system in (a) is

$$\mathbf{X}(t) = c_1 \begin{pmatrix} 3 \\ 1 \end{pmatrix} e^{-3t} + c_2 \left[\begin{pmatrix} 3 \\ 1 \end{pmatrix} te^{-3t} + \begin{pmatrix} \frac{1}{2} \\ 0 \end{pmatrix} e^{-3t} \right].$$

The initial condition gives $c_1 = 0$ and $c_2 = 2$, and so

$$x = (6t + 1)e^{-3t}, \quad y = 2te^{-3t}$$

are parametric equations for the solution.

The general solution of the system in (b) is

$$\mathbf{X}(t) = c_1 \begin{pmatrix} \cos t + \sin t \\ \cos t \end{pmatrix} + c_2 \begin{pmatrix} \cos t - \sin t \\ -\sin t \end{pmatrix}.$$

The initial condition gives $c_1 = 0$ and $c_2 = 1$, and so

$$x = \cos t - \sin t, \quad y = -\sin t$$

are parametric equations for the ellipse. Note that $y < 0$ for small positive values of t, and therefore the ellipse is traversed in the clockwise direction.

The solutions of (a) and (b) are shown in Figures 10.17(a) and (b), respectively.

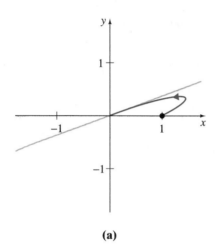

(a)

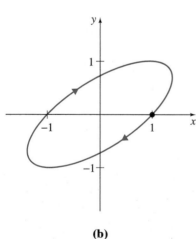

(b)

Figure 10.17

Figure 10.18 conveniently summarizes the results of this section. The general geometric nature of the solutions can be determined by computing the trace and determinant of $\mathbf{A}$. In practice, graphs of the solutions are most easily obtained *not* by constructing explicit eigenvalue-eigenvector

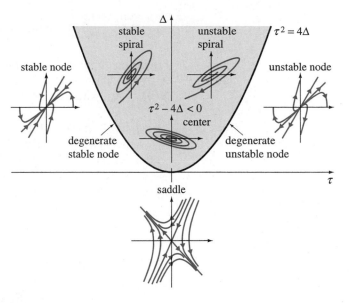

Figure 10.18

solutions but rather by generating the solutions using a numerical solver and the Runge-Kutta method for first-order systems.

EXAMPLE 4 **Classifying Critical Points**

Classify the critical point $(0, 0)$ of each of the following linear systems $\mathbf{X}' = \mathbf{AX}$:

(a) $\mathbf{A} = \begin{pmatrix} 1.01 & 3.10 \\ -1.10 & -1.02 \end{pmatrix}$ **(b)** $\mathbf{A} = \begin{pmatrix} -a\hat{x} & -ab\hat{x} \\ -cd\hat{y} & -d\hat{y} \end{pmatrix}$

for positive constants a, b, c, d, $\hat{x}$, and $\hat{y}$.

Solution

(a) For this matrix $\tau = -0.01$, $\Delta = 2.3798$, and so $\tau^2 - 4\Delta < 0$. Using Figure 10.18, we see that $(0, 0)$ is a stable spiral point.

(b) This matrix arises from the Lotka-Volterra competition model, which we will study in Section 10.4. Since $\tau = -(a\hat{x} + d\hat{y})$ and all constants in the matrix are positive, $\tau < 0$. The determinant may be written as $\Delta = ad\hat{x}\hat{y}(1 - bc)$. If $bc > 1$, then $\Delta < 0$ and the critical point is a saddle point. If $bc < 1$, $\Delta > 0$ and the critical point is either a stable node, a degenerate stable node, or a stable spiral point. In all three cases $\lim_{t \to \infty} \mathbf{X}(t) = 0$. ∎

The answers to the questions posed at the beginning of this section for the linear plane autonomous system (1) with $ad - bc \neq 0$ are summarized in the next theorem.

> **THEOREM 10.1   Stability Criteria for Linear Systems**
>
> For a linear plane autonomous system $\mathbf{X}' = \mathbf{AX}$ with det $\mathbf{A} \neq 0$, let $\mathbf{X} = \mathbf{X}(t)$ denote the solution that satisfies the initial condition $\mathbf{X}(0) = \mathbf{X}_0$, where $\mathbf{X}_0 \neq \mathbf{0}$.
>
>   (a) $\lim_{t \to \infty} \mathbf{X}(t) = \mathbf{0}$ if and only if the eigenvalues of $\mathbf{A}$ have negative real parts. This will occur when $\Delta > 0$ and $\tau < 0$.
>   (b) $\mathbf{X}(t)$ is periodic if and only if the eigenvalues of $\mathbf{A}$ are pure imaginary. This will occur when $\Delta > 0$ and $\tau = 0$.
>   (c) In all other cases, given any neighborhood of the origin, there is at least one $\mathbf{X}_0$ in the neighborhood for which $\mathbf{X}(t)$ becomes unbounded as t increases.

Remarks The terminology used to describe the types of critical points varies from text to text. The following table lists many of the alternative terms that you may encounter in your reading.

Term	Alternative Terms
critical point	equilibrium point, singular point, stationary point, rest point
spiral point	focus, focal point, vortex point
stable node or spiral point	attractor, sink
unstable node or spiral point	repeller, source

EXERCISES 10.2

Answers to odd-numbered problems begin on page AN-15.

In Problems 1–8 the general solution of the linear system $\mathbf{X}' = \mathbf{AX}$ is given.
(a) In each case discuss the nature of the solutions in a neighborhood of $(0, 0)$.

(b) With the aid of a graphics calculator or computer software sketch the solution that satisfies $\mathbf{X}(0) = (1, 1)$.

1. $\mathbf{A} = \begin{pmatrix} -2 & -2 \\ -2 & -5 \end{pmatrix}$, $\mathbf{X}(t) = c_1 \begin{pmatrix} 2 \\ -1 \end{pmatrix} e^{-t} + c_2 \begin{pmatrix} 1 \\ 2 \end{pmatrix} e^{-6t}$

2. $\mathbf{A} = \begin{pmatrix} -1 & -2 \\ 3 & 4 \end{pmatrix}$, $\mathbf{X}(t) = c_1 \begin{pmatrix} 1 \\ -1 \end{pmatrix} e^{t} + c_2 \begin{pmatrix} -4 \\ 6 \end{pmatrix} e^{2t}$

3. $\mathbf{A} = \begin{pmatrix} 1 & -1 \\ 1 & 1 \end{pmatrix}$, $\mathbf{X}(t) = e^{t} \left[c_1 \begin{pmatrix} -\sin t \\ \cos t \end{pmatrix} + c_2 \begin{pmatrix} \cos t \\ \sin t \end{pmatrix} \right]$

4. $\mathbf{A} = \begin{pmatrix} -1 & -4 \\ 1 & -1 \end{pmatrix}$, $\mathbf{X}(t) = e^{-t} \left[c_1 \begin{pmatrix} 2 \cos 2t \\ \sin 2t \end{pmatrix} + c_2 \begin{pmatrix} -2 \sin 2t \\ \cos 2t \end{pmatrix} \right]$

5. $\mathbf{A} = \begin{pmatrix} -6 & 5 \\ -5 & 4 \end{pmatrix}$, $\mathbf{X}(t) = c_1 \begin{pmatrix} 1 \\ 1 \end{pmatrix} e^{-t} + c_2 \left[\begin{pmatrix} 1 \\ 1 \end{pmatrix} te^{-t} + \begin{pmatrix} 0 \\ \frac{1}{5} \end{pmatrix} e^{-t} \right]$

6. $\mathbf{A} = \begin{pmatrix} 2 & 4 \\ -1 & 6 \end{pmatrix}$, $\mathbf{X}(t) = c_1 \begin{pmatrix} 2 \\ 1 \end{pmatrix} e^{4t} + c_2 \left[\begin{pmatrix} 2 \\ 1 \end{pmatrix} te^{4t} + \begin{pmatrix} 1 \\ 1 \end{pmatrix} e^{4t} \right]$

7. $\mathbf{A} = \begin{pmatrix} 2 & -1 \\ 3 & -2 \end{pmatrix}$, $\mathbf{X}(t) = c_1 \begin{pmatrix} 1 \\ 1 \end{pmatrix} e^{t} + c_2 \begin{pmatrix} 1 \\ 3 \end{pmatrix} e^{-t}$

8. $\mathbf{A} = \begin{pmatrix} -1 & 5 \\ -1 & 1 \end{pmatrix}$, $\mathbf{X}(t) = c_1 \begin{pmatrix} 5 \cos 2t \\ \cos 2t - 2\sin 2t \end{pmatrix} + c_2 \begin{pmatrix} 5 \sin 2t \\ 2\cos 2t + \sin 2t \end{pmatrix}$

In Problems 9–16 classify the critical point $(0, 0)$ of the given linear system by computing the trace τ and determinant Δ and using Figure 10.18.

9. $x' = -5x + 3y$
$\quad y' = 2x + 7y$

10. $x' = -5x + 3y$
$\quad\ \ y' = 2x - 7y$

11. $x' = -5x + 3y$
$\quad\ \ y' = -2x + 5y$

12. $x' = -5x + 3y$
$\quad\ \ y' = -7x + 4y$

13. $x' = -\frac{3}{2}x + \frac{1}{4}y$
$\quad\ \ y' = -x - \frac{1}{2}y$

14. $x' = \frac{3}{2}x + \frac{1}{4}y$
$\quad\ \ y' = -x + \frac{1}{2}y$

15. $x' = 0.02x - 0.11y$
$\quad\ \ y' = 0.10x - 0.05y$

16. $x' = 0.03x + 0.01y$
$\quad\ \ y' = -0.01x + 0.05y$

17. Determine conditions on the real constant μ so that $(0, 0)$ is a center for the linear system

$$x' = -\mu x + y$$
$$y' = -x + \mu y.$$

18. Determine a condition on the real constant μ so that $(0, 0)$ is a stable spiral point of the linear system

$$x' = y$$
$$y' = -x + \mu y.$$

19. Show that $(0, 0)$ is always an unstable critical point of the linear system

$$x' = \mu x + y$$
$$y' = -x + y,$$

where μ is a real constant and $\mu \neq -1$. When is $(0, 0)$ an unstable saddle point? When is $(0, 0)$ an unstable spiral point?

20. Let $\mathbf{X} = \mathbf{X}(t)$ be the response of the linear dynamical system

$$x' = \alpha x - \beta y$$
$$y' = \beta x + \alpha y$$

that satisfies the initial condition $\mathbf{X}(0) = \mathbf{X}_0$. Determine conditions on the real constants α and β that will ensure $\lim_{t \to \infty} \mathbf{X}(t) = (0, 0)$. Can $(0, 0)$ be a node or saddle point?

21. Show that the nonhomogeneous linear system $\mathbf{X}' = \mathbf{AX} + \mathbf{F}$ has a unique critical point $\mathbf{X}_1$ when $\Delta = \det \mathbf{A} \neq 0$. Conclude that if $\mathbf{X} = \mathbf{X}(t)$ is a solution to the nonhomogeneous system, $\tau < 0$ and $\Delta > 0$, then $\lim_{t \to \infty} \mathbf{X}(t) = \mathbf{X}_1$. [*Hint:* $\mathbf{X}(t) = \mathbf{X}_c(t) + \mathbf{X}_1$.]

22. In Example 4(b) show that $(0, 0)$ is a stable node when $bc < 1$.

In Problems 23–26 a nonhomogeneous linear system $\mathbf{X}' = \mathbf{AX} + \mathbf{F}$ is given.
(a) In each case determine the unique critical point $\mathbf{X}_1$.
(b) Use a numerical solver to determine the nature of the critical point in (a).
(c) Investigate the relationship between $\mathbf{X}_1$ and the critical point $(0, 0)$ of the homogeneous linear system $\mathbf{X}' = \mathbf{AX}$.

23. $x' = 2x + 3y - 6$
$\quad\; y' = -x - 2y + 5$

24. $x' = -5x + 9y + 13$
$\quad\; y' = -x - 11y - 23$

25. $x' = 0.1x - 0.2y + 0.35$
$\quad\; y' = 0.1x + 0.1y - 0.25$

26. $x' = 3x - 2y - 1$
$\quad\; y' = 5x - 3y - 2$

10.3 LINEARIZATION AND LOCAL STABILITY

• Stable and unstable critical points • Linearization • Jacobian matrix of a nonlinear system • Predicting stability using linearization • Classifying critical points using linearization • Phase-plane analysis

The key idea in this section is that of linearization. Recall from calculus that a local linear approximation, or **linearization**, of a differentiable function $f(x)$ at a point $(x_1, f(x_1))$ is the equation of the tangent line to the graph of f at that point: $y = f(x_1) + f'(x_1)(x - x_1)$. For x close to x_1 the values $y(x)$ obtained from the equation of the tangent line are approximations to the corresponding values $f(x)$. We shall see that linearization can be used to approximate a nonlinear differential equation by a linear equation and to approximate a nonlinear system of first-order differential equations by a linear system.

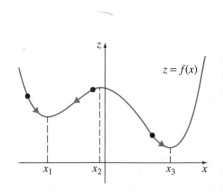

Figure 10.19

The Sliding Bead   We start this section by refining the stability concepts introduced in Section 10.2 in such a way that they will apply to nonlinear autonomous systems as well. Although the linear system $\mathbf{X}' = \mathbf{AX}$ had only one critical point when det $\mathbf{A} \neq 0$, we saw in Section 10.1 that a nonlinear system may have many critical points. We therefore cannot expect that a particle placed initially at a $\mathbf{X}_0$ will remain near a given critical point $\mathbf{X}_1$ unless $\mathbf{X}_0$ has been placed sufficiently close to $\mathbf{X}_1$ to begin with. The particle might well be driven to a second critical point. To emphasize this idea, consider the physical system shown in Figure 10.19 in which a bead slides along the curve $z = f(x)$ under the influence of gravity alone. We will show in Section 10.4 that the x-coordinate of the bead satisfies a nonlinear second-order differential equation $x'' = g(x, x')$, and therefore letting $y = x'$ satisfies the nonlinear autonomous system

$$x' = y$$
$$y' = g(x, y).$$

If the bead is positioned at $P = (x, f(x))$ and given zero initial velocity, the bead will remain at P provided $f'(x) = 0$. If the bead is placed near the critical point located at $x = x_1$, it will remain near $x = x_1$ only if its initial velocity does not drive it over the "hump" at $x = x_2$ toward the critical point located at $x = x_3$. Therefore $\mathbf{X}(0) = (x(0), x'(0))$ must be near $(x_1, 0)$.

In the next definition we will denote the distance between two points $\mathbf{X}$ and $\mathbf{Y}$ by $|\mathbf{X} - \mathbf{Y}|$. Recall that if $\mathbf{X} = (x_1, x_2, \ldots, x_n)$ and $\mathbf{Y} = (y_1, y_2, \ldots, y_n)$, then

$$|\mathbf{X} - \mathbf{Y}| = \sqrt{(x_1 - y_1)^2 + (x_2 - y_2)^2 + \cdots + (x_n - y_n)^2}.$$

DEFINITION 10.1 Stable Critical Points

Let $\mathbf{X}_1$ be a critical point of an autonomous system, and let $\mathbf{X} = \mathbf{X}(t)$ denote the solution that satisfies the initial condition $\mathbf{X}(0) = \mathbf{X}_0$, where $\mathbf{X}_0 \neq \mathbf{X}_1$. We say that $\mathbf{X}_1$ is a **stable critical point** when, given any radius $\rho > 0$, there is a corresponding radius $r > 0$ such that if the initial position $\mathbf{X}_0$ satisfies $|\mathbf{X}_0 - \mathbf{X}_1| < r$, then the corresponding solution $\mathbf{X}(t)$ satisfies $|\mathbf{X}(t) - \mathbf{X}_1| < \rho$ for all $t > 0$. If, in addition, $\lim_{t \to \infty} \mathbf{X}(t) = \mathbf{X}_1$ whenever $|\mathbf{X}_0 - \mathbf{X}_1| < r$, we call $\mathbf{X}_1$ an **asymptotically stable critical point.**

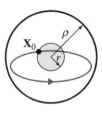

(a)

This definition is illustrated in Figure 10.20(a). Given any disk of radius ρ about the critical point $\mathbf{X}_1$, a solution will remain inside this disk provided $\mathbf{X}(0) = \mathbf{X}_0$ is selected sufficiently close to $\mathbf{X}_1$. It is *not* necessary that a solution approach the critical point in order for $\mathbf{X}_1$ to be stable. Stable nodes, stable spiral points, and centers are all examples of stable critical points for linear systems. To emphasize that $\mathbf{X}_0$ must be selected close to $\mathbf{X}_1$, the terminology **locally stable critical point** is also used.

By negating Definition 10.1 we obtain the definition of an unstable critical point.

DEFINITION 10.2 Unstable Critical Point

Let $\mathbf{X}_1$ be a critical point of an autonomous system, and let $\mathbf{X} = \mathbf{X}(t)$ denote the solution that satisfies the initial condition $\mathbf{X}(0) = \mathbf{X}_0$, where $\mathbf{X}_0 \neq \mathbf{X}_1$. We say that $\mathbf{X}_1$ is an **unstable critical point** if there is a disk of radius $\rho > 0$ with the property that, for any $r > 0$, there is at least one initial position $\mathbf{X}_0$ that satisfies $|\mathbf{X}_0 - \mathbf{X}_1| < r$ yet the corresponding solution $\mathbf{X}(t)$ satisfies $|\mathbf{X}(t) - \mathbf{X}_1| \geq \rho$ for at least one $t > 0$.

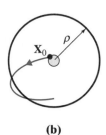

(b)

Figure 10.20

If a critical point $\mathbf{X}_1$ is unstable, no matter how small the neighborhood about $\mathbf{X}_1$, an initial position $\mathbf{X}_0$ can always be found that results in the solution leaving some disk of radius ρ at some future time t. See Figure 10.20(b). Therefore unstable nodes, unstable spiral points, and saddle points are all examples of unstable critical points for linear systems. In Figure 10.19 the critical point $(x_2, 0)$ is unstable. The slightest displacement or initial velocity results in the bead sliding away from the point $(x_2, f(x_2))$.

EXAMPLE 1   **A Stable Critical Point**

Show that $(0, 0)$ is a stable critical point of the nonlinear plane autonomous system

$$x' = -y - x\sqrt{x^2 + y^2}$$
$$y' = x - y\sqrt{x^2 + y^2}$$

considered in Example 6 of Section 10.1.

Solution   In Example 6 of Section 10.1 we showed that in polar coordinates $r = 1/(r + c_1)$, $\theta = t + c_2$ is the solution of the system. If $\mathbf{X}(0) = (r_0, \theta_0)$ is the initial condition in polar coordinates, then

$$r = \frac{r_0}{r_0 t + 1}, \quad \theta = t + \theta_0.$$

Note that $r \leq r_0$ for $t \geq 0$, and r approaches $(0, 0)$ as t increases. Therefore, given $\rho > 0$, a solution that starts less than ρ units from $(0, 0)$ remains within ρ units of the origin for all $t \geq 0$. Hence the critical point $(0, 0)$ is stable and is in fact asymptotically stable. A typical solution is shown in Figure 10.21.   ∎

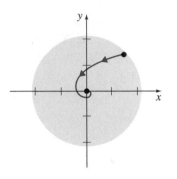

Figure 10.21

EXAMPLE 2   **An Unstable Critical Point**

When expressed in polar coordinates, a plane autonomous system takes the form

$$\frac{dr}{dt} = 0.05r(3 - r)$$

$$\frac{d\theta}{dt} = -1.$$

Show that $(x, y) = (0, 0)$ is an unstable critical point.

Solution   Since $x = r\cos\theta$ and $y = r\sin\theta$, we have

$$\frac{dx}{dt} = -r\sin\theta\frac{d\theta}{dt} + \frac{dr}{dt}\cos\theta$$

$$\frac{dy}{dt} = r\cos\theta\frac{d\theta}{dt} + \frac{dr}{dt}\sin\theta.$$

From $dr/dt = 0.05r(3 - r)$ we see that $dr/dt = 0$ when $r = 0$ and can conclude that $(x, y) = (0, 0)$ is a critical point by substituting $r = 0$ into the new system.

   The differential equation $dr/dt = 0.05r(3 - r)$ is a logistic equation that can be solved using either separation of variables or equation (5) in Section 3.2. If $r(0) = r_0$ and $r_0 \neq 0$, then

$$r = \frac{3}{1 + c_0 e^{-0.15t}},$$

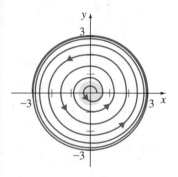

Figure 10.22

where $c_0 = (3 - r_0)/r_0$. Since

$$\lim_{t \to \infty} \frac{3}{1 + c_0 e^{-0.15t}} = 3,$$

it follows that no matter how close to $(0, 0)$ a solution starts, the solution will leave a disk of radius 1 about the origin. Therefore $(0, 0)$ is an unstable critical point. A typical solution that starts near $(0, 0)$ is shown in Figure 10.22. ∎

Linearization   It is rarely possible to determine the stability of a critical point of a nonlinear system by finding explicit solutions, as in Examples 1 and 2. Instead we replace the term $\mathbf{g}(\mathbf{X})$ in the original autonomous system $\mathbf{X}' = \mathbf{g}(\mathbf{X})$ by a linear term $\mathbf{A}(\mathbf{X} - \mathbf{X}_1)$ that most closely approximates $\mathbf{g}(\mathbf{X})$ in a neighborhood of $\mathbf{X}_1$. This replacement process, called **linearization,** will be illustrated first for the first-order differential equation $x' = g(x)$.

An equation of the tangent line to the curve $y = g(x)$ at $x = x_1$ is $y = g(x_1) + g'(x_1)(x - x_1)$, and if x_1 is a critical point of $x' = g(x)$, we have $x' = g(x) \approx g'(x_1)(x - x_1)$. The general solution to the linear differential equation

$$x' = g'(x_1)(x - x_1)$$

is $x = x_1 + ce^{\lambda_1 t}$, where $\lambda_1 = g'(x_1)$. Thus if $g'(x_1) < 0$, then $x(t)$ approaches x_1. Theorem 10.2 asserts that the same behavior occurs in the original differential equation provided $x(0) = x_0$ is selected close enough to x_1.

THEOREM 10.2   **Stability Criteria for $x' = g(x)$**

Let x_1 be a critical point of the autonomous differential equation $x' = g(x)$, where g is differentiable at x_1.

(a) If $g'(x_1) < 0$, then x_1 is an asymptotically stable critical point.
(b) If $g'(x_1) > 0$, then x_1 is an unstable critical point.

EXAMPLE 3   **Stability in a Nonlinear First-Order DE**

Both $x = \pi/4$ and $x = 5\pi/4$ are critical points of the autonomous differential equation $x' = \cos x - \sin x$. This differential equation is difficult to solve explicitly, but we can use Theorem 10.2 to predict the behavior of solutions near these two critical points.

Since $g'(x) = -\sin x - \cos x$, $g'(\pi/4) = -\sqrt{2} < 0$ and $g'(5\pi/4) = \sqrt{2} > 0$. Therefore $x = \pi/4$ is an asymptotically stable critical point, but $x = 5\pi/4$ is unstable. In Figure 10.23 we used a numerical solver to investigate solutions that start near $(0, \pi/4)$ and $(0, 5\pi/4)$. Observe that solution curves that start close to $(0, 5\pi/4)$ quickly move away from the line $x = 5\pi/4$, as predicted. ∎

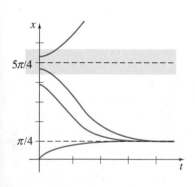

Figure 10.23

EXAMPLE 4   **Stability Analysis of the Logistic DE**

Without solving explicitly, analyze the critical points of the logistic differential equation (see Section 3.2) $x' = \dfrac{r}{K} x(K - x)$, where r and K are positive constants.

Solution   The two critical points are $x = 0$ and $x = K$, and so from $g'(x) = r(K - 2x)/K$ we get $g'(0) = r$ and $g'(K) = -r$. By Theorem 10.2 we conclude that $x = 0$ is an unstable critical point and $x = K$ is an asymptotically stable critical point. ∎

Jacobian Matrix   A similar analysis may be carried out for a plane autonomous system. An equation of the tangent plane to the surface $z = g(x, y)$ at $\mathbf{X}_1 = (x_1, y_1)$ is

$$z = g(x_1, y_1) + \frac{\partial g}{\partial x}\bigg|_{(x_1, y_1)} (x - x_1) + \frac{\partial g}{\partial y}\bigg|_{(x_1, y_1)} (y - y_1),$$

and $g(x, y)$ may be approximated by its tangent plane in a neighborhood of $\mathbf{X}_1$.

When $\mathbf{X}_1$ is a critical point of a plane autonomous system, $P(x_1, y_1) = Q(x_1, y_1) = 0$ and we have

$$x' = P(x, y) \approx \frac{\partial P}{\partial x}\bigg|_{(x_1, y_1)} (x - x_1) + \frac{\partial P}{\partial y}\bigg|_{(x_1, y_1)} (y - y_1)$$

$$y' = Q(x, y) \approx \frac{\partial Q}{\partial x}\bigg|_{(x_1, y_1)} (x - x_1) + \frac{\partial Q}{\partial y}\bigg|_{(x_1, y_1)} (y - y_1).$$

The original system $\mathbf{X}' = \mathbf{g}(\mathbf{X})$ may be approximated in a neighborhood of the critical point $\mathbf{X}_1$ by the linear system $\mathbf{X}' = \mathbf{A}(\mathbf{X} - \mathbf{X}_1)$, where

$$\mathbf{A} = \begin{pmatrix} \dfrac{\partial P}{\partial x}\bigg|_{(x_1, y_1)} & \dfrac{\partial P}{\partial y}\bigg|_{(x_1, y_1)} \\ \dfrac{\partial Q}{\partial x}\bigg|_{(x_1, y_1)} & \dfrac{\partial Q}{\partial y}\bigg|_{(x_1, y_1)} \end{pmatrix}.$$

This matrix is called the **Jacobian matrix** at $\mathbf{X}_1$ and is denoted by $\mathbf{g}'(\mathbf{X}_1)$. If we let $\mathbf{H} = \mathbf{X} - \mathbf{X}_1$, then the linear system $\mathbf{X}' = \mathbf{A}(\mathbf{X} - \mathbf{X}_1)$ becomes $\mathbf{H}' = \mathbf{AH}$, which is the form of the linear system analyzed in Section 10.2. The critical point $\mathbf{X} = \mathbf{X}_1$ for $\mathbf{X}' = \mathbf{A}(\mathbf{X} - \mathbf{X}_1)$ now corresponds to the critical point $\mathbf{H} = \mathbf{0}$ for $\mathbf{H}' = \mathbf{AH}$. If the eigenvalues of $\mathbf{A}$ have negative real parts, then, by Theorem 10.1, $\mathbf{0}$ is an asymptotically stable critical point for $\mathbf{H}' = \mathbf{AH}$. If there is an eigenvalue with positive real part, $\mathbf{H} = \mathbf{0}$ is an unstable critical point. Theorem 10.3 asserts that the same conclusions can be made for the critical point $\mathbf{X}_1$ of the original system.

THEOREM 10.3   **Stability Criteria for Plane Autonomous Systems**

Let $\mathbf{X}_1$ be a critical point of the plane autonomous system $\mathbf{X}' = \mathbf{g}(\mathbf{X})$, where $P(x, y)$ and $Q(x, y)$ have continuous first partials in a neighborhood of $\mathbf{X}_1$.

(a) If the eigenvalues of $\mathbf{A} = \mathbf{g}'(\mathbf{X}_1)$ have negative real part, then $\mathbf{X}_1$ is an asymptotically stable critical point.
(b) If $\mathbf{A} = \mathbf{g}'(\mathbf{X}_1)$ has an eigenvalue with positive real part, then $\mathbf{X}_1$ is an unstable critical point.

EXAMPLE 5   **Stability Analysis of Nonlinear Systems**

Classify (if possible) the critical points of each of the following plane autonomous systems as stable or unstable:

(a) $x' = x^2 + y^2 - 6$ (b) $x' = 0.01x(100 - x - y)$
     $y' = x^2 - y$ $y' = 0.05y(60 - y - 0.2x)$

Solution   The critical points of each system were determined in Example 4 of Section 10.1.

(a) The critical points are $(\sqrt{2}, 2)$ and $(-\sqrt{2}, 2)$, the Jacobian matrix is

$$\mathbf{g}'(\mathbf{X}) = \begin{pmatrix} 2x & 2y \\ 2x & -1 \end{pmatrix},$$

and so

$$\mathbf{A}_1 = \mathbf{g}'((\sqrt{2}, 2)) = \begin{pmatrix} 2\sqrt{2} & 4 \\ 2\sqrt{2} & -1 \end{pmatrix} \quad \text{and} \quad \mathbf{A}_2 = \mathbf{g}'((-\sqrt{2}, 2)) = \begin{pmatrix} -2\sqrt{2} & 4 \\ -2\sqrt{2} & -1 \end{pmatrix}.$$

Since the determinant of $\mathbf{A}_1$ is negative, $\mathbf{A}_1$ has a positive real eigenvalue. Therefore $(\sqrt{2}, 2)$ is an unstable critical point. Matrix $\mathbf{A}_2$ has a positive determinant and a negative trace, and so both eigenvalues have negative real parts. It follows that $(-\sqrt{2}, 2)$ is a stable critical point.

(b) The critical points are $(0, 0)$, $(0, 60)$, $(100, 0)$, and $(50, 50)$, the Jacobian matrix is

$$\mathbf{g}'(\mathbf{X}) = \begin{pmatrix} 0.01(100 - 2x - y) & -0.01x \\ -0.01y & 0.05(60 - 2y - 0.2x) \end{pmatrix},$$

and so

$$\mathbf{A}_1 = \mathbf{g}'((0, 0)) = \begin{pmatrix} 1 & 0 \\ 0 & 3 \end{pmatrix} \qquad \mathbf{A}_2 = \mathbf{g}'((0, 60)) = \begin{pmatrix} 0.4 & 0 \\ -0.6 & -3 \end{pmatrix}$$

$$\mathbf{A}_3 = \mathbf{g}'((100, 0)) = \begin{pmatrix} -1 & -1 \\ 0 & 2 \end{pmatrix} \quad \mathbf{A}_4 = \mathbf{g}'((50, 50)) = \begin{pmatrix} -0.5 & -0.5 \\ -0.5 & -2.5 \end{pmatrix}.$$

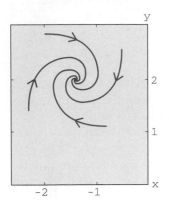

Figure 10.24

Since the matrix $\mathbf{A}_1$ has a positive determinant and a positive trace, both eigenvalues have positive real parts. Therefore $(0, 0)$ is an unstable critical point. The determinants of matrices $\mathbf{A}_2$ and $\mathbf{A}_3$ are negative, and so in each case one of the eigenvalues is positive. Therefore both $(0, 60)$ and $(100, 0)$ are unstable critical points. Since the matrix $\mathbf{A}_4$ has a positive determinant and a negative trace, $(50, 50)$ is a stable critical point. ∎

In Example 5 we did not compute $\tau^2 - 4\Delta$ (as in Section 10.2) and attempt to further classify the critical points as stable nodes, stable spiral points, saddle points, and so on. For example, for $\mathbf{X}_1 = (-\sqrt{2}, 2)$ in Example 5(a), $\tau^2 - 4\Delta < 0$, and if the system were linear, we would be able to conclude that $\mathbf{X}_1$ was a stable spiral point. Figure 10.24 shows several solution curves near $\mathbf{X}_1$ that were obtained with a numerical solver, and each solution does *appear* to spiral in toward the critical point.

Classifying Critical Points It is natural to ask whether we can infer more geometric information about the solutions near a critical point $\mathbf{X}_1$ of a nonlinear autonomous system from an analysis of the critical point of the corresponding linear system. The answer is summarized in Figure 10.25, but you should note the following comments.

- (*i*) In five separate cases (stable node, stable spiral point, unstable spiral point, unstable node, and saddle) the critical point may be categorized like the critical point in the corresponding linear system. The solutions have the same general geometric features as the solutions to the linear system, and the smaller the neighborhood about $\mathbf{X}_1$, the closer the resemblance.
- (*ii*) If $\tau^2 = 4\Delta$ and $\tau > 0$, the critical point $\mathbf{X}_1$ is unstable, but in this borderline case *we are not yet able to decide whether $\mathbf{X}_1$ is*

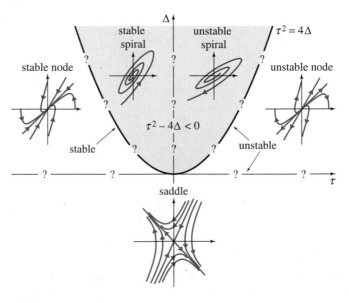

Figure 10.25

an unstable spiral, unstable node, or degenerate unstable node. Likewise, if $\tau^2 = 4\Delta$ and $\tau < 0$, the critical point $\mathbf{X}_1$ is stable but may be either a stable spiral, a stable node, or a degenerate stable node.

(*iii*) If $\tau = 0$ and $\Delta > 0$, the eigenvalues of $\mathbf{A} = \mathbf{g}'(\mathbf{X})$ are pure imaginary, and in this borderline case $\mathbf{X}_1$ may be either a stable spiral, an unstable spiral, or a center. *It is therefore not yet possible to determine whether $\mathbf{X}_1$ is stable or unstable.*

EXAMPLE 6 **Classifying Critical Points of a Nonlinear System**

Classify each critical point of the plane autonomous system in Example 5(b) as a stable node, a stable spiral point, an unstable spiral point, an unstable node, or a saddle point.

Solution   For the matrix $\mathbf{A}_1$ corresponding to $(0, 0)$, $\Delta = 3$, $\tau = 4$, and so $\tau^2 - 4\Delta = 4$. Therefore $(0, 0)$ is an unstable node. The critical points $(0, 60)$ and $(100, 0)$ are saddles since $\Delta < 0$ in both cases. For matrix $\mathbf{A}_4$, $\Delta > 0$, $\tau < 0$, and $\tau^2 - 4\Delta > 0$. It follows that $(50, 50)$ is a stable node. Experiment with a numerical solver to verify these conclusions. ■

EXAMPLE 7 **Stability Analysis for a Soft Spring**

Recall from Section 5.3 that the second-order differential equation $mx'' + kx + k_1 x^3 = 0$, for $k > 0$, represents a general model for the free, undamped oscillations of a mass m attached to a nonlinear spring. If $k = 1$ and $k_1 = -1$, the spring is called *soft* and the plane autonomous system corresponding to the nonlinear second-order differential equation $x'' + x - x^3 = 0$ is

$$x' = y$$
$$y' = x^3 - x.$$

Find and classify (if possible) the critical points.

Solution   Since $x^3 - x = x(x^2 - 1)$, the critical points are $(0, 0)$, $(1, 0)$, and $(-1, 0)$. The corresponding Jacobian matrices are

$$\mathbf{A}_1 = \mathbf{g}'((0,0)) = \begin{pmatrix} 0 & 1 \\ -1 & 0 \end{pmatrix}, \quad \mathbf{A}_2 = \mathbf{g}'((1,0)) = \mathbf{g}'((-1,0)) = \begin{pmatrix} 0 & 1 \\ 2 & 0 \end{pmatrix}.$$

Since det $\mathbf{A}_2 < 0$, critical points $(1, 0)$ and $(-1, 0)$ are both saddle points. The eigenvalues of matrix $\mathbf{A}_1$ are $\pm i$, and according to comment (*iii*), the status of the critical point at $(0, 0)$ remains in doubt. It may be either a stable spiral, an unstable spiral, or a center. ■

The Phase-Plane Method   The linearization method, when successful, can provide useful information on the local behavior of solutions near critical points. It is of little help if we are interested in solutions whose initial position $\mathbf{X}(0) = \mathbf{X}_0$ is not close to a critical point or if we wish to obtain a global view of the family of solution curves. The **phase-plane method** is based on the fact that

$$\frac{dy}{dx} = \frac{dy/dt}{dx/dt} = \frac{Q(x, y)}{P(x, y)}$$

and attempts to find y as a function of x using one of the methods available for solving first-order differential equations (Chapter 2). As we show in Examples 8 and 9, the method can sometimes be used to decide whether a critical point such as $(0, 0)$ in Example 7 is a stable spiral, an unstable spiral, or a center.

EXAMPLE 8   Phase-Plane Method

Use the phase-plane method to classify the sole critical point $(0, 0)$ of the plane autonomous system

$$x' = y^2$$
$$y' = x^2.$$

Solution   The determinant of the Jacobian matrix

$$\mathbf{g}'(\mathbf{X}) = \begin{pmatrix} 0 & 2y \\ 2x & 0 \end{pmatrix}.$$

is 0 at $(0, 0)$, and so the nature of the critical point $(0, 0)$ remains in doubt. Using the phase-plane method, we obtain the first-order differential equation

$$\frac{dy}{dx} = \frac{dy/dt}{dx/dt} = \frac{x^2}{y^2},$$

which can be easily solved by separation of variables:

$$\int y^2 \, dy = \int x^2 \, dx \quad \text{or} \quad y^3 = x^3 + c.$$

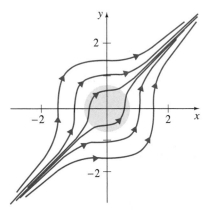

Figure 10.26

If $\mathbf{X}(0) = (0, y_0)$, it follows that $y^3 = x^3 + y_0^3$ or $y = \sqrt[3]{x^3 + y_0^3}$. Figure 10.26 shows a collection of solution curves corresponding to various choices for y_0. The nature of the critical point is clear from this phase portrait: No matter how close to $(0, 0)$ the solution starts, $\mathbf{X}(t)$ moves away from the origin as t increases. The critical point at $(0, 0)$ is therefore unstable. ∎

EXAMPLE 9   Phase-Plane Analysis of a Soft Spring

Use the phase-plane method to determine the nature of the solutions to $x'' + x - x^3 = 0$ in a neighborhood of $(0, 0)$.

Solution   If we let $dx/dt = y$, then $dy/dt = x^3 - x$. From this we obtain the first-order differential equation

$$\frac{dy}{dx} = \frac{dy/dt}{dx/dt} = \frac{x^3 - x}{y},$$

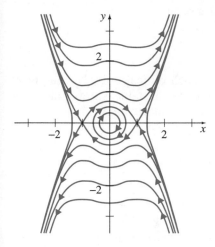

Figure 10.27

which can be solved by separation of variables. Integrating

$$\int y\, dy = \int (x^3 - x)\, dx \quad \text{gives} \quad \frac{y^2}{2} = \frac{x^4}{4} - \frac{x^2}{2} + c.$$

After completing the square, we can write the solution as $y^2 = (x^2 - 1)^2/2 + c_0$. If $\mathbf{X}(0) = (x_0, 0)$, where $0 < x_0 < 1$, then $c_0 = -(x_0^2 - 1)^2/2$, and so

$$y^2 = \frac{(x^2 - 1)^2}{2} - \frac{(x_0^2 - 1)^2}{2} = \frac{(2 - x^2 - x_0^2)(x_0^2 - x^2)}{2}.$$

Note that $y = 0$ when $x = -x_0$. In addition, the right-hand side is positive when $-x_0 < x < x_0$, and so each x has *two* corresponding values of y. The solution $\mathbf{X} = \mathbf{X}(t)$ that satisfies $\mathbf{X}(0) = (x_0, 0)$ is therefore periodic, and so $(0, 0)$ is a center.

Figure 10.27 shows a family of solution curves, or phase portrait, of the original system. We used the original plane autonomous system to determine the directions indicated on each trajectory. ∎

EXERCISES 10.3

Answers to odd-numbered problems begin on page AN-16.

1. Show that $(0, 0)$ is an asymptotically stable critical point of the nonlinear autonomous system

$$x' = \alpha x - \beta y + y^2$$
$$y' = \beta x + \alpha y - xy$$

when $\alpha < 0$ and an unstable critical point when $\alpha > 0$. [*Hint*: Switch to polar coordinates.]

2. When expressed in polar coordinates, a plane autonomous system takes the form

$$\frac{dr}{dt} = \alpha r(5 - r)$$
$$\frac{d\theta}{dt} = -1.$$

Show that $(0, 0)$ is an asymptotically stable critical point if and only if $\alpha < 0$.

In Problems 3–10, without solving explicitly, classify the critical points of the given first-order autonomous differential equation as either asymptotically stable or unstable. All constants are assumed to be positive.

3. $\dfrac{dx}{dt} = kx(n + 1 - x)$

4. $\dfrac{dx}{dt} = -kx \ln \dfrac{x}{K}, \quad x > 0$

5. $\dfrac{dT}{dt} = k(T - T_0)$

6. $m\dfrac{dv}{dt} = mg - kv$

7. $\dfrac{dx}{dt} = k(\alpha - x)(\beta - x), \quad \alpha > \beta$

8. $\dfrac{dx}{dt} = k(\alpha - x)(\beta - x)(\gamma - x), \quad \alpha > \beta > \gamma$

9. $\dfrac{dP}{dt} = P(a - bP)(1 - cP^{-1}), \quad P > 0, a < bc$

10. $\dfrac{dA}{dt} = k\sqrt{A}\,(K - \sqrt{A}), \quad A > 0$

In Problems 11–20 classify (if possible) each critical point of the given plane autonomous system as a stable node, a stable spiral point, an unstable spiral point, an unstable node, or a saddle point.

11. $x' = 1 - 2xy$
  $y' = 2xy - y$

12. $x' = x^2 - y^2 - 1$
  $y' = 2y$

13. $x' = y - x^2 + 2$
  $y' = x^2 - xy$

14. $x' = 2x - y^2$
  $y' = -y + xy$

15. $x' = -3x + y^2 + 2$
  $y' = x^2 - y^2$

16. $x' = xy - 3y - 4$
  $y' = y^2 - x^2$

17. $x' = -2xy$
  $y' = y - x + xy - y^3$

18. $x' = x(1 - x^2 - 3y^2)$
  $y' = y(3 - x^2 - 3y^2)$

19. $x' = x(10 - x - \frac{1}{2}y)$
  $y' = y(16 - y - x)$

20. $x' = -2x + y + 10$
  $y' = 2x - y - 15\,\dfrac{y}{y + 5}$

In Problems 21–26 classify (if possible) each critical point of the given second-order differential equation as a stable node, a stable spiral point, an unstable spiral point, an unstable node, or a saddle point.

21. $\theta'' = (\cos\theta - 0.5)\sin\theta, \quad |\theta| < \pi$

22. $x'' + x = (\frac{1}{2} - 3(x')^2)x' - x^2$

23. $x'' + x'(1 - x^3) - x^2 = 0$

24. $x'' + 4\dfrac{x}{1 + x^2} + 2x' = 0$

25. $x'' + x = \epsilon x^3$ for $\epsilon > 0$

26. $x'' + x - \epsilon x|x| = 0$ for $\epsilon > 0$ $\left[Hint: \dfrac{d}{dx}\,x\,|x| = 2\,|x|. \right]$

27. Show that the nonlinear second-order differential equation

$$(1 + \alpha^2 x^2)x'' + (\beta + \alpha^2(x')^2)x = 0$$

has a saddle point at $(0, 0)$ when $\beta < 0$.

28. Show that the dynamical system

$$x' = -\alpha x + xy$$
$$y' = 1 - \beta y - x^2$$

has a unique critical point when $\alpha\beta > 1$ and that this critical point is stable when $\beta > 0$.

29. **(a)** Show that the plane autonomous system

$$x' = -x + y - x^3$$
$$y' = -x - y + y^2$$

has two critical points by sketching the graphs of $-x + y - x^3 = 0$ and $-x - y + y^2 = 0$. Classify the critical point at $(0, 0)$.

(b) Show that the second critical point $X_1 = (0.88054, 1.56327)$ is a saddle point.

30. (a) Show that $(0, 0)$ is the only critical point of the **Raleigh differential equation**

$$x'' + \epsilon(\tfrac{1}{3}(x')^3 - x') + x = 0.$$

(b) Show that $(0, 0)$ is unstable when $\epsilon > 0$. When is $(0, 0)$ an unstable spiral point?

(c) Show that $(0, 0)$ is stable when $\epsilon < 0$. When is $(0, 0)$ a stable spiral point?

(d) Show that $(0, 0)$ is a center when $\epsilon = 0$.

31. Use the phase-plane method to show that $(0, 0)$ is a center of the nonlinear second-order differential equation $x'' + 2x^3 = 0$.

32. Use the phase-plane method to show that the solution to the nonlinear second-order differential equation $x'' + 2x - x^2 = 0$ that satisfies $x(0) = 1$ and $x'(0) = 0$ is periodic.

33. (a) Find the critical points of the plane autonomous system

$$x' = 2xy$$
$$y' = 1 - x^2 + y^2,$$

and show that linearization gives no information about the nature of these critical points.

(b) Use the phase-plane method to show that the critical points in (a) are both centers.

[*Hint*: Let $u = y^2/x$, and show that $(x - c)^2 + y^2 = c^2 - 1$.]

34. The origin is the only critical point of the nonlinear second-order differential equation $x'' + (x')^2 + x = 0$.

(a) Show that the phase-plane method leads to the Bernoulli differential equation $dy/dx = -y - xy^{-1}$.

(b) Show that the solution satisfying $x(0) = \tfrac{1}{2}$ and $x'(0) = 0$ is not periodic.

35. A solution of the nonlinear second-order differential equation $x'' + x - x^3 = 0$ satisfies $x(0) = 0$ and $x'(0) = v_0$. Use the phase-plane method to determine when the resulting solution is periodic. [*Hint*: See Example 9.]

36. The nonlinear differential equation $x'' + x = 1 + \epsilon x^2$ arises in the analysis of planetary motion using relativity theory. Classify (if possible) all critical points of the corresponding plane autonomous system.

37. When a nonlinear capacitor is present in an LCR circuit, the voltage drop is no longer given by q/C but is more accurately described by $\alpha q + \beta q^3$, where α and β are constants and $\alpha > 0$. Differential equation (34) of Section 5.1 for the free circuit is then replaced by

$$L\frac{d^2q}{dt^2} + R\frac{dq}{dt} + \alpha q + \beta q^3 = 0.$$

Find and classify all critical points of this nonlinear differential equation. [*Hint*: Divide into the two cases $\beta > 0$ and $\beta < 0$.]

38. The nonlinear equation $mx'' + kx + k_1x^3 = 0$, for $k > 0$, represents a general model for the free, undamped oscillations of a mass m attached to a spring. If $k_1 > 0$, the spring is called *hard* (see Example 1 in Section 5.3). Determine the nature of the solutions to $x'' + x + x^3 = 0$ in a neighborhood of $(0, 0)$.

39. The nonlinear equation $\theta'' + \sin \theta = \frac{1}{2}$ can be interpreted as a model for a certain pendulum with a constant driving function.

 (a) Show that $(\pi/6, 0)$ and $(5\pi/6, 0)$ are critical points of the corresponding plane autonomous system.

 (b) Classify the critical point $(5\pi/6, 0)$ using linearization.

 (c) Use the phase-plane method to classify the critical point $(\pi/6, 0)$.

Discussion Problems

40. (a) Show that $(0, 0)$ is an isolated critical point of the plane autonomous system

$$x' = x^4 - 2xy^3$$
$$y' = 2x^3y - y^4$$

but that linearization gives no useful information about the nature of this critical point.

 (b) Use the phase-plane method to show that $x^3 + y^3 = 3cxy$. This classic curve is called a *folium of Descartes*. Parametric equations for a folium are

$$x = \frac{3ct}{1 + t^3}, \quad y = \frac{3ct^2}{1 + t^3}.$$

 [*Hint:* The differential equation in x and y is homogeneous.]

 (c) Use graphing software or a numerical solver to sketch solution curves. Based on your sketch, would you classify the critical point as stable or unstable? Would you classify the critical point as a node, saddle point, center, or spiral point? Explain.

10.4 MODELING USING AUTONOMOUS SYSTEMS

• Stability analysis of a nonlinear pendulum • Nonlinear oscillations and stability • Predator-prey interactions and population cycles • Competitive interactions and coexistence

Many applications from physics give rise to nonlinear second-order differential equations of the form $x'' = g(x, x')$. For example, in the analysis of free, damped motion in Section 5.1 we assumed that the damping force was proportional to the velocity x'; the resulting model $mx'' = -\beta x' - kx$ is a linear differential equation. But if the magnitude of the damping force is proportional to the square of the velocity, the new differential equation $mx'' = -\beta x'|x'| - kx$ is nonlinear. The corresponding plane autonomous system is nonlinear:

$$x' = y$$
$$y' = -\frac{\beta}{m}y|y| - \frac{k}{m}x.$$

In this section we will use the results of Section 10.3 to analyze the nonlinear pendulum, the motion of a bead on a curve, the Lotka-Volterra predator-prey models, and the Lotka-Volterra competition model. Additional models are presented in the exercises.

The Nonlinear Pendulum In (6) of Section 5.3 we showed that the displacement angle θ for a simple pendulum satisfies the nonlinear second-order differential equation

$$\frac{d^2\theta}{dt^2} + \frac{g}{l}\sin\theta = 0.$$

When we let $x = \theta$ and $y = \theta'$, this second-order differential equation may be rewritten as the dynamical system

$$x' = y$$

$$y' = -\frac{g}{l}\sin x.$$

The critical points are $(\pm k\pi, 0)$, and the Jacobian matrix is easily shown to be

$$\mathbf{g}'((\pm k\pi, 0)) = \begin{pmatrix} 0 & 1 \\ (-1)^{k+1}\frac{g}{l} & 0 \end{pmatrix}.$$

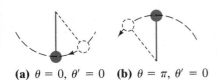

(a) $\theta = 0$, $\theta' = 0$ **(b)** $\theta = \pi$, $\theta' = 0$

Figure 10.28

If $k = 2n + 1$, $\Delta < 0$, and so all critical points $(\pm(2n + 1)\pi, 0)$ are saddle points. In particular, the critical point at $(\pi, 0)$ is unstable as expected. See Figure 10.28. When $k = 2n$, the eigenvalues are pure imaginary, and so the nature of these critical points remains in doubt. Since we have assumed that there are no damping forces acting on the pendulum, we expect that all of the critical points $(\pm 2n\pi, 0)$ are centers. This can be verified using the phase-plane method. From

$$\frac{dy}{dx} = \frac{dy/dt}{dx/dt} = -\frac{g}{l}\frac{\sin x}{y}$$

it follows that $y^2 = (2g/l)\cos x + c$. If $\mathbf{X}(0) = (x_0, 0)$, then

$$y^2 = \frac{2g}{l}(\cos x - \cos x_0).$$

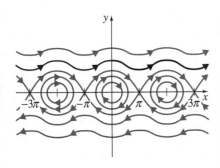

Figure 10.29

Note that $y = 0$ when $x = -x_0$, and that $(2g/l)(\cos x - \cos x_0) > 0$ for $|x| < |x_0| < \pi$. Thus each such x has two corresponding values of y, and so the solution $\mathbf{X} = \mathbf{X}(t)$ that satisfies $\mathbf{X}(0) = (x_0, 0)$ is periodic. We may conclude that $(0, 0)$ is a center. Observe that $x = \theta$ increases for solutions that correspond to large initial velocities, such as the one drawn in black in Figure 10.29. In this case the pendulum spins in complete circles about its pivot.

EXAMPLE 1   **Periodic Solutions of the Pendulum DE**

A pendulum in an equilibrium position with $\theta = 0$ is given an initial angular velocity of ω_0 rad/s. Determine under what conditions the resulting motion is periodic.

Solution   We are asked to examine the solution of the plane autonomous system that satisfies $\mathbf{X}(0) = (0, \omega_0)$. From $y^2 = (2g/l)\cos x + c$ it follows that

$$y^2 = \frac{2g}{l}\left(\cos x - 1 + \frac{l}{2g}\omega_0^2\right).$$

To establish that the solution $\mathbf{X}(t)$ is periodic it is sufficient to show that there are two x-intercepts $x = \pm x_0$ between $-\pi$ and π and that the right-hand side is positive for $|x| < |x_0|$. Each such x then has two corresponding values of y.

If $y = 0$, $\cos x = 1 - (l/2g)\omega_0^2$, and this equation has two solutions $x = \pm x_0$ between $-\pi$ and π, provided $1 - (l/2g)\omega_0^2 > -1$. Note that $(2g/l)(\cos x - \cos x_0)$ is then positive for $|x| < |x_0|$. This restriction on the initial angular velocity may be written as $|\omega_0| < 2\sqrt{g/l}$. ∎

Nonlinear Oscillations: The Sliding Bead   Suppose, as shown in Figure 10.30, a bead with mass m slides along a thin wire whose shape is described by the function $z = f(x)$. A wide variety of nonlinear oscillations can be obtained by changing the shape of the wire and by making different assumptions about the forces acting on the bead.

The tangential force $\mathbf{F}$ due to the weight $W = mg$ has magnitude $mg \sin \theta$, and therefore the x-component of $\mathbf{F}$ is $F_x = -mg \sin \theta \cos \theta$. Since $\tan \theta = f'(x)$, we may use the identities $1 + \tan^2\theta = \sec^2\theta$ and $\sin^2\theta = 1 - \cos^2\theta$ to conclude that

$$F_x = -mg \sin \theta \cos \theta = -mg \frac{f'(x)}{1 + [f'(x)]^2}.$$

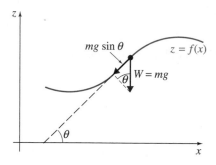

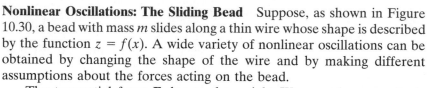

Figure 10.30

We assume (as in Section 5.1) that a damping force $\mathbf{D}$, acting in the direction opposite to the motion, is a constant multiple of the velocity of the bead. The x-component of $\mathbf{D}$ is therefore $D_x = -\beta x'$. If we ignore the frictional force between the wire and the bead and assume that no other external forces are impressed on the system, it follows from Newton's second law that

$$mx'' = -mg \frac{f'(x)}{1 + [f'(x)]^2} - \beta x',$$

and the corresponding plane autonomous system is

$$x' = y$$
$$y' = -g \frac{f'(x)}{1 + [f'(x)]^2} - \frac{\beta}{m} y.$$

If $\mathbf{X}_1 = (x_1, y_1)$ is a critical point of the system, $y_1 = 0$ and therefore $f'(x_1) = 0$. The bead must therefore be at rest at a point on the wire where the tangent line is horizontal. When f is twice differentiable, the Jacobian matrix at $\mathbf{X}_1$ is

$$\mathbf{g}'(\mathbf{X}_1) = \begin{pmatrix} 0 & 1 \\ -gf''(x_1) & -\beta/m \end{pmatrix},$$

and so $\tau = -\beta/m$, $\Delta = gf''(x_1)$, and $\tau^2 - 4\Delta = \beta^2/m^2 - 4gf''(x_1)$. Using the results of Section 10.3, we can make the following conclusions:

(i) $f''(x_1) < 0$
A relative maximum therefore occurs at $x = x_1$, and since $\Delta < 0$, an *unstable saddle point* occurs at $\mathbf{X}_1 = (x_1, 0)$.
(ii) $f''(x_1) > 0$ and $\beta > 0$
A relative minimum therefore occurs at $x = x_1$, and since $\tau < 0$ and $\Delta > 0$, $\mathbf{X}_1 = (x_1, 0)$ is a *stable critical point*. If

$\beta^2 > 4gm^2 f''(x_1)$, the system is **overdamped** and the critical point is a *stable node*. If $\beta^2 < 4gm^2 f''(x_1)$, the system is **underdamped** and the critical point is a *stable spiral point*. The exact nature of the stable critical point is still in doubt if $\beta^2 = 4gm^2 f''(x_1)$.

(*iii*) $f''(x_1) > 0$ and the system is undamped ($\beta = 0$)

In this case the eigenvalues are pure imaginary, but the phase-plane method can be used to show that the critical point is a *center*. Therefore solutions with $\mathbf{X}(0) = (x(0), x'(0))$ near $\mathbf{X}_1 = (x_1, 0)$ are periodic.

EXAMPLE 2   **Bead Sliding along a Sine Wave**

A 10-gram bead slides along the graph of $z = \sin x$. According to conclusion (*ii*), the relative minima at $x_1 = -\pi/2$ and $3\pi/2$ give rise to stable critical points (see Figure 10.31). Since $f''(-\pi/2) = f''(3\pi/2) = 1$, the system will be underdamped provided $\beta^2 < 4gm^2$. If we use SI units, $m = 0.01$ kg and $g = 9.8$ m/s^2, and so the condition for an underdamped system becomes $\beta^2 < 3.92 \times 10^{-3}$.

If $\beta = 0.01$ is the damping constant, both of these critical points are stable spiral points. The two solutions corresponding to initial conditions $\mathbf{X}(0) = (x(0), x'(0)) = (-2\pi, 10)$ and $\mathbf{X}(0) = (-2\pi, 15)$, respectively, were obtained using a numerical solver and are shown in Figure 10.32. When $x'(0) = 10$, the bead has enough momentum to make it over the hill at $x = -3\pi/2$ but not over the hill at $x = \pi/2$. The bead then approaches the relative minimum based at $x = -\pi/2$. If $x'(0) = 15$, the bead has the momentum to make it over both hills, but then it rocks back and forth in the valley based at $x = 3\pi/2$ and approaches the point $(3\pi/2, -1)$ on the wire. Experiment with other initial conditions using your numerical solver.

Figure 10.33 shows a collection of solution curves obtained from a numerical solver for the undamped case. Since $\beta = 0$, the critical points corresponding to $x_1 = -\pi/2$ and $3\pi/2$ are now centers. When $\mathbf{X}(0) = (-2\pi, 10)$, the bead has sufficient momentum to move over *all* hills. The figure also indicates that when the bead is released from rest at a position on the wire between $x = -3\pi/2$ and $x = \pi/2$, the resulting motion is periodic. ∎

Lotka-Volterra Predator-Prey Model   A **predator-prey interaction** between two species occurs when one species (the predator) feeds on a second species (the prey). For example, the snowy owl feeds almost exclusively on a common arctic rodent called a lemming, while a lemming uses arctic tundra plants as its food supply. Interest in using mathematics to help explain predator-prey interactions has been stimulated by the observation of population cycles in many arctic mammals. In the MacKenzie River district of Canada, for example, the principal prey of the lynx is the snowshoe hare, and both populations cycle with a period of about 10 years.

There are many predator-prey models that lead to plane autonomous systems with at least one periodic solution. The first such model was constructed independently by pioneer biomathematicians A. Lotka (1925)

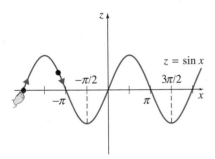

Figure 10.31

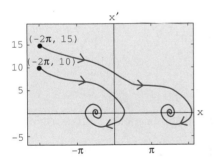

Figure 10.32 $\beta = 0.01$

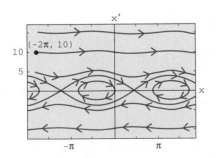

Figure 10.33 $\beta = 0$

and V. Volterra (1926). If x denotes the number of predators and y denotes the number of prey, then the Lotka-Volterra model takes the form

$$x' = -ax + bxy = x(-a + by)$$
$$y' = -cxy + dy = y(-cx + d),$$

where a, b, c, and d are positive constants.

Note that in the absence of predators ($x = 0$), $y' = dy$, and so the number of prey grows exponentially. In the absence of prey, $x' = -ax$, and so the predator population becomes extinct. The term $-cxy$ represents the death rate due to predation. The model therefore assumes that this death rate is directly proportional to the number of possible encounters xy between predator and prey at a particular time t, and the term bxy represents the resulting positive contribution to the predator population.

The critical points of this plane autonomous system are $(0, 0)$ and $(d/c, a/b)$, and the corresponding Jacobian matrices are

$$\mathbf{A}_1 = \mathbf{g}'((0,0)) = \begin{pmatrix} -a & 0 \\ 0 & d \end{pmatrix} \quad \text{and} \quad \mathbf{A}_2 = \mathbf{g}'((d/c, a/b)) = \begin{pmatrix} 0 & bd/c \\ -ac/b & 0 \end{pmatrix}.$$

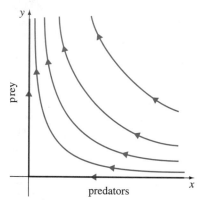

prey

predators

Figure 10.34 Solutions near $(0, 0)$

The critical point at $(0, 0)$ is a saddle point, and Figure 10.34 shows a typical profile of solutions that are in the first quadrant and near $(0, 0)$.

Since the matrix $\mathbf{A}_2$ has pure imaginary eigenvalues $\lambda = \pm\sqrt{ad}\, i$, the critical point $(d/c, a/b)$ *may* be a center. This possibility can be investigated using the phase-plane method. Since

$$\frac{dy}{dx} = \frac{y(-cx + d)}{x(-a + by)},$$

we separate variables and obtain

$$\int \frac{-a + by}{y}\, dy = \int \frac{-cx + d}{x}\, dx$$

so that

$$-a \ln y + by = -cx + d \ln x + c_1 \quad \text{or} \quad (x^d e^{-cx})(y^a e^{-by}) = c_0.$$

The following argument establishes that all solution curves that originate in the first quadrant are periodic.

Typical graphs of the nonnegative functions $F(x) = x^d e^{-cx}$ and $G(y) = y^a e^{-by}$ are shown in Figure 10.35. It is not hard to show that $F(x)$ has an absolute maximum at $x = d/c$, whereas $G(y)$ has an absolute maximum at $y = a/b$. Note that, with the exception of 0 and the absolute maximum, F and G each take on all values in their range precisely twice.

These graphs can be used to establish the following properties of a solution curve that originates at a noncritical point (x_0, y_0) in the first quadrant.

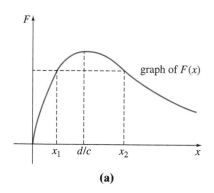

graph of $F(x)$

x_1 d/c x_2

(a)

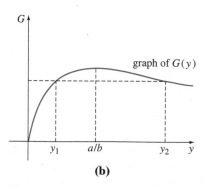

graph of $G(y)$

y_1 a/b y_2

(b)

Figure 10.35

(1) If $y = a/b$, the equation $F(x)G(y) = c_0$ has exactly two solutions x_m and x_M that satisfy $x_m < d/c < x_M$.
(2) If $x_m < x_1 < x_M$ and $x = x_1$, then $F(x)G(y) = c_0$ has exactly two solutions y_1 and y_2 that satisfy $y_1 < a/b < y_2$.
(3) If x is outside the interval $[x_m, x_M]$, then $F(x)G(y) = c_0$ has no solutions.

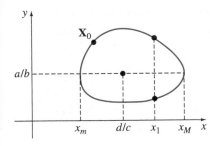

Figure 10.36

We will give the demonstration of (1) and outline parts (2) and (3) in the exercises. Since $(x_0, y_0) \neq (d/c, a/b)$, $F(x_0)G(y_0) < F(d/c)G(a/b)$. If $y = a/b$, then

$$0 < \frac{c_0}{G(a/b)} = \frac{F(x_0)G(y_0)}{G(a/b)} < \frac{F(d/c)G(a/b)}{G(a/b)} = F(d/c).$$

Therefore $F(x) = c_0/G(a/b)$ has precisely two solutions x_m and x_M that satisfy $x_m < d/c < x_M$. The graph of a typical periodic solution is shown in Figure 10.36.

> ### EXAMPLE 3   Predator-Prey Population Cycles

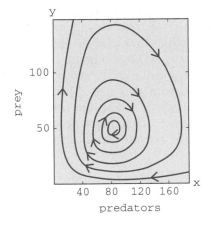

Figure 10.37

If we let $a = 0.1$, $b = 0.002$, $c = 0.0025$, and $d = 0.2$ in the Lotka-Volterra predator-prey model, the critical point in the first quadrant is $(d/c, a/b) = (80, 50)$, and we know that this critical point is a center. See Figure 10.37 in which we have used a numerical solver to generate these cycles. The closer the initial condition $\mathbf{X}_0$ is to $(80, 50)$, the more the periodic solutions resemble the elliptical solutions to the corresponding linear system. The eigenvalues of $\mathbf{g}'((80, 50))$ are $\lambda = \pm\sqrt{ad}\,i = \pm\sqrt{2}/10\,i$, and so the solutions near the critical point have period $p \approx 10\sqrt{2}\,\pi$, or about 44.4. ∎

Lotka-Volterra Competition Model   A **competitive interaction** occurs when two or more species compete for the food, water, light, and space resources of an ecosystem. The use of one of these resources by one population therefore inhibits the ability of another population to survive and grow. Under what conditions can two competing species coexist? A number of mathematical models have been constructed that offer insights into conditions that permit coexistence. If x denotes the number in species I and y denotes the number in species II, then the Lotka-Volterra model takes the form

$$x' = \frac{r_1}{K_1}x(K_1 - x - \alpha_{12}y)$$

$$y' = \frac{r_2}{K_2}y(K_2 - y - \alpha_{21}x). \tag{1}$$

Note that in the absence of species II ($y = 0$), $x' = (r_1/K_1)x(K_1 - x)$, and so the first population grows logistically and approaches the steady-state population K_1 (see Section 3.3 and Example 4 in Section 10.3). A similar statement holds for species II growing in the absence of species I. The term $-\alpha_{21}xy$ in the second equation stems from the competitive effect of species I on species II. The model therefore assumes that this rate of inhibition is directly proportional to the number of possible competitive pairs xy at a particular time t.

This plane autonomous system has critical points at $(0, 0)$, $(K_1, 0)$, and $(0, K_2)$. When $\alpha_{12}\alpha_{21} \neq 0$, the lines $K_1 - x - \alpha_{12}y = 0$ and $K_2 - y - \alpha_{21}x = 0$ intersect to produce a fourth critical point $\hat{\mathbf{X}} = (\hat{x}, \hat{y})$. Figure

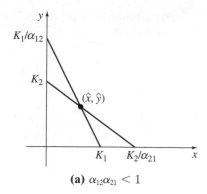

(a) $\alpha_{12}\alpha_{21} < 1$

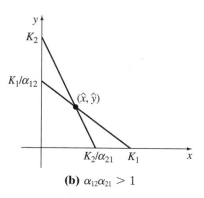

(b) $\alpha_{12}\alpha_{21} > 1$

Figure 10.38

10.38 shows the two conditions under which $(\hat{x}, \hat{y})$ is in the first quadrant. The trace and determinant of the Jacobian matrix at $(\hat{x}, \hat{y})$ are, respectively,

$$\tau = -\hat{x}\frac{r_1}{K_1} - \hat{y}\frac{r_2}{K_2} \quad \text{and} \quad \Delta = (1 - \alpha_{12}\alpha_{21})\hat{x}\hat{y}\frac{r_1 r_2}{K_1 K_2}.$$

In case (a) of Figure 10.38, $K_1/\alpha_{12} > K_2$ and $K_2/\alpha_{21} > K_1$. It follows that $\alpha_{12}\alpha_{21} < 1$, $\tau < 0$, and $\Delta > 0$. Since

$$\tau^2 - 4\Delta = \left(\hat{x}\frac{r_1}{K_1} + \hat{y}\frac{r_2}{K_2}\right)^2 + 4(\alpha_{12}\alpha_{21} - 1)\hat{x}\hat{y}\frac{r_1 r_2}{K_1 K_2}$$

$$= \left(\hat{x}\frac{r_1}{K_1} - \hat{y}\frac{r_2}{K_2}\right)^2 + 4\alpha_{12}\alpha_{21}\hat{x}\hat{y}\frac{r_1 r_2}{K_1 K_2},$$

$\tau^2 - 4\Delta > 0$, and so $(\hat{x}, \hat{y})$ is a stable node. Therefore if $\mathbf{X}(0) = \mathbf{X}_0$ is sufficiently close to $\hat{\mathbf{X}} = (\hat{x}, \hat{y})$, $\lim_{t \to \infty} \mathbf{X}(t) = \hat{\mathbf{X}}$, and we may conclude that coexistence is possible. The demonstration that case (b) leads to a saddle point and the investigation of the nature of critical points at $(0, 0)$, $(K_1, 0)$, and $(0, K_2)$ are left to the exercises.

When the competitive interactions between two species are weak, both of the coefficients α_{12} and α_{21} will be small, and so the conditions $K_1/\alpha_{12} > K_2$ and $K_2/\alpha_{21} > K_1$ may be satisfied. This might occur when there is a small overlap in the ranges of two predator species that hunt for a common prey.

EXAMPLE 4 A Lotka-Volterra Competition Model

A competitive interaction is described by the Lotka-Volterra competition model

$$x' = 0.004x(50 - x - 0.75y)$$
$$y' = 0.001y(100 - y - 3.0x).$$

Find and classify all critical points of the system.

Solution Critical points occur at $(0, 0)$, $(50, 0)$, $(0, 100)$ and at the solution $(20, 40)$ of the system

$$x + 0.75y = 50$$
$$3.0x + y = 100.$$

Since $\alpha_{12}\alpha_{21} = 2.25 > 1$, we have case (b) in Figure 10.38, and so the critical point at $(20, 40)$ is a saddle point. The Jacobian matrix is

$$\mathbf{g}'(\mathbf{X}) = \begin{pmatrix} 0.2 - 0.008x - 0.003y & -0.003x \\ -0.003y & 0.1 - 0.002y - 0.003x \end{pmatrix},$$

and we obtain

$$\mathbf{g}'((0,0)) = \begin{pmatrix} 0.2 & 0 \\ 0 & 0.1 \end{pmatrix}, \quad \mathbf{g}'((50,0)) = \begin{pmatrix} -0.2 & -0.15 \\ 0 & -0.05 \end{pmatrix}, \quad \mathbf{g}'((0,100)) = \begin{pmatrix} -0.1 & 0 \\ -0.3 & -0.1 \end{pmatrix}.$$

Therefore $(0, 0)$ is an unstable node, whereas both $(50, 0)$ and $(0, 100)$ are stable nodes. (Check this!) ∎

Coexistence can also occur in the Lotka-Volterra competition model if there is at least one periodic solution lying entirely in the first quadrant. It is possible to show, however, that this model has no periodic solutions.

EXERCISES 10.4

Answers to odd-numbered problems begin on page AN-16.

Nonlinear Pendulum

1. A pendulum is released at $\theta = \pi/3$ and is given an initial angular velocity of ω_0 rad/s. Determine under what conditions the resulting motion is periodic.

2. (a) If a pendulum is released from rest at $\theta = \theta_0$, show that the angular velocity is again 0 when $\theta = -\theta_0$.
   (b) The period T of the pendulum is the amount of time needed for θ to change from θ_0 to $-\theta_0$ and back to θ_0. Show that

$$T = \sqrt{\frac{2L}{g}} \int_{-\theta_0}^{\theta_0} \frac{1}{\sqrt{\cos \theta - \cos \theta_0}} \, d\theta.$$

Sliding Bead

3. A bead with mass m slides along a thin wire whose shape is described by the function $z = f(x)$. If $\mathbf{X}_1 = (x_1, y_1)$ is a critical point of the plane autonomous system associated with the sliding bead, verify that the Jacobian matrix at $\mathbf{X}_1$ is

$$\mathbf{g}'(\mathbf{X}_1) = \begin{pmatrix} 0 & 1 \\ -gf''(x_1) & -\beta/m \end{pmatrix}.$$

4. A bead with mass m slides along a thin wire whose shape is described by the function $z = f(x)$. When $f'(x_1) = 0, f''(x_1) > 0$, and the system is undamped, the critical point $\mathbf{X}_1 = (x_1, 0)$ is a center. Estimate the period of the bead when $x(0)$ is near x_1 and $x'(0) = 0$.

5. A bead is released from the position $x(0) = x_0$ on the curve $z = x^2/2$ with initial velocity $x'(0) = v_0$ cm/s.
   (a) Use the phase-plane method to show that the resulting solution is periodic when the system is undamped.
   (b) Show that the maximum height z_{max} to which the bead rises is given by $z_{max} = \frac{1}{2}[e^{v_0^2/g}(1 + x_0^2) - 1]$.

6. Rework Problem 5 with $z = \cosh x$.

Predator-Prey Models

7. (Refer to Figure 10.36.) If $x_m < x_1 < x_M$ and $x = x_1$, show that $F(x)G(y) = c_0$ has exactly two solutions y_1 and y_2 that satisfy $y_1 < a/b < y_2$. [*Hint:* First show that $G(y) = c_0/F(x_1) < G(a/b)$.]

8. From (1) and (3) on page 474, conclude that the maximum number of predators occurs when $y = a/b$.

9. In many fishery science models the rate at which a species is caught is assumed to be directly proportional to its abundance. If both predator and prey are being exploited in this manner, the Lotka-Volterra differential equations take the form

$$x' = -ax + bxy - \epsilon_1 x$$
$$y' = -cxy + dy - \epsilon_2 y,$$

where ϵ_1 and ϵ_2 are positive constants.

(a) When $\epsilon_2 < d$, show that there is a new critical point in the first quadrant that is a center.

(b) **Volterra's principle** states that a moderate amount of exploitation increases the average number of prey and decreases the average number of predators. Is this fisheries model consistent with Volterra's principle?

10. A predator-prey interaction is described by the Lotka-Volterra model

$$x' = -0.1x + 0.02xy$$
$$y' = 0.2y - 0.025xy.$$

(a) Find the critical point in the first quadrant, and use a numerical solver to sketch some population cycles.

(b) Estimate the period of the periodic solutions that are close to the critical point in (a).

Competition Models

11. A competitive interaction is described by the Lotka-Volterra competition model

$$x' = 0.08x(20 - 0.4x - 0.3y)$$
$$y' = 0.06y(10 - 0.1y - 0.3x).$$

Find and classify all critical points of the system.

12. In (1) show that $(0, 0)$ is always an unstable node.

13. In (1) show that $(K_1, 0)$ is a stable node when $K_1 > K_2/\alpha_{21}$ and a saddle point when $K_1 < K_2/\alpha_{21}$.

14. Use Problems 12 and 13 to establish that $(0, 0)$, $(K_1, 0)$, and $(0, K_2)$ are unstable when $\hat{\mathbf{X}} = (\hat{x}, \hat{y})$ is a stable node.

15. In (1) show that $\hat{\mathbf{X}} = (\hat{x}, \hat{y})$ is a saddle point when

$$\frac{K_1}{\alpha_{12}} < K_2 \quad \text{and} \quad \frac{K_2}{\alpha_{21}} < K_1.$$

Miscellaneous Nonlinear Models

16. If we assume that a damping force acts in a direction opposite to the motion of a pendulum and with a magnitude directly proportional to the angular velocity $d\theta/dt$, the displacement angle θ for the pendulum satisfies the nonlinear second-order differential equation

$$ml\frac{d^2\theta}{dt^2} = -mg\sin\theta - \beta\frac{d\theta}{dt}.$$

(a) Write the second-order differential equation as a plane autonomous system, and find all critical points.

(b) Find a condition on m, l, and β that will make $(0, 0)$ a stable spiral point.

17. In the analysis of free, damped motion in Section 5.1 we assumed that the damping force was proportional to the velocity x'. Frequently the magnitude of this damping force is proportional to the square of the velocity, and the new differential equation becomes

$$x'' = -\frac{\beta}{m}x'|x'| - \frac{k}{m}x.$$

(a) Write the second-order differential equation as a plane autonomous system, and find all critical points.

(b) The system is called *overdamped* when $(0, 0)$ is a stable node and is called *underdamped* when $(0, 0)$ is a stable spiral point. Physical considerations suggest that $(0, 0)$ must be an asymptotically stable critical point. Show that the system is necessarily underdamped.

$$\left[Hint: \frac{d}{dy}(y|y|) = 2|y|. \right]$$

Discussion Problems

18. A bead with mass m slides along a thin wire whose shape may be described by the function $z = f(x)$. Small stretches of the wire act like an inclined plane, and in mechanics it is assumed that the magnitude of the frictional force between the bead and wire is directly proportional to $mg\cos\theta$ (see Figure 10.30).

(a) Explain why the new differential equation for the x-coordinate of the bead is

$$x'' = g\frac{\mu - f'(x)}{1 + [f'(x)]^2} - \frac{\beta}{m}x'$$

for some positive constant μ.

(b) Investigate the critical points of the corresponding plane autonomous system. Under what conditions is a critical point a saddle point? A stable spiral point?

19. An undamped oscillation satisfies a nonlinear second-order differential equation of the form $x'' + f(x) = 0$, where $f(0) = 0$ and $xf(x) > 0$ for $x \neq 0$ and $-d < x < d$. Use the phase-plane method to investigate whether it is possible for the critical point $(0, 0)$ to be a stable spiral point. $\left[Hint: \text{Let } F(x) = \int_0^x f(u)\,du, \text{ and show that } y^2 + 2F(x) = c. \right]$

20. The Lotka-Volterra predator-prey model assumes that, in the absence of predators, the number of prey grows exponentially. If we make the alternative assumption that the prey population grows logistically, the new system is

$$x' = -ax + bxy$$

$$y' = -cxy + \frac{r}{K}y(K - y),$$

where a, b, c, r, and K are positive and $K > a/b$.

(a) Show that the system has critical points at $(0, 0)$, $(0, K)$, and $(\hat{x}, \hat{y})$, where $\hat{y} = a/b$ and $c\hat{x} = \frac{r}{K}(K - \hat{y})$.

(b) Show that the critical points at $(0, 0)$ and $(0, K)$ are saddle points, whereas the critical point at $(\hat{x}, \hat{y})$ is either a stable node or a stable spiral point.

(c) Show that $(\hat{x}, \hat{y})$ is a stable spiral point if $\hat{y} < \dfrac{4bK^2}{r + 4bK}$. Explain why this case will occur when the carrying capacity K of the prey is large.

21. The dynamical system

$$x' = \alpha \frac{y}{1 + y}x - x$$

$$y' = -\frac{y}{1 + y}x - y + \beta$$

arises in a model for the growth of microorganisms in a chemostat, a simple laboratory device in which a nutrient from a supply source flows into a growth chamber. In the system, x denotes the concentration of the microorganisms in the growth chamber, y denotes the concentration of nutrients, and $\alpha > 1$ and $\beta > 0$ are constants that can be adjusted by the experimenter. Find conditions on α and β that ensure that the system has a single critical point $(\hat{x}, \hat{y})$ in the first quadrant, and investigate the stability of this critical point.

22. Use the methods of this chapter together with a numerical solver to investigate stability in the nonlinear spring/mass system modeled by

$$x'' + 8x - 6x^3 + x^5 = 0.$$

(See Problem 8 in Exercises 5.3.)

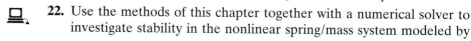

CHAPTER 10 IN REVIEW

Answers to odd-numbered problems begin on page AN-16.

Answer Problems 1–10 without referring back to the text. Fill in the blank or answer true or false.

1. The second-order differential equation $x'' + f(x') + g(x) = 0$ can be written as a plane autonomous system. _____

2. If $\mathbf{X} = \mathbf{X}(t)$ is a solution to a plane autonomous system and $\mathbf{X}(t_1) = \mathbf{X}(t_2)$ for $t_1 \neq t_2$, then $\mathbf{X}(t)$ is a periodic solution. _____

3. If the trace of the matrix $\mathbf{A}$ is 0 and $\det \mathbf{A} \neq 0$, then the critical point $(0, 0)$ of the linear system $\mathbf{X}' = \mathbf{AX}$ may be classified as _____.

4. If the critical point $(0, 0)$ of the linear system $\mathbf{X}' = \mathbf{AX}$ is a stable spiral point, then the eigenvalues of $\mathbf{A}$ are _____.

5. If the critical point $(0,0)$ of the linear system $\mathbf{X}' = \mathbf{AX}$ is a saddle point and $\mathbf{X} = \mathbf{X}(t)$ is a solution, then $\lim_{t\to\infty} \mathbf{X}(t)$ does not exist. _____

6. If the Jacobian matrix $\mathbf{A} = \mathbf{g}'(\mathbf{X}_1)$ at a critical point of a plane autonomous system has positive trace and determinant, then the critical point $\mathbf{X}_1$ is unstable. _____

7. It is possible to show, using linearization, that a nonlinear plane autonomous system has periodic solutions. _____

8. All solutions to the pendulum equation $\dfrac{d^2\theta}{dt^2} + \dfrac{g}{l}\sin\theta = 0$ are periodic. _____

9. For what value(s) of α does the plane autonomous system

$$x' = \alpha x - 2y$$
$$y' = -\alpha x + y$$

possess periodic solutions? _____

10. For what values of n is $x = n\pi$ an asymptotically stable critical point of the autonomous first-order differential equation $x' = \sin x$? _____

11. Solve the following nonlinear plane autonomous system by switching to polar coordinates, and describe the geometric behavior of the solution that satisfies the given initial condition:

$$x' = -y - x(\sqrt{x^2 + y^2})^3$$
$$y' = x - y(\sqrt{x^2 + y^2})^3,$$
$$\mathbf{X}(0) = (1, 0).$$

12. Discuss the geometric nature of the solutions to the linear system $\mathbf{X}' = \mathbf{AX}$ given that the general solution is

(a) $\mathbf{X}(t) = c_1\begin{pmatrix}1\\1\end{pmatrix}e^{-t} + c_2\begin{pmatrix}1\\-2\end{pmatrix}e^{-2t}$

(b) $\mathbf{X}(t) = c_1\begin{pmatrix}1\\-1\end{pmatrix}e^{-t} + c_2\begin{pmatrix}1\\2\end{pmatrix}e^{2t}$

13. Classify the critical point $(0,0)$ of the given linear system by computing the trace τ and determinant Δ.

(a) $x' = -3x + 4y$
$\quad\ y' = -5x + 3y$

(b) $x' = -3x + 2y$
$\quad\ y' = -2x + y$

14. Find and classify (if possible) the critical points of the plane autonomous system

$$x' = x + xy - 3x^2$$
$$y' = 4y - 2xy - y^2.$$

15. Determine the value(s) of α for which $(0, 0)$ is a stable critical point for the plane autonomous system (in polar coordinates)

$$r' = \alpha r$$
$$\theta' = 1.$$

16. Classify the critical point $(0, 0)$ of the plane autonomous system corresponding to the nonlinear second-order differential equation

$$x'' + \mu(x^2 - 1)x' + x = 0,$$

where μ is a real constant.

17. Without solving explicitly, classify (if possible) the critical points of the autonomous first-order differential equation $x' = (x^2 - 1)e^{-x/2}$ as asymptotically stable or unstable.

18. Use the phase-plane method to show that the solutions to the nonlinear second-order differential equation $x'' = -2x\sqrt{(x')^2 + 1}$ that satisfy $x(0) = x_0$ and $x'(0) = 0$ are periodic.

19. In Section 5.1 we assumed that the restoring force F of the spring satisfied Hooke's law $F = ks$, where s is the elongation of the spring and k is a positive constant of proportionality. If we replace this assumption with the nonlinear law $F = ks^3$, the new differential equation for damped motion of the hard spring becomes

$$mx'' = -\beta x' - k(s + x)^3 + mg,$$

where $ks^3 = mg$. The system is called overdamped when $(0, 0)$ is a stable node and is called underdamped when $(0, 0)$ is a stable spiral point. Find new conditions on m, k, and β that will lead to overdamping and underdamping.

20. The rod of a pendulum is attached to a movable joint at a point P and rotates at an angular speed of ω (rad/s) in the plane perpendicular to the rod. See Figure 10.39. As a result the bob of the rotating pendulum experiences an additional centripetal force, and the new differential equation for θ becomes

$$ml\frac{d^2\theta}{dt^2} = \omega^2 ml \sin\theta\cos\theta - mg\sin\theta - \beta\frac{d\theta}{dt}.$$

(a) If $\omega^2 < g/l$, show that $(0, 0)$ is a stable critical point and is the only critical point in the domain $-\pi < \theta < \pi$. Describe what occurs physically when $\theta(0) = \theta_0$, $\theta'(0) = 0$, and θ_0 is small.

(b) If $\omega^2 > g/l$, show that $(0, 0)$ is unstable and there are two additional stable critical points $(\pm\hat{\theta}, 0)$ in the domain $-\pi < \theta < \pi$. Describe what occurs physically when $\theta(0) = \theta_0$, $\theta'(0) = 0$, and θ_0 is small.

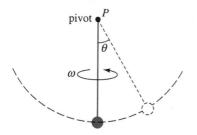

Figure 10.39

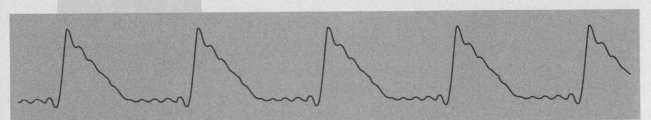

Partial sums of a Fourier series; see page 493.

11 ORTHOGONAL FUNCTIONS AND FOURIER SERIES

INTRODUCTION   In calculus you studied vectors in 2- and 3-space and saw that two non-zero vectors are orthogonal when their dot, or inner, product is zero. Beyond the level of calculus, the notions of vectors, orthogonality, and inner product often lose their geometric interpretation. These concepts have been generalized; it is perfectly common to think of a function as a vector. We can then say that two different functions are orthogonal when their inner product is zero, but we shall see that the inner product of these vectors (functions) is actually a definite integral. The concept of orthogonal functions is fundamental to the material covered in Chapters 12 and 13.

Another concept encountered in calculus is the expansion of a given function f in a power series. In this chapter we shall see how to expand a function f in terms of an infinite set of orthogonal functions.

11.1 ORTHOGONAL FUNCTIONS

• Inner product • Orthogonal functions • Orthogonal set
• Norm • Square norm • Orthonormal set • Orthogonality with respect to a weight function • Orthogonal series expansion

In advanced mathematics a function is considered to be a generalization of a vector. In this section we shall see how the two vector concepts of inner, or dot, product and orthogonality can be extended to functions. Recall that if $\mathbf{u}$ and $\mathbf{v}$ are vectors in 3-space, then the inner product $(\mathbf{u}, \mathbf{v})$ (also written as $\mathbf{u} \cdot \mathbf{v}$) possesses the following properties:

    (*i*) $(\mathbf{u}, \mathbf{v}) = (\mathbf{v}, \mathbf{u})$
    (*ii*) $(k\mathbf{u}, \mathbf{v}) = k(\mathbf{u}, \mathbf{v})$, k a scalar
    (*iii*) $(\mathbf{u}, \mathbf{u}) = 0$ if $\mathbf{u} = \mathbf{0}$ and $(\mathbf{u}, \mathbf{u}) > 0$ if $\mathbf{u} \neq \mathbf{0}$
    (*iv*) $(\mathbf{u} + \mathbf{v}, \mathbf{w}) = (\mathbf{u}, \mathbf{w}) + (\mathbf{v}, \mathbf{w})$.

We expect that a generalization of the inner product concept should have these same properties.

Inner Product of Two Functions Suppose that f_1 and f_2 are functions defined on an interval $[a, b]$.* Since a definite integral on the interval of the product $f_1(x)f_2(x)$ possesses properties (*i*)–(*iv*) of the inner product of vectors whenever the integral exists, we are prompted to make the following definition.

DEFINITION 11.1   Inner Product of Functions

The **inner product** of two functions f_1 and f_2 on an interval $[a, b]$ is the number

$$(f_1, f_2) = \int_a^b f_1(x)f_2(x)\, dx.$$

Orthogonal Functions Motivated by the fact that two vectors $\mathbf{u}$ and $\mathbf{v}$ are orthogonal whenever their inner product is zero, we define **orthogonal functions** in a similar manner.

DEFINITION 11.2   Orthogonal Functions

Two functions f_1 and f_2 are said to be **orthogonal** on an interval $[a, b]$ if

$$(f_1, f_2) - \int_a^b f_1(x)f_2(x)\, dx = 0. \tag{1}$$

---

*The interval could also be $(-\infty, \infty)$, $[0, \infty)$, and so on.

For example, the functions $f_1(x) = x^2$ and $f_2(x) = x^3$ are orthogonal on the interval $[-1, 1]$ since

$$(f_1, f_2) = \int_{-1}^{1} x^2 \cdot x^3 \, dx = \tfrac{1}{6} x^6 \big|_{-1}^{1} = 0.$$

Unlike in vector analysis, where the word *orthogonal* is a synonym for *perpendicular,* in this present context the term *orthogonal* and condition (1) have no geometric significance.

Orthogonal Sets   We are primarily interested in infinite sets of orthogonal functions.

DEFINITION 11.3   **Orthogonal Set**

A set of real-valued functions $\{\phi_0(x), \phi_1(x), \phi_2(x), \ldots\}$ is said to be **orthogonal** on an interval $[a, b]$ if

$$(\phi_m, \phi_n) = \int_{a}^{b} \phi_m(x)\phi_n(x) \, dx = 0, \quad m \neq n. \tag{2}$$

Orthonormal Sets   The norm, or length $\|\mathbf{u}\|$, of a vector $\mathbf{u}$ can be expressed in terms of the inner product. The expression $(\mathbf{u}, \mathbf{u}) = \|\mathbf{u}\|^2$ is called the square norm, and so the norm is $\|\mathbf{u}\| = \sqrt{(\mathbf{u}, \mathbf{u})}$. Similarly, the **square norm** of a function ϕ_n is $\|\phi_n(x)\|^2 = (\phi_n, \phi_n)$, and so the **norm,** or its generalized length, is $\|\phi_n(x)\| = \sqrt{(\phi_n, \phi_n)}$. In other words, the square norm and norm of a function ϕ_n in an orthogonal set $\{\phi_n(x)\}$ are, respectively,

$$\|\phi_n(x)\|^2 = \int_{a}^{b} \phi_n^2(x) \, dx \quad \text{and} \quad \|\phi_n(x)\| = \sqrt{\int_{a}^{b} \phi_n^2(x) \, dx}. \tag{3}$$

If $\{\phi_n(x)\}$ is an orthogonal set of functions on the interval $[a, b]$ with the property that $\|\phi_n(x)\| = 1$ for $n = 0, 1, 2, \ldots$, then $\{\phi_n(x)\}$ is said to be an **orthonormal set** on the interval.

EXAMPLE 1   **Orthogonal Set of Functions**

Show that the set $\{1, \cos x, \cos 2x, \ldots\}$ is orthogonal on the interval $[-\pi, \pi]$.

Solution   If we make the identification $\phi_0(x) = 1$ and $\phi_n(x) = \cos nx$, we must then show that $\int_{-\pi}^{\pi} \phi_0(x)\phi_n(x) \, dx = 0, n \neq 0$, and $\int_{-\pi}^{\pi} \phi_m(x)\phi_n(x) \, dx = 0, m \neq n$. We have, in the first case,

$$(\phi_0, \phi_n) = \int_{-\pi}^{\pi} \phi_0(x)\phi_n(x) \, dx = \int_{-\pi}^{\pi} \cos nx \, dx$$

$$= \frac{1}{n} \sin nx \bigg|_{-\pi}^{\pi} = \frac{1}{n} [\sin n\pi - \sin(-n\pi)] = 0, \quad n \neq 0,$$

and in the second,

$$(\phi_m, \phi_n) = \int_{-\pi}^{\pi} \phi_m(x)\phi_n(x)\, dx$$

$$= \int_{-\pi}^{\pi} \cos mx \cos nx\, dx$$

$$= \frac{1}{2} \int_{-\pi}^{\pi} [\cos(m + n)x + \cos(m - n)x]\, dx \qquad \leftarrow \text{trig identity}$$

$$= \frac{1}{2} \left[\frac{\sin(m + n)x}{m + n} + \frac{\sin(m - n)x}{m - n} \right]_{-\pi}^{\pi} = 0, \quad m \neq n. \qquad ∎$$

EXAMPLE 2   Norms

Find the norm of each function in the orthogonal set given in Example 1.

Solution   For $\phi_0(x) = 1$ we have, from (3),

$$\|\phi_0(x)\|^2 = \int_{-\pi}^{\pi} dx = 2\pi,$$

so $\|\phi_0(x)\| = \sqrt{2\pi}$. For $\phi_n(x) = \cos nx$, $n > 0$, it follows that

$$\|\phi_n(x)\|^2 = \int_{-\pi}^{\pi} \cos^2 nx\, dx = \frac{1}{2} \int_{-\pi}^{\pi} [1 + \cos 2nx]\, dx = \pi.$$

Thus for $n > 0$, $\|\phi_n(x)\| = \sqrt{\pi}$. ∎

Any orthogonal set of nonzero functions $\{\phi_n(x)\}$, $n = 0, 1, 2, \ldots$ can be *normalized*—that is, made into an orthonormal set—by dividing each function by its norm. It follows from Examples 1 and 2 that the set

$$\left\{ \frac{1}{\sqrt{2\pi}}, \frac{\cos x}{\sqrt{\pi}}, \frac{\cos 2x}{\sqrt{\pi}}, \ldots \right\}$$

is orthonormal on $[-\pi, \pi]$.

We shall make one more analogy between vectors and functions. Suppose $\mathbf{v}_1$, $\mathbf{v}_2$, and $\mathbf{v}_3$ are three mutually orthogonal nonzero vectors in 3-space. Such an orthogonal set can be used as a basis for 3-space; that is, any three-dimensional vector can be written as a linear combination

$$\mathbf{u} = c_1\mathbf{v}_1 + c_2\mathbf{v}_2 + c_3\mathbf{v}_3, \tag{4}$$

where the c_i, $i = 1, 2, 3$ are scalars called the components of the vector. Each component c_i can be expressed in terms of $\mathbf{u}$ and the corresponding vector $\mathbf{v}_i$. To see this we take the inner product of (4) with $\mathbf{v}_1$:

$$(\mathbf{u}, \mathbf{v}_1) = c_1(\mathbf{v}_1, \mathbf{v}_1) + c_2(\mathbf{v}_2, \mathbf{v}_1) + c_3(\mathbf{v}_3, \mathbf{v}_1) = c_1\|\mathbf{v}_1\|^2 + c_2 \cdot 0 + c_3 \cdot 0.$$

Hence $c_1 = \dfrac{(\mathbf{u}, \mathbf{v}_1)}{\|\mathbf{v}_1\|^2}.$

In like manner we find that the components c_2 and c_3 are given by

$$c_2 = \frac{(\mathbf{u}, \mathbf{v}_2)}{\|\mathbf{v}_2\|^2} \quad \text{and} \quad c_3 = \frac{(\mathbf{u}, \mathbf{v}_3)}{\|\mathbf{v}_3\|^2}.$$

Hence (4) can be expressed as

$$\mathbf{u} = \frac{(\mathbf{u}, \mathbf{v}_1)}{\|\mathbf{v}_1\|^2}\mathbf{v}_1 + \frac{(\mathbf{u}, \mathbf{v}_2)}{\|\mathbf{v}_2\|^2}\mathbf{v}_2 + \frac{(\mathbf{u}, \mathbf{v}_3)}{\|\mathbf{v}_3\|^2}\mathbf{v}_3 = \sum_{n=1}^{3} \frac{(\mathbf{u}, \mathbf{v}_n)}{\|\mathbf{v}_n\|^2}\mathbf{v}_n. \tag{5}$$

Orthogonal Series Expansion   Suppose $\{\phi_n(x)\}$ is an infinite orthogonal set of functions on an interval $[a, b]$. We ask: If $y = f(x)$ is a function defined on the interval $[a, b]$, is it possible to determine a set of coefficients c_n, $n = 0, 1, 2, \ldots$ for which

$$f(x) = c_0\phi_0(x) + c_1\phi_1(x) + \cdots + c_n\phi_n(x) + \cdots? \tag{6}$$

As in the foregoing discussion on finding components of a vector, we can find the coefficients c_n by utilizing the inner product. Multiplying (6) by $\phi_m(x)$ and integrating over the interval $[a, b]$ gives

$$\int_a^b f(x)\phi_m(x)\,dx = c_0\int_a^b \phi_0(x)\phi_m(x)\,dx + c_1\int_a^b \phi_1(x)\phi_m(x)\,dx + \cdots + c_n\int_a^b \phi_n(x)\phi_m(x)\,dx + \cdots$$

$$= c_0(\phi_0, \phi_m) + c_1(\phi_1, \phi_m) + \cdots + c_n(\phi_n, \phi_m) + \cdots.$$

By orthogonality each term on the right-hand side of the last equation is zero *except* when $m = n$. In this case we have

$$\int_a^b f(x)\phi_n(x)\,dx = c_n\int_a^b \phi_n^2(x)\,dx.$$

It follows that the required coefficients are

$$c_n = \frac{\int_a^b f(x)\phi_n(x)\,dx}{\int_a^b \phi_n^2(x)\,dx}, \quad n = 0, 1, 2, \ldots.$$

In other words,

$$f(x) = \sum_{n=0}^{\infty} c_n\phi_n(x), \tag{7}$$

where

$$c_n = \frac{\int_a^b f(x)\phi_n(x)\,dx}{\|\phi_n(x)\|^2}. \tag{8}$$

With inner product notation, (7) becomes

$$f(x) = \sum_{n=0}^{\infty} \frac{(f, \phi_n)}{\|\phi_n(x)\|^2}\phi_n(x). \tag{9}$$

Thus (9) is seen to be the functional analogue of the vector result given in (5).

DEFINITION 11.4   **Orthogonal Set/Weight Function**

A set of real-valued functions $\{\phi_0(x), \phi_1(x), \phi_2(x), \ldots\}$ is said to be **orthogonal with respect to a weight function** $w(x)$ on an interval $[a, b]$ if

$$\int_a^b w(x)\phi_m(x)\phi_n(x)\,dx = 0, \quad m \neq n.$$

The usual assumption is that $w(x) > 0$ on the interval of orthogonality $[a, b]$. The set $\{1, \cos x, \cos 2x, \ldots\}$ in Example 1 is orthogonal with respect to the weight function $w(x) = 1$ on the interval $[-\pi, \pi]$.

If $\{\phi_n(x)\}$ is orthogonal with respect to a weight function $w(x)$ on the interval $[a, b]$, then multiplying (6) by $w(x)\phi_n(x)$ and integrating yields

$$c_n = \frac{\int_a^b f(x)w(x)\phi_n(x)\,dx}{\|\phi_n(x)\|^2}, \tag{10}$$

where
$$\|\phi_n(x)\|^2 = \int_a^b w(x)\,\phi_n^2(x)\,dx. \tag{11}$$

The series (7) with coefficients given by either (8) or (10) is said to be an **orthogonal series expansion** of f or a **generalized Fourier series.**

Complete Sets The procedure outlined for determining the coefficients c_n was *formal*; that is, basic questions on whether or not an orthogonal series expansion such as (7) is actually possible were ignored. Also, to expand f in a series of orthogonal functions, it is certainly necessary that f not be orthogonal to each ϕ_n of the orthogonal set $\{\phi_n(x)\}$. (If f were orthogonal to every ϕ_n, then $c_n = 0$, $n = 0, 1, 2, \ldots$.) To avoid the latter problem we shall assume, for the remainder of the discussion, that an orthogonal set is **complete.** This means that the only function orthogonal to each member of the set is the zero function.

EXERCISES 11.1

Answers to odd-numbered problems begin on page AN-16.

In Problems 1–6 show that the given functions are orthogonal on the indicated interval.

1. $f_1(x) = x$, $f_2(x) = x^2$; $[-2, 2]$

2. $f_1(x) = x^3$, $f_2(x) = x^2 + 1$; $[-1, 1]$

3. $f_1(x) = e^x$, $f_2(x) = xe^{-x} - e^{-x}$; $[0, 2]$

4. $f_1(x) = \cos x$, $f_2(x) = \sin^2 x$; $[0, \pi]$

5. $f_1(x) = x$, $f_2(x) = \cos 2x$; $[-\pi/2, \pi/2]$

6. $f_1(x) = e^x$, $f_2(x) = \sin x$; $[\pi/4, 5\pi/4]$

In Problems 7–12 show that the given set of functions is orthogonal on the indicated interval. Find the norm of each function in the set.

7. $\{\sin x, \sin 3x, \sin 5x, \ldots\}$; $[0, \pi/2]$

8. $\{\cos x, \cos 3x, \cos 5x, \ldots\}$; $[0, \pi/2]$

9. $\{\sin nx\}$, $n = 1, 2, 3, \ldots$; $[0, \pi]$

10. $\left\{\sin \dfrac{n\pi}{p}x\right\}$, $n = 1, 2, 3, \ldots$; $[0, p]$

11. $\left\{1, \cos \dfrac{n\pi}{p}x\right\}$, $n = 1, 2, 3, \ldots$; $[0, p]$

12. $\left\{1, \cos \dfrac{n\pi}{p}x, \sin \dfrac{m\pi}{p}x\right\}$, $n = 1, 2, 3, \ldots, m = 1, 2, 3, \ldots$; $[-p, p]$

In Problems 13 and 14 verify by direct integration that the functions are orthogonal with respect to the indicated weight function on the given interval.

13. $H_0(x) = 1$, $H_1(x) = 2x$, $H_2(x) = 4x^2 - 2$; $w(x) = e^{-x^2}$, $(-\infty, \infty)$

14. $L_0(x) = 1$, $L_1(x) = -x + 1$, $L_2(x) = \dfrac{1}{2}x^2 - 2x + 1$; $w(x) = e^{-x}$, $[0, \infty)$

15. Let $\{\phi_n(x)\}$ be an orthogonal set of functions on $[a, b]$ such that $\phi_0(x) = 1$. Show that $\int_a^b \phi_n(x)\, dx = 0$ for $n = 1, 2, \ldots$.

16. Let $\{\phi_n(x)\}$ be an orthogonal set of functions on $[a, b]$ such that $\phi_0(x) = 1$ and $\phi_1(x) = x$. Show that $\int_a^b (\alpha x + \beta)\phi_n(x)\, dx = 0$ for $n = 2, 3, \ldots$ and any constants α and β.

17. Let $\{\phi_n(x)\}$ be an orthogonal set of functions on $[a, b]$. Show that $\|\phi_m(x) + \phi_n(x)\|^2 = \|\phi_m(x)\|^2 + \|\phi_n(x)\|^2$, $m \neq n$.

18. From Problem 1 we know that $f_1(x) = x$ and $f_2(x) = x^2$ are orthogonal on $[-2, 2]$. Find constants c_1 and c_2 such that $f_3(x) = x + c_1 x^2 + c_2 x^3$ is orthogonal to both f_1 and f_2 on the same interval.

19. The set of functions $\{\sin nx\}$, $n = 1, 2, 3, \ldots$ is orthogonal on the interval $[-\pi, \pi]$. Show that the set is not complete.

20. Suppose f_1, f_2, and f_3 are functions continuous on the interval $[a, b]$. Show that $(f_1 + f_2, f_3) = (f_1, f_3) + (f_2, f_3)$.

Discussion Problems

21. A real-valued function f is said to be **periodic** with period T if $f(x + T) = f(x)$. For example, 4π is a period of $\sin x$ since $\sin(x + 4\pi) = \sin x$. The smallest value of T for which $f(x + T) = f(x)$ holds is called the **fundamental period** of f. For example, the fundamental period of $f(x) = \sin x$ is $T = 2\pi$. What is the fundamental period of each of the following functions?

(a) $f(x) = \cos 2\pi x$ **(b)** $f(x) = \sin \dfrac{4}{L} x$

(c) $f(x) = \sin x + \sin 2x$ **(d)** $f(x) = \sin 2x + \cos 4x$

(e) $f(x) = \sin 3x + \cos 2x$

(f) $f(x) = A_0 + \displaystyle\sum_{n=1}^{\infty}\left(A_n \cos \dfrac{n\pi}{p} x + B_n \sin \dfrac{n\pi}{p} x\right)$, A_n and B_n depend only on n

11.2 FOURIER SERIES

- *Trigonometric series*   • *Fourier series*   • *Fourier coefficients*
- *Convergence of a Fourier series*   • *Periodic extension*

We have just seen that if $\{\phi_0(x), \phi_1(x), \phi_2(x), \ldots\}$ is an orthogonal set on an interval $[a, b]$ and if f is a function defined on the same interval then we can formally expand f in an orthogonal series $c_0\phi_0(x) + c_1\phi_1(x) + c_2\phi_2(x) + \cdots$, where the coefficients c_n are determined using the inner product concept. The orthogonal set of trigonometric functions

$$\left\{1, \cos\frac{\pi}{p}x, \cos\frac{2\pi}{p}x, \cos\frac{3\pi}{p}x, \ldots, \sin\frac{\pi}{p}x, \sin\frac{2\pi}{p}x, \sin\frac{3\pi}{p}x, \ldots\right\} \qquad \textbf{(1)}$$

will be of particular importance later on in the solution of certain kinds of boundary-value problems involving linear partial differential equations. The set (1) is orthogonal on the interval $[-p, p]$ (see Problem 12 in Exercises 11.1).

A Trigonometric Series Suppose that f is a function defined on the interval $[-p, p]$ and can be expanded in an orthogonal series consisting of the trigonometric functions in the orthogonal set (1); that is,

$$f(x) = \frac{a_0}{2} + \sum_{n=1}^{\infty} \left(a_n \cos \frac{n\pi}{p} x + b_n \sin \frac{n\pi}{p} x \right). \tag{2}$$

The coefficients $a_0, a_1, a_2, \ldots, b_1, b_2, \ldots$ can be determined in exactly the same manner as in the general discussion of orthogonal series expansions on page 487. Before proceeding, note that we have chosen to write the coefficient of 1 in the set (1) as $a_0/2$ rather than a_0. This is for convenience only; the formula of a_n will then reduce to a_0 for $n = 0$.

Now integrating both sides of (2) from $-p$ to p gives

$$\int_{-p}^{p} f(x)\, dx = \frac{a_0}{2} \int_{-p}^{p} dx + \sum_{n=1}^{\infty} \left(a_n \int_{-p}^{p} \cos \frac{n\pi}{p} x\, dx + b_n \int_{-p}^{p} \sin \frac{n\pi}{p} x\, dx \right). \tag{3}$$

Since $\cos(n\pi x/p)$ and $\sin(n\pi x/p), n \geq 1$ are orthogonal to 1 on the interval, the right side of (3) reduces to a single term:

$$\int_{-p}^{p} f(x)\, dx = \frac{a_0}{2} \int_{-p}^{p} dx = \frac{a_0}{2} x \Big|_{-p}^{p} = pa_0.$$

Solving for a_0 yields

$$a_0 = \frac{1}{p} \int_{-p}^{p} f(x)\, dx. \tag{4}$$

Now we multiply (2) by $\cos(m\pi x/p)$ and integrate:

$$\int_{-p}^{p} f(x) \cos \frac{m\pi}{p} x\, dx = \frac{a_0}{2} \int_{-p}^{p} \cos \frac{m\pi}{p} x\, dx$$

$$+ \sum_{n=1}^{\infty} \left(a_n \int_{-p}^{p} \cos \frac{m\pi}{p} x \cos \frac{n\pi}{p} x\, dx + b_n \int_{-p}^{p} \cos \frac{m\pi}{p} x \sin \frac{n\pi}{p} x\, dx \right). \tag{5}$$

By orthogonality we have

$$\int_{-p}^{p} \cos \frac{m\pi}{p} x\, dx = 0, \quad m > 0, \qquad \int_{-p}^{p} \cos \frac{m\pi}{p} x \sin \frac{n\pi}{p} x\, dx = 0,$$

and

$$\int_{-p}^{p} \cos \frac{m\pi}{p} x \cos \frac{n\pi}{p} x\, dx = \begin{cases} 0, & m \neq n \\ p, & m = n. \end{cases}$$

Thus (5) reduces to $\quad \int_{-p}^{p} f(x) \cos \frac{n\pi}{p} x\, dx = a_n p,$

and so $\quad a_n = \frac{1}{p} \int_{-p}^{p} f(x) \cos \frac{n\pi}{p} x\, dx. \tag{6}$

Finally, if we multiply (2) by $\sin(m\pi x/p)$, integrate, and make use of the results

$$\int_{-p}^{p} \sin \frac{m\pi}{p} x\, dx = 0, \quad m > 0, \qquad \int_{-p}^{p} \sin \frac{m\pi}{p} x \sin \frac{n\pi}{p} x\, dx = 0,$$

and

$$\int_{-p}^{p} \sin \frac{m\pi}{p} x \sin \frac{n\pi}{p} x\, dx = \begin{cases} 0, & m \neq n \\ p, & m = n \end{cases}$$

we find that $\quad b_n = \frac{1}{p} \int_{-p}^{p} f(x) \sin \frac{n\pi}{p} x\, dx. \tag{7}$

The trigonometric series (2) with coefficients a_0, a_n, and b_n defined by (4), (6), and (7), respectively, is said to be the **Fourier series** of the function f. The coefficients obtained from (4), (6), and (7) are referred to as **Fourier coefficients** of f.

In finding the coefficients a_0, a_n, and b_n we assumed that f was integrable on the interval and that (2), as well as the series obtained by multiplying (2) by $\cos(m\pi x/p)$, converged in such a manner as to permit term-by-term integration. Until (2) is shown to be convergent for a given function f, the equality sign is not to be taken in a strict or literal sense. Some texts use the symbol $\sim$ in place of $=$. In view of the fact that most functions in applications are of a type that guarantees convergence of the series, we shall use the equality symbol. We summarize the results:

DEFINITION 11.5 **Fourier Series**

The **Fourier series** of a function f defined on the interval $(-p, p)$ is given by

$$f(x) = \frac{a_0}{2} + \sum_{n=1}^{\infty}\left(a_n \cos\frac{n\pi}{p}x + b_n \sin\frac{n\pi}{p}x\right), \tag{8}$$

where
$$a_0 = \frac{1}{p}\int_{-p}^{p} f(x)\,dx \tag{9}$$

$$a_n = \frac{1}{p}\int_{-p}^{p} f(x)\cos\frac{n\pi}{p}x\,dx \tag{10}$$

$$b_n = \frac{1}{p}\int_{-p}^{p} f(x)\sin\frac{n\pi}{p}x\,dx. \tag{11}$$

EXAMPLE 1 **Expansion in a Fourier Series**

Expand
$$f(x) = \begin{cases} 0, & -\pi < x < 0 \\ \pi - x, & 0 \le x < \pi \end{cases} \tag{12}$$

in a Fourier series.

Solution The graph of f is given in Figure 11.1. With $p = \pi$ we have from (9) and (10) that

$$a_0 = \frac{1}{\pi}\int_{-\pi}^{\pi} f(x)\,dx = \frac{1}{\pi}\left[\int_{-\pi}^{0} 0\,dx + \int_{0}^{\pi} (\pi - x)\,dx\right] = \frac{1}{\pi}\left[\pi x - \frac{x^2}{2}\right]_0^{\pi} = \frac{\pi}{2}$$

$$a_n = \frac{1}{\pi}\int_{-\pi}^{\pi} f(x)\cos nx\,dx = \frac{1}{\pi}\left[\int_{-\pi}^{0} 0\,dx + \int_{0}^{\pi} (\pi - x)\cos nx\,dx\right]$$

$$= \frac{1}{\pi}\left[(\pi - x)\frac{\sin nx}{n}\Big|_0^{\pi} + \frac{1}{n}\int_{0}^{\pi} \sin nx\,dx\right]$$

$$= -\frac{1}{n\pi}\frac{\cos nx}{n}\Big|_0^{\pi} = \frac{1 - (-1)^n}{n^2\pi},$$

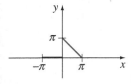

y

π

$-\pi$ $\quad$ π $\quad$ x

Figure 11.1

where we have used $\cos n\pi = (-1)^n$. In like manner we find from (11) that

$$b_n = \frac{1}{\pi} \int_0^\pi (\pi - x) \sin nx\, dx = \frac{1}{n}.$$

Therefore $f(x) = \frac{\pi}{4} + \sum_{n=1}^\infty \left\{ \frac{1 - (-1)^n}{n^2 \pi} \cos nx + \frac{1}{n} \sin nx \right\}.$ **(13)**   ∎

Note that a_n defined by (10) reduces to a_0 given by (9) when we set $n = 0$. But as Example 1 shows, this may not be the case *after* the integral for a_n is evaluated.

Convergence of a Fourier Series   The following theorem gives sufficient conditions for convergence of a Fourier series at a point.

THEOREM 11.1   **Conditions for Convergence**

Let f and f' be piecewise continuous on the interval $(-p, p)$; that is, let f and f' be continuous except at a finite number of points in the interval and have only finite discontinuities at these points. Then the Fourier series of f on the interval converges to $f(x)$ at a point of continuity. At a point of discontinuity the Fourier series converges to the average

$$\frac{f(x+) + f(x-)}{2},$$

where $f(x+)$ and $f(x-)$ denote the limit of f at x from the right and from the left, respectively.*

For a proof of this theorem you are referred to the classic text by Churchill and Brown.†

EXAMPLE 2   **Convergence of a Point of Discontinuity**

The function (12) in Example 1 satisfies the conditions of Theorem 11.1. Thus for every x in the interval $(-\pi, \pi)$, except at $x = 0$, the series (13) will converge to $f(x)$. At $x = 0$ the function is discontinuous and so the series (13) will converge to

$$\frac{f(0+) + f(0-)}{2} = \frac{\pi + 0}{2} = \frac{\pi}{2}.$$ ∎

---

*In other words, for x a point in the interval and $h > 0$,

$$f(x+) = \lim_{h \to 0} f(x + h), \quad f(x-) = \lim_{h \to 0} f(x - h).$$

†Ruel V. Churchill and James Ward Brown, *Fourier Series and Boundary Value Problems* (New York: McGraw-Hill).

Periodic Extension   Observe that each of the functions in the basic set (1) has a different fundamental period*—namely, $2p/n$, $n \geq 1$—but since a positive integer multiple of a period is also a period, we see that all of the functions have in common the period $2p$. (Verify.) Hence the right-hand side of (2) is $2p$-periodic; indeed, $2p$ is the **fundamental period** of the sum. We conclude that a Fourier series not only represents the function on the interval $(-p, p)$ but also gives the **periodic extension** of f outside this interval. We can now apply Theorem 11.1 to the periodic extension of f, or we may assume from the outset that the given function is periodic with period $2p$; that is, $f(x + 2p) = f(x)$. When f is piecewise continuous and the right- and left-hand derivatives exist at $x = -p$ and $x = p$, respectively, then the series (8) converges to the average

$$[f(p-) + f(-p+)]/2$$

at these endpoints and to this value extended periodically to $\pm 3p$, $\pm 5p$, $\pm 7p$, and so on.

The Fourier series in (13) converges to the periodic extension of (12) on the entire x-axis. At 0, $\pm 2\pi$, $\pm 4\pi$, ... and at $\pm \pi$, $\pm 3\pi$, $\pm 5\pi$, ... the series converges to the values

$$\frac{f(0+) + f(0-)}{2} = \frac{\pi}{2} \quad \text{and} \quad \frac{f(\pi+) + f(\pi-)}{2} = 0,$$

respectively. The solid dots in Figure 11.2 represent the value $\pi/2$.

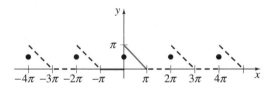

Figure 11.2

Sequence of Partial Sums   It is interesting to see how the sequence of partial sums $\{S_N(x)\}$ of a Fourier series approximates a function. For example, the first three partial sums of (13) are

$$S_1(x) = \frac{\pi}{4}, \quad S_2(x) = \frac{\pi}{4} + \frac{2}{\pi}\cos x + \sin x, \quad \text{and} \quad S_3(x) = \frac{\pi}{4} + \frac{2}{\pi}\cos x + \sin x + \frac{1}{2}\sin 2x.$$

In Figure 11.3 we have used a CAS to graph the partial sums $S_3(x)$, $S_5(x)$, $S_8(x)$, and $S_{15}(x)$ of (13) on the interval $(-\pi, \pi)$. Figure 11.3(e) shows the periodic extension using $S_{15}(x)$ on $(-4\pi, 4\pi)$.

---

*See Problem 21 in Exercises 11.1.

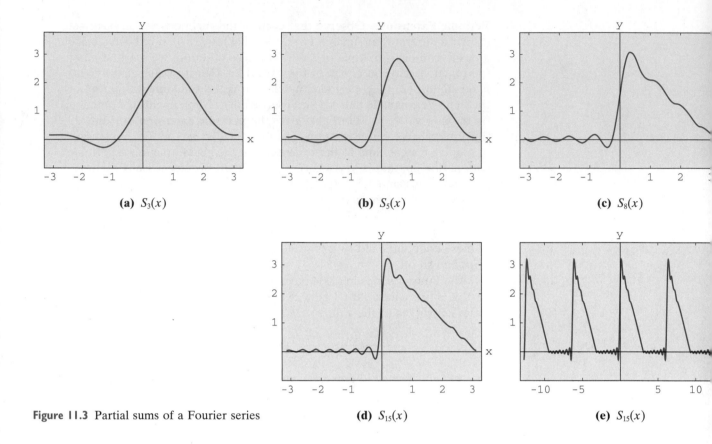

(a) $S_3(x)$ **(b)** $S_5(x)$ **(c)** $S_8(x)$

Figure 11.3 Partial sums of a Fourier series   **(d)** $S_{15}(x)$ **(e)** $S_{15}(x)$

EXERCISES 11.2

Answers to odd-numbered problems begin on page AN-16.

In Problems 1–16 find the Fourier series of f on the given interval.

1. $f(x) = \begin{cases} 0, & -\pi < x < 0 \\ 1, & 0 \le x < \pi \end{cases}$

2. $f(x) = \begin{cases} -1, & -\pi < x < 0 \\ 2, & 0 \le x < \pi \end{cases}$

3. $f(x) = \begin{cases} 1, & -1 < x < 0 \\ x, & 0 \le x < 1 \end{cases}$

4. $f(x) = \begin{cases} 0, & -1 < x < 0 \\ x, & 0 \le x < 1 \end{cases}$

5. $f(x) = \begin{cases} 0, & -\pi < x < 0 \\ x^2, & 0 \le x < \pi \end{cases}$

6. $f(x) = \begin{cases} \pi^2, & -\pi < x < 0 \\ \pi^2 - x^2, & 0 \le x < \pi \end{cases}$

7. $f(x) = x + \pi, \quad -\pi < x < \pi$

8. $f(x) = 3 - 2x, \quad -\pi < x < \pi$

9. $f(x) = \begin{cases} 0, & -\pi < x < 0 \\ \sin x, & 0 \le x < \pi \end{cases}$

10. $f(x) = \begin{cases} 0, & -\pi/2 < x < 0 \\ \cos x, & 0 \le x < \pi/2 \end{cases}$

11. $f(x) = \begin{cases} 0, & -2 < x < -1 \\ -2, & -1 \le x < 0 \\ 1, & 0 \le x < 1 \\ 0, & 1 \le x < 2 \end{cases}$

12. $f(x) = \begin{cases} 0, & -2 < x < 0 \\ x, & 0 \le x < 1 \\ 1, & 1 \le x < 2 \end{cases}$

13. $f(x) = \begin{cases} 1, & -5 < x < 0 \\ 1 + x, & 0 \le x < 5 \end{cases}$

14. $f(x) = \begin{cases} 2 + x, & -2 < x < 0 \\ 2, & 0 \le x < 2 \end{cases}$

15. $f(x) = e^x$, $-\pi < x < \pi$

16. $f(x) = \begin{cases} 0, & -\pi < x < 0 \\ e^x - 1, & 0 \le x < \pi \end{cases}$

17. Use the result of Problem 5 to show

$$\frac{\pi^2}{6} = 1 + \frac{1}{2^2} + \frac{1}{3^2} + \frac{1}{4^2} + \cdots \quad \text{and} \quad \frac{\pi^2}{12} = 1 - \frac{1}{2^2} + \frac{1}{3^2} - \frac{1}{4^2} + \cdots.$$

18. Use Problem 17 to find a series that gives the numerical value of $\pi^2/8$.

19. Use the result of Problem 7 to show

$$\frac{\pi}{4} = 1 - \frac{1}{3} + \frac{1}{5} - \frac{1}{7} + \cdots.$$

20. Use the result of Problem 9 to show

$$\frac{\pi}{4} = \frac{1}{2} + \frac{1}{1 \cdot 3} - \frac{1}{3 \cdot 5} + \frac{1}{5 \cdot 7} - \frac{1}{7 \cdot 9} + \cdots.$$

21. (a) Use the complex exponential form of the cosine and sine,

$$\cos \frac{n\pi}{p} x = \frac{e^{in\pi x/p} + e^{-in\pi x/p}}{2}, \quad \sin \frac{n\pi}{p} x = \frac{e^{in\pi x/p} - e^{-in\pi x/p}}{2i},$$

to show that (8) can be written in the **complex form**

$$f(x) = \sum_{n=-\infty}^{\infty} c_n e^{in\pi x/p},$$

where $c_0 = a_0/2$, $c_n = (a_n - ib_n)/2$, and $c_{-n} = (a_n + ib_n)/2$, where $n = 1, 2, 3, \ldots$.

(b) Show that c_0, c_n, and c_{-n} of part (a) can be written as one integral

$$c_n = \frac{1}{2p} \int_{-p}^{p} f(x) e^{-in\pi x/p} \, dx, \quad n = 0, \pm 1, \pm 2, \ldots.$$

22. Use the results of Problem 21 to find the complex form of the Fourier series of $f(x) = e^{-x}$ on the interval $-\pi < x < \pi$.

11.3 FOURIER COSINE AND SINE SERIES

- *Even and odd functions* • *Properties of even and odd functions*
- *Fourier cosine series* • *Fourier sine series* • *Sequence of partial sums* • *Gibbs phenomenon* • *Half-range expansions*

The effort expended in the evaluation of coefficients a_0, a_n, and b_n in expanding a function f in a Fourier series is reduced significantly when f is either an even or an odd function. A function f is said to be

even if $f(-x) = f(x)$ and **odd** if $f(-x) = -f(x)$.

On a symmetric interval such as $(-p, p)$, the graph of an even function possesses symmetry with respect to the y-axis whereas the graph of an odd function possesses symmetry with respect to the origin.

Some Examples It is likely the origin of the words *even* and *odd* derives from the fact that the graphs of polynomial functions that consist of all even powers of x are symmetric with respect to the y-axis whereas graphs

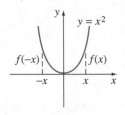

Figure 11.4

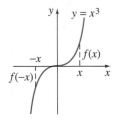

Figure 11.5

of polynomials that consist of all odd powers of x are symmetric with respect to the origin. For example,

$$\overset{\downarrow \text{ even integer}}{f(x) = x^2 \text{ is even since } f(-x) = (-x)^2 = x^2 = f(x)}$$

$$\overset{\downarrow \text{ odd integer}}{f(x) = x^3 \text{ is odd since } f(-x) = (-x)^3 = -x^3 = -f(x).}$$

See Figures 11.4 and 11.5. The trigonometric cosine and sine functions are even and odd functions, respectively, since $\cos(-x) = \cos x$ and $\sin(-x) = -\sin x$. The exponential functions $f(x) = e^x$ and $f(x) = e^{-x}$ are neither odd nor even.

Properties of Even and Odd Functions   The following theorem lists some properties of even and odd functions.

THEOREM 11.2   **Properties of Even/Odd Functions**

   (a) The product of two even functions is even.
   (b) The product of two odd functions is even.
   (c) The product of an even function and an odd function is odd.
   (d) The sum (difference) of two even functions is even.
   (e) The sum (difference) of two odd functions is odd.
   (f) If f is even, then $\int_{-a}^{a} f(x)\,dx = 2\int_{0}^{a} f(x)\,dx$.
   (g) If f is odd, then $\int_{-a}^{a} f(x)\,dx = 0$.

Proof of (b)   Let us suppose that f and g are odd functions. Then we have $f(-x) = -f(x)$ and $g(-x) = -g(x)$. If we define the product of f and g as $F(x) = f(x)g(x)$, then

$$F(-x) = f(-x)g(-x) = (-f(x))(-g(x)) = f(x)g(x) = F(x).$$

This shows that the product F of two odd functions is an even function. The proofs of the remaining properties are left as exercises. See Problem 49 in Exercises 11.3.   ∎

Cosine and Sine Series   If f is an even function on $(-p, p)$, then in view of the foregoing properties the coefficients (9), (10), and (11) of Section 11.2 become

$$a_0 = \frac{1}{p}\int_{-p}^{p} f(x)\,dx = \frac{2}{p}\int_{0}^{p} f(x)\,dx$$

$$a_n = \frac{1}{p}\int_{-p}^{p} \underbrace{f(x)\cos\frac{n\pi}{p}x}_{\text{even}}\,dx = \frac{2}{p}\int_{0}^{p} f(x)\cos\frac{n\pi}{p}x\,dx$$

$$b_n = \frac{1}{p}\int_{-p}^{p} \underbrace{f(x)\sin\frac{n\pi}{p}x}_{\text{odd}}\,dx = 0.$$

Similarly, when f is odd on the interval $(-p, p)$,

$$a_n = 0, \quad n = 0, 1, 2, \ldots, \quad b_n = \frac{2}{p} \int_0^p f(x) \sin \frac{n\pi}{p} x \, dx.$$

We summarize the results in the following definition.

DEFINITION 11.6 Fourier Cosine and Sine Series

(*i*) The Fourier series of an even function on the interval $(-p, p)$ is the **cosine series**

$$f(x) = \frac{a_0}{2} + \sum_{n=1}^{\infty} a_n \cos \frac{n\pi}{p} x, \tag{1}$$

where

$$a_0 = \frac{2}{p} \int_0^p f(x) \, dx \tag{2}$$

$$a_n = \frac{2}{p} \int_0^p f(x) \cos \frac{n\pi}{p} x \, dx. \tag{3}$$

(*ii*) The Fourier series of an odd function on the interval $(-p, p)$ is the **sine series**

$$f(x) = \sum_{n=1}^{\infty} b_n \sin \frac{n\pi}{p} x, \tag{4}$$

where

$$b_n = \frac{2}{p} \int_0^p f(x) \sin \frac{n\pi}{p} x \, dx. \tag{5}$$

EXAMPLE 1 Expansion in a Sine Series

Expand $f(x) = x$, $-2 < x < 2$ in a Fourier series.

Solution Inspection of Figure 11.6 shows that the given function is odd on the interval $(-2, 2)$, and so we expand f in a sine series. With the identification $2p = 4$ we have $p = 2$. Thus (5), after integration by parts, is

$$b_n = \int_0^2 x \sin \frac{n\pi}{2} x \, dx = \frac{4(-1)^{n+1}}{n\pi}.$$

Therefore

$$f(x) = \frac{4}{\pi} \sum_{n=1}^{\infty} \frac{(-1)^{n+1}}{n} \sin \frac{n\pi}{2} x. \tag{6} \blacksquare$$

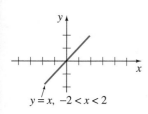

$y = x, \ -2 < x < 2$

Figure 11.6

The function in Example 1 satisfies the conditions of Theorem 11.1. Hence the series (6) converges to the function on $(-2, 2)$ and the periodic extension (of period 4) given in Figure 11.7.

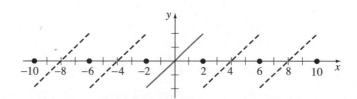

Figure 11.7

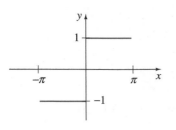

Figure 11.8

EXAMPLE 2   Expansion in a Sine Series

The function $f(x) = \begin{cases} -1, & -\pi < x < 0 \\ 1, & 0 \le x < \pi \end{cases}$ shown in Figure 11.8 is odd on the interval $(-\pi, \pi)$. With $p = \pi$ we have, from (5),

$$b_n = \frac{2}{\pi} \int_0^\pi (1) \sin nx \, dx = \frac{2}{\pi} \frac{1 - (-1)^n}{n},$$

and so

$$f(x) = \frac{2}{\pi} \sum_{n=1}^\infty \frac{1 - (-1)^n}{n} \sin nx. \qquad (7) \quad \blacksquare$$

Gibbs Phenomenon   With the aid of a CAS we have plotted the graphs $S_1(x)$, $S_2(x)$, $S_3(x)$, and $S_{15}(x)$ of the partial sums of nonzero terms of (7) in Figure 11.9. As seen in Figure 11.9(d), the graph of $S_{15}(x)$ has pronounced spikes near the discontinuities at $x = 0$, $x = \pi$, $x = -\pi$, and so on. This "overshooting" by the partial sums S_N from the functional values near a point of discontinuity does not smooth out but remains fairly constant, even when the value N is taken to be large. This behavior of a Fourier series near a point at which f is discontinuous is known as the **Gibbs phenomenon**.

The periodic extension of f in Example 2 onto the entire x-axis is a meander function (see page 347).

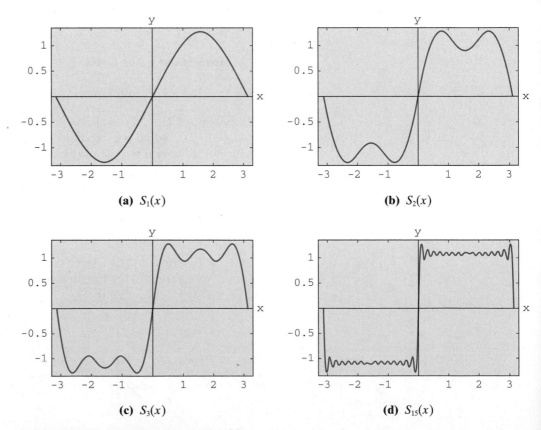

Figure 11.9 Partial sums of sine series (7)

Half-Range Expansions Throughout the preceding discussion it was understood that a function f was defined on an interval with the origin as midpoint—that is, $-p < x < p$. However, in many instances we are interested in representing a function that is defined only for $0 < x < L$ by a trigonometric series. This can be done in many different ways by supplying an arbitrary *definition* of the function on the interval $-L < x < 0$. For brevity we consider the three most important cases. If $y = f(x)$ is defined on the interval $0 < x < L$,

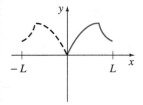

Figure 11.10

(*i*) reflect the graph of the function about the *y*-axis onto $-L < x < 0$; the function is now even on $-L < x < L$ (see Figure 11.10); or

(*ii*) reflect the graph of the function through the origin onto $-L < x < 0$; the function is now odd on $-L < x < L$ (see Figure 11.11); or

(*iii*) define f on $-L < x < 0$ by $f(x) = f(x + L)$ (see Figure 11.12).

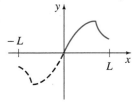

Figure 11.11

Note that the coefficients of the series (1) and (4) utilize only the definition of the function on $0 < x < p$ (that is, half of the interval $-p < x < p$). Hence in practice there is no actual need to make the reflections described in (*i*) and (*ii*). If f is defined on $0 < x < L$, we simply identify the half-period as the length of the interval $p = L$. The coefficient formulas (2), (3), and (5) and the corresponding series yield either an even or an odd periodic extension of period $2L$ of the original function. The cosine and sine series obtained in this manner are known as **half-range expansions.** Lastly, in case (*iii*) we are defining the functional values on the interval $-L < x < 0$ to be the same as the values on $0 < x < L$. As in the previous two cases, there is no real need to do this. It can be shown that the set of functions in (1) of Section 11.2 is orthogonal on $a \leq x \leq a + 2p$ for any real number a. Choosing $a = -p$, we obtain the limits of integration in (9), (10), and (11) of that section. But for $a = 0$ the limits of integration are from $x = 0$ to $x = 2p$. Thus if f is defined over the interval $0 < x < L$, we identify $2p = L$ or $p = L/2$. The resulting Fourier series will give the periodic extension of f with period L. In this manner the values to which the series converges will be the same on $-L < x < 0$ as on $0 < x < L$.

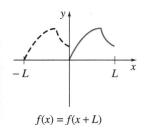

$f(x) = f(x + L)$

Figure 11.12

| EXAMPLE 3 | **Expansion in Three Series** |

Expand $f(x) = x^2$, $0 < x < L$,
(a) in a cosine series   **(b)** in a sine series   **(c)** in a Fourier series.

Solution The graph of the function is given in Figure 11.13.
(a) We have

$$a_n = \frac{2}{L}\int_0^L x^2\, dx = \frac{2}{3}L^2, \quad a_n = \frac{2}{L}\int_0^L x^2 \cos\frac{n\pi}{L}x\, dx = \frac{4L^2(-1)^n}{n^2\pi^2},$$

where integration by parts was used twice in the evaluation of a_n.

Thus $\qquad\qquad f(x) = \frac{L^2}{3} + \frac{4L^2}{\pi^2}\sum_{n=1}^{\infty}\frac{(-1)^n}{n^2}\cos\frac{n\pi}{L}x.$ **(8)**

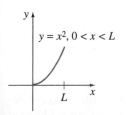

$y = x^2$, $0 < x < L$

Figure 11.13

(b) In this case we must again integrate by parts twice:

$$b_n = \frac{2}{L} \int_0^L x^2 \sin \frac{n\pi}{L} x \, dx = \frac{2L^2(-1)^{n+1}}{n\pi} + \frac{4L^2}{n^3\pi^3} [(-1)^n - 1].$$

Hence $f(x) = \frac{2L^2}{\pi} \sum_{n=1}^{\infty} \left\{ \frac{(-1)^{n+1}}{n} + \frac{2}{n^3\pi^2} [(-1)^n - 1] \right\} \sin \frac{n\pi}{L} x.$ **(9)**

(c) With $p = L/2$, $1/p = 2/L$, and $n\pi/p = 2n\pi/L$, we have

$$a_0 = \frac{2}{L} \int_0^L x^2 \, dx = \frac{2}{3} L^2, \quad a_n = \frac{2}{L} \int_0^L x^2 \cos \frac{2n\pi}{L} x \, dx = \frac{L^2}{n^2\pi^2},$$

and $$b_n = \frac{2}{L} \int_0^L x^2 \sin \frac{2n\pi}{L} x \, dx = -\frac{L^2}{n\pi}.$$

Therefore $f(x) = \frac{L^2}{3} + \frac{L^2}{\pi} \sum_{n=1}^{\infty} \left\{ \frac{1}{n^2\pi} \cos \frac{2n\pi}{L} x - \frac{1}{n} \sin \frac{2n\pi}{L} x \right\}.$ **(10)**

The series (8), (9), and (10) converge to the 2L-periodic even extension of f, the 2L-periodic odd extension of f, and the L-periodic extension of f, respectively. The graphs of these periodic extensions are shown in Figure 11.14.

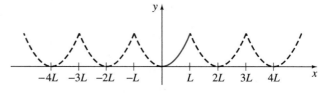

(a) cosine series

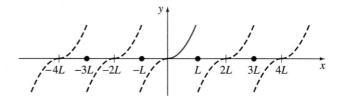

(b) sine series

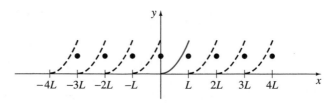

(c) Fourier series

Figure 11.14

Periodic Driving Force   Fourier series are sometimes useful in determining a particular solution of a differential equation describing a physical system in which the input or driving force $f(t)$ is periodic. In the next example we find a particular solution of the differential equation

$$m\frac{d^2x}{dt^2} + kx = f(t) \tag{11}$$

by first representing f by a half-range sine expansion and then assuming a particular solution of the form

$$x_p(t) = \sum_{n=1}^{\infty} B_n \sin\frac{n\pi}{p}t. \tag{12}$$

EXAMPLE 4 **Particular Solution of a DE**

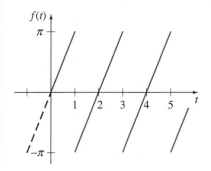

Figure 11.15

An undamped spring-mass system, in which the mass $m = \frac{1}{16}$ slug and the spring constant $k = 4$ lb/ft, is driven by the 2-periodic external force $f(t)$ shown in Figure 11.15. Although the force $f(t)$ acts on the system for $t > 0$, note that if we extend the graph of the function in a 2-periodic manner to the negative t-axis we obtain an odd function. In practical terms this means that we need only find the half-range sine expansion of $f(t) = \pi t$, $0 < t < 1$. With $p = 1$ it follows from (5) and integration by parts that

$$b_n = 2\int_0^1 \pi t \sin n\pi t\, dt = \frac{2(-1)^{n+1}}{n}.$$

From (11) the differential equation of motion is seen to be

$$\frac{1}{16}\frac{d^2x}{dt^2} + 4x = \sum_{n=1}^{\infty}\frac{2(-1)^{n+1}}{n}\sin n\pi t. \tag{13}$$

To find a particular solution $x_p(t)$ of (13) we substitute (12) into the equation and equate coefficients of $\sin n\pi t$. This yields

$$\left(-\frac{1}{16}n^2\pi^2 + 4\right)B_n = \frac{2(-1)^{n+1}}{n} \quad\text{or}\quad B_n = \frac{32(-1)^{n+1}}{n(64 - n^2\pi^2)}.$$

Thus
$$x_p(t) = \sum_{n=1}^{\infty}\frac{32(-1)^{n+1}}{n(64 - n^2\pi^2)}\sin n\pi t. \tag{14}$$

Observe in the solution (14) that there is no integer $n \geq 1$ for which the denominator $64 - n^2\pi^2$ of B_n is zero. In general, if there *is* a value of n, say N, for which $N\pi/p = \omega$, where $\omega = \sqrt{k/m}$, then the system described by (11) is in a state of pure resonance. In other words, we have pure resonance if the Fourier series expansion of the driving force $f(t)$ contains a term $\sin(N\pi/L)t$ (or $\cos(N\pi/L)t$) that has the same frequency as the free vibrations.

Of course, if the $2p$-periodic extension of the driving force f onto the negative t-axis yields an even function, then we expand f in a cosine series.

SECTION 11.3 EXERCISES

Answers to odd-numbered problems begin on page AN-16.

In Problems 1–10 determine whether the function is even, odd, or neither.

1. $f(x) = \sin 3x$ **2.** $f(x) = x\cos x$

3. $f(x) = x^2 + x$ **4.** $f(x) = x^3 - 4x$

5. $f(x) = e^{|x|}$

6. $f(x) = e^x - e^{-x}$

7. $f(x) = \begin{cases} x^2, & -1 < x < 0 \\ -x^2, & 0 \le x < 1 \end{cases}$

8. $f(x) = \begin{cases} x + 5, & -2 < x < 0 \\ -x + 5, & 0 \le x < 2 \end{cases}$

9. $f(x) = x^3, 0 \le x \le 2$

10. $f(x) = |x^5|$

In Problems 11–24 expand the given function in an appropriate cosine or sine series.

11. $f(x) = \begin{cases} -1, & -\pi < x < 0 \\ 1, & 0 \le x < \pi \end{cases}$

12. $f(x) = \begin{cases} 1, & -2 < x < -1 \\ 0, & -1 < x < 1 \\ 1, & 1 < x < 2 \end{cases}$

13. $f(x) = |x|, -\pi < x < \pi$

14. $f(x) = x, -\pi < x < \pi$

15. $f(x) = x^2, -1 < x < 1$

16. $f(x) = x|x|, -1 < x < 1$

17. $f(x) = \pi^2 - x^2, -\pi < x < \pi$

18. $f(x) = x^3, -\pi < x < \pi$

19. $f(x) = \begin{cases} x - 1, & -\pi < x < 0 \\ x + 1, & 0 \le x < \pi \end{cases}$

20. $f(x) = \begin{cases} x + 1, & -1 < x < 0 \\ x - 1, & 0 \le x < 1 \end{cases}$

21. $f(x) = \begin{cases} 1, & -2 < x < -1 \\ -x, & -1 \le x < 0 \\ x, & 0 \le x < 1 \\ 1, & 1 \le x < 2 \end{cases}$

22. $f(x) = \begin{cases} -\pi, & -2\pi < x < -\pi \\ x, & -\pi \le x < \pi \\ \pi, & \pi \le x < 2\pi \end{cases}$

23. $f(x) = |\sin x|, -\pi < x < \pi$

24. $f(x) = \cos x, -\pi/2 < x < \pi/2$

In Problems 25–34 find the half-range cosine and sine expansions of the given function.

25. $f(x) = \begin{cases} 1, & 0 < x < \frac{1}{2} \\ 0, & \frac{1}{2} \le x < 1 \end{cases}$

26. $f(x) = \begin{cases} 0, & 0 < x < \frac{1}{2} \\ 1, & \frac{1}{2} \le x < 1 \end{cases}$

27. $f(x) = \cos x, 0 < x < \pi/2$

28. $f(x) = \sin x, 0 < x < \pi$

29. $f(x) = \begin{cases} x, & 0 < x < \pi/2 \\ \pi - x, & \pi/2 \le x < \pi \end{cases}$

30. $f(x) = \begin{cases} 0, & 0 < x < \pi \\ x - \pi, & \pi \le x < 2\pi \end{cases}$

31. $f(x) = \begin{cases} x, & 0 < x < 1 \\ 1, & 1 \le x < 2 \end{cases}$

32. $f(x) = \begin{cases} 1, & 0 < x < 1 \\ 2 - x, & 1 \le x < 2 \end{cases}$

33. $f(x) = x^2 + x, 0 < x < 1$

34. $f(x) = x(2 - x), 0 < x < 2$

In Problems 35–38 expand the given function in a Fourier series.

35. $f(x) = x^2, \quad 0 < x < 2\pi$

36. $f(x) = x, \quad 0 < x < \pi$

37. $f(x) = x + 1, \quad 0 < x < 1$

38. $f(x) = 2 - x, \quad 0 < x < 2$

In Problems 39 and 40 proceed as in Example 4 to find a particular solution $x_p(t)$ of equation (11) when $m = 1$, $k = 10$, and the driving force $f(t)$ is as given. Assume that when $f(t)$ is extended to the negative t-axis in a periodic manner, the resulting function is odd.

39. $f(t) = \begin{cases} 5, & 0 < t < \pi \\ -5, & \pi < t < 2\pi \end{cases}; \quad f(t + 2\pi) = f(t)$

40. $f(t) = 1 - t, 0 < t < 2; \quad f(t + 2) = f(t)$

In Problems 41 and 42 proceed as in Example 4 to find a particular solution $x_p(t)$ of equation (11) when $m = \frac{1}{4}$, $k = 12$, and the driving force $f(t)$ is as given. Assume that when $f(t)$ is extended to the negative t-axis in a periodic manner, the resulting function is even.

41. $f(t) = 2\pi t - t^2, 0 < t < 2\pi;\quad f(t + 2\pi) = f(t)$

42. $f(t) = \begin{cases} t, & 0 < t < \frac{1}{2} \\ 1 - t, & \frac{1}{2} < t < 1 \end{cases};\quad f(t + 1) = f(t)$

43. (a) Solve the differential equation in Problem 39, $x'' + 10x = f(t)$, subject to the initial conditions $x(0) = 0$, $x'(0) = 0$.
  (b) Use a CAS to plot the graph of the solution $x(t)$ in part (a).

44. (a) Solve the differential equation in Problem 41, $\frac{1}{4}x'' + 12x = f(t)$, subject to the initial conditions $x(0) = 1$, $x'(0) = 0$.
  (b) Use a CAS to plot the graph of the solution $x(t)$ in part (a).

45. Suppose a uniform beam of length L is simply supported at $x = 0$ and at $x = L$. If the load per unit length is given by $w(x) = w_0 x/L$, $0 < x < L$, then the differential equation for the deflection $y(x)$ is

$$EI\frac{d^4y}{dx^4} = \frac{w_0 x}{L},$$

where E, I, and w_0 are constants. (See (4) in Section 5.2.)
  (a) Expand $w(x)$ in a half-range sine series.
  (b) Use the method of Example 4 to find a particular solution $y_p(x)$ of the differential equation.

46. Proceed as in Problem 45 to find a particular solution $y_p(x)$ when the load per unit length is as given in Figure 11.16.

47. When a uniform beam is supported by an elastic foundation and subject to a load per unit length $w(x)$, the differential equation for its deflection $y(x)$ is

$$EI\frac{d^4y}{dx^4} + ky = w(x),$$

where k is the modulus of the foundation. Suppose that the beam and elastic foundation are infinite in length (that is, $-\infty < x < \infty$) and that the load per unit length is the periodic function

$$w(x) = \begin{cases} 0, & -\pi < x < -\pi/2 \\ w_0, & -\pi/2 \leq x \leq \pi/2, \\ 0 & \pi/2 < x < \pi \end{cases} \quad w(x + 2\pi) = w(x).$$

Use the method of Example 4 to find a particular solution $y_p(x)$ of the differential equation.

Discussion Problems

48. Prove properties (a), (c), (d), (f), and (g) in Theorem 11.2.

49. There is only one function that is both even and odd. What is it?

50. As we know from Chapter 4, the general solution of the differential equation in Problem 47 is $y = y_c + y_p$. Discuss why we can argue on physical grounds that the solution of Problem 47 is simply y_p. [*Hint:* Consider $y = y_c + y_p$ as $x \rightarrow \pm\infty$.]

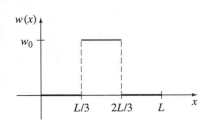

Figure 11.16

Computer Lab Assignments

In Problems 51 and 52 use a CAS to plot graphs of partial sums $\{S_N(x)\}$ of the given trigonometric series. Experiment with different values of N and graphs on different intervals of the x-axis. Use your graphs to conjecture a closed-form expression for a function f defined for $0 < x < L$ that is represented by the series.

51. $f(x) = -\dfrac{\pi}{4} + \displaystyle\sum_{n=1}^{\infty} \left[\dfrac{(-1)^n - 1}{n^2 \pi} \cos nx + \dfrac{1 - 2(-1)^n}{n} \sin nx \right]$

52. $f(x) = \dfrac{1}{4} + \dfrac{4}{\pi^2} \displaystyle\sum_{n=1}^{\infty} \dfrac{1}{n^2} \left(1 - \cos \dfrac{n\pi}{2} \right) \cos \dfrac{n\pi}{2} x$

53. Is your answer in Problem 51 or in Problem 52 unique? Given a function f defined on a symmetric interval about the origin $-a < x < a$ that has the same trigonometric series
   (a) as in Problem 51   **(b)** as in Problem 52.

11.4 STURM-LIOUVILLE PROBLEM

• Eigenvalues • Eigenfunctions • Regular Sturm-Liouville problem • Singular Sturm-Liouville problem • Properties of Sturm-Liouville problems • Self-adjoint form • Orthogonality relation

For convenience we present here a brief review of some of the ordinary differential equations that will be of importance in the sections and chapters that follow.

Linear equations	General solutions
$y' + \lambda y = 0$	$y = c_1 e^{-\lambda x}$
$y'' + \lambda y = 0, \quad \lambda > 0$	$y = c_1 \cos \sqrt{\lambda}\, x + c_2 \sin \sqrt{\lambda}\, x$
$y'' - \lambda y = 0, \quad \lambda > 0$	$y = c_1 e^{-\sqrt{\lambda} x} + c_2 e^{\sqrt{\lambda} x}$ or
	$y = c_1 \cosh \sqrt{\lambda}\, x + c_2 \sinh \sqrt{\lambda}\, x$

Cauchy-Euler equation	General solutions, $x > 0$
$x^2 y'' + xy' - \lambda^2 y = 0$	$y = c_1 x^{-\lambda} + c_2 x^{\lambda}, \quad \lambda \neq 0$
	$y = c_1 + c_2 \ln x, \quad \lambda = 0$

Parametric Bessel equation ($\nu = 0$)	General solution, $x > 0$
$xy'' + y' + \lambda^2 xy = 0$	$y = c_1 J_0(\lambda x) + c_2 Y_0(\lambda x)$

Legendre's equation ($n = 0, 1, 2, \ldots$)	Particular solutions are polynomials
$(1 - x^2)y'' - 2xy' + n(n+1)y = 0$	$P_0(x) = 1,\ P_1(x) = x,\quad P_2(x) = \tfrac{1}{2}(3x^2 - 1),\ \ \ldots$

As a rule, the exponential form $y = c_1 e^{-\sqrt{\lambda} x} + c_2 e^{\sqrt{\lambda} x}$ of the general solution of $y'' - \lambda y = 0, \lambda > 0$ is used when the domain of x is an infinite or semi-infinite interval; the hyperbolic form $y = c_1 \cosh \sqrt{\lambda}\, x + c_2 \sinh \sqrt{\lambda}\, x$ is used when the domain of x is a finite interval.

Eigenvalues and Eigenfunctions Orthogonal functions arise in the solution of differential equations. More to the point, an orthogonal set of functions can be generated by solving a two-point boundary-value problem involving a linear second-order differential equation containing a parameter λ. In Example 2 of Section 5.2 we saw that the boundary-value problem

$$y'' + \lambda y = 0, \quad y(0) = 0, \quad y(L) = 0 \tag{1}$$

possessed nontrivial solutions only when the parameter λ took on the values $\lambda = n^2\pi^2/L^2, n = 1, 2, 3, \ldots$, called **eigenvalues.** The corresponding nontrivial solutions $y = c_2 \sin(n\pi x/L)$ or simply $y = \sin(n\pi x/L)$ are called the **eigenfunctions** of the problem. For example, for the boundary-value problem (1), $\lambda = -2$ *is not an eigenvalue* whereas $\lambda = 9\pi^2/L^2$ $(n = 3)$ *is an eigenvalue,* and so for

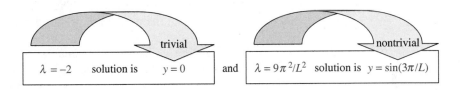

$\lambda = -2$ solution is $y = 0$ and $\lambda = 9\pi^2/L^2$ solution is $y = \sin(3\pi/L)$

For our purposes in this chapter it is important to recognize that the set $\{\sin(n\pi x/L)\}$, $n = 1, 2, 3, \ldots$ is the orthogonal set of functions on the interval $[0, L]$ used as the basis for the Fourier sine series.

EXAMPLE 1   **Eigenvalues and Eigenfunctions**

It is left as an exercise to verify that the eigenvalues and eigenfunctions for the boundary-value problem

$$y'' + \lambda y = 0, \quad y'(0) = 0, \quad y'(L) = 0 \tag{2}$$

are, respectively, $\lambda = n^2\pi^2/L^2$, $n = 0, 1, 2, \ldots$ and $y = c_1 \cos(n\pi x/L)$, $c_1 \neq 0$. In contrast to (1), $\lambda = 0$ is an eigenvalue of this problem and $y = 1$ is the corresponding eigenfunction. The latter comes from solving $y'' = 0$ subject to the same boundary conditions, $y'(0) = 0$, $y'(L) = 0$, but can be incorporated in $y = \cos(n\pi x/L)$ by permitting $n = 0$ and $c_1 = 1$. The set $\{\cos(n\pi x/L)\}$, $n = 0, 1, 2, 3, \ldots$ is orthogonal on the interval $[0, L]$. See Problem 3 in Exercises 11.4. ▪

Regular Sturm-Liouville Problem The problems in (1) and (2) are special cases of an important two-point boundary-value problem. Let p, q, r, and r' be real-valued functions continuous on an interval $[a, b]$, and let $r(x) > 0$ and $p(x) > 0$ for every x in the interval. Then

$$\textit{Solve:} \qquad \frac{d}{dx}[r(x)y'] + (q(x) + \lambda p(x))y = 0 \tag{3}$$

$$\textit{Subject to:} \qquad \alpha_1 y(a) + \beta_1 y'(a) = 0 \tag{4}$$

$$\alpha_2 y(a) + \beta_2 y'(a) = 0 \tag{5}$$

is said to be a **regular Sturm-Liouville problem.** The coefficients in (4) and (5) are assumed to be real and independent of λ. In addition α_1 and β_1 are not both zero, and α_2 and β_2 are not both zero.

Since the differential equation (3) is homogeneous, the Sturm-Liouville problem always has the trivial solution $y = 0$. However, this solution is of no interest to us. As in Example 1, in solving such a problem we seek to find numbers λ (eigenvalues) and nontrivial solutions y that depend on λ (eigenfunctions).

The boundary conditions in (4) and (5) (a linear combination of y and y' equal to zero at a point) are also **homogeneous.** The two boundary conditions in Example 1 and in Example 2, which follows, are homogeneous; a boundary condition such as $\alpha_1 y(a) + \beta_1 y'(a) = c$, where the constant c is nonzero, is **nonhomogeneous.** A boundary-value problem that consists of a homogeneous linear differential equation and homogeneous boundary conditions is, of course, said to be homogeneous; otherwise it is nonhomogeneous. The Sturm-Liouville problem given in (3)–(5) is a homogeneous boundary-value problem.

Properties   Theorem 11.3 is a list of some of the more important of the many properties of the regular Sturm-Liouville problem. We shall prove only the last property.

THEOREM 11.3   **Properties of the Regular Sturm-Liouville Problem**

(a) There exist an infinite number of real eigenvalues that can be arranged in increasing order $\lambda_1 < \lambda_2 < \lambda_3 < \cdots \lambda_n < \cdots$ such that $\lambda_n \to \infty$ as $n \to \infty$.
(b) For each eigenvalue there is only one eigenfunction (except for nonzero multiples).
(c) Eigenfunctions corresponding to different eigenvalues are linearly independent.
(d) The set of eigenfunctions corresponding to the set of eigenvalues is orthogonal with respect to the weight function $p(x)$ on the interval $[a, b]$.

Proof of (d)   Let y_m and y_n be eigenfunctions corresponding to eigenvalues λ_m and λ_n, respectively. Then

$$\frac{d}{dx}[r(x)y_m'] + (q(x) + \lambda_m p(x))y_m = 0 \tag{6}$$

$$\frac{d}{dx}[r(x)y_n'] + (q(x) + \lambda_n p(x))y_n = 0. \tag{7}$$

Multiplying (6) by y_n and (7) by y_m and subtracting the two equations gives

$$(\lambda_m - \lambda_n)p(x)y_m y_n = y_m \frac{d}{dx}[r(x)y_n'] - y_n \frac{d}{dx}[r(x)y_m'].$$

Integrating this last result by parts from $x = a$ to $x = b$ then yields

$$(\lambda_m - \lambda_n)\int_a^b p(x)y_m y_n \, dx = r(b)[y_m(b)y_n'(b) - y_n(b)y_m'(b)] - r(a)[y_m(a)y_n'(a) - y_n(a)y_m'(a)]. \tag{8}$$

Now the eigenfunctions y_m and y_n must both satisfy the boundary conditions (4) and (5). In particular, from (4) we have

$$\alpha_1 y_m(a) + \beta_1 y_m'(a) = 0$$
$$\alpha_1 y_n(a) + \beta_1 y_n'(a) = 0.$$

In order for this system to be satisfied by α_1 and β_1, not both zero, the determinant of the coefficients must be zero:

$$y_m(a)y_n'(a) - y_n(a)y_m'(a) = 0.$$

A similar argument applied to (5) also gives

$$y_m(b)y_n'(b) - y_n(b)y_m'(b) = 0.$$

Since both members of the right-hand side of (8) are zero, we have established the orthogonality relation

$$\int_a^b p(x)y_m(x)y_n(x)\, dx = 0, \quad \lambda_m \neq \lambda_n. \tag{9} \quad \blacksquare$$

EXAMPLE 2   A Regular Sturm-Liouville Problem

Solve the boundary-value problem

$$y'' + \lambda y = 0, \quad y(0) = 0, \quad y(1) + y'(1) = 0. \tag{10}$$

Solution   You should verify that for $\lambda < 0$ and for $\lambda = 0$ the problem in (10) possesses only the trivial solution $y = 0$. For $\lambda > 0$ the general solution of the differential equation is $y = c_1 \cos \sqrt{\lambda}x + c_2 \sin \sqrt{\lambda}x$. Now the condition $y(0) = 0$ implies $c_1 = 0$ in this solution. When $y = c_2 \sin \sqrt{\lambda}x$, the second boundary condition $y(1) + y'(1) = 0$ is satisfied if

$$c_2 \sin \sqrt{\lambda} + c_2 \sqrt{\lambda} \cos \sqrt{\lambda} = 0.$$

Choosing $c_2 \neq 0$, we see that the last equation is equivalent to

$$\tan \sqrt{\lambda} = -\sqrt{\lambda}. \tag{11}$$

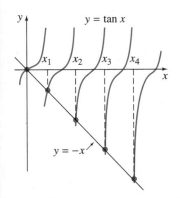

Figure 11.17

If we let $x = \sqrt{\lambda}$, then Figure 11.17 shows the plausibility that there exist an infinite number of roots of the equation $\tan x = -x$. The eigenvalues of problem (10) are then $\lambda_n = x_n^2$, where x_n, $n = 1, 2, 3, \ldots$ are the consecutive *positive* roots. With the aid of a CAS it is easily shown that, to four rounded decimal places, $x_1 = 2.0288$, $x_2 = 4.9132$, $x_3 = 7.9787$, and $x_4 = 11.0855$, and the corresponding solutions are $y_1 = \sin 2.0288x$, $y_2 = \sin 4.9132x$, $y_3 = \sin 7.9787x$, and $y_4 = \sin 11.0855x$. In general, the eigenfunctions of the problem are $\{\sin \sqrt{\lambda_n}x\}$, $n = 1, 2, 3, \ldots$.

With identifications $r(x) = 1$, $q(x) = 0$, $p(x) = 1$, $\alpha_1 = 1$, $\beta_1 = 0$, $\alpha_2 = 1$, $\beta_2 = 1$ we see that (10) is a regular Sturm-Liouville problem. Thus $\{\sin \sqrt{\lambda_n}x\}$, $n = 1, 2, 3, \ldots$ is an orthogonal set with respect to the weight function $p(x) = 1$ on the interval $[0, 1]$.   ▬

In some circumstances we can prove the orthogonality of the solutions of (3) without the necessity of specifying a boundary condition at $x = a$ and at $x = b$.

Singular Sturm-Liouville Problem If $r(a) = 0$, or $r(b) = 0$, or $r(a) = r(b) = 0$, the Sturm-Liouville problem consisting of the differential equation (3) and boundary conditions (4) and (5) is said to be **singular.** In the first case $x = a$ may be a singular point of the differential equation, and consequently a solution of (3) may become unbounded as $x \to a$. However, we see from (8) that if $r(a) = 0$ then no boundary condition is required at $x = a$ to prove orthogonality of the eigenfunctions, provided these solutions are bounded at that point. This latter requirement guarantees the existence of the integrals involved. By assuming the solutions of (3) are bounded on the closed interval $[a, b]$ we can say that

- (*i*) if $r(a) = 0$, then the orthogonality relation (9) holds with no boundary condition at $x = a$;
- (*ii*) if $r(b) = 0$, then the orthogonality relation (9) holds with no boundary condition at $x = b$;* and
- (*iii*) if $r(a) = r(b) = 0$, then the orthogonality relation (9) holds with no boundary condition at either $x = a$ or $x = b$.

We note in passing that a Sturm-Liouville problem is singular when the interval under consideration is infinite. See Problems 11 and 12 in Exercises 11.4.

Self-Adjoint Form If the coefficients are continuous and $a(x) \neq 0$ for all x in some interval, then any second-order differential equation

$$a(x)y'' + b(x)y' + (c(x) + \lambda d(x))y = 0 \tag{12}$$

can be put into the so-called **self-adjoint form** (3) by multiplying the equation by the integrating factor

$$\frac{1}{a(x)} e^{\int (b(x)/a(x))dx}. \tag{13}$$

To see this we observe that the differential equation

$$e^{\int(b/a)dx}y'' + \frac{b(x)}{a(x)}e^{\int(b/a)dx}y' + \left(\frac{c(x)}{a(x)}e^{\int(b/a)dx} + \lambda\frac{d(x)}{a(x)}e^{\int(b/a)dx}\right)y = 0$$

is the same as

$$\frac{d}{dx}[\underbrace{e^{\int(b/a)dx}}_{r(x)}y'] + \left(\underbrace{\frac{c(x)}{a(x)}e^{\int(b/a)dx}}_{q(x)} + \lambda\underbrace{\frac{d(x)}{a(x)}e^{\int(b/a)dx}}_{p(x)}\right)y = 0. \tag{14}$$

It is certainly not necessary to put a second-order differential equation (12) into the self-adjoint form (3) in order to solve the DE. For our

*Conditions (*i*) and (*ii*) are equivalent to choosing $\alpha_1 = 0$, $\beta_1 = 0$ and $\alpha_2 = 0$, $\beta_2 = 0$, respectively.

purposes we use the form given in (14) to determine the weight function $p(x)$ needed in the orthogonality relation (9). The next two examples illustrate orthogonality relations for Bessel functions and for Legendre polynomials.

EXAMPLE 3   Parametric Bessel Equation

In Section 6.3 we saw that the general solution of the parametric Bessel equation $x^2y'' + xy' + (\lambda^2 x^2 - n^2)y = 0, n = 0, 1, 2, \ldots$ was $y = c_1 J_n(\lambda x) + c_2 Y_n(\lambda x)$. From (13) we find that $1/x$ is an integrating factor. Multiplying Bessel's equation by this factor yields the self-adjoint form

$$\frac{d}{dx}[xy'] + \left(\lambda^2 x - \frac{n^2}{x}\right)y = 0,$$

where we identify $r(x) = x, q(x) = -n^2/x$, and $p(x) = x$ and λ is replaced by λ^2. Now $r(0) = 0$, and of the two solutions $J_n(\lambda x)$ and $Y_n(\lambda x)$ only $J_n(\lambda x)$ is bounded at $x = 0$. Thus in view of (i), the set $\{J_n(\lambda_i x)\}, i = 1,$ 2, 3, $\ldots$ is orthogonal with respect to the weight function $p(x) = x$ on an interval $[0, b]$. The orthogonality relation is

$$\int_0^b x J_n(\lambda_i x) J_n(\lambda_j x)\, dx = 0, \quad \lambda_i \neq \lambda_j,$$

provided the eigenvalues $\lambda_i, i = 1, 2, 3, \ldots$ are defined by means of a boundary condition at $x = b$ of the type

$$\alpha_2 J_n(\lambda b) + \beta_2 \lambda J_n'(\lambda b) = 0.^* \tag{15}$$

For any choice of α_2 and β_2, not both zero, it is known that (15) has an infinite number of roots $x_i = \lambda_i b$ from which we get $\lambda_i = x_i/b$. More will be said about the eigenvalues in the next chapter. ∎

EXAMPLE 4   Legendre's Equation

The Legendre polynomials $P_n(x)$ are bounded solutions of Legendre's differential equation $(1 - x^2)y'' - 2xy' + n(n + 1)y = 0, n = 0, 1, 2, \ldots$ on the interval $[-1, 1]$. From the self-adjoint form

$$\frac{d}{dx}[(1 - x^2)y'] + n(n + 1)y = 0$$

we read off $r(x) = 1 - x^2, q(x) = 0, p(x) = 1$, and $\lambda = n(n + 1)$. Observing that $r(-1) = r(1) = 0$, we find from (iii) that the set $\{P_n(x)\}, n = 0, 1, 2,$ $\ldots$ is orthogonal with respect to weight function $p(x) = 1$ on $[-1, 1]$. The orthogonality relation is

$$\int_{-1}^1 P_m(x) P_n(x)\, dx = 0, \quad m \neq n. \quad ∎$$

---

*The extra factor of λ comes from the Chain Rule: $(d/dx)J_n(\lambda x) = \lambda J_n'(\lambda x)$.

EXERCISES 11.4

Answers to odd-numbered problems begin on page AN-17.

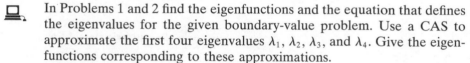

In Problems 1 and 2 find the eigenfunctions and the equation that defines the eigenvalues for the given boundary-value problem. Use a CAS to approximate the first four eigenvalues λ_1, λ_2, λ_3, and λ_4. Give the eigenfunctions corresponding to these approximations.

1. $y'' + \lambda y = 0$, $y'(0) = 0$, $y(1) + y'(1) = 0$

2. $y'' + \lambda y = 0$, $y(0) + y'(0) = 0$, $y(1) = 0$

3. Consider $y'' + \lambda y = 0$ subject to $y'(0) = 0$, $y'(L) = 0$. Show that the eigenfunctions are $\left\{1, \cos\dfrac{\pi}{L}x, \cos\dfrac{2\pi}{L}x, \ldots\right\}$. This set, which is orthogonal on $[0, L]$, is the basis for the Fourier cosine series.

4. Consider $y'' + \lambda y = 0$ subject to the periodic boundary conditions $y(-L) = y(L)$, $y'(-L) = y'(L)$. Show that the eigenfunctions are

$$\left\{1, \cos\frac{\pi}{L}x, \cos\frac{2\pi}{L}x, \ldots, \sin\frac{\pi}{L}x, \sin\frac{2\pi}{L}x, \sin\frac{3\pi}{L}x, \ldots\right\}.$$

This set, which is orthogonal on $[-L, L]$, is the basis for the Fourier series.

5. Find the square norm of each eigenfunction in Problem 1.

6. Show that for the eigenfunctions in Example 2,
$$\|\sin\sqrt{\lambda_n}x\|^2 = \tfrac{1}{2}[1 + \cos^2\sqrt{\lambda_n}].$$

7. (a) Find the eigenvalues and eigenfunctions of the boundary-value problem
$$x^2y'' + xy' + \lambda y = 0, \quad y(1) = 0, \quad y(5) = 0.$$
(b) Put the differential equation in self-adjoint form.
(c) Give an orthogonality relation.

8. (a) Find the eigenvalues and eigenfunctions of the boundary-value problem
$$y'' + y' + \lambda y = 0, \quad y(0) = 0, \quad y(2) = 0.$$
(b) Put the differential equation in self-adjoint form.
(c) Give an orthogonality relation.

9. Laguerre's differential equation $xy'' + (1 - x)y' + ny = 0, n = 0, 1, 2, \ldots$ has polynomial solutions $L_n(x)$. Put the equation in self-adjoint form, and give an orthogonality relation.

10. Hermite's differential equation $y'' - 2xy' + 2ny = 0, n = 0, 1, 2, \ldots$ has polynomial solutions $H_n(x)$. Put the equation in self-adjoint form and give an orthogonality relation.

11. Consider the regular Sturm-Liouville problem:
$$\frac{d}{dx}[(1 + x^2)y'] + \frac{1}{1 + x^2}y = 0, \quad y(0) = 0, \quad y(1) = 0.$$

(a) Find the eigenvalues and eigenfunctions of the boundary-value problem. [*Hint:* Let $x = \tan\theta$ and then use the Chain Rule.]
(b) Give an orthogonality relation.

12. **(a)** Find the eigenfunctions and the equation that defines the eigenvalues for the boundary-value problem

$$x^2y'' + xy' + (\lambda^2 x^2 - 1)y = 0, \quad y \text{ is bounded at } x = 0, \quad y(3) = 0.$$

   (b) Use Table 6.2 to find the approximate values of the first four eigenvalues λ_1, λ_2, λ_3, and λ_4.

Discussion Problems

13. Consider the special case of the regular Sturm-Liouville problem on the interval $[a, b]$:

$$\frac{d}{dx}[r(x)y'] + \lambda p(x)y = 0, \quad y'(a) = 0, \quad y'(b) = 0.$$

   Is $\lambda = 0$ an eigenvalue of the problem? Defend your answer.

Computer Lab Assignments

14. **(a)** Give an orthogonality relation for the Sturm-Liouville problem in Problem 1.
   (b) Use a CAS as an aid in verifying the orthogonality relation for the eigenfunctions y_1 and y_2 that correspond to the first two eigenvalues λ_1 and λ_2, respectively.

15. **(a)** Give an orthogonality relation for the Sturm-Liouville problem in Problem 2.
   (b) Use a CAS as an aid in verifying the orthogonality relation for the eigenfunctions y_1 and y_2 that correspond to the first two eigenvalues λ_1 and λ_2, respectively.

11.5 BESSEL AND LEGENDRE SERIES

- *Orthogonal set of Bessel functions* • *Fourier-Bessel series* • *Differential recurrence relations* • *Forms of Fourier-Bessel series* • *Convergence of a Fourier-Bessel series* • *Orthogonal set of Legendre polynomial* • *Fourier-Legendre series* • *Convergence of a Fourier-Legendre series*

Fourier series, Fourier cosine series, and Fourier sine series are three ways of expanding a function in terms of an orthogonal set of functions. But such expansions are by no means limited to orthogonal sets of trigonometric functions. We saw in Section 11.1 that a function f defined on an interval (a, b) could be expanded, at least formally, in terms of *any* set of functions $\{\phi_n(x)\}$ that is orthogonal with respect to a weight function on $[a, b]$. Many of these orthogonal series expansions or generalized Fourier series stem from Sturm-Liouville problems which, in turn, arise from attempts to solve linear partial differential equations serving as models for physical systems. Fourier series and orthogonal series expansions, as well as the two series considered in this section, will appear in the subsequent consideration of these applications in Chapters 12 and 13.

11.5.1 FOURIER-BESSEL SERIES

We saw in Example 3 of Section 11.4 that the set of Bessel functions $\{J_n(\lambda_i x)\}$, $i = 1, 2, 3, \ldots$ is orthogonal with respect to the weight function $p(x) = x$ on an interval $[0, b]$ when the eigenvalues λ_i are defined by means of a boundary condition of the form

$$\alpha_2 J_n(\lambda b) + \beta_2 \lambda J_n'(\lambda b) = 0. \tag{1}$$

From (7) and (8) of Section 11.1 the generalized Fourier expansion of a function f defined on $(0, b)$ in terms of this orthogonal set is

$$f(x) = \sum_{i=1}^{\infty} c_i J_n(\lambda_i x), \tag{2}$$

where

$$c_i = \frac{\int_0^b x J_n(\lambda_i x) f(x) \, dx}{\int_0^b x J_n^2(\lambda_i x) \, dx}. \tag{3}$$

The series (2) with coefficients (3) is called a **Fourier-Bessel series.**

Differential Recurrence Relations   The **differential recurrence relations** that were given in (15) and (14) of Section 6.3,

$$\frac{d}{dx}[x^n J_n(x)] = x^n J_{n-1}(x) \tag{4}$$

$$\frac{d}{dx}[x^{-n} J_n(x)] = -x^{-n} J_{n+1}(x), \tag{5}$$

are often useful in the evaluation of the coefficients in (3).

Square Norm The value of the square norm $\|J_n(\lambda_i x)\|^2 = \int_0^b x J_n^2(\lambda_i x) \, dx$ depends on how the eigenvalues λ_i are defined. If $y = J_n(\lambda x)$, then we know from Example 3 of Section 11.4 that

$$\frac{d}{dx}[xy'] + \left(\lambda^2 x - \frac{n^2}{x}\right) y = 0.$$

After we multiply by $2xy'$, this equation can be written as

$$\frac{d}{dx}[xy']^2 + (\lambda^2 x^2 - n^2)\frac{d}{dx}[y]^2 = 0.$$

Integrating the last result by parts on $[0, b]$ then gives

$$2\lambda^2 \int_0^b xy^2 \, dx = \left([xy']^2 + (\lambda^2 x^2 - n^2)y^2\right)\Big|_0^b.$$

Since $y = J_n(\lambda x)$, the lower limit is zero for $n > 0$ because $J_n(0) = 0$. For $n = 0$ the quantity $[xy']^2 + \lambda^2 x^2 y^2$ is zero at $x = 0$. Thus

$$2\lambda^2 \int_0^b x J_n^2(\lambda x) \, dx = \lambda^2 b^2 [J_n'(\lambda b)]^2 + (\lambda^2 b^2 - n^2)[J_n(\lambda b)]^2, \tag{6}$$

where we have used the Chain Rule to write $y' = \lambda J_n'(\lambda x)$.

We now consider three cases of (1).

Case I If we choose $\alpha_2 = 1$ and $\beta_2 = 0$, then (1) is

$$J_n(\lambda b) = 0. \tag{7}$$

There are an infinite number of positive roots x_i of (7) (see Figure 6.3), and so the eigenvalues are positive and are given by $\lambda_i = x_i/b$. No new eigenvalues result from the negative roots of (7) since $J_n(-x) = (-1)^n J_n(x)$. (See page 295.) The number 0 is not an eigenvalue for any n since $J_n(0) = 0$ for $n = 1, 2, 3, \ldots$ and $J_0(0) = 1$. In other words, if $\lambda = 0$, we get the trivial function (which is never an eigenfunction) for $n = 1, 2, 3, \ldots$, and for $n = 0$, $\lambda = 0$ does not satisfy equation (7). When (5) is written in the form $xJ_n'(x) = nJ_n(x) - xJ_{n+1}(x)$, it follows from (6) and (7) that the square norm of $J_n(\lambda_i x)$ is

$$\|J_n(\lambda_i x)\|^2 = \frac{b^2}{2} J_{n+1}^2(\lambda_i b). \tag{8}$$

Case II If we choose $\alpha_2 = h \geq 0$, $\beta_2 = b$, then (1) is

$$hJ_n(\lambda b) + \lambda b J_n'(\lambda b) = 0. \tag{9}$$

Equation (9) has an infinite number of positive roots x_i for each positive integer $n = 1, 2, 3, \ldots$. As before, the eigenvalues are obtained from $\lambda_i = x_i/b$. $\lambda = 0$ is not an eigenvalue for $n = 1, 2, 3, \ldots$. Substituting $\lambda_i b J_n'(\lambda_i b) = -hJ_n(\lambda_i b)$ into (6), we find that the square norm of $J_n(\lambda_i x)$ is now

$$\|J_n(\lambda_i x)\|^2 = \frac{\lambda_i^2 b^2 - n^2 + h^2}{2\lambda_i^2} J_n^2(\lambda_i b). \tag{10}$$

Case III If $h = 0$ and $n = 0$ in (9), the eigenvalues λ_i are defined from the roots of

$$J_0'(\lambda b) = 0. \tag{11}$$

Even though (11) is just a special case of (9), it is the only situation for which $\lambda = 0$ is an eigenvalue. To see this observe that for $n = 0$ the result in (5) implies that

$$J_0'(\lambda b) = 0 \quad \text{is equivalent to} \quad J_1(\lambda b) = 0. \tag{12}$$

Since $x_1 = 0$ is a root of this equation and since $J_0(0) = 1$ is nontrivial, we conclude that $\lambda_1 = 0$ is an eigenvalue. But obviously we cannot use (10) when $h = 0$, $n = 0$, and $\lambda_1 = 0$. However, by definition of the square norm we have

$$\|1\|^2 = \int_0^b x\, dx = \frac{b^2}{2}. \tag{13}$$

For $\lambda_i > 0$ we can use (10) with $h = 0$ and $n = 0$:

$$\|J_0(\lambda_i x)\|^2 = \frac{b^2}{2} J_0^2(\lambda_i b). \tag{14}$$

The following summary gives the corresponding three forms of the series (2).

DEFINITION 11.7   Fourier-Bessel Series

The **Fourier-Bessel series** of a function f defined on the interval $(0, b)$ is given by

$$(i) \qquad\qquad f(x) = \sum_{i=1}^{\infty} c_i J_n(\lambda_i x) \qquad\qquad (15)$$

$$c_i = \frac{2}{b^2 J_{n+1}^2(\lambda_i b)} \int_0^b x J_n(\lambda_i x) f(x)\, dx, \qquad (16)$$

where the λ_i are defined by $J_n(\lambda b) = 0$.

$$(ii) \qquad\qquad f(x) = \sum_{i=1}^{\infty} c_i J_n(\lambda_i x) \qquad\qquad (17)$$

$$c_i = \frac{2\lambda_i^2}{(\lambda_i^2 b^2 - n^2 + h^2) J_n^2(\lambda_i b)} \int_0^b x J_n(\lambda_i x) f(x)\, dx, \qquad (18)$$

where the λ_i are defined by $h J_n(\lambda b) + \lambda b J_n'(\lambda b) = 0$.

$$(iii) \qquad\qquad f(x) = c_1 + \sum_{i=2}^{\infty} c_i J_0(\lambda_i x) \qquad\qquad (19)$$

$$c_1 = \frac{2}{b^2} \int_0^b x f(x)\, dx, \quad c_i = \frac{2}{b^2 J_0^2(\lambda_i b)} \int_0^b x J_0(\lambda_i x) f(x)\, dx, \qquad (20)$$

where the λ_i are defined by $J_0'(\lambda b) = 0$.

Convergence of a Fourier-Bessel Series   Sufficient conditions for the convergence of a Fourier-Bessel series are not particularly restrictive.

THEOREM 11.4   **Conditions for Convergence**

If f and f' are piecewise continuous on the open interval $(0, b)$, then a Fourier-Bessel expansion of f converges to $f(x)$ at any point where f is continuous and to the average $[f(x-) + f(x+)]/2$ at a point where f is discontinuous.

EXAMPLE 1   **Expansion in a Fourier-Bessel Series**

Expand $f(x) = x$, $0 < x < 3$ in a Fourier-Bessel series, using Bessel functions of order one that satisfy the boundary condition $J_1(3\lambda) = 0$.

Solution   We use (15) where the coefficients c_i are given by (16):

$$c_i = \frac{2}{3^2 J_2^2(3\lambda_i)} \int_0^3 x^2 J_1(\lambda_i x)\, dx.$$

To evaluate this integral we let $t = \lambda_i x$ and use (4) in the form $(d/dt)[t^2 J_2(t)] = t^2 J_1(t)$:

$$c_i = \frac{2}{9\lambda_i^3 J_2^2(3\lambda_i)} \int_0^{3\lambda_i} \frac{d}{dt}[t^2 J_2(t)]\, dt = \frac{2}{\lambda_i J_2(3\lambda_i)}.$$

Therefore the desired expansion is

$$f(x) = 2 \sum_{i=1}^{\infty} \frac{1}{\lambda_i J_2(3\lambda_i)} J_1(\lambda_i x).$$

■

EXAMPLE 2   **Expansion in a Fourier-Bessel Series**

If eigenvalues λ_i in Example 1 are defined by $J_1(3\lambda) + \lambda J_1'(3\lambda) = 0$, then the only thing that changes in the expansion is the value of the square norm. Multiplying the boundary condition by 3 gives $3J_1(3\lambda) + 3\lambda J_1'(3\lambda) = 0$, which now matches (9) when $h = 3$, $b = 3$, and $n = 1$. Thus (18) and (17) yield, in turn,

$$c_i = \frac{18\lambda_i J_2(3\lambda_i)}{(9\lambda_i^2 + 8) J_1^2(3\lambda_i)}$$

and

$$f(x) = 18 \sum_{i=1}^{\infty} \frac{\lambda_i J_2(3\lambda_i)}{(9\lambda_i^2 + 8) J_1^2(3\lambda_i)} J_1(\lambda_i x).$$

■

Use of Computers   Since Bessel functions are "built-in functions" in a CAS, it is a straightforward task to find the approximate values of the eigenvalues λ_i and the coefficients c_i in a Fourier-Bessel series. For example, in (9) we can think of $x_i = \lambda_i b$ as a positive root of the equation $h J_n(x) + x J_n'(x) = 0$. Thus in Example 2 we have used a CAS to find the first five positive roots x_i of $3 J_1(x) + x J_1'(x) = 0$, and from these roots we obtain the first five eigenvalues: $\lambda_1 = x_1/3 = 0.98320$, $\lambda_2 = x_2/3 = 1.94704$, $\lambda_3 = x_3/3 = 2.95758$, $\lambda_4 = x_4/3 = 3.98538$, and $\lambda_5 = x_5/3 = 5.02078$. Knowing the roots $x_i = 3\lambda_i$ and the eigenvalues, we again use a CAS to calculate the numerical values of $J_2(3\lambda_i)$, $J_1^2(3\lambda_i)$, and finally the coefficients c_i. In this manner we find that the fifth partial sum $S_5(x)$ for the Fourier-Bessel series representation of $f(x) = x$, $0 < x < 3$ in Example 2 is

$$S_5(x) = 4.01844 J_1(0.98320x) - 1.86937 J_1(1.94704x)$$
$$+ 1.07106 J_1(2.95758x) - 0.70306 J_1(3.98538x)$$
$$+ 0.50343 J_1(5.02078x).$$

The graph of $S_5(x)$ on the interval $0 < x < 3$ is shown in Figure 11.18(a). In Figure 11.18(b) we have graphed $S_{10}(x)$ on the interval $0 < x < 50$. Notice that outside the interval of definition $0 < x < 3$ the series does not converge to a periodic extension of f because Bessel functions are not periodic functions. See Problems 11 and 12 in Exercises 11.5.

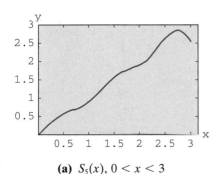

(a) $S_5(x)$, $0 < x < 3$

(b) $S_{10}(x)$, $0 < x < 50$

Figure 11.18 Partial sums of a Fourier-Bessel series

11.5.2   FOURIER-LEGENDRE SERIES

From Example 4 of Section 11.4 we know that the set of Legendre polynomials $\{P_n(x)\}$, $n = 0, 1, 2, \ldots$ is orthogonal with respect to the weight function $p(x) = 1$ on $[-1, 1]$. Furthermore, it can be proved that the square norm of a polynomial $P_n(x)$ depends on n in the following manner:

$$\|P_n(x)\|^2 = \int_{-1}^{1} P_n^2(x)\, dx = \frac{2}{2n + 1}.$$

The orthogonal series expansion of a function in terms of the Legendre polynomials is summarized in the next definition.

DEFINITION 11.8   Fourier-Legendre Series

The **Fourier-Legendre series** of a function f on an interval $(-1, 1)$ is given by

$$f(x) = \sum_{n=0}^{\infty} c_n P_n(x), \tag{21}$$

where
$$c_n = \frac{2n + 1}{2} \int_{-1}^{1} f(x) P_n(x)\, dx. \tag{22}$$

Convergence of a Fourier-Legendre Series   Sufficient conditions for convergence of a Fourier-Legendre series are given in the next theorem.

THEOREM 11.5   Conditions for Convergence

If f and f' are piecewise continuous on $(-1, 1)$, then the Fourier-Legendre series (21) converges to $f(x)$ at a point of continuity and to $[f(x+) + f(x-)]/2$ at a point of discontinuity.

EXAMPLE 3   Expansion in a Fourier-Legendre Series

Write out the first four nonzero terms in the Fourier-Legendre expansion of

$$f(x) = \begin{cases} 0, & -1 < x < 0 \\ 1, & 0 \le x < 1. \end{cases}$$

Solution   The first several Legendre polynomials are listed on page 300. From these and (22) we find

$$c_0 = \frac{1}{2} \int_{-1}^{1} f(x) P_0(x)\, dx = \frac{1}{2} \int_{0}^{1} 1 \cdot 1\, dx = \frac{1}{2}$$

$$c_1 = \frac{3}{2} \int_{-1}^{1} f(x) P_1(x)\, dx = \frac{3}{2} \int_{0}^{1} 1 \cdot x\, dx = \frac{3}{4}$$

$$c_2 = \frac{5}{2} \int_{-1}^{1} f(x) P_2(x)\, dx = \frac{5}{2} \int_{0}^{1} 1 \cdot \frac{1}{2}(3x^2 - 1)\, dx = 0$$

$$c_3 = \frac{7}{2} \int_{-1}^{1} f(x) P_3(x)\, dx = \frac{7}{2} \int_{0}^{1} 1 \cdot \frac{1}{2}(5x^3 - 3x)\, dx = -\frac{7}{16}$$

$$c_4 = \frac{9}{2} \int_{-1}^{1} f(x) P_4(x)\, dx = \frac{9}{2} \int_{0}^{1} 1 \cdot \frac{1}{8}(35x^4 - 30x^2 + 3)\, dx = 0$$

$$c_5 = \frac{11}{2} \int_{-1}^{1} f(x) P_5(x)\, dx = \frac{11}{2} \int_{0}^{1} 1 \cdot \frac{1}{8}(63x^5 - 70x^3 + 15x)\, dx = \frac{11}{32}.$$

Hence $\quad f(x) = \frac{1}{2} P_0(x) + \frac{3}{4} P_1(x) - \frac{7}{16} P_3(x) + \frac{11}{32} P_5(x) + \cdots.$ ■

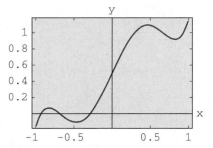

Figure 11.19 Partial sum $S_5(x)$ of Fourier-Legendre series

Like the Bessel functions, Legendre polynomials are built-in functions in computer algebra systems such as *Maple* and *Mathematica,* and so each of the coefficients just listed can be found using the integration application of such a program. Indeed, using a CAS we further find that $c_6 = 0$ and $c_7 = -\frac{65}{256}$. The fifth partial sum of the Fourier-Legendre series representation of the function f defined in Example 3 is then

$$S_5(x) = \frac{1}{2} P_0(x) + \frac{3}{4} P_1(x) - \frac{7}{16} P_3(x) + \frac{11}{32} P_5(x) - \frac{65}{256} P_7(x).$$

The graph of $S_5(x)$ on the interval $-1 < x < 1$ is given in Figure 11.19.

Alternative Form of Series In applications the Fourier-Legendre series appears in an alternative form. If we let $x = \cos\theta$, then $x = 1$ implies $\theta = 0$ whereas $x = -1$ implies $\theta = \pi$. Since $dx = -\sin\theta\, d\theta$, (21) and (22) become, respectively,

$$F(\theta) = \sum_{n=0}^{\infty} c_n P_n(\cos\theta) \tag{23}$$

$$c_n = \frac{2n+1}{2} \int_{0}^{\pi} F(\theta) P_n(\cos\theta) \sin\theta\, d\theta, \tag{24}$$

where $f(\cos\theta)$ has been replaced by $F(\theta)$.

EXERCISES 11.5

Answers to odd-numbered problems begin on page AN-17.

11.5.1 Fourier-Bessel Series

1. Find the first four eigenvalues λ_k defined by $J_1(3\lambda) = 0$. [*Hint:* See Table 6.2 in Section 6.3.]
2. Find the first four eigenvalues λ_k defined by $J_0'(2\lambda) = 0$.

In Problems 3–6 expand $f(x) = 1, 0 < x < 2$ in a Fourier-Bessel series, using Bessel functions of order zero that satisfy the given boundary condition.

3. $J_0(2\lambda) = 0$ **4.** $J_0'(2\lambda) = 0$
5. $J_0(2\lambda) + 2\lambda J_0'(2\lambda) = 0$ **6.** $J_0(2\lambda) + \lambda J_0'(2\lambda) = 0$

In Problems 7–10 expand the given function in a Fourier-Bessel series, using Bessel functions of the same order as in the indicated boundary condition.

7. $f(x) = 5x, \quad 0 < x < 4$
$3J_1(4\lambda) + 4\lambda J_1'(4\lambda) = 0$

8. $f(x) = x^2, \quad 0 < x < 1$
$J_2(\lambda) = 0$

9. $f(x) = x^2, \quad 0 < x < 3$
$J_0'(3\lambda) = 0 \quad [Hint: t^3 = t^2 \cdot t.]$

10. $f(x) = 1 - x^2, \quad 0 < x < 1$
$J_0(\lambda) = 0$

Computer Lab Assignments

11. (a) Use a CAS to plot the graph of $y = 3J_1(x) + xJ_1'(x)$ on an interval so that the first five positive x-intercepts of the graph are shown.

(b) Use the root-finding capability of your CAS to approximate the first five roots x_i of the equation $3J_1(x) + xJ_1'(x) = 0$.

(c) Use the data obtained in part (b) to find the first five positive eigenvalues λ_i that satisfy $3J_1(4\lambda) + 4\lambda J_1'(4\lambda) = 0$. (See Problem 7.)

(d) If instructed, find the first ten positive eigenvalues.

12. (a) Use the eigenvalues in part (c) of Problem 11 and a CAS to approximate the values of the first five coefficients c_i of the Fourier-Bessel series obtained in Problem 7.

(b) Use a CAS to plot the graphs of the partial sums $S_N(x)$, $N = 1$, 2, 3, 4, 5 of the Fourier-Bessel series in Problem 7.

(c) If instructed, plot the graph of the partial sum $S_{10}(x)$ on $0 < x < 4$ and on $0 < x < 50$.

11.5.2 Fourier-Legendre Series

In Problems 13 and 14 write out the first five nonzero terms in the Fourier-Legendre expansion of the given function. If instructed, use a CAS as an aid in evaluating the coefficients. Use a CAS to plot the graph of the partial sum $S_5(x)$.

13. $f(x) = \begin{cases} 0, & -1 < x < 0 \\ x, & 0 < x < 1 \end{cases}$

14. $f(x) = e^x, \quad -1 < x < 1$

15. The first three Legendre polynomials are $P_0(x) = 1$, $P_1(x) = x$, and $P_2(x) = \frac{1}{2}(3x^2 - 1)$. If $x = \cos \theta$, then $P_0(\cos \theta) = 1$ and $P_1(\cos \theta) = \cos \theta$. Show that $P_2(\cos \theta) = \frac{1}{4}(3 \cos 2\theta + 1)$.

16. Use the results given in Problem 15 to find a Fourier-Legendre expansion (23) of $F(\theta) = 1 - \cos 2\theta$.

17. A Legendre polynomial $P_n(x)$ is an even or odd function, depending on whether n is even or odd. Show that if f is an even function on $(-1, 1)$, then (21) and (22) become, respectively,

$$f(x) = \sum_{n=0}^{\infty} c_{2n} P_{2n}(x) \tag{25}$$

$$c_{2n} = (4n + 1) \int_0^1 f(x) P_{2n}(x) \, dx. \tag{26}$$

18. Show that if f is an odd function on $(-1, 1)$, then (21) and (22) become, respectively,

$$f(x) = \sum_{n=0}^{\infty} c_{2n+1} P_{2n+1}(x) \tag{27}$$

$$c_{2n+1} = (4n + 3) \int_0^1 f(x) P_{2n+1}(x) \, dx. \tag{28}$$

The series (25) and (26) can also be used when f is defined on only the interval $(0, 1)$. Both series represent f on $(0, 1)$; but on the interval $(-1, 0)$, (25) represents an even extension whereas (27) represents an odd extension. In Problems 19 and 20 write out the first four nonzero terms in the indicated expansion of the given function. What function does the series represent on $(-1, 1)$? Use a CAS to plot the graph of the partial sum $S_4(x)$.

19. $f(x) = x, 0 < x < 1$;   (25)   **20.** $f(x) = 1, 0 < x < 1$;   (27)

Discussion Problems

21. If the partial sums in Problem 12 were plotted on a symmetric interval such as $-30 < x < 30$, would the graphs possess any symmetry? Explain.

22. (a) Sketch, by hand, a graph of what you think the Fourier-Bessel series in Problem 3 converges to on the interval $-2 < x < 2$.

  (b) Sketch, by hand, a graph of what you think the Fourier-Bessel series would converge to on the interval $-1 < x < 1$ if the eigenvalues in Problem 7 were defined by $3J_2(2\lambda) + 4\lambda J_2'(2\lambda) = 0$.

23. Discuss why a Fourier-Legendre expansion of a polynomial function that is defined on the interval $(-1, 1)$ is necessarily a finite series. Find the finite Fourier-Legendre series of
  (a) $f(x) = x^2$ **(b)** $f(x) = x^3$.

CHAPTER 11 IN REVIEW

Answers to odd-numbered problems begin on page AN-17.

In Problems 1–6 fill in the blank or answer true or false without referring back to the text.

1. The functions $f(x) = x^2 - 1$ and $g(x) = x^5$ are orthogonal on $[-\pi, \pi]$. _____

2. The product of an odd function f with an odd function g is _____.

3. To expand $f(x) = |x| + 1, -\pi < x < \pi$ in an appropriate trigonometric series we would use a _____ series.

4. $y = 0$ is never an eigenfunction of a Sturm-Liouville problem. _____

5. $\lambda = 0$ is never an eigenvalue of a Sturm-Liouville problem. _____

6. If the function $f(x) = \begin{cases} x + 1, & -1 < x < 0 \\ -x, & 0 < x < 1 \end{cases}$ is expanded in a Fourier series, the series will converge to _____ at $x = -1$, to _____ at $x = 0$, and to _____ at $x = 1$.

7. Suppose the function $f(x) = x^2 + 1$, $0 < x < 3$ is expanded in a Fourier series, a cosine series, and a sine series. Give the value to which each series will converge at $x = 0$.

8. What is the corresponding eigenfunction for the boundary-value problem $y'' + \lambda y = 0$, $y'(0) = 0$, $y(\pi/2) = 0$ for $\lambda = 25$?

9. **Chebyshev's differential equation** $(1 - x^2)y'' - xy' + n^2y = 0$ has a polynomial solution $y = T_n(x)$ for $n = 0, 1, 2, \ldots$. Specify the weight function $w(x)$ and the interval over which the set of Chebyshev polynomials $\{T_n(x)\}$ is orthogonal. Give an orthogonality relation.

10. The set of Legendre polynomials $\{P_n(x)\}$, where $P_0(x) = 1$, $P_1(x) = x, \ldots$, is orthogonal with respect to the weight function $w(x) = 1$ on the interval $[-1, 1]$. Explain why $\int_{-1}^{1} P_n(x)\, dx = 0$ for $n > 0$.

11. Without doing any work, explain why the cosine series of $f(x) = \cos^2 x$, $0 < x < \pi$ is the finite series $f(x) = \frac{1}{2} + \frac{1}{2}\cos 2x$.

12. **(a)** Show that the set

$$\left\{ \sin \frac{\pi}{2L} x, \sin \frac{3\pi}{2L} x, \sin \frac{5\pi}{2L} x, \ldots \right\}$$

is orthogonal on the interval $0 \leq x \leq L$.

(b) Find the norm of each function in part (a). Construct an orthonormal set.

13. Expand $f(x) = |x| - x$, $-1 < x < 1$ in a Fourier series.

14. Expand $f(x) = 2x^2 - 1$, $-1 < x < 1$ in a Fourier series.

15. Expand $f(x) = e^x$, $0 < x < 1$
   (a) in a cosine series   **(b)** in a sine series.

16. In Problems 13, 14, and 15, sketch the periodic extension of f to which each series converges.

17. Find the eigenvalues and eigenfunctions of the boundary-value problem

$$x^2 y'' + xy' + 9\lambda y = 0, \quad y'(1) = 0, \quad y(e) = 0.$$

18. Give an orthogonality relation for the eigenfunctions in Problem 17.

19. Expand $f(x) = \begin{cases} 1, & 0 < x < 2 \\ 0, & 2 < x < 4 \end{cases}$ in a Fourier-Bessel series, using Bessel functions of order zero that satisfy the boundary-condition $J_0(4\lambda) = 0$.

20. Expand $f(x) = x^4$, $-1 < x < 1$ in a Fourier-Legendre series.

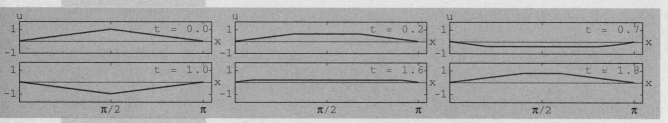

The plucked string; see Problem 21, page 542.

12 PARTIAL DIFFERENTIAL EQUATIONS AND BOUNDARY-VALUE PROBLEMS IN RECTANGULAR COORDINATES

INTRODUCTION In this and the next two chapters the emphasis will be on two procedures used in solving partial differential equations that occur frequently in problems involving temperature distributions, vibrations, and potentials. These problems, called boundary-value problems, are described by relatively simple linear second-order partial differential equations. The thrust of both of these procedures is to find solutions of a partial differential equation by reducing it to one or more ordinary differential equations.

We begin with the method of separation of variables for linear partial differential equations. The application of this method leads us back to the important concepts in Chapter 11—namely, eigenvalues, eigenfunctions, and the expansion of a function in an infinite series of orthogonal functions.

521

12.1 SEPARABLE PARTIAL DIFFERENTIAL EQUATIONS

- *Linear second-order PDE* • *Homogeneous equation*
- *Nonhomogeneous equation* • *Separable equations* • *Solution*
- *Separation constant* • *Superposition principle* • *Classification of linear second-order PDEs*

Partial differential equations (PDEs), like ordinary differential equations, are classified as either linear or nonlinear. Analogously to a linear ODE (see (6) of Section 1.1), the dependent variable and its partial derivatives appear only to the first power in a linear PDE. In this and the chapters that follow we shall be interested only in *linear* PDEs, which will be, for the most part, *second-order* equations.

Linear Partial Differential Equation If we let u denote the dependent variable and x and y the independent variables, then the general form of a **linear second-order partial differential equation** is given by

$$A\frac{\partial^2 u}{\partial x^2} + B\frac{\partial^2 u}{\partial x\,\partial y} + C\frac{\partial^2 u}{\partial y^2} + D\frac{\partial u}{\partial x} + E\frac{\partial u}{\partial y} + Fu = G, \qquad \textbf{(1)}$$

where A, B, C, ..., G are functions of x and y. When $G(x, y) = 0$, equation (1) is said to be **homogeneous;** otherwise, it is **nonhomogeneous.** For example, the equations

$$\frac{\partial^2 u}{\partial x^2} + \frac{\partial^2 u}{\partial y^2} = 0 \quad \text{and} \quad \frac{\partial^2 u}{\partial x^2} - \frac{\partial u}{\partial y} = xy$$

are homogeneous and nonhomogeneous, respectively.

Solution of a PDE A **solution** of a linear partial differential equation (1) is a function $u(x, y)$ of two independent variables that possesses all partial derivatives occurring in the equation and that satisfies the equation in some region of the xy-plane.

It is not our intention to examine procedures for finding *general solutions* of linear partial differential equations. Not only is a general solution of a linear second-order PDE often difficult to obtain; it also is usually not all that useful in applications. Thus our focus throughout will be on finding *particular solutions* of some of the more important linear PDEs—that is, equations that appear in many applications.

Separation of Variables There are several methods that can be tried to find particular solutions of a linear PDE. In the **method of separation of variables** we seek to find a particular solution in the form of a *product* of a function of x and a function of y:

$$u(x, y) = X(x)Y(y).$$

With this assumption, it is sometimes possible to reduce a linear PDE in two variables to two ODEs. To this end we note that

$$\frac{\partial u}{\partial x} = X'Y, \quad \frac{\partial u}{\partial y} = XY'$$

and
$$\frac{\partial^2 u}{\partial x^2} = X''Y, \quad \frac{\partial^2 u}{\partial y^2} = XY'',$$

where the primes denote ordinary differentiation.

EXAMPLE 1 **Using Separation of Variables**

Find product solutions of $\dfrac{\partial^2 u}{\partial x^2} = 4 \dfrac{\partial u}{\partial y}$.

Solution If $u(x, y) = X(x)Y(y)$, the equation becomes
$$X''Y = 4XY'.$$

After dividing both sides by $4XY$, we have separated the variables:
$$\frac{X''}{4X} = \frac{Y'}{Y}.$$

Since the left-hand side of the last equation is independent of y and is equal to the right-hand side, which is independent of x, we conclude that both sides of the equation are independent of x *and* y. In other words, each side of the equation must be a constant. In practice it is convenient to write this real *separation constant* as either λ^2 or $-\lambda^2$. We distinguish the following three cases.

Case I If $\lambda^2 > 0$, the two equalities
$$\frac{X''}{4X} = \frac{Y'}{Y} = \lambda^2$$

lead to $\qquad X'' - 4\lambda^2 X = 0 \quad \text{and} \quad Y' - \lambda^2 Y = 0.$

These latter equations have the solutions
$$X = c_1 \cosh 2\lambda x + c_2 \sinh 2\lambda x \quad \text{and} \quad Y = c_3 e^{\lambda^2 y},$$

respectively. Thus a particular solution of the partial differential equation is
$$\begin{aligned}
u &= XY \\
&= (c_1 \cosh 2\lambda x + c_2 \sinh 2\lambda x)(c_3 e^{\lambda^2 y}) \\
&= A_1 e^{\lambda^2 y} \cosh 2\lambda x + B_1 e^{\lambda^2 y} \sinh 2\lambda x, \tag{2}
\end{aligned}$$

where $A_1 = c_1 c_3$ and $B_1 = c_2 c_3$.

Case II If $-\lambda^2 < 0$, the two equalities
$$\frac{X''}{4X} = \frac{Y'}{Y} = -\lambda^2$$

give $\qquad X'' + 4\lambda^2 X = 0 \quad \text{and} \quad Y' + \lambda^2 Y = 0.$

Since the solutions of these equations are
$$X = c_4 \cos 2\lambda x + c_5 \sin 2\lambda x \quad \text{and} \quad Y = c_6 e^{-\lambda^2 y},$$

respectively, another particular solution is

$$u = A_2 e^{-\lambda^2 y} \cos 2\lambda x + B_2 e^{-\lambda^2 y} \sin 2\lambda x, \tag{3}$$

where $A_2 = c_4 c_6$ and $B_2 = c_5 c_6$.

Case III   If $\lambda^2 = 0$, it follows that

$$X'' = 0 \text{ and } Y' = 0.$$

In this case $X = c_7 x + c_8$ and $Y = c_9$, so

$$u = A_3 x + B_3, \tag{4}$$

where $A_3 = c_7 c_9$ and $B_3 = c_8 c_9$.   ■

It is left as an exercise to verify that (2), (3), and (4) satisfy the given equation. See Problem 29 in Exercises 12.1.

Separation of variables is not a general method for finding particular solutions; some linear partial differential equations are simply not separable. You should verify that the assumption $u = XY$ does not lead to a solution for $\partial^2 u / \partial x^2 - \partial u / \partial y = x$.

Superposition Principle   The following theorem is analogous to Theorem 4.2 and is known as the **superposition principle.**

> **THEOREM 12.1**   **Superposition Principle**
>
> If $u_1, u_2, \ldots, u_k$ are solutions of a homogeneous linear partial differential equation, then the linear combination
>
> $$u = c_1 u_1 + c_2 u_2 + \cdots + c_k u_k,$$
>
> where the c_i, $i = 1, 2, \ldots, k$ are constants, is also a solution.

Throughout the remainder of the chapter we shall assume that whenever we have an infinite set $u_1, u_2, u_3, \ldots$ of solutions of a homogeneous linear equation, we can construct yet another solution u by forming the infinite series

$$u = \sum_{k=1}^{\infty} c_k u_k,$$

where the c_i, $i = 1, 2, \ldots$ are constants.

Classification of Equations   A linear second-order partial differential equation in two independent variables with constant coefficients can be classified as one of three types. This classification depends only on the coefficients of the second-order derivatives. Of course, we assume that at least one of the coefficients A, B, and C is not zero.

> **DEFINITION 12.1** **Classification of Equations**
>
> The linear second-order partial differential equation
>
> $$A\frac{\partial^2 u}{\partial x^2} + B\frac{\partial^2 u}{\partial x\,\partial y} + C\frac{\partial^2 u}{\partial y^2} + D\frac{\partial u}{\partial x} + E\frac{\partial u}{\partial y} + Fu = 0,$$
>
> where A, B, C, D, E, and F are real constants, is said to be
>
> > **hyperbolic** if $\quad B^2 - 4AC > 0,$
> > **parabolic** if $\quad B^2 - 4AC = 0,$
> > **elliptic** if $\quad\quad B^2 - 4AC < 0.$

EXAMPLE 2 **Classifying Linear Second-Order PDEs**

Classify the following equations:

(a) $3\dfrac{\partial^2 u}{\partial x^2} = \dfrac{\partial u}{\partial y}$ **(b)** $\dfrac{\partial^2 u}{\partial x^2} = \dfrac{\partial^2 u}{\partial y^2}$ **(c)** $\dfrac{\partial^2 u}{\partial x^2} + \dfrac{\partial^2 u}{\partial y^2} = 0$

Solution

(a) By rewriting the given equation as

$$3\frac{\partial^2 u}{\partial x^2} - \frac{\partial u}{\partial y} = 0$$

we can make the identifications $A = 3$, $B = 0$, and $C = 0$. Since $B^2 - 4AC = 0$, the equation is parabolic.

(b) By rewriting the equation as

$$\frac{\partial^2 u}{\partial x^2} - \frac{\partial^2 u}{\partial y^2} = 0$$

we see that $A = 1$, $B = 0$, $C = -1$, and $B^2 - 4AC = -4(1)(-1) > 0$. The equation is hyperbolic.

(c) With $A = 1$, $B = 0$, $C = 1$, and $B^2 - 4AC = -4(1)(1) < 0$, the equation is elliptic. ■

A detailed explanation of why we would want to classify a second-order partial differential equation is beyond the scope of this text. But the answer lies in the fact that we wish to solve partial differential equations subject to certain side conditions known as boundary and initial conditions. The kinds of side conditions appropriate for a given equation depend on whether the equation is hyperbolic, parabolic, or elliptic.

EXERCISES 12.1 ————————————————————————

Answers to odd-numbered problems begin on page AN-17.

In Problems 1–16 use separation of variables to find, if possible, product solutions for the given partial differential equation.

1. $\dfrac{\partial u}{\partial x} = \dfrac{\partial u}{\partial y}$

2. $\dfrac{\partial u}{\partial x} + 3\dfrac{\partial u}{\partial y} = 0$

3. $u_x + u_y = u$

4. $u_x = u_y + u$

5. $x \dfrac{\partial u}{\partial x} = y \dfrac{\partial u}{\partial y}$

6. $y \dfrac{\partial u}{\partial x} + x \dfrac{\partial u}{\partial y} = 0$

7. $\dfrac{\partial^2 u}{\partial x^2} + \dfrac{\partial^2 u}{\partial x \partial y} + \dfrac{\partial^2 u}{\partial y^2} = 0$

8. $y \dfrac{\partial^2 u}{\partial x \partial y} + u = 0$

9. $k \dfrac{\partial^2 u}{\partial x^2} - u = \dfrac{\partial u}{\partial t}, \quad k > 0$

10. $k \dfrac{\partial^2 u}{\partial x^2} = \dfrac{\partial u}{\partial t}, \quad k > 0$

11. $a^2 \dfrac{\partial^2 u}{\partial x^2} = \dfrac{\partial^2 u}{\partial t^2}$

12. $a^2 \dfrac{\partial^2 u}{\partial x^2} = \dfrac{\partial^2 u}{\partial t^2} + 2k \dfrac{\partial u}{\partial t}, \quad k > 0$

13. $\dfrac{\partial^2 u}{\partial x^2} + \dfrac{\partial^2 u}{\partial y^2} = 0$

14. $x^2 \dfrac{\partial^2 u}{\partial x^2} + \dfrac{\partial^2 u}{\partial y^2} = 0$

15. $u_{xx} + u_{yy} = u$

16. $a^2 u_{xx} - g = u_{tt}, \quad g$ a constant

In Problems 17–26 classify the given partial differential equation as hyperbolic, parabolic, or elliptic.

17. $\dfrac{\partial^2 u}{\partial x^2} + \dfrac{\partial^2 u}{\partial x \partial y} + \dfrac{\partial^2 u}{\partial y^2} = 0$

18. $3 \dfrac{\partial^2 u}{\partial x^2} + 5 \dfrac{\partial^2 u}{\partial x \partial y} + \dfrac{\partial^2 u}{\partial y^2} = 0$

19. $\dfrac{\partial^2 u}{\partial x^2} + 6 \dfrac{\partial^2 u}{\partial x \partial y} + 9 \dfrac{\partial^2 u}{\partial y^2} = 0$

20. $\dfrac{\partial^2 u}{\partial x^2} - \dfrac{\partial^2 u}{\partial x \partial y} - 3 \dfrac{\partial^2 u}{\partial y^2} = 0$

21. $\dfrac{\partial^2 u}{\partial x^2} = 9 \dfrac{\partial^2 u}{\partial x \partial y}$

22. $\dfrac{\partial^2 u}{\partial x \partial y} - \dfrac{\partial^2 u}{\partial y^2} + 2 \dfrac{\partial u}{\partial x} = 0$

23. $\dfrac{\partial^2 u}{\partial x^2} + 2 \dfrac{\partial^2 u}{\partial x \partial y} + \dfrac{\partial^2 u}{\partial y^2} + \dfrac{\partial u}{\partial x} - 6 \dfrac{\partial u}{\partial y} = 0$

24. $\dfrac{\partial^2 u}{\partial x^2} + \dfrac{\partial^2 u}{\partial y^2} = u$

25. $a^2 \dfrac{\partial^2 u}{\partial x^2} = \dfrac{\partial^2 u}{\partial t^2}$

26. $k \dfrac{\partial^2 u}{\partial x^2} = \dfrac{\partial u}{\partial t}, \quad k > 0$

27. Show that the equation

$$k \left(\dfrac{\partial^2 u}{\partial r^2} + \dfrac{1}{r} \dfrac{\partial u}{\partial r} \right) = \dfrac{\partial u}{\partial t}$$

possesses the product solution

$$u = e^{-k\lambda^2 t}(AJ_0(\lambda r) + BY_0(\lambda r)).$$

28. Show that the equation

$$\dfrac{\partial^2 u}{\partial r^2} + \dfrac{1}{r} \dfrac{\partial u}{\partial r} + \dfrac{1}{r^2} \dfrac{\partial^2 u}{\partial \theta^2} = 0$$

possesses the product solution

$$u = (c_1 \cos\lambda\theta + c_2 \sin\lambda\theta)(c_3 r^\lambda + c_4 r^{-\lambda}).$$

29. Verify that the product solutions (2), (3), and (4) satisfy the first-order equation in Example 1.

30. Definition 12.1 generalizes to linear PDEs with coefficients that are functions of x and y. Determine the regions in the xy-plane for which the equation

$$(xy + 1)\frac{\partial^2 u}{\partial x^2} + (x + 2y)\frac{\partial^2 u}{\partial x \partial y} + \frac{\partial^2 u}{\partial y^2} + xy^2 u = 0$$

is hyperbolic, parabolic, or elliptic.

12.2 CLASSICAL EQUATIONS AND BOUNDARY-VALUE PROBLEMS

• One-dimensional heat equation • One-dimensional wave equation • Laplace's equation in two variables • Laplacian • Initial conditions • Boundary conditions • Types of boundary conditions • Boundary-value problems • Classical PDEs

In the remainder of this chapter and Chapters 13–15 we shall be concerned with finding solutions of the following classical equations of mathematical physics:

$$k\frac{\partial^2 u}{\partial x^2} = \frac{\partial u}{\partial t}, \quad k > 0 \tag{1}$$

$$a^2\frac{\partial^2 u}{\partial x^2} = \frac{\partial^2 u}{\partial t^2} \tag{2}$$

$$\frac{\partial^2 u}{\partial x^2} + \frac{\partial^2 u}{\partial y^2} = 0 \tag{3}$$

or slight variations of these equations. The PDEs (1), (2), and (3) are known, respectively, as the **one-dimensional heat equation,** the **one-dimensional wave equation,** and **Laplace's equation in two dimensions.** "One-dimensional" refers to the fact that x denotes a spatial variable whereas t represents time. Note from Theorem 12.1 that the heat equation (1) is parabolic, the wave equation (2) is hyperbolic, and Laplace's equation is elliptic. This observation will be important in Chapter 15.

Heat Equation   Equation (1) occurs in the theory of heat flow—that is, heat transferred by conduction in a rod or in a thin wire. The function $u(x, t)$ represents temperature at a point x along the rod at some time t. Problems in mechanical vibrations often lead to the wave equation (2). For purposes of discussion, a solution $u(x, t)$ of (2) will represent the displacement of an idealized string. Finally, a solution $u(x, y)$ of Laplace's equation (3) can be interpreted as the steady-state (that is, time-independent) temperature distribution throughout a thin, two-dimensional plate.

Even though we have to make many simplifying assumptions, it is worthwhile to see how equations such as (1) and (2) arise.

Suppose a thin circular rod of length L has a cross-sectional area A and coincides with the x-axis on the interval $[0, L]$. See Figure 12.1. Let us suppose:

cross-section of area A

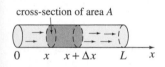

Figure 12.1

- The flow of heat within the rod takes place only in the x-direction.
- The lateral, or curved, surface of the rod is insulated; that is, no heat escapes from this surface.

- No heat is being generated within the rod.
- The rod is homogeneous; that is, its mass per unit volume ρ is a constant.
- The specific heat γ and thermal conductivity K of the material of the rod are constants.

To derive the partial differential equation satisfied by the temperature $u(x, t)$ we need two empirical laws of heat conduction:

(*i*) *The quantity of heat Q in an element of mass m is*

$$Q = \gamma m u, \tag{4}$$

where u is the temperature of the element.

(*ii*) *The rate of heat flow Q_t through the cross-section indicated in Figure 12.1 is proportional to the area A of the cross-section and the partial derivative with respect to x of the temperature:*

$$Q_t = -KAu_x. \tag{5}$$

Since heat flows in the direction of decreasing temperature, the minus sign in (5) is used to ensure that Q_t is positive for $u_x < 0$ (heat flow to the right) and negative for $u_x > 0$ (heat flow to the left). If the circular slice of the rod shown in Figure 12.1 between x and $x + \Delta x$ is very thin, then $u(x, t)$ can be taken as the approximate temperature at each point in the interval. Now the mass of the slice is $m = \rho(A \, \Delta x)$, and so it follows from (4) that the quantity of heat in it is

$$Q = \gamma \rho A \, \Delta x \, u. \tag{6}$$

Furthermore, when heat flows in the positive x-direction, we see from (5) that heat builds up in the slice at the net rate

$$-KAu_x(x, t) - [-KAu_x(x + \Delta x, t)] = KA[u_x(x + \Delta x, t) - u_x(x, t)]. \tag{7}$$

By differentiating (6) with respect to t we see that this net rate is also given by

$$Q_t = \gamma \rho A \, \Delta x \, u_t. \tag{8}$$

Equating (7) and (8) gives

$$\frac{K}{\gamma \rho} \frac{u_x(x + \Delta x, t) - u_x(x, t)}{\Delta x} = u_t. \tag{9}$$

Taking the limit of (9) as $\Delta x \to 0$ finally yields (1) in the form*

$$\frac{K}{\gamma \rho} u_{xx} = u_t.$$

It is customary to let $k = K/\gamma \rho$ and call this positive constant the **thermal diffusivity.**

Wave Equation   Consider a string of length L, such as a guitar string, stretched taut between two points on the x-axis—say, $x = 0$ and $x = L$.

---

*Recall from calculus that $u_{xx} = \displaystyle\lim_{\Delta x \to 0} \frac{u_x(x + \Delta x, t) - u_x(x, t)}{\Delta x}$.

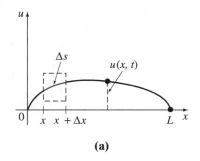

(a)

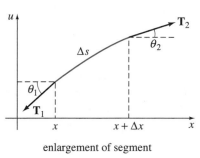

enlargement of segment

(b)

Figure 12.2

When the string starts to vibrate, assume that the motion takes place in the xu-plane in such a manner that each point on the string moves in a direction perpendicular to the x-axis (transverse vibrations). As shown in Figure 12.2(a), let $u(x, t)$ denote the vertical displacement of any point on the string measured from the x-axis for $t > 0$. We further assume:

- The string is perfectly flexible.
- The string is homogeneous; that is, its mass per unit length ρ is a constant.
- The displacements u are small compared to the length of the string.
- The slope of the curve is small at all points.
- The tension **T** acts tangent to the string, and its magnitude T is the same at all points.
- The tension is large compared with the force of gravity.
- No other external forces act on the string.

Now in Figure 12.2(b) the tensions $\mathbf{T}_1$ and $\mathbf{T}_2$ are tangent to the ends of the curve on the interval $[x, x + \Delta x]$. For small θ_1 and θ_2 the net vertical force acting on the corresponding element Δs of the string is then

$$T \sin \theta_2 - T \sin \theta_1 \approx T \tan \theta_2 - T \tan \theta_1$$
$$= T[u_x(x + \Delta x, t) - u_x(x, t)],^*$$

where $T = |\mathbf{T}_1| = |\mathbf{T}_2|$. Now $\rho \Delta s \approx \rho \Delta x$ is the mass of the string on $[x, x + \Delta x]$, and so Newton's second law gives

$$T[u_x(x + \Delta x, t) - u_x(x, t)] = \rho \Delta x \, u_{tt}$$

or

$$\frac{u_x(x + \Delta x, t) - u_x(x, t)}{\Delta x} = \frac{\rho}{T} u_{tt}.$$

If the limit is taken as $\Delta x \to 0$, the last equation becomes $u_{xx} = (\rho/T)u_{tt}$. This of course is (2) with $a^2 = T/\rho$.

Laplace's Equation   Although we shall not present its derivation, Laplace's equation in two and three dimensions occurs in time-independent problems involving potentials such as electrostatic, gravitational, and velocity in fluid mechanics. Moreover, a solution of Laplace's equation can also be interpreted as a steady-state temperature distribution. As illustrated in Figure 12.3, a solution $u(x, y)$ of (3) could represent the temperature that varies from point to point—but not with time—of a rectangular plate. Laplace's equation in two dimensions and in three dimensions is abbreviated as $\nabla^2 u = 0$, where

$$\nabla^2 u = \frac{\partial^2 u}{\partial x^2} + \frac{\partial^2 u}{\partial y^2} \quad \text{and} \quad \nabla^2 u = \frac{\partial^2 u}{\partial x^2} + \frac{\partial^2 u}{\partial y^2} + \frac{\partial^2 u}{\partial z^2}$$

are called, respectively, the two-dimensional **Laplacian** and the three-dimensional Laplacian of a function u.

We often wish to find solutions of equations (1), (2), and (3) that satisfy certain side conditions.

temperature as a function of position on the hot plate

thermometer

Figure 12.3

---

*$\tan \theta_2 = u_x(x + \Delta x, t)$ and $\tan \theta_1 = u_x(x, t)$ are equivalent expressions for slope.

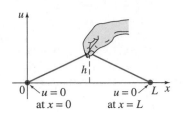

Figure 12.4

Initial Conditions Since solutions of (1) and (2) depend on time t, we can prescribe what happens at $t = 0$; that is, we can give **initial conditions (IC)**. If $f(x)$ denotes the initial temperature distribution throughout the rod in Figure 12.1, then a solution $u(x, t)$ of (1) must satisfy the single initial condition $u(x, 0) = f(x)$, $0 < x < L$. On the other hand, for a vibrating string, we can specify its initial displacement (or shape) $f(x)$ as well as its initial velocity $g(x)$. In mathematical terms we seek a function $u(x, t)$ satisfying (2) and the two initial conditions:

$$u(x, 0) = f(x), \quad \left.\frac{\partial u}{\partial t}\right|_{t=0} = g(x), \quad 0 < x < L. \tag{10}$$

For example, the string could be plucked, as shown in Figure 12.4, and released from rest ($g(x) = 0$).

Boundary Conditions The string in Figure 12.4 is secured to the x-axis at $x = 0$ and $x = L$ for all time. We interpret this by the two **boundary conditions (BC):**

$$u(0, t) = 0, \quad u(L, t) = 0, \quad t > 0.$$

Note that in this context the function f in (10) is continuous, and consequently $f(0) = 0$ and $f(L) = 0$. In general, there are three types of boundary conditions associated with equations (1), (2), and (3). On a boundary we can specify the values of *one* of the following:

$$(i) \ u, \quad (ii) \ \frac{\partial u}{\partial n}, \quad \text{or} \quad (iii) \ \frac{\partial u}{\partial n} + hu, \quad h \text{ a constant.}$$

Here $\partial u/\partial n$ denotes the normal derivative of u (the directional derivative of u in the direction perpendicular to the boundary). A boundary condition of the first type (i) is called a **Dirichlet condition;** a boundary condition of the second type (ii) is called a **Neumann condition;** and a boundary condition of the third type (iii) is known as a **Robin condition.** For example, for $t > 0$ a typical condition at the right-hand end of the rod in Figure 12.1 can be

$$(i)' \ u(L, t) = u_0, \quad u_0 \text{ a constant,}$$

$$(ii)' \ \left.\frac{\partial u}{\partial x}\right|_{x=L} = 0, \text{ or}$$

$$(iii)' \ \left.\frac{\partial u}{\partial x}\right|_{x=L} = -h(u(L, t) - u_m), \quad h > 0 \text{ and } u_m \text{ constants.}$$

Condition $(i)'$ simply states that the boundary $x = L$ is held by some means at a constant *temperature* u_0 for all time $t > 0$. Condition $(ii)'$ indicates that the boundary $x = L$ is *insulated*. From the empirical law of heat transfer, the flux of heat across a boundary (that is, the amount of heat per unit area per unit time conducted across the boundary) is proportional to the value of the normal derivative $\partial u/\partial n$ of the temperature u. Thus when the boundary $x = L$ is thermally insulated, no heat flows into or out of the rod and so

$$\left.\frac{\partial u}{\partial x}\right|_{x=L} = 0.$$

We can interpret *(iii)'* to mean that *heat is lost* from the right-hand end of the rod by being in contact with a medium, such as air or water, that is held at a constant temperature. From Newton's law of cooling, the outward flux of heat from the rod is proportional to the difference between the temperature $u(L, t)$ at the boundary and the temperature u_m of the surrounding medium. We note that if heat is lost from the left-hand end of the rod, the boundary condition is

$$\frac{\partial u}{\partial x}\bigg|_{x=0} = h(u(0, t) - u_m).$$

The change in algebraic sign is consistent with the assumption that the rod is at a higher temperature than the medium surrounding the ends so that $u(0, t) > u_m$ and $u(L, t) > u_m$. At $x = 0$ and $x = L$, the slopes $u_x(0, t)$ and $u_x(L, t)$ must be positive and negative, respectively.

Of course, at the ends of the rod we can specify different conditions at the same time. For example, we could have

$$\frac{\partial u}{\partial x}\bigg|_{x=0} = 0 \quad \text{and} \quad u(L, t) = u_0, \quad t > 0.$$

We note that the boundary condition in *(i)'* is homogeneous if $u_0 = 0$; if $u_0 \neq 0$, the boundary condition is nonhomogeneous. The boundary condition *(ii)'* is homogeneous; *(iii)'* is homogeneous if $u_m = 0$ and nonhomogeneous if $u_m \neq 0$.

Boundary-Value Problems   Problems such as

Solve: $\qquad a^2 \dfrac{\partial^2 u}{\partial x^2} = \dfrac{\partial^2 u}{\partial t^2}, \quad 0 < x < L, \quad t > 0$

Subject to: (BC) $\quad u(0, t) = 0, \qquad u(L, t) = 0, \qquad t > 0$ $\qquad$ **(11)**

$\qquad\qquad$ (IC) $\quad u(x, 0) = f(x), \quad \dfrac{\partial u}{\partial t}\bigg|_{t=0} = g(x), \quad 0 < x < L$

and

Solve: $\qquad \dfrac{\partial^2 u}{\partial x^2} + \dfrac{\partial^2 u}{\partial y^2} = 0, \quad 0 < x < a, \quad 0 < y < b$

Subject to: (BC) $\begin{cases} \dfrac{\partial u}{\partial x}\bigg|_{x=0} = 0, \quad \dfrac{\partial u}{\partial x}\bigg|_{x=a} = 0, \qquad 0 < y < b \\[2mm] u(x, 0) = 0, \quad u(x, b) = f(x), \quad 0 < x < a \end{cases}$ $\qquad$ **(12)**

are called **boundary-value problems.**

Modifications   The partial differential equations (1), (2), and (3) must be modified to take into consideration internal or external influences acting on the physical system. More general forms of the one-dimensional heat and wave equations are, respectively,

$$k\frac{\partial^2 u}{\partial x^2} + G(x, t, u, u_x) = \frac{\partial u}{\partial t} \qquad \textbf{(13)}$$

and $\qquad\qquad$ $a^2 \dfrac{\partial^2 u}{\partial x^2} + F(x, t, u, u_t) = \dfrac{\partial^2 u}{\partial t^2}.$ $\qquad$ **(14)**

For example, if there is heat transfer from the lateral surface of a rod into a surrounding medium that is held at a constant temperature u_m, then the heat equation (13) is

$$k\frac{\partial^2 u}{\partial x^2} - h(u - u_m) = \frac{\partial u}{\partial t}.$$

In (14) the function F could represent the various forces acting on the string. For example, when external, damping, and elastic restoring forces are taken into account, (14) assumes the form

$$a^2\frac{\partial^2 u}{\partial x^2} + f(x, t) = \frac{\partial^2 u}{\partial t^2} + c\frac{\partial u}{\partial t} + ku. \qquad \textbf{(15)}$$

↑ external force   ↑ damping force   ↑ restoring force

Remarks The analysis of a wide variety of diverse phenomena yields mathematical models (1), (2), or (3) or their generalizations involving a greater number of spatial variables. For example, (1) is sometimes called the **diffusion equation** since the diffusion of dissolved substances in solution is analogous to the flow of heat in a solid. The function $u(x, t)$ satisfying the partial differential equation in this case represents the concentration of the dissolved substance. Similarly, equation (1) arises in the study of the flow of electricity in a long cable or transmission line. In this setting (1) is known as the **telegraph equation.** It can be shown that under certain assumptions the current and the voltage in the line are functions satisfying two equations identical with (1). The wave equation (2) also appears in the theory of high-frequency transmission lines, fluid mechanics, acoustics, and elasticity. Laplace's equation (3) is encountered in the static displacement of membranes.

EXERCISES 12.2

Answers to odd-numbered problems begin on page AN-18.

In Problems 1–4 a rod of length L coincides with the interval $[0, L]$ on the x-axis. Set up the boundary-value problem for the temperature $u(x, t)$.

1. The left end is held at temperature zero, and the right end is insulated. The initial temperature is $f(x)$ throughout.

2. The left end is held at temperature u_0, and the right end is held at temperature u_1. The initial temperature is zero throughout.

3. The left end is held at temperature 100, and there is heat transfer from the right end into the surrounding medium at temperature zero. The initial temperature is $f(x)$ throughout.

4. The ends are insulated, and there is heat transfer from the lateral surface into the surrounding medium at temperature 50. The initial temperature is 100 throughout.

In Problems 5–8 a string of length L coincides with the interval $[0, L]$ on the x-axis. Set up the boundary-value problem for the displacement $u(x, t)$.

5. The ends are secured to the x-axis. The string is released from rest from the initial displacement $x(L - x)$.

6. The ends are secured to the x-axis. Initially the string is undisplaced but has the initial velocity $\sin(\pi x/L)$.

7. The left end is secured to the x-axis, but the right end moves in a transverse manner according to $\sin \pi t$. The string is released from rest from the initial displacement $f(x)$. For $t > 0$ the transverse vibrations are damped with a force proportional to the instantaneous velocity.

8. The ends are secured to the x-axis, and the string is initially at rest on that axis. An external vertical force proportional to the horizontal distance from the left end acts on the string for $t > 0$.

In Problems 9 and 10 set up the boundary-value problem for the steady-state temperature $u(x, y)$.

9. A thin rectangular plate coincides with the region defined by $0 \leq x \leq 4$, $0 \leq y \leq 2$. The left end and the bottom of the plate are insulated. The top of the plate is held at temperature zero, and the right end of the plate is held at temperature $f(y)$.

10. A semi-infinite plate coincides with the region defined by $0 \leq x \leq \pi$, $y \geq 0$. The left end is held at temperature e^{-y}, and the right end is held at temperature 100 for $0 < y \leq 1$ and temperature zero for $y > 1$. The bottom of the plate is held at temperature $f(x)$.

12.3 HEAT EQUATION

- *Solution of a boundary-value problem by separation of variables*
- *Eigenvalues*   • *Eigenfunctions*

Consider a thin rod of length L with an initial temperature $f(x)$ throughout and whose ends are held at temperature zero for all time $t > 0$. If the rod shown in Figure 12.5 satisfies the assumptions given on pages 527–528, then the temperature $u(x, t)$ in the rod is determined from the boundary-value problem

$$k\frac{\partial^2 u}{\partial x^2} = \frac{\partial u}{\partial t}, \quad 0 < x < L, \quad t > 0 \tag{1}$$

$$u(0, t) = 0, \quad u(L, t) = 0, \quad t > 0 \tag{2}$$

$$u(x, 0) = f(x), \quad 0 < x < L. \tag{3}$$

Solution of the BVP   Using the product $u(x, t) = X(x)T(t)$ and $-\lambda^2$ as a separation constant leads to

$$\frac{X''}{X} = \frac{T'}{kT} = -\lambda^2 \tag{4}$$

$u = 0$ $u = 0$

0 L x

Figure 12.5

and
$$X'' + \lambda^2 X = 0 \tag{5}$$
$$T' + k\lambda^2 T = 0$$
$$X = c_1 \cos \lambda x + c_2 \sin \lambda x \tag{6}$$
$$T = c_3 e^{-k\lambda^2 t}. \tag{7}$$

From $u(0, t) = X(0)T(t) = 0$ and $u(L, t) = X(L)T(t) = 0$

we must have $X(0) = 0$ and $X(L) = 0$. These homogeneous boundary conditions together with the homogeneous equation (5) constitute a regular Sturm-Liouville problem. Applying the first of these conditions in (6) immediately gives $c_1 = 0$. Therefore

$$X = c_2 \sin \lambda x.$$

The second boundary condition now implies $X(L) = c_2 \sin \lambda L = 0$. If $c_2 = 0$ then $X = 0$, so $u = 0$. To obtain a nontrivial solution u we must have $c_2 \neq 0$, and so the last equation is satisfied when

$$\sin \lambda L = 0.$$

This implies that $\lambda L = n\pi$ or $\lambda = n\pi/L$, where $n = 1, 2, 3, \ldots$. The values

$$\lambda = \frac{n\pi}{L}, \quad n = 1, 2, 3, \ldots \tag{8}$$

and the corresponding solutions

$$X = c_2 \sin \frac{n\pi}{L} x, \quad n = 1, 2, 3, \ldots \tag{9}$$

are the **eigenvalues** and **eigenfunctions,** respectively, of the problem. From (7) we have $T = c_3 e^{-k(n^2\pi^2/L^2)t}$, and so

$$u_n = XT = A_n e^{-k(n^2\pi^2/L^2)t} \sin \frac{n\pi}{L} x, \tag{10}$$

where we have replaced the constant $c_2 c_3$ by A_n. The products $u_n(x, t)$ satisfy the partial differential equation (1) and the boundary conditions (2) for each value of the positive integer n.* However, in order for the functions given in (10) to satisfy the initial condition (3), we would have to choose the coefficient A_n in such a manner that

$$u_n(x, 0) = f(x) = A_n \sin \frac{n\pi}{L} x. \tag{11}$$

In general, we would not expect condition (11) to be satisfied for an arbitrary, but reasonable, choice of f. Therefore we are forced to admit that $u_n(x, t)$ is *not a solution of the given problem.* Now by the superposition principle the function

$$u(x, t) = \sum_{n=1}^{\infty} u_n = \sum_{n=1}^{\infty} A_n e^{-k(n^2\pi^2/L^2)t} \sin \frac{n\pi}{L} x \tag{12}$$

---

*You are urged to verify that if the separation constant is chosen as either $\lambda^2 = 0$ or $\lambda^2 > 0$, then the only solution of (1) satisfying (2) is $u = 0$.

must also, though formally, satisfy (1) and (2). Substituting $t = 0$ into (12) implies

$$u(x, 0) = f(x) = \sum_{n=1}^{\infty} A_n \sin \frac{n\pi}{L} x.$$

This last expression is recognized as the half-range expansion of f in a sine series. Thus if we make the identification $A_n = b_n$, $n = 1, 2, 3, \ldots$, it follows from (5) of Section 11.3 that

$$A_n = \frac{2}{L} \int_0^L f(x) \sin \frac{n\pi}{L} x \, dx.$$

We conclude that a solution of the boundary-value problem described in (1), (2), and (3) is given by the infinite series

$$u(x, t) = \frac{2}{L} \sum_{n=1}^{\infty} \left(\int_0^L f(x) \sin \frac{n\pi}{L} x \, dx \right) e^{-k(n^2\pi^2/L^2)t} \sin \frac{n\pi}{L} x.$$

In the special case when $u(x, 0) = 100$, $L = \pi$, and $k = 1$, you should verify that the coefficients A_n are given by

$$A_n = \frac{200}{\pi} \left[\frac{1 - (-1)^n}{n} \right],$$

and so

$$u(x, t) = \frac{200}{\pi} \sum_{n=1}^{\infty} \left[\frac{1 - (-1)^n}{n} \right] e^{-n^2 t} \sin nx. \tag{13}$$

The graph of the solution (13) is a surface in 3-space. We could use the 3D-plot application of a computer algebra system to approximate this surface by graphing partial sums $S_n(x, t)$ over a rectangular region defined by $0 \le x \le \pi$, $0 \le t \le T$. Alternatively, with the aid of the 2D-plot application of a CAS we can plot the solution $u(x, t)$ on the x-interval $[0, \pi]$ for increasing values of time t. See Figure 12.6(a). In Figure 12.6(b) the solution $u(x, t)$ is graphed on the t-interval $[0, 6]$ for increasing values of x ($x = 0$ is the left end and $x = \pi/2$ is the midpoint of the rod of length $L = \pi$). Both sets of graphs verify that which is apparent in (13)—namely, $u(x, t) \to 0$ as $t \to \infty$.

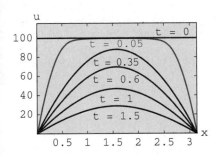

(a) $u(x, t)$ graphed as a function of x for various fixed times

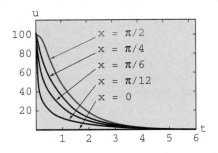

(b) $u(x, t)$ graphed as a function of t for various fixed positions

Figure 12.6

EXERCISES 12.3

Answers to odd-numbered problems begin on page AN-18.

In Problems 1 and 2 solve the heat equation (1) subject to the given conditions. Assume a rod of length L.

1. $u(0, t) = 0$, $u(L, t) = 0$

$$u(x, 0) = \begin{cases} 1, & 0 < x < L/2 \\ 0, & L/2 < x < L \end{cases}$$

2. $u(0, t) = 0$, $u(L, t) = 0$

$$u(x, 0) = x(L - x)$$

3. Find the temperature $u(x, t)$ in a rod of length L if the initial temperature is $f(x)$ throughout and if the ends $x = 0$ and $x = L$ are insulated.

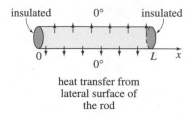

heat transfer from
lateral surface of
the rod

Figure 12.7

4. Solve Problem 3 if $L = 2$ and $f(x) = \begin{cases} x, & 0 < x < 1 \\ 0, & 1 < x < 2. \end{cases}$

5. Suppose heat is lost from the lateral surface of a thin rod of length L into a surrounding medium at temperature zero. If the linear law of heat transfer applies, then the heat equation takes on the form $k\partial^2 u/\partial x^2 - hu = \partial u/\partial t, \ 0 < x < L, \ t > 0, \ h$ a constant. Find the temperature $u(x, t)$ if the initial temperature is $f(x)$ throughout and the ends $x = 0$ and $x = L$ are insulated. See Figure 12.7.

6. Solve Problem 5 if the ends $x = 0$ and $x = L$ are held at temperature zero.

Computer Lab Assignments

7. (a) Solve the heat equation (1) subject to

$$u(0, t) = 0, \quad u(100, t) = 0, \quad t > 0$$

$$u(x, 0) = \begin{cases} 0.8x, & 0 \le x \le 50 \\ 0.8(100 - x), & 50 < x \le 100. \end{cases}$$

(b) Use the 3D-plot application of your CAS to graph the partial sum $S_5(x, t)$ consisting of the first five nonzero terms of the solution in part (a) for $0 \le x \le 100, 0 \le t \le 200$. Assume that $k = 1.6352$. Experiment with various three-dimensional viewing perspectives of the surface (called the **ViewPoint** option in *Mathematica*).

Discussion Problems

8. In Figure 12.6(b) we have the graphs of $u(x, t)$ on the interval $0 \le t \le 6$ for $x = 0, x = \pi/12, x = \pi/6, x = \pi/4$, and $x = \pi/2$. Describe or sketch the graphs of $u(x, t)$ on the same time interval but for the fixed values $x = 3\pi/4, x = 5\pi/6, x = 11\pi/12$, and $x = \pi$.

12.4 WAVE EQUATION

• Solution of a boundary-value problem by separation of variables • Standing waves • Normal modes • First normal mode • Fundamental frequency • Overtones

We are now in a position to solve the boundary-value problem (11) discussed in Section 12.2. The vertical displacement $u(x, t)$ of the vibrating string of length L shown in Figure 12.2(a) is determined from

$$a^2 \frac{\partial^2 u}{\partial x^2} = \frac{\partial^2 u}{\partial t^2}, \quad 0 < x < L, \quad t > 0 \tag{1}$$

$$u(0, t) = 0, \quad u(L, t) = 0, \quad t > 0 \tag{2}$$

$$u(x, 0) = f(x), \quad \left.\frac{\partial u}{\partial t}\right|_{t=0} = g(x). \tag{3}$$

Solution of the BVP   With the usual assumption that $u(x, t) = X(x)T(t)$, separating variables in (1) gives

$$\frac{X''}{X} = \frac{T''}{a^2 T} = -\lambda^2$$

so that

$$X'' + \lambda^2 X = 0 \quad \text{and} \quad T'' + \lambda^2 a^2 T = 0$$

and therefore

$$X = c_1 \cos \lambda x + c_2 \sin \lambda x$$
$$T = c_3 \cos \lambda a t + c_4 \sin \lambda a t.$$

As before, the boundary conditions (2) translate into $X(0) = 0$ and $X(L) = 0$. In turn we find

$$c_1 = 0 \quad \text{and} \quad c_2 \sin \lambda L = 0.$$

This last equation yields the eigenvalues $\lambda = n\pi/L$, where $n = 1, 2, 3, \ldots$. The corresponding eigenfunctions are

$$X = c_2 \sin \frac{n\pi}{L} x, \quad n = 1, 2, 3, \ldots.$$

Thus solutions of equation (1) satisfying the boundary conditions (2) are

$$u_n = \left(A_n \cos \frac{n\pi a}{L} t + B_n \sin \frac{n\pi a}{L} t \right) \sin \frac{n\pi}{L} x \qquad \textbf{(4)}$$

and

$$u(x, t) = \sum_{n=1}^{\infty} \left(A_n \cos \frac{n\pi a}{L} t + B_n \sin \frac{n\pi a}{L} t \right) \sin \frac{n\pi}{L} x. \qquad \textbf{(5)}$$

Setting $t = 0$ in (5) gives

$$u(x, 0) = f(x) = \sum_{n=1}^{\infty} A_n \sin \frac{n\pi}{L} x,$$

which is a half-range expansion for f in a sine series. As in the discussion of the heat equation, we can write $A_n = b_n$:

$$A_n = \frac{2}{L} \int_0^L f(x) \sin \frac{n\pi}{L} x \, dx. \qquad \textbf{(6)}$$

To determine B_n we differentiate (5) with respect to t and then set $t = 0$:

$$\frac{\partial u}{\partial t} = \sum_{n=1}^{\infty} \left(-A_n \frac{n\pi a}{L} \sin \frac{n\pi a}{L} t + B_n \frac{n\pi a}{L} \cos \frac{n\pi a}{L} t \right) \sin \frac{n\pi}{L} x$$

$$\left. \frac{\partial u}{\partial t} \right|_{t=0} = g(x) = \sum_{n=1}^{\infty} \left(B_n \frac{n\pi a}{L} \right) \sin \frac{n\pi}{L} x.$$

In order for this last series to be the half-range sine expansion of g on the interval, the *total* coefficient $B_n n\pi a/L$ must be given by the form of (5) in Section 11.3—that is,

$$B_n \frac{n\pi a}{L} = \frac{2}{L} \int_0^L g(x) \sin \frac{n\pi}{L} x \, dx,$$

from which we obtain

$$B_n = \frac{2}{n\pi a} \int_0^L g(x) \sin \frac{n\pi}{L} x \, dx. \tag{7}$$

The solution of the problem consists of the series (5) with A_n and B_n defined by (6) and (7), respectively.

We note that when the string is released from *rest*, then $g(x) = 0$ for every x in $0 \le x \le L$ and consequently $B_n = 0$.

Standing Waves   Recall from the derivation of the wave equation in Section 12.2 that the constant a appearing in the solution of the boundary-value problem in (1), (2), and (3) is given by $\sqrt{T/\rho}$, where ρ is mass per unit length and T is the magnitude of the tension in the string. When T is large enough, the vibrating string produces a musical sound. This sound is the result of standing waves. The solution (5) is a superposition of product solutions called **standing waves** or **normal modes**:

$$u(x, t) = u_1(x, t) + u_2(x, t) + u_3(x, t) + \cdots.$$

In view of (6) and (7) of Section 5.1, the product solutions (4) can be written as

$$u_n(x, t) = C_n \sin\left(\frac{n\pi a}{L} t + \phi_n\right) \sin \frac{n\pi}{L} x, \tag{8}$$

where $C_n = \sqrt{A_n^2 + B_n^2}$ and ϕ_n is defined by $\sin \phi_n = A_n/C_n$ and $\cos \phi_n = B_n/C_n$. For $n = 1, 2, 3, \ldots$ the standing waves are essentially the graphs of $\sin(n\pi x/L)$, with a time-varying amplitude given by

$$C_n \sin\left(\frac{n\pi a}{L} t + \phi_n\right).$$

(a) first standing wave

Alternatively, we see from (8) that at a fixed value of x each product function $u_n(x, t)$ represents simple harmonic motion with amplitude $C_n|\sin(n\pi x/L)|$ and frequency $f_n = na/2L$. In other words, each point on a standing wave vibrates with a different amplitude but with the same frequency. When $n = 1$,

$$u_1(x, t) = C_1 \sin\left(\frac{\pi a}{L} t + \phi_1\right) \sin \frac{\pi}{L} x$$

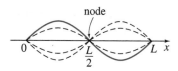

(b) second standing wave

is called the **first standing wave**, the **first normal mode**, or the **fundamental mode of vibration**. The first three standing waves, or normal modes, are shown in Figure 12.8. The dashed graphs represent the standing waves at various values of time. The points in the interval $(0, L)$, for which $\sin(n\pi/L)x = 0$, correspond to points on a standing wave where there is no motion. These points are called **nodes**. For example, in Figures 12.8(b) and (c) we see that the second standing wave has one node at $L/2$ and the third standing wave has two nodes at $L/3$ and $2L/3$. In general, the nth normal mode of vibration has $n - 1$ nodes.

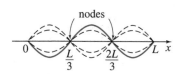

(c) third standing wave

Figure 12.8

The frequency

$$f_1 = \frac{a}{2L} = \frac{1}{2L}\sqrt{\frac{T}{\rho}}$$

of the first normal mode is called the **fundamental frequency** or **first harmonic** and is directly related to the pitch produced by a stringed instrument. It is apparent that the greater the tension on the string, the higher the pitch of the sound. The frequencies f_n of the other normal modes, which are integer multiples of the fundamental frequency, are called **overtones**. The second harmonic is the first overtone, and so on.

EXERCISES 12.4

Answers to odd-numbered problems begin on page AN-18.

In Problems 1–8 solve the wave equation (1) subject to the given conditions.

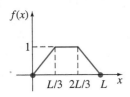

$f(x)$

Figure 12.9

1. $u(0, t) = 0, \quad u(L, t) = 0$

$u(x, 0) = \frac{1}{4}x(L - x), \quad \left.\dfrac{\partial u}{\partial t}\right|_{t=0} = 0$

2. $u(0, t) = 0, \quad u(L, t) = 0$

$u(x, 0) = 0, \quad \left.\dfrac{\partial u}{\partial t}\right|_{t=0} = x(L - x)$

3. $u(0, t) = 0, \quad u(L, t) = 0$
$u(x, 0),$ as specified in

Figure 12.9, $\left.\dfrac{\partial u}{\partial t}\right|_{t=0} = 0$

4. $u(0, t) = 0, \quad u(\pi, t) = 0$
$u(x, 0) = \frac{1}{6}x(\pi^2 - x^2),$

$\left.\dfrac{\partial u}{\partial t}\right|_{t=0} = 0$

5. $u(0, t) = 0, \quad u(\pi, t) = 0$

$u(x, 0) = 0, \quad \left.\dfrac{\partial u}{\partial t}\right|_{t=0} = \sin x$

6. $u(0, t) = 0, \quad u(1, t) = 0$
$u(x, 0) = 0.01 \sin 3\pi x,$

$\left.\dfrac{\partial u}{\partial t}\right|_{t=0} = 0$

7. $u(0, t) = 0, \quad u(L, t) = 0$

$u(x, 0) = \begin{cases} \dfrac{2hx}{L}, & 0 < x < \dfrac{L}{2} \\ 2h\left(1 - \dfrac{x}{L}\right), & \dfrac{L}{2} \le x < L \end{cases}, \quad \left.\dfrac{\partial u}{\partial t}\right|_{t=0} = 0$

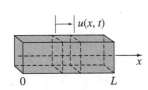

$u(x, t)$

Figure 12.10

8. $\left.\dfrac{\partial u}{\partial x}\right|_{x=0} = 0, \quad \left.\dfrac{\partial u}{\partial x}\right|_{x=L} = 0$

$u(x, 0) = x, \quad \left.\dfrac{\partial u}{\partial t}\right|_{t=0} = 0$

This problem could describe the longitudinal displacement $u(x, t)$ of a vibrating elastic bar. The boundary conditions at $x = 0$ and $x = L$ are called **free-end conditions**. See Figure 12.10.

9. A string is stretched and secured on the x-axis at $x = 0$ and $x = \pi$ for $t > 0$. If the transverse vibrations take place in a medium that imparts a resistance proportional to the instantaneous velocity, then the wave equation takes on the form

$$\frac{\partial^2 u}{\partial x^2} = \frac{\partial^2 u}{\partial t^2} + 2\beta \frac{\partial u}{\partial t}, \quad 0 < \beta < 1, \quad t > 0.$$

Find the displacement $u(x, t)$ if the string starts from rest from the initial displacement $f(x)$.

10. Show that a solution of the boundary-value problem

$$\frac{\partial^2 u}{\partial x^2} = \frac{\partial^2 u}{\partial t^2} + u, \quad 0 < x < \pi, \quad t > 0$$

$$u(0, t) = 0, \quad u(\pi, t) = 0, \quad t > 0$$

$$u(x, 0) = \begin{cases} x, & 0 < x < \pi/2 \\ \pi - x, & \pi/2 \le x < \pi \end{cases}$$

$$\frac{\partial u}{\partial t}\bigg|_{t=0} = 0, \quad 0 < x < \pi$$

is
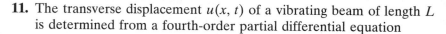
$$u(x, t) = \frac{4}{\pi} \sum_{k=1}^{\infty} \frac{(-1)^{k+1}}{(2k-1)^2} \sin(2k-1)x \cos \sqrt{(2k-1)^2 + 1}\, t.$$

11. The transverse displacement $u(x, t)$ of a vibrating beam of length L is determined from a fourth-order partial differential equation

$$a^2 \frac{\partial^4 u}{\partial x^4} + \frac{\partial^2 u}{\partial t^2} = 0, \quad 0 < x < L, \quad t > 0.$$

Figure 12.11 Simply supported beam

If the beam is **simply supported,** as shown in Figure 12.11, the boundary and initial conditions are

$$u(0, t) = 0, \quad u(L, t) = 0, \quad t > 0$$

$$\frac{\partial^2 u}{\partial x^2}\bigg|_{x=0} = 0, \quad \frac{\partial^2 u}{\partial x^2}\bigg|_{x=L} = 0, \quad t > 0$$

$$u(x, 0) = f(x), \quad \frac{\partial u}{\partial t}\bigg|_{t=0} = g(x), \quad 0 < x < L.$$

Solve for $u(x, t)$. [*Hint*: For convenience use λ^4 instead of λ^2 when separating variables.]

12. If the ends of the beam in Problem 11 are **embedded** at $x = 0$ and $x = L$, the boundary conditions become, for $t > 0$,

$$u(0, t) = 0, \quad u(L, t) = 0$$

$$\frac{\partial u}{\partial x}\bigg|_{x=0} = 0, \quad \frac{\partial u}{\partial x}\bigg|_{x=L} = 0.$$

(a) Show that the eigenvalues of the problem are $\lambda = x_n/L$, where x_n, $n = 1, 2, 3, \ldots$ are the positive roots of the equation $\cosh x \cos x = 1$.

(b) Show graphically that the equation in part (a) has an infinite number of roots.

(c) Use a calculator or a computer to find approximations to the first four eigenvalues. Use four decimal places.

13. Consider the boundary-value problem given in (1), (2), and (3) of this section. If $g(x) = 0$ on $0 < x < L$, show that the solution of the problem can be written as

$$u(x, t) = \frac{1}{2}[f(x + at) + f(x - at)].$$

[*Hint*: Use the identity $2 \sin \theta_1 \cos \theta_2 = \sin(\theta_1 + \theta_2) + \sin(\theta_1 - \theta_2)$.]

14. The vertical displacement $u(x, t)$ of an infinitely long string is determined from the initial-value problem

$$a^2 \frac{\partial^2 u}{\partial x^2} = \frac{\partial^2 u}{\partial t^2}, \quad -\infty < x < \infty, \quad t > 0$$

$$u(x, 0) = f(x), \quad \left.\frac{\partial u}{\partial t}\right|_{t=0} = g(x). \tag{9}$$

This problem can be solved without separating variables.
(a) Show that the wave equation can be put into the form $\partial^2 u / \partial \eta \partial \xi = 0$ by means of the substitutions $\xi = x + at$ and $\eta = x - at$.
(b) Integrate the partial differential equation in part (a), first with respect to η and then with respect to ξ, to show that $u(x, t) = F(x + at) + G(x - at)$, where F and G are arbitrary twice differentiable functions, is a solution of the wave equation. Use this solution and the given initial conditions to show that

$$F(x) = \frac{1}{2}f(x) + \frac{1}{2a}\int_{x_0}^{x} g(s)\, ds + c$$

and

$$G(x) = \frac{1}{2}f(x) - \frac{1}{2a}\int_{x_0}^{x} g(s)\, ds - c,$$

where x_0 is arbitrary and c is a constant of integration.
(c) Use the results in part (b) to show that

$$u(x, t) = \frac{1}{2}[f(x + at) + f(x - at)] + \frac{1}{2a}\int_{x-at}^{x+at} g(s)\, ds. \tag{10}$$

Note that when the initial velocity $g(x) = 0$ we obtain

$$u(x, t) = \frac{1}{2}[f(x + at) + f(x - at)], \quad -\infty < x < \infty.$$

The last solution can be interpreted as a superposition of two **traveling waves**, one moving to the right (that is, $\frac{1}{2}f(x - at)$) and one moving to the left ($\frac{1}{2}f(x + at)$). Both waves travel with speed a and have the same basic shape as the initial displacement $f(x)$. The form of $u(x, t)$ given in (10) is called **d'Alembert's solution**.

In Problems 15–17 use d'Alembert's solution (10) to solve the initial-value problem in Problem 14 subject to the given initial conditions.

15. $f(x) = \sin x, \quad g(x) = 1$
16. $f(x) = \sin x, \quad g(x) = \cos x$
17. $f(x) = 0, \quad g(x) = \sin 2x$
18. Suppose $f(x) = 1/(1 + x^2)$, $g(x) = 0$, and $a = 1$ for the initial-value problem stated in Problem 14. Graph d'Alembert's solution in this case at the times $t = 0$, $t = 1$, and $t = 3$.

Computer Lab Assignments

19. A model for an infinitely long string that is initially held at the three points $(-1, 0)$, $(1, 0)$, and $(0, 1)$ and then simultaneously released at all three points at time $t = 0$ is given by (9) with

$$f(x) = \begin{cases} 1 - |x|, & |x| \le 1 \\ 0, & |x| > 1 \end{cases} \quad \text{and} \quad g(x) = 0.$$

(a) Plot the initial position of the string on the interval $[-6, 6]$.

(b) Use a CAS to plot d'Alembert's solution (10) on $[-6, 6]$ for $t = 0.2k$, $k = 0, 1, 2, \ldots, 25$.

(c) Use the animation feature of your computer algebra system to make a movie of the solution. Describe the motion of the string over time.

20. An infinitely long string coinciding with the x-axis is struck at the origin with a hammer whose head is 0.2 inch in diameter. A model for the motion of the string is given by (9) with

$$f(x) = 0 \quad \text{and} \quad g(x) = \begin{cases} 1, & |x| \leq 0.1 \\ 0, & |x| > 0.1 \end{cases}$$

(a) Use a CAS to plot d'Alembert's solution (10) on $[-6, 6]$ for $t = 0.2k$, $k = 0, 1, 2, \ldots, 25$.

(b) Use the animation feature of your computer algebra system to make a movie of the solution. Describe the motion of the string over time.

21. The model of the vibrating string in Problem 7 is called the **plucked string.** The string is tied to the x-axis at $x = 0$ and $x = L$ and is held at $x = L/2$ at h units above the x-axis. See Figure 12.4. Starting at $t = 0$ the string is released from rest.

(a) Use a CAS to plot the partial sum $S_6(x, t)$—that is, the first six nonzero terms of your solution—for $t = 0.1k$, $k = 0, 1, 2, \ldots, 20$. Assume that $a = 1$, $h = 1$, and $L = \pi$.

(b) Use the animation feature of your computer algebra system to make a movie of the solution to Problem 7.

12.5 LAPLACE'S EQUATION

• *Solution of a boundary-value problem by separation of variables* • *Dirichlet problem* • *Superposition principle*

Suppose we wish to find the steady-state temperature $u(x, y)$ in a rectangular plate whose vertical edges are insulated, as shown in Figure 12.12. When no heat escapes from the lateral faces of the plate, we solve the following boundary-value problem:

$$\frac{\partial^2 u}{\partial x^2} + \frac{\partial^2 u}{\partial y^2} = 0, \quad 0 < x < a, \quad 0 < y < b \tag{1}$$

$$\left.\frac{\partial u}{\partial x}\right|_{x=0} = 0, \quad \left.\frac{\partial u}{\partial x}\right|_{x=a} = 0, \quad 0 < y < b \tag{2}$$

$$u(x, 0) = 0, \quad u(x, b) = f(x), \quad 0 < x < a. \tag{3}$$

Solution of the BVP With $u(x, y) = X(x)Y(y)$ separation of variables in (1) leads to

$$\frac{X''}{X} = -\frac{Y''}{Y} = -\lambda^2$$

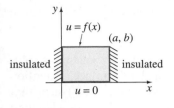

Figure 12.12

$$X'' + \lambda^2 X = 0 \tag{4}$$

$$Y'' - \lambda^2 Y = 0 \tag{5}$$

$$X = c_1 \cos \lambda x + c_2 \sin \lambda x, \tag{6}$$

and since $0 < y < b$ is a finite interval, we use the solution

$$Y = c_3 \cosh \lambda y + c_4 \sinh \lambda y. \tag{7}$$

Now the first three boundary conditions translate into $X'(0) = 0$, $X'(a) = 0$, and $Y(0) = 0$. Differentiating X and setting $x = 0$ implies $c_2 = 0$ and, therefore, $X = c_1 \cos \lambda x$. Differentiating this last expression and then setting $x = a$ gives $-c_1 \lambda \sin \lambda a = 0$. This last condition is satisfied when $\lambda = 0$ or when $\lambda a = n\pi$ or $\lambda = n\pi/a$, $n = 1, 2, \ldots$ Observe that $\lambda = 0$ implies that (4) is $X'' = 0$. The general solution of this equation is given by the linear function $X = c_1 + c_2 x$ and *not* by (6). In this case the boundary conditions $X'(0) = 0$, $X'(a) = 0$ demand that $X = c_1$. In this example, unlike the previous two examples, we are forced to conclude that $\lambda = 0$ is an eigenvalue. Corresponding $\lambda = 0$ with $n = 0$, we get the eigenfunctions

$$X = c_1, \quad n = 0 \quad \text{and} \quad X = c_1 \cos \frac{n\pi}{a} x, \quad n = 1, 2, \ldots.$$

Finally, the condition $Y(0) = 0$ dictates that $c_3 = 0$ in (7) when $\lambda > 0$. However, when $\lambda = 0$, equation (5) becomes $Y'' = 0$, and thus the solution is given by $Y = c_3 + c_4 y$ rather than by (7). But $Y(0) = 0$ implies again that $c_3 = 0$, and so $Y = c_4 y$. Thus product solutions of the equation satisfying the first three boundary conditions are

$$A_0 y, \quad n = 0 \quad \text{and} \quad A_n \sinh \frac{n\pi}{a} y \cos \frac{n\pi}{a} x, \quad n = 1, 2, \ldots.$$

The superposition principle yields another solution

$$u(x, y) = A_0 y + \sum_{n=1}^{\infty} A_n \sinh \frac{n\pi}{a} y \cos \frac{n\pi}{a} x. \tag{8}$$

Substituting $y = b$ in (8) gives

$$u(x, b) = f(x) = A_0 b + \sum_{n=1}^{\infty} \left(A_n \sinh \frac{n\pi}{a} b \right) \cos \frac{n\pi}{a} x,$$

which in this case is a half-range expansion of f in a cosine series. If we make the identifications $A_0 b = a_0/2$ and $A_n \sinh(n\pi b/a) = a_n$, $n = 1, 2, 3, \ldots$, it follows from (2) and (3) of Section 11.3 that

$$2A_0 b = \frac{2}{a} \int_0^a f(x) \, dx$$

$$A_0 = \frac{1}{ab} \int_0^a f(x) \, dx \tag{9}$$

and

$$A_n \sinh \frac{n\pi}{a} b = \frac{2}{a} \int_0^a f(x) \cos \frac{n\pi}{a} x \, dx$$

$$A_n = \frac{2}{a \sinh \frac{n\pi}{a} b} \int_0^a f(x) \cos \frac{n\pi}{a} x \, dx. \tag{10}$$

The solution of this problem consists of the series given in (8), where A_0 and A_n are defined by (9) and (10), respectively.

Dirichlet Problem A boundary-value problem in which we seek a solution to an elliptic partial differential equation, such as Laplace's equation $\nabla^2 u = 0$, within a bounded region R (in the plane or 3-space) such that u takes on prescribed values on the entire boundary of the region is called a **Dirichlet problem**. In Problem 1 in Exercises 12.5 you are asked to show that the solution of the Dirichlet problem for a rectangular region

$$\frac{\partial^2 u}{\partial x^2} + \frac{\partial^2 u}{\partial y^2} = 0, \quad 0 < x < a, \quad 0 < y < b$$

$$u(0, y) = 0, \quad u(a, y) = 0, \qquad 0 < y < b$$

$$u(x, 0) = 0, \quad u(x, b) = f(x), \quad 0 < x < a$$

is

$$u(x, y) = \sum_{n=1}^{\infty} A_n \sinh \frac{n\pi}{a} y \sin \frac{n\pi}{a} x, \text{ where } A_n = \frac{2}{a \sinh \frac{n\pi b}{a}} \int_0^a f(x) \sin \frac{n\pi}{a} x \, dx.$$

In the special case when $f(x) = 100$, $a = 1$, and $b = 1$, the coefficients A_n are given by $A_n = 200 \frac{1 - (-1)^n}{n\pi \sinh n\pi}$. With the help of a CAS we plotted the surface defined by $u(x, y)$ over the region $R: 0 \leq x \leq 1, 0 \leq y \leq 1$ in Figure 12.13(a). You can see in the figure that the boundary conditions are satisfied; especially note that along $y = 1$, $u = 100$ for $0 \leq x \leq 1$. The **isotherms,** or curves in the rectangular region along which the temperature $u(x, y)$ is constant, can be obtained using the contour-plot application of a CAS and are illustrated in Figure 12.13(b). The isotherms can also be visualized as the curves of intersection (projected into the xy-plane) of the horizontal planes $u = 80$, $u = 60$, and so on, with the surface in Figure 12.13(a). Notice that throughout the region the maximum temperature is $u = 100$ and occurs on the portion of the boundary corresponding to $y = 1$. This is no coincidence. There is a **maximum principle** that states that a solution u of Laplace's equation within a bounded region R with boundary B (such as a rectangle, circle, sphere, and so on) takes on its maximum and minimum values on B. In addition, it can be proved that u can have no relative extrema (maxima or minima) in the interior of R. This last statement is clearly borne out by the surface shown in Figure 12.13(a).

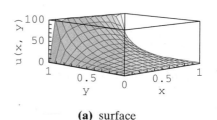

(a) surface

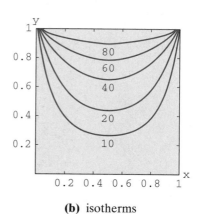

(b) isotherms

Figure 12.13

Superposition Principle A Dirichlet problem for a rectangle can readily be solved by separation of variables when homogeneous boundary conditions are specified on two *parallel* boundaries. However, the method of separation of variables is not applicable to a Dirichlet problem when the boundary conditions on all four sides of the rectangle are nonhomogeneous. To get around this difficulty we break the problem

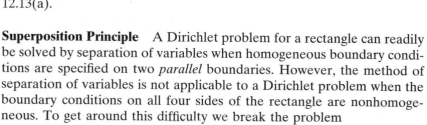

$$\frac{\partial^2 u}{\partial x^2} + \frac{\partial^2 u}{\partial y^2} = 0, \quad 0 < x < a, \quad 0 < y < b$$

$$u(0, y) = F(y), \quad u(a, y) = G(y), \quad 0 < y < b \tag{11}$$

$$u(x, 0) = f(x), \quad u(x, b) = g(x), \quad 0 < x < a$$

into two problems, each of which has homogeneous boundary conditions on parallel boundaries, as shown:

Problem 1

$$\frac{\partial^2 u_1}{\partial x^2} + \frac{\partial^2 u_1}{\partial y^2} = 0, \quad 0 < x < a, \quad 0 < y < b$$

$u_1(0, y) = 0, \qquad u_1(a, y) = 0, \quad 0 < y < b$

$u_1(x, 0) = f(x), \qquad u_1(x, b) = g(x), \quad 0 < x < a$

Problem 2

$$\frac{\partial^2 u_2}{\partial x^2} + \frac{\partial^2 u_2}{\partial y^2} = 0, \quad 0 < x < a, \quad 0 < y < b$$

$u_2(0, y) = F(y), \quad u_2(a, y) = G(y), \quad 0 < y < b$

$u_2(x, 0) = 0, \qquad u_2(x, b) = 0, \quad 0 < x < a$

Suppose u_1 and u_2 are the solutions of Problems 1 and 2, respectively. If we define $u(x, y) = u_1(x, y) + u_2(x, y)$, it is seen that u satisfies all boundary conditions in the original problem (11). For example,

$$u(0, y) = u_1(0, y) + u_2(0, y) = 0 + F(y) = F(y)$$
$$u(x, b) = u_1(x, b) + u_2(x, b) = g(x) + 0 = g(x)$$

and so on. Furthermore, u is a solution of Laplace's equation by Theorem 12.1. In other words, by solving Problems 1 and 2 and adding their solutions we have solved the original problem. This additive property of solutions is known as the superposition principle. See Figure 12.14.

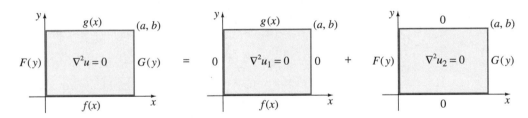

Figure 12.14 Solution u = Solution u_1 of Problem 1 + Solution u_2 of Problem 2

We leave as exercises (see Problems 13 and 14 in Exercises 12.5) to show that a solution of Problem 1 is

$$u_1(x, y) = \sum_{n=1}^{\infty} \left\{ A_n \cosh \frac{n\pi}{a} y + B_n \sinh \frac{n\pi}{a} y \right\} \sin \frac{n\pi}{a} x,$$

where $\quad A_n = \dfrac{2}{a} \displaystyle\int_0^a f(x) \sin \frac{n\pi}{a} x \, dx$

$$B_n = \frac{1}{\sinh \dfrac{n\pi}{a} b} \left(\frac{2}{a} \int_0^a g(x) \sin \frac{n\pi}{a} x \, dx - A_n \cosh \frac{n\pi}{a} b \right),$$

and that a solution of Problem 2 is

$$u_2(x, y) = \sum_{n=1}^{\infty} \left\{ A_n \cosh \frac{n\pi}{b} x + B_n \sinh \frac{n\pi}{b} x \right\} \sin \frac{n\pi}{b} y,$$

where $\quad A_n = \dfrac{2}{b} \displaystyle\int_0^b F(y) \sin \frac{n\pi}{b} y \, dy$

$$B_n = \frac{1}{\sinh \dfrac{n\pi}{b} a} \left(\frac{2}{b} \int_0^b G(y) \sin \frac{n\pi}{b} y \, dy - A_n \cosh \frac{n\pi}{b} a \right).$$

EXERCISES 12.5

Answers to odd-numbered problems begin on page AN-18.

In Problems 1–10 solve Laplace's equation (1) for a rectangular plate subject to the given boundary conditions.

1. $u(0, y) = 0, \quad u(a, y) = 0$
$u(x, 0) = 0, \quad u(x, b) = f(x)$

2. $u(0, y) = 0, \quad u(a, y) = 0$
$\left.\dfrac{\partial u}{\partial y}\right|_{y=0} = 0, \quad u(x, b) = f(x)$

3. $u(0, y) = 0, \quad u(a, y) = 0$
$u(x, 0) = f(x), \quad u(x, b) = 0$

4. $\left.\dfrac{\partial u}{\partial x}\right|_{x=0} = 0, \quad \left.\dfrac{\partial u}{\partial x}\right|_{x=a} = 0$
$u(x, 0) = x, \quad u(x, b) = 0$

5. $u(0, y) = 0, \quad u(1, y) = 1 - y$
$\left.\dfrac{\partial u}{\partial y}\right|_{y=0} = 0, \quad \left.\dfrac{\partial u}{\partial y}\right|_{y=1} = 0$

6. $u(0, y) = g(y), \quad \left.\dfrac{\partial u}{\partial x}\right|_{x=1} = 0$
$\left.\dfrac{\partial u}{\partial y}\right|_{y=0} = 0, \quad \left.\dfrac{\partial u}{\partial y}\right|_{y=\pi} = 0$

7. $\left.\dfrac{\partial u}{\partial x}\right|_{x=0} = u(0, y), \quad u(\pi, y) = 1$
$u(x, 0) = 0, \quad u(x, \pi) = 0$

8. $u(0, y) = 0, \quad u(1, y) = 0$
$\left.\dfrac{\partial u}{\partial y}\right|_{y=0} = u(x,0), \quad u(x,1) = f(x)$

9. $u(0, y) = 0, \quad u(1, y) = 0$
$u(x, 0) = 100, \quad u(x, 1) = 200$

10. $u(0, y) = 10y, \quad \left.\dfrac{\partial u}{\partial x}\right|_{x=1} = -1$
$u(x, 0) = 0, \quad u(x, 1) = 0$

In Problems 11 and 12 solve Laplace's equation (1) for the given semi-infinite plate extending in the positive y-direction. In each case assume $u(x, y)$ is bounded as $y \to \infty$.

11.

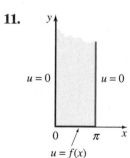

12.

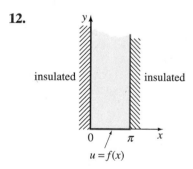

Figure 12.15

Figure 12.16

In Problems 13 and 14 solve Laplace's equation (1) for a rectangular plate subject to the given boundary conditions.

13. $u(0, y) = 0, \quad u(a, y) = 0$
$u(x, 0) = f(x), \quad u(x, b) = g(x)$

14. $u(0, y) = F(y), \quad u(a, y) = G(y)$
$u(x, 0) = 0, \quad u(x, b) = 0$

In Problems 15 and 16 use the superposition principle to solve Laplace's equation (1) for a square plate subject to the given boundary conditions.

15. $u(0, y) = 1, \quad u(\pi, y) = 1$
$u(x, 0) = 0, \quad u(x, \pi) = 1$

16. $u(0, y) = 0, \quad u(2, y) = y(2 - y)$
$u(x, 0) = 0,$
$u(x, 2) = \begin{cases} x, & 0 < x < 1 \\ 2 - x, & 1 \le x < 2 \end{cases}$

Computer Lab Assignments

17. (a) Use the contour-plot application of your CAS to graph the isotherms $u = 170, 140, 110, 80, 60, 30$ for the solution of Problem 9. Use the partial sum $S_5(x, y)$ consisting of the first five nonzero terms of the solution.

   (b) Use the 3D-plot application of your CAS to graph the partial sum $S_5(x, y)$.

18. Use the contour-plot application of your CAS to graph the isotherms $u = 2, 1, 0.5, 0.2, 0.1, 0.05, 0, -0.05$ for the solution of Problem 10. Use the partial sum $S_5(x, y)$ consisting of the first five nonzero terms of the solution.

Discussion Problems

19. (a) In Problem 1 suppose that $a = b = \pi$ and $f(x) = 100x(\pi - x)$. Without using the solution $u(x, y)$, sketch, by hand, what the surface would look like over the rectangular region defined by $0 \le x \le \pi, 0 \le y \le \pi$.

   (b) What is the maximum value of the temperature u for $0 \le x \le \pi$, $0 \le y \le \pi$?

   (c) Use the information in part (a) to compute the coefficients for your answer in Problem 1. Then use the 3D-plot application of your CAS to graph the partial sum $S_5(x, t)$ consisting of the first five nonzero terms of the solution in part (a) for $0 \le x \le \pi$, $0 \le y \le \pi$. Use different perspectives and then compare with your sketch from part (a).

20. In Problem 16 what is the maximum value of the temperature u for $0 \le x \le 2, 0 \le y \le 2$?

12.6 NONHOMOGENEOUS EQUATIONS AND BOUNDARY CONDITIONS

- *Change of dependent variable* • *Steady-state solution*
- *Transient solution*

The method of separation of variables may not be applicable to a boundary-value problem when the partial differential equation or boundary conditions are nonhomogeneous. For example, when heat is generated at a constant rate r within a rod of finite length, the heat equation takes on the form

$$k \frac{\partial^2 u}{\partial x^2} + r = \frac{\partial u}{\partial t}. \tag{1}$$

Equation (1) is nonhomogeneous and is readily shown not to be separable. On the other hand, suppose we wish to solve the usual heat equation $ku_{xx} = u_t$ when the boundaries $x = 0$ and $x = L$ are held at nonzero temperatures k_1 and k_2. Even though the substitution $u(x, t) = X(x)T(t)$ separates the PDE, we quickly find ourselves at an impasse in determining eigenvalues and eigenfunctions since no conclusion can be drawn from $u(0, t) = X(0)T(t) = k_1$ and $u(L, t) = X(L)T(t) = k_2$.

Change of Dependent Variable A few problems involving nonhomogeneous equations and nonhomogeneous boundary conditions can be solved by changing the dependent variable u to a new dependent variable v by means of the substitution $u = v + \psi$. The basic idea is to determine ψ, a function of *one* variable, in such a manner that v, a function of *two* variables, is made to satisfy a homogeneous partial differential equation and homogeneous boundary conditions. The following example illustrates the procedure.

EXAMPLE 1 **Nonhomogeneous Equation and Boundary Condition**

Solve equation (1) subject to

$$u(0, t) = 0, \quad u(1, t) = u_0, \quad t > 0$$
$$u(x, 0) = f(x), \quad 0 < x < 1.$$

Solution Both the equation and the boundary condition at $x = 1$ are nonhomogeneous. If we let $u(x, t) = v(x, t) + \psi(x)$, then

$$\frac{\partial^2 u}{\partial x^2} = \frac{\partial^2 v}{\partial x^2} + \psi'' \quad \text{and} \quad \frac{\partial u}{\partial t} = \frac{\partial v}{\partial t}.$$

Substituting these results into (1) gives

$$k\frac{\partial^2 v}{\partial x^2} + k\psi'' + r = \frac{\partial v}{\partial t}. \tag{2}$$

Equation (2) reduces to a homogeneous equation if we demand that ψ satisfy

$$k\psi'' + r = 0 \quad \text{or} \quad \psi'' = -\frac{r}{k}.$$

Integrating the last equation twice reveals that

$$\psi(x) = -\frac{r}{2k}x^2 + c_1 x + c_2. \tag{3}$$

Furthermore

$$u(0, t) = v(0, t) + \psi(0) = 0$$
$$u(1, t) = v(1, t) + \psi(1) = u_0.$$

We have $v(0, t) = 0$ and $v(1, t) = 0$, provided

$$\psi(0) = 0 \quad \text{and} \quad \psi(1) = u_0.$$

Applying the latter two conditions to (3) gives, in turn, $c_2 = 0$ and $c_1 = r/2k + u_0$. Consequently

$$\psi(x) = -\frac{r}{2k}x^2 + \left(\frac{r}{2k} + u_0\right)x.$$

Last, the initial condition $u(x, 0) = v(x, 0) + \psi(x)$ implies $v(x, 0) = u(x, 0) - \psi(x) = f(x) - \psi(x)$. Thus to determine $v(x, t)$ we solve the *new* boundary-value problem

$$k\frac{\partial^2 v}{\partial x^2} = \frac{\partial v}{\partial t}, \quad 0 < x < 1, \quad t > 0$$

$$v(0, t) = 0, \quad v(1, t) = 0, \quad t > 0$$

$$v(x, 0) = f(x) + \frac{r}{2k}x^2 - \left(\frac{r}{2k} + u_0\right)x$$

by separation of variables. In the usual manner we find

$$v(x, t) = \sum_{n=1}^{\infty} A_n e^{-kn^2\pi^2 t} \sin n\pi x,$$

where $\qquad A_n = 2\int_0^1 \left[f(x) + \frac{r}{2k}x^2 - \left(\frac{r}{2k} + u_0\right)x\right] \sin n\pi x \, dx.$ $\qquad$ **(4)**

Finally, a solution of the original problem is obtained by adding $\psi(x)$ and $v(x, t)$:

$$u(x, t) = -\frac{r}{2k}x^2 + \left(\frac{r}{2k} + u_0\right)x + \sum_{n=1}^{\infty} A_n e^{-kn^2\pi^2 t} \sin n\pi x, \qquad \textbf{(5)}$$

where the A_n are defined in (4). ■

Observe in (5) that $u(x, t) \to \psi(x)$ as $t \to \infty$. In the context of solving forms of the heat equation, ψ is called a **steady-state solution.** Since $v(x, t) \to 0$ as $t \to \infty$, v is called a **transient solution.**

The substitution $u = v + \psi$ can also be used on problems involving forms of the wave equation as well as Laplace's equation.

EXERCISES 12.6

Answers to odd-numbered problems begin on page AN-19.

In Problems 1 and 2 solve the heat equation $ku_{xx} = u_t, 0 < x < 1, t > 0$ subject to the given conditions.

1. $u(0, t) = 100, \quad u(1, t) = 100$
$\quad u(x, 0) = 0$

2. $u(0, t) = u_0, \quad u(1, t) = 0$
$\quad u(x, 0) = f(x)$

In Problems 3 and 4 solve the partial differential equation (1) subject to the given conditions.

3. $u(0, t) = u_0, \quad u(1, t) = u_0$
$\quad u(x, 0) = 0$

4. $u(0, t) = u_0, \quad u(1, t) = u_1$
$\quad u(x, 0) = f(x)$

5. Solve the boundary-value problem

$$k\frac{\partial^2 u}{\partial x^2} + Ae^{-\beta x} = \frac{\partial u}{\partial t}, \quad \beta > 0, \quad 0 < x < 1, \quad t > 0$$

$$u(0, t) = 0, \quad u(1, t) = 0, \quad t > 0$$

$$u(x, 0) = f(x), \quad 0 < x < 1.$$

The partial differential equation is a form of the heat equation when heat is generated within a thin rod from radioactive decay of the material.

6. Solve the boundary-value problem

$$k\frac{\partial^2 u}{\partial x^2} - hu = \frac{\partial u}{\partial t}, \quad 0 < x < \pi, \quad t > 0$$

$$u(0, t) = 0, \quad u(\pi, t) = u_0, \quad t > 0$$

$$u(x, 0) = 0, \quad 0 < x < \pi.$$

7. Find a steady-state solution $\psi(x)$ of the boundary-value problem

$$k\frac{\partial^2 u}{\partial x^2} - h(u - u_0) = \frac{\partial u}{\partial t}, \quad 0 < x < 1, \quad t > 0$$

$$u(0, t) = u_0, \quad u(1, t) = 0, \quad t > 0$$

$$u(x, 0) = f(x), \quad 0 < x < 1.$$

8. Find a steady-state solution $\psi(x)$ if the rod in Problem 7 is semi-infinite extending in the positive x-direction, it radiates from its lateral surface into a medium at temperature zero, and

$$u(0, t) = u_0, \quad \lim_{x \to \infty} u(x, t) = 0, \quad t > 0$$

$$u(x, 0) = f(x), \quad x > 0.$$

9. When a vibrating string is subjected to an external vertical force that varies with the horizontal distance from the left end, the wave equation takes on the form

$$a^2 \frac{\partial^2 u}{\partial x^2} + Ax = \frac{\partial^2 u}{\partial t^2},$$

where A is a constant. Solve this partial differential equation subject to

$$u(0, t) = 0, \quad u(1, t) = 0, \quad t > 0$$

$$u(x, 0) = 0, \quad \frac{\partial u}{\partial t}\bigg|_{t=0} = 0, \quad 0 < x < 1.$$

10. A string initially at rest on the x-axis is secured on the x-axis at $x = 0$ and $x = 1$. If the string is allowed to fall under its own weight for $t > 0$, the displacement $u(x, t)$ satisfies

$$a^2 \frac{\partial^2 u}{\partial x^2} - g = \frac{\partial^2 u}{\partial t^2}, \quad 0 < x < 1, \quad t > 0,$$

where g is the acceleration of gravity. Solve for $u(x, t)$.

11. Find the steady-state temperature $u(x, y)$ in the semi-infinite plate shown in Figure 12.17. Assume that the temperature is bounded as $x \to \infty$. [*Hint:* Try $u(x, y) = v(x, y) + \psi(y)$.]

12. Poisson's equation

$$\frac{\partial^2 u}{\partial x^2} + \frac{\partial^2 u}{\partial y^2} = -h, \quad h > 0$$

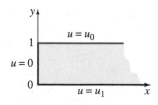

Figure 12.17

occurs in many problems involving electric potential. Solve the above equation subject to the conditions

$$u(0, y) = 0, \quad u(\pi, y) = 1, \quad y > 0$$
$$u(x, 0) = 0, \quad 0 < x < \pi.$$

12.7 ORTHOGONAL SERIES EXPANSIONS

- *Boundary-value problems that do not lead to Fourier series*
- *Using orthogonal series expansions*

For certain types of boundary conditions, the method of separation of variables and the superposition principle lead to an expansion of a function in a trigonometric series that is *not* a Fourier series. To solve the problems in this section we shall utilize the concept of orthogonal series expansions or generalized Fourier series developed in Section 11.1.

EXAMPLE 1 **Using Orthogonal Series Expansions**

The temperature in a rod of unit length in which there is heat transfer from its right boundary into a surrounding medium kept at a constant temperature zero is determined from

$$k \frac{\partial^2 u}{\partial x^2} = \frac{\partial u}{\partial t}, \quad 0 < x < 1, \quad t > 0$$

$$u(0, t) = 0, \quad \frac{\partial u}{\partial x}\bigg|_{x=1} = -hu(1, t), \quad h > 0, \quad t > 0$$

$$u(x, 0) = 1, \quad 0 < x < 1.$$

Solve for $u(x, t)$.

Solution The method of separation of variables gives

$$X'' + \lambda^2 X = 0, \quad T' + k\lambda^2 T = 0 \tag{1}$$

$$X(x) = c_1 \cos \lambda x + c_2 \sin \lambda x \quad \text{and} \quad T(t) = c_3 e^{-k\lambda^2 t}.$$

Since $u = XT$, we find that the boundary conditions become

$$X(0) = 0 \quad \text{and} \quad X'(1) = -hX(1). \tag{2}$$

The first condition in (2) immediately gives $c_1 = 0$. Applying the second condition in (2) to $X(x) = c_2 \sin \lambda x$ yields

$$\lambda \cos \lambda = -h \sin \lambda \quad \text{or} \quad \tan \lambda = -\frac{\lambda}{h}. \tag{3}$$

From the analysis in Example 2 of Section 11.4 we know that the last equation in (3) has an infinite number of roots. The consecutive positive roots λ_n, $n = 1, 2, 3, \ldots$ are the eigenvalues of the problem, and the corresponding eigenfunctions are $X(x) = c_2 \sin \lambda_n x$, $n = 1, 2, 3, \ldots$. Thus

$$u_n = XT = A_n e^{-k\lambda_n^2 t} \sin \lambda_n x \quad \text{and} \quad u(x, t) = \sum_{n=1}^{\infty} A_n e^{-k\lambda_n^2 t} \sin \lambda_n x.$$

Now at $t = 0$, $u(x, 0) = 1$, $0 < x < 1$, so

$$1 = \sum_{n=1}^{\infty} A_n \sin \lambda_n x. \tag{4}$$

The series in (4) is not a Fourier sine series; it is an expansion in terms of the orthogonal functions arising from the regular Sturm-Liouville problem consisting of the first differential equation in (1) and boundary conditions (2). It follows that the set of eigenfunctions $\{\sin \lambda_n x\}$, $n = 1, 2, 3, \ldots$, where the λ's are defined by $\tan \lambda = -\lambda/h$, is orthogonal with respect to the weight function $p(x) = 1$ on the interval $[0, 1]$. With $f(x) = 1$ in (8) of Section 11.1 we can write

$$A_n = \frac{\int_0^1 \sin \lambda_n x \, dx}{\int_0^1 \sin^2 \lambda_n x \, dx}. \tag{5}$$

To evaluate the square norm of each of the eigenfunctions we use a trigonometric identity:

$$\int_0^1 \sin^2 \lambda_n x \, dx = \frac{1}{2} \int_0^1 [1 - \cos 2\lambda_n x] \, dx = \frac{1}{2} \left[1 - \frac{1}{2\lambda_n} \sin 2\lambda_n \right]. \tag{6}$$

Using $\sin 2\lambda_n = 2 \sin \lambda_n \cos \lambda_n$ and $\lambda_n \cos \lambda_n = -h \sin \lambda_n$, we simplify (6) to

$$\int_0^1 \sin^2 \lambda_n x \, dx = \frac{1}{2h} [h + \cos^2 \lambda_n].$$

Also

$$\int_0^1 \sin \lambda_n x \, dx = -\frac{1}{\lambda_n} \cos \lambda_n x \Big|_0^1 = \frac{1}{\lambda_n} [1 - \cos \lambda_n].$$

Consequently (5) becomes

$$A_n = \frac{2h(1 - \cos \lambda_n)}{\lambda_n(h + \cos^2 \lambda_n)}.$$

Finally, a solution of the boundary-value problem is

$$u(x, t) = 2h \sum_{n=1}^{\infty} \frac{1 - \cos \lambda_n}{\lambda_n(h + \cos^2 \lambda_n)} e^{-k\lambda_n^2 t} \sin \lambda_n x. \qquad \blacksquare$$

EXAMPLE 2   Using Orthogonal Series Expansions

The twist angle $\theta(x, t)$ of a torsionally vibrating shaft of unit length is determined from

$$a^2 \frac{\partial^2 \theta}{\partial x^2} = \frac{\partial^2 \theta}{\partial t^2}, \quad 0 < x < 1, \quad t > 0$$

$$\theta(0, t) = 0, \quad \frac{\partial \theta}{\partial x}\Big|_{x=1} = 0, \quad t > 0$$

$$\theta(x, 0) = x, \quad \frac{\partial \theta}{\partial t}\Big|_{t=0} = 0, \quad 0 < x < 1.$$

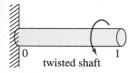

Figure 12.18

twisted shaft

See Figure 12.18. The boundary condition at $x = 1$ is called a free-end condition. Solve for $\theta(x, t)$.

Solution Using $\theta = XT$, we find

$$X'' + \lambda^2 X = 0, \quad T'' + a^2\lambda^2 T = 0$$

$$X(x) = c_1 \cos \lambda x + c_2 \sin \lambda x \quad \text{and} \quad T(t) = c_3 \cos a\lambda t + c_4 \sin a\lambda t.$$

The boundary conditions $X(0) = 0$ and $X'(1) = 0$ give $c_1 = 0$ and $c_2 \cos \lambda = 0$, respectively. Since the cosine function is zero at odd multiples of $\pi/2$, the eigenvalues of the problem are $\lambda = (2n - 1)(\pi/2)$, $n = 1, 2, 3, \ldots$. The initial condition $T'(0) = 0$ gives $c_4 = 0$, so

$$\theta_n = XT = A_n \cos a\left(\frac{2n - 1}{2}\right)\pi t \, \sin\left(\frac{2n - 1}{2}\right)\pi x.$$

In order to satisfy the remaining initial condition we form

$$\theta(x, t) = \sum_{n=1}^{\infty} A_n \cos a\left(\frac{2n - 1}{2}\right)\pi t \, \sin\left(\frac{2n - 1}{2}\right)\pi x. \tag{7}$$

When $t = 0$, we must have, for $0 < x < 1$,

$$\theta(x, 0) = x = \sum_{n=1}^{\infty} A_n \sin\left(\frac{2n - 1}{2}\right)\pi x. \tag{8}$$

As in Example 1, the set of eigenfunctions $\left\{\sin\left(\frac{2n - 1}{2}\right)\pi x\right\}$, $n = 1, 2, 3, \ldots$ is orthogonal with respect to the weight function $p(x) = 1$ on the interval $[0, 1]$. The series $\sum_{n=1}^{\infty} A_n \sin\left(\frac{2n - 1}{2}\right)\pi x$ is not a Fourier sine series since the argument of the sine is not an integer multiple of $\pi x/L$ ($L = 1$ in this case). The series is again a generalized Fourier series. Hence from (8) of Section 11.1 the coefficients in (7) are

$$A_n = \frac{\int_0^1 x \sin\left(\frac{2n - 1}{2}\right)\pi x \, dx}{\int_0^1 \sin^2\left(\frac{2n - 1}{2}\right)\pi x \, dx}.$$

Carrying out the two integrations, we arrive at

$$A_n = \frac{8(-1)^{n+1}}{(2n - 1)^2\pi^2}.$$

The twist angle is then

$$\theta(x, t) = \frac{8}{\pi^2} \sum_{n=1}^{\infty} \frac{(-1)^{n+1}}{(2n - 1)^2} \cos a\left(\frac{2n - 1}{2}\right)\pi t \, \sin\left(\frac{2n - 1}{2}\right)\pi x. \quad \blacksquare$$

Answers to odd-numbered problems begin on page AN-19.

1. In Example 1 find the temperature $u(x, t)$ when the left end of the rod is insulated.

2. Solve the boundary-value problem

$$k \frac{\partial^2 u}{\partial x^2} = \frac{\partial u}{\partial t}, \quad 0 < x < 1, \quad t > 0$$

$$u(0, t) = 0, \quad \frac{\partial u}{\partial x}\Big|_{x=1} = -h(u(1, t) - u_0), \quad h > 0, \quad t > 0$$

$$u(x, 0) = f(x), \quad 0 < x < 1.$$

3. Find the steady-state temperature for a rectangular plate for which the boundary conditions are

$$u(0, y) = 0, \quad \frac{\partial u}{\partial x}\Big|_{x=a} = -hu(a, y), \quad 0 < y < b$$

$$u(x, 0) = 0, \quad u(x, b) = f(x), \quad 0 < x < a.$$

4. Solve the boundary-value problem

$$\frac{\partial^2 u}{\partial x^2} + \frac{\partial^2 u}{\partial y^2} = 0, \quad 0 < y < 1, \quad x > 0$$

$$u(0, y) = u_0, \quad \lim_{x \to \infty} u(x, y) = 0, \quad 0 < y < 1$$

$$\frac{\partial u}{\partial y}\Big|_{y=0} = 0, \quad \frac{\partial u}{\partial y}\Big|_{y=1} = -hu(x, 1), \quad h > 0, \quad x > 0.$$

5. Find the temperature $u(x, t)$ in a rod of length L if the initial temperature is $f(x)$ throughout and if the end $x = 0$ is kept at temperature zero and the end $x = L$ is insulated.

6. Solve the boundary-value problem

$$a^2 \frac{\partial^2 u}{\partial x^2} = \frac{\partial^2 u}{\partial t^2}, \quad 0 < x < L, \quad t > 0$$

$$u(0, t) = 0, \quad E\frac{\partial u}{\partial x}\Big|_{x=L} = F_0, \quad t > 0$$

$$u(x, 0) = 0, \quad \frac{\partial u}{\partial t}\Big|_{t=0} = 0, \quad 0 < x < L.$$

The solution $u(x, t)$ represents the longitudinal displacement of a vibrating elastic bar that is anchored at its left end and is subjected to a constant force of magnitude F_0 at its right end. See Figure 12.10 on page 539. E is a constant called the modulus of elasticity.

7. Solve the boundary-value problem

$$\frac{\partial^2 u}{\partial x^2} + \frac{\partial^2 u}{\partial y^2} = 0, \quad 0 < x < 1, \quad 0 < y < 1$$

$$\frac{\partial u}{\partial x}\Big|_{x=0} = 0, \quad u(1, y) = u_0, \quad 0 < y < 1$$

$$u(x, 0) = 0, \quad \frac{\partial u}{\partial y}\Big|_{y=1} = 0, \quad 0 < x < 1.$$

8. The initial temperature in a rod of unit length is $f(x)$ throughout. There is heat transfer from both ends, $x = 0$ and $x = 1$, into a surrounding medium kept at a constant temperature zero. Show that

$$u(x, t) = \sum_{n=1}^{\infty} A_n e^{-k\lambda_n^2 t}(\lambda_n \cos \lambda_n x + h \sin \lambda_n x),$$

where $A_n = \dfrac{2}{(\lambda_n^2 + 2h + h^2)} \displaystyle\int_0^1 f(x)(\lambda_n \cos \lambda_n x + h \sin \lambda_n x) \, dx$

and the λ_n, $n = 1, 2, 3, \ldots$ are the consecutive positive roots of $\tan \lambda = 2\lambda h/(\lambda^2 - h^2)$.

9. A vibrating cantilever beam is embedded at its left end ($x = 0$) and free at its right end ($x = 1$). See Figure 12.19. The transverse displacement $u(x, t)$ of the beam is determined from the boundary-value problem

$$\frac{\partial^4 u}{\partial x^4} + \frac{\partial^2 u}{\partial t^2} = 0, \quad 0 < x < 1, \quad t > 0$$

$$u(0, t) = 0, \quad \left.\frac{\partial u}{\partial x}\right|_{x=0} = 0, \quad t > 0$$

$$\left.\frac{\partial^2 u}{\partial x^2}\right|_{x=1} = 0, \quad \left.\frac{\partial^3 u}{\partial x^3}\right|_{x=1} = 0, \quad t > 0$$

$$u(x, 0) = f(x), \quad \left.\frac{\partial u}{\partial t}\right|_{t=0} = g(x), \quad x > 0.$$

(a) Using the separation constant λ^4, show that the eigenvalues of the problem are determined from the equation $\cos \lambda \cosh \lambda = -1$.

(b) Use a calculator or a computer to find approximations to the first two positive eigenvalues.

10. (a) Find an equation that defines the eigenvalues when the ends of the beam in Problem 9 are embedded at $x = 0$ and $x = 1$.

(b) Use a calculator or a computer to find approximations to the first two positive eigenvalues.

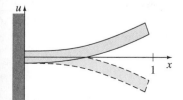

Figure 12.19

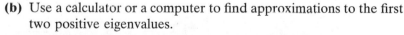

12.8 BOUNDARY-VALUE PROBLEMS INVOLVING FOURIER SERIES IN TWO VARIABLES

- *Two-dimensional heat equation*   • *Two-dimensional wave equation*
- *Sine series in two variables*   • *Cosine series in two variables*

Up to now we have solved problems involving only the one-dimensional heat and wave equations. In this section we will show how to extend the method of separation of variables to certain problems involving the two-dimensional versions of these partial differential equations. Since the unknown function u is a function of three independent variables x, y, and t, the assumption of a particular solution of the form $X(x)Y(y)T(t)$ leads naturally to the notion of a Fourier series in two variables x and y.

In the exercises we will also consider Laplace's equation in three spatial variables.

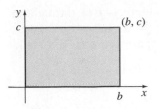

Figure 12.20

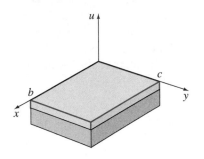

Figure 12.21

Heat and Wave Equations in Two Dimensions Suppose that the rectangular region shown in Figure 12.20 is a thin plate in which the temperature u is a function of time t and position (x, y). Then, under suitable conditions, $u(x, y, t)$ can be shown to satisfy the **two-dimensional heat equation**

$$k\left(\frac{\partial^2 u}{\partial x^2} + \frac{\partial^2 u}{\partial y^2}\right) = \frac{\partial u}{\partial t}. \tag{1}$$

On the other hand, suppose that Figure 12.21 represents a rectangular frame over which a thin flexible membrane has been stretched (a rectangular drum). If the membrane is set in motion, then its displacement u, measured from the xy-plane (transverse vibrations), is also a function of t and position (x, y). When the vibrations are small, free, and undamped, $u(x, y, t)$ satisfies the **two-dimensional wave equation**

$$a^2\left(\frac{\partial^2 u}{\partial x^2} + \frac{\partial^2 u}{\partial y^2}\right) = \frac{\partial^2 u}{\partial t^2}. \tag{2}$$

In the next example, we apply the method of separation of variables to a boundary-value problem involving (1).

EXAMPLE 1 **Heat Flow in Two-Dimensions**

Find the temperature $u(x, y, t)$ in the plate shown in Figure 12.20 if the initial temperature is $f(x, y)$ throughout and if the boundaries are held at temperature zero.

Solution We must solve

$$k\left(\frac{\partial^2 u}{\partial x^2} + \frac{\partial^2 u}{\partial y^2}\right) = \frac{\partial u}{\partial t}, \quad 0 < x < b, \quad 0 < y < c, \quad t > 0$$

subject to

$$u(0, y, t) = 0, \qquad u(b, y, t) = 0, \quad 0 < y < c, \quad t > 0$$
$$u(x, 0, t) = 0, \qquad u(x, c, t) = 0, \quad 0 < x < b, \quad t > 0$$
$$u(x, y, 0) = f(x, y), \quad 0 < x < b, \quad 0 < y < c.$$

To separate variables in the partial differential equation in three independent variables we try to find a product solution $u(x, y, t) = X(x)Y(y)T(t)$. After substitution this assumption yields

$$k(X''YT + XY''T) = XYT'$$

or

$$\frac{X''}{X} = -\frac{Y''}{Y} + \frac{T'}{kT}. \tag{3}$$

Since the left-hand side of (3) depends only on x and the right-hand side depends only on y and t, we must have both sides equal to a constant $-\lambda^2$:

$$\frac{X''}{X} = -\frac{Y''}{Y} + \frac{T'}{kT} = -\lambda^2$$

and so

$$X'' + \lambda^2 X = 0 \tag{4}$$

$$\frac{Y''}{Y} = \frac{T'}{kT} + \lambda^2. \tag{5}$$

By the same reasoning, if we introduce another separation constant $-\mu^2$ in (5), then

$$\frac{Y''}{Y} = -\mu^2 \quad \text{and} \quad \frac{T'}{kT} + \lambda^2 = -\mu^2$$

$$Y'' + \mu^2 Y = 0 \quad \text{and} \quad T' + k(\lambda^2 + \mu^2)T = 0. \tag{6}$$

Now the solutions of the equations in (4) and (6) are, respectively,

$$X(x) = c_1 \cos \lambda x + c_2 \sin \lambda x \tag{7}$$
$$Y(y) = c_3 \cos \mu y + c_4 \sin \mu y \tag{8}$$
$$T(t) = c_5 e^{-k(\lambda^2 + \mu^2)t}. \tag{9}$$

But the boundary conditions

$$\left.\begin{array}{ll} u(0, y, t) = 0, & u(b, y, t) = 0 \\ u(x, 0, t) = 0, & u(x, c, t) = 0 \end{array}\right\} \quad \text{imply} \quad \begin{cases} X(0) = 0, & X(b) = 0 \\ Y(0) = 0, & Y(c) = 0. \end{cases}$$

Applying these conditions to (7) and (8) gives $c_1 = 0$, $c_3 = 0$ and $c_2 \sin \lambda b = 0$, $c_4 \sin \mu c = 0$. The latter equations in turn imply

$$\lambda = \frac{m\pi}{b}, \quad m = 1, 2, 3, \dots; \quad \mu = \frac{n\pi}{c}, \quad n = 1, 2, 3, \dots.$$

Thus a product solution of the two-dimensional heat equation that satisfies the boundary conditions is

$$u_{mn}(x, y, t) = A_{mn} e^{-k[(m\pi/b)^2 + (n\pi/c)^2]t} \sin \frac{m\pi}{b} x \sin \frac{n\pi}{c} y,$$

where A_{mn} is an arbitrary constant. Because we have two independent sets of eigenvalues, we are prompted to try the superposition principle in the form of a double sum

$$u(x, y, t) = \sum_{m=1}^{\infty} \sum_{n=1}^{\infty} A_{mn} e^{-k[(m\pi/b)^2 + (n\pi/c)^2]t} \sin \frac{m\pi}{b} x \sin \frac{n\pi}{c} y. \tag{10}$$

Now at $t = 0$ we must have

$$u(x, y, 0) = f(x, y) = \sum_{m=1}^{\infty} \sum_{n=1}^{\infty} A_{mn} \sin \frac{m\pi}{b} x \sin \frac{n\pi}{c} y. \tag{11}$$

We can find the coefficients A_{mn} by multiplying the double sum (11) by the product $\sin(m\pi x/b) \sin(n\pi y/c)$ and integrating over the rectangle $0 \le x \le b$, $0 \le y \le c$. It follows that

$$A_{mn} = \frac{4}{bc} \int_0^c \int_0^b f(x, y) \sin \frac{m\pi}{b} x \sin \frac{n\pi}{c} y \, dx \, dy. \tag{12}$$

Thus the solution of the boundary-value problem consists of (10) with the A_{mn} defined by (12). ▬

The series (11) with coefficients (12) is called a **sine series in two variables** or a **double sine series**. We summarize next the **cosine series in two variables**.

The **double cosine series** of a function $f(x, y)$ defined over a rectangular region $0 \leq x \leq b, 0 \leq y \leq c$ is given by

$$f(x, y) = A_{00} + \sum_{m=1}^{\infty} A_{m0} \cos \frac{m\pi}{b} x + \sum_{n=1}^{\infty} A_{0n} \cos \frac{n\pi}{c} y$$

$$+ \sum_{m=1}^{\infty} \sum_{n=1}^{\infty} A_{mn} \cos \frac{m\pi}{b} x \cos \frac{n\pi}{c} y,$$

where

$$A_{00} = \frac{1}{bc} \int_0^c \int_0^b f(x, y) \, dx \, dy$$

$$A_{m0} = \frac{2}{bc} \int_0^c \int_0^b f(x, y) \cos \frac{m\pi}{b} x \, dx \, dy$$

$$A_{0n} = \frac{2}{bc} \int_0^c \int_0^b f(x, y) \cos \frac{n\pi}{c} y \, dx \, dy$$

$$A_{mn} = \frac{4}{bc} \int_0^c \int_0^b f(x, y) \cos \frac{m\pi}{b} x \cos \frac{n\pi}{c} y \, dx \, dy.$$

See Problem 2 in Exercises 12.8.

EXERCISES 12.8

Answers to odd-numbered problems begin on page AN-19.

In Problems 1 and 2 solve the heat equation (1) subject to the given conditions.

1. $u(0, y, t) = 0, \quad u(\pi, y, t) = 0$
$u(x, 0, t) = 0, \quad u(x, \pi, t) = 0$
$u(x, y, 0) = u_0$

2. $\dfrac{\partial u}{\partial x}\Big|_{x=0} = 0, \quad \dfrac{\partial u}{\partial x}\Big|_{x=1} = 0$
$\dfrac{\partial u}{\partial y}\Big|_{y=0} = 0, \quad \dfrac{\partial u}{\partial y}\Big|_{y=1} = 0$
$u(x, y, 0) = xy$

In Problems 3 and 4 solve the wave equation (2) subject to the given conditions.

3. $u(0, y, t) = 0, \quad u(\pi, y, t) = 0$
$u(x, 0, t) = 0, \quad u(x, \pi, t) = 0$
$u(x, y, 0) = xy(x - \pi)(y - \pi)$
$\dfrac{\partial u}{\partial t}\Big|_{t=0} = 0$

4. $u(0, y, t) = 0, \quad u(b, y, t) = 0$
$u(x, 0, t) = 0, \quad u(x, c, t) = 0$
$u(x, y, 0) = f(x, y)$
$\dfrac{\partial u}{\partial t}\Big|_{t=0} = g(x, y)$

The steady-state temperature $u(x, y, z)$ in the rectangular parallelepiped shown in Figure 12.22 satisfies Laplace's equation in three dimensions:

$$\frac{\partial^2 u}{\partial x^2} + \frac{\partial^2 u}{\partial y^2} + \frac{\partial^2 u}{\partial z^2} = 0. \tag{13}$$

5. Solve Laplace's equation (13) if the top ($z = c$) of the parallelepiped is kept at temperature $f(x, y)$ and the remaining sides are kept at temperature zero.

6. Solve Laplace's equation (13) if the bottom ($z = 0$) of the parallelepiped is kept at temperature $f(x, y)$ and the remaining sides are kept at temperature zero.

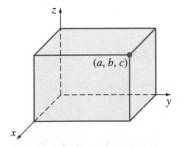

Figure 12.22

CHAPTER 12 IN REVIEW

Answers to odd-numbered problems begin on page AN-19.

1. Use separation of variables to find product solutions of

$$\frac{\partial^2 u}{\partial x\, \partial y} = u.$$

2. Use separation of variables to find product solutions of

$$\frac{\partial^2 u}{\partial x^2} + \frac{\partial^2 u}{\partial y^2} + 2\frac{\partial u}{\partial x} + 2\frac{\partial u}{\partial y} = 0.$$

Is it possible to choose a separation constant so that both X and Y are oscillatory functions?

3. Find a steady-state solution $\psi(x)$ of the boundary-value problem

$$k\frac{\partial^2 u}{\partial x^2} = \frac{\partial u}{\partial t}, \quad 0 < x < \pi, \quad t > 0$$

$$u(0, t) = u_0, \quad -\frac{\partial u}{\partial x}\Big|_{x=\pi} = u(\pi, t) - u_1, \quad t > 0$$

$$u(x, 0) = 0, \quad 0 < x < \pi.$$

4. Give a physical interpretation for the boundary conditions in Problem 3.

5. At $t = 0$ a string of unit length is stretched on the positive x-axis. The ends of the string $x = 0$ and $x = 1$ are secured on the x-axis for $t > 0$. Find the displacement $u(x, t)$ if the initial velocity $g(x)$ is as given in Figure 12.23.

6. The partial differential equation

$$\frac{\partial^2 u}{\partial x^2} + x^2 = \frac{\partial^2 u}{\partial t^2}$$

is a form of the wave equation when an external vertical force proportional to the square of the horizontal distance from the left end is applied to the string. The string is secured at $x = 0$ one unit above the x-axis and on the x-axis at $x = 1$ for $t > 0$. Find the displacement $u(x, t)$ if the string starts from rest from the initial displacement $f(x)$.

7. Find the steady-state temperature $u(x, y)$ in the square plate shown in Figure 12.24.

8. Find the steady-state temperature $u(x, y)$ in the semi-infinite plate shown in Figure 12.25.

9. Solve Problem 8 if the boundaries $y = 0$ and $y = \pi$ are held at temperature zero for all time.

10. Find the temperature $u(x, t)$ in the infinite plate of width $2L$ shown in Figure 12.26 if the initial temperature is u_0 throughout. [*Hint:* $u(x, 0) = u_0, -L < x < L$ is an even function of x.]

11. Solve the boundary-value problem

$$\frac{\partial^2 u}{\partial x^2} = \frac{\partial u}{\partial t}, \quad 0 < x < \pi, \quad t > 0$$

$$u(0, t) = 0, \quad u(\pi, t) = 0, \quad t > 0$$

$$u(x, 0) = \sin x, \quad 0 < x < \pi.$$

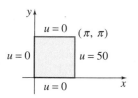

Figure 12.23

Figure 12.24

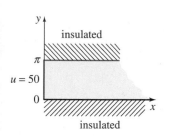

Figure 12.25

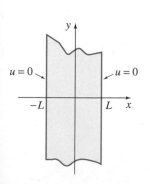

Figure 12.26

12. Solve the boundary-value problem

$$k\frac{\partial^2 u}{\partial x^2} + \sin 2\pi x = \frac{\partial u}{\partial t}, \quad 0 < x < 1, \quad t > 0$$

$$u(0, t) = 0, \quad u(1, t) = 0, \quad t > 0$$

$$u(x, 0) = \sin \pi x, \quad 0 < x < 1.$$

13. Find a formal series solution of the problem

$$\frac{\partial^2 u}{\partial x^2} + 2\frac{\partial u}{\partial x} = \frac{\partial^2 u}{\partial t^2} + 2\frac{\partial u}{\partial t} + u, \quad 0 < x < \pi, \quad t > 0$$

$$u(0, t) = 0, \quad u(\pi, t) = 0, \quad t > 0$$

$$\left.\frac{\partial u}{\partial t}\right|_{t=0} = 0, \quad 0 < x < \pi.$$

Do not attempt to evaluate the coefficients in the series.

14. The concentration $c(x, t)$ of a substance that both diffuses in a medium and is convected by currents in the medium satisfies the partial differential equation

$$k\frac{\partial^2 c}{\partial x^2} - h\frac{\partial c}{\partial x} = \frac{\partial c}{\partial t}, \quad h \text{ a constant.}$$

Solve the equation subject to

$$c(0, t) = 0, \quad c(1, t) = 0, \quad t > 0$$

$$c(x, 0) = c_0, \quad 0 < x < 1,$$

where c_0 is a constant.

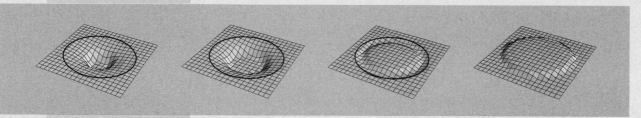

The shape of a drumbeat; see Problem 13, page 574.

13

BOUNDARY-VALUE PROBLEMS IN OTHER COORDINATE SYSTEMS

INTRODUCTION   Up to this point all the boundary-value problems that we have considered have been expressed in terms of a rectangular coordinate system. If, however, we wished to find, say, temperatures in a circular disk, in a circular cylinder, or in a sphere, we would naturally try to describe the problem in terms of polar coordinates, cylindrical coordinates, or spherical coordinates, respectively. It is essential then that we express the Laplacian $\nabla^2 u$ in terms of these different coordinate systems. In Sections 13.2 and 13.3 we put the theory of Fourier-Bessel series and Fourier-Legendre series to practical use.

13.1 PROBLEMS INVOLVING LAPLACE'S EQUATION IN POLAR COORDINATES

• Laplacian in polar coordinates   • Laplace's equation in polar coordinates   • Steady-state temperatures   • Double eigenvalues

The relationship between polar coordinates in the plane and rectangular coordinates is given by

$$x = r \cos \theta, \quad y = r \sin \theta \quad \text{and} \quad r^2 = x^2 + y^2, \quad \tan \theta = \frac{y}{x}.$$

See Figure 13.1. The first pair of equations transforms polar coordinates (r, θ) into rectangular coordinates (x, y); the second pair of equations enables us to transform rectangular coordinates into polar coordinates. These equations also make it possible to convert the two-dimensional Laplacian of a function $u(x, y)$,

$$\nabla^2 u = \frac{\partial^2 u}{\partial x^2} + \frac{\partial^2 u}{\partial y^2},$$

into polar coordinates.

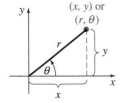

Figure 13.1

Laplacian in Polar Coordinates   You are encouraged to work through the Chain Rule for partial derivatives and show that

$$\frac{\partial u}{\partial x} = \frac{\partial u}{\partial r}\frac{\partial r}{\partial x} + \frac{\partial u}{\partial \theta}\frac{\partial \theta}{\partial x} = \cos \theta \frac{\partial u}{\partial r} - \frac{\sin \theta}{r}\frac{\partial u}{\partial \theta}$$

$$\frac{\partial u}{\partial y} = \frac{\partial u}{\partial r}\frac{\partial r}{\partial y} + \frac{\partial u}{\partial \theta}\frac{\partial \theta}{\partial y} = \sin \theta \frac{\partial u}{\partial r} + \frac{\cos \theta}{r}\frac{\partial u}{\partial \theta}$$

$$\frac{\partial^2 u}{\partial x^2} = \cos^2 \theta \frac{\partial^2 u}{\partial r^2} - \frac{2 \sin \theta \cos \theta}{r}\frac{\partial^2 u}{\partial r \, \partial \theta} + \frac{\sin^2 \theta}{r^2}\frac{\partial^2 u}{\partial \theta^2} + \frac{\sin^2 \theta}{r}\frac{\partial u}{\partial r} + \frac{2 \sin \theta \cos \theta}{r}\frac{\partial u}{\partial \theta} \quad (1)$$

$$\frac{\partial^2 u}{\partial y^2} = \sin^2 \theta \frac{\partial^2 u}{\partial r^2} + \frac{2 \sin \theta \cos \theta}{r}\frac{\partial^2 u}{\partial r \, \partial \theta} + \frac{\cos^2 \theta}{r^2}\frac{\partial^2 u}{\partial \theta^2} + \frac{\cos^2 \theta}{r}\frac{\partial u}{\partial r} - \frac{2 \sin \theta \cos \theta}{r}\frac{\partial u}{\partial \theta}. \quad (2)$$

Adding (1) and (2) and simplifying yields the Laplacian of u in polar coordinates:

$$\nabla^2 u = \frac{\partial^2 u}{\partial x^2} + \frac{\partial^2 u}{\partial y^2} = \frac{\partial^2 u}{\partial r^2} + \frac{1}{r}\frac{\partial u}{\partial r} + \frac{1}{r^2}\frac{\partial^2 u}{\partial \theta^2}.$$

In this section we shall focus only on problems involving Laplace's equation in polar coordinates:

$$\frac{\partial^2 u}{\partial r^2} + \frac{1}{r}\frac{\partial u}{\partial r} + \frac{1}{r^2}\frac{\partial^2 u}{\partial \theta^2} = 0.$$

Our first example is the Dirichlet problem for a circular disk.

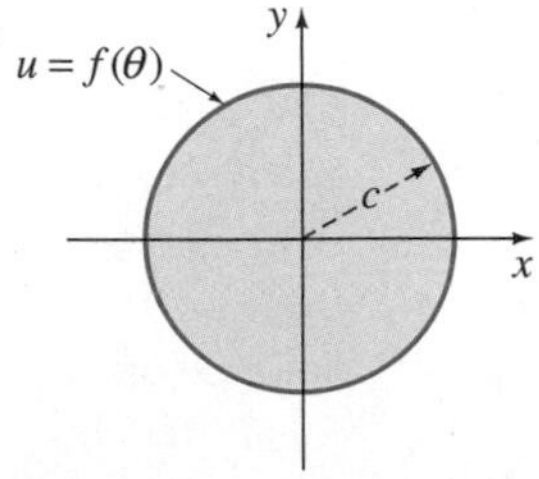

EXAMPLE 1   **Steady Temperatures in a Circular Plate**

Find the steady-state temperature $u(r, \theta)$ in a circular plate of radius c if the temperature of the circumference is $u(c, \theta) = f(\theta)$, $0 < \theta < 2\pi$. See Figure 13.2.

$u = f(\theta)$

Figure 13.2

Solution   We must solve Laplace's equation

$$\frac{\partial^2 u}{\partial r^2} + \frac{1}{r}\frac{\partial u}{\partial r} + \frac{1}{r^2}\frac{\partial^2 u}{\partial \theta^2} = 0, \quad 0 < \theta < 2\pi, \quad 0 < r < c$$

subject to $u(c, \theta) = f(\theta), 0 < \theta < 2\pi$.

If we define $u = R(r)\Theta(\theta)$, then separation of variables gives

$$\frac{r^2 R'' + rR'}{R} = -\frac{\Theta''}{\Theta} = \lambda^2$$

and
$$r^2 R'' + rR' - \lambda^2 R = 0 \tag{3}$$
$$\Theta'' + \lambda^2 \Theta = 0. \tag{4}$$

The solution of (4) is

$$\Theta = c_1 \cos \lambda\theta + c_2 \sin \lambda\theta. \tag{5}$$

Equation (3) is recognized as a Cauchy-Euler equation. Its general solution is

$$R = c_3 r^\lambda + c_4 r^{-\lambda}. \tag{6}$$

Now there are no explicit conditions in the statement of this problem that enable us to determine any of the coefficients or the eigenvalues. However, there are some implicit conditions.

First, our physical intuition leads us to expect that the temperature $u(r, \theta)$ should be bounded inside the circle $r = c$. Moreover, the temperature u should be the same at a specified point in the circle regardless of the polar description of that point. Since $(r, \theta + 2\pi)$ is an equivalent description of the point (r, θ), we must have $u(r, \theta) = u(r, \theta + 2\pi)$. In other words, the temperature $u(r, \theta)$ must be periodic with period 2π. But the only way the solutions of (4) can be 2π-periodic is if we take $\lambda = n$, where $n = 0, 1, 2, \dots$. Thus for $n > 0$, (5) and (6) become

$$\Theta = c_1 \cos n\theta + c_2 \sin n\theta \tag{7}$$
$$R = c_3 r^n + c_4 r^{-n}. \tag{8}$$

Now observe in (8) that $r^{-n} = 1/r^n$. In order for the solution to be bounded at the center of the plate (that is, at $r = 0$) we must define $c_4 = 0$. Hence for $\lambda = n > 0$,

$$u_n = R\Theta = r^n(A_n \cos n\theta + B_n \sin n\theta), \tag{9}$$

where $c_3 c_1$ and $c_3 c_2$ have been replaced by A_n and B_n, respectively.

There is one last item of consideration—namely, $n = 0$ is an eigenvalue. For this value the solutions of (3) and (4) are not given by (5) and (6). The solutions of the differential equations $\Theta'' = 0$ and $r^2 R'' + rR' = 0$ are

$$\Theta = c_5\theta + c_6 \quad \text{and} \quad R = c_7 + c_8 \ln r.$$

Periodicity demands that we take $c_5 = 0$, whereas the expectation that the temperature is finite at $r = 0$ demands that we take $c_8 = 0$. In this case a product solution of the partial differential equation is simply

$$u_0 = A_0, \tag{10}$$

where A_0 represents $c_6 c_7$.

The superposition principle then gives $u(r, \theta) = \sum_{n=0}^{\infty} u_n$ or

$$u(r, \theta) = A_0 + \sum_{n=1}^{\infty} r^n(A_n \cos n\theta + B_n \sin n\theta). \tag{11}$$

Finally, by applying the boundary condition at $r = c$ to the result in (11) we recognize

$$f(\theta) = A_0 + \sum_{n=1}^{\infty} c^n(A_n \cos n\theta + B_n \sin n\theta)$$

as an expansion of f in a full Fourier series. Consequently we can make the identifications

$$A_0 = \frac{a_0}{2}, \quad c^n A_n = a_n, \quad \text{and} \quad c^n B_n = b_n.$$

That is,
$$A_0 = \frac{1}{2\pi} \int_0^{2\pi} f(\theta) \, d\theta \tag{12}$$

$$A_n = \frac{1}{c^n \pi} \int_0^{2\pi} f(\theta) \cos n\theta \, d\theta \tag{13}$$

$$B_n = \frac{1}{c^n \pi} \int_0^{2\pi} f(\theta) \sin n\theta \, d\theta. \tag{14}$$

The solution of the problem consists of the series given in (11), where A_0, A_n, and B_n are defined in (12), (13), and (14). ∎

Observe in Example 1 that corresponding to each positive eigenvalue, $n = 1, 2, 3, \ldots$, there are two different eigenfunctions—namely, $\cos n\theta$ and $\sin n\theta$. In this situation the eigenvalues are sometimes called **double eigenvalues.**

EXAMPLE 2   **Steady Temperatures in a Semicircular Plate**

Find the steady-state temperature $u(r, \theta)$ in the semicircular plate shown in Figure 13.3.

Solution   The boundary-value problem is

$$\frac{\partial^2 u}{\partial r^2} + \frac{1}{r}\frac{\partial u}{\partial r} + \frac{1}{r^2}\frac{\partial^2 u}{\partial \theta^2} = 0, \quad 0 < \theta < \pi, \quad 0 < r < c$$

$$u(c, \theta) = u_0, \quad 0 < \theta < \pi$$

$$u(r, 0) = 0, \quad u(r, \pi) = 0, \quad 0 < r < c.$$

Defining $u = R(r)\Theta(\theta)$ and separating variables gives

$$\frac{r^2 R'' + rR'}{R} = -\frac{\Theta''}{\Theta} = \lambda^2$$

and
$$r^2 R'' + rR' - \lambda^2 R = 0 \tag{15}$$

$$\Theta'' + \lambda^2 \Theta = 0. \tag{16}$$

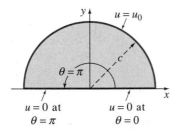

Figure 13.3

$u = u_0$
$\theta = \pi$
$u = 0$ at $\theta = \pi$
$u = 0$ at $\theta = 0$

Applying the boundary conditions $\Theta(0) = 0$ and $\Theta(\pi) = 0$ to the solution $\Theta = c_1 \cos \lambda\theta + c_2 \sin \lambda\theta$ of (16) gives, in turn, $c_1 = 0$ and $\lambda = n$, $n = 1$, $2, 3, \ldots$. Hence $\Theta = c_2 \sin n\theta$. In this problem, unlike the problem in Example 1, $n = 0$ is not an eigenvalue. With $\lambda = n$ the solution of (15) is $R = c_3 r^n + c_4 r^{-n}$. But the assumption that $u(r, \theta)$ is bounded at $r = 0$ prompts us to define $c_4 = 0$. Therefore $u_n = R(r)\Theta(\theta) = A_n r^n \sin n\theta$ and

$$u(r, \theta) = \sum_{n=1}^{\infty} A_n r^n \sin n\theta.$$

The remaining boundary condition at $r = c$ gives the sine series

$$u_0 = \sum_{n=1}^{\infty} A_n c^n \sin n\theta.$$

Consequently
$$A_n c^n = \frac{2}{\pi} \int_0^{\pi} u_0 \sin n\theta \, d\theta,$$

and so
$$A_n = \frac{2u_0}{\pi c^n} \frac{1 - (-1)^n}{n}.$$

Hence the solution of the problem is given by

$$u(r, \theta) = \frac{2u_0}{\pi} \sum_{n=1}^{\infty} \frac{1 - (-1)^n}{n} \left(\frac{r}{c}\right)^n \sin n\theta. \qquad \blacksquare$$

EXERCISES 13.1

Answers to odd-numbered problems begin on page AN-19.

In Problems 1–4 find the steady-state temperature $u(r, \theta)$ in a circular plate of radius $r = 1$ if the temperature on the circumference is as given.

1. $u(1, \theta) = \begin{cases} u_0, & 0 < \theta < \pi \\ 0, & \pi < \theta < 2\pi \end{cases}$ **2.** $u(1, \theta) = \begin{cases} \theta, & 0 < \theta < \pi \\ \pi - \theta, & \pi < \theta < 2\pi \end{cases}$

3. $u(1, \theta) = 2\pi\theta - \theta^2$, $0 < \theta < 2\pi$ **4.** $u(1, \theta) = \theta$, $0 < \theta < 2\pi$

5. Solve the exterior Dirichlet problem for a circular disk of radius c if $u(c, \theta) = f(\theta)$, $0 < \theta < 2\pi$. In other words, find the steady-state temperature $u(r, \theta)$ in a plate that coincides with the entire xy-plane in which a circular hole of radius c has been cut out around the origin and the temperature on the circumference of the hole is $f(\theta)$. [*Hint:* Assume that the temperature is bounded as $r \to \infty$.]

6. Find the steady-state temperature in the quarter-circular plate shown in Figure 13.4.

7. If the boundaries $\theta = 0$ and $\theta = \pi/2$ in Figure 13.4 are insulated, we then have, respectively,

$$\left.\frac{\partial u}{\partial \theta}\right|_{\theta=0} = 0, \quad \left.\frac{\partial u}{\partial \theta}\right|_{\theta=\pi/2} = 0.$$

Find the steady-state temperature if $u(c, \theta) = \begin{cases} 1, & 0 < \theta < \pi/4 \\ 0, & \pi/4 < \theta < \pi/2. \end{cases}$

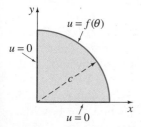

$u = f(\theta)$

$u = 0$

$u = 0$

Figure 13.4

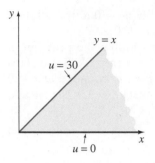

Figure 13.5

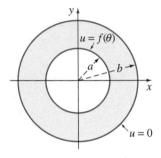

Figure 13.6

8. Find the steady-state temperature in the infinite wedge-shaped plate shown in Figure 13.5. [*Hint:* Assume that the temperature is bounded as $r \to 0$ and as $r \to \infty$.]

9. Find the steady-state temperature $u(r, \theta)$ in the circular ring shown in Figure 13.6. [*Hint:* Proceed as in Example 1.]

10. If the boundary conditions for the circular ring in Figure 13.6 are $u(a, \theta) = u_0, u(b, \theta) = u_1, 0 < \theta < 2\pi, u_0$ and u_1 constants, show that the steady-state temperature is given by

$$u(r, \theta) = \frac{u_0 \ln(r/b) - u_1 \ln(r/a)}{\ln(a/b)}.$$

[*Hint:* Try a solution of the form $u(r, \theta) = v(r, \theta) + \psi(r)$.]

11. Find the steady-state temperature $u(r, \theta)$ in a semicircular ring if

$$u(a, \theta) = \theta(\pi - \theta), \quad u(b, \theta) = 0, \quad 0 < \theta < \pi$$
$$u(r, 0) = 0, \qquad\qquad u(r, \pi) = 0, \quad a < r < b.$$

12. Find the steady-state temperature $u(r, \theta)$ in a semicircular plate of radius $r = 1$ if

$$u(1, \theta) = u_0, \quad 0 < \theta < \pi$$
$$u(r, 0) = 0, \quad u(r, \pi) = u_0, \quad 0 < r < 1,$$

u_0 a constant.

13. Find the steady-state temperature $u(r, \theta)$ in a semicircular plate of radius $r = 2$ if

$$u(2, \theta) = \begin{cases} u_0, & 0 < \theta < \pi/2 \\ 0, & \pi/2 < \theta < \pi, \end{cases}$$

u_0 a constant, and the edges $\theta = 0$ and $\theta = \pi$ are insulated.

Discussion Problems

14. **(a)** Find the series solution for $u(r, \theta)$ in Example 1 when

$$u(1, \theta) = \begin{cases} 100, & 0 < \theta < \pi \\ 0, & \pi < \theta < 2\pi. \end{cases}$$

(See Problem 1.)

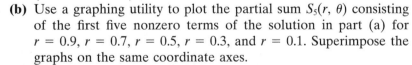

(b) Use a graphing utility to plot the partial sum $S_5(r, \theta)$ consisting of the first five nonzero terms of the solution in part (a) for $r = 0.9, r = 0.7, r = 0.5, r = 0.3$, and $r = 0.1$. Superimpose the graphs on the same coordinate axes.

(c) Approximate the temperatures $u(0.9, 1.3), u(0.7, 2), u(0.5, 3.5), u(0.3, 4), u(0.1, 5.5)$. Then approximate $u(0.9, 2\pi - 1.3), u(0.7, 2\pi - 2), u(0.5, 2\pi - 3.5), u(0.3, 2\pi - 4), u(0.1, 2\pi - 5.5)$.

(d) What is the temperature at the center of the circular plate? Discuss why it is appropriate to call this value the average temperature in the plate. [*Hint:* Look at the graphs in part (b) and look at the numbers in part (c).]

15. Consider the circular ring shown in Figure 13.6. Discuss how the steady-state temperature $u(r, \theta)$ can be found when the boundary conditions are $u(a, \theta) = f(\theta), u(b, \theta) = g(\theta), 0 \le \theta \le 2\pi$.

13.2 PROBLEMS IN POLAR AND CYLINDRICAL COORDINATES: BESSEL FUNCTIONS

• Heat equation in polar coordinates • Wave equation in polar coordinates • Problems with radial symmetry • Radial vibrations • Using Fourier-Bessel series • Standing waves • Nodal line • Laplacian in cylindrical coordinates

The two-dimensional heat and wave equations

$$k\left(\frac{\partial^2 u}{\partial x^2} + \frac{\partial^2 u}{\partial y^2}\right) = \frac{\partial u}{\partial t} \quad \text{and} \quad a^2\left(\frac{\partial^2 u}{\partial x^2} + \frac{\partial^2 u}{\partial y^2}\right) = \frac{\partial^2 u}{\partial t^2}$$

expressed in polar coordinates are, in turn,

$$k\left(\frac{\partial^2 u}{\partial r^2} + \frac{1}{r}\frac{\partial u}{\partial r} + \frac{1}{r^2}\frac{\partial^2 u}{\partial \theta^2}\right) = \frac{\partial u}{\partial t} \quad \text{and} \quad a^2\left(\frac{\partial^2 u}{\partial r^2} + \frac{1}{r}\frac{\partial u}{\partial r} + \frac{1}{r^2}\frac{\partial^2 u}{\partial \theta^2}\right) = \frac{\partial^2 u}{\partial t^2},$$

where $u = u(r, \theta, t)$. To solve a boundary-value problem involving either of these equations by separation of variables we define $u = R(r)\,\Theta(\theta)\,T(t)$. As in Section 12.8, this assumption leads to multiple infinite series. See Problem 12 in Exercises 13.2. In the discussion that follows we shall consider a simpler, but still important, class of problems.

Radial Symmetry Boundary-value problems in polar coordinates in which the unknown function u is independent of the angular coordinate θ are said to possess **radial symmetry.** In this case the heat and wave equations take, respectively, the forms

$$k\left(\frac{\partial^2 u}{\partial r^2} + \frac{1}{r}\frac{\partial u}{\partial r}\right) = \frac{\partial u}{\partial t} \quad \text{and} \quad a^2\left(\frac{\partial^2 u}{\partial r^2} + \frac{1}{r}\frac{\partial u}{\partial r}\right) = \frac{\partial^2 u}{\partial t^2}, \qquad \textbf{(1)}$$

where $u = u(r, t)$. Vibrations described by the second equation in (1) are said to be **radial vibrations.**

The first example deals with the free radial vibrations of a thin circular membrane. We assume that the displacements are small and that the motion is such that each point on the membrane moves in a direction perpendicular to the xy-plane (transverse vibrations)—that is, the u-axis is perpendicular to the xy-plane. A physical model to keep in mind while reading this example is a vibrating drumhead.

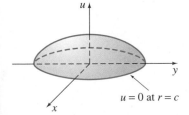

$u = 0$ at $r = c$

Figure 13.7

EXAMPLE 1   **Radial Vibrations of a Circular Membrane**

Find the displacement $u(r, t)$ of a circular membrane of radius c clamped along its circumference if its initial displacement is $f(r)$ and its initial velocity is $g(r)$. See Figure 13.7.

Solution   The boundary-value problem to be solved is

$$a^2 \left(\frac{\partial^2 u}{\partial r^2} + \frac{1}{r} \frac{\partial u}{\partial r} \right) = \frac{\partial^2 u}{\partial t^2}, \quad 0 < r < c, \quad t > 0$$

$$u(c, t) = 0, \quad t > 0$$

$$u(r, 0) = f(r), \quad \left. \frac{\partial u}{\partial t} \right|_{t=0} = g(r), \quad 0 < r < c.$$

Substituting $u = R(r)T(t)$ into the partial differential equation and separating variables gives

$$\frac{R'' + \frac{1}{r} R'}{R} = \frac{T''}{a^2 T} = -\lambda^2$$

and
$$rR'' + R' + \lambda^2 rR = 0 \qquad \text{(2)}$$
$$T'' + a^2 \lambda^2 T = 0. \qquad \text{(3)}$$

Now (2) is not a Cauchy-Euler equation but the parametric Bessel differential equation with order $\nu = 0$. Its general solution is

$$R = c_1 J_0(\lambda r) + c_2 Y_0(\lambda r).$$

The general solution of the familiar equation (3) is

$$T = c_3 \cos a\lambda t + c_4 \sin a\lambda t.$$

Now recall $Y_0(\lambda r) \to -\infty$ as $r \to 0^+$, and so the implicit assumption that the displacement $u(r, t)$ should be bounded at $r = 0$ forces us to define $c_2 = 0$. Thus $R = c_1 J_0(\lambda r)$.

Since the boundary condition $u(c, t) = 0$ is equivalent to $R(c) = 0$, we must have $c_1 J_0(\lambda c) = 0$. We rule out $c_1 = 0$ (this would lead to a trivial solution of the partial differential equation), so consequently

$$J_0(\lambda c) = 0. \qquad \text{(4)}$$

This equation defines the positive eigenvalues λ_n of the problem: If x_n are the positive roots of (4), then $\lambda_n = x_n/c$. Product solutions that satisfy the partial differential equation and the boundary condition are

$$u_n = RT = (A_n \cos a\lambda_n t + B_n \sin a\lambda_n t) J_0(\lambda_n r), \qquad \text{(5)}$$

where we have done the usual relabeling of the constants. The superposition principle then gives

$$u(r, t) = \sum_{n=1}^{\infty} (A_n \cos a\lambda_n t + B_n \sin a\lambda_n t) J_0(\lambda_n r). \qquad \text{(6)}$$

The given initial conditions determine the coefficients A_n and B_n.
Setting $t = 0$ in (6) and using $u(r, 0) = f(r)$ gives

$$f(r) = \sum_{n=1}^{\infty} A_n J_0(\lambda_n r). \qquad \text{(7)}$$

This last result is recognized as the Fourier-Bessel expansion of the function f on the interval $(0, c)$. Hence by a direct comparison of (4) and (7)

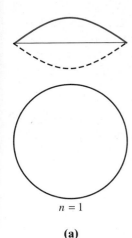

$n = 1$

(a)

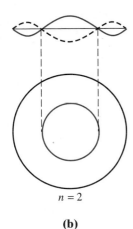

$n = 2$

(b)

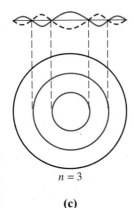

$n = 3$

(c)

Figure 13.8

with (7) and (15) of Section 11.5 we can identify the coefficients A_n with those given in (16) of Section 11.5:

$$A_n = \frac{2}{c^2 J_1^2(\lambda_n c)} \int_0^c r J_0(\lambda_n r) f(r) \, dr. \tag{8}$$

Next, we differentiate (6) with respect to t, set $t = 0$, and use $u_t(r, 0) = g(r)$:

$$g(r) = \sum_{n=1}^{\infty} a \lambda_n B_n J_0(\lambda_n r).$$

This is now a Fourier-Bessel expansion of the function g. By identifying the total coefficient $a\lambda_n B_n$ with (16) of Section 11.5 we can write

$$B_n = \frac{2}{a \lambda_n c^2 J_1^2(\lambda_n c)} \int_0^c r J_0(\lambda_n r) g(r) \, dr. \tag{9}$$

Finally, the solution of the given boundary-value problem is the series (6) with A_n and B_n as defined in (8) and (9). ∎

Analogous to (8) of Section 12.4, the product solutions (5) are called **standing waves.** For $n = 1, 2, 3, \ldots$, the standing waves are basically the graph of $J_0(\lambda_n r)$ with the time-varying amplitude

$$A_n \cos a\lambda_n t + B_n \sin a\lambda_n t.$$

The standing waves at different values of time are represented by the dashed graphs in Figure 13.8. The zeros of each standing wave in the interval $(0, c)$ are the roots of $J_0(\lambda_n r) = 0$ and correspond to the set of points on a standing wave where there is no motion. This set of points is called a **nodal line.** If (as in Example 1) the positive roots of $J_0(\lambda_n c) = 0$ are denoted by x_n, then $\lambda_n c = x_n$ implies $\lambda_n = x_n/c$ and consequently the zeros of the standing waves are determined from

$$J_0(\lambda_n r) = J_0\left(\frac{x_n}{c} r\right) = 0.$$

Now the first three positive zeros of J_0 are (approximately) $x_1 = 2.4$, $x_2 = 5.5$, and $x_3 = 8.7$. Thus for $n = 1$, the first positive root of

$$J_0\left(\frac{x_1}{c} r\right) = 0 \quad \text{is} \quad \frac{2.4}{c} r = 2.4 \quad \text{or} \quad r = c.$$

Since we are seeking zeros of the standing waves in the open interval $(0, c)$, the last result means that the first standing wave has no nodal line. For $n = 2$, the first two positive roots of

$$J_0\left(\frac{x_2}{c} r\right) = 0 \quad \text{are determined from} \quad \frac{5.5}{c} r = 2.4 \quad \text{and} \quad \frac{5.5}{c} r = 5.5.$$

Thus the second standing wave has one nodal line defined by $r = x_1 c/x_2 = 2.4c/5.5$. Note that $r \approx 0.44c < c$. For $n = 3$, a similar analysis

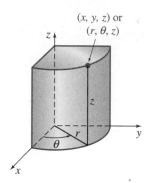

Figure 13.9

shows that there are two nodal lines defined by $r = x_1 c / x_3 = 2.4c/8.7$ and $r = x_2 c / x_3 = 5.5c/8.7$. In general, the nth standing wave has $n - 1$ nodal lines $r = x_1 c / x_n$, $r = x_2 c / x_n$, $\ldots$, $r = x_{n-1} c / x_n$. Since $r = constant$ is an equation of a circle in polar coordinates, we see in Figure 13.8 that the nodal lines of a standing wave are concentric circles.

Laplacian in Cylindrical Coordinates   From Figure 13.9 we see that the relationship between the cylindrical coordinates of a point in space and its rectangular coordinates is given by

$$x = r \cos \theta, \quad y = r \sin \theta, \quad z = z.$$

It follows immediately from the derivation of the Laplacian in polar coordinates (page 562) that in cylindrical coordinates the three-dimensional Laplacian

$$\nabla^2 u = \frac{\partial^2 u}{\partial x^2} + \frac{\partial^2 u}{\partial y^2} + \frac{\partial^2 u}{\partial z^2} \quad \text{becomes} \quad \nabla^2 u = \frac{\partial^2 u}{\partial r^2} + \frac{1}{r}\frac{\partial u}{\partial r} + \frac{1}{r^2}\frac{\partial^2 u}{\partial \theta^2} + \frac{\partial^2 u}{\partial z^2}.$$

EXAMPLE 2   **Steady Temperatures in a Circular Cylinder**

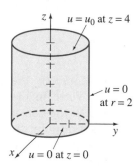

Figure 13.10

Find the steady-state temperature in the circular cylinder shown in Figure 13.10.

Solution   The prescribed boundary conditions suggest that the temperature u has radial symmetry. Accordingly, $u(r, z)$ is determined from

$$\frac{\partial^2 u}{\partial r^2} + \frac{1}{r}\frac{\partial u}{\partial r} + \frac{\partial^2 u}{\partial z^2} = 0, \quad 0 < r < 2, \quad 0 < z < 4$$

$$u(2, z) = 0, \quad 0 < z < 4$$

$$u(r, 0) = 0, \quad u(r, 4) = u_0, \quad 0 < r < 2.$$

Using $u = R(r)Z(z)$ and separating variables gives

$$\frac{R'' + \dfrac{1}{r}R'}{R} = -\frac{Z''}{Z} = -\lambda^2$$

and

$$rR'' + R' + \lambda^2 rR = 0 \tag{10}$$

$$Z'' - \lambda^2 Z = 0. \tag{11}$$

A negative separation constant is used here since there is no reason to expect the solution $u(r, z)$ to be periodic in z. The solution of (10) is

$$R = c_1 J_0(\lambda r) + c_2 Y_0(\lambda r),$$

and since the solution of (11) is defined on the finite interval $[0, 2]$, we write its general solution as

$$Z - c_3 \cosh \lambda z + c_4 \sinh \lambda z.$$

As in Example 1, the assumption that the function u is bounded at $r = 0$ demands that $c_2 = 0$. The condition $u(2, z) = 0$ implies

$R(2) = 0$. This equation,

$$J_0(2\lambda) = 0, \tag{12}$$

defines the positive eigenvalues λ_n of the problem. Last, $Z(0) = 0$ implies $c_3 = 0$. Hence we have $R = c_1 J_0(\lambda_n r)$, $Z = c_4 \sinh \lambda_n z$, and

$$u_n = RZ = A_n \sinh \lambda_n z J_0(\lambda_n r)$$

$$u(r, z) = \sum_{n=1}^{\infty} A_n \sinh \lambda_n z J_0(\lambda_n r).$$

The remaining boundary condition at $z = 4$ then gives the Fourier-Bessel series

$$u_0 = \sum_{n=1}^{\infty} A_n \sinh 4\lambda_n J_0(\lambda_n r),$$

so that in view of (12) the coefficients are defined by (16) of Section 11.5,

$$A_n \sinh 4\lambda_n = \frac{2u_0}{2^2 J_1^2(2\lambda_n)} \int_0^2 r J_0(\lambda_n r)\, dr.$$

To evaluate the last integral we first use the substitution $t = \lambda_n r$, followed by $\dfrac{d}{dt}[tJ_1(t)] = tJ_0(t)$:

$$A_n \sinh 4\lambda_n = \frac{u_0}{2\lambda_n^2 J_1^2(2\lambda_n)} \int_0^{2\lambda_n} \frac{d}{dt}[tJ_1(t)]\, dt = \frac{u_0}{\lambda_n J_1(2\lambda_n)}.$$

Finally we arrive at $A_n = \dfrac{u_0}{\lambda_n \sinh 4\lambda_n J_1(2\lambda_n)}.$

Thus the temperature in the cylinder is given by

$$u(r, z) = u_0 \sum_{n=1}^{\infty} \frac{\sinh \lambda_n z J_0(\lambda_n r)}{\lambda_n \sinh 4\lambda_n J_1(2\lambda_n)}. \qquad \blacksquare$$

EXERCISES 13.2

Answers to odd-numbered problems begin on page AN-20.

1. Find the displacement $u(r, t)$ in Example 1 if $f(r) = 0$ and the circular membrane is given an initial unit velocity in the upward direction.

2. A circular membrane of unit radius 1 is clamped along its circumference. Find the displacement $u(r, t)$ if the membrane starts from rest from the initial displacement $f(r) = 1 - r^2$, $0 < r < 1$. [*Hint:* See Problem 10 in Exercises 11.5.]

3. Find the steady-state temperature $u(r, z)$ in the cylinder in Example 2 if the boundary conditions are $u(2, z) = 0$, $0 < z < 4$, $u(r, 0) = u_0$, $u(r, 4) = 0$, $0 < r < 2$.

4. If the lateral side of the cylinder in Example 2 is insulated, then

$$\frac{\partial u}{\partial r}\bigg|_{r=2} = 0, \quad 0 < z < 4.$$

(a) Find the steady-state temperature $u(r, z)$ when $u(r, 4) = f(r)$, $0 < r < 2$.

(b) Show that the steady-state temperature in part (a) reduces to $u(r, z) = u_0 z/4$ when $f(r) = u_0$. [*Hint:* Use (11) of Section 11.5.]

5. The temperature in a circular plate of radius c is determined from the boundary-value problem

$$k\left(\frac{\partial^2 u}{\partial r^2} + \frac{1}{r}\frac{\partial u}{\partial r}\right) = \frac{\partial u}{\partial t}, \quad 0 < r < c, \quad t > 0$$

$$u(c, t) = 0, \quad t > 0$$

$$u(r, 0) = f(r), \quad 0 < r < c.$$

Solve for $u(r, t)$.

6. Solve Problem 5 if the edge $r = c$ of the plate is insulated.

7. When there is heat transfer from the lateral side of an infinite circular cylinder of unit radius (see Figure 13.11) into a surrounding medium at temperature zero, the temperature inside the cylinder is determined from

$$k\left(\frac{\partial^2 u}{\partial r^2} + \frac{1}{r}\frac{\partial u}{\partial r}\right) = \frac{\partial u}{\partial t}, \quad 0 < r < 1, \quad t > 0$$

$$\frac{\partial u}{\partial r}\bigg|_{r=1} = -hu(1, t), \quad h > 0, \quad t > 0$$

$$u(r, 0) = f(r), \quad 0 < r < 1.$$

Solve for $u(r, t)$.

8. Find the steady-state temperature $u(r, z)$ in a semi-infinite cylinder of unit radius ($z \geq 0$) if there is heat transfer from its lateral side into a surrounding medium at temperature zero and if the temperature of the base $z = 0$ is held at a constant temperature u_0.

9. A circular plate is a composite of two different materials in the form of concentric circles. See Figure 13.12. The temperature in the plate is determined from the boundary-value problem

$$\frac{\partial^2 u}{\partial r^2} + \frac{1}{r}\frac{\partial u}{\partial r} = \frac{\partial u}{\partial t}, \quad 0 < r < 2, \quad t > 0$$

$$u(2, t) = 100, \quad t > 0$$

$$u(r, 0) = \begin{cases} 200, & 0 < r < 1 \\ 100, & 1 < r < 2 \end{cases}$$

Solve for $u(r, t)$. [*Hint:* Let $u(r, t) = v(r, t) + \psi(r)$.]

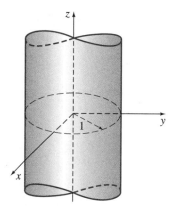

Figure 13.11

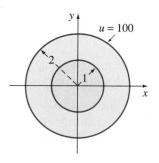

Figure 13.12

10. Solve the boundary-value problem

$$\frac{\partial^2 u}{\partial r^2} + \frac{1}{r}\frac{\partial u}{\partial r} + \beta = \frac{\partial u}{\partial t}, \quad 0 < r < 1, \quad t > 0, \quad \beta \text{ a constant}$$

$$u(1, t) = 0, \quad t > 0$$

$$u(r, 0) = 0, \quad 0 < r < 1.$$

11. The horizontal displacement $u(x, t)$ of a heavy chain of length L oscillating in a vertical plane satisfies the partial differential equation

$$g\frac{\partial}{\partial x}\left(x\frac{\partial u}{\partial x}\right) = \frac{\partial^2 u}{\partial t^2}, \quad 0 < x < L, \quad t > 0.$$

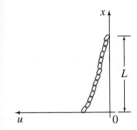

Figure 13.13

See Figure 13.13.

(a) Using $-\lambda^2$ as a separation constant, show that the ordinary differential equation in the spatial variable x is $xX'' + X' + \lambda^2 X = 0$. Solve this equation by means of the substitution $x = \tau^2/4$.

(b) Use the result of part (a) to solve the given partial differential equation subject to

$$u(L, t) = 0, \quad t > 0$$

$$u(x, 0) = f(x), \quad \frac{\partial u}{\partial t}\bigg|_{t=0} = 0, \quad 0 < x < L.$$

[*Hint:* Assume the oscillations at the free end $x = 0$ are finite.]

12. In this problem we consider the general case—that is, with θ dependence—of the vibrating circular membrane of radius c:

$$a^2\left(\frac{\partial^2 u}{\partial r^2} + \frac{1}{r}\frac{\partial u}{\partial r} + \frac{1}{r^2}\frac{\partial^2 u}{\partial \theta^2}\right) = \frac{\partial^2 u}{\partial t^2}, \quad 0 < r < c, \quad t > 0$$

$$u(c, \theta, t) = 0, \quad 0 < \theta < 2\pi, \quad t > 0$$

$$u(r, \theta, 0) = f(r, \theta), \quad 0 < r < c, \quad 0 < \theta < 2\pi$$

$$\frac{\partial u}{\partial t}\bigg|_{t=0} = g(r, \theta), \quad 0 < r < c, \quad 0 < \theta < 2\pi.$$

(a) Assume that $u = R(r)\Theta(\theta)T(t)$ and that the separation constants are $-\lambda^2$ and $-\nu^2$. Show that the separated differential equations are

$$T'' + a^2\lambda^2 T = 0, \quad \Theta'' + \nu^2\Theta = 0$$

$$r^2 R'' + rR' + (\lambda^2 r^2 - \nu^2)R = 0.$$

(b) Solve the separated equations.

(c) Show that the eigenvalues and eigenfunctions of the problem are as follows:

Eigenvalues: $\nu = n, n = 0, 1, 2, \ldots$; eigenfunctions: 1, $\cos n\theta$, $\sin n\theta$.

Eigenvalues: $\lambda_{ni} = x_{ni}/c, i = 1, 2, \ldots$, where, for each n, x_{ni} are the positive roots of $J_n(\lambda c) = 0$; eigenfunctions: $J_n(\lambda_{ni}r) = 0$.

(d) Use the superposition principle to determine a multiple series solution. Do not attempt to evaluate the coefficients.

Computer Lab Assignments

13. Consider an idealized drum consisting of a thin membrane stretched over a circular frame of unit radius. When such a drum is struck at its center, one hears a sound that is frequently described as a dull thud rather than a melodic tone. We can model a single drumbeat using the boundary-value problem solved in Example 1.

(a) Find the solution $u(r, t)$ given in (6) when $c = 1$, $f(r) = 0$, and

$$g(r) = \begin{cases} -v_0, & 0 \le r < b \\ 0, & b \le r < 1. \end{cases}$$

(b) Show that the frequency of the standing wave $u_n(r, t)$ is $f_n = a\lambda_n / 2\pi$, where λ_n is the nth positive zero of $J_0(x)$. Unlike the solution of the one-dimensional wave equation in Section 12.4, the frequencies are not integer multiples of the fundamental frequency f_1. Show that $f_2 \approx 2.295 f_1$ and $f_3 \approx 3.598 f_1$. We say that the drumbeat produces **anharmonic overtones.** As a result, the displacement function $u(r, t)$ is not periodic, and so our ideal drum cannot produce a sustained tone.

(c) Let $a = 1$, $b = \frac{1}{4}$, and $v_0 = 1$ in your solution in part (a). Use a CAS to graph the fifth partial sum $S_5(r, t)$ for the times $t = 0, 0.1, 0.2, 0.3, \ldots, 5.9, 6.0$ on the interval $-1 \le r \le 1$. Use the animation capabilities of your CAS to produce a movie of these vibrations.

(d) For a greater challenge, use the 3D-plot application of your CAS to make a movie of the motion of the circular drum head that is shown in cross-section in part (c). [*Hint:* There are several ways of proceeding. For a fixed time, either graph u as a function of x and y using $r = \sqrt{x^2 + y^2}$ or use the equivalent of *Mathematica*'s **CylindricalPlot3D**.]

14. (a) Consider Example 1 with $a = 1$, $c = 10$, $g(r) = 0$, and $f(r) = 1 - r/10$, $0 < r < 10$. Use a CAS as an aid in finding the numerical values of the first three eigenvalues $\lambda_1, \lambda_2, \lambda_3$ of the boundary-value problem and the first three coefficients A_1, A_2, A_3 of the solution $u(r, t)$ given in (6). Write the third partial sum $S_3(r, t)$ of the series solution.

(b) Use a CAS to plot the graph of $S_3(r, t)$ for $t = 0, 4, 10, 12, 20$.

15. Solve Problem 5 with boundary conditions $u(c, t) = 200$, $u(r, 0) = 0$. With these imposed conditions, one would expect intuitively that at any interior point of the plate, $u(r, t) \to 200$ as $t \to \infty$. Assume that $c = 10$ and that the plate is cast iron so that $k = 0.1$ (approximately). Use a CAS as an aid in finding the numerical values of the first five eigenvalues $\lambda_1, \lambda_2, \lambda_3, \lambda_4, \lambda_5$ of the boundary-value problem and the five coefficients A_1, A_2, A_3, A_4, A_5 in the solution $u(r, t)$. Let the corresponding approximate solution be denoted by $S_5(r, t)$. Plot $S_5(5, t)$ and $S_5(0, t)$ on a sufficiently large time interval $0 \le t \le T$. Use the plots of $S_5(5, t)$ and $S_5(0, t)$ to estimate the times (in seconds) for which $u(5, t) \approx 100$ and $u(0, t) \approx 100$. Repeat for $u(5, t) \approx 200$ and $u(0, t) \approx 200$.

13.3 PROBLEMS IN SPHERICAL COORDINATES: LEGENDRE POLYNOMIALS

• Laplacian in spherical coordinates • Laplace's equation in spherical coordinates • Steady-state temperatures • Using Fourier-Legendre series

The relationship between spherical coordinates in 3-space and rectangular coordinates is given by

$$x = r \sin \theta \cos \phi, \quad y = r \sin \theta \sin \phi, \quad z = r \cos \theta. \tag{1}$$

The angle θ is called the polar angle and the angle ϕ is called the azimuthal angle.* See Figure 13.14. Although r, θ, and ϕ can take on any real values, in applications involving partial differential equations in spherical coordinates it usually is assumed that $r \geq 0$, $0 \leq \theta \leq \pi$, and $0 \leq \phi \leq 2\pi$.

We use the equations in (1) to convert the three-dimensional Laplacian of a function $u(x, y, z)$,

$$\nabla^2 u = \frac{\partial^2 u}{\partial x^2} + \frac{\partial^2 u}{\partial y^2} + \frac{\partial^2 u}{\partial z^2},$$

into spherical coordinates.

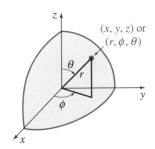

Figure 13.14

Laplacian in Spherical Coordinates If we use the Chain Rule as we did in Section 13.1, the three-dimensional Laplacian becomes

$$\nabla^2 u = \frac{\partial^2 u}{\partial r^2} + \frac{2}{r} \frac{\partial u}{\partial r} + \frac{1}{r^2 \sin^2 \theta} \frac{\partial^2 u}{\partial \phi^2} + \frac{1}{r^2} \frac{\partial^2 u}{\partial \theta^2} + \frac{\cot \theta}{r^2} \frac{\partial u}{\partial \theta}, \tag{2}$$

where $u = u(r, \theta, \phi)$. As you might imagine, problems involving (2) can be quite formidable. Consequently we shall consider only a few of the simpler problems that are independent of the azimuthal angle ϕ.

EXAMPLE 1 **Steady Temperatures in a Sphere**

Find the steady-state temperature $u(r, \theta)$ in the sphere shown in Figure 13.15.

Solution The temperature is determined from

$$\frac{\partial^2 u}{\partial r^2} + \frac{2}{r} \frac{\partial u}{\partial r} + \frac{1}{r^2} \frac{\partial^2 u}{\partial \theta^2} + \frac{\cot \theta}{r^2} \frac{\partial u}{\partial \theta} = 0, \quad 0 < r < c, \quad 0 < \theta < \pi$$

$$u(c, \theta) = f(\theta), \quad 0 < \theta < \pi.$$

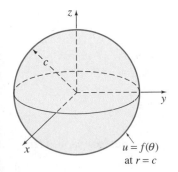

Figure 13.15

---

*The reader is cautioned that there is no agreement about the use of symbols to denote the two angles in Figure 13.14. So if you consult other texts, you should be aware that often the symbol ϕ is used for the polar angle and the symbol θ for the azimuthal angle. Also, some authors use ρ instead of r.

If $u = R(r)\Theta(\theta)$, the partial differential equation separates as

$$\frac{r^2 R'' + 2rR'}{R} = -\frac{\Theta'' + \cot\theta\,\Theta'}{\Theta} = \lambda^2,$$

and so

$$r^2 R'' + 2rR' - \lambda^2 R = 0 \tag{3}$$

$$\sin\theta\,\Theta'' + \cos\theta\,\Theta' + \lambda^2 \sin\theta\,\Theta = 0. \tag{4}$$

After we substitute $x = \cos\theta$, $0 \le \theta \le \pi$, (4) becomes

$$(1 - x^2)\frac{d^2\Theta}{dx^2} - 2x\frac{d\Theta}{dx} + \lambda^2\Theta = 0, \quad -1 \le x \le 1. \tag{5}$$

The latter equation is a form of Legendre's equation (see Problem 31 in Exercises 6.3). Now the only solutions of (5) that are continuous and have continuous derivatives on the closed interval $[-1, 1]$ are the Legendre polynomials $P_n(x)$ corresponding to $\lambda^2 = n(n + 1)$, $n = 0, 1, 2, \ldots$. Thus we take the solutions of (4) to be

$$\Theta = P_n(\cos\theta).$$

Furthermore, when $\lambda^2 = n(n + 1)$, the general solution of the Cauchy-Euler equation (3) is

$$R = c_1 r^n + c_2 r^{-(n+1)}.$$

Since we again expect $u(r, \theta)$ to be bounded at $r = 0$, we define $c_2 = 0$. Hence $u_n = A_n r^n P_n(\cos\theta)$, and

$$u(r, \theta) = \sum_{n=0}^{\infty} A_n r^n P_n(\cos\theta).$$

At $r = c$,

$$f(\theta) = \sum_{n=0}^{\infty} A_n c^n P_n(\cos\theta).$$

Therefore $A_n c^n$ are the coefficients of the Fourier-Legendre series (23) of Section 11.5:

$$A_n = \frac{2n + 1}{2c^n}\int_0^\pi f(\theta) P_n(\cos\theta)\sin\theta\,d\theta.$$

It follows that the solution is

$$u(r, \theta) = \sum_{n=0}^{\infty} \left(\frac{2n + 1}{2}\int_0^\pi f(\theta) P_n(\cos\theta)\sin\theta\,d\theta\right)\left(\frac{r}{c}\right)^n P_n(\cos\theta). \quad \blacksquare$$

EXERCISES 13.3

Answers to odd-numbered problems begin on page AN-20.

1. Solve the problem in Example 1 if

$$f(\theta) = \begin{cases} 50, & 0 < \theta < \pi/2 \\ 0, & \pi/2 < \theta < \pi. \end{cases}$$

Write out the first four nonzero terms of the series solution. [*Hint:* See Example 3 in Section 11.5.]

2. The solution $u(r, \theta)$ in Example 1 could also be interpreted as the potential inside the sphere due to a charge distribution $f(\theta)$ on its surface. Find the potential outside the sphere.

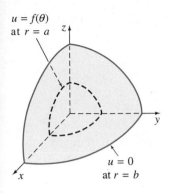

$u = f(\theta)$
at $r = a$

$u = 0$
at $r = b$

Figure 13.16

3. Find the solution of the problem in Example 1 if $f(\theta) = \cos\theta$, $0 < \theta < \pi$. [*Hint:* $P_1(\cos\theta) = \cos\theta$. Use orthogonality.]

4. Find the solution of the problem in Example 1 if $f(\theta) = 1 - \cos 2\theta$, $0 < \theta < \pi$. [*Hint:* See Problem 16 in Exercises 11.5.]

5. Find the steady-state temperature $u(r, \theta)$ within a hollow sphere $a < r < b$ if its inner surface $r = a$ is kept at temperature $f(\theta)$ and its outer surface $r = b$ is kept at temperature zero. The sphere in the first octant is shown in Figure 13.16.

6. The steady-state temperature in a hemisphere of radius $r = c$ is determined from

$$\frac{\partial^2 u}{\partial r^2} + \frac{2}{r}\frac{\partial u}{\partial r} + \frac{1}{r^2}\frac{\partial^2 u}{\partial \theta^2} + \frac{\cot\theta}{r^2}\frac{\partial u}{\partial \theta} = 0, \quad 0 < r < c, \quad 0 < \theta < \frac{\pi}{2}$$

$$u\left(r, \frac{\pi}{2}\right) = 0, \quad 0 < r < c$$

$$u(c, \theta) = f(\theta), \quad 0 < \theta < \frac{\pi}{2}.$$

Solve for $u(r, \theta)$. [*Hint:* $P_n(0) = 0$ only if n is odd. Also see Problem 18 in Exercises 11.5.]

7. Solve Problem 6 when the base of the hemisphere is insulated; that is,

$$\left.\frac{\partial u}{\partial \theta}\right|_{\theta = \pi/2} = 0, \quad 0 < r < c.$$

8. Solve Problem 6 for $r > c$.

9. The time-dependent temperature within a sphere of unit radius is determined from

$$\frac{\partial^2 u}{\partial r^2} + \frac{2}{r}\frac{\partial u}{\partial r} = \frac{\partial u}{\partial t}, \quad 0 < r < 1, \quad t > 0$$

$$u(1, t) = 100, \quad t > 0$$

$$u(r, 0) = 0, \quad 0 < r < 1.$$

Solve for $u(r, t)$. [*Hint:* Verify that the left side of the partial differential equation can be written as $\dfrac{1}{r}\dfrac{\partial^2}{\partial r^2}(ru)$. Let $ru(r, t) = v(r, t) + \psi(r)$. Use only functions that are bounded as $r \to 0$.]

10. A uniform solid sphere of radius 1 at an initial constant temperature u_0 throughout is dropped into a large container of fluid that is kept at a constant temperature u_1 ($u_1 > u_0$) for all time. See Figure 13.17. Since there is heat transfer across the boundary $r = 1$, the temperature $u(r, t)$ in the sphere is determined from the boundary-value problem

$$\frac{\partial^2 u}{\partial r^2} + \frac{2}{r}\frac{\partial u}{\partial r} = \frac{\partial u}{\partial t}, \quad 0 < r < 1, \quad t > 0$$

$$\left.\frac{\partial u}{\partial r}\right|_{r=1} = -h(u(1, t) - u_1), \quad 0 < h < 1$$

$$u(r, 0) = u_0, \quad 0 < r < 1.$$

Solve for $u(r, t)$. [*Hint:* Proceed as in Problem 9.]

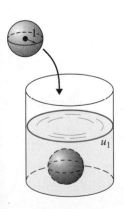

u_1

Figure 13.17

11. Solve the boundary-value problem involving spherical vibrations:

$$a^2\left(\frac{\partial^2 u}{\partial r^2} + \frac{2}{r}\frac{\partial u}{\partial r}\right) = \frac{\partial^2 u}{\partial t^2}, \quad 0 < r < c, \quad t > 0$$

$$u(c, t) = 0, \quad t > 0$$

$$u(r, 0) = f(r), \quad \frac{\partial u}{\partial t}\bigg|_{t=0} = g(r), \quad 0 < r < c.$$

[*Hint:* Write the left side of the partial differential equation as $a^2 \frac{1}{r}\frac{\partial^2}{\partial r^2}(ru)$. Let $v(r, t) = ru(r, t)$.]

12. A conducting sphere of radius $r = c$ is grounded and placed in a uniform electric field that has intensity E in the z-direction. The potential $u(r, \theta)$ outside the sphere is determined from the boundary-value problem

$$\frac{\partial^2 u}{\partial r^2} + \frac{2}{r}\frac{\partial u}{\partial r} + \frac{1}{r^2}\frac{\partial^2 u}{\partial \theta^2} + \frac{\cot\theta}{r^2}\frac{\partial u}{\partial \theta} = 0, \quad r > c, \quad 0 < \theta < \pi$$

$$u(c, \theta) = 0, \quad 0 < \theta < \pi$$

$$\lim_{r \to \infty} u(r, \theta) = -Ez = -Er\cos\theta.$$

Show that $\qquad u(r, \theta) = -Er\cos\theta + E\frac{c^3}{r^2}\cos\theta.$

[*Hint:* Explain why $\int_0^\pi \cos\theta\, P_n(\cos\theta)\sin\theta\, d\theta = 0$ for all nonnegative integers except $n = 1$. See (24) of Section 11.5.]

CHAPTER 13 IN REVIEW

Answers to odd-numbered problems begin on page AN-20.

1. Find the steady-state temperature $u(r, \theta)$ in a circular plate of radius c if the temperature on the circumference is given by

$$u(c, \theta) = \begin{cases} u_0, & 0 < \theta < \pi \\ -u_0, & \pi < \theta < 2\pi. \end{cases}$$

2. Find the steady-state temperature in the circular plate in Problem 1 if

$$u(c, \theta) = \begin{cases} 1, & 0 < \theta < \pi/2 \\ 0, & \pi/2 < \theta < 3\pi/2 \\ 1, & 3\pi/2 < \theta < 2\pi. \end{cases}$$

3. Find the steady-state temperature $u(r, \theta)$ in a semicircular plate of radius 1 if

$$u(1, \theta) = u_0(\pi\theta - \theta^2), \quad 0 < \theta < \pi$$

$$u(r, 0) = 0, \quad u(r, \pi) = 0, \quad 0 < r < 1.$$

4. Find the steady-state temperature $u(r, \theta)$ in the semicircular plate in Problem 3 if $u(1, \theta) = \sin\theta$, $0 < \theta < \pi$.

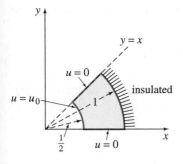

Figure 13.18

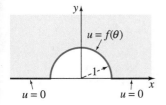

Figure 13.19

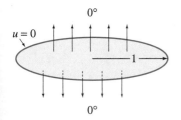

Figure 13.20

5. Find the steady-state temperature $u(r, \theta)$ in the plate shown in Figure 13.18.

6. Find the steady-state temperature $u(r, \theta)$ in the infinite plate shown in Figure 13.19.

7. Suppose heat is lost from the flat surfaces of a very thin circular unit disk into a surrounding medium at temperature zero. If the linear law of heat transfer applies, the heat equation assumes the form

$$\frac{\partial^2 u}{\partial r^2} + \frac{1}{r}\frac{\partial u}{\partial r} - hu = \frac{\partial u}{\partial t}, \quad h > 0, \quad 0 < r < 1, \quad t > 0.$$

See Figure 13.20. Find the temperature $u(r, t)$ if the edge $r = 1$ is kept at temperature zero and if initially the temperature of the plate is unity throughout.

8. Suppose x_k is a positive zero of J_0. Show that a solution of the boundary-value problem

$$a^2\left(\frac{\partial^2 u}{\partial r^2} + \frac{1}{r}\frac{\partial u}{\partial r}\right) = \frac{\partial^2 u}{\partial t^2}, \quad 0 < r < 1, \quad t > 0$$

$$u(1, t) = 0, \quad t > 0$$

$$u(r, 0) = u_0 J_0(x_k r), \quad \frac{\partial u}{\partial t}\bigg|_{t=0} = 0, \quad 0 < r < 1$$

is $u(r, t) = u_0 J_0(x_k r) \cos ax_k t$.

9. Find the steady-state temperature $u(r, z)$ in the cylinder in Figure 13.10 if the lateral side is kept at temperature zero, the top $z = 4$ is kept at temperature 50, and the base $z = 0$ is insulated.

10. Solve the boundary-value problem

$$\frac{\partial^2 u}{\partial r^2} + \frac{1}{r}\frac{\partial u}{\partial r} + \frac{\partial^2 u}{\partial z^2} = 0, \quad 0 < r < 1, \quad 0 < z < 1$$

$$\frac{\partial u}{\partial r}\bigg|_{r=1} = 0, \quad 0 < z < 1$$

$$u(r, 0) = f(r), \quad u(r, 1) = g(r), \quad 0 < r < 1.$$

11. Find the steady-state temperature $u(r, \theta)$ in a sphere of unit radius if the surface is kept at

$$u(1, \theta) = \begin{cases} 100, & 0 < \theta < \pi/2 \\ -100, & \pi/2 < \theta < \pi. \end{cases}$$

[*Hint:* See Problem 20 in Exercises 11.5.]

12. Solve the boundary-value problem

$$\frac{\partial^2 u}{\partial r^2} + \frac{2}{r}\frac{\partial u}{\partial r} = \frac{\partial^2 u}{\partial t^2}, \quad 0 < r < 1, \quad t > 0$$

$$\frac{\partial u}{\partial r}\bigg|_{r=1} = 0, \quad t > 0$$

$$u(r, 0) = f(r), \quad \frac{\partial u}{\partial t}\bigg|_{t=0} = g(r), \quad 0 < r < 1.$$

[*Hint:* Proceed as in Problems 9 and 10 in Exercises 13.3, but let $v(r, t) = ru(r, t)$. See Section 12.7.]

13. The function $u(x) = Y_0(\lambda a)J_0(\lambda x) - J_0(\lambda a)Y_0(\lambda x)$, $a > 0$ is a solution of the parametric Bessel equation

$$x^2 \frac{d^2u}{dx^2} + x\frac{du}{dx} + \lambda^2 x^2 u = 0$$

on the interval $a \leq x \leq b$. If the eigenvalues λ_n are defined by the positive roots of the equation $Y_0(\lambda a)J_0(\lambda b) - J_0(\lambda a)Y_0(\lambda b) = 0$, show that the functions

$$u_m(x) = Y_0(\lambda_m a)J_0(\lambda_m x) - J_0(\lambda_m a)Y_0(\lambda_m x)$$
$$u_n(x) = Y_0(\lambda_n a)J_0(\lambda_n x) - J_0(\lambda_n a)Y_0(\lambda_n x)$$

are orthogonal with respect to the weight function $p(x) = x$ on the interval $[a, b]$; that is,

$$\int_a^b xu_m(x)u_n(x)\, dx = 0, \quad m \neq n.$$

[*Hint:* Follow the procedure on page 506.]

14. Use the results of Problem 13 to solve the following boundary-value problem for the temperature $u(r, t)$ in a circular ring:

$$\frac{\partial^2 u}{\partial r^2} + \frac{1}{r}\frac{\partial u}{\partial r} = \frac{\partial u}{\partial t}, \quad a < r < b, \quad t > 0$$
$$u(a, t) = 0, \quad u(b, t) = 0, \quad t > 0$$
$$u(r, 0) = f(r), \quad a < r < b.$$

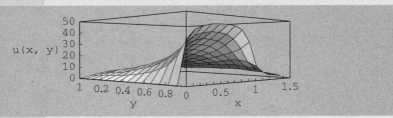

Solution of Laplace's equation; see Problem 22, page 607.

14 INTEGRAL TRANSFORM METHOD

INTRODUCTION The method of separation of variables that we employed in Chapters 12 and 13 is a powerful, but not universally applicable, method for solving boundary-value problems. If the partial differential equation in question is nonhomogeneous, or if the boundary conditions are time dependent, or if the domain of the spatial variable is infinite $(-\infty, \infty)$ or semi-infinite (a, ∞), we may be able to use an integral transform to solve the problem. In Section 14.2 we shall solve problems that involve the heat and wave equations by means of the familiar Laplace transform. In Section 14.4 three new integral transforms—Fourier transforms—will be introduced and used.

14.1   ERROR FUNCTION

- *Two integral definitions*   • *Complementary error function*
- *Probability integral*

There are many functions in mathematics that are defined by means of an integral. For example, in many traditional calculus texts the natural logarithm is defined as $\ln x = \int_1^x \frac{1}{t}\, dt, x > 0$. In earlier chapters we saw, albeit briefly, the error function erf(x), the complementary error function erfc(x), the sine integral function Si(x), the Fresnel sine integral $S(x)$, and the gamma function $\Gamma(\alpha)$; all of these functions are defined in terms of an integral. Before applying the Laplace transform to boundary-value problems we need to know a little more about the error function and the complementary error function. In this section we examine the graphs and a few of the obvious properties of erf(x) and erfc(x).

Properties and Graphs   Recall from (14) of Section 2.3 that the definitions of the **error function** erf(x) and **complementary error function** erfc(x) are, respectively,

$$\text{erf}(x) = \frac{2}{\sqrt{\pi}} \int_0^x e^{-u^2}\, du \quad \text{and} \quad \text{erfc}(x) = \frac{2}{\sqrt{\pi}} \int_x^\infty e^{-u^2}\, du. \qquad (1)$$

With the aid of polar coordinates it can be demonstrated that

$$\int_0^\infty e^{-u^2}\, du = \frac{\sqrt{\pi}}{2} \quad \text{or} \quad \frac{2}{\sqrt{\pi}} \int_0^\infty e^{-u^2}\, du = 1.$$

Thus from the additive interval property of definite integrals the last result is the same as

$$\frac{2}{\sqrt{\pi}} \left[\int_0^x e^{-u^2}\, du + \int_x^\infty e^{-u^2}\, du \right] = 1.$$

This shows that erf(x) and erfc(x) are related by the identity

$$\text{erf}(x) + \text{erfc}(x) = 1. \qquad (2)$$

The graphs of erf(x) and erfc(x) for $x \geq 0$ are given in Figure 14.1. Note that erf(0) = 0, erfc(0) = 1 and that erf(x) → 1, erfc(x) → 0 as $x \to \infty$. Other numerical values of erf(x) and erfc(x) can be obtained from a CAS or tables. In tables the error function is often referred to as the **probability integral**. The domain of erf(x) and of erfc(x) is $(-\infty, \infty)$. In Problem 11 in Exercises 14.1 you are asked to obtain the graph of each function on this interval and to deduce a few additional properties.

Table 14.1, of Laplace transforms, will be useful in the exercises in the next section. The proofs of these results are complicated and will not be given.

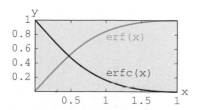

Figure 14.1 erf(x) and erfc(x), $x \geq 0$

TABLE 14.1

$f(t),\ a > 0$	$\mathscr{L}\{f(t)\} = F(s)$	$f(t),\ a > 0$	$\mathscr{L}\{f(t)\} = F(s)$
1. $\dfrac{1}{\sqrt{\pi t}}\,e^{-a^2/4t}$	$\dfrac{e^{-a\sqrt{s}}}{\sqrt{s}}$	**4.** $2\sqrt{\dfrac{t}{\pi}}\,e^{-a^2/4t} - a\,\mathrm{erfc}\left(\dfrac{a}{2\sqrt{t}}\right)$	$\dfrac{e^{-a\sqrt{s}}}{s\sqrt{s}}$
2. $\dfrac{a}{2\sqrt{\pi t^3}}\,e^{-a^2/4t}$	$e^{-a\sqrt{s}}$	**5.** $e^{ab}e^{b^2 t}\,\mathrm{erfc}\left(b\sqrt{t} + \dfrac{a}{2\sqrt{t}}\right)$	$\dfrac{e^{-a\sqrt{s}}}{\sqrt{s}(\sqrt{s} + b)}$
3. $\mathrm{erfc}\left(\dfrac{a}{2\sqrt{t}}\right)$	$\dfrac{e^{-a\sqrt{s}}}{s}$	**6.** $-e^{ab}e^{b^2 t}\,\mathrm{erfc}\left(b\sqrt{t} + \dfrac{a}{2\sqrt{t}}\right) + \mathrm{erfc}\left(\dfrac{a}{2\sqrt{t}}\right)$	$\dfrac{be^{-a\sqrt{s}}}{s(\sqrt{s} + b)}$

EXERCISES 14.1

Answers to odd-numbered problems begin on page AN-20.

1. (a) Show that $\mathrm{erf}(\sqrt{t}) = \dfrac{1}{\sqrt{\pi}}\displaystyle\int_0^t \dfrac{e^{-\tau}}{\sqrt{\tau}}\,d\tau$.

(b) Use the convolution theorem and the results of Problems 39 and 40 in Exercises 7.1 to show that

$$\mathscr{L}\{\mathrm{erf}(\sqrt{t})\} = \dfrac{1}{s\sqrt{s + 1}}.$$

2. Use the result of Problem 1 to show that

$$\mathscr{L}\{\mathrm{erfc}(\sqrt{t})\} = \dfrac{1}{s}\left[1 - \dfrac{1}{\sqrt{s + 1}}\right].$$

3. Use the result of Problem 1 to show that

$$\mathscr{L}\{e^t\,\mathrm{erf}(\sqrt{t})\} = \dfrac{1}{\sqrt{s}(s - 1)}.$$

4. Use the result of Problem 2 to show that

$$\mathscr{L}\{e^t\,\mathrm{erfc}(\sqrt{t})\} = \dfrac{1}{\sqrt{s}(\sqrt{s} + 1)}.$$

5. Let C, G, R, and x be constants. Use Table 14.1 to show that

$$\mathscr{L}^{-1}\left\{\dfrac{C}{Cs + G}\left(1 - e^{-x\sqrt{RCs + RG}}\right)\right\} = e^{-Gt/C}\,\mathrm{erf}\left(\dfrac{x}{2}\sqrt{\dfrac{RC}{t}}\right).$$

6. Let a be a constant. Show that

$$\mathscr{L}^{-1}\left\{\dfrac{\sinh a\sqrt{s}}{s\sinh\sqrt{s}}\right\} = \sum_{n=0}^{\infty}\left[\mathrm{erf}\left(\dfrac{2n + 1 + a}{2\sqrt{t}}\right) - \mathrm{erf}\left(\dfrac{2n + 1 - a}{2\sqrt{t}}\right)\right].$$

[*Hint*: Use the exponential definition of the hyperbolic sine. Expand $1/(1 - e^{-2\sqrt{s}})$ in a geometric series.]

7. Use the Laplace transform and Table 14.1 to solve the integral equation

$$y(t) = 1 - \int_0^t \frac{y(\tau)}{\sqrt{t-\tau}} \, d\tau.$$

8. Use the third and fifth entries in Table 14.1 to derive the sixth entry.

9. Show that $\int_a^b e^{-u^2} \, du = \frac{\sqrt{\pi}}{2} [\operatorname{erf}(b) - \operatorname{erf}(a)]$.

10. Show that $\int_{-a}^a e^{-u^2} \, du = \sqrt{\pi} \operatorname{erf}(a)$.

Computer Lab Assignments

11. The functions $\operatorname{erf}(x)$ and $\operatorname{erfc}(x)$ are defined for $x < 0$. Use a CAS to superimpose the graphs of $\operatorname{erf}(x)$ and $\operatorname{erfc}(x)$ on the same axes for $-10 \leq x \leq 10$. Do the graphs possess any symmetry? What are $\lim_{x \to -\infty} \operatorname{erf}(x)$ and $\lim_{x \to -\infty} \operatorname{erfc}(x)$?

14.2 APPLICATION OF THE LAPLACE TRANSFORM

> • *Transforms of partial derivatives*   • *Solutions of boundary-value problems*

Recall that the Laplace transform of a function $f(t)$ is given by

$$\mathcal{L}\{f(t)\} = \int_0^\infty e^{-st} f(t) \, dt$$

and is denoted by $F(s)$. The Laplace transform of a function of two variables, $u(x, t)$, with respect to the variable t is defined by

$$\mathcal{L}\{u(x, t)\} = \int_0^\infty e^{-st} u(x, t) \, dt,$$

where x is treated as a parameter. We continue the convention of using capital letters to denote the Laplace transform of a function by writing $\mathcal{L}\{u(x, t)\} = U(x, s)$. Throughout this section we shall assume that the operational properties developed in Sections 7.2–7.4 apply to functions of two variables.

Transform of Partial Derivatives   The transform of the partial derivative $\partial u/\partial t$ follows analogously from the result given in (6) of Section 7.2:

$$\mathcal{L}\left\{\frac{\partial u}{\partial t}\right\} = s\mathcal{L}\{u(x, t)\} - u(x, 0);$$

that is,

$$\mathcal{L}\left\{\frac{\partial u}{\partial t}\right\} = sU(x, s) - u(x, 0). \tag{1}$$

Similarly,

$$\mathcal{L}\left\{\frac{\partial^2 u}{\partial t^2}\right\} = s^2 U(x, s) - su(x, 0) - u_t(x, 0). \tag{2}$$

Since we are transforming with respect to t, we further suppose that it is legitimate to interchange integration and differentiation in the transform of $\partial^2 u / \partial x^2$:

$$\mathcal{L}\left\{\frac{\partial^2 u}{\partial x^2}\right\} = \int_0^\infty e^{-st} \frac{\partial^2 u}{\partial x^2}\, dt = \int_0^\infty \frac{\partial^2}{\partial x^2}\left[e^{-st} u(x, t)\right] dt$$

$$= \frac{d^2}{dx^2}\int_0^\infty e^{-st} u(x, t)\, dt = \frac{d^2}{dx^2} \mathcal{L}\{u(x, t)\};$$

that is,
$$\mathcal{L}\left\{\frac{\partial^2 u}{\partial x^2}\right\} = \frac{d^2 U}{dx^2}. \tag{3}$$

In view of (1) and (2) we see that the Laplace transform is suited to problems with initial conditions—namely, those problems associated with the heat equation or the wave equation.

EXAMPLE 1 Laplace Transform of a PDE

Find the Laplace transform of the wave equation $a^2 \dfrac{\partial^2 u}{\partial x^2} = \dfrac{\partial^2 u}{\partial t^2}$, $t > 0$.

Solution From (2) and (3),

$$\mathcal{L}\left\{a^2 \frac{\partial^2 u}{\partial x^2}\right\} = \mathcal{L}\left\{\frac{\partial^2 u}{\partial t^2}\right\}$$

becomes

$$a^2 \frac{d^2}{dx^2} \mathcal{L}\{u(x, t)\} = s^2 \mathcal{L}\{u(x, t)\} - s u(x, 0) - u_t(x, 0)$$

or
$$a^2 \frac{d^2 U}{dx^2} - s^2 U = -s u(x, 0) - u_t(x, 0). \tag{4} \quad\blacksquare$$

The Laplace transform with respect to t of either the wave equation or the heat equation eliminates that variable, and for the one-dimensional equations the transformed equations are then *ordinary differential equations* in the spatial variable x. In solving a transformed equation we treat s as a parameter.

EXAMPLE 2 Using the Laplace Transform to Solve a BVP

Solve
$$\frac{\partial^2 u}{\partial x^2} = \frac{\partial^2 u}{\partial t^2}, \quad 0 < x < 1, \quad t > 0$$

subject to
$$u(0, t) = 0, \quad u(1, t) = 0, \quad t > 0$$

$$u(x, 0) = 0, \quad \frac{\partial u}{\partial t}\Big|_{t=0} = \sin \pi x, \quad 0 < x < 1.$$

Solution The partial differential equation is recognized as the wave equation with $a = 1$. From (4) and the given initial conditions, the transformed equation is

$$\frac{d^2U}{dx^2} - s^2U = -\sin \pi x, \tag{5}$$

where $U(x, s) = \mathcal{L}\{u(x, t)\}$. Since the boundary conditions are functions of t, we must also find their Laplace transforms:

$$\mathcal{L}\{u(0, t)\} = U(0, s) = 0 \quad \text{and} \quad \mathcal{L}\{u(1, t)\} = U(1, s) = 0. \tag{6}$$

The results in (6) are boundary conditions for the ordinary differential equation (5). Since (5) is defined over a finite interval, its complementary function is

$$U_c(x, s) = c_1 \cosh sx + c_2 \sinh sx.$$

The method of undetermined coefficients yields a particular solution

$$U_p(x, s) = \frac{1}{s^2 + \pi^2} \sin \pi x.$$

Hence $$U(x, s) = c_1 \cosh sx + c_2 \sinh sx + \frac{1}{s^2 + \pi^2} \sin \pi x.$$

But the conditions $U(0, s) = 0$ and $U(1, s) = 0$ yield, in turn, $c_1 = 0$ and $c_2 = 0$. We conclude that

$$U(x, s) = \frac{1}{s^2 + \pi^2} \sin \pi x$$

$$u(x, t) = \mathcal{L}^{-1}\left\{\frac{1}{s^2 + \pi^2} \sin \pi x\right\} = \frac{1}{\pi} \sin \pi x \, \mathcal{L}^{-1}\left\{\frac{\pi}{s^2 + \pi^2}\right\}.$$

Therefore $$u(x, t) = \frac{1}{\pi} \sin \pi x \sin \pi t. \qquad \blacksquare$$

EXAMPLE 3   Using the Laplace Transform to Solve a BVP

A very long string is initially at rest on the nonnegative x-axis. The string is secured at $x = 0$, and its distant right end slides down a frictionless vertical support. The string is set in motion by letting it fall under its own weight. Find the displacement $u(x, t)$.

Solution   Since the force of gravity is taken into consideration, it can be shown that the wave equation has the form

$$a^2 \frac{\partial^2 u}{\partial x^2} - g = \frac{\partial^2 u}{\partial t^2}, \quad x > 0, \quad t > 0.$$

The boundary and initial conditions are, respectively,

$$u(0, t) = 0, \quad \lim_{x \to \infty} \frac{\partial u}{\partial x} = 0, \quad t > 0$$

$$u(x, 0) = 0, \quad \left.\frac{\partial u}{\partial t}\right|_{t=0} = 0, \quad x > 0.$$

The second boundary condition $\lim_{x \to \infty} \partial u/\partial x = 0$ indicates that the string is horizontal at a great distance from the left end. Now from (2) and (3),

$$\mathscr{L}\left\{ a^2 \frac{\partial^2 u}{\partial x^2} \right\} - \mathscr{L}\{g\} = \mathscr{L}\left\{ \frac{\partial^2 u}{\partial t^2} \right\}$$

becomes
$$a^2 \frac{d^2 U}{dx^2} - \frac{g}{s} = s^2 U - su(x, 0) - u_t(x, 0)$$

or, in view of the initial conditions,

$$\frac{d^2 U}{dx^2} - \frac{s^2}{a^2} U = \frac{g}{a^2 s}.$$

The transforms of the boundary conditions are

$$\mathscr{L}\{u(0, t)\} = U(0, s) = 0 \quad \text{and} \quad \mathscr{L}\left\{ \lim_{x \to \infty} \frac{\partial u}{\partial x} \right\} = \lim_{x \to \infty} \frac{dU}{dx} = 0.$$

With the aid of undetermined coefficients, the general solution of the transformed equation is found to be

$$U(x, s) = c_1 e^{-(x/a)s} + c_2 e^{(x/a)s} - \frac{g}{s^3}.$$

The boundary condition $\lim_{x \to \infty} dU/dx = 0$ implies $c_2 = 0$, and $U(0, s) = 0$ gives $c_1 = g/s^3$. Therefore

$$U(x, s) = \frac{g}{s^3} e^{-(x/a)s} - \frac{g}{s^3}.$$

Now by the second translation theorem we have

$$u(x, t) = \mathscr{L}^{-1}\left\{ \frac{g}{s^3} e^{-(x/a)s} - \frac{g}{s^3} \right\} = \frac{1}{2} g \left(t - \frac{x}{a} \right)^2 \mathscr{U}\left(t - \frac{x}{a} \right) - \frac{1}{2} gt^2$$

or
$$u(x, t) = \begin{cases} -\dfrac{1}{2} gt^2, & 0 \le t < \dfrac{x}{a} \\[2mm] -\dfrac{g}{2a^2} (2axt - x^2), & t \ge \dfrac{x}{a}. \end{cases}$$

To interpret the solution let us suppose that $t > 0$ is fixed. For $0 \le x \le at$ the string is the shape of a parabola passing through $(0, 0)$ and $(at, -\frac{1}{2}gt^2)$. For $x > at$ the string is described by the horizontal line $u = -\frac{1}{2}gt^2$. See Figure 14.2. ∎

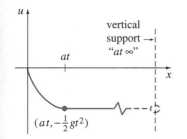

Figure 14.2

Observe that the problem in the next example could be solved by the procedure in Section 12.6. The Laplace transform provides an alternative solution.

EXAMPLE 4 **A Solution in Terms of erf(x)**

Solve the heat equation

$$\frac{\partial^2 u}{\partial x^2} = \frac{\partial u}{\partial t}, \quad 0 < x < 1, \quad t > 0$$

subject to $\quad u(0, t) = 0, \quad u(1, t) = u_0, \quad t > 0$
$$u(x, 0) = 0, \quad 0 < x < 1.$$

Solution   From (1) and (3) and the given initial condition,

$$\mathscr{L}\left\{\frac{\partial^2 u}{\partial x^2}\right\} = \mathscr{L}\left\{\frac{\partial u}{\partial t}\right\}$$

becomes $\qquad\qquad \dfrac{d^2 U}{dx^2} - sU = 0.$ $\qquad\qquad$ (7)

The transforms of the boundary conditions are

$$U(0, s) = 0 \quad \text{and} \quad U(1, s) = \frac{u_0}{s}.$$ $\qquad$ (8)

Since we are concerned with a finite interval on the x-axis, we choose to write the general solution of (7) as

$$U(x, s) = c_1 \cosh(\sqrt{s}\,x) + c_2 \sinh(\sqrt{s}\,x).$$

Applying the two boundary conditions in (8) yields, respectively, $c_1 = 0$ and $c_2 = u_0/(s \sinh \sqrt{s})$. Thus

$$U(x, s) = u_0 \frac{\sinh(\sqrt{s}\,x)}{s \sinh \sqrt{s}}.$$

Now the inverse transform of the latter function cannot be found in most tables. However, by writing

$$\frac{\sinh(\sqrt{s}\,x)}{s \sinh \sqrt{s}} = \frac{e^{\sqrt{s}\,x} - e^{-\sqrt{s}\,x}}{s(e^{\sqrt{s}} - e^{-\sqrt{s}})} = \frac{e^{(x-1)\sqrt{s}} - e^{-(x+1)\sqrt{s}}}{s(1 - e^{-2\sqrt{s}})}$$

and using the geometric series

$$\frac{1}{1 - e^{-2\sqrt{s}}} = \sum_{n=0}^{\infty} e^{-2n\sqrt{s}}$$

we find

$$\frac{\sinh(\sqrt{s}\,x)}{s \sinh \sqrt{s}} = \sum_{n=0}^{\infty}\left[\frac{e^{-(2n+1-x)\sqrt{s}}}{s} - \frac{e^{-(2n+1+x)\sqrt{s}}}{s}\right].$$

If we assume that the inverse Laplace transform can be done term by term, it follows from entry 3 of Table 14.1 that

$$u(x, t) = u_0 \mathscr{L}^{-1}\left\{\frac{\sinh(\sqrt{s}\,x)}{s \sinh \sqrt{s}}\right\}$$

$$= u_0 \sum_{n=0}^{\infty}\left[\mathscr{L}^{-1}\left\{\frac{e^{-(2n+1-x)\sqrt{s}}}{s}\right\} - \mathscr{L}^{-1}\left\{\frac{e^{-(2n+1+x)\sqrt{s}}}{s}\right\}\right]$$

$$= u_0 \sum_{n=0}^{\infty}\left[\operatorname{erfc}\left(\frac{2n+1-x}{2\sqrt{t}}\right) - \operatorname{erfc}\left(\frac{2n+1+x}{2\sqrt{t}}\right)\right].$$ $\qquad$ (9)

The solution (9) can be rewritten in terms of the error function using $\operatorname{erfc}(x) = 1 - \operatorname{erf}(x)$:

$$u(x, t) = u_0 \sum_{n=0}^{\infty} \left[\operatorname{erf}\left(\frac{2n + 1 + x}{2\sqrt{t}}\right) - \operatorname{erf}\left(\frac{2n + 1 - x}{2\sqrt{t}}\right) \right]. \quad (10) \quad \blacksquare$$

Figure 14.3(a), obtained with the aid of the 3D-plot application in a CAS, shows the surface over the rectangular region $0 \le x \le 1, 0 \le t \le 6$ defined by the partial sum $S_{10}(x, t)$ of the solution (10) with $u_0 = 100$. It is apparent from the surface and the accompanying two-dimensional graphs that at a fixed value of x (the curve of intersection of a plane slicing the surface perpendicular to the x-axis on the interval $0 \le x \le 1$)

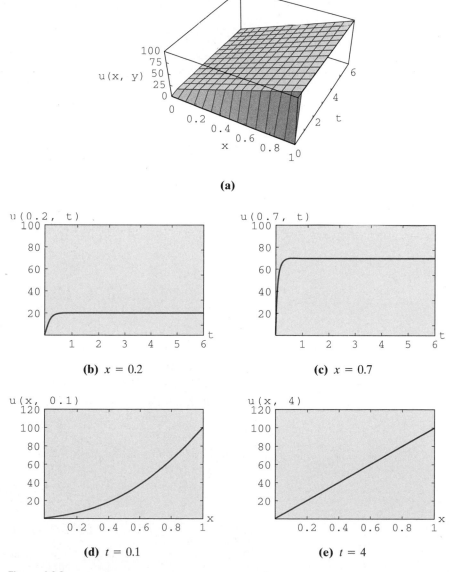

(a)

(b) $x = 0.2$

(c) $x = 0.7$

(d) $t = 0.1$

(e) $t = 4$

Figure 14.3

the temperature $u(x, t)$ increases rapidly to a constant value as time increases. See Figures 14.3(b) and 14.3(c). For a fixed time (the curve of intersection of a plane slicing the surface perpendicular to the t-axis) the temperature $u(x, t)$ naturally increases from 0 to 100. See Figures 14.3(d) and 14.3(e).

EXERCISES 14.2

Answers to odd-numbered problems begin on page AN-20.

In the following problems use tables as necessary.

1. A string is stretched along the x-axis between $(0, 0)$ and $(L, 0)$. Find the displacement $u(x, t)$ if the string starts from rest in the initial position $A \sin(\pi x/L)$.

2. Solve the boundary-value problem

$$\frac{\partial^2 u}{\partial x^2} = \frac{\partial^2 u}{\partial t^2}, \quad 0 < x < 1, \quad t > 0$$

$$u(0, t) = 0, \quad u(1, t) = 0$$

$$u(x, 0) = 0, \quad \left.\frac{\partial u}{\partial t}\right|_{t=0} = 2 \sin \pi x + 4 \sin 3\pi x.$$

3. The displacement of a semi-infinite elastic string is determined from

$$a^2 \frac{\partial^2 u}{\partial x^2} = \frac{\partial^2 u}{\partial t^2}, \quad x > 0, \quad t > 0$$

$$u(0, t) = f(t), \quad \lim_{x \to \infty} u(x, t) = 0, \quad t > 0$$

$$u(x, 0) = 0, \quad \left.\frac{\partial u}{\partial t}\right|_{t=0} = 0, \quad x > 0.$$

Solve for $u(x, t)$.

4. Solve the boundary-value problem in Problem 3 when

$$f(t) = \begin{cases} \sin \pi t, & 0 \le t \le 1 \\ 0, & t > 1 \end{cases}$$

Sketch the displacement $u(x, t)$ for $t > 1$.

5. In Example 3 find the displacement $u(x, t)$ when the left end of the string at $x = 0$ is given an oscillatory motion described by $f(t) = A \sin \omega t$.

6. The displacement $u(x, t)$ of a string that is driven by an external force is determined from

$$\frac{\partial^2 u}{\partial x^2} + \sin \pi x \sin \omega t = \frac{\partial^2 u}{\partial t^2}, \quad 0 < x < 1, \quad t > 0$$

$$u(0, t) = 0, \quad u(1, t) = 0, \quad t > 0$$

$$u(x, 0) = 0, \quad \left.\frac{\partial u}{\partial t}\right|_{t=0} = 0, \quad 0 < x < 1.$$

Solve for $u(x, t)$.

7. A uniform bar is clamped at $x = 0$ and is initially at rest. If a constant force F_0 is applied to the free end at $x = L$, the longitudinal displacement $u(x, t)$ of a cross-section of the bar is determined from

$$a^2 \frac{\partial^2 u}{\partial x^2} = \frac{\partial^2 u}{\partial t^2}, \quad 0 < x < L, \quad t > 0$$

$$u(0, t) = 0, \quad E \frac{\partial u}{\partial x}\Big|_{x=L} = F_0, \quad E \text{ a constant}, \quad t > 0$$

$$u(x, 0) = 0, \quad \frac{\partial u}{\partial t}\Big|_{t=0} = 0, \quad 0 < x < L.$$

Solve for $u(x, t)$. [*Hint*: Expand $1/(1 + e^{-2sL/a})$ in a geometric series.]

8. A uniform semi-infinite elastic beam moving along the x-axis with a constant velocity $-v_0$ is brought to a stop by hitting a wall at time $t = 0$. See Figure 14.4. The longitudinal displacement $u(x, t)$ is determined from

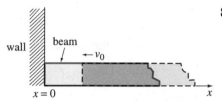

$$a^2 \frac{\partial^2 u}{\partial x^2} = \frac{\partial^2 u}{\partial t^2}, \quad x > 0, \quad t > 0$$

$$u(0, t) = 0, \quad \lim_{x \to \infty} \frac{\partial u}{\partial x} = 0, \quad t > 0$$

$$u(x, 0) = 0, \quad \frac{\partial u}{\partial t}\Big|_{t=0} = -v_0, \quad x > 0.$$

Figure 14.4

Solve for $u(x, t)$.

9. Solve the boundary-value problem

$$\frac{\partial^2 u}{\partial x^2} = \frac{\partial^2 u}{\partial t^2}, \quad x > 0, \quad t > 0$$

$$u(0, t) = 0, \quad \lim_{x \to \infty} u(x, t) = 0, \quad t > 0$$

$$u(x, 0) = xe^{-x}, \quad \frac{\partial u}{\partial t}\Big|_{t=0} = 0, \quad x > 0.$$

10. Solve the boundary-value problem

$$\frac{\partial^2 u}{\partial x^2} = \frac{\partial^2 u}{\partial t^2}, \quad x > 0, \quad t > 0$$

$$u(0, t) = 1, \quad \lim_{x \to \infty} u(x, t) = 0, \quad t > 0$$

$$u(x, 0) = e^{-x}, \quad \frac{\partial u}{\partial t}\Big|_{t=0} = 0, \quad x > 0.$$

11. (a) The temperature in a semi-infinite solid is modeled by the boundary-value problem

$$k \frac{\partial^2 u}{\partial x^2} = \frac{\partial u}{\partial t}, \quad x > 0, \quad t > 0$$

$$u(0, t) = u_0 \quad \lim_{x \to \infty} u(x, t) = 0, \quad t > 0$$

$$u(x, 0) = 0, \quad x > 0.$$

Solve for $u(x, t)$. Use the solution to determine analytically the value of $\lim_{t \to \infty} u(x, t), \; x > 0$.

(b) Use a CAS to obtain the graph of $u(x, t)$ over the rectangular region $0 \leq x \leq 10$, $0 \leq t \leq 15$. Assume $u_0 = 100$ and $k = 1$. Indicate the two boundary conditions and initial condition on your graph. Use 2D and 3D plots of $u(x, t)$ to verify your answer to part (a).

12. (a) In Problem 11 if there is a constant flux of heat into the solid at its left-hand boundary, then the boundary condition is $\dfrac{\partial u}{\partial x}\Big|_{x=0} = -A$, $A > 0$, $t > 0$. Solve for $u(x, t)$. Use the solution to determine analytically the value of $\lim_{t \to \infty} u(x, t)$, $x > 0$.

(b) Use a CAS to obtain the graph of $u(x, t)$ over the rectangular region $0 \leq x \leq 10$, $0 \leq t \leq 15$. Assume $u_0 = 100$ and $k = 1$. Use 2D and 3D plots of $u(x, t)$ to verify your answer to part (a).

In Problems 13–20 use the Laplace transform to solve the heat equation $u_{xx} = u_t$, $x > 0$, $t > 0$ subject to the given conditions.

13. $u(0, t) = u_0$, $\lim\limits_{x \to \infty} u(x, t) = u_1$
$u(x, 0) = u_1$

14. $u(0, t) = u_0$, $\lim\limits_{x \to \infty} \dfrac{u(x, t)}{x} = u_1$
$u(x, 0) = u_1 x$

15. $\dfrac{\partial u}{\partial x}\Big|_{x=0} = u(0, t)$, $\lim\limits_{x \to \infty} u(x, t) = u_0$
$u(x, 0) = u_0$

16. $\dfrac{\partial u}{\partial x}\Big|_{x=0} = u(0, t) - 50$, $\lim\limits_{x \to \infty} u(x, t) = 0$
$u(x, 0) = 0$

17. $u(0, t) = f(t)$, $\lim\limits_{x \to \infty} u(x, t) = 0$
$u(x, 0) = 0$
[*Hint:* Use the convolution theorem.]

18. $\dfrac{\partial u}{\partial x}\Big|_{x=0} = -f(t)$, $\lim\limits_{x \to \infty} u(x, t) = 0$
$u(x, 0) = 0$

19. $u(0, t) = 60 + 40\,\mathcal{U}(t - 2)$, $\lim\limits_{x \to \infty} u(x, t) = 60$
$u(x, 0) = 60$

20. $u(0, t) = \begin{cases} 20, & 0 < t < 1 \\ 0, & t \geq 1 \end{cases}$, $\lim\limits_{x \to \infty} u(x, t) = 100$
$u(x, 0) = 100$

21. Solve the boundary-value problem

$$\frac{\partial^2 u}{\partial x^2} = \frac{\partial u}{\partial t}, \quad -\infty < x < 1, \quad t > 0$$

$$\frac{\partial u}{\partial x}\Big|_{x=1} = 100 - u(1, t), \quad \lim_{x \to -\infty} u(x, t) = 0, \quad t > 0$$

$$u(x, 0) = 0, \quad -\infty < x < 1.$$

22. Show that a solution of the boundary-value problem

$$k \frac{\partial^2 u}{\partial x^2} + r = \frac{\partial u}{\partial t}, \quad x > 0, \quad t > 0$$

$$u(0, t) = 0, \quad \lim_{x \to \infty} \frac{\partial u}{\partial x} = 0, \quad t > 0$$

$$u(x, 0) = 0, \quad x > 0,$$

where r is a constant, is given by

$$u(x, t) = rt - r \int_0^t \operatorname{erfc}\left(\frac{x}{2\sqrt{k\tau}}\right) d\tau.$$

23. A rod of length L is held at a constant temperature u_0 at its ends $x = 0$ and $x = L$. If the rod's initial temperature is $u_0 + u_0 \sin(x\pi/L)$, solve the heat equation $u_{xx} = u_t$, $0 < x < L$, $t > 0$ for the temperature $u(x, t)$.

24. If there is a heat transfer from the lateral surface of a thin wire of length L into a medium at constant temperature u_m, then the heat equation takes on the form

$$k \frac{\partial^2 u}{\partial x^2} - h(u - u_m) = \frac{\partial u}{\partial t}, \quad 0 < x < L, \quad t > 0,$$

where h is a constant. Find the temperature $u(x, t)$ if the initial temperature is a constant u_0 throughout and the ends $x = 0$ and $x = L$ are insulated.

25. A rod of unit length is insulated at $x = 0$ and is kept at temperature zero at $x = 1$. If the initial temperature of the rod is a constant u_0, solve $ku_{xx} = u_t$, $0 < x < 1$, $t > 0$ for the temperature $u(x, t)$. [*Hint:* Expand $1/(1 + e^{-2\sqrt{s/k}})$ in a geometric series.]

26. An infinite porous slab of unit width is immersed in a solution of constant concentration c_0. A dissolved substance in the solution diffuses into the slab. The concentration $c(x, t)$ in the slab is determined from

$$D \frac{\partial^2 c}{\partial x^2} = \frac{\partial c}{\partial t}, \quad 0 < x < 1, \quad t > 0$$

$$c(0, t) = c_0, \quad c(1, t) = c_0, \quad t > 0$$

$$c(x, 0) = 0, \quad 0 < x < 1,$$

where D is a constant. Solve for $c(x, t)$.

27. A very long telephone transmission line is initially at a constant potential u_0. If the line is grounded at $x = 0$ and insulated at the distant right end, then the potential $u(x, t)$ at a point x along the line at time t is determined from

$$\frac{\partial^2 u}{\partial x^2} - RC \frac{\partial u}{\partial t} - RGu = 0, \quad x > 0, \quad t > 0$$

$$u(0, t) = 0, \quad \lim_{x \to \infty} \frac{\partial u}{\partial x} = 0, \quad t > 0$$

$$u(x, 0) = u_0, \quad x > 0,$$

where R, C, and G are constants known as resistance, capacitance, and conductance, respectively. Solve for $u(x, t)$. [*Hint:* See Problem 5 in Exercises 14.1.]

28. Show that a solution of the boundary-value problem

$$\frac{\partial^2 u}{\partial x^2} - hu = \frac{\partial u}{\partial t}, \quad x > 0, \quad t > 0$$

$$u(0, t) = u_0, \quad \lim_{x \to \infty} u(x, t) = 0, \quad t > 0$$

$$u(x, 0) = 0, \quad x > 0$$

is

$$u(x, t) = \frac{u_0 x}{2\sqrt{\pi}} \int_0^t \frac{e^{-h\tau - x^2/4\tau}}{\tau^{3/2}} \, d\tau.$$

29. Humans gather most of their information on the outside world through sight and sound. But many creatures use chemical signals as their primary means of communication; for example, honey bees, when alarmed, emit a substance and fan their wings feverishly to relay the warning signal to the bees that attend to the queen. These molecular messages between members of the same species are called pheromones. The signals may be carried by moving air or water or by a *diffusion process* in which the random movement of gas molecules transports the chemical away from its source. Figure 14.5 shows an ant emitting an alarm chemical into the still air of a tunnel. If $c(x, t)$ denotes the concentration of the chemical x centimeters from the source at time t, then $c(x, t)$ satisfies

$$k\frac{\partial^2 c}{\partial x^2} = \frac{\partial c}{\partial t}, \quad x > 0, \quad t > 0$$

and k is a positive constant. The emission of pheromones as a discrete pulse gives rise to a boundary condition of the form

$$\frac{\partial c}{\partial x}\bigg|_{x=0} = -A\delta(t).$$

(a) Solve the boundary-value problem if it is further known that $c(x, 0) = 0$, $x > 0$ and $\lim_{x \to \infty} c(x, t) = 0$, $t > 0$.

(b) Use a CAS to plot the graph of the solution in part (a) for $x \geq 0$ at the fixed times $t = 0.1$, $t = 0.5$, $t = 1$, $t = 2$, and $t = 5$.

(c) For any fixed time t, show that $\int_0^\infty c(x, t) \, dx = Ak$. Thus Ak represents the total amount of chemical discharged.

30. Starting at $t = 0$, a concentrated load of magnitude F_0 moves with a constant velocity v_0 along a semi-infinite string. In this case the wave equation becomes

$$a^2\frac{\partial^2 u}{\partial x^2} = \frac{\partial^2 u}{\partial t^2} + F_0\delta\left(t - \frac{x}{v_0}\right).$$

Figure 14.5

Solve the above partial differential equation subject to

$$u(0, t) = 0, \quad \lim_{x \to \infty} u(x, t) = 0, \quad t > 0$$

$$u(x, 0) = 0, \quad \left. \frac{\partial u}{\partial t} \right|_{t=0} = 0, \quad x > 0$$

(a) when $v_0 \neq a$ **(b)** when $v_0 = a$.

14.3 FOURIER INTEGRAL

• Definition of the Fourier integral • Convergence of the Fourier integral • Fourier cosine and sine integrals • Complex form of the Fourier integral

We used Fourier series in Chapters 11 and 12 to represent a function f defined on a *finite* interval $(-p, p)$ or $(0, L)$. When f and f' are piecewise continuous on such an interval, a Fourier series represents the function on the interval and converges to a periodic extension of f outside the interval. In this way we are justified in saying that Fourier series are associated only with periodic functions. We shall now derive, in a nonrigorous fashion, a means of representing certain kinds of nonperiodic functions that are defined on either an *infinite* interval $(-\infty, \infty)$ or a *semi-infinite* interval $(0, \infty)$.

From Fourier Series to Fourier Integral Suppose a function f is defined on $(-p, p)$. If we use the integral definitions of the coefficients (9), (10), and (11) of Section 11.2 in (8) of that section, then the Fourier series of f on the interval is

$$f(x) = \frac{1}{2p} \int_{-p}^{p} f(t)\, dt + \frac{1}{p} \sum_{n=1}^{\infty} \left[\left(\int_{-p}^{p} f(t) \cos \frac{n\pi}{p} t\, dt \right) \cos \frac{n\pi}{p} x + \left(\int_{-p}^{p} f(t) \sin \frac{n\pi}{p} t\, dt \right) \sin \frac{n\pi}{p} x \right]. \quad \textbf{(1)}$$

If we let $\alpha_n = n\pi/p$, $\Delta\alpha = \alpha_{n+1} - \alpha_n = \pi/p$, then (1) becomes

$$f(x) = \frac{1}{2\pi} \left(\int_{-p}^{p} f(t)\, dt \right) \Delta\alpha + \frac{1}{\pi} \sum_{n=1}^{\infty} \left[\left(\int_{-p}^{p} f(t) \cos \alpha_n t\, dt \right) \cos \alpha_n x + \left(\int_{-p}^{p} f(t) \sin \alpha_n t\, dt \right) \sin \alpha_n x \right] \Delta\alpha. \quad \textbf{(2)}$$

We now expand the interval $(-p, p)$ by letting $p \to \infty$. Since $p \to \infty$ implies that $\Delta\alpha \to 0$, the limit of (2) has the form $\lim_{\Delta\alpha \to 0} \sum_{n=1}^{\infty} F(\alpha_n)\, \Delta\alpha$, which is suggestive of the definition of the integral $\int_0^\infty F(\alpha)\, d\alpha$. Thus if $\int_{-\infty}^{\infty} f(t)\, dt$ exists, the limit of the first term in (2) is zero and the limit of the sum becomes

$$f(x) = \frac{1}{\pi} \int_0^\infty \left[\left(\int_{-\infty}^{\infty} f(t) \cos \alpha t\, dt \right) \cos \alpha x + \left(\int_{-\infty}^{\infty} f(t) \sin \alpha t\, dt \right) \sin \alpha x \right] d\alpha. \quad \textbf{(3)}$$

The result given in (3) is called the **Fourier integral** of f on $(-\infty, \infty)$. As the following summary shows, the basic structure of the Fourier integral is reminiscent of that of a Fourier series.

DEFINITION 14.2 Fourier Integral

The **Fourier integral** of a function f defined on the interval $(-\infty, \infty)$ is given by

$$f(x) = \frac{1}{\pi} \int_0^\infty [A(\alpha) \cos \alpha x + B(\alpha) \sin \alpha x] \, d\alpha, \qquad (4)$$

where
$$A(\alpha) = \int_{-\infty}^\infty f(x) \cos \alpha x \, dx \qquad (5)$$

$$B(\alpha) = \int_{-\infty}^\infty f(x) \sin \alpha x \, dx. \qquad (6)$$

Convergence of a Fourier Integral Sufficient conditions under which a Fourier integral converges to $f(x)$ are similar to, but slightly more restrictive than, the conditions for a Fourier series.

THEOREM 14.1 Conditions for Convergence

Let f and f' be piecewise continuous on every finite interval, and let f be absolutely integrable on $(-\infty, \infty)$.* Then the Fourier integral of f on the interval converges to $f(x)$ at a point of continuity. At a point of discontinuity the Fourier integral will converge to the average

$$\frac{f(x+) + f(x-)}{2},$$

where $f(x+)$ and $f(x-)$ denote the limit of f at x from the right and from the left, respectively.

EXAMPLE 1 Fourier Integral Representation

Find the Fourier integral representation of the function

$$f(x) = \begin{cases} 0, & x < 0 \\ 1, & 0 < x < 2 \\ 0, & x > 2 \end{cases}$$

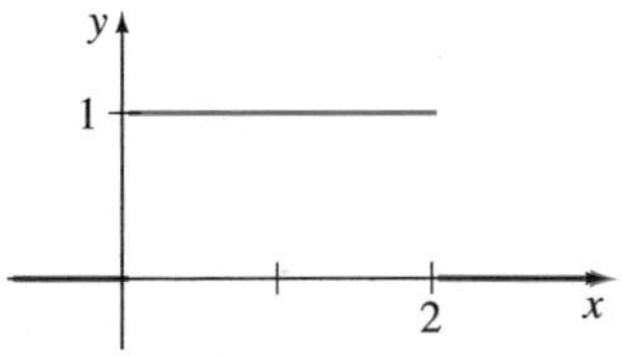

Figure 14.6

Solution The function, whose graph is shown in Figure 14.6, satisfies the hypotheses of Theorem 14.1. Hence from (5) and (6) we have

---

*This means that the integral $\int_{-\infty}^\infty |f(x)| \, dx$ converges.

at once

$$A(\alpha) = \int_{-\infty}^{\infty} f(x) \cos \alpha x \, dx$$

$$= \int_{-\infty}^{0} f(x) \cos \alpha x \, dx + \int_{0}^{2} f(x) \cos \alpha x \, dx + \int_{2}^{\infty} f(x) \cos \alpha x \, dx$$

$$= \int_{0}^{2} \cos \alpha x \, dx = \frac{\sin 2\alpha}{\alpha}$$

$$B(\alpha) = \int_{-\infty}^{\infty} f(x) \sin \alpha x \, dx = \int_{0}^{2} \sin \alpha x \, dx = \frac{1 - \cos 2\alpha}{\alpha}.$$

Substituting these coefficients into (4) then gives

$$f(x) = \frac{1}{\pi} \int_{0}^{\infty} \left[\left(\frac{\sin 2\alpha}{\alpha} \right) \cos \alpha x + \left(\frac{1 - \cos 2\alpha}{\alpha} \right) \sin \alpha x \right] d\alpha.$$

When we use trigonometric identities, the last integral simplifies to

$$f(x) = \frac{2}{\pi} \int_{0}^{\infty} \frac{\sin \alpha \cos \alpha (x - 1)}{\alpha} \, d\alpha. \qquad (7) \quad \blacksquare$$

The Fourier integral can be used to evaluate integrals. For example, at $x = 1$ it follows from Theorem 14.1 that (7) converges to $f(1)$; that is,

$$\int_{0}^{\infty} \frac{\sin \alpha}{\alpha} \, d\alpha = \frac{\pi}{2}.$$

The latter result is worthy of special note since it cannot be obtained in the "usual" manner; the integrand $(\sin x)/x$ does not possess an antiderivative that is an elementary function.

Cosine and Sine Integrals When f is an even function on the interval $(-\infty, \infty)$, then the product $f(x) \cos \alpha x$ is also an even function whereas $f(x) \sin \alpha x$ is an odd function. As a consequence of property (g) of Theorem 11.2, $B(\alpha) = 0$, and so (4) becomes

$$f(x) = \frac{2}{\pi} \int_{0}^{\infty} \left(\int_{0}^{\infty} f(t) \cos \alpha t \, dt \right) \cos \alpha x \, d\alpha.$$

Here we have also used property (f) of Theorem 11.2 to write

$$\int_{-\infty}^{\infty} f(t) \cos \alpha t \, dt = 2 \int_{0}^{\infty} f(t) \cos \alpha t \, dt.$$

Similarly, when f is an odd function on $(-\infty, \infty)$, products $f(x) \cos \alpha x$ and $f(x) \sin \alpha x$ are odd and even functions, respectively. Therefore $A(\alpha) = 0$ and

$$f(x) = \frac{2}{\pi} \int_{0}^{\infty} \left(\int_{0}^{\infty} f(t) \sin \alpha t \, dt \right) \sin \alpha x \, d\alpha.$$

We summarize in the following definition.

DEFINITION 14.3   Fourier Cosine and Sine Integrals

(*i*) The Fourier integral of an even function on the interval $(-\infty, \infty)$ is the **cosine integral**

$$f(x) = \frac{2}{\pi} \int_0^\infty A(\alpha) \cos \alpha x \, d\alpha, \qquad (8)$$

where

$$A(\alpha) = \int_0^\infty f(x) \cos \alpha x \, dx. \qquad (9)$$

(*ii*) The Fourier integral of an odd function on the interval $(-\infty, \infty)$ is the **sine integral**

$$f(x) = \frac{2}{\pi} \int_0^\infty B(\alpha) \sin \alpha x \, d\alpha, \qquad (10)$$

where

$$B(\alpha) = \int_0^\infty f(x) \sin \alpha x \, dx. \qquad (11)$$

EXAMPLE 2   Cosine Integral Representation

Find the Fourier integral representation of the function

$$f(x) = \begin{cases} 1, & |x| < a \\ 0, & |x| > a. \end{cases}$$

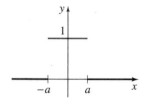

Figure 14.7

Solution   It is apparent from Figure 14.7 that f is an even function. Hence we represent f by the Fourier cosine integral (8). From (9) we obtain

$$A(\alpha) = \int_0^\infty f(x) \cos \alpha x \, dx = \int_0^a f(x) \cos \alpha x \, dx + \int_a^\infty f(x) \cos \alpha x \, dx$$

$$= \int_0^a \cos \alpha x \, dx = \frac{\sin a\alpha}{\alpha},$$

and so

$$f(x) = \frac{2}{\pi} \int_0^\infty \frac{\sin a\alpha \cos \alpha x}{\alpha} \, d\alpha. \qquad \blacksquare$$

The integrals (8) and (10) can be used when f is neither odd nor even and defined only on the half-line $(0, \infty)$. In this case (8) represents f on the interval $(0, \infty)$ and its even (but not periodic) extension to $(-\infty, 0)$, whereas (10) represents f on $(0, \infty)$ and its odd extension to the interval $(-\infty, 0)$. The next example illustrates this concept.

EXAMPLE 3   Cosine and Sine Integral Representations

Represent $f(x) = e^{-x}, x > 0$

(**a**) by a cosine integral   (**b**) by a sine integral.

Figure 14.8

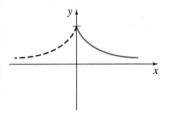

(a) cosine integral

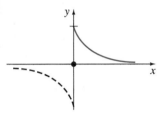

(b) sine integral

Figure 14.9

Solution   The graph of the function is given in Figure 14.8.

(a) Using integration by parts, we find

$$A(\alpha) = \int_0^\infty e^{-x} \cos \alpha x \, dx = \frac{1}{1 + \alpha^2}.$$

Therefore the cosine integral of f is

$$f(x) = \frac{2}{\pi} \int_0^\infty \frac{\cos \alpha x}{1 + \alpha^2} \, d\alpha.$$

(b) Similarly, we have

$$B(\alpha) = \int_0^\infty e^{-x} \sin \alpha x \, dx = \frac{\alpha}{1 + \alpha^2}.$$

The sine integral of f is then

$$f(x) = \frac{2}{\pi} \int_0^\infty \frac{\alpha \sin \alpha x}{1 + \alpha^2} \, d\alpha.$$

Figure 14.9 shows the graphs of the functions and their extensions represented by the two integrals.   ■

Complex Form   The Fourier integral (4) also possesses an equivalent **complex form**, or **exponential form**, that is analogous to the complex form of a Fourier series (see Problem 21 in Exercises 11.2). If (5) and (6) are substituted into (4), then

$$f(x) = \frac{1}{\pi} \int_0^\infty \int_{-\infty}^\infty f(t)[\cos \alpha t \cos \alpha x + \sin \alpha t \sin \alpha x] \, dt \, d\alpha$$

$$= \frac{1}{\pi} \int_0^\infty \int_{-\infty}^\infty f(t) \cos \alpha(t - x) \, dt \, d\alpha$$

$$= \frac{1}{2\pi} \int_{-\infty}^\infty \int_{-\infty}^\infty f(t) \cos \alpha(t - x) \, dt \, d\alpha \qquad (12)$$

$$= \frac{1}{2\pi} \int_{-\infty}^\infty \int_{-\infty}^\infty f(t)[\cos \alpha(t - x) + i \sin \alpha(t - x)] \, dt \, d\alpha \qquad (13)$$

$$= \frac{1}{2\pi} \int_{-\infty}^\infty \int_{-\infty}^\infty f(t) e^{i\alpha(t-x)} \, dt \, d\alpha$$

$$= \frac{1}{2\pi} \int_{-\infty}^\infty \left(\int_{-\infty}^\infty f(t) e^{i\alpha t} \, dt \right) e^{-i\alpha x} \, d\alpha. \qquad (14)$$

We note that (12) follows from the fact that the integrand is an even function of α. In (13) we have simply added zero to the integrand;

$$i \int_{-\infty}^\infty \int_{-\infty}^\infty f(t) \sin \alpha(t - x) \, dt \, d\alpha = 0$$

because the integrand is an odd function of α. The integral in (14) can be expressed as

$$f(x) = \frac{1}{2\pi} \int_{-\infty}^\infty C(\alpha) e^{-i\alpha x} \, d\alpha, \qquad (15)$$

where

$$C(\alpha) = \int_{-\infty}^\infty f(x) e^{i\alpha x} \, dx. \qquad (16)$$

This latter form of the Fourier integral will be put to use in the next section when we return to the solution of boundary-value problems.

EXERCISES 14.3

Answers to odd-numbered problems begin on page AN-21.

In Problems 1–6 find the Fourier integral representation of the given function.

1. $f(x) = \begin{cases} 0, & x < -1 \\ -1, & -1 < x < 0 \\ 2, & 0 < x < 1 \\ 0, & x > 1 \end{cases}$

2. $f(x) = \begin{cases} 0, & x < \pi \\ 4, & \pi < x < 2\pi \\ 0, & x > 2\pi \end{cases}$

3. $f(x) = \begin{cases} 0, & x < 0 \\ x, & 0 < x < 3 \\ 0, & x > 3 \end{cases}$

4. $f(x) = \begin{cases} 0, & x < 0 \\ \sin x, & 0 \le x \le \pi \\ 0, & x > \pi \end{cases}$

5. $f(x) = \begin{cases} 0, & x < 0 \\ e^{-x}, & x > 0 \end{cases}$

6. $f(x) = \begin{cases} e^x, & |x| < 1 \\ 0, & |x| > 1 \end{cases}$

In Problems 7–12 represent the given function by an appropriate cosine or sine integral.

7. $f(x) = \begin{cases} 0, & x < -1 \\ -5, & -1 < x < 0 \\ 5, & 0 < x < 1 \\ 0, & x > 1 \end{cases}$

8. $f(x) = \begin{cases} 0, & |x| < 1 \\ \pi, & 1 < |x| < 2 \\ 0, & |x| > 2 \end{cases}$

9. $f(x) = \begin{cases} |x|, & |x| < \pi \\ 0, & |x| > \pi \end{cases}$

10. $f(x) = \begin{cases} x, & |x| < \pi \\ 0, & |x| > \pi \end{cases}$

11. $f(x) = e^{-|x|} \sin x$

12. $f(x) = xe^{-|x|}$

In Problems 13–16 find the cosine and sine integral representations of the given function.

13. $f(x) = e^{-kx}, \quad k > 0, x > 0$

14. $f(x) = e^{-x} - e^{-3x}, \quad x > 0$

15. $f(x) = xe^{-2x}, \quad x > 0$

16. $f(x) = e^{-x} \cos x, \quad x > 0$

In Problems 17 and 18 solve the given integral equation for the function f.

17. $\displaystyle\int_0^\infty f(x) \cos \alpha x \, dx = e^{-\alpha}$

18. $\displaystyle\int_0^\infty f(x) \sin \alpha x \, dx = \begin{cases} 1, & 0 < \alpha < 1 \\ 0, & \alpha > 1 \end{cases}$

19. (a) Use (7) to show that

$$\int_0^\infty \frac{\sin 2x}{x} \, dx = \frac{\pi}{2}.$$

[*Hint:* α is a dummy variable of integration.]

(b) Show in general that, for $k > 0$,

$$\int_0^\infty \frac{\sin kx}{x} \, dx = \frac{\pi}{2}.$$

20. Use the complex form (15) to find the Fourier integral representation of $f(x) = e^{-|x|}$. Show that the result is the same as that obtained from (8).

14.4 FOURIER TRANSFORMS

• Integral transform pairs • Inverse transform • Kernel of an integral transform • Fourier transform pairs • Existence of Fourier transforms • Operational properties • Solution of BVPs

The Laplace transform $F(s)$ of a function $f(t)$ is defined by an integral, but up to now we have been using the symbolic representation $f(t) = \mathcal{L}^{-1}\{F(s)\}$ to denote the inverse Laplace transform of $F(s)$. Actually, the inverse Laplace transform is also an integral. If

$$\mathcal{L}\{f(t)\} = \int_0^\infty e^{-st}f(t)\,dt = F(s) \quad \text{then} \quad \mathcal{L}^{-1}\{F(s)\} = \frac{1}{2\pi i}\int_{\gamma-i\infty}^{\gamma+i\infty} e^{st}F(s)\,dx = f(t).$$

The last integral is called a **contour integral**; its evaluation requires the use of complex variables and is beyond the scope of this text. The point is this: integral transforms appear in **transform pairs**. If $f(x)$ is transformed into $F(\alpha)$ by an **integral transform** $F(\alpha) = \int_a^b f(x)K(\alpha, x)\,dx$, then the function f can be recovered by another integral transform $f(x) = \int_c^d F(\alpha)H(\alpha, x)\,d\alpha$, called the **inverse transform**. The functions K and H are called the **kernels** of their respective transforms. We identify $K(s, t) = e^{-st}$ as the kernel of the Laplace transform and $H(s, t) = e^{st}/2\pi i$ as the kernel of the inverse Laplace transform.

Fourier Transform Pairs The Fourier integral is the source of three new integral transforms. From (16)–(15), (11)–(10), and (9)–(8) of the preceding section we are prompted to define the following **Fourier transform pairs**.

DEFINITION 14.4 Fourier Transform Pairs

(*i*) Fourier transform:
$$\mathcal{F}\{f(x)\} = \int_{-\infty}^\infty f(x)e^{i\alpha x}\,dx = F(\alpha) \tag{1}$$

Inverse Fourier transform:
$$\mathcal{F}^{-1}\{F(\alpha)\} = \frac{1}{2\pi}\int_{-\infty}^\infty F(\alpha)e^{-i\alpha x}\,d\alpha = f(x) \tag{2}$$

(*ii*) Fourier sine transform:
$$\mathcal{F}_s\{f(x)\} = \int_0^\infty f(x)\sin \alpha x\,dx = F(\alpha) \tag{3}$$

Inverse Fourier sine transform:
$$\mathcal{F}_s^{-1}\{F(\alpha)\} = \frac{2}{\pi}\int_0^\infty F(\alpha)\sin \alpha x\,d\alpha = f(x) \tag{4}$$

(*iii*) Fourier cosine transform:
$$\mathcal{F}_c\{f(x)\} = \int_0^\infty f(x)\cos \alpha x\,dx = F(\alpha) \tag{5}$$

Inverse Fourier cosine transform:
$$\mathcal{F}_c^{-1}\{F(\alpha)\} = \frac{2}{\pi}\int_0^\infty F(\alpha)\cos \alpha x\,d\alpha = f(x) \tag{6}$$

Existence The conditions under which (1), (3), and (5) exist are more stringent than those for the Laplace transform. For example, you should verify that $\mathscr{F}\{1\}$, $\mathscr{F}_s\{1\}$, and $\mathscr{F}_c\{1\}$ do not exist. Sufficient conditions for existence are that f be absolutely integrable on the appropriate interval and that f and f' be piecewise continuous on every finite interval.

Operational Properties   Since our immediate goal is to apply these new transforms to boundary-value problems, we need to examine the transforms of derivatives.

Fourier Transform   Suppose that f is continuous and absolutely integrable on the interval $(-\infty, \infty)$ and f' is piecewise continuous on every finite interval. If $f(x) \to 0$ as $x \to \pm\infty$, then integration by parts gives

$$\mathscr{F}\{f'(x)\} = \int_{-\infty}^{\infty} f'(x)e^{i\alpha x}\,dx$$

$$= f(x)e^{i\alpha x}\Big|_{-\infty}^{\infty} - i\alpha \int_{-\infty}^{\infty} f(x)e^{i\alpha x}\,dx$$

$$= -i\alpha \int_{-\infty}^{\infty} f(x)e^{i\alpha x}\,dx;$$

that is, $$\mathscr{F}\{f'(x)\} = -i\alpha F(\alpha). \tag{7}$$

Similarly, under the added assumptions that f' is continuous on $(-\infty, \infty)$, $f''(x)$ is piecewise continuous on every finite interval, and $f'(x) \to 0$ as $x \to \pm\infty$, we have

$$\mathscr{F}\{f''(x)\} = (-i\alpha)^2 \mathscr{F}\{f(x)\} = -\alpha^2 F(\alpha). \tag{8}$$

It is important to be aware that the sine and cosine transforms are not suitable for transforming the first derivative (or, for that matter, any derivative of *odd* order). It is readily shown that

$$\mathscr{F}_s\{f'(x)\} = -\alpha\mathscr{F}_c\{f(x)\} \quad \text{and} \quad \mathscr{F}_c\{f'(x)\} = \alpha\mathscr{F}_s\{f(x)\} - f(0).$$

The difficulty is apparent; the transform of $f'(x)$ is not expressed in terms of the original integral transform.

Fourier Sine Transform   Suppose that f and f' are continuous, f is absolutely integrable on the interval $[0, \infty)$, and f'' is piecewise continuous on every finite interval. If $f \to 0$ and $f' \to 0$ as $x \to \infty$, then

$$\mathscr{F}_s\{f''(x)\} = \int_0^{\infty} f''(x)\sin\alpha x\,dx$$

$$= f'(x)\sin\alpha x\Big|_0^{\infty} - \alpha \int_0^{\infty} f'(x)\cos\alpha x\,dx$$

$$= -\alpha\left[f(x)\cos\alpha x\Big|_0^{\infty} + \alpha \int_0^{\infty} f(x)\sin\alpha x\,dx \right]$$

$$= \alpha f(0) - \alpha^2\mathscr{F}_s\{f(x)\};$$

that is, $$\mathscr{F}_s\{f''(x)\} = -\alpha^2 F(\alpha) + \alpha f(0). \tag{9}$$

Fourier Cosine Transform   Under the same assumptions that lead to (9) we find the Fourier cosine transform of $f''(x)$ to be

$$\mathscr{F}_c\{f''(x)\} = -\alpha^2 F(\alpha) - f'(0). \tag{10}$$

A natural question is "How do we know which transform to use on a given boundary-value problem?" Clearly, to use a Fourier transform the domain of the variable to be eliminated must be $(-\infty, \infty)$. To utilize a sine or cosine transform the domain of at least one of the variables in the problem must be $[0, \infty)$. But the determining factor in choosing between the sine transform and the cosine transform is the type of boundary condition specified at zero.

In the examples that follow we shall assume without further mention that both u and $\partial u/\partial x$ (or $\partial u/\partial y$) approach zero as $x \to \pm\infty$. This is not a major restriction since these conditions hold in most applications.

EXAMPLE 1   Using the Fourier Transform

Solve the heat equation $k\dfrac{\partial^2 u}{\partial x^2} = \dfrac{\partial u}{\partial t}$, $-\infty < x < \infty$, $t > 0$ subject to

$$u(x, 0) = f(x), \quad \text{where} \quad f(x) = \begin{cases} u_0, & |x| < 1 \\ 0, & |x| > 1. \end{cases}$$

Solution   The problem can be interpreted as finding the temperature $u(x, t)$ in an infinite rod. Since the domain of x is the infinite interval $(-\infty, \infty)$, we use the Fourier transform (1) and define

$$\mathscr{F}\{u(x, t)\} = \int_{-\infty}^{\infty} u(x, t)e^{i\alpha x}\, dx = U(\alpha, t).$$

If we transform the partial differential equation and use (8),

$$\mathscr{F}\left\{k\frac{\partial^2 u}{\partial x^2}\right\} = \mathscr{F}\left\{\frac{\partial u}{\partial t}\right\}$$

yields $\quad -k\alpha^2 U(\alpha, t) = \dfrac{dU}{dt} \quad \text{or} \quad \dfrac{dU}{dt} + k\alpha^2 U(\alpha, t) = 0.$

Solving the last equation gives $U(\alpha, t) = ce^{-k\alpha^2 t}$. Now the transform of the initial condition is

$$\mathscr{F}\{u(x, 0)\} = \int_{-\infty}^{\infty} f(x)e^{i\alpha x}\, dx = \int_{-1}^{1} u_0 e^{i\alpha x}\, dx = u_0 \frac{e^{i\alpha} - e^{-i\alpha}}{i\alpha}.$$

This result is the same as $U(\alpha, 0) = 2u_0\dfrac{\sin\alpha}{\alpha}$. Applying this condition to the solution $U(\alpha, t)$ gives $U(\alpha, 0) = c = (2u_0 \sin\alpha)/\alpha$, and so

$$U(\alpha, t) = 2u_0 \frac{\sin\alpha}{\alpha} e^{-k\alpha^2 t}.$$

It then follows from the inversion integral (2) that

$$u(x, t) = \frac{u_0}{\pi} \int_{-\infty}^{\infty} \frac{\sin\alpha}{\alpha} e^{-k\alpha^2 t} e^{-i\alpha x}\, d\alpha.$$

The last expression can be simplified somewhat by Euler's formula $e^{-i\alpha x} = \cos\alpha x - i\sin\alpha x$ and noting that $\displaystyle\int_{-\infty}^{\infty} \frac{\sin\alpha}{\alpha} e^{-k\alpha^2 t} \sin\alpha x\, d\alpha = 0$

since the integrand is an odd function of α. Hence we finally have

$$u(x, t) = \frac{u_0}{\pi} \int_{-\infty}^{\infty} \frac{\sin \alpha \cos \alpha x}{\alpha} e^{-k\alpha^2 t} \, d\alpha. \tag{11} \blacksquare$$

It is left to the reader to show that the solution (11) can be expressed in terms of the error function. See Problem 4 in Exercises 14.4.

EXAMPLE 2   **Using the Cosine Transform**

The steady-state temperature in a semi-infinite plate is determined from

$$\frac{\partial^2 u}{\partial x^2} + \frac{\partial^2 u}{\partial y^2} = 0, \quad 0 < x < \pi, \quad y > 0$$

$$u(0, y) = 0, \quad u(\pi, u) = e^{-y}, \quad y > 0$$

$$\left. \frac{\partial u}{\partial y} \right|_{y=0} = 0, \quad 0 < x < \pi.$$

Solve for $u(x, y)$.

Solution   The domain of the variable y and the prescribed condition at $y = 0$ indicate that the Fourier cosine transform is suitable for the problem. We define

$$\mathscr{F}_c\{u(x, y)\} = \int_0^{\infty} u(x, y) \cos \alpha y \, dy = U(x, \alpha).$$

In view of (10), $\mathscr{F}_c\left\{\dfrac{\partial^2 u}{\partial x^2}\right\} + \mathscr{F}_c\left\{\dfrac{\partial^2 u}{\partial y^2}\right\} = \mathscr{F}_c\{0\}$

becomes

$$\frac{d^2 U}{dx^2} - \alpha^2 U(x, \alpha) - u_y(x, 0) = 0 \quad \text{or} \quad \frac{d^2 U}{dx^2} - \alpha^2 U = 0.$$

Since the domain of x is a finite interval, we choose to write the solution of the ordinary differential equation as

$$U(x, \alpha) = c_1 \cosh \alpha x + c_2 \sinh \alpha x. \tag{12}$$

Now $\mathscr{F}_c\{u(0, y)\} = \mathscr{F}_c\{0\}$ and $\mathscr{F}_c\{u(\pi, y)\} = \mathscr{F}_c\{e^{-y}\}$ are in turn equivalent to

$$U(0, \alpha) = 0 \quad \text{and} \quad U(\pi, \alpha) = \frac{1}{1 + \alpha^2}.$$

When we apply these latter conditions, the solution (12) gives $c_1 = 0$ and $c_2 = 1/[(1 + \alpha^2) \sinh \alpha\pi]$. Therefore

$$U(x, \alpha) = \frac{\sinh \alpha x}{(1 + \alpha^2) \sinh \alpha\pi},$$

and so from (6) we arrive at

$$u(x, y) = \frac{2}{\pi} \int_0^{\infty} \frac{\sinh \alpha x}{(1 + \alpha^2) \sinh \alpha\pi} \cos \alpha y \, d\alpha. \tag{13} \blacksquare$$

Had $u(x, 0)$ been given in Example 2 rather than $u_y(x, 0)$, then the sine transform would have been appropriate.

EXERCISES 14.4

Answers to odd-numbered problems begin on page AN-21.

In Problems 1–21 use the Fourier integral transforms of this section to solve the given boundary-value problem. Make assumptions about boundedness where necessary.

1. $k\dfrac{\partial^2 u}{\partial x^2} = \dfrac{\partial u}{\partial t}$, $-\infty < x < \infty$, $t > 0$

$u(x, 0) = e^{-|x|}$, $-\infty < x < \infty$

2. $k\dfrac{\partial^2 u}{\partial x^2} = \dfrac{\partial u}{\partial t}$, $-\infty < x < \infty$, $t > 0$

$$u(x, 0) = \begin{cases} 0, & x < -1 \\ -100, & -1 < x < 0 \\ 100, & 0 < x < 1 \\ 0, & x > 1 \end{cases}$$

3. Use the result $\mathscr{F}\{e^{-x^2/4p^2}\} = 2\sqrt{\pi}\,pe^{-p^2\alpha^2}$ to solve the boundary-value problem

$$k\frac{\partial^2 u}{\partial x^2} = \frac{\partial u}{\partial t}, \quad -\infty < x < \infty, \quad t > 0$$

$$u(x, 0) = e^{-x^2}, \quad -\infty < x < \infty.$$

4. (a) If $\mathscr{F}\{f(x)\} = F(\alpha)$ and $\mathscr{F}\{g(x)\} = G(\alpha)$, then the **convolution theorem** for the Fourier transform is given by

$$\int_{-\infty}^{\infty} f(\tau)g(x - \tau)\, d\tau = \mathscr{F}^{-1}\{F(\alpha)G(\alpha)\}.$$

Use this result and $\mathscr{F}\{e^{-x^2/4p^2}\} = 2\sqrt{\pi}\,pe^{-p^2\alpha^2}$ to show that a solution of the boundary-value problem

$$k\frac{\partial^2 u}{\partial x^2} = \frac{\partial u}{\partial t}, \quad -\infty < x < \infty, \quad t > 0$$

$$u(x, 0) = f(x), \quad -\infty < x < \infty$$

is

$$u(x, t) = \frac{1}{2\sqrt{k\pi t}}\int_{-\infty}^{\infty} f(\tau)e^{-(x-\tau)^2/4kt}\, d\tau.$$

(b) Use the change of variables $u = (x - \tau)/2\sqrt{kt}$ and Problem 9 in Exercises 14.1 to show that the solution of Example 1 is

$$u(x, t) = \frac{u_0}{2}\left[\operatorname{erf}\left(\frac{x + 1}{2\sqrt{kt}}\right) - \operatorname{erf}\left(\frac{x - 1}{2\sqrt{kt}}\right)\right].$$

(c) Assume $u_0 = 100$ and $k = 1$. Use a CAS to obtain the graph of $u(x, t)$ over the rectangular region $-4 \le x \le 4, 0 \le t \le 6$. Use a 2D-plot application to superimpose the graphs of $u(x, t)$ for $t = 0.05, 0.125, 0.5, 1, 2, 4, 6$, and 15 on the interval $-4 \le x \le 4$. Use the graphs to conjecture the values of $\lim_{t\to\infty} u(x, t)$ and $\lim_{x\to\infty} u(x, t)$. Then prove these results analytically using the properties of $\operatorname{erf}(x)$.

5. Find the temperature $u(x, t)$ in a semi-infinite rod if $u(0, t) = u_0, t > 0$ and $u(x, 0) = 0, x > 0$.

6. Use the result $\int_0^\infty \dfrac{\sin \alpha x}{\alpha} d\alpha = \dfrac{\pi}{2}, x > 0$ to show that the solution of Problem 5 can be written as

$$u(x, t) = u_0 - \frac{2u_0}{\pi} \int_0^\infty \frac{\sin \alpha x}{\alpha} e^{-k\alpha^2 t} d\alpha.$$

7. Find the temperature $u(x, t)$ in a semi-infinite rod if $u(0, t) = 0, t > 0$ and

$$u(x, 0) = \begin{cases} 1, & 0 < x < 1 \\ 0, & x > 1. \end{cases}$$

8. Solve Problem 5 if the condition at the left boundary is

$$\frac{\partial u}{\partial x}\bigg|_{x=0} = -A, \quad t > 0,$$

where A is a constant.

9. Solve Problem 7 if the end $x = 0$ is insulated.

10. Find the temperature $u(x, t)$ in a semi-infinite rod if $u(0, t) = 1, t > 0$ and $u(x, 0) = e^{-x}, x > 0$.

11. (a) $a^2 \dfrac{\partial^2 u}{\partial x^2} = \dfrac{\partial^2 u}{\partial t^2}, \quad -\infty < x < \infty, \quad t > 0$

$$u(x, 0) = f(x), \quad \frac{\partial u}{\partial t}\bigg|_{t=0} = g(x), \quad -\infty < x < \infty$$

(b) If $g(x) = 0$, show that the solution of part (a) can be written as $u(x, t) = \frac{1}{2}[f(x + at) + f(x - at)]$.

12. Find the displacement $u(x, t)$ of a semi-infinite string if

$$u(0, t) = 0, \quad t > 0$$
$$u(x, 0) = xe^{-x}, \quad \frac{\partial u}{\partial t}\bigg|_{t=0} = 0, \quad x > 0.$$

13. Solve the problem in Example 2 if the boundary conditions at $x = 0$ and $x = \pi$ are reversed: $u(0, y) = e^{-y}, u(\pi, y) = 0, y > 0$.

14. Solve the problem in Example 2 if the boundary condition at $y = 0$ is $u(x, 0) = 1, 0 < x < \pi$.

15. Find the steady-state temperature $u(x, y)$ in a plate defined by $x \geq 0$, $y \geq 0$ if the boundary $x = 0$ is insulated and, at $y = 0$,

$$u(x, 0) = \begin{cases} 50, & 0 < x < 1 \\ 0, & x > 1 \end{cases}$$

16. Solve Problem 15 if the boundary condition at $x = 0$ is $u(0, y) = 0$, $y > 0$.

17. $\dfrac{\partial^2 u}{\partial x^2} + \dfrac{\partial^2 u}{\partial y^2} = 0, \quad x > 0, \quad 0 < y < 2$

$u(0, y) = 0, \quad 0 < y < 2$

$u(x, 0) = f(x), \quad u(x, 2) = 0, \quad x > 0$

18. $\dfrac{\partial^2 u}{\partial x^2} + \dfrac{\partial^2 u}{\partial y^2} = 0, \quad 0 < x < \pi, \quad y > 0$

$u(0, y) = f(y), \quad \dfrac{\partial u}{\partial x}\bigg|_{x=\pi} = 0, \quad y > 0$

$\dfrac{\partial u}{\partial y}\bigg|_{y=0} = 0, \quad 0 < x < \pi$

In Problems 19 and 20 find the steady-state temperature in the plate given in the figure. [*Hint:* One way of proceeding is to express Problems 19 and 20 as two- and three-boundary-value problems, respectively. Use the superposition principle (see Section 12.5).]

19.

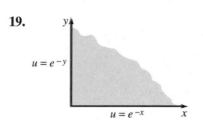

Figure 14.10

20.

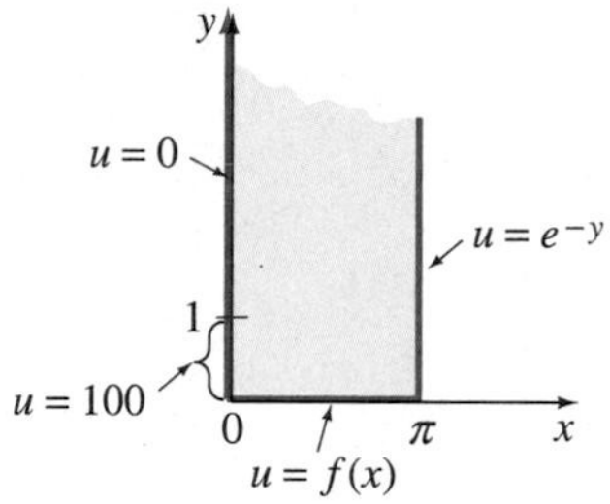

Figure 14.11

21. Use the transform $\mathscr{F}\{e^{-x^2/4p^2}\}$ given in Problem 3 to find the steady-state temperature in the infinite strip shown in Figure 14.12.

22. The solution of Problem 16 can be integrated. Use entries 42 and 43 of the table in Appendix III to show that

$$u(x, y) = \frac{100}{\pi}\left[\arctan\frac{x}{y} - \frac{1}{2}\arctan\frac{x+1}{y} - \frac{1}{2}\arctan\frac{x-1}{y}\right].$$

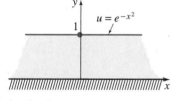

insulated

Figure 14.12

CHAPTER 14 IN REVIEW

Answers to odd-numbered problems begin on page AN-21.

In Problems 1–13 solve the given boundary-value problem by an appropriate integral transform. Make assumptions about boundedness where necessary.

1. $\dfrac{\partial^2 u}{\partial x^2} + \dfrac{\partial^2 u}{\partial y^2} = 0, \quad x > 0, \quad 0 < y < \pi$

$\dfrac{\partial u}{\partial x}\bigg|_{x=0} = 0, \quad 0 < y < \pi$

$u(x, 0) = 0, \quad \dfrac{\partial u}{\partial y}\bigg|_{y=\pi} = e^{-x}, \quad x > 0$

2. $\dfrac{\partial^2 u}{\partial x^2} = \dfrac{\partial u}{\partial t}, \quad 0 < x < 1, \quad t > 0$

$u(0, t) = 0, \quad u(1, t) = 0, \quad t > 0$

$u(x, 0) = 50\sin 2\pi x, \quad 0 < x < 1$

3. $\dfrac{\partial^2 u}{\partial x^2} - hu = \dfrac{\partial u}{\partial t}, \quad h > 0, \quad x > 0, \quad t > 0$

$u(0, t) = 0, \quad \lim\limits_{x \to \infty} \dfrac{\partial u}{\partial x} = 0, \quad t > 0$

$u(x, 0) = u_0, \quad x > 0$

4. $\dfrac{\partial u}{\partial t} - \dfrac{\partial^2 u}{\partial x^2} = e^{-|x|}, \quad -\infty < x < \infty, \quad t > 0$

$u(x, 0) = 0, \quad -\infty < x < \infty$

5. $\dfrac{\partial^2 u}{\partial x^2} = \dfrac{\partial u}{\partial t}, \quad x > 0, \quad t > 0$

$u(0, t) = t, \quad \lim\limits_{x \to \infty} u(x, t) = 0$

$u(x, 0) = 0, \quad x > 0 \quad$ [*Hint:* Use Theorem 7.9.]

6. $\dfrac{\partial^2 u}{\partial x^2} = \dfrac{\partial^2 u}{\partial t^2}, \quad 0 < x < 1, \quad t > 0$

$u(0, t) = 0, \quad u(1, t) = 0, \quad t > 0$

$u(x, 0) = \sin \pi x, \quad \left.\dfrac{\partial u}{\partial t}\right|_{t=0} = -\sin \pi x, \quad 0 < x < 1$

7. $k \dfrac{\partial^2 u}{\partial x^2} = \dfrac{\partial u}{\partial t}, \quad -\infty < x < \infty, \quad t > 0$

$u(x, 0) = \begin{cases} 0, & x < 0 \\ u_0, & 0 < x < \pi \\ 0, & x > \pi \end{cases}$

8. $\dfrac{\partial^2 u}{\partial x^2} + \dfrac{\partial^2 u}{\partial y^2} = 0, \quad 0 < x < \pi, \quad y > 0$

$u(0, y) = 0, \quad u(\pi, y) = \begin{cases} 0, & 0 < y < 1 \\ 1, & 1 < y < 2 \\ 0, & y > 2 \end{cases}$

$\left.\dfrac{\partial u}{\partial y}\right|_{y=0} = 0, \quad 0 < x < \pi$

9. $\dfrac{\partial^2 u}{\partial x^2} + \dfrac{\partial^2 u}{\partial y^2} = 0, \quad x > 0, \quad y > 0$

$u(0, y) = \begin{cases} 50, & 0 < y < 1 \\ 0, & y > 1 \end{cases}$

$u(x, 0) = \begin{cases} 100, & 0 < x < 1 \\ 0, & x > 1 \end{cases}$

10. $\dfrac{\partial^2 u}{\partial x^2} + r = \dfrac{\partial u}{\partial t}, \quad 0 < x < 1, \quad t > 0$

$\left.\dfrac{\partial u}{\partial x}\right|_{x=0} = 0, \quad u(1, t) = 0, \quad t > 0$

$u(x, 0) = 0, \quad 0 < x < 1$

11. $\dfrac{\partial^2 u}{\partial x^2} + \dfrac{\partial^2 u}{\partial y^2} = 0, \quad x > 0, \quad 0 < y < \pi$

$u(0, y) = A, \quad 0 < y < \pi$

$\dfrac{\partial u}{\partial y}\Big|_{y=0} = 0, \quad \dfrac{\partial u}{\partial y}\Big|_{y=\pi} = Be^{-x}, \quad x > 0$

12. $\dfrac{\partial^2 u}{\partial x^2} = \dfrac{\partial u}{\partial t}, \quad 0 < x < 1, \quad t > 0$

$u(0, t) = u_0, \quad u(1, t) = u_0, \quad t > 0$
$u(x, 0) = 0, \quad 0 < x < 1$

[*Hint:* Use the identity $\sinh(x - y) = \sinh x \cosh y - \cosh x \sinh y$, and then use Problem 6 in Exercises 14.1.]

13. $k\dfrac{\partial^2 u}{\partial x^2} = \dfrac{\partial u}{\partial t}, \quad -\infty < x < \infty, \quad t > 0$

$u(x, 0) = \begin{cases} 0, & x < 0 \\ e^{-x}, & x > 0 \end{cases}$

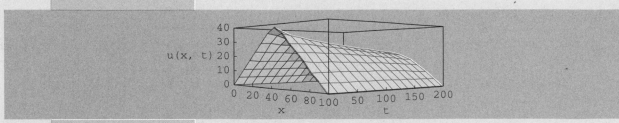

$u(x, t)$

Solution of the heat equation; see Problem 6(d), page 624.

15 NUMERICAL SOLUTIONS OF PARTIAL DIFFERENTIAL EQUATIONS

15.1 Elliptic Equations
15.2 Parabolic Equations
15.3 Hyperbolic Equations

Chapter 15 in Review

INTRODUCTION In Section 9.5 we saw that one way of approximating a solution of a second-order boundary-value problem was to work with a finite difference equation replacement of the ordinary differential equation. The difference equation was constructed by replacing the ordinary derivatives y'' and y' by difference quotients. The same idea carries over to BVPs involving partial differential equations. In the respective sections of this chapter, we shall form a difference equation replacement for Laplace's equation, the heat equation, and the wave equation by replacing the partial derivatives u_{xx}, u_{yy}, u_{tt}, and u_t by difference quotients.

15.1 ELLIPTIC EQUATIONS

• Difference equation replacement for Laplace's equation   • Five-point approximation   • Mesh size   • Interior point   • Boundary point   • Dirichlet problem   • Sparse matrix   • Banded matrix   • Gauss-Seidel iteration   • Diagonally dominant matrix

Using the difference quotients derived in Section 9.5, we can easily replace a linear second-order partial differential equation by a difference equation. In the discussion that follows we shall confine our attention to elliptic partial differential equations such as Laplace's equation,

$$\frac{\partial^2 u}{\partial x^2} + \frac{\partial^2 u}{\partial y^2} = 0.$$

A Difference Equation Replacement for Laplace's Equation   If u is a function of two variables x and y, we can form two central differences

$$u(x + h, y) - 2u(x, y) + u(x - h, y) \quad \text{and} \quad u(x, y + h) - 2u(x, y) + u(x, y - h).$$

It then follows from (6) of Section 9.5 that difference quotient approximations for the second partial derivatives u_{xx} and u_{yy} are given by

$$\frac{\partial^2 u}{\partial x^2} \approx \frac{1}{h^2}[u(x + h, y) - 2u(x, y) + u(x - h, y)] \tag{1}$$

$$\frac{\partial^2 u}{\partial y^2} \approx \frac{1}{h^2}[u(x, y + h) - 2u(x, y) + u(x, y - h)]. \tag{2}$$

Now by adding (1) and (2) we obtain a **five-point approximation** to the Laplacian:

$$\frac{\partial^2 u}{\partial x^2} + \frac{\partial^2 u}{\partial y^2} \approx \frac{1}{h^2}[u(x + h, y) + u(x, y + h) + u(x - h, y) + u(x, y - h) - 4u(x, y)].$$

Hence we can replace Laplace's equation $\dfrac{\partial^2 u}{\partial x^2} + \dfrac{\partial^2 u}{\partial y^2} = 0$ by the difference equation

$$u(x + h, y) + u(x, y + h) + u(x - h, y) + u(x, y - h) - 4u(x, y) = 0. \tag{3}$$

If we adopt the notation $u(x, y) = u_{ij}$ and

$$u(x + h, y) = u_{i+1,j}, \quad u(x, y + h) = u_{i,j+1}$$
$$u(x - h, y) = u_{i-1,j}, \quad u(x, y - h) = u_{i,j-1},$$

then (3) becomes

$$u_{i+1,j} + u_{i,j+1} + u_{i-1,j} + u_{i,j-1} - 4u_{ij} = 0. \tag{4}$$

To understand (4) a little better, suppose a rectangular grid consisting of horizontal lines spaced h units apart and vertical lines spaced h units apart is placed over a region R bounded by a curve C in which we are seeking a solution of Laplace's equation. The number h is called the **mesh**

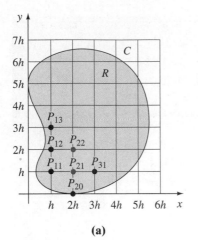

(a)

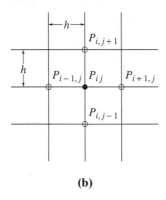

(b)

Figure 15.1

size. See Figure 15.1(a). The points of intersection $P_{ij} = P(ih, jh)$, i and j integers, of the lines are called **mesh points** or **lattice points.** A mesh point is an **interior point** if its four nearest neighboring mesh points are points of R. Points in R or on C that are not interior points are called **boundary points.** For example, in Figure 15.1(a) we have

$$P_{20} = P(2h, 0), \quad P_{11} = P(h, h), \quad P_{21} = P(2h, h), \quad P_{22} = P(2h, 2h),$$

and so on. Of the points listed, P_{21} and P_{22} are interior points, P_{20} and P_{11} are boundary points. In Figure 15.1(a) the interior points are the colored dots and the boundary points are the black dots. Now from (4) we see that

$$u_{ij} = \frac{1}{4}[u_{i+1,j} + u_{i,j+1} + u_{i-1,j} + u_{i,j-1}], \tag{5}$$

and so, as shown in Figure 15.1(b), the value of u_{ij} at an interior mesh point of R is the average of the values of u at four neighboring mesh points. The neighboring points $P_{i+1,j}$, $P_{i,j+1}$, $P_{i-1,j}$, and $P_{i,j-1}$ correspond, respectively, to the four points on a compass E, N, W, and S.

Dirichlet Problem Recall that in the **Dirichlet problem** for Laplace's equation $\nabla^2 u = 0$ the values of $u(x, y)$ are prescribed on the boundary of a region R. The basic idea is to find an approximate solution to Laplace's equation at interior mesh points by replacing the partial differential equation at these points by the difference equation (4). Hence the approximate values of u at the mesh points—namely, the u_{ij}—are related to each other and, possibly, to known values of u if a mesh point lies on the boundary. In this manner we obtain a system of linear algebraic equations that we solve for the unknown u_{ij}. The following example illustrates the method for a square region.

EXAMPLE 1 **A BVP Revisited**

In Problem 16 of Exercises 12.5 you were asked to solve the boundary-value problem

$$\frac{\partial^2 u}{\partial x^2} + \frac{\partial^2 u}{\partial y^2} = 0, \quad 0 < x < 2, \quad 0 < y < 2$$

$$u(0, y) = 0, \quad u(2, y) = y(2 - y), \quad 0 < y < 2$$

$$u(x, 0) = 0, \quad u(x, 2) = \begin{cases} x, & 0 < x < 1 \\ 2 - x, & 1 \le x < 2 \end{cases}$$

utilizing the superposition principle. To apply the present numerical method let us start with a mesh size of $h = \frac{2}{3}$. As we see in Figure 15.2, that choice yields four interior points and eight boundary points. The numbers listed next to the boundary points are the exact values of u obtained from the specified condition along that boundary. For example, at $P_{31} = P(3h, h) = P(2, \frac{2}{3})$ we have $x = 2$ and $y = \frac{2}{3}$, and so the condition $u(2, y)$ gives $u(2, \frac{2}{3}) = \frac{2}{3}(2 - \frac{2}{3}) = \frac{8}{9}$. Similarly, at $P_{13} = P(\frac{2}{3}, 2)$ the condition $u(x, 2)$ gives $u(\frac{2}{3}, 2) = \frac{2}{3}$. We now apply (4) at each interior point. For example, at P_{11} we have $i = 1$ and $j = 1$, so (4) becomes

$$u_{21} + u_{12} + u_{01} + u_{10} - 4u_{11} = 0.$$

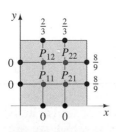

Figure 15.2

Since $u_{01} = u(0, \frac{2}{3}) = 0$ and $u_{10} = u(\frac{2}{3}, 0) = 0$, the foregoing equation becomes $-4u_{11} + u_{21} + u_{12} = 0$. Repeating this, in turn, at P_{21}, P_{12}, and P_{22}, we get three additional equations:

$$
\begin{aligned}
-4u_{11} + \ u_{21} + \ u_{12} \qquad\qquad &= 0 \\
u_{11} - 4u_{21} + \qquad\qquad u_{22} &= -\frac{8}{9} \\
u_{11} \qquad\qquad - 4u_{12} + \ u_{22} &= -\frac{2}{3} \\
u_{21} + \ u_{12} - 4u_{22} &= -\frac{14}{9}.
\end{aligned}
$$

(6)

Using a computer algebra system to solve the system, we find the approximate values at the four interior points to be

$$
u_{11} = \frac{7}{36} = 0.1944, \quad u_{21} = \frac{5}{12} = 0.4167, \quad u_{12} = \frac{13}{36} = 0.3611, \quad u_{22} = \frac{7}{12} = 0.5833. \quad\blacksquare
$$

As in the discussion of ordinary differential equations, we expect that a smaller value of h will improve the accuracy of the approximation. However, using a smaller mesh size means, of course, that there are more interior mesh points, and correspondingly there is a much larger system of equations to be solved. For a *square* region whose length of side is L, a mesh size of $h = L/n$ will yield a total of $(n - 1)^2$ interior mesh points. In Example 1, for $n = 8$, the mesh size is a reasonable $h = \frac{2}{8} = \frac{1}{4}$, but the number of interior points is $(8 - 1)^2 = 49$. Thus we have 49 equations in 49 unknowns. In the next example we use a mesh size of $h = \frac{1}{2}$.

<table>
<tr><td>EXAMPLE 2</td><td>**Example 1 with More Mesh Points**</td></tr>
</table>

As we see in Figure 15.3, with $n = 4$, a mesh size $h = \frac{2}{4} = \frac{1}{2}$ for the square in Example 1 gives $3^2 = 9$ interior mesh points. Applying (4) at these points and using the indicated boundary conditions, we get nine equations in nine unknowns. So that you can verify the results, we give the system in an unsimplified form:

$$
\begin{aligned}
u_{21} + u_{12} + \ 0 + \ 0 - 4u_{11} &= 0 \\
u_{31} + u_{22} + u_{11} + \ 0 - 4u_{21} &= 0 \\
\frac{3}{4} + u_{32} + u_{21} + \ 0 - 4u_{31} &= 0 \\
u_{22} + u_{13} + u_{11} + \ 0 - 4u_{12} &= 0 \\
u_{32} + u_{23} + u_{12} + u_{21} - 4u_{22} &= 0 \\
1 + u_{33} + u_{22} + u_{31} - 4u_{32} &= 0 \\
u_{23} + \frac{1}{2} + \ 0 + u_{12} - 4u_{13} &= 0 \\
u_{33} + \ 1 + u_{13} + u_{22} - 4u_{23} &= 0 \\
\frac{3}{4} + \frac{1}{2} + u_{23} + u_{32} - 4u_{33} &= 0.
\end{aligned}
$$

(7)

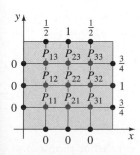

Figure 15.3

In this case a CAS yields

$$u_{11} = \frac{7}{64} = 0.1094, \quad u_{21} = \frac{51}{224} = 0.2277, \quad u_{31} = \frac{177}{448} = 0.3951$$

$$u_{12} = \frac{47}{224} = 0.2098, \quad u_{22} = \frac{13}{32} = 0.4063, \quad u_{32} = \frac{135}{224} = 0.6027$$

$$u_{13} = \frac{145}{448} = 0.3237, \quad u_{23} = \frac{131}{224} = 0.5848, \quad u_{33} = \frac{39}{64} = 0.6094. \quad \blacksquare$$

After we simplify (7), it is interesting to note that the 9×9 matrix of coefficients is

$$\begin{pmatrix} -4 & 1 & 0 & 1 & 0 & 0 & 0 & 0 & 0 \\ 1 & -4 & 1 & 0 & 1 & 0 & 0 & 0 & 0 \\ 0 & 1 & -4 & 0 & 0 & 1 & 0 & 0 & 0 \\ 1 & 0 & 0 & -4 & 1 & 0 & 1 & 0 & 0 \\ 0 & 1 & 0 & 1 & -4 & 1 & 0 & 1 & 0 \\ 0 & 0 & 1 & 0 & 1 & -4 & 0 & 0 & 1 \\ 0 & 0 & 0 & 1 & 0 & 0 & -4 & 1 & 0 \\ 0 & 0 & 0 & 0 & 1 & 0 & 1 & -4 & 1 \\ 0 & 0 & 0 & 0 & 0 & 1 & 0 & 1 & -4 \end{pmatrix}. \quad \textbf{(8)}$$

This is an example of a **sparse matrix** in that a large percentage of the entries are zeros. The matrix (8) is also an example of a **banded matrix.** These kinds of matrices are characterized by the properties that the entries on the main diagonal and on diagonals (or bands) parallel to the main diagonal are all nonzero.

Gauss-Seidel Iteration   Problems requiring approximations to solutions of partial differential equations invariably lead to large systems of linear algebraic equations. It is not uncommon to have to solve systems involving hundreds of equations. Although a direct method of solution such as Gaussian elimination leaves unchanged the zero entries outside the bands in a matrix such as (8), it does fill in the positions between the bands with nonzeros. Since storing very large matrices uses up a large portion of computer memory, it is usual practice to solve a large system in an indirect manner. One popular indirect method is called **Gauss-Seidel iteration.**

We shall illustrate this method for the system in (6). For the sake of simplicity we replace the double-subscripted variables u_{11}, u_{21}, u_{12}, and u_{22} by x_1, x_2, x_3, and x_4, respectively.

EXAMPLE 3   **Gauss-Seidel Iteration**

Step 1:   *Solve each equation for the variables on the main diagonal of the system.*   That is, in (6) solve the first equation for x_1, the second equation for x_2, and so on:

$$\begin{aligned} x_1 &= 0.25x_2 + 0.25x_3 \\ x_2 &= 0.25x_1 + 0.25x_4 + 0.2222 \\ x_3 &= 0.25x_1 + 0.25x_4 + 0.1667 \\ x_4 &= 0.25x_2 + 0.25x_3 + 0.3889. \end{aligned} \quad \textbf{(9)}$$

These equations can be obtained directly by using (5) rather than (4) at the interior points.

Step 2: *Iterations.* We start by making an initial guess for the values of x_1, x_2, x_3, and x_4. If this were simply a system of linear equations and we knew nothing about the solution, we could start with $x_1 = 0$, $x_2 = 0$, $x_3 = 0$, $x_4 = 0$. But since the solution of (9) represents approximations to a solution of a boundary-value problem, it would seem reasonable to use as the initial guess for the values of $x_1 = u_{11}$, $x_2 = u_{21}$, $x_3 = u_{12}$, and $x_4 = u_{22}$ the average of all the boundary conditions. In this case the average of the numbers at the eight boundary points shown in Figure 15.2 is approximately 0.4. Thus our initial guess is $x_1 = 0.4$, $x_2 = 0.4$, $x_3 = 0.4$, and $x_4 = 0.4$. Iterations of the Gauss-Seidel method use the x values as soon as they are computed. Note that the first equation in (9) depends only on x_2 and x_3; thus substituting $x_2 = 0.4$ and $x_3 = 0.4$ gives $x_1 = 0.2$. Since the second and third equations depend on x_1 and x_4, we use the newly calculated values $x_1 = 0.2$ and $x_4 = 0.4$ to obtain $x_2 = 0.3722$ and $x_3 = 0.3167$. The fourth equation depends on x_2 and x_3, so we use the new values $x_2 = 0.3722$ and $x_3 = 0.3167$ to get $x_4 = 0.5611$. In summary, the first iteration has given the values

$$x_1 = 0.2, \quad x_2 = 0.3722, \quad x_3 = 0.3167, \quad x_4 = 0.5611.$$

Note how close these numbers are already to the actual values given at the end of Example 1.

The second iteration starts with substituting $x_2 = 0.3722$ and $x_3 = 0.3167$ into the first equation. This gives $x_1 = 0.1722$. From $x_1 = 0.1722$ and the last computed value of x_4 (namely, $x_4 = 0.5611$), the second and third equations give, in turn, $x_2 = 0.4055$ and $x_3 = 0.3500$. Using these two values, we find from the fourth equation that $x_4 = 0.5678$. At the end of the second iteration we have

$$x_1 = 0.1722, \quad x_2 = 0.4055, \quad x_3 = 0.3500, \quad x_4 = 0.5678.$$

The third through seventh iterations are summarized in Table 15.1.

TABLE 15.1

Iteration	3rd	4th	5th	6th	7th
x_1	0.1889	0.1931	0.1941	0.1944	0.1944
x_2	0.4139	0.4160	0.4165	0.4166	0.4166
x_3	0.3584	0.3605	0.3610	0.3611	0.3611
x_4	0.5820	0.5830	0.5833	0.5833	0.5833

Note To apply Gauss-Seidel iteration to a general system of n linear equations in n unknowns, the variable x_i must actually appear in the ith equation of the system. Moreover, after each equation is solved for x_i, $i = 1, 2, \ldots, n$, the resulting system has the form $\mathbf{X} = \mathbf{AX} + \mathbf{B}$, where all the entries on the main diagonal of $\mathbf{A}$ are zero.

Remarks (*i*) In the examples given in this section the values of u_{ij} were determined using known values of u at boundary points. But what do we do if the region is such that boundary points do not coincide with the actual boundary C of the region R? In this case the required values can be obtained by interpolation.

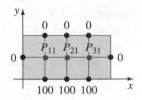

Figure 15.4

(*ii*) It is sometimes possible to cut down the number of equations to solve by using symmetry. Consider the rectangular region shown in Figure 15.4. The boundary conditions are $u = 0$ along $x = 0$, $x = 2$, $y = 1$ and $u = 100$ along $y = 0$. The region is symmetric about the lines $x = 1$ and $y = \frac{1}{2}$, and the interior points P_{11} and P_{31} are equidistant from the neighboring boundary points at which the specified values of u are the same. Consequently we assume that $u_{11} = u_{31}$, and so the system of three equations in three unknowns reduces to two equations in two unknowns. See Problem 2.

(*iii*) Although on a computer this may not be noticeable, convergence of Gauss-Seidel iteration may not be particularly fast. Also, in a more general setting Gauss-Seidel iteration may not converge at all. If a system of n linear equations in n unknowns is written in the standard form $\mathbf{AX} = \mathbf{B}$, then a sufficient condition for convergence is that the coefficient matrix $\mathbf{A}$ be **diagonally dominant.** This means that the absolute value of each entry on the main diagonal of $\mathbf{A}$ is greater than the sum of the absolute values of the remaining entries in the same row. For example, the coefficient matrix (8) for the system (7) is clearly diagonally dominant since in each row $|-4|$ is greater than the sum of the absolute values of the other entries. You should verify that the coefficient matrix for the system (6) has the same property. Diagonal dominance guarantees that Gauss-Seidel iteration converges for any initial values.

EXERCISES 15.1

Answers to odd-numbered problems begin on page AN-22.

 In Problems 1–8 use a computer as a computation aid.

In Problems 1–4 use (4) to approximate the solution of Laplace's equation at the interior points of the given region. Use symmetry when possible.

1. $u(0, y) = 0$, $u(3, y) = y(2 - y)$, $0 < y < 2$
   $u(x, 0) = 0$, $u(x, 2) = x(3 - x)$, $0 < x < 3$
   mesh size: $h = 1$

2. $u(0, y) = 0$, $u(2, y) = 0$, $0 < y < 1$
   $u(x, 0) = 100$, $u(x, 1) = 0$, $0 < x < 2$
   mesh size: $h = \frac{1}{2}$

3. $u(0, y) = 0$, $u(1, y) = 0$, $0 < y < 1$
   $u(x, 0) = 0$, $u(x, 1) = \sin \pi x$, $0 < x < 1$
   mesh size: $h = \frac{1}{3}$

4. $u(0, y) = 108y^2(1 - y)$, $u(1, y) = 0$, $0 < y < 1$
   $u(x, 0) = 0$, $u(x, 1) = 0$, $0 < x < 1$
   mesh size: $h = \frac{1}{3}$

In Problems 5 and 6 use (5) and Gauss-Seidel iteration to approximate the solution of Laplace's equation at the interior points of a unit square. Use the mesh size $h = \frac{1}{4}$. In Problem 5 the boundary conditions are given; in Problem 6 the values of u at boundary points are given in Figure 15.5.

5. $u(0, y) = 0$, $u(1, y) = 100y$, $0 < y < 1$
$u(x, 0) = 0$, $u(x, 1) = 100x$, $0 < x < 1$

6.

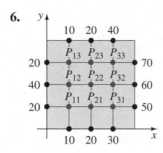

Figure 15.5

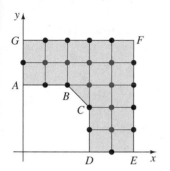

Figure 15.6

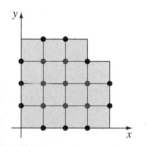

Figure 15.7

7. (a) In Problem 12 of Exercises 12.6 you solved a potential problem using a special form of Poisson's equation $\dfrac{\partial^2 u}{\partial x^2} + \dfrac{\partial^2 u}{\partial y^2} = f(x, y)$. Show that the difference equation replacement for Poisson's equation is

$$u_{i+1,j} + u_{i,j+1} + u_{i-1,j} + u_{i,j-1} - 4u_{ij} = h^2 f(x, y).$$

(b) Use the result in part (a) to approximate the solution of the Poisson equation $\dfrac{\partial^2 u}{\partial x^2} + \dfrac{\partial^2 u}{\partial y^2} = -2$ at the interior points of the region in Figure 15.6. The mesh size is $h = \frac{1}{2}$, $u = 1$ at every point along $ABCD$, and $u = 0$ at every point along $DEFGA$. Use symmetry and, if necessary, Gauss-Seidel iteration.

8. Use the result in part (a) of Problem 7 to approximate the solution of the Poisson equation $\dfrac{\partial^2 u}{\partial x^2} + \dfrac{\partial^2 u}{\partial y^2} = -64$ at the interior points of the region in Figure 15.7. The mesh size is $h = \frac{1}{8}$, and $u = 0$ at every point on the boundary of the region. If necessary, use Gauss-Seidel iteration.

15.2 PARABOLIC EQUATIONS

- *Difference equation replacement for the heat equation*
- *Explicit finite difference method* • *Instability of the finite difference method* • *Implicit finite difference method*
- *Crank-Nicholson method* • *Tridiagonal matrix*

In this section we examine two methods for approximating the solution of a boundary-value problem of the type

$$c\frac{\partial^2 u}{\partial x^2} = \frac{\partial u}{\partial t}, \quad 0 < x < a, \quad t > 0$$

$$u(0, t) = T_1, \quad u(a, t) = T_2, \quad t > 0 \tag{1}$$

$$u(x, 0) = f(x), \quad 0 < x < a.$$

The equation in this problem is the one-dimensional heat equation and is, as we know from Section 12.1, a parabolic PDE. Recall that the function f can be interpreted as the initial temperature distribution in a homogeneous rod extending from $x = 0$ to $x = a$. The constants T_1 and T_2 represent temperatures at the endpoints of the rod. Although we shall not prove it, the boundary-value problem has a unique solution when f is continuous on the closed interval $[0, a]$. This latter condition will be assumed, and so we replace the initial condition in (1) by $u(x, 0) = f(x), 0 \le x \le a$.

A Difference Equation Replacement for the Heat Equation To approximate the solution of (1) we first replace the heat equation by a difference equation. Using the central difference approximation (1) of Section 15.1,

$$\frac{\partial^2 u}{\partial x^2} \approx \frac{1}{h^2}[u(x + h, t) - 2u(x, t) + u(x - h, t)],$$

and the forward difference approximation (3) of Section 9.5,

$$\frac{\partial u}{\partial t} \approx \frac{1}{k}[u(x, t + k) - u(x, t)],$$

we can replace the heat equation $c\dfrac{\partial^2 u}{\partial x^2} = \dfrac{\partial u}{\partial t}$ by the following difference equation:

$$\frac{c}{h^2}[u(x + h, t) - 2u(x, t) + u(x - h, t)] = \frac{1}{k}[u(x, t + k) - u(x, t)]. \quad (2)$$

If we let $\lambda = ck/h^2$ and

$$u(x, t) = u_{ij}, \quad u(x + h, t) = u_{i+1,j}, \quad u(x - h, t) = u_{i-1,j}, \quad u(x, t + k) = u_{i,j+1},$$

then (2) becomes, after simplifying,

$$u_{i,j+1} = \lambda u_{i+1,j} + (1 - 2\lambda)u_{ij} + \lambda u_{i-1,j}. \quad (3)$$

We wish to approximate a solution of a problem such as (1) on a rectangular region in the xt-plane defined by the inequalities $0 \le x \le a$, $0 \le t \le T$ for some value of time T. Over this region we place a rectangular grid consisting of vertical lines h units apart and horizontal lines k units apart. If we choose two positive integers n and m and define

$$h = \frac{a}{n} \quad \text{and} \quad k = \frac{T}{m},$$

then the vertical and horizontal grid lines are defined by

$$x_i = ih, \quad i = 0, 1, 2, \ldots, n \quad \text{and} \quad t_j = jk, \quad j = 0, 1, 2, \ldots, m.$$

See Figure 15.8.

As illustrated in Figure 15.9, the idea here is to use formula (3) to estimate the values of the solution $u(x, t)$ at the points on the $(j + 1)$st time line using only values from the jth time line. For example, the values on the first time line ($j = 1$) depend on the initial condition $u_{i,0} = u(x_i, 0) = f(x_i)$ given on the zeroth time line ($j = 0$). This kind of numerical procedure is called an **explicit finite difference method**.

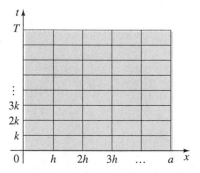

Figure 15.8

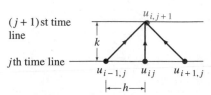

Figure 15.9

EXAMPLE 1 **Using the Finite Difference Method**

Consider the boundary-value problem

$$\frac{\partial^2 u}{\partial x^2} = \frac{\partial u}{\partial t}, \quad 0 < x < 1, \quad 0 < t < 0.5$$

$$u(0, t) = 0, \quad u(1, t) = 0, \quad 0 \le t \le 0.5$$

$$u(x, 0) = \sin \pi x, \quad 0 \le x \le 1.$$

First we identify $c = 1$, $a = 1$, and $T = 0.5$. If we choose, say, $n = 5$ and $m = 50$, then $h = 1/5 = 0.2$, $k = 0.5/50 = 0.01$, $\lambda = 0.25$,

$$x_i = i\frac{1}{5}, \quad i = 0, 1, 2, 3, 4, 5 \quad \text{and} \quad t_j = j\frac{1}{100}, \quad j = 0, 1, 2, \dots, 50.$$

Thus (3) becomes

$$u_{i,j+1} = 0.25(u_{i+1,j} + 2u_{ij} + u_{i-1,j}).$$

By setting $j = 0$ in this formula we get a formula for the approximations to the temperature u on the first time line:

$$u_{i,1} = 0.25(u_{i+1,0} + 2u_{i,0} + u_{i-1,0}).$$

If we then let $i = 1, \dots, 4$ in the last equation, we obtain, in turn,

$$u_{11} = 0.25(u_{20} + 2u_{10} + u_{00})$$
$$u_{21} = 0.25(u_{30} + 2u_{20} + u_{10})$$
$$u_{31} = 0.25(u_{40} + 2u_{30} + u_{20})$$
$$u_{41} = 0.25(u_{50} + 2u_{40} + u_{30}).$$

The first equation in this list is interpreted as

$$u_{11} = 0.25(u(x_2, 0) + 2u(x_1, 0) + u(0, 0))$$
$$= 0.25(u(0.4, 0) + 2u(0.2, 0) + u(0, 0)).$$

From the initial condition $u(x, 0) = \sin \pi x$ the last line becomes

$$u_{11} = 0.25(0.951056516 + 2(0.587785252) + 0) = 0.531656755.$$

This number represents an approximation to the temperature $u(0.2, 0.01)$.

The remaining calculations are summarized in Table 15.2 on page 620. ▪

You should verify, using the methods of Chapter 12, that the exact solution of the boundary-value problem in Example 1 is given by $u(x, t) = e^{-\pi^2 t} \sin \pi x$. Some sample values are given in Table 15.3.

TABLE 15.2 Explicit Difference Equation Approximation with $h = 0.2$, $k = 0.01$, $\lambda = 0.25$

Time	$x = 0.20$	$x = 0.40$	$x = 0.60$	$x = 0.80$	Time	$x = 0.20$	$x = 0.40$	$x = 0.60$	$x = 0.80$
0.00	0.5878	0.9511	0.9511	0.5878	0.26	0.0432	0.0700	0.0700	0.0432
0.01	0.5317	0.8602	0.8602	0.5317	0.27	0.0391	0.0633	0.0633	0.0391
0.02	0.4809	0.7781	0.7781	0.4809	0.28	0.0354	0.0572	0.0572	0.0354
0.03	0.4350	0.7038	0.7038	0.4350	0.29	0.0320	0.0518	0.0518	0.0320
0.04	0.3934	0.6366	0.6366	0.3934	0.30	0.0289	0.0468	0.0468	0.0289
0.05	0.3559	0.5758	0.5758	0.3559	0.31	0.0262	0.0424	0.0424	0.0262
0.06	0.3219	0.5208	0.5208	0.3219	0.32	0.0237	0.0383	0.0383	0.0237
0.07	0.2911	0.4711	0.4711	0.2911	0.33	0.0214	0.0347	0.0347	0.0214
0.08	0.2633	0.4261	0.4261	0.2633	0.34	0.0194	0.0313	0.0313	0.0194
0.09	0.2382	0.3854	0.3854	0.2382	0.35	0.0175	0.0284	0.0284	0.0175
0.10	0.2154	0.3486	0.3486	0.2154	0.36	0.0159	0.0256	0.0256	0.0159
0.11	0.1949	0.3153	0.3153	0.1949	0.37	0.0143	0.0232	0.0232	0.0143
0.12	0.1763	0.2852	0.2852	0.1763	0.38	0.0130	0.0210	0.0210	0.0130
0.13	0.1594	0.2580	0.2580	0.1594	0.39	0.0117	0.0190	0.0190	0.0117
0.14	0.1442	0.2333	0.2333	0.1442	0.40	0.0106	0.0172	0.0172	0.0106
0.15	0.1304	0.2111	0.2111	0.1304	0.41	0.0096	0.0155	0.0155	0.0096
0.16	0.1180	0.1909	0.1909	0.1180	0.42	0.0087	0.0140	0.0140	0.0087
0.17	0.1067	0.1727	0.1727	0.1067	0.43	0.0079	0.0127	0.0127	0.0079
0.18	0.0965	0.1562	0.1562	0.0965	0.44	0.0071	0.0115	0.0115	0.0071
0.19	0.0873	0.1413	0.1413	0.0873	0.45	0.0064	0.0104	0.0104	0.0064
0.20	0.0790	0.1278	0.1278	0.0790	0.46	0.0058	0.0094	0.0094	0.0058
0.21	0.0714	0.1156	0.1156	0.0714	0.47	0.0053	0.0085	0.0085	0.0053
0.22	0.0646	0.1045	0.1045	0.0646	0.48	0.0048	0.0077	0.0077	0.0048
0.23	0.0584	0.0946	0.0946	0.0584	0.49	0.0043	0.0070	0.0070	0.0043
0.24	0.0529	0.0855	0.0855	0.0529	0.50	0.0039	0.0063	0.0063	0.0039
0.25	0.0478	0.0774	0.0774	0.0478					

TABLE 15.3

Exact	Approx.
$u(0.4, 0.05) = 0.5806$	$u_{25} = 0.5758$
$u(0.6, 0.06) = 0.5261$	$u_{36} = 0.5208$
$u(0.2, 0.10) = 0.2191$	$u_{1,10} = 0.2154$
$u(0.8, 0.14) = 0.1476$	$u_{4,14} = 0.1442$

Stability   These approximations are comparable to the exact values and are accurate enough for some purposes. But there is a difficulty with the foregoing method. Recall that a numerical method is **unstable** if round-off errors or any other errors grow too rapidly as the computations proceed. The numerical procedure illustrated in Example 1 can exhibit this kind of behavior. It can be proved that the procedure is stable if λ is less than or equal to 0.5 but unstable otherwise. To obtain $\lambda = 0.25 \leq 0.5$ in Example 1 we had to choose the value $k = 0.01$; the necessity of using very small step sizes in the time direction is the principal fault of this method. You are urged to work Problem 12 in Exercises 15.2 and witness the predictable instability when $\lambda = 1$.

Crank-Nicholson Method   There are **implicit finite difference methods** for solving parabolic partial differential equations. These methods require that we solve a system of equations to determine the approximate values of u on the $(j + 1)$st time line. However, implicit methods do not suffer from instability problems.

The algorithm introduced by J. Crank and P. Nicholson in 1947 is used mostly for solving the heat equation. The algorithm consists of replacing the second partial derivative in $c\dfrac{\partial^2 u}{\partial x^2} = \dfrac{\partial u}{\partial t}$ by an average of two central difference quotients, one evaluated at t and the other at $t + k$:

$$\frac{c}{2}\left[\frac{u(x+h,t)-2u(x,t)+u(x-h,t)}{h^2} + \frac{u(x+h,t+k)-2u(x,t+k)+u(x-h,t+k)}{h^2}\right] \tag{4}$$
$$= \frac{1}{k}[u(x,t+k)-u(x,t)].$$

If we again define $\lambda = ck/h^2$, then after rearranging terms we can write (4) as

$$-u_{i-1,j+1} + \alpha u_{i,j+1} - u_{i+1,j+1} = u_{i+1,j} - \beta u_{ij} + u_{i-1,j}, \tag{5}$$

where $\alpha = 2(1 + 1/\lambda)$ and $\beta = 2(1 - 1/\lambda)$, $j = 0, 1, \ldots, m - 1$, and $i = 1, 2, \ldots, n - 1$.

For each choice of j the difference equation (5) for $i = 1, 2, \ldots, n - 1$ gives $n - 1$ equations in $n - 1$ unknowns $u_{i,j+1}$. Because of the prescribed boundary conditions, the values of $u_{i,j+1}$ are known for $i = 0$ and for $i = n$. For example, in the case $n = 4$ the system of equations for determining the approximate values of u on the $(j + 1)$st time line is

$$-u_{0,j+1} + \alpha u_{1,j+1} - u_{2,j+1} = u_{2,j} - \beta u_{1,j} + u_{0,j}$$
$$-u_{1,j+1} + \alpha u_{2,j+1} - u_{3,j+1} = u_{3,j} - \beta u_{2,j} + u_{1,j}$$
$$-u_{2,j+1} + \alpha u_{3,j+1} - u_{4,j+1} = u_{4,j} - \beta u_{3,j} + u_{2,j}$$

or

$$\alpha u_{1,j+1} - u_{2,j+1} \qquad\qquad = b_1$$
$$-u_{1,j+1} + \alpha u_{2,j+1} - u_{3,j+1} = b_2 \tag{6}$$
$$- u_{2,j+1} + \alpha u_{3,j+1} = b_3,$$

where

$$b_1 = u_{2,j} - \beta u_{1,j} + u_{0,j} + u_{0,j+1}$$
$$b_2 = u_{3,j} - \beta u_{2,j} + u_{1,j}$$
$$b_3 = u_{4,j} - \beta u_{3,j} + u_{2,j} + u_{4,j+1}.$$

In general, if we use the difference equation (5) to determine values of u on the $(j + 1)$st time line, we need to solve a linear system $\mathbf{AX} = \mathbf{B}$, where the coefficient matrix $\mathbf{A}$ is a **tridiagonal matrix,**

$$\mathbf{A} = \begin{pmatrix} \alpha & -1 & 0 & 0 & 0 & \cdots & & 0 \\ -1 & \alpha & -1 & 0 & 0 & & & 0 \\ 0 & -1 & \alpha & -1 & 0 & & & 0 \\ 0 & 0 & -1 & \alpha & -1 & & & 0 \\ \vdots & & & & & \ddots & & \vdots \\ 0 & 0 & 0 & 0 & 0 & & \alpha & -1 \\ 0 & 0 & 0 & 0 & 0 & \cdots & -1 & \alpha \end{pmatrix},$$

and the entries of the column matrix $\mathbf{B}$ are

$$b_1 = u_{2,j} - \beta u_{1,j} + u_{0,j} + u_{0,j+1}$$
$$b_2 = u_{3,j} - \beta u_{2,j} + u_{1,j}$$
$$b_3 = u_{4,j} - \beta u_{3,j} + u_{2,j}$$
$$\vdots$$
$$b_{n-1} = u_{n,j} - \beta u_{n-1,j} + u_{n-2,j} + u_{n,j+1}.$$

EXAMPLE 2   Using the Crank-Nicholson Method

Use the Crank-Nicholson method to approximate the solution of the boundary-value problem

$$0.25 \frac{\partial^2 u}{\partial x^2} = \frac{\partial u}{\partial t}, \quad 0 < x < 2, \quad 0 < t < 0.3$$

$$u(0, t) = 0, \quad u(2, t) = 0, \quad 0 \le t \le 0.3$$

$$u(x, 0) = \sin \pi x, \quad 0 \le x \le 2,$$

using $n = 8$ and $m = 30$.

Solution   From the identifications $a = 2$, $T = 0.3$, $h = \frac{1}{4} = 0.25$, $k = \frac{1}{100} = 0.01$, and $c = 0.25$ we get $\lambda = 0.04$. With the aid of a computer we get the results in Table 15.4 on page 623. ∎

You should verify that the function satisfying all the conditions in Example 2 is $u(x, t) = e^{-\pi^2 t/4} \sin \pi x$. The sample comparisons listed in Table 15.5 show that the absolute errors are of the order 10^{-2} or 10^{-3}. Smaller errors can be obtained by decreasing either h or k.

TABLE 15.5

Exact	Approx.
$u(0.75, 0.05) = 0.6250$	$u_{35} = 0.6289$
$u(0.50, 0.20) = 0.6105$	$u_{2,20} = 0.6259$
$u(0.25, 0.10) = 0.5525$	$u_{1,10} = 0.5594$

TABLE 15.4   Crank-Nicholson Method with $h = 0.25$, $k = 0.01$, $\lambda = 0.25$

Time	$x = 0.25$	$x = 0.50$	$x = 0.75$	$x = 1.00$	$x = 1.25$	$x = 1.50$	$x = 1.75$
0.00	0.7071	1.0000	0.7071	0.0000	−0.7071	−1.0000	−0.7071
0.01	0.6907	0.9768	0.6907	0.0000	−0.6907	−0.9768	−0.6907
0.02	0.6747	0.9542	0.6747	0.0000	−0.6747	−0.9542	−0.6747
0.03	0.6591	0.9321	0.6591	0.0000	−0.6591	−0.9321	−0.6591
0.04	0.6438	0.9105	0.6438	0.0000	−0.6438	−0.9105	−0.6438
0.05	0.6289	0.8894	0.6289	0.0000	−0.6289	−0.8894	−0.6289
0.06	0.6144	0.8688	0.6144	0.0000	−0.6144	−0.8688	−0.6144
0.07	0.6001	0.8487	0.6001	0.0000	−0.6001	−0.8487	−0.6001
0.08	0.5862	0.8291	0.5862	0.0000	−0.5862	−0.8291	−0.5862
0.09	0.5727	0.8099	0.5727	0.0000	−0.5727	−0.8099	−0.5727
0.10	0.5594	0.7911	0.5594	0.0000	−0.5594	−0.7911	−0.5594
0.11	0.5464	0.7728	0.5464	0.0000	−0.5464	−0.7728	−0.5464
0.12	0.5338	0.7549	0.5338	0.0000	−0.5338	−0.7549	−0.5338
0.13	0.5214	0.7374	0.5214	0.0000	−0.5214	−0.7374	−0.5214
0.14	0.5093	0.7203	0.5093	0.0000	−0.5093	−0.7203	−0.5093
0.15	0.4975	0.7036	0.4975	0.0000	−0.4975	−0.7036	−0.4975
0.16	0.4860	0.6873	0.4860	0.0000	−0.4860	−0.6873	−0.4860
0.17	0.4748	0.6714	0.4748	0.0000	−0.4748	−0.6714	−0.4748
0.18	0.4638	0.6559	0.4638	0.0000	−0.4638	−0.6559	−0.4638
0.19	0.4530	0.6407	0.4530	0.0000	−0.4530	−0.6407	−0.4530
0.20	0.4425	0.6258	0.4425	0.0000	−0.4425	−0.6258	−0.4425
0.21	0.4323	0.6114	0.4323	0.0000	−0.4323	−0.6114	−0.4323
0.22	0.4223	0.5972	0.4223	0.0000	−0.4223	−0.5972	−0.4223
0.23	0.4125	0.5834	0.4125	0.0000	−0.4125	−0.5834	−0.4125
0.24	0.4029	0.5699	0.4029	0.0000	−0.4029	−0.5699	−0.4029
0.25	0.3936	0.5567	0.3936	0.0000	−0.3936	−0.5567	−0.3936
0.26	0.3845	0.5438	0.3845	0.0000	−0.3845	−0.5438	−0.3845
0.27	0.3756	0.5312	0.3756	0.0000	−0.3756	−0.5312	− 0.3756
0.28	0.3669	0.5189	0.3669	0.0000	−0.3669	−0.5189	−0.3669
0.29	0.3584	0.5068	0.3584	0.0000	−0.3584	−0.5068	−0.3584
0.30	0.3501	0.4951	0.3501	0.0000	−0.3501	−0.4951	−0.3501

EXERCISES 15.2

Answers to odd-numbered problems begin on page AN-22.

In Problems 1–12 use a computer as a computation aid.

1. Use the difference equation (3) to approximate the solution of the boundary-value problem

$$\frac{\partial^2 u}{\partial x^2} = \frac{\partial u}{\partial t}, \quad 0 < x < 2, \quad 0 < t < 1$$

$$u(0, t) = 0, \quad u(2, t) = 0, \quad 0 \le t \le 1$$

$$u(x, 0) = \begin{cases} 1, & 0 \le x \le 1 \\ 0, & 1 < x \le 2. \end{cases}$$

Use $n = 8$ and $m = 40$.

2. Using the Fourier series solution obtained in Problem 1 of Exercises 12.3, with $L = 2$, one can sum the first 20 terms to estimate the values for $u(0.25, 0.1)$, $u(1, 0.5)$, and $u(1.5, 0.8)$ for the solution $u(x, t)$ of Problem 1 above. A student wrote a computer program to do this and obtained the results $u(0.25, 0.1) = 0.3794$, $u(1, 0.5) = 0.1854$, and $u(1.5, 0.8) = 0.0623$. Assume these results are accurate for all digits given. Compare these values with the approximations obtained in Problem 1 above. Find the absolute errors in each case.

3. Solve Problem 1 by the Crank-Nicholson method with $n = 8$ and $m = 40$. Use the values for $u(0.25, 0.1)$, $u(1, 0.5)$, and $u(1.5, 0.8)$ given in Problem 2 to compute the absolute errors.

4. Repeat Problem 1 using $n = 8$ and $m = 20$. Use the values for $u(0.25, 0.1)$, $u(1, 0.5)$, and $u(1.5, 0.8)$ given in Problem 2 to compute the absolute errors. Why are the approximations so inaccurate in this case?

5. Solve Problem 1 by the Crank-Nicholson method with $n = 8$ and $m = 20$. Use the values for $u(0.25, 0.1)$, $u(1, 0.5)$, and $u(1.5, 0.8)$ given in Problem 2 to compute the absolute errors. Compare the absolute errors with those obtained in Problem 4.

6. It was shown in Section 12.2 that if a rod of length L is made of a material with thermal conductivity K, specific heat γ, and density ρ, the temperature $u(x, t)$ satisfies the partial differential equation

$$\frac{K}{\gamma\rho}\frac{\partial^2 u}{\partial x^2} = \frac{\partial u}{\partial t}, \quad 0 < x < L.$$

Consider the boundary-value problem consisting of the foregoing equation and the following conditions:

$$u(0, t) = 0, \quad u(L, t) = 0, \quad 0 \leq t \leq 10$$
$$u(x, 0) = f(x), \quad 0 \leq x \leq L.$$

Use the difference equation (3) in this section with $n = 10$ and $m = 10$ to approximate the solution of the boundary-value problem when

(a) $L = 20$, $K = 0.15$, $\rho = 8.0$, $\gamma = 0.11$, $f(x) = 30$
(b) $L = 50$, $K = 0.15$, $\rho = 8.0$, $\gamma = 0.11$, $f(x) = 30$
(c) $L = 20$, $K = 1.10$, $\rho = 2.7$, $\gamma = 0.22$, $f(x) = 0.5x(20 - x)$
(d) $L = 100$, $K = 1.04$, $\rho = 10.6$, $\gamma = 0.06$,

$$f(x) = \begin{cases} 0.8x, & 0 \leq x \leq 50 \\ 0.8(100 - x), & 50 < x \leq 100 \end{cases}$$

7. Solve Problem 6 by the Crank-Nicholson method with $n = 10$ and $m = 10$.

8. Repeat Problem 6 if the endpoint temperatures are $u(0, t) = 0$, $u(L, t) = 20$, $0 \leq t \leq 10$.

9. Solve Problem 8 by the Crank-Nicholson method.

10. Consider the boundary-value problem in Example 2. Assume that $n = 4$.
(a) Find the new value of λ.
(b) Use the Crank-Nicholson difference equation (5) to find the sys-

tem of equations for u_{11}, u_{21}, and u_{31}—that is, the approximate values of u on the first time line. [*Hint:* Set $j = 0$ in (5), and let i take on the values 1, 2, 3.]

(c) Solve the system of three equations without the aid of a computer program. Compare your results with the corresponding entries in Table 15.4.

11. Consider a rod whose length is $L = 20$ for which $K = 1.05$, $\rho = 10.6$, and $\gamma = 0.056$. Suppose

$$u(0, t) = 20, \quad u(20, t) = 30$$
$$u(x, 0) = 50.$$

(a) Use the method outlined in Section 12.6 to find the steady-state solution $\psi(x)$.

(b) Use the Crank-Nicholson method to approximate the temperatures $u(x, t)$ for $0 \le t \le T_{max}$. Select T_{max} large enough to allow the temperatures to approach the steady-state values. Compare the approximations for $t = T_{max}$ with the values of $\psi(x)$ found in part (a).

12. Use the difference equation (3) to approximate the solution of the boundary-value problem

$$\frac{\partial^2 u}{\partial x^2} = \frac{\partial u}{\partial t}, \quad 0 < x < 1, \quad 0 < t < 1$$

$$u(0, t) = 0, \quad u(1, t) = 0, \quad 0 \le t \le 1$$
$$u(x, 0) = \sin \pi x, \quad 0 \le x \le 1.$$

Use $n = 5$ and $m = 25$.

15.3 HYPERBOLIC EQUATIONS
- *Difference equation replacement for the wave equation*
- *Explicit finite difference method* • *Instability of the finite difference method* • *Discretization and round-off errors*

We turn now to a numerical method for approximating a solution of a boundary-value problem involving the one-dimensional wave equation:

$$c^2 \frac{\partial^2 u}{\partial x^2} = \frac{\partial^2 u}{\partial t^2}, \quad 0 < x < a, \quad t > 0$$

$$u(0, t) = 0, \quad u(a, t) = 0, \quad t > 0 \tag{1}$$

$$u(x, 0) = f(x), \quad \left.\frac{\partial u}{\partial t}\right|_{t=0} = g(x), \quad 0 \le x \le a.$$

The wave equation is an example of a hyperbolic partial differential equation. This problem has a unique solution if the functions f and g have continuous second derivatives on the interval $(0, a)$ and $f(a) = f(0) = 0$.

A Difference Equation Replacement for the Wave Equation As in the preceding two sections, our first objective is to replace the partial differential equation in (1) by a difference equation. In this case we replace the two second-order partial derivatives by the central difference quotients

$$\frac{\partial^2 u}{\partial x^2} \approx \frac{1}{h^2}[u(x+h,t) - 2u(x,t) + u(x-h,t)]$$

and

$$\frac{\partial^2 u}{\partial t^2} \approx \frac{1}{k^2}[u(x,t+k) - 2u(x,t) + u(x,t-k)].$$

Thus we replace $c^2 \dfrac{\partial^2 u}{\partial x^2} = \dfrac{\partial^2 u}{\partial t^2}$ by

$$\frac{c^2}{h^2}[u(x+h,t) - 2u(x,t) + u(x-h,t)] = \frac{1}{k^2}[u(x,t+k) - 2u(x,t) + u(x,t-k)]. \qquad \textbf{(2)}$$

If we let $\lambda = ck/h$, then (2) can be expressed as

$$u_{i,j+1} = \lambda^2 u_{i+1,j} + 2(1 - \lambda^2)u_{ij} + \lambda^2 u_{i-1,j} - u_{i,j-1} \qquad \textbf{(3)}$$

for $i = 1, 2, \ldots, n-1$ and $j = 1, 2, \ldots, m-1$.

The numerical method based on equation (3), like the first method considered in Section 15.2, is an explicit finite difference method. As before, we use the difference equation to approximate the solution $u(x, t)$ of (1) over a rectangular region in the xt-plane defined by the inequalities $0 \le x \le a$, $0 \le t \le T$. If n and m are positive integers and

$$h = \frac{a}{n} \quad \text{and} \quad k = \frac{T}{m},$$

the vertical and horizontal grid lines on this region are defined by

$$x_i = ih, \quad i = 0, 1, 2, \ldots, n \quad \text{and} \quad t_j = jk, \quad j = 0, 1, 2, \ldots, m.$$

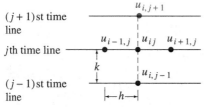

(j + 1)st time line

jth time line

(j − 1)st time line

Figure 15.10

As shown in Figure 15.10, (3) enables us to obtain the approximation $u_{i,j+1}$ on the $(j + 1)$st time line from the values indicated on the jth and $(j − 1)$st time lines. Moreover, we use

$$u_{0,j} = u(0, jk) = 0, \quad u_{n,j} = u(a, jk) = 0 \qquad \leftarrow \text{boundary conditions}$$

and

$$u_{i,0} = u(x_i, 0) = f(x_i). \qquad \leftarrow \text{initial condition}$$

There is one minor problem in getting started. You can see from (3) that for $j = 1$ we need to know the values of $u_{i,1}$ (that is, the estimates of u on the first time line) in order to find $u_{i,2}$. But from Figure 15.10, with $j = 0$, we see that the values of $u_{i,1}$ on the first time line depend on the values of $u_{i,0}$ on the zeroth time line and on the values of $u_{i,-1}$. To compute these latter values we make use of the initial-velocity condition $u_t(x, 0) = g(x)$. At $t = 0$ it follows from (5) of Section 9.5 that

$$g(x_i) = u_t(x_i, 0) \approx \frac{u(x_i, k) - u(x_i, -k)}{2k}. \qquad \textbf{(4)}$$

In order to make sense of the term $u(x_i, -k) = u_{i,-1}$ in (4) we have to imagine $u(x, t)$ extended backward in time. It follows from (4) that

$$u(x_i, -k) \approx u(x_i, k) - 2kg(x_i).$$

This last result suggests that we define

$$u_{i,-1} = u_{i,1} - 2kg(x_i) \qquad \textbf{(5)}$$

in the iteration of (3). By substituting (5) into (3) when $j = 0$ we get the special case

$$u_{i,1} = \frac{\lambda^2}{2}(u_{i+1,0} + u_{i-1,0}) + (1 - \lambda^2)u_{i,0} + kg(x_i). \tag{6}$$

EXAMPLE 1 Using the Finite Difference Method

Approximate the solution of the boundary-value problem

$$4\frac{\partial^2 u}{\partial x^2} = \frac{\partial^2 u}{\partial t^2}, \quad 0 < x < 1, \quad 0 < t < 1$$

$$u(0, t) = 0, \qquad u(1, t) = 0, \quad 0 \le t \le 1$$

$$u(x, 0) = \sin \pi x, \quad \frac{\partial u}{\partial t}\bigg|_{t=0} = 0, \quad 0 \le x \le 1,$$

using (3) with $n = 5$ and $m = 20$.

Solution   We make the identifications $c = 2$, $a = 1$, and $T = 1$. With $n = 5$ and $m = 20$ we get $h = \frac{1}{5} = 0.2$, $k = \frac{1}{20} = 0.05$, and $\lambda = 0.5$. Thus, with $g(x) = 0$, equations (6) and (3) become, respectively,

$$u_{i,1} = 0.125(u_{i+1,0} + u_{i-1,0}) + 0.75u_{i,0} \tag{7}$$

$$u_{i,j+1} = 0.25u_{i+1,j} + 1.5u_{ij} + 0.25u_{i-1,j} - u_{i,j-1}. \tag{8}$$

For $i = 1, 2, 3, 4$, equation (7) yields the following values for the $u_{i,1}$ on the first time line:

$$\begin{aligned}
u_{11} &= 0.125(u_{20} + u_{00}) + 0.75u_{10} = 0.55972100 \\
u_{21} &= 0.125(u_{30} + u_{10}) + 0.75u_{20} = 0.90564761 \\
u_{31} &= 0.125(u_{40} + u_{20}) + 0.75u_{30} = 0.90564761 \\
u_{41} &= 0.125(u_{50} + u_{30}) + 0.75u_{40} = 0.55972100.
\end{aligned} \tag{9}$$

Note that the results given in (9) were obtained from the initial condition $u(x, 0) = \sin \pi x$. For example, $u_{20} = \sin(0.2\pi)$, and so on. Now $j = 1$ in (8) gives

$$u_{i,2} = 0.25u_{i+1,1} + 1.5u_{i,1} + 0.25u_{i-1,1} - u_{i,0},$$

and so for $i = 1, 2, 3, 4$ we get

$$\begin{aligned}
u_{12} &= 0.25u_{21} + 1.5u_{11} + 0.25u_{01} - u_{10} \\
u_{22} &= 0.25u_{31} + 1.5u_{21} + 0.25u_{11} - u_{20} \\
u_{32} &= 0.25u_{41} + 1.5u_{31} + 0.25u_{21} - u_{30} \\
u_{42} &= 0.25u_{51} + 1.5u_{41} + 0.25u_{31} - u_{40}.
\end{aligned}$$

Using the boundary conditions, the initial conditions, and the data obtained in (9), we get from these equations the approximations for u on the second time line. These results and the remaining calculations are summarized in Table 15.6.

TABLE 15.6 Explicit Difference Equation Approximation with $h = 0.2$, $k = 0.05$, $\lambda = 0.5$

Time	$x = 0.20$	$x = 0.40$	$x = 0.60$	$x = 0.80$	Time	$x = 0.20$	$x = 0.40$	$x = 0.60$	$x = 0.80$
0.00	0.5878	0.9511	0.9511	0.5878	0.55	−0.5663	−0.9163	−0.9163	−0.5663
0.05	0.5597	0.9056	0.9056	0.5597	0.60	−0.4912	−0.7947	−0.7947	−0.4912
0.10	0.4782	0.7738	0.7738	0.4782	0.65	−0.3691	−0.5973	−0.5973	−0.3691
0.15	0.3510	0.5680	0.5680	0.3510	0.70	−0.2119	−0.3428	−0.3428	−0.2119
0.20	0.1903	0.3080	0.3080	0.1903	0.75	−0.0344	−0.0556	−0.0556	−0.0344
0.25	0.0115	0.0185	0.0185	0.0115	0.80	0.1464	0.2369	0.2369	0.1464
0.30	−0.1685	−0.2727	−0.2727	−0.1685	0.85	0.3132	0.5068	0.5068	0.3132
0.35	−0.3324	−0.5378	−0.5378	−0.3324	0.90	0.4501	0.7283	0.7283	0.4501
0.40	−0.4645	−0.7516	−0.7516	−0.4645	0.95	0.5440	0.8803	0.8803	0.5440
0.45	−0.5523	−0.8936	−0.8936	−0.5523	1.00	0.5860	0.9482	0.9482	0.5860
0.50	−0.5873	−0.9503	−0.9503	−0.5873					

It is readily verified that the exact solution of the problem in Example 1 is $u(x, t) = \sin \pi x \cos 2\pi t$. With this function we can compare the exact results with the approximations given in Table 15.6. For example, some sample values are listed in Table 15.7.

TABLE 15.7

Exact	Approximation
$u(0.4, 0.25) = 0$	$u_{25} = 0.0185$
$u(0.6, 0.3) = -0.2939$	$u_{36} = -0.2727$
$u(0.2, 0.5) = -0.5878$	$u_{1,10} = -0.5873$
$u(0.8, 0.7) = -0.1816$	$u_{4,14} = -0.2119$

As you can see, the approximations are in the same "ball park" as the exact values, but the accuracy is not particularly impressive. We can, however, obtain more accurate results. The accuracy of this algorithm varies with the choice of λ. Of course, λ is determined by the choice of the integers n and m, which in turn determine the values of the step sizes h and k. It can be proved that the best accuracy is always obtained from this method when the ratio $\lambda = kc/h$ is equal to one—in other words, when the step in the time direction is $k = h/c$. For example, the choice $n = 8$ and $m = 16$ yields $h = \frac{1}{8}$, $k = \frac{1}{16}$, and $\lambda = 1$. The sample values listed in Table 15.8 clearly show the improved accuracy.

TABLE 15.8

Exact	Approximation
$u(0.25, 0.3125) = -0.2706$	$u_{25} = -0.2706$
$u(0.375, 0.375) = -0.6533$	$u_{36} = -0.6533$
$u(0.125, 0.625) = -0.2706$	$u_{1,10} = -0.2706$

Stability   We note in conclusion that this explicit finite difference method for the wave equation is stable when $\lambda \leq 1$ and unstable when $\lambda > 1$.

Answers to odd-numbered problems begin on page AN-27.

In Problems 1, 3, 5, and 6 use a computer as a computation aid.

1. Use the difference equation (3) to approximate the solution of the boundary-value problem

$$c^2 \frac{\partial^2 u}{\partial x^2} = \frac{\partial^2 u}{\partial t^2}, \quad 0 < x < a, \quad 0 < t < T$$

$$u(0, t) = 0, \qquad u(a, t) = 0, \quad 0 \leq t \leq T$$

$$u(x, 0) = f(x), \quad \frac{\partial u}{\partial t}\Big|_{t=0} = 0, \quad 0 \leq x \leq a$$

when

   (a) $c = 1, a = 1, T = 1, f(x) = x(1 - x); \quad n = 4$ and $m = 10$
   (b) $c = 1, a = 2, T = 1, f(x) = e^{-16(x-1)^2}; \quad n = 5$ and $m = 10$
   (c) $c = \sqrt{2}, a = 1, T = 1, f(x) = \begin{cases} 0, & 0 \leq x \leq 0.5 \\ 0.5, & 0.5 < x \leq 1 \end{cases}; \quad n = 10$ and $m = 25$.

2. Consider the boundary-value problem

$$\frac{\partial^2 u}{\partial x^2} = \frac{\partial^2 u}{\partial t^2}, \quad 0 < x < 1, \quad 0 < t < 0.5$$

$$u(0, t) = 0, \qquad u(1, t) = 0, \quad 0 \leq t \leq 0.5$$

$$u(x, 0) = \sin \pi x, \quad \frac{\partial u}{\partial t}\Big|_{t=0} = 0, \quad 0 \leq x \leq 1.$$

   (a) Use the methods of Chapter 12 to verify that the solution of the problem is $u(x, t) = \sin \pi x \cos \pi t$.
   (b) Use the method of this section to approximate the solution of the problem without the aid of a computer program. Use $n = 4$ and $m = 5$.
   (c) Compute the absolute error at each interior grid point.

3. Approximate the solution of the boundary-value problem in Problem 2 using a computer program with
   (a) $n = 5, m = 10$ **(b)** $n = 5, m = 20$.

4. Given the boundary-value problem

$$\frac{\partial^2 u}{\partial x^2} = \frac{\partial^2 u}{\partial t^2}, \quad 0 < x < 1, \quad 0 < t < 1$$

$$u(0, t) = 0, \qquad u(1, t) = 0, \quad 0 \leq t \leq 1$$

$$u(x, 0) = x(1 - x), \quad \frac{\partial u}{\partial t}\Big|_{t=0} = 0, \quad 0 \leq x \leq 1,$$

use $h = k = \frac{1}{5}$ in equation (6) to compute the values of $u_{i,1}$ by hand.

5. It was shown in Section 12.2 that the equation of a vibrating string is

$$\frac{T}{\rho}\frac{\partial^2 u}{\partial x^2} = \frac{\partial^2 u}{\partial t^2},$$

where T is the constant magnitude of the tension in the string and ρ is its mass per unit length. Suppose a string of length 60 centimeters is secured to the x-axis at its ends and is released from rest from the initial displacement

$$f(x) = \begin{cases} 0.01x, & 0 \leq x \leq 30 \\ 0.30 - \dfrac{x-30}{100}, & 30 < x \leq 60. \end{cases}$$

Use the difference equation (3) in this section to approximate the solution of the boundary-value problem when $h = 10$, $k = 5\sqrt{\rho/T}$ and where $\rho = 0.0225$ g/cm, $T = 1.4 \times 10^7$ dynes. Use $m = 50$.

6. Repeat Problem 5 using

$$f(x) = \begin{cases} 0.2x, & 0 \leq x \leq 15 \\ 0.30 - \dfrac{x-15}{150}, & 15 < x \leq 60. \end{cases}$$

and $h = 10$, $k = 2.5\sqrt{\rho/T}$. Use $m = 50$.

CHAPTER 15 IN REVIEW

Answers to odd-numbered problems begin on page AN-30.

1. Consider the boundary-value problem

$$\frac{\partial^2 u}{\partial x^2} + \frac{\partial^2 u}{\partial y^2} = 0, \quad 0 < x < 2, \quad 0 < y < 1$$

$$u(0, y) = 0, \quad u(2, y) = 50, \quad 0 < y < 1$$

$$u(x, 0) = 0, \quad u(x, 1) = 0, \quad 0 < x < 2.$$

Approximate the solution of the differential equation at the interior points of the region with mesh size $h = \frac{1}{2}$. Use Gaussian elimination or Gauss-Seidel iteration.

2. Solve Problem 1 using mesh size $h = \frac{1}{4}$. Use Gauss-Seidel iteration.

3. Consider the boundary-value problem

$$\frac{\partial^2 u}{\partial x^2} = \frac{\partial u}{\partial t}, \quad 0 < x < 1, \quad 0 < t < 0.05$$

$$u(0, t) = 0, \quad u(1, t) = 0, \quad t > 0$$

$$u(x, 0) = x, \quad 0 < x < 1.$$

(a) Note that the initial temperature $u(x, 0) = x$ indicates that the temperature at the right boundary $x = 1$ should be $u(1, 0) = 1$, whereas the boundary conditions imply that $u(1, 0) = 0$. Write a computer program for the explicit finite difference method so that the boundary conditions prevail for all times considered, including $t = 0$. Use the program to complete Table 15.9.

TABLE 15.9

Time	x = 0.00	x = 0.20	x = 0.40	x = 0.60	x = 0.80	x = 1.00
0.00	0.0000	0.2000	0.4000	0.6000	0.8000	0.0000
0.01	0.0000					0.0000
0.02	0.0000					0.0000
0.03	0.0000					0.0000
0.04	0.0000					0.0000
0.05	0.0000					0.0000

(b) Modify your computer program so that the initial condition prevails at the boundaries at $t = 0$. Use this program to complete Table 15.10.

TABLE 15.10

Time	x = 0.00	x = 0.20	x = 0.40	x = 0.60	x = 0.80	x = 1.00
0.00	0.0000	0.2000	0.4000	0.6000	0.8000	1.0000
0.01	0.0000					0.0000
0.02	0.0000					0.0000
0.03	0.0000					0.0000
0.04	0.0000					0.0000
0.05	0.0000					0.0000

(c) Are Tables 15.9 and 15.10 related in any way? Use a larger time interval if necessary.

APPENDIX

I

GAMMA FUNCTION

Euler's integral definition of the **gamma function** is

$$\Gamma(x) = \int_0^\infty t^{x-1}e^{-t}\,dt. \tag{1}$$

Convergence of the integral requires that $x - 1 > -1$ or $x > 0$. The recurrence relation

$$\Gamma(x + 1) = x\Gamma(x), \tag{2}$$

which we saw in Section 6.3, can be obtained from (1) with integration by parts. Now when $x = 1$, $\Gamma(1) = \int_0^\infty e^{-t}\,dt = 1$, and thus (2) gives

$$\Gamma(2) = 1\Gamma(1) = 1$$
$$\Gamma(3) = 2\Gamma(2) = 2 \cdot 1$$
$$\Gamma(4) = 3\Gamma(3) = 3 \cdot 2 \cdot 1$$

and so on. In this manner it is seen that when n is a positive integer,

$$\Gamma(n + 1) = n!.$$

For this reason the gamma function is often called the **generalized factorial function.**

   Although the integral form (1) does not converge for $x < 0$, it can be shown by means of alternative definitions that the gamma function is defined for all real and complex numbers *except $x = -n, n = 0, 1, 2, \ldots$*. As a consequence, (2) is actually valid for $x \neq -n$. The graph of $\Gamma(x)$, considered as a function of a real variable x, is as given in Figure I.1. Observe that the nonpositive integers correspond to vertical asymptotes of the graph.

   In Problems 23 and 24 of Exercises 6.3 we utilized the fact that $\Gamma(\frac{1}{2}) = \sqrt{\pi}$. This result can be derived from (1) by setting $x = \frac{1}{2}$:

$$\Gamma(\tfrac{1}{2}) = \int_0^\infty t^{-1/2}e^{-t}\,dt. \tag{3}$$

When we let $t = u^2$, (3) can be written as $\Gamma(\frac{1}{2}) = 2\int_0^\infty e^{-u^2}\,du$. But

$$\int_0^\infty e^{-u^2}\,du = \int_0^\infty e^{-v^2}\,dv,$$

and so

$$[\Gamma(\tfrac{1}{2})]^2 = \left(2\int_0^\infty e^{-u^2}\,du\right)\left(2\int_0^\infty e^{-v^2}\,dv\right)$$
$$= 4\int_0^\infty\int_0^\infty e^{-(u^2+v^2)}\,du\,dv.$$

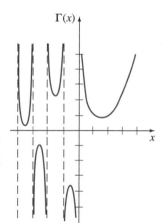

$\Gamma(x)$

Figure I.1

Switching to polar coordinates $u = r\cos\theta$, $v = r\sin\theta$ enables us to evaluate the double integral:

$$4\int_0^\infty \int_0^\infty e^{-(u^2+v^2)}\,du\,dv = 4\int_0^{\pi/2}\int_0^\infty e^{-r^2}r\,dr\,d\theta = \pi.$$

Hence $[\Gamma(\tfrac{1}{2})]^2 = \pi$ or $\Gamma(\tfrac{1}{2}) = \sqrt{\pi}.$ **(4)**

EXAMPLE 1   **Value of $\Gamma(-\tfrac{1}{2})$**

Evaluate $\Gamma(-\tfrac{1}{2})$.

Solution   In view of (2) and (4) it follows that, with $x = -\tfrac{1}{2}$,

$$\Gamma(\tfrac{1}{2}) = -\tfrac{1}{2}\Gamma(-\tfrac{1}{2}).$$

Therefore $\Gamma(-\tfrac{1}{2}) = -2\Gamma(\tfrac{1}{2}) = -2\sqrt{\pi}.$ ■

EXERCISES FOR APPENDIX I

Answers to odd-numbered problems begin on page AN-30.

1. Evaluate.
   (a) $\Gamma(5)$ **(b)** $\Gamma(7)$ **(c)** $\Gamma(-\tfrac{3}{2})$ **(d)** $\Gamma(-\tfrac{5}{2})$

2. Use (1) and the fact that $\Gamma(\tfrac{6}{5}) = 0.92$ to evaluate $\displaystyle\int_0^\infty x^5 e^{-x^5}\,dx$.
   [*Hint:* Let $t = x^5$.]

3. Use (1) and the fact that $\Gamma(\tfrac{5}{3}) = 0.89$ to evaluate $\displaystyle\int_0^\infty x^4 e^{-x^3}\,dx$.

4. Evaluate $\displaystyle\int_0^1 x^3 \left(\ln\frac{1}{x}\right)^3 dx$. [*Hint:* Let $t = -\ln x$.]

5. Use the fact that $\Gamma(x) > \displaystyle\int_0^1 t^{x-1}e^{-t}\,dt$ to show that $\Gamma(x)$ is unbounded as $x \to 0^+$.

6. Use (1) to derive (2) for $x > 0$.

APPENDIX

II

INTRODUCTION TO MATRICES

II.1   BASIC DEFINITIONS AND THEORY

DEFINITION II.1   Matrix

A **matrix A** is any rectangular array of numbers or functions:

$$\mathbf{A} = \begin{pmatrix} a_{11} & a_{12} & \cdots & a_{1n} \\ a_{21} & a_{22} & \cdots & a_{2n} \\ \vdots & & & \vdots \\ a_{m1} & a_{m2} & \cdots & a_{mn} \end{pmatrix}. \tag{1}$$

If a matrix has m rows and n columns, we say that its **size** is m by n (written $m \times n$). An $n \times n$ matrix is called a **square matrix** of order n.

The element, or entry, in the ith row and jth column of an $m \times n$ matrix **A** is written a_{ij}. An $m \times n$ matrix **A** is then abbreviated as $\mathbf{A} = (a_{ij})_{m \times n}$ or simply $\mathbf{A} = (a_{ij})$. A 1×1 matrix is simply one constant or function.

DEFINITION II.2   Equality of Matrices

Two $m \times n$ matrices **A** and **B** are **equal** if $a_{ij} = b_{ij}$ for each i and j.

DEFINITION II.3   Column Matrix

A **column matrix X** is any matrix having n rows and one column:

$$\mathbf{X} = \begin{pmatrix} b_{11} \\ b_{21} \\ \vdots \\ b_{n1} \end{pmatrix} = (b_{i1})_{n \times 1}.$$

A column matrix is also called a **column vector** or simply a **vector.**

DEFINITION II.4   **Multiples of Matrices**

A **multiple** of a matrix $\mathbf{A}$ is defined to be

$$k\mathbf{A} = \begin{pmatrix} ka_{11} & ka_{12} & \cdots & ka_{1n} \\ ka_{21} & ka_{22} & \cdots & ka_{2n} \\ \vdots & & & \vdots \\ ka_{m1} & ka_{m2} & \cdots & ka_{mn} \end{pmatrix} = (ka_{ij})_{m \times n},$$

where k is a constant or a function.

EXAMPLE 1   **Multiples of Matrices**

(a) $5\begin{pmatrix} 2 & -3 \\ 4 & -1 \\ \frac{1}{5} & 6 \end{pmatrix} = \begin{pmatrix} 10 & -15 \\ 20 & -5 \\ 1 & 30 \end{pmatrix}$ **(b)** $e^t\begin{pmatrix} 1 \\ -2 \\ 4 \end{pmatrix} = \begin{pmatrix} e^t \\ -2e^t \\ 4e^t \end{pmatrix}$ ▬

We note in passing that for any matrix $\mathbf{A}$ the product $k\mathbf{A}$ is the same as $\mathbf{A}k$. For example,

$$e^{-3t}\begin{pmatrix} 2 \\ 5 \end{pmatrix} = \begin{pmatrix} 2e^{-3t} \\ 5e^{-3t} \end{pmatrix} = \begin{pmatrix} 2 \\ 5 \end{pmatrix}e^{-3t}.$$

DEFINITION II.5   **Addition of Matrices**

The **sum** of two $m \times n$ matrices $\mathbf{A}$ and $\mathbf{B}$ is defined to be the matrix

$$\mathbf{A} + \mathbf{B} = (a_{ij} + b_{ij})_{m \times n}.$$

In other words, when adding two matrices of the same size, we add the corresponding elements.

EXAMPLE 2   **Matrix Addition**

The sum of $\mathbf{A} = \begin{pmatrix} 2 & -1 & 3 \\ 0 & 4 & 6 \\ -6 & 10 & -5 \end{pmatrix}$ and $\mathbf{B} = \begin{pmatrix} 4 & 7 & -8 \\ 9 & 3 & 5 \\ 1 & -1 & 2 \end{pmatrix}$ is

$$\mathbf{A} + \mathbf{B} = \begin{pmatrix} 2+4 & -1+7 & 3+(-8) \\ 0+9 & 4+3 & 6+5 \\ -6+1 & 10+(-1) & -5+2 \end{pmatrix} = \begin{pmatrix} 6 & 6 & -5 \\ 9 & 7 & 11 \\ -5 & 9 & -3 \end{pmatrix}.$$ ▬

| EXAMPLE 3 | **A Matrix Written as a Sum of Column Matrices** |

The single matrix $\begin{pmatrix} 3t^2 - 2e^t \\ t^2 + 7t \\ 5t \end{pmatrix}$ can be written as the sum of three column vectors:

$$\begin{pmatrix} 3t^2 - 2e^t \\ t^2 + 7t \\ 5t \end{pmatrix} = \begin{pmatrix} 3t^2 \\ t^2 \\ 0 \end{pmatrix} + \begin{pmatrix} 0 \\ 7t \\ 5t \end{pmatrix} + \begin{pmatrix} -2e^t \\ 0 \\ 0 \end{pmatrix} = \begin{pmatrix} 3 \\ 1 \\ 0 \end{pmatrix} t^2 + \begin{pmatrix} 0 \\ 7 \\ 5 \end{pmatrix} t + \begin{pmatrix} -2 \\ 0 \\ 0 \end{pmatrix} e^t. \qquad \blacksquare$$

The **difference** of two $m \times n$ matrices is defined in the usual manner: $\mathbf{A} - \mathbf{B} = \mathbf{A} + (-\mathbf{B})$, where $-\mathbf{B} = (-1)\mathbf{B}$.

| DEFINITION II.6 | **Multiplication of Matrices** |

Let $\mathbf{A}$ be a matrix having m rows and n columns and $\mathbf{B}$ be a matrix having n rows and p columns. We define the **product AB** to be the $m \times p$ matrix

$$\mathbf{AB} = \begin{pmatrix} a_{11} & a_{12} & \cdots & a_{1n} \\ a_{21} & a_{22} & \cdots & a_{2n} \\ \vdots & \vdots & & \vdots \\ a_{m1} & a_{m2} & \cdots & a_{mn} \end{pmatrix} \begin{pmatrix} b_{11} & b_{12} & \cdots & b_{1p} \\ b_{21} & b_{22} & \cdots & b_{2p} \\ \vdots & \vdots & & \vdots \\ b_{n1} & b_{n2} & \cdots & b_{np} \end{pmatrix}$$

$$= \begin{pmatrix} a_{11}b_{11} + a_{12}b_{21} + \cdots + a_{1n}b_{n1} & \cdots & a_{11}b_{1p} + a_{12}b_{2p} + \cdots + a_{1n}b_{np} \\ a_{21}b_{11} + a_{22}b_{21} + \cdots + a_{2n}b_{n1} & \cdots & a_{21}b_{1p} + a_{22}b_{2p} + \cdots + a_{2n}b_{np} \\ \vdots & & \vdots \\ a_{m1}b_{11} + a_{m2}b_{21} + \cdots + a_{mn}b_{n1} & \cdots & a_{m1}b_{1p} + a_{m2}b_{2p} + \cdots + a_{mn}b_{np} \end{pmatrix}$$

$$= \left(\sum_{k=1}^{n} a_{ik}b_{kj} \right)_{m \times p}.$$

Note carefully in Definition II.6 that the product $\mathbf{AB} = \mathbf{C}$ is defined only when the number of columns in the matrix $\mathbf{A}$ is the same as the number of rows in $\mathbf{B}$. The size of the product can be determined from

$$\mathbf{A}_{m \times n} \mathbf{B}_{n \times p} = \mathbf{C}_{m \times p}.$$

Also, you might recognize that the entries in, say, the ith row of the final matrix $\mathbf{AB}$ are formed by using the component definition of the inner, or dot, product of the ith row of $\mathbf{A}$ with each of the columns of $\mathbf{B}$.

EXAMPLE 4   **Multiplication of Matrices**

(a) For $\mathbf{A} = \begin{pmatrix} 4 & 7 \\ 3 & 5 \end{pmatrix}$ and $\mathbf{B} = \begin{pmatrix} 9 & -2 \\ 6 & 8 \end{pmatrix}$,

$$\mathbf{AB} = \begin{pmatrix} 4 \cdot 9 + 7 \cdot 6 & 4 \cdot (-2) + 7 \cdot 8 \\ 3 \cdot 9 + 5 \cdot 6 & 3 \cdot (-2) + 5 \cdot 8 \end{pmatrix} = \begin{pmatrix} 78 & 48 \\ 57 & 34 \end{pmatrix}.$$

(b) For $\mathbf{A} = \begin{pmatrix} 5 & 8 \\ 1 & 0 \\ 2 & 7 \end{pmatrix}$ and $\mathbf{B} = \begin{pmatrix} -4 & -3 \\ 2 & 0 \end{pmatrix}$,

$$\mathbf{AB} = \begin{pmatrix} 5 \cdot (-4) + 8 \cdot 2 & 5 \cdot (-3) + 8 \cdot 0 \\ 1 \cdot (-4) + 0 \cdot 2 & 1 \cdot (-3) + 0 \cdot 0 \\ 2 \cdot (-4) + 7 \cdot 2 & 2 \cdot (-3) + 7 \cdot 0 \end{pmatrix} = \begin{pmatrix} -4 & -15 \\ -4 & -3 \\ 6 & -6 \end{pmatrix}.$$ ■

In general, *matrix multiplication is not commutative*; that is, $\mathbf{AB} \neq \mathbf{BA}$. Observe in part (a) of Example 4 that $\mathbf{BA} = \begin{pmatrix} 30 & 53 \\ 48 & 82 \end{pmatrix}$, whereas in part (b) the product $\mathbf{BA}$ is not defined since Definition II.6 requires that the first matrix (in this case $\mathbf{B}$) have the same number of columns as the second matrix has rows.

We are particularly interested in the product of a square matrix and a column vector.

EXAMPLE 5   **Multiplication of Matrices**

(a) $\begin{pmatrix} 2 & -1 & 3 \\ 0 & 4 & 5 \\ 1 & -7 & 9 \end{pmatrix} \begin{pmatrix} -3 \\ 6 \\ 4 \end{pmatrix} = \begin{pmatrix} 2 \cdot (-3) + (-1) \cdot 6 + 3 \cdot 4 \\ 0 \cdot (-3) + \quad 4 \ \cdot 6 + 5 \cdot 4 \\ 1 \cdot (-3) + (-7) \cdot 6 + 9 \cdot 4 \end{pmatrix} = \begin{pmatrix} 0 \\ 44 \\ -9 \end{pmatrix}$

(b) $\begin{pmatrix} -4 & 2 \\ 3 & 8 \end{pmatrix} \begin{pmatrix} x \\ y \end{pmatrix} = \begin{pmatrix} -4x + 2y \\ 3x + 8y \end{pmatrix}$ ■

Multiplicative Identity   For a given positive integer n, the $n \times n$ matrix

$$\mathbf{I} = \begin{pmatrix} 1 & 0 & 0 & \cdots & 0 \\ 0 & 1 & 0 & \cdots & 0 \\ \vdots & & & & \vdots \\ 0 & 0 & 0 & \cdots & 1 \end{pmatrix}$$

is called the **multiplicative identity matrix.** It follows from Definition II.6 that for any $n \times n$ matrix $\mathbf{A}$,

$$\mathbf{AI} = \mathbf{IA} = \mathbf{A}.$$

Also, it is readily verified that if **X** is an $n \times 1$ column matrix, then **IX = X**.

Zero Matrix   A matrix consisting of all zero entries is called a **zero matrix** and is denoted by **0**. For example,

$$\mathbf{0} = \begin{pmatrix} 0 \\ 0 \end{pmatrix}, \quad \mathbf{0} = \begin{pmatrix} 0 & 0 \\ 0 & 0 \end{pmatrix}, \quad \mathbf{0} = \begin{pmatrix} 0 & 0 \\ 0 & 0 \\ 0 & 0 \end{pmatrix},$$

and so on. If **A** and **0** are $m \times n$ matrices, then

$$\mathbf{A} + \mathbf{0} = \mathbf{0} + \mathbf{A} = \mathbf{A}.$$

Associative Law   Although we shall not prove it, matrix multiplication is **associative.** If **A** is an $m \times p$ matrix, **B** a $p \times r$ matrix, and **C** an $r \times n$ matrix, then

$$\mathbf{A}(\mathbf{BC}) = (\mathbf{AB})\mathbf{C}$$

is an $m \times n$ matrix.

Distributive Law   If all products are defined, multiplication is **distributive** over addition:

$$\mathbf{A}(\mathbf{B} + \mathbf{C}) = \mathbf{AB} + \mathbf{AC} \quad \text{and} \quad (\mathbf{B} + \mathbf{C})\mathbf{A} = \mathbf{BA} + \mathbf{CA}.$$

Determinant of a Matrix   Associated with every *square* matrix **A** of constants is a number called the **determinant of the matrix,** which is denoted by det **A**.

EXAMPLE 6   **Determinant of a Square Matrix**

For $\mathbf{A} = \begin{pmatrix} 3 & 6 & 2 \\ 2 & 5 & 1 \\ -1 & 2 & 4 \end{pmatrix}$ we expand det **A** by cofactors of the first row:

$$\det \mathbf{A} = \begin{vmatrix} 3 & 6 & 2 \\ 2 & 5 & 1 \\ -1 & 2 & 4 \end{vmatrix} = 3 \begin{vmatrix} 5 & 1 \\ 2 & 4 \end{vmatrix} - 6 \begin{vmatrix} 2 & 1 \\ -1 & 4 \end{vmatrix} + 2 \begin{vmatrix} 2 & 5 \\ -1 & 2 \end{vmatrix}$$

$$= 3(20 - 2) - 6(8 + 1) + 2(4 + 5) = 18. \quad \blacksquare$$

It can be proved that a determinant det **A** can be expanded by cofactors using any row or column. If det **A** has a row (or a column) containing many zero entries, then wisdom dictates that we expand the determinant by that row (or column).

DEFINITION II.7   **Transpose of a Matrix**

The **transpose** of the $m \times n$ matrix (1) is the $n \times m$ matrix $\mathbf{A}^T$ given by

$$\mathbf{A}^T = \begin{pmatrix} a_{11} & a_{21} & \cdots & a_{m1} \\ a_{12} & a_{22} & \cdots & a_{m2} \\ \vdots & & & \vdots \\ a_{1n} & a_{2n} & \cdots & a_{mn} \end{pmatrix}.$$

In other words, the rows of a matrix $\mathbf{A}$ become the columns of its transpose $\mathbf{A}^T$.

EXAMPLE 7   **Transpose of a Matrix**

(a) The transpose of $\mathbf{A} = \begin{pmatrix} 3 & 6 & 2 \\ 2 & 5 & 1 \\ -1 & 2 & 4 \end{pmatrix}$ is $\mathbf{A}^T = \begin{pmatrix} 3 & 2 & -1 \\ 6 & 5 & 2 \\ 2 & 1 & 4 \end{pmatrix}$.

(b) If $\mathbf{X} = \begin{pmatrix} 5 \\ 0 \\ 3 \end{pmatrix}$, then $\mathbf{X}^T = (5 \quad 0 \quad 3)$.

DEFINITION II.8   **Multiplicative Inverse of a Matrix**

Let $\mathbf{A}$ be an $n \times n$ matrix. If there exists an $n \times n$ matrix $\mathbf{B}$ such that

$$\mathbf{AB} = \mathbf{BA} = \mathbf{I},$$

where $\mathbf{I}$ is the multiplicative identity, then $\mathbf{B}$ is said to be the **multiplicative inverse of $\mathbf{A}$** and is denoted by $\mathbf{B} = \mathbf{A}^{-1}$.

DEFINITION II.9   **Nonsingular/Singular Matrices**

Let $\mathbf{A}$ be an $n \times n$ matrix. If det $\mathbf{A} \neq 0$, then $\mathbf{A}$ is said to be **nonsingular.** If det $\mathbf{A} = 0$, then $\mathbf{A}$ is said to be **singular.**

The following theorem gives a necessary and sufficient condition for a square matrix to have a multiplicative inverse.

THEOREM II.1	**Nonsingularity Implies A Has an Inverse**

An $n \times n$ matrix $\mathbf{A}$ has a multiplicative inverse $\mathbf{A}^{-1}$ if and only if $\mathbf{A}$ is nonsingular.

The following theorem gives one way of finding the multiplicative inverse for a nonsingular matrix.

THEOREM II.2	**A Formula for the Inverse of a Matrix**

Let $\mathbf{A}$ be an $n \times n$ nonsingular matrix and let $C_{ij} = (-1)^{i+j}M_{ij}$, where M_{ij} is the determinant of the $(n-1) \times (n-1)$ matrix obtained by deleting the ith row and jth column from $\mathbf{A}$. Then

$$\mathbf{A}^{-1} = \frac{1}{\det \mathbf{A}}(C_{ij})^T. \tag{2}$$

Each C_{ij} in Theorem II.2 is simply the **cofactor** (signed minor) of the corresponding entry a_{ij} in $\mathbf{A}$. Note that the transpose is utilized in formula (2).

For future reference we observe in the case of a 2×2 nonsingular matrix

$$\mathbf{A} = \begin{pmatrix} a_{11} & a_{12} \\ a_{21} & a_{22} \end{pmatrix}$$

that $C_{11} = a_{22}$, $C_{12} = -a_{21}$, $C_{21} = -a_{12}$, and $C_{22} = a_{11}$. Thus

$$\mathbf{A}^{-1} = \frac{1}{\det \mathbf{A}} \begin{pmatrix} a_{22} & -a_{21} \\ -a_{12} & a_{11} \end{pmatrix}^T = \frac{1}{\det \mathbf{A}} \begin{pmatrix} a_{22} & -a_{12} \\ -a_{21} & a_{11} \end{pmatrix}. \tag{3}$$

For a 3×3 nonsingular matrix

$$\mathbf{A} = \begin{pmatrix} a_{11} & a_{12} & a_{13} \\ a_{21} & a_{22} & a_{23} \\ a_{31} & a_{32} & a_{33} \end{pmatrix},$$

$$C_{11} = \begin{vmatrix} a_{22} & a_{23} \\ a_{32} & a_{33} \end{vmatrix}, \quad C_{12} = -\begin{vmatrix} a_{21} & a_{23} \\ a_{31} & a_{33} \end{vmatrix}, \quad C_{13} = \begin{vmatrix} a_{21} & a_{22} \\ a_{31} & a_{32} \end{vmatrix},$$

and so on. Carrying out the transposition gives

$$\mathbf{A}^{-1} = \frac{1}{\det \mathbf{A}} \begin{pmatrix} C_{11} & C_{21} & C_{31} \\ C_{12} & C_{22} & C_{32} \\ C_{13} & C_{23} & C_{33} \end{pmatrix}. \tag{4}$$

EXAMPLE 8   **Inverse of a 2 × 2 Matrix**

Find the multiplicative inverse for $\mathbf{A} = \begin{pmatrix} 1 & 4 \\ 2 & 10 \end{pmatrix}$.

Solution   Since det $\mathbf{A} = 10 - 8 = 2 \neq 0$, $\mathbf{A}$ is nonsingular. It follows from Theorem II.1 that $\mathbf{A}^{-1}$ exists. From (3) we find

$$\mathbf{A}^{-1} = \frac{1}{2}\begin{pmatrix} 10 & -4 \\ -2 & 1 \end{pmatrix} = \begin{pmatrix} 5 & -2 \\ -1 & \frac{1}{2} \end{pmatrix}.$$

Not every square matrix has a multiplicative inverse. The matrix $\mathbf{A} = \begin{pmatrix} 2 & 2 \\ 3 & 3 \end{pmatrix}$ is singular since det $\mathbf{A} = 0$. Hence $\mathbf{A}^{-1}$ does not exist.

EXAMPLE 9   **Inverse of a 3 × 3 Matrix**

Find the multiplicative inverse for $\mathbf{A} = \begin{pmatrix} 2 & 2 & 0 \\ -2 & 1 & 1 \\ 3 & 0 & 1 \end{pmatrix}$.

Solution   Since det $\mathbf{A} = 12 \neq 0$, the given matrix is nonsingular. The cofactors corresponding to the entries in each row of det $\mathbf{A}$ are

$$C_{11} = \begin{vmatrix} 1 & 1 \\ 0 & 1 \end{vmatrix} = 1 \qquad C_{12} = -\begin{vmatrix} -2 & 1 \\ 3 & 1 \end{vmatrix} = 5 \qquad C_{13} = \begin{vmatrix} -2 & 1 \\ 3 & 0 \end{vmatrix} = -3$$

$$C_{21} = -\begin{vmatrix} 2 & 0 \\ 0 & 1 \end{vmatrix} = -2 \qquad C_{22} = \begin{vmatrix} 2 & 0 \\ 3 & 1 \end{vmatrix} = 2 \qquad C_{23} = -\begin{vmatrix} 2 & 2 \\ 3 & 0 \end{vmatrix} = 6$$

$$C_{31} = \begin{vmatrix} 2 & 0 \\ 1 & 1 \end{vmatrix} = 2 \qquad C_{32} = -\begin{vmatrix} 2 & 0 \\ -2 & 1 \end{vmatrix} = -2 \qquad C_{33} = \begin{vmatrix} 2 & 2 \\ -2 & 1 \end{vmatrix} = 6.$$

It follows from (4) that

$$\mathbf{A}^{-1} = \frac{1}{12}\begin{pmatrix} 1 & -2 & 2 \\ 5 & 2 & -2 \\ -3 & 6 & 6 \end{pmatrix} = \begin{pmatrix} \frac{1}{12} & -\frac{1}{6} & \frac{1}{6} \\ \frac{5}{12} & \frac{1}{6} & -\frac{1}{6} \\ -\frac{1}{4} & \frac{1}{2} & \frac{1}{2} \end{pmatrix}.$$

You are urged to verify that $\mathbf{A}^{-1}\mathbf{A} = \mathbf{A}\mathbf{A}^{-1} = \mathbf{I}$.

Formula (2) presents obvious difficulties for nonsingular matrices larger than 3 × 3. For example, to apply (2) to a 4 × 4 matrix we would have to calculate *sixteen* 3 × 3 determinants.* In the case of a large matrix there are more efficient ways of finding $\mathbf{A}^{-1}$. The curious reader is referred to any text in linear algebra.

Since our goal is to apply the concept of a matrix to systems of linear first-order differential equations, we need the following definitions.

---

*Strictly speaking, a determinant is a number, but it is sometimes convenient to refer to a determinant as if it were an array.

DEFINITION II.10 **Derivative of a Matrix of Functions**

If $\mathbf{A}(t) = (a_{ij}(t))_{m \times n}$ is a matrix whose entries are functions differentiable on a common interval, then

$$\frac{d\mathbf{A}}{dt} = \left(\frac{d}{dt} a_{ij} \right)_{m \times n}.$$

DEFINITION II.11 **Integral of a Matrix of Functions**

If $\mathbf{A}(t) = (a_{ij}(t))_{m \times n}$ is a matrix whose entries are functions continuous on a common interval containing t and t_0, then

$$\int_{t_0}^{t} \mathbf{A}(s)\, ds = \left(\int_{t_0}^{t} a_{ij}(s)\, ds \right)_{m \times n}.$$

To differentiate (integrate) a matrix of functions we simply differentiate (integrate) each entry. The derivative of a matrix is also denoted by $\mathbf{A}'(t)$.

EXAMPLE 10 **Derivative/Integral of a Matrix**

If

$$\mathbf{X}(t) = \begin{pmatrix} \sin 2t \\ e^{3t} \\ 8t - 1 \end{pmatrix}, \quad \text{then} \quad \mathbf{X}'(t) = \begin{pmatrix} \dfrac{d}{dt} \sin 2t \\[6pt] \dfrac{d}{dt} e^{3t} \\[6pt] \dfrac{d}{dt}(8t - 1) \end{pmatrix} = \begin{pmatrix} 2\cos 2t \\ 3e^{3t} \\ 8 \end{pmatrix}$$

and

$$\int_{0}^{t} \mathbf{X}(s)\, ds = \begin{pmatrix} \int_{0}^{t} \sin 2s\, ds \\[6pt] \int_{0}^{t} e^{3s}\, ds \\[6pt] \int_{0}^{t} (8s - 1)\, ds \end{pmatrix} = \begin{pmatrix} -\frac{1}{2}\cos 2t + \frac{1}{2} \\[4pt] \frac{1}{3}e^{3t} - \frac{1}{3} \\[4pt] 4t^2 - t \end{pmatrix}.$$

$\blacksquare$

II.2 GAUSSIAN AND GAUSS-JORDAN ELIMINATION

Matrices are an invaluable aid in solving algebraic systems of n linear equations in n unknowns,

$$\begin{aligned}
a_{11}x_1 + a_{12}x_2 + \cdots + a_{1n}x_n &= b_1 \\
a_{21}x_1 + a_{22}x_2 + \cdots + a_{2n}x_n &= b_2 \\
&\;\;\vdots \\
a_{n1}x_1 + a_{n2}x_2 + \cdots + a_{nn}x_n &= b_n.
\end{aligned} \tag{5}$$

If **A** denotes the matrix of coefficients in (5), we know that Cramer's rule could be used to solve the system whenever det $\mathbf{A} \neq 0$. However, that rule requires a herculean effort if **A** is larger than 3×3. The procedure that we shall now consider has the distinct advantage of being not only an efficient way of handling large systems but also a means of solving consistent systems (5) in which det $\mathbf{A} = 0$ and a means of solving m linear equations in n unknowns.

DEFINITION II.12   Augmented Matrix

The **augmented matrix** of the system (5) is the $n \times (n + 1)$ matrix

$$\begin{pmatrix} a_{11} & a_{12} & \cdots & a_{1n} & b_1 \\ a_{21} & a_{22} & \cdots & a_{2n} & b_2 \\ \vdots & & & & \vdots \\ a_{n1} & a_{n2} & \cdots & a_{nn} & b_n \end{pmatrix}.$$

If **B** is the column matrix of the b_i, $i = 1, 2, \ldots, n$, the augmented matrix of (5) is denoted by $(\mathbf{A} \mid \mathbf{B})$.

Elementary Row Operations   Recall from algebra that we can transform an algebraic system of equations into an equivalent system (that is, one having the same solution) by multiplying an equation by a nonzero constant, interchanging the positions of any two equations in a system, and adding a nonzero constant multiple of an equation to another equation. These operations on equations in a system are, in turn, equivalent to **elementary row operations** on an augmented matrix:

   (*i*) Multiply a row by a nonzero constant.
   (*ii*) Interchange any two rows.
   (*iii*) Add a nonzero constant multiple of one row to any other row.

Elimination Methods   To solve a system such as (5) using an augmented matrix we use either **Gaussian elimination** or the **Gauss-Jordan elimination method.** In the former method we carry out a succession of elementary row operations until we arrive at an augmented matrix in **row-echelon form:**

   (*i*) The first nonzero entry in a nonzero row is 1.
   (*ii*) In consecutive nonzero rows, the first entry 1 in the lower row appears to the right of the first 1 in the higher row.
   (*iii*) Rows consisting of all 0's are at the bottom of the matrix.

In the Gauss-Jordan method the row operations are continued until we obtain an augmented matrix that is in **reduced row-echelon form.** A reduced row-echelon matrix has the same three properties listed above in addition to the following one:

   (*iv*) A column containing a first entry 1 has 0's everywhere else.

EXAMPLE 11 **Row-Echelon/Reduced Row-Echelon Form**

(a) The augmented matrices

$$\begin{pmatrix} 1 & 5 & 0 & | & 2 \\ 0 & 1 & 0 & | & -1 \\ 0 & 0 & 0 & | & 0 \end{pmatrix} \text{ and } \begin{pmatrix} 0 & 0 & 1 & -6 & 2 & | & 2 \\ 0 & 0 & 0 & 0 & 1 & | & 4 \end{pmatrix}$$

are in row-echelon form. You should verify that the three criteria are satisfied.

(b) The augmented matrices

$$\begin{pmatrix} 1 & 0 & 0 & | & 7 \\ 0 & 1 & 0 & | & -1 \\ 0 & 0 & 0 & | & 0 \end{pmatrix} \text{ and } \begin{pmatrix} 0 & 0 & 1 & -6 & 0 & | & -6 \\ 0 & 0 & 0 & 0 & 1 & | & 4 \end{pmatrix}$$

are in reduced row-echelon form. Note that the remaining entries in the columns containing a leading entry 1 are all 0's. ▪

Note that in Gaussian elimination we stop once we have obtained *an* augmented matrix in row-echelon form. In other words, by using different sequences of row operations we may arrive at different row-echelon forms. This method then requires the use of back-substitution. In Gauss-Jordan elimination we stop when we have obtained *the* augmented matrix in reduced row-echelon form. Any sequence of row operations will lead to the same augmented matrix in reduced row-echelon form. This method does not require back-substitution; the solution of the system will be apparent by inspection of the final matrix. In terms of the equations of the original system, our goal in both methods is simply to make the coefficient of x_1 in the first equation* equal to 1 and then use multiples of that equation to eliminate x_1 from other equations. The process is repeated on the other variables.

To keep track of the row operations on an augmented matrix, we utilize the following notation:

Symbol	Meaning
R_{ij}	Interchange rows i and j
cR_i	Multiply the ith row by the nonzero constant c
$cR_i + R_j$	Multiply the ith row by c and add to the jth row

EXAMPLE 12 **Solution by Elimination**

Solve
$$2x_1 + 6x_2 + x_3 = 7$$
$$x_1 + 2x_2 - x_3 = -1$$
$$5x_1 + 7x_2 - 4x_3 = 9$$

using **(a)** Gaussian elimination and **(b)** Gauss-Jordan elimination.

---

*We can always interchange equations so that the first equation contains the variable x_1.

Solution   **(a)** Using row operations on the augmented matrix of the system, we obtain

$$\begin{pmatrix} 2 & 6 & 1 & 7 \\ 1 & 2 & -1 & -1 \\ 5 & 7 & -4 & 9 \end{pmatrix} \xrightarrow{R_{12}} \begin{pmatrix} 1 & 2 & -1 & -1 \\ 2 & 6 & 1 & 7 \\ 5 & 7 & -4 & 9 \end{pmatrix} \xrightarrow[-5R_1 + R_3]{-2R_1 + R_2} \begin{pmatrix} 1 & 2 & -1 & -1 \\ 0 & 2 & 3 & 9 \\ 0 & -3 & 1 & 14 \end{pmatrix}$$

$$\xrightarrow{\frac{1}{2}R_2} \begin{pmatrix} 1 & 2 & -1 & -1 \\ 0 & 1 & \frac{3}{2} & \frac{9}{2} \\ 0 & -3 & 1 & 14 \end{pmatrix} \xrightarrow{3R_2 + R_3} \begin{pmatrix} 1 & 2 & -1 & -1 \\ 0 & 1 & \frac{3}{2} & \frac{9}{2} \\ 0 & 0 & \frac{11}{2} & \frac{55}{2} \end{pmatrix} \xrightarrow{\frac{2}{11}R_3} \begin{pmatrix} 1 & 2 & -1 & -1 \\ 0 & 1 & \frac{3}{2} & \frac{9}{2} \\ 0 & 0 & 1 & 5 \end{pmatrix}.$$

The last matrix is in row-echelon form and represents the system

$$x_1 + 2x_2 - x_3 = -1$$
$$x_2 + \frac{3}{2}x_3 = \frac{9}{2}$$
$$x_3 = 5.$$

Substituting $x_3 = 5$ into the second equation then gives $x_2 = -3$. Substituting both these values back into the first equation finally yields $x_1 = 10$.

(b) We start with the last matrix above. Since the first entries in the second and third rows are 1's, we must, in turn, make the remaining entries in the second and third columns 0's:

$$\begin{pmatrix} 1 & 2 & -1 & -1 \\ 0 & 1 & \frac{3}{2} & \frac{9}{2} \\ 0 & 0 & 1 & 5 \end{pmatrix} \xrightarrow{-2R_2 + R_1} \begin{pmatrix} 1 & 0 & -4 & -10 \\ 0 & 1 & \frac{3}{2} & \frac{9}{2} \\ 0 & 0 & 1 & 5 \end{pmatrix} \xrightarrow[-\frac{3}{2}R_3 + R_2]{4R_3 + R_1} \begin{pmatrix} 1 & 0 & 0 & 10 \\ 0 & 1 & 0 & -3 \\ 0 & 0 & 1 & 5 \end{pmatrix}.$$

The last matrix is now in reduced row-echelon form. Because of what the matrix means in terms of equations, it is evident that the solution of the system is $x_1 = 10$, $x_2 = -3$, $x_3 = 5$.   ∎

EXAMPLE 13   Gauss-Jordan Elimination

Solve
$$x + 3y - 2z = -7$$
$$4x + y + 3z = 5$$
$$2x - 5y + 7z = 19.$$

Solution   We solve the system using Gauss-Jordan elimination:

$$\begin{pmatrix} 1 & 3 & -2 & -7 \\ 4 & 1 & 3 & 5 \\ 2 & -5 & 7 & 19 \end{pmatrix} \xrightarrow[-2R_1 + R_3]{-4R_1 + R_2} \begin{pmatrix} 1 & 3 & -2 & -7 \\ 0 & -11 & 11 & 33 \\ 0 & -11 & 11 & 33 \end{pmatrix}$$

$$\xrightarrow[-\frac{1}{11}R_3]{-\frac{1}{11}R_2} \begin{pmatrix} 1 & 3 & -2 & -7 \\ 0 & 1 & -1 & -3 \\ 0 & 1 & -1 & -3 \end{pmatrix} \xrightarrow[-R_2 + R_3]{-3R_2 + R_1} \begin{pmatrix} 1 & 0 & 1 & 2 \\ 0 & 1 & -1 & -3 \\ 0 & 0 & 0 & 0 \end{pmatrix}.$$

In this case the last matrix in reduced row-echelon form implies that the original system of three equations in three unknowns is really equivalent to two equations in three unknowns. Since only z is common to both equations (the nonzero rows), we can assign its values arbitrarily. If we let $z = t$, where t represents any real number, then we see that the system has infinitely many solutions: $x = 2 - t$, $y = -3 + t$, $z = t$. Geometrically, these equations are the parametric equations for the line of intersection of the planes $x + 0y + z = 2$ and $0x + y - z = 3$. ∎

Using Row Operations to Find an Inverse   Because of the number of determinants that must be evaluated, formula (2) in Theorem II.2 is seldom used to find the inverse when the matrix **A** is large. In the case of 3×3 or larger matrices, the method described in the next theorem is a particular efficient means for finding $\mathbf{A}^{-1}$.

THEOREM II.3   **Finding $\mathbf{A}^{-1}$ Using Elementary Row Operations**

If an $n \times n$ matrix **A** can be transformed into the $n \times n$ identity **I** by a sequence of elementary row operations, then **A** is nonsingular. The same sequence of operations that transforms **A** into the identity **I** will also transform **I** into $\mathbf{A}^{-1}$.

It is convenient to carry out these row operations on **A** and **I** simultaneously by means of an $n \times 2n$ matrix obtained by augmenting **A** with the identity **I** as shown here:

$$(\mathbf{A} \mid \mathbf{I}) = \begin{pmatrix} a_{11} & a_{12} & \cdots & a_{1n} & 1 & 0 & \cdots & 0 \\ a_{21} & a_{22} & \cdots & a_{2n} & 1 & 0 & \cdots & 0 \\ \vdots & & & \vdots & \vdots & & & \vdots \\ a_{n1} & a_{n2} & \cdots & a_{nn} & 0 & 0 & \cdots & 1 \end{pmatrix}.$$

The procedure for finding $\mathbf{A}^{-1}$ is outlined in the following diagram:

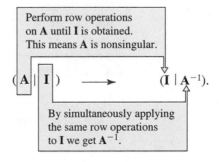

$$(\mathbf{A} \mid \mathbf{I}) \longrightarrow (\mathbf{I} \mid \mathbf{A}^{-1}).$$

Perform row operations on **A** until **I** is obtained. This means **A** is nonsingular.

By simultaneously applying the same row operations to **I** we get $\mathbf{A}^{-1}$.

EXAMPLE 14   **Inverse by Elementary Row Operations**

Find the multiplicative inverse for $\mathbf{A} = \begin{pmatrix} 2 & 0 & 1 \\ -2 & 3 & 4 \\ -5 & 5 & 6 \end{pmatrix}$.

Solution   We shall use the same notation as we did when we reduced an augmented matrix to reduced row-echelon form:

$$\begin{pmatrix} 2 & 0 & 1 & | & 1 & 0 & 0 \\ -2 & 3 & 4 & | & 0 & 1 & 0 \\ -5 & 5 & 6 & | & 0 & 0 & 1 \end{pmatrix} \xrightarrow{\frac{1}{2}R_1} \begin{pmatrix} 1 & 0 & \frac{1}{2} & | & \frac{1}{2} & 0 & 0 \\ -2 & 3 & 4 & | & 0 & 1 & 0 \\ -5 & 5 & 6 & | & 0 & 0 & 1 \end{pmatrix} \xrightarrow[5R_1 + R_3]{2R_1 + R_2} \begin{pmatrix} 1 & 0 & \frac{1}{2} & | & \frac{1}{2} & 0 & 0 \\ 0 & 3 & 5 & | & 1 & 1 & 0 \\ 0 & 5 & \frac{17}{2} & | & \frac{5}{2} & 0 & 1 \end{pmatrix}$$

$$\xrightarrow[\frac{1}{5}R_3]{\frac{1}{3}R_2} \begin{pmatrix} 1 & 0 & \frac{1}{2} & | & \frac{1}{2} & 0 & 0 \\ 0 & 1 & \frac{5}{3} & | & \frac{1}{3} & \frac{1}{3} & 0 \\ 0 & 1 & \frac{17}{10} & | & \frac{1}{2} & 0 & \frac{1}{5} \end{pmatrix} \xrightarrow{-R_2 + R_3} \begin{pmatrix} 1 & 0 & \frac{1}{2} & | & \frac{1}{2} & 0 & 0 \\ 0 & 1 & \frac{5}{3} & | & \frac{1}{3} & \frac{1}{3} & 0 \\ 0 & 0 & \frac{1}{30} & | & \frac{1}{6} & -\frac{1}{3} & \frac{1}{5} \end{pmatrix}$$

$$\xrightarrow{30R_3} \begin{pmatrix} 1 & 0 & \frac{1}{2} & | & \frac{1}{2} & 0 & 0 \\ 0 & 1 & \frac{5}{3} & | & \frac{1}{3} & \frac{1}{3} & 0 \\ 0 & 0 & 1 & | & 5 & -10 & 6 \end{pmatrix} \xrightarrow[-\frac{5}{3}R_3 + R_2]{-\frac{1}{2}R_3 + R_1} \begin{pmatrix} 1 & 0 & 0 & | & -2 & 5 & -3 \\ 0 & 1 & 0 & | & -8 & 17 & -10 \\ 0 & 0 & 1 & | & 5 & -10 & 6 \end{pmatrix}.$$

Since **I** appears to the left of the vertical line, we conclude that the matrix to the right of the line is

$$\mathbf{A}^{-1} = \begin{pmatrix} -2 & 5 & -3 \\ -8 & 17 & -10 \\ 5 & -10 & 6 \end{pmatrix}.$$

■

If row reduction of $(\mathbf{A} \mid \mathbf{I})$ leads to the situation

$$(\mathbf{A} \mid \mathbf{I}) \xrightarrow[\text{operations}]{\text{row}} (\mathbf{B} \mid \mathbf{C}),$$

where the matrix **B** contains a row of zeros, then necessarily **A** is singular. Since further reduction of **B** always yields another matrix with a row of zeros, we can never transform **A** into **I**.

II.3   THE EIGENVALUE PROBLEM

Gauss-Jordan elimination can be used to find the **eigenvectors** of a square matrix.

DEFINITION II.13   **Eigenvalues and Eigenvectors**

Let **A** be an $n \times n$ matrix. A number λ is said to be an **eigenvalue** of **A** if there exists a *nonzero* solution vector **K** of the linear system

$$\mathbf{AK} = \lambda\mathbf{K}. \tag{6}$$

The solution vector **K** is said to be an **eigenvector** corresponding to the eigenvalue λ.

The word *eigenvalue* is a combination of German and English terms adapted from the German word *eigenwert,* which, translated literally, is "proper value." Eigenvalues and eigenvectors are also called **characteristic values** and **characteristic vectors,** respectively.

EXAMPLE 15   **Eigenvector of a Matrix**

Verify that $\mathbf{K} = \begin{pmatrix} 1 \\ -1 \\ 1 \end{pmatrix}$ is an eigenvector of the matrix

$$\mathbf{A} = \begin{pmatrix} 0 & -1 & -3 \\ 2 & 3 & 3 \\ -2 & 1 & 1 \end{pmatrix}.$$

Solution   By carrying out the multiplication $\mathbf{AK}$ we see that

$$\mathbf{AK} = \begin{pmatrix} 0 & -1 & -3 \\ 2 & 3 & 3 \\ -2 & 1 & 1 \end{pmatrix}\begin{pmatrix} 1 \\ -1 \\ 1 \end{pmatrix} = \begin{pmatrix} -2 \\ 2 \\ -2 \end{pmatrix} = (-2)\begin{pmatrix} 1 \\ -1 \\ 1 \end{pmatrix} = \overset{\text{eigenvalue}}{\overset{\downarrow}{(-2)}}\mathbf{K}.$$

We see from the preceding line and Definition II.13 that $\lambda = -2$ is an eigenvalue of $\mathbf{A}$. ∎

Using properties of matrix algebra, we can write (6) in the alternative form

$$(\mathbf{A} - \lambda\mathbf{I})\mathbf{K} = \mathbf{0}, \tag{7}$$

where $\mathbf{I}$ is the multiplicative identity. If we let

$$\mathbf{K} = \begin{pmatrix} k_1 \\ k_2 \\ \vdots \\ k_n \end{pmatrix},$$

then (7) is the same as

$$\begin{aligned}
(a_{11} - \lambda)k_1 + \quad a_{12}k_2 + \cdots + \quad a_{1n}k_n &= 0 \\
a_{21}k_1 + (a_{22} - \lambda)k_2 + \cdots + \quad a_{2n}k_n &= 0 \\
&\;\;\vdots \\
a_{n1}k_1 + \quad a_{n2}k_2 + \cdots + (a_{nn} - \lambda)k_n &= 0.
\end{aligned} \tag{8}$$

Although an obvious solution of (8) is $k_1 = 0$, $k_2 = 0$, ..., $k_n = 0$, we are seeking only nontrivial solutions. It is known that a homogeneous system of n linear equations in n unknowns (that is, $b_i = 0$, $i = 1, 2, \ldots, n$ in (5)) has a nontrivial solution if and only if the determinant of the coefficient matrix is equal to zero. Thus to find a nonzero solution $\mathbf{K}$ for (7) we must have

$$\det(\mathbf{A} - \lambda\mathbf{I}) = 0. \tag{9}$$

Inspection of (8) shows that the expansion of $\det(\mathbf{A} - \lambda\mathbf{I})$ by cofactors results in an nth-degree polynomial in λ. The equation (9) is called the **characteristic equation** of $\mathbf{A}$. Thus *the eigenvalues of $\mathbf{A}$ are the roots of the characteristic equation.* To find an eigenvector corresponding to an eigenvalue λ we simply solve the system of equations $(\mathbf{A} - \lambda\mathbf{I})\mathbf{K} = \mathbf{0}$ by applying Gauss-Jordan elimination to the augmented matrix $(\mathbf{A} - \lambda\mathbf{I}|\mathbf{0})$.

EXAMPLE 16   Eigenvalues/Eigenvectors

Find the eigenvalues and eigenvectors of $\mathbf{A} = \begin{pmatrix} 1 & 2 & 1 \\ 6 & -1 & 0 \\ -1 & -2 & -1 \end{pmatrix}$.

Solution   To expand the determinant in the characteristic equation we use the cofactors of the second row:

$$\det(\mathbf{A} - \lambda\mathbf{I}) = \begin{vmatrix} 1-\lambda & 2 & 1 \\ 6 & -1-\lambda & 0 \\ -1 & -2 & -1-\lambda \end{vmatrix} = -\lambda^3 - \lambda^2 + 12\lambda = 0.$$

From $-\lambda^3 - \lambda^2 + 12\lambda = -\lambda(\lambda + 4)(\lambda - 3) = 0$ we see that the eigenvalues are $\lambda_1 = 0$, $\lambda_2 = -4$, and $\lambda_3 = 3$. To find the eigenvectors we must now reduce $(\mathbf{A} - \lambda\mathbf{I}|\mathbf{0})$ three times corresponding to the three distinct eigenvalues.

For $\lambda_1 = 0$ we have

$$(\mathbf{A} - 0\mathbf{I}|\mathbf{0}) = \begin{pmatrix} 1 & 2 & 1 & | & 0 \\ 6 & -1 & 0 & | & 0 \\ -1 & -2 & -1 & | & 0 \end{pmatrix} \xrightarrow[\substack{R_1 + R_3}]{-6R_1 + R_2} \begin{pmatrix} 1 & 2 & 1 & | & 0 \\ 0 & -13 & -6 & | & 0 \\ 0 & 0 & 0 & | & 0 \end{pmatrix}$$

$$\xrightarrow{-\frac{1}{13}R_2} \begin{pmatrix} 1 & 2 & 1 & | & 0 \\ 0 & 1 & \frac{6}{13} & | & 0 \\ 0 & 0 & 0 & | & 0 \end{pmatrix} \xrightarrow{-2R_2 + R_1} \begin{pmatrix} 1 & 0 & \frac{1}{13} & | & 0 \\ 0 & 1 & \frac{6}{13} & | & 0 \\ 0 & 0 & 0 & | & 0 \end{pmatrix}.$$

Thus we see that $k_1 = -\frac{1}{13}k_3$ and $k_2 = -\frac{6}{13}k_3$. Choosing $k_3 = -13$, we get the eigenvector*

$$\mathbf{K}_1 = \begin{pmatrix} 1 \\ 6 \\ -13 \end{pmatrix}.$$

For $\lambda_2 = -4$,

$$(\mathbf{A} + 4\mathbf{I}|\mathbf{0}) = \begin{pmatrix} 5 & 2 & 1 & | & 0 \\ 6 & 3 & 0 & | & 0 \\ -1 & -2 & 3 & | & 0 \end{pmatrix} \xrightarrow[R_{31}]{-R_3} \begin{pmatrix} 1 & 2 & -3 & | & 0 \\ 6 & 3 & 0 & | & 0 \\ 5 & 2 & 1 & | & 0 \end{pmatrix}$$

$$\xrightarrow[\substack{-5R_1 + R_3}]{-6R_1 + R_2} \begin{pmatrix} 1 & 2 & -3 & | & 0 \\ 0 & -9 & 18 & | & 0 \\ 0 & -8 & 16 & | & 0 \end{pmatrix} \xrightarrow[\substack{-\frac{1}{8}R_3}]{-\frac{1}{9}R_2} \begin{pmatrix} 1 & 2 & -3 & | & 0 \\ 0 & 1 & -2 & | & 0 \\ 0 & 1 & -2 & | & 0 \end{pmatrix} \xrightarrow[\substack{-R_2 + R_3}]{-2R_2 + R_1} \begin{pmatrix} 1 & 0 & 1 & | & 0 \\ 0 & 1 & -2 & | & 0 \\ 0 & 0 & 0 & | & 0 \end{pmatrix}$$

---

*Of course k_3 could be chosen as any nonzero number. In other words, a nonzero constant multiple of an eigenvector is also an eigenvector.

implies $k_1 = -k_3$ and $k_2 = 2k_3$. Choosing $k_3 = 1$ then yields the second eigenvector

$$\mathbf{K}_2 = \begin{pmatrix} -1 \\ 2 \\ 1 \end{pmatrix}.$$

Finally, for $\lambda_3 = 3$ Gauss-Jordan elimination gives

$$(\mathbf{A} - 3\mathbf{I}|\mathbf{0}) = \begin{pmatrix} -2 & 2 & 1 & | & 0 \\ 6 & -4 & 0 & | & 0 \\ -1 & -2 & -4 & | & 0 \end{pmatrix} \xrightarrow[\text{operations}]{\text{row}} \begin{pmatrix} 1 & 0 & 1 & | & 0 \\ 0 & 1 & \frac{3}{2} & | & 0 \\ 0 & 0 & 0 & | & 0 \end{pmatrix},$$

and so $k_1 = -k_3$ and $k_2 = -\frac{3}{2}k_3$. The choice of $k_3 = -2$ leads to the third eigenvector:

$$\mathbf{K}_3 = \begin{pmatrix} 2 \\ 3 \\ -2 \end{pmatrix}.$$

∎

When an $n \times n$ matrix $\mathbf{A}$ possesses n distinct eigenvalues $\lambda_1, \lambda_2, \ldots, \lambda_n$, it can be proved that a set of n linearly independent* eigenvectors $\mathbf{K}_1, \mathbf{K}_2, \ldots, \mathbf{K}_n$ can be found. However, when the characteristic equation has repeated roots, it may not be possible to find n linearly independent eigenvectors for $\mathbf{A}$.

EXAMPLE 17 **Eigenvalues/Eigenvectors**

Find the eigenvalues and eigenvectors of $\mathbf{A} = \begin{pmatrix} 3 & 4 \\ -1 & 7 \end{pmatrix}$.

Solution   From the characteristic equation

$$\det(\mathbf{A} - \lambda\mathbf{I}) = \begin{vmatrix} 3 - \lambda & 4 \\ -1 & 7 - \lambda \end{vmatrix} = (\lambda - 5)^2 = 0$$

we see that $\lambda_1 = \lambda_2 = 5$ is an eigenvalue of multiplicity two. In the case of a 2×2 matrix there is no need to use Gauss-Jordan elimination. To find the eigenvector(s) corresponding to $\lambda_1 = 5$ we resort to the system $(\mathbf{A} - 5\mathbf{I}|\mathbf{0})$ in its equivalent form

$$-2k_1 + 4k_2 = 0$$
$$-k_1 + 2k_2 = 0.$$

It is apparent from this system that $k_1 = 2k_2$. Thus if we choose $k_2 = 1$, we find the single eigenvector

$$\mathbf{K}_1 = \begin{pmatrix} 2 \\ 1 \end{pmatrix}.$$

∎

*Linear independence of column vectors is defined in exactly the same manner as for functions.

EXAMPLE 18   Eigenvalues/Eigenvectors

Find the eigenvalues and eigenvectors of $\mathbf{A} = \begin{pmatrix} 9 & 1 & 1 \\ 1 & 9 & 1 \\ 1 & 1 & 9 \end{pmatrix}$.

Solution   The characteristic equation

$$\det(\mathbf{A} - \lambda\mathbf{I}) = \begin{vmatrix} 9-\lambda & 1 & 1 \\ 1 & 9-\lambda & 1 \\ 1 & 1 & 9-\lambda \end{vmatrix} = -(\lambda - 11)(\lambda - 8)^2 = 0$$

shows that $\lambda_1 = 11$ and that $\lambda_2 = \lambda_3 = 8$ is an eigenvalue of multiplicity two. For $\lambda_1 = 11$ Gauss-Jordan elimination gives

$$(\mathbf{A} - 11\mathbf{I}|\mathbf{0}) = \begin{pmatrix} -2 & 1 & 1 & | & 0 \\ 1 & -2 & 1 & | & 0 \\ 1 & 1 & -2 & | & 0 \end{pmatrix} \xrightarrow[\text{operations}]{\text{row}} \begin{pmatrix} 1 & 0 & -1 & | & 0 \\ 0 & 1 & -1 & | & 0 \\ 0 & 0 & 0 & | & 0 \end{pmatrix}.$$

Hence $k_1 = k_3$ and $k_2 = k_3$. If $k_3 = 1$, then

$$\mathbf{K}_1 = \begin{pmatrix} 1 \\ 1 \\ 1 \end{pmatrix}.$$

Now for $\lambda_2 = 8$ we have

$$(\mathbf{A} - 8\mathbf{I}|\mathbf{0}) = \begin{pmatrix} 1 & 1 & 1 & | & 0 \\ 1 & 1 & 1 & | & 0 \\ 1 & 1 & 1 & | & 0 \end{pmatrix} \xrightarrow[\text{operations}]{\text{row}} \begin{pmatrix} 1 & 1 & 1 & | & 0 \\ 0 & 0 & 0 & | & 0 \\ 0 & 0 & 0 & | & 0 \end{pmatrix}.$$

In the equation $k_1 + k_2 + k_3 = 0$ we are free to select two of the variables arbitrarily. Choosing, on the one hand, $k_2 = 1$, $k_3 = 0$ and, on the other, $k_2 = 0$, $k_3 = 1$, we obtain two linearly independent eigenvectors

$$\mathbf{K}_2 = \begin{pmatrix} -1 \\ 1 \\ 0 \end{pmatrix} \quad \text{and} \quad \mathbf{K}_3 = \begin{pmatrix} -1 \\ 0 \\ 1 \end{pmatrix}.$$

EXERCISES FOR APPENDIX II

Answers to odd-numbered problems begin on page AN-30.

II.I   Basic Definitions and Theory

1. If $\mathbf{A} = \begin{pmatrix} 4 & 5 \\ -6 & 9 \end{pmatrix}$ and $\mathbf{B} = \begin{pmatrix} -2 & 6 \\ 8 & -10 \end{pmatrix}$, find

(a) $\mathbf{A} + \mathbf{B}$ (b) $\mathbf{B} - \mathbf{A}$ (c) $2\mathbf{A} + 3\mathbf{B}$

2. If $\mathbf{A} = \begin{pmatrix} -2 & 0 \\ 4 & 1 \\ 7 & 3 \end{pmatrix}$ and $\mathbf{B} = \begin{pmatrix} 3 & -1 \\ 0 & 2 \\ -4 & -2 \end{pmatrix}$, find

    (a) $\mathbf{A} - \mathbf{B}$ **(b)** $\mathbf{B} - \mathbf{A}$ **(c)** $2(\mathbf{A} + \mathbf{B})$

3. If $\mathbf{A} = \begin{pmatrix} 2 & -3 \\ -5 & 4 \end{pmatrix}$ and $\mathbf{B} = \begin{pmatrix} -1 & 6 \\ 3 & 2 \end{pmatrix}$, find

    (a) $\mathbf{AB}$ **(b)** $\mathbf{BA}$ **(c)** $\mathbf{A}^2 = \mathbf{AA}$ **(d)** $\mathbf{B}^2 = \mathbf{BB}$

4. If $\mathbf{A} = \begin{pmatrix} 1 & 4 \\ 5 & 10 \\ 8 & 12 \end{pmatrix}$ and $\mathbf{B} = \begin{pmatrix} -4 & 6 & -3 \\ 1 & -3 & 2 \end{pmatrix}$, find

    (a) $\mathbf{AB}$ **(b)** $\mathbf{BA}$

5. If $\mathbf{A} = \begin{pmatrix} 1 & -2 \\ -2 & 4 \end{pmatrix}$, $\mathbf{B} = \begin{pmatrix} 6 & 3 \\ 2 & 1 \end{pmatrix}$, and $\mathbf{C} = \begin{pmatrix} 0 & 2 \\ 3 & 4 \end{pmatrix}$, find

    (a) $\mathbf{BC}$ **(b)** $\mathbf{A}(\mathbf{BC})$ **(c)** $\mathbf{C}(\mathbf{BA})$ **(d)** $\mathbf{A}(\mathbf{B} + \mathbf{C})$

6. If $\mathbf{A} = (5 \quad -6 \quad 7)$, $\mathbf{B} = \begin{pmatrix} 3 \\ 4 \\ -1 \end{pmatrix}$, and $\mathbf{C} = \begin{pmatrix} 1 & 2 & 4 \\ 0 & 1 & -1 \\ 3 & 2 & 1 \end{pmatrix}$, find

    (a) $\mathbf{AB}$ **(b)** $\mathbf{BA}$ **(c)** $(\mathbf{BA})\mathbf{C}$ **(d)** $(\mathbf{AB})\mathbf{C}$

7. If $\mathbf{A} = \begin{pmatrix} 4 \\ 8 \\ -10 \end{pmatrix}$ and $\mathbf{B} = (2 \quad 4 \quad 5)$, find

    (a) $\mathbf{A}^T\mathbf{A}$ **(b)** $\mathbf{B}^T\mathbf{B}$ **(c)** $\mathbf{A} + \mathbf{B}^T$

8. If $\mathbf{A} = \begin{pmatrix} 1 & 2 \\ 2 & 4 \end{pmatrix}$ and $\mathbf{B} = \begin{pmatrix} -2 & 3 \\ 5 & 7 \end{pmatrix}$, find

    (a) $\mathbf{A} + \mathbf{B}^T$ **(b)** $2\mathbf{A}^T - \mathbf{B}^T$ **(c)** $\mathbf{A}^T(\mathbf{A} - \mathbf{B})$

9. If $\mathbf{A} = \begin{pmatrix} 3 & 4 \\ 8 & 1 \end{pmatrix}$ and $\mathbf{B} = \begin{pmatrix} 5 & 10 \\ -2 & -5 \end{pmatrix}$, find

    (a) $(\mathbf{AB})^T$ **(b)** $\mathbf{B}^T\mathbf{A}^T$

10. If $\mathbf{A} = \begin{pmatrix} 5 & 9 \\ -4 & 6 \end{pmatrix}$ and $\mathbf{B} = \begin{pmatrix} -3 & 11 \\ -7 & 2 \end{pmatrix}$, find

    (a) $\mathbf{A}^T + \mathbf{B}^T$ **(b)** $(\mathbf{A} + \mathbf{B})^T$

In Problems 11–14 write the given sum as a single column matrix.

11. $4\begin{pmatrix} -1 \\ 2 \end{pmatrix} - 2\begin{pmatrix} 2 \\ 8 \end{pmatrix} + 3\begin{pmatrix} -2 \\ 3 \end{pmatrix}$

12. $3t\begin{pmatrix} 2 \\ t \\ -1 \end{pmatrix} + (t - 1)\begin{pmatrix} -1 \\ -t \\ 3 \end{pmatrix} - 2\begin{pmatrix} 3t \\ 4 \\ -5t \end{pmatrix}$

13. $\begin{pmatrix} 2 & -3 \\ 1 & 4 \end{pmatrix}\begin{pmatrix} -2 \\ 5 \end{pmatrix} - \begin{pmatrix} -1 & 6 \\ -2 & 3 \end{pmatrix}\begin{pmatrix} -7 \\ 2 \end{pmatrix}$

14. $\begin{pmatrix} 1 & -3 & 4 \\ 2 & 5 & -1 \\ 0 & -4 & -2 \end{pmatrix}\begin{pmatrix} t \\ 2t - 1 \\ -t \end{pmatrix} + \begin{pmatrix} -t \\ 1 \\ 4 \end{pmatrix} - \begin{pmatrix} 2 \\ 8 \\ -6 \end{pmatrix}$

In Problems 15–22 determine whether the given matrix is singular or nonsingular. If it is nonsingular, find $\mathbf{A}^{-1}$ using Theorem II.2.

15. $\mathbf{A} = \begin{pmatrix} -3 & 6 \\ -2 & 4 \end{pmatrix}$

16. $\mathbf{A} = \begin{pmatrix} 2 & 5 \\ 1 & 4 \end{pmatrix}$

17. $\mathbf{A} = \begin{pmatrix} 4 & 8 \\ -3 & -5 \end{pmatrix}$

18. $\mathbf{A} = \begin{pmatrix} 7 & 10 \\ 2 & 2 \end{pmatrix}$

19. $\mathbf{A} = \begin{pmatrix} 2 & 1 & 0 \\ -1 & 2 & 1 \\ 1 & 2 & 1 \end{pmatrix}$

20. $\mathbf{A} = \begin{pmatrix} 3 & 2 & 1 \\ 4 & 1 & 0 \\ -2 & 5 & -1 \end{pmatrix}$

21. $\mathbf{A} = \begin{pmatrix} 2 & 1 & 1 \\ 1 & -2 & -3 \\ 3 & 2 & 4 \end{pmatrix}$

22. $\mathbf{A} = \begin{pmatrix} 4 & 1 & -1 \\ 6 & 2 & -3 \\ -2 & -1 & 2 \end{pmatrix}$

In Problems 23 and 24 show that the given matrix is nonsingular for every real value of t. Find $\mathbf{A}^{-1}(t)$ using Theorem II.2.

23. $\mathbf{A}(t) = \begin{pmatrix} 2e^{-t} & e^{4t} \\ 4e^{-t} & 3e^{4t} \end{pmatrix}$

24. $\mathbf{A}(t) = \begin{pmatrix} 2e^{t} \sin t & -2e^{t} \cos t \\ e^{t} \cos t & e^{t} \sin t \end{pmatrix}$

In Problems 25–28 find $d\mathbf{X}/dt$.

25. $\mathbf{X} = \begin{pmatrix} 5e^{-t} \\ 2e^{-t} \\ -7e^{-t} \end{pmatrix}$

26. $\mathbf{X} = \begin{pmatrix} \frac{1}{2} \sin 2t - 4 \cos 2t \\ -3 \sin 2t + 5 \cos 2t \end{pmatrix}$

27. $\mathbf{X} = 2 \begin{pmatrix} 1 \\ -1 \end{pmatrix} e^{2t} + 4 \begin{pmatrix} 2 \\ 1 \end{pmatrix} e^{-3t}$

28. $\mathbf{X} = \begin{pmatrix} 5te^{2t} \\ t \sin 3t \end{pmatrix}$

29. Let $\mathbf{A}(t) = \begin{pmatrix} e^{4t} & \cos \pi t \\ 2t & 3t^2 - 1 \end{pmatrix}$. Find

    (a) $\dfrac{d\mathbf{A}}{dt}$
    (b) $\displaystyle\int_{0}^{2} \mathbf{A}(t)\, dt$
    (c) $\displaystyle\int_{0}^{t} \mathbf{A}(s)\, ds$

30. Let $\mathbf{A}(t) = \begin{pmatrix} \dfrac{1}{t^2 + 1} & 3t \\ t^2 & t \end{pmatrix}$ and $\mathbf{B}(t) = \begin{pmatrix} 6t & 2 \\ 1/t & 4t \end{pmatrix}$. Find

    (a) $\dfrac{d\mathbf{A}}{dt}$
    (b) $\dfrac{d\mathbf{B}}{dt}$
    (c) $\displaystyle\int_{0}^{1} \mathbf{A}(t)\, dt$
    (d) $\displaystyle\int_{1}^{2} \mathbf{B}(t)\, dt$

    (e) $\mathbf{A}(t)\mathbf{B}(t)$
    (f) $\dfrac{d}{dt}\mathbf{A}(t)\mathbf{B}(t)$
    (g) $\displaystyle\int_{1}^{t} \mathbf{A}(s)\mathbf{B}(s)\, ds$

II.2   Gaussian and Gauss-Jordan Elimination

In Problems 31–38 solve the given system of equations by either Gaussian elimination or Gauss-Jordan elimination.

31. $\begin{aligned} x + y - 2z &= 14 \\ 2x - y + z &= 0 \\ 6x + 3y + 4z &= 1 \end{aligned}$

32. $\begin{aligned} 5x - 2y + 4z &= 10 \\ x + y + z &= 9 \\ 4x - 3y + 3z &= 1 \end{aligned}$

33.
$$y + z = -5$$
$$5x + 4y - 16z = -10$$
$$x - y - 5z = 7$$

34.
$$3x + y + z = 4$$
$$4x + 2y - z = 7$$
$$x + y - 3z = 6$$

35.
$$2x + y + z = 4$$
$$10x - 2y + 2z = -1$$
$$6x - 2y + 4z = 8$$

36.
$$x + 2z = 8$$
$$x + 2y - 2z = 4$$
$$2x + 5y - 6z = 6$$

37.
$$x_1 + x_2 - x_3 - x_4 = -1$$
$$x_1 + x_2 + x_3 + x_4 = 3$$
$$x_1 - x_2 + x_3 - x_4 = 3$$
$$4x_1 + x_2 - 2x_3 + x_4 = 0$$

38.
$$2x_1 + x_2 + x_3 = 0$$
$$x_1 + 3x_2 + x_3 = 0$$
$$7x_1 + x_2 + 3x_3 = 0$$

In Problems 39 and 40 use Gauss-Jordan elimination to demonstrate that the given system of equations has no solution.

39.
$$x + 2y + 4z = 2$$
$$2x + 4y + 3z = 1$$
$$x + 2y - z = 7$$

40.
$$x_1 + x_2 - x_3 + 3x_4 = 1$$
$$x_2 - x_3 - 4x_4 = 0$$
$$x_1 + 2x_2 - 2x_3 - x_4 = 6$$
$$4x_1 + 7x_2 - 7x_3 = 9$$

In Problems 41–46 use Theorem II.3 to find $\mathbf{A}^{-1}$ for the given matrix or show that no inverse exists.

41. $\mathbf{A} = \begin{pmatrix} 4 & 2 & 3 \\ 2 & 1 & 0 \\ -1 & -2 & 0 \end{pmatrix}$

42. $\mathbf{A} = \begin{pmatrix} 2 & 4 & -2 \\ 4 & 2 & -2 \\ 8 & 10 & -6 \end{pmatrix}$

43. $\mathbf{A} = \begin{pmatrix} -1 & 3 & 0 \\ 1 & -2 & 1 \\ 0 & 1 & 2 \end{pmatrix}$

44. $\mathbf{A} = \begin{pmatrix} 1 & 2 & 3 \\ 0 & 1 & 4 \\ 0 & 0 & 8 \end{pmatrix}$

45. $\mathbf{A} = \begin{pmatrix} 1 & 2 & 3 & 1 \\ -1 & 0 & 2 & 1 \\ 2 & 1 & -3 & 0 \\ 1 & 1 & 2 & 1 \end{pmatrix}$

46. $\mathbf{A} = \begin{pmatrix} 1 & 0 & 0 & 0 \\ 0 & 0 & 1 & 0 \\ 0 & 0 & 0 & 1 \\ 0 & 1 & 0 & 0 \end{pmatrix}$

II.3   The Eigenvalue Problem

In Problems 47–54 find the eigenvalues and eigenvectors of the given matrix.

47. $\begin{pmatrix} -1 & 2 \\ -7 & 8 \end{pmatrix}$

48. $\begin{pmatrix} 2 & 1 \\ 2 & 1 \end{pmatrix}$

49. $\begin{pmatrix} -8 & -1 \\ 16 & 0 \end{pmatrix}$

50. $\begin{pmatrix} 1 & 1 \\ \frac{1}{4} & 1 \end{pmatrix}$

51. $\begin{pmatrix} 5 & -1 & 0 \\ 0 & -5 & 9 \\ 5 & -1 & 0 \end{pmatrix}$

52. $\begin{pmatrix} 3 & 0 & 0 \\ 0 & 2 & 0 \\ 4 & 0 & 1 \end{pmatrix}$

53. $\begin{pmatrix} 0 & 4 & 0 \\ -1 & -4 & 0 \\ 0 & 0 & -2 \end{pmatrix}$

54. $\begin{pmatrix} 1 & 6 & 0 \\ 0 & 2 & 1 \\ 0 & 1 & 2 \end{pmatrix}$

In Problems 55 and 56 show that the given matrix has complex eigenvalues. Find the eigenvectors of the matrix.

55. $\begin{pmatrix} -1 & 2 \\ -5 & 1 \end{pmatrix}$

56. $\begin{pmatrix} 2 & -1 & 0 \\ 5 & 2 & 4 \\ 0 & 1 & 2 \end{pmatrix}$

Miscellaneous Problems

57. If $\mathbf{A}(t)$ is a 2×2 matrix of differentiable functions and $\mathbf{X}(t)$ is a 2×1 column matrix of differentiable functions, prove the product rule

$$\frac{d}{dt}[\mathbf{A}(t)\mathbf{X}(t)] = \mathbf{A}(t)\mathbf{X}'(t) + \mathbf{A}'(t)\mathbf{X}(t).$$

58. Derive formula (3). [*Hint:* Find a matrix $\mathbf{B} = \begin{pmatrix} b_{11} & b_{12} \\ b_{21} & b_{22} \end{pmatrix}$ for which $\mathbf{AB} = \mathbf{I}$. Solve for b_{11}, b_{12}, b_{21}, and b_{22}. Then show that $\mathbf{BA} = \mathbf{I}$.]

59. If $\mathbf{A}$ is nonsingular and $\mathbf{AB} = \mathbf{AC}$, show that $\mathbf{B} = \mathbf{C}$.

60. If $\mathbf{A}$ and $\mathbf{B}$ are nonsingular, show that $(\mathbf{AB})^{-1} = \mathbf{B}^{-1}\mathbf{A}^{-1}$.

61. Let $\mathbf{A}$ and $\mathbf{B}$ be $n \times n$ matrices. In general, is

$$(\mathbf{A} + \mathbf{B})^2 = \mathbf{A}^2 + 2\mathbf{AB} + \mathbf{B}^2?$$

62. A square matrix $\mathbf{A}$ is said to be a **diagonal matrix** if all its entries off the main diagonal are zero—that is, $a_{ij} = 0$, $i \neq j$. The entries a_{ii} on the main diagonal may or may not be zero. The multiplicative identity matrix $\mathbf{I}$ is an example of a diagonal matrix.
(a) Find the inverse of the 2×2 diagonal matrix

$$\mathbf{A} = \begin{pmatrix} a_{11} & 0 \\ 0 & a_{22} \end{pmatrix}$$

when $a_{11} \neq 0$, $a_{22} \neq 0$.
(b) Find the inverse of a 3×3 diagonal matrix $\mathbf{A}$ whose main diagonal entries a_{ii} are all nonzero.
(c) In general, what is the inverse of an $n \times n$ diagonal matrix $\mathbf{A}$ whose main diagonal entries a_{ii} are all nonzero?

APPENDIX

III

LAPLACE TRANSFORMS

$f(t)$	$\mathcal{L}\{f(t)\} = F(s)$
1. 1	$\dfrac{1}{s}$
2. t	$\dfrac{1}{s^2}$
3. t^n	$\dfrac{n!}{s^{n+1}}, \quad n$ a positive integer
4. $t^{-1/2}$	$\sqrt{\dfrac{\pi}{s}}$
5. $t^{1/2}$	$\dfrac{\sqrt{\pi}}{2s^{3/2}}$
6. t^α	$\dfrac{\Gamma(\alpha+1)}{s^{\alpha+1}}, \quad \alpha > -1$
7. $\sin kt$	$\dfrac{k}{s^2 + k^2}$
8. $\cos kt$	$\dfrac{s}{s^2 + k^2}$
9. $\sin^2 kt$	$\dfrac{2k^2}{s(s^2 + 4k^2)}$
10. $\cos^2 kt$	$\dfrac{s^2 + 2k^2}{s(s^2 + 4k^2)}$
11. e^{at}	$\dfrac{1}{s - a}$
12. $\sinh kt$	$\dfrac{k}{s^2 - k^2}$
13. $\cosh kt$	$\dfrac{s}{s^2 - k^2}$
14. $\sinh^2 kt$	$\dfrac{2k^2}{s(s^2 - 4k^2)}$
15. $\cosh^2 kt$	$\dfrac{s^2 - 2k^2}{s(s^2 - 4k^2)}$
16. te^{at}	$\dfrac{1}{(s - a)^2}$

$f(t)$	$\mathscr{L}\{f(t)\} = F(s)$
17. $t^n e^{at}$	$\dfrac{n!}{(s-a)^{n+1}}$, n a positive integer
18. $e^{at} \sin kt$	$\dfrac{k}{(s-a)^2 + k^2}$
19. $e^{at} \cos kt$	$\dfrac{s-a}{(s-a)^2 + k^2}$
20. $e^{at} \sinh kt$	$\dfrac{k}{(s-a)^2 - k^2}$
21. $e^{at} \cosh kt$	$\dfrac{s-a}{(s-a)^2 - k^2}$
22. $t \sin kt$	$\dfrac{2ks}{(s^2 + k^2)^2}$
23. $t \cos kt$	$\dfrac{s^2 - k^2}{(s^2 + k^2)^2}$
24. $\sin kt + kt \cos kt$	$\dfrac{2ks^2}{(s^2 + k^2)^2}$
25. $\sin kt - kt \cos kt$	$\dfrac{2k^3}{(s^2 + k^2)^2}$
26. $t \sinh kt$	$\dfrac{2ks}{(s^2 - k^2)^2}$
27. $t \cosh kt$	$\dfrac{s^2 + k^2}{(s^2 - k^2)^2}$
28. $\dfrac{e^{at} - e^{bt}}{a - b}$	$\dfrac{1}{(s-a)(s-b)}$
29. $\dfrac{ae^{at} - be^{bt}}{a - b}$	$\dfrac{s}{(s-a)(s-b)}$
30. $1 - \cos kt$	$\dfrac{k^2}{s(s^2 + k^2)}$
31. $kt - \sin kt$	$\dfrac{k^3}{s^2(s^2 + k^2)}$
32. $\dfrac{a \sin bt - b \sin at}{ab(a^2 - b^2)}$	$\dfrac{1}{(s^2 + a^2)(s^2 + b^2)}$
33. $\dfrac{\cos bt - \cos at}{a^2 - b^2}$	$\dfrac{s}{(s^2 + a^2)(s^2 + b^2)}$
34. $\sin kt \sinh kt$	$\dfrac{2k^2 s}{s^4 + 4k^4}$
35. $\sin kt \cosh kt$	$\dfrac{k(s^2 + 2k^2)}{s^4 + 4k^4}$
36. $\cos kt \sinh kt$	$\dfrac{k(s^2 - 2k^2)}{s^4 + 4k^4}$
37. $\cos kt \cosh kt$	$\dfrac{s^3}{s^4 + 4k^4}$

$f(t)$	$\mathcal{L}\{f(t)\} = F(s)$
38. $J_0(kt)$	$\dfrac{1}{\sqrt{s^2 + k^2}}$
39. $\dfrac{e^{bt} - e^{at}}{t}$	$\ln \dfrac{s - a}{s - b}$
40. $\dfrac{2(1 - \cos kt)}{t}$	$\ln \dfrac{s^2 + k^2}{s^2}$
41. $\dfrac{2(1 - \cosh kt)}{t}$	$\ln \dfrac{s^2 - k^2}{s^2}$
42. $\dfrac{\sin at}{t}$	$\arctan\left(\dfrac{a}{s}\right)$
43. $\dfrac{\sin at \cos bt}{t}$	$\dfrac{1}{2} \arctan \dfrac{a + b}{s} + \dfrac{1}{2} \arctan \dfrac{a - b}{s}$
44. $\dfrac{1}{\sqrt{\pi t}} e^{-a^2/4t}$	$\dfrac{e^{-a\sqrt{s}}}{\sqrt{s}}$
45. $\dfrac{a}{2\sqrt{\pi t^3}} e^{-a^2/4t}$	$e^{-a\sqrt{s}}$
46. $\operatorname{erfc}\left(\dfrac{a}{2\sqrt{t}}\right)$	$\dfrac{e^{-a\sqrt{s}}}{s}$
47. $2\sqrt{\dfrac{t}{\pi}} e^{-a^2/4t} - a\operatorname{erfc}\left(\dfrac{a}{2\sqrt{t}}\right)$	$\dfrac{e^{-a\sqrt{s}}}{s\sqrt{s}}$
48. $e^{ab}e^{b^2 t}\operatorname{erfc}\left(b\sqrt{t} + \dfrac{a}{2\sqrt{t}}\right)$	$\dfrac{e^{-a\sqrt{s}}}{\sqrt{s}(\sqrt{s} + b)}$
49. $-e^{ab}e^{b^2 t}\operatorname{erfc}\left(b\sqrt{t} + \dfrac{a}{2\sqrt{t}}\right)$ $\quad + \operatorname{erfc}\left(\dfrac{a}{2\sqrt{t}}\right)$	$\dfrac{be^{-a\sqrt{s}}}{s(\sqrt{s} + b)}$
50. $\delta(t)$	1
51. $\delta(t - t_0)$	e^{-st_0}
52. $e^{at}f(t)$	$F(s - a)$
53. $f(t - a)\,\mathcal{U}(t - a)$	$e^{-as}F(s)$
54. $\mathcal{U}(t - a)$	$\dfrac{e^{-as}}{s}$
55. $f^{(n)}(t)$	$s^n F(s) - s^{(n-1)}f(0) - \cdots - f^{(n-1)}(0)$
56. $t^n f(t)$	$(-1)^n \dfrac{d^n}{ds^n} F(s)$
57. $\displaystyle\int_0^t f(\tau)g(t - \tau)\,d\tau$	$F(s)G(s)$

SELECTED ANSWERS FOR
ODD-NUMBERED PROBLEMS

EXERCISES 1.1 (PAGE 10)

1. second-order linear
3. first-order, linear in x but nonlinear in y
5. fourth-order linear   **7.** second-order nonlinear
9. third-order linear
15. $X = \dfrac{e^t - 1}{e^t - 2}$ defined on $(-\infty, \ln 2)$ or on $(\ln 2, \infty)$
23. (a) $-3 \le x < \infty$ **(b)** $-3 < x < \infty$
25. (a) $m = -2$ **(b)** $m = 2, m = 3$

EXERCISES 1.2 (PAGE 19)

1. $y = 1/(1 - 4e^{-x})$ **3.** $x = -\cos t + 8 \sin t$
5. $x = \dfrac{\sqrt{3}}{4} \cos t + \dfrac{1}{4} \sin t$ **7.** $y = \frac{3}{2}e^x - \frac{1}{2}e^{-x}$
9. $y = 5e^{-x-1}$ **11.** $y = 0, y = x^3$
13. half-planes defined by either $y > 0$ or $y < 0$
15. half-planes defined by either $x > 0$ or $x < 0$
17. the regions defined by $y > 2, y < -2$, or $-2 < y < 2$
19. any region not containing $(0, 0)$
21. yes   **23.** no
25. (a) $y = cx$
  (b) any rectangular region not touching the y-axis
  (c) No, the function is not differentiable at $x = 0$.
27. (b) $y = 1/(1 - x)$ on $(-\infty, 1); y = -1/(x + 1)$ on $(-1, \infty)$
  (c) $y = y_0/(1 - y_0 x), y_0 \ne 0$. Consider the cases $y_0 < 0$ and $y_0 > 0$.

EXERCISES 1.3 (PAGE 31)

1. $\dfrac{dP}{dt} = kP + r; \dfrac{dP}{dt} = kP - r$ **3.** $\dfrac{dP}{dt} = k_1 P - k_2 P^2$
7. $\dfrac{dx}{dt} = kx(1000 - x)$ **9.** $\dfrac{dA}{dt} = -\dfrac{A}{100}$
11. $\dfrac{dh}{dt} = -\dfrac{c\pi}{450}\sqrt{h}$ **13.** $L\dfrac{di}{dt} + Ri = E(t)$
15. $m\dfrac{dv}{dt} = mg - kv^2$ **17.** $m\dfrac{d^2x}{dt^2} = -kx$
19. $x\dfrac{d^2x}{dt^2} + \left(\dfrac{dx}{dt}\right)^2 + 32x = 160$ **21.** $\dfrac{d^2r}{dt^2} + \dfrac{gR^2}{r^2} = 0$
23. $\dfrac{dA}{dt} = k(M - A), k > 0$ **25.** $\dfrac{dx}{dt} + kx = r, k > 0$
27. $\dfrac{dy}{dx} = \dfrac{-x + \sqrt{x^2 + y^2}}{y}$

CHAPTER 1 IN REVIEW (PAGE 37)

1. $\dfrac{dy}{dx} = -\dfrac{y}{x}$ **3.** $y'' + k^2 y = 0$ **5.** $y'' - 2y' + y = 0$
7. (a), (d)   **9.** (b)   **11.** (b)

13. $y = c_1$ and $y = c_2 e^x, c_1$ and c_2 constants   **15.** $y' = x^2 + y^2$
17. (a) The domain is all real numbers.
  (b) $(-\infty, 0)$ or $(0, \infty)$
19. (c) $y = \begin{cases} -x^2, & x < 0 \\ x^2, & x \ge 0 \end{cases}$ **21.** $(0, \infty)$
25. $\dfrac{dv}{dt} = 16(1 - \sqrt{3}\,\mu)$

EXERCISES 2.1 (PAGE 47)

17. 0 is asymptotically stable (attractor); 3 is unstable (repeller)
19. 2 is semistable
21. -2 is unstable (repeller); 0 is semistable; 2 is asymptotically stable (attractor)
23. -1 is asymptotically stable (attractor); 0 is unstable (repeller)
25. (a) mg/k **(b)** $\sqrt{mg/k}$

EXERCISES 2.2 (PAGE 57)

1. $y = -\frac{1}{5}\cos 5x + c$ **3.** $y = \frac{1}{3}e^{-3x} + c$
5. $y = cx^4$ **7.** $-3e^{-2y} = 2e^{3x} + c$
9. $\frac{1}{3}x^3 \ln x - \frac{1}{9}x^3 = \frac{1}{2}y^2 + 2y + \ln|y| + c$
11. $4\cos y = 2x + \sin 2x + c$
13. $(e^x + 1)^{-2} + 2(e^y + 1)^{-1} = c$
15. $S = ce^{kr}$ **17.** $P = \dfrac{ce^t}{1 + ce^t}$
19. $(y + 3)^5 e^x = c(x + 4)^5 e^y$ **21.** $y = \sin(\frac{1}{2}x^2 + c)$
23. $x = \tan\left(4y - \dfrac{3\pi}{4}\right)$ **25.** $xy = e^{-(1 + 1/x)}$
27. $y = \dfrac{\sqrt{3}}{2}\sqrt{1 - x^2} + \frac{1}{2}x$
29. (a) $y = 2, y = -2, y = 2\dfrac{3 - e^{4x-1}}{3 + e^{4x-1}}$
33. $y = 1$ **35.** $y = 1 + \dfrac{1}{10}\tan\dfrac{x}{10}$
39. (a) $y = -\sqrt{x^2 + x - 1}$ **(c)** $\left(-\infty, -\dfrac{1}{2} - \dfrac{\sqrt{5}}{2}\right)$

EXERCISES 2.3 (PAGE 69)

1. $y = ce^{5x}, -\infty < x < \infty$
3. $y = \frac{1}{4}e^{3x} + ce^{-x}, -\infty < x < \infty$
5. $y = \frac{1}{3} + ce^{-x^3}, -\infty < x < \infty$
7. $y = x^{-1}\ln x + cx^{-1}, x > 0$
9. $y = cx - x\cos x, x > 0$
11. $y = \frac{1}{7}x^3 - \frac{1}{5}x + cx^{-4}, x > 0$
13. $y = \dfrac{1}{2x^2}e^x + \dfrac{c}{x^2}e^{-x}, x > 0$
15. $x = 2y^6 + cy^4, y > 0$
17. $y = \sin x + c\cos x, -\pi/2 < x < \pi/2$

19. $(x + 1)e^x y = x^2 + c, x > -1$

21. $(\sec\theta + \tan\theta)r = \theta - \cos\theta + c, -\pi/2 < \theta < \pi/2$

23. $y = e^{-3x} + \dfrac{c}{x}e^{-3x}, 0 < x < \infty$

25. $y = \dfrac{e^x}{x} + \dfrac{2 - e}{x}, x > 0$

27. $i = \dfrac{E}{R} + \left(i_0 - \dfrac{E}{R}\right)e^{-Rt/L}, -\infty < t < \infty$

29. $(x + 1)y = x\ln x - x + 21, x > 0$

31. $y = \begin{cases} \frac{1}{2}(1 - e^{-2x}), & 0 \le x \le 3 \\ \frac{1}{2}(e^6 - 1)e^{-2x}, & x > 3 \end{cases}$

33. $y = \begin{cases} \frac{1}{2} + \frac{3}{2}e^{-x^2}, & 0 \le x < 1 \\ (\frac{1}{2}e + \frac{3}{2})e^{-x^2}, & x \ge 1 \end{cases}$

35. $y = \begin{cases} 2x - 1 + 4e^{-2x}, & 0 \le x \le 1 \\ 4x^2\ln x + (1 + 4e^{-2})x^2, & x > 1 \end{cases}$

37. $y = e^{-e^x}\displaystyle\int_0^x e^{e^t}\,dt + e^{1-e^x}; y = e^{1-e^x}; y = 1$

EXERCISES 2.4 (PAGE 78)

1. $x^2 - x + \frac{3}{2}y^2 + 7y = c$ **3.** $\frac{5}{2}x^2 + 4xy - 2y^4 = c$

5. $x^2y^2 - 3x + 4y = c$ **7.** not exact

9. $xy^3 + y^2\cos x - \frac{1}{2}x^2 = c$ **11.** not exact

13. $xy - 2xe^x + 2e^x - 2x^3 = c$ **15.** $x^3y^3 - \tan^{-1}3x = c$

17. $-\ln|\cos x| + \cos x\sin y = c$

19. $t^4y - 5t^3 - ty + y^3 = c$ **21.** $\frac{1}{3}x^3 + x^2y + xy^2 - y = \frac{4}{3}$

23. $4ty + t^2 - 5t + 3y^2 - y = 8$

25. $y^2\sin x - x^3y - x^2 + y\ln y - y = 0$

27. $k = 10$ **29.** $x^2y^2\cos x = c$

31. $x^2y^2 + x^3 = c$ **33.** $3x^2y^3 + y^4 = c$

35. $-2ye^{3x} + \frac{10}{3}e^{3x} + x = c$ **37.** $e^{y^2}(x^2 + 4) = 20$

39. (c) $y_1(x) = -x^2 - \sqrt{x^4 - x^3 + 4}$,

$y_2(x) = -x^2 + \sqrt{x^4 - x^3 + 4}$

EXERCISES 2.5 (PAGE 84)

1. $x\ln|x| + y = cx$

3. $(x - y)\ln|x - y| = y + c(x - y)$

5. $x + y\ln|x| = cy$ **7.** $\ln(x^2 + y^2) + 2\tan^{-1}\left(\dfrac{y}{x}\right) = c$

9. $4x = y(\ln|y| - c)^2$ **11.** $y^3 + 3x^3\ln|x| = 8x^3$

13. $\ln|x| = e^{y/x} - 1$ **15.** $y^3 = 1 + cx^{-3}$

17. $y^{-3} = x + \frac{1}{3} + ce^{3x}$

19. $e^{t/y} = ct$

21. $y^{-3} = -\frac{9}{5}x^{-1} + \frac{49}{5}x^{-6}$

23. $y = -x - 1 + \tan(x + c)$

25. $2y - 2x + \sin 2(x + y) = c$

27. $4(y - 2x + 3) = (x + c)^2$

29. $-\cot(x + y) + \csc(x + y) = x + \sqrt{2} - 1$

EXERCISES 2.6 (PAGE 91)

1. $y_2 = 2.9800, y_4 = 3.1151$

3. $y_{10} = 2.5937, y_{20} = 2.6533; y = e^x$

5. $y_5 = 0.4198, y_{10} = 0.4124$

7. $y_5 = 0.5639, y_{10} = 0.5565$

9. $y_5 = 1.2194, y_{10} = 1.2696$

CHAPTER 2 IN REVIEW (PAGE 92)

1. $-A/k$, repeller, attractor

3. $\dfrac{dy}{dx} = (y - 1)^2(y - 3)^3$

5. semistable for n even and unstable for n odd; semistable for n even and asymptotically stable for n odd

9. $2x + \sin 2x = 2\ln(y^2 + 1) + c$

11. $(6x + 1)y^3 = -3x^3 + c$

13. $Q = \dfrac{c}{t} + \dfrac{t^4}{25}(-1 + 5\ln t)$ **15.** $y = \frac{1}{4} + c(x^2 + 4)^{-4}$

17. $y = \csc x, \pi < x < 2\pi$

19. (b) $y = \frac{1}{4}(x + 2\sqrt{y_0} - x_0)^2, x \ge x_0 - 2\sqrt{y_0}$

EXERCISES 3.1 (PAGE 103)

1. 7.9 yr; 10 yr **3.** 760

5. $I(15) = 0.00098I_0$ or approximately 0.1% of I_0

7. 11 h **9.** 136.5 h

11. 15,600 yr **13.** $T(1) = 36.76°F$; approximately 3.06 min

15. approximately 82.1 s; approximately 145.7 s

17. $A(t) = 200 - 170e^{-t/50}$

19. $A(t) = 1000 - 1000e^{-t/100}$; 0.0975 lb/gal

21. 64.38 lb **23.** $i(t) = \frac{3}{5} - \frac{3}{5}e^{-500t}$; $i \to \frac{3}{5}$ as $t \to \infty$

25. $q(t) = \frac{1}{100} - \frac{1}{100}e^{-50t}; i(t) = \frac{1}{2}e^{-50t}$

27. $i(t) = \begin{cases} 60 - 60e^{-t/10}, & 0 \le t \le 20 \\ 60(e^2 - 1)e^{-t/10}, & t > 20 \end{cases}$

29. (a) $v(t) = \dfrac{mg}{k} + \left(v_0 - \dfrac{mg}{k}\right)e^{-kt/m}$

(b) $v \to \dfrac{mg}{k}$ as $t \to \infty$

(c) $s(t) = \dfrac{mg}{k}t - \dfrac{m}{k}\left(v_0 - \dfrac{mg}{k}\right)e^{-kt/m} + \dfrac{m}{k}\left(v_0 - \dfrac{mg}{k}\right) + s_0$

33. (a) $v(t) = \dfrac{\rho g}{4k}\left(\dfrac{k}{\rho}t + r_0\right) - \dfrac{\rho g r_0}{4k}\left(\dfrac{r_0}{\frac{k}{\rho}t + r_0}\right)^3$

(b) $33\frac{1}{3}$ min

35. (a) $P(t) = P_0 e^{(k_1 - k_2)t}$

37. (a) As $t \to \infty, A \to \dfrac{k_1 M}{k_1 + k_2}$. Observe that for $k_1 > 0$

and $k_2 > 0$, then $\dfrac{k_1 M}{k_1 + k_2} < M$.

(b) $A(t) = \dfrac{k_1 M}{k_1 + k_2}(1 - e^{-(k_1 + k_2)t})$

EXERCISES 3.2 (PAGE 115)

1. (a) $N = 2000$

(b) $N(t) = \dfrac{1.0005e^t}{1 + 0.005e^t}; N(10) = 1834$

3. 1,000,000; 5.29 mo

7. 29.3 g; $X \to 60$ as $t \to \infty$; 0 g of A and 30 g of B

9. (a) $h(t) = \left(\sqrt{H} - \dfrac{4A_h}{A_w}t\right)^2; I$ is $0 \le t \le \dfrac{\sqrt{H}A_w}{4A_h}$

(b) $576\sqrt{10}$ s or 30.36 min

11. (a) approximately 856.65 s or 14.31 min

(b) 243 s or 4.05 min

13. (a) $v(t) = \sqrt{\dfrac{mg}{k}} \tanh\left(\sqrt{\dfrac{kg}{m}} t + c_1\right),$

where $c_1 = \tanh^{-1}\left(\sqrt{\dfrac{k}{mg}}\, v_0\right)$

(b) $\sqrt{\dfrac{mg}{k}}$

(c) $s(t) = \dfrac{m}{k} \ln \cosh\left(\sqrt{\dfrac{kg}{m}} t + c_1\right) + c_2,$

where $c_2 = -\left(\dfrac{m}{k}\right) \ln \cosh c_1$

15. (a) $m\dfrac{dv}{dt} = mg - kv^2 - \rho V,$

where ρ is the weight density of water

(b) $v(t) = \sqrt{\dfrac{mg - \rho V}{k}} \tanh\left(\dfrac{\sqrt{kmg - k\rho V}}{m} t + c_1\right)$

(c) $\sqrt{\dfrac{mg - \rho V}{k}}$

17. (a) $W = 0$ and $W = 2$
(b) $W(x) = 2 \operatorname{sech}^2(x - c_1)$
(c) $W(x) = 2 \operatorname{sech}^2 x$

EXERCISES 3.3 (PAGE 126)

1. $x(t) = x_0 e^{-\lambda_1 t}$

$y(t) = \dfrac{x_0 \lambda_1}{\lambda_2 - \lambda_1}\left(e^{-\lambda_1 t} - e^{-\lambda_2 t}\right)$

$z(t) = x_0\left(1 - \dfrac{\lambda_2}{\lambda_2 - \lambda_1} e^{-\lambda_1 t} + \dfrac{\lambda_1}{\lambda_2 - \lambda_1} e^{-\lambda_2 t}\right)$

3. 5, 20, 147 days. The time when $y(t)$ and $z(t)$ are the same makes sense because most of A and half of B are gone, so half of C should have been formed.

5. $\dfrac{dx_1}{dt} = 6 - \dfrac{2}{25}x_1 + \dfrac{1}{50}x_2$

$\dfrac{dx_2}{dt} = \dfrac{2}{25}x_1 - \dfrac{2}{25}x_2$

7. (a) $\dfrac{dx_1}{dt} = 3\dfrac{x_2}{100 - t} - 2\dfrac{x_1}{100 + t}$

$\dfrac{dx_2}{dt} = 2\dfrac{x_1}{100 + t} - 3\dfrac{x_2}{100 - t}$

(b) $x_1(t) + x_2(t) = 150; x_2(30) \approx 47.4$ lb

13. $L_1\dfrac{di_2}{dt} + (R_1 + R_2)i_2 + R_1 i_3 = E(t)$

$L_2\dfrac{di_3}{dt} + R_1 i_2 + (R_1 + R_3)i_3 = E(t)$

15. $i(0) = i_0, s(0) = n - i_0, r(0) = 0$

CHAPTER 3 IN REVIEW (PAGE 130)

1. $P(45) = 8.99$ billion

3. (a) $E(t) = \begin{cases} 0, & 0 \le t < 4 \\ 12e^{-(t-4)/RC}, & 4 \le t < 6 \\ 0, & 6 \le t < 10 \\ 12e^{-(t-10)/RC}, & 10 \le t < 12 \\ 0, & 12 \le t < 16 \\ 12e^{-(t-16)/RC} & 16 \le t < 18 \end{cases}$

5. (a) $T(t) = \dfrac{T_2 + BT_1}{1 + B} + \dfrac{T_1 - T_2}{1 + B} e^{k(1+B)t}$

(b) $\dfrac{T_2 + BT_1}{1 + B}$

(c) $\dfrac{T_2 + BT_1}{1 + B}$

7. $x(\theta) = k\theta - \dfrac{k}{2}\sin 2\theta + c, y(\theta) = k \sin^2 \theta$

9. $x(t) = \dfrac{\alpha c_1 e^{\alpha k_1 t}}{1 + c_1 e^{\alpha k_1 t}}, y(t) = c_2(1 + c_1 e^{\alpha k_1 t})^{k_2/k_1}$

EXERCISES 4.1 (PAGE 151)

1. $y = \tfrac{1}{2}e^x - \tfrac{1}{2}e^{-x}$ **3.** $y = 3x - 4x \ln x$

9. $(-\infty, 2)$ **11.** $y = \dfrac{e}{e^2 - 1}(e^x - e^{-x})$

13. (a) $y = e^x \cos x - e^x \sin x$
(b) no solution
(c) $y = e^x \cos x + e^{-\pi/2} e^x \sin x$
(d) $y = c_2 e^x \sin x$, where c_2 is arbitrary

15. dependent **17.** dependent
19. dependent **21.** independent
23. The functions satisfy the DE and are linearly independent on the interval since
$W(e^{-3x}, e^{4x}) = 7e^x \ne 0; y = c_1 e^{-3x} + c_2 e^{4x}$.
25. The functions satisfy the DE and are linearly independent on the interval since
$W(e^x \cos 2x, e^x \sin 2x) = 2e^{2x} \ne 0;$
$y = c_1 e^x \cos 2x + c_2 e^x \sin 2x$.
27. The functions satisfy the DE and are linearly independent on the interval since
$W(x^3, x^4) = x^6 \ne 0; y = c_1 x^3 + c_2 x^4$.
29. The functions satisfy the DE and are linearly independent on the interval since
$W(x, x^{-2}, x^{-2} \ln x) = 9x^{-6} \ne 0;$
$y = c_1 x + c_2 x^{-2} + c_3 x^{-2} \ln x$.
35. (b) $y_p = x^2 + 3x + 3e^{2x}; y_p = -2x^2 - 6x - \tfrac{1}{3}e^{2x}$

EXERCISES 4.2 (PAGE 157)

1. $y_2 = xe^{2x}$ **3.** $y_2 = \sin 4x$
5. $y_2 = \sinh x$ **7.** $y_2 = xe^{2x/3}$
9. $y_2 = x^4 \ln|x|$ **11.** $y_2 = 1$
13. $y_2 = x \cos(\ln x)$ **15.** $y_2 = x^2 + x + 2$
17. $y_2 = e^{2x}, y_p = -\tfrac{1}{2}$ **19.** $y_2 = e^{2x}, y_p = \tfrac{5}{2}e^{3x}$

EXERCISES 4.3 (PAGE 163)

1. $y = c_1 + c_2 e^{-x/4}$ **3.** $y = c_1 e^{3x} + c_2 e^{-2x}$
5. $y = c_1 e^{-4x} + c_2 xe^{-4x}$ **7.** $y = c_1 e^{2x/3} + c_2 xe^{-x/4}$
9. $y = c_1 \cos 3x + c_2 \sin 3x$ **11.** $y = e^{2x}(c_1 \cos x + c_2 \sin x)$

13. $y = e^{-x/3}\left(c_1 \cos \dfrac{\sqrt{2}}{3}x + c_2 \sin \dfrac{\sqrt{2}}{3}x\right)$

15. $y = c_1 + c_2 e^{-x} + c_3 e^{5x}$ **17.** $y = c_1 e^{-x} + c_2 e^{3x} + c_3 xe^{3x}$
19. $u = c_1 e^t + e^{-t}(c_2 \cos t + c_3 \sin t)$
21. $y = c_1 e^{-x} + c_2 xe^{-x} + c_3 x^2 e^{-x}$

23. $y = c_1 + c_2 x + e^{-x/2}\left(c_3 \cos \dfrac{\sqrt{3}}{2}x + c_4 \sin \dfrac{\sqrt{3}}{2}x\right)$

25. $y = c_1 \cos \dfrac{\sqrt{3}}{2} x + c_2 \sin \dfrac{\sqrt{3}}{2} x$
$\qquad + c_3 x \cos \dfrac{\sqrt{3}}{2} x + c_4 x \sin \dfrac{\sqrt{3}}{2} x$

27. $u = c_1 e^r + c_2 r e^r + c_3 e^{-r} + c_4 r e^{-r} + c_5 e^{-5r}$

29. $y = 2 \cos 4x - \frac{1}{2} \sin 4x$ **31.** $y = -\frac{1}{3} e^{-(t-1)} + \frac{1}{3} e^{5(t-1)}$

33. $y = 0$ **35.** $y = \frac{5}{36} - \frac{5}{36} e^{-6x} + \frac{1}{6} x e^{-6x}$

37. $y = e^{5x} - x e^{5x}$ **39.** $y = -2 \cos x$

41. $y = \dfrac{1}{2}\left(1 - \dfrac{5}{\sqrt{3}}\right) e^{-\sqrt{3}x} - \dfrac{1}{2}\left(1 + \dfrac{5}{\sqrt{3}}\right) e^{-\sqrt{3}x}$;

$\qquad y = \cosh \sqrt{3}x + \dfrac{5}{\sqrt{3}} \sinh \sqrt{3}x$

49. (a) $x(t) = c_1 \cosh \dfrac{8}{\sqrt{L}} t + c_2 \sinh \dfrac{8}{\sqrt{L}} t$

$\qquad$ **(b)** $x(t) = x_0 \cosh \dfrac{8}{\sqrt{L}} t$ **(c)** 17.69 ft/s

EXERCISES 4.4 (PAGE 176)

1. $y = c_1 e^{-x} + c_2 e^{-2x} + 3$

3. $y = c_1 e^{5x} + c_2 x e^{5x} + \frac{6}{5} x + \frac{3}{5}$

5. $y = c_1 e^{-2x} + c_2 x e^{-2x} + x^2 - 4x + \frac{7}{2}$

7. $y = c_1 \cos \sqrt{3}x + c_2 \sin \sqrt{3}x + (-4x^2 + 4x - \frac{4}{3})e^{3x}$

9. $y = c_1 + c_2 e^x + 3x$

11. $y = c_1 e^{x/2} + c_2 x e^{x/2} + 12 + \frac{1}{2} x^2 e^{x/2}$

13. $y = c_1 \cos 2x + c_2 \sin 2x - \frac{3}{4} x \cos 2x$

15. $y = c_1 \cos x + c_2 \sin x - \frac{1}{2} x^2 \cos x + \frac{1}{2} x \sin x$

17. $y = c_1 e^x \cos 2x + c_2 e^x \sin 2x + \frac{1}{4} x e^x \sin 2x$

19. $y = c_1 e^{-x} + c_2 x e^{-x} - \frac{1}{2} \cos x + \frac{12}{25} \sin 2x - \frac{9}{25} \cos 2x$

21. $y = c_1 + c_2 x + c_3 e^{6x} - \frac{1}{4} x^2 - \frac{6}{37} \cos x + \frac{1}{37} \sin x$

23. $y = c_1 e^x + c_2 x e^x + c_3 x^2 e^x - x - 3 - \frac{2}{3} x^3 e^x$

25. $y = c_1 \cos x + c_2 \sin x + c_3 x \cos x$
$\qquad + c_4 x \sin x + x^2 - 2x - 3$

27. $y = \sqrt{2} \sin 2x - \frac{1}{2}$

29. $y = -200 + 200 e^{-x/5} - 3x^2 + 30x$

31. $y = -10 e^{-2x} \cos x + 9 e^{-2x} \sin x + 7 e^{-4x}$

33. $x = \dfrac{F_0}{2\omega^2} \sin \omega t - \dfrac{F_0}{2\omega} t \cos \omega t$

35. $y = 11 - 11 e^x + 9x e^x + 2x - 12 x^2 e^x + \frac{1}{2} e^{5x}$

37. $y = 6 \cos x - 6(\cot 1) \sin x + x^2 - 1$

39. $y = \begin{cases} \cos 2x + \frac{5}{6} \sin 2x + \frac{1}{3} \sin x, & 0 \le x \le \pi/2 \\ \frac{2}{3} \cos 2x + \frac{5}{6} \sin 2x, & x > \pi/2 \end{cases}$

EXERCISES 4.5 (PAGE 186)

1. $(3D - 2)(3D + 2)y = \sin x$

3. $(D - 6)(D + 2)y = x - 6$ **5.** $D(D + 5)^2 y = e^x$

7. $(D - 1)(D - 2)(D + 5)y = x e^{-x}$

9. $D(D + 2)(D^2 - 2D + 4)y = 4$ **15.** D^4 **17.** $D(D - 2)$

19. $D^2 + 4$ **21.** $D^3(D^2 + 16)$ **23.** $(D + 1)(D - 1)^3$

25. $D(D^2 - 2D + 5)$ **27.** $1, x, x^2, x^3, x^4$ **29.** $e^{6x}, e^{-3x/2}$

31. $\cos \sqrt{5}x, \sin \sqrt{5}x$ **33.** $1, e^{5x}, x e^{5x}$

35. $y = c_1 e^{-3x} + c_2 e^{3x} - 6$ **37.** $y = c_1 + c_2 e^{-x} + 3x$

39. $y = c_1 e^{-2x} + c_2 x e^{-2x} + \frac{1}{2} x + 1$

41. $y = c_1 + c_2 x + c_3 e^{-x} + \frac{2}{3} x^4 - \frac{8}{3} x^3 + 8x^2$

43. $y = c_1 e^{-3x} + c_2 e^{4x} + \frac{1}{7} x e^{4x}$

45. $y = c_1 e^{-x} + c_2 e^{3x} - e^x + 3$

47. $y = c_1 \cos 5x + c_2 \sin 5x + \frac{1}{4} \sin x$

49. $y = c_1 e^{-3x} + c_2 x e^{-3x} - \frac{1}{49} x e^{4x} + \frac{2}{343} e^{4x}$

51. $y = c_1 e^{-x} + c_2 e^x + \frac{1}{6} x^3 e^x - \frac{1}{4} x^2 e^x + \frac{1}{4} x e^x - 5$

53. $y = e^x(c_1 \cos 2x + c_2 \sin 2x) + \frac{1}{3} e^x \sin x$

55. $y = c_1 \cos 5x + c_2 \sin 5x - 2x \cos 5x$

57. $y = e^{-x/2}\left(c_1 \cos \dfrac{\sqrt{3}}{2} x + c_2 \sin \dfrac{\sqrt{3}}{2} x\right)$
$\qquad + \sin x + 2 \cos x - x \cos x$

59. $y = c_1 + c_2 x + c_3 e^{-8x} + \frac{11}{256} x^2 + \frac{7}{32} x^3 - \frac{1}{16} x^4$

61. $y = c_1 e^x + c_2 x e^x + c_3 x^2 e^x + \frac{1}{6} x^3 e^x + x - 13$

63. $y = c_1 + c_2 x + c_3 e^x + c_4 x e^x + \frac{1}{2} x^2 e^x + \frac{1}{2} x^2$

65. $y = \frac{5}{8} e^{-8x} + \frac{5}{8} e^{8x} - \frac{1}{4}$

67. $y = -\frac{41}{125} + \frac{41}{125} e^{5x} - \frac{1}{10} x^2 + \frac{9}{25} x$

69. $y = -\pi \cos x - \frac{11}{3} \sin x - \frac{8}{3} \cos 2x + 2x \cos x$

71. $y = 2 e^{2x} \cos 2x - \frac{3}{64} e^{2x} \sin 2x + \frac{1}{8} x^3 + \frac{3}{16} x^2 + \frac{3}{32} x$

EXERCISES 4.6 (PAGE 192)

1. $y = c_1 \cos x + c_2 \sin x + x \sin x + \cos x \ln|\cos x|$

3. $y = c_1 \cos x + c_2 \sin x - \frac{1}{2} x \cos x$

5. $y = c_1 \cos x + c_2 \sin x + \frac{1}{2} - \frac{1}{6} \cos 2x$

7. $y = c_1 e^x + c_2 e^{-x} + \frac{1}{2} x \sinh x$

9. $y = c_1 e^{2x} + c_2 e^{-2x} + \dfrac{1}{4}\left(e^{2x} \ln|x| - e^{-2x} \displaystyle\int_{x_0}^{x} \dfrac{e^{4t}}{t} dt\right), x_0 > 0$

11. $y = c_1 e^{-x} + c_2 e^{-2x} + (e^{-x} + e^{-2x}) \ln(1 + e^x)$

13. $y = c_1 e^{-2x} + c_2 e^{-x} - e^{-2x} \sin e^x$

15. $y = c_1 e^{-t} + c_2 t e^{-t} + \frac{1}{2} t^2 e^{-t} \ln t - \frac{3}{4} t^2 e^{-t}$

17. $y = c_1 e^x \sin x + c_2 e^x \cos x + \frac{1}{3} x e^x \sin x$
$\qquad + \frac{1}{3} e^x \cos x \ln|\cos x|$

19. $y = \frac{1}{4} e^{-x/2} + \frac{3}{4} e^{x/2} + \frac{1}{8} x^2 e^{x/2} - \frac{1}{4} x e^{x/2}$

21. $y = \frac{4}{9} e^{-4x} + \frac{25}{36} e^{2x} - \frac{1}{4} e^{-2x} + \frac{1}{9} e^{-x}$

23. $y = c_1 x^{-1/2} \cos x + c_2 x^{-1/2} \sin x + x^{-1/2}$

25. $y = c_1 + c_2 \cos x + c_3 \sin x - \ln|\cos x|$
$\qquad - \sin x \ln|\sec x + \tan x|$

EXERCISES 4.7 (PAGE 199)

1. $y = c_1 x^{-1} + c_2 x^2$ **3.** $y = c_1 + c_2 \ln x$

5. $y = c_1 \cos(2 \ln x) + c_2 \sin(2 \ln x)$

7. $y = c_1 x^{(2-\sqrt{6})} + c_2 x^{(2+\sqrt{6})}$

9. $y = c_1 \cos(\frac{1}{5} \ln x) + c_2 \sin(\frac{1}{5} \ln x)$

11. $y = c_1 x^{-2} + c_2 x^{-2} \ln x$

13. $y = x^{-1/2}\left[c_1 \cos\left(\dfrac{\sqrt{3}}{6} \ln x\right) + c_2 \sin\left(\dfrac{\sqrt{3}}{6} \ln x\right)\right]$

15. $y = c_1 x^3 + c_2 \cos(\sqrt{2} \ln x) + c_3 \sin(\sqrt{2} \ln x)$

17. $y = c_1 + c_2 x + c_3 x^2 + c_4 x^{-3}$

19. $y = c_1 + c_2 x^5 + \frac{1}{5} x^5 \ln x$

21. $y = c_1 x + c_2 x \ln x + x(\ln x)^2$ **23.** $y = 2 - 2x^{-2}$

25. $y = \cos(\ln x) + 2 \sin(\ln x)$ **27.** $y = \frac{3}{4} - \ln x + \frac{1}{4} x^2$

29. $y = c_1 x^{-1} + c_2 x^{-8} + \frac{1}{30} x^2$

31. $y = x^2[c_1 \cos(3 \ln x) + c_2 \sin(3 \ln x)] + \frac{4}{13} + \frac{3}{10} x$

39. $y = 2(-x)^{1/2} - 5(-x)^{1/2} \ln(-x), x < 0$

EXERCISES 4.8 (PAGE 205)

1. $x = c_1 e^t + c_2 t e^t$
$\qquad y = (c_1 - c_2) e^t + c_2 t e^t$

3. $x = c_1 \cos t + c_2 \sin t + t + 1$
$\qquad y = c_1 \sin t - c_2 \cos t + t - 1$

5. $x = \frac{1}{2}c_1 \sin t + \frac{1}{2}c_2 \cos t - 2c_3 \sin \sqrt{6}t - 2c_4 \cos \sqrt{6}t$
$y = c_1 \sin t + c_2 \cos t + c_3 \sin \sqrt{6}t + c_4 \cos \sqrt{6}t$

7. $x = c_1 e^{2t} + c_2 e^{-2t} + c_3 \sin 2t + c_4 \cos 2t + \frac{1}{5}e^t$
$y = c_1 e^{2t} + c_2 e^{-2t} - c_3 \sin 2t - c_4 \cos 2t - \frac{1}{5}e^t$

9. $x = c_1 - c_2 \cos t + c_3 \sin t + \frac{17}{15}e^{3t}$
$y = c_1 + c_2 \sin t + c_3 \cos t - \frac{4}{15}e^{3t}$

11. $x = c_1 e^t + c_2 e^{-t/2} \cos \frac{\sqrt{3}}{2}t + c_3 e^{-t/2} \sin \frac{\sqrt{3}}{2}t$

$y = \left(-\frac{3}{2}c_2 - \frac{\sqrt{3}}{2}c_3\right)e^{-t/2}\cos\frac{\sqrt{3}}{2}t$
$\quad + \left(\frac{\sqrt{3}}{2}c_2 - \frac{3}{2}c_3\right)e^{-t/2}\sin\frac{\sqrt{3}}{2}t$

13. $x = c_1 e^{4t} + \frac{4}{3}e^t$
$y = -\frac{3}{4}c_1 e^{4t} + c_2 + 5e^t$

15. $x = c_1 + c_2 t + c_3 e^t + c_4 e^{-t} - \frac{1}{2}t^2$
$y = (c_1 - c_2 + 2) + (c_2 + 1)t + c_4 e^{-t} - \frac{1}{2}t^2$

17. $x = c_1 e^t + c_2 e^{-t/2} \sin \frac{\sqrt{3}}{2}t + c_3 e^{-t/2} \cos \frac{\sqrt{3}}{2}t$

$y = c_1 e^t + \left(-\frac{1}{2}c_2 - \frac{\sqrt{3}}{2}c_3\right)e^{-t/2}\sin\frac{\sqrt{3}}{2}t$
$\quad + \left(\frac{\sqrt{3}}{2}c_2 - \frac{1}{2}c_3\right)e^{-t/2}\cos\frac{\sqrt{3}}{2}t$

$z = c_1 e^t + \left(-\frac{1}{2}c_2 + \frac{\sqrt{3}}{2}c_3\right)e^{-t/2}\sin\frac{\sqrt{3}}{2}t$
$\quad + \left(-\frac{\sqrt{3}}{2}c_2 - \frac{1}{2}c_3\right)e^{-t/2}\cos\frac{\sqrt{3}}{2}t$

19. $x = -6c_1 e^{-t} - 3c_2 e^{-2t} + 2c_3 e^{3t}$
$y = c_1 e^{-t} + c_2 e^{-2t} + c_3 e^{3t}$
$z = 5c_1 e^{-t} + c_2 e^{-2t} + c_3 e^{3t}$

21. $x = e^{-3t+3} - te^{-3t+3}$
$y = -e^{-3t+3} + 2te^{-3t+3}$

23. $m\dfrac{d^2x}{dt^2} = 0$

$m\dfrac{d^2y}{dt^2} = -mg$

$x = c_1 t + c_2$
$y = -\frac{1}{2}gt^2 + c_3 t + c_4$

EXERCISES 4.9 (PAGE 211)

3. $y = \ln|\cos(c_1 - x)| + c_2$

5. $y = \dfrac{1}{c_1^2}\ln|c_1 x + 1| - \dfrac{1}{c_1}x + c_2$

7. $\frac{1}{3}y^3 - c_1 y = x + c_2$

9. $y = \tan\left(\dfrac{\pi}{4} - \dfrac{x}{2}\right), -\dfrac{\pi}{2} < x < \dfrac{3\pi}{2}$

11. $y = -\dfrac{1}{c_1}\sqrt{1 - c_1^2 x^2} + c_2$

13. $y = 1 + x + \frac{1}{2}x^2 + \frac{1}{2}x^3 + \frac{1}{6}x^4 + \frac{1}{10}x^5 + \cdots$

15. $y = 1 + x - \frac{1}{2}x^2 + \frac{2}{3}x^3 - \frac{1}{4}x^4 + \frac{7}{60}x^5 + \cdots$

17. $y = -\sqrt{1-x^2}$

CHAPTER 4 IN REVIEW (PAGE 212)

1. $y = 0$ **3.** false

5. $(-\infty, 0); (0, \infty)$

7. $y = c_1 e^{3x} + c_2 e^{-5x} + c_3 xe^{-5x} + c_4 e^x + c_5 xe^x + c_6 x^2 e^x$;
$y = c_1 x^3 + c_2 x^{-5} + c_3 x^{-5}\ln x + c_4 x + c_5 x \ln x + c_6 x(\ln x)^2$

9. $y = c_1 e^{(1+\sqrt{3})x} + c_2 e^{(1-\sqrt{3})x}$ **11.** $y = c_1 + c_2 e^{-5x} + c_3 xe^{-5x}$

13. $y = c_1 e^{-x/3} + e^{-3x/2}\left(c_2 \cos\dfrac{\sqrt{7}}{2}x + c_3 \sin\dfrac{\sqrt{7}}{2}x\right)$

15. $y = e^{3x/2}\left(c_2 \cos\dfrac{\sqrt{11}}{2}x + c_3 \sin\dfrac{\sqrt{11}}{2}x\right)$

$\quad + \dfrac{4}{5}x^3 + \dfrac{36}{25}x^2 + \dfrac{46}{125}x - \dfrac{222}{625}$

17. $y = c_1 + c_2 e^{2x} + c_3 e^{3x} + \frac{1}{5}\sin x - \frac{1}{5}\cos x + \frac{4}{3}x$

19. $y = e^x(c_1 \cos x + c_2 \sin x) - e^x \cos x \ln|\sec x + \tan x|$

21. $y = c_1 x^{-1/3} + c_2 x^{1/2}$ **23.** $y = c_1 x^2 + c_2 x^3 + x^4 - x^2 \ln x$

25. (a) $y = c_1 \cos \omega x + c_2 \sin \omega x + A \cos \alpha x + B \sin \alpha x,$
$\quad \omega \neq \alpha;$
$\quad y = c_1 \cos \omega x + c_2 \sin \omega x + Ax \cos \omega x + Bx \sin \omega x,$
$\quad \omega = \alpha$

(b) $y = c_1 e^{-\omega x} + c_2 e^{\omega x} + Ae^{\alpha x}, \omega \neq \alpha;$
$\quad y = c_1 e^{-\omega x} + c_2 e^{\omega x} + Axe^{\omega x}, \omega = \alpha$

27. (a) $y = c_1 \cosh x + c_2 \sinh x + c_3 x \cosh x + c_4 x \sinh x$
(b) $y_p = Ax^2 \cosh x + Bx^2 \sinh x$

29. $y = e^{x-\pi}\cos x$ **31.** $y = \frac{13}{4}e^x - \frac{5}{4}e^{-x} - x - \frac{1}{2}\sin x$

33. $y = x^2 + 4$

37. $x = -c_1 e^t - \frac{3}{2}c_2 e^{2t} + \frac{5}{2}$
$y = c_1 e^t + c_2 e^{2t} - 3$

39. $x = c_1 e^t + c_2 e^{5t} + te^t$
$y = -c_1 e^t + 3c_2 e^{5t} - te^t + 2e^t$

EXERCISES 5.1 (PAGE 229)

1. $\dfrac{\sqrt{2}\pi}{8}$ **3.** $x(t) = -\frac{1}{4}\cos 4\sqrt{6}\,t$

5. (a) $x\left(\dfrac{\pi}{12}\right) = -\dfrac{1}{4}; x\left(\dfrac{\pi}{8}\right) = -\dfrac{1}{2}; x\left(\dfrac{\pi}{6}\right) = -\dfrac{1}{4};$

$\quad x\left(\dfrac{\pi}{4}\right) = \dfrac{1}{2}; x\left(\dfrac{9\pi}{32}\right) = \dfrac{\sqrt{2}}{4}$

(b) 4 ft/s; downward

(c) $t = \dfrac{(2n+1)\pi}{16}, n = 0, 1, 2, \ldots$

7. (a) the 20-kg mass
(b) the 20-kg mass; the 50-kg mass
(c) $t = n\pi, n = 0, 1, 2, \ldots$; at the equilibrium position; the 50-kg mass is moving upward whereas the 20-kg mass is moving upward when n is even and downward when n is odd.

9. $x(t) = \dfrac{1}{2}\cos 2t + \dfrac{3}{4}\sin 2t = \dfrac{\sqrt{13}}{4}\sin(2t + 0.5880)$

11. (a) $x(t) = -\frac{2}{3}\cos 10t + \frac{1}{2}\sin 10t = \frac{5}{6}\sin(10t - 0.927)$

(b) $\dfrac{5}{6}$ ft; $\dfrac{\pi}{5}$

(c) 15 cycles

(d) 0.721 s

(e) $\dfrac{(2n+1)\pi}{20} + 0.0927, n = 0, 1, 2, \ldots$

(f) $x(3) = -0.597$ ft

(g) $x'(3) = -5.814$ ft/s

(h) $x''(3) = 59.702$ ft/s^2

(i) $\pm 8\frac{1}{3}$ ft/s

(j) $0.1451 + \dfrac{n\pi}{5}; 0.3545 + \dfrac{n\pi}{5}, n = 0, 1, 2, \ldots$

(k) $0.3545 + \dfrac{n\pi}{5}, n = 0, 1, 2, \ldots$

13. 120 lb/ft; $x(t) = \dfrac{\sqrt{3}}{12} \sin 8\sqrt{3}\,t$

17. (a) above **(b)** heading upward

19. (a) below **(b)** heading upward

21. $\frac{1}{4}$ s; $\frac{1}{2}$ s, $x(\frac{1}{2}) = e^{-2}$; that is, the weight is approximately 0.14 ft below the equilibrium position.

23. (a) $x(t) = \frac{4}{3}e^{-2t} - \frac{1}{3}e^{-8t}$
   (b) $x(t) = -\frac{2}{3}e^{-2t} + \frac{5}{3}e^{-8t}$

25. (a) $x(t) = e^{-2t}(-\cos 4t - \frac{1}{2}\sin 4t)$
   (b) $x(t) = \dfrac{\sqrt{5}}{2} e^{-2t} \sin(4t + 4.249)$
   (c) $t = 1.294$ s

27. (a) $\beta > \frac{5}{2}$ **(b)** $\beta = \frac{5}{2}$ **(c)** $0 < \beta < \frac{5}{2}$

29. $x(t) = e^{-t/2}\left(-\dfrac{4}{3}\cos\dfrac{\sqrt{47}}{2}t - \dfrac{64}{3\sqrt{47}}\sin\dfrac{\sqrt{47}}{2}t\right)$
   $+ \dfrac{10}{3}(\cos 3t + \sin 3t)$

31. $x(t) = \frac{1}{4}e^{-4t} + te^{-4t} - \frac{1}{4}\cos 4t$

33. $x(t) = -\frac{1}{2}\cos 4t + \frac{9}{4}\sin 4t + \frac{1}{2}e^{-2t}\cos 4t$
   $- 2e^{-2t}\sin 4t$

35. (a) $m\dfrac{d^2x}{dt^2} = -k(x-h) - \beta\dfrac{dx}{dt}$ or
   $\dfrac{d^2x}{dt^2} + 2\lambda\dfrac{dx}{dt} + \omega^2 x = \omega^2 h(t)$,
   where $2\lambda = \beta/m$ and $\omega^2 = k/m$
   (b) $x(t) = e^{-2t}(-\frac{56}{13}\cos 2t - \frac{72}{13}\sin 2t) + \frac{56}{13}\cos t + \frac{32}{13}\sin t$

37. $x(t) = -\cos 2t - \frac{1}{8}\sin 2t + \frac{3}{4}t\sin 2t + \frac{5}{4}t\cos 2t$

39. (b) $\dfrac{F_0}{2\omega}t\sin\omega t$

45. 4.568 C; 0.0509 s

47. $q(t) = 10 - 10e^{-3t}(\cos 3t + \sin 3t)$
   $i(t) = 60e^{-3t}\sin 3t$; 10.432 C

49. $q_p = \frac{100}{13}\sin t + \frac{150}{13}\cos t$
   $i_p = \frac{100}{13}\cos t - \frac{150}{13}\sin t$

53. $q(t) = -\frac{1}{2}e^{-10t}(\cos 10t + \sin 10t) + \frac{3}{2}; \frac{3}{2}$ C

57. $q(t) = \left(q_0 - \dfrac{E_0 C}{1 - \gamma^2 LC}\right)\cos\dfrac{t}{\sqrt{LC}}$
   $+ \sqrt{LC}\,i_0\sin\dfrac{t}{\sqrt{LC}} + \dfrac{E_0 C}{1 - \gamma^2 LC}\cos\gamma t$

   $i(t) = i_0\cos\dfrac{t}{\sqrt{LC}}$
   $- \dfrac{1}{\sqrt{LC}}\left(q_0 - \dfrac{E_0 C}{1 - \gamma^2 LC}\right)\sin\dfrac{t}{\sqrt{LC}}$
   $- \dfrac{E_0 C\gamma}{1 - \gamma^2 LC}\sin\gamma t$

EXERCISES 5.2 (PAGE 243)

1. (a) $y(x) = \dfrac{w_0}{24EI}(6L^2x^2 - 4Lx^3 + x^4)$

3. (a) $y(x) = \dfrac{w_0}{48EI}(3L^2x^2 - 5Lx^3 + 2x^4)$

5. (a) $y(x) = \dfrac{w_0}{360EI}(7L^4x - 10L^2x^3 + 3x^5)$
   (c) $x \approx 0.51933$, $y_{max} \approx 0.234799$

7. $y(x) = -\dfrac{w_0 EI}{P^2}\cosh\sqrt{\dfrac{P}{EI}}\,x$
   $+ \left(\dfrac{w_0 EI}{P^2}\sinh\sqrt{\dfrac{P}{EI}}\,L - \dfrac{w_0 L\sqrt{EI}}{P\sqrt{P}}\right)\dfrac{\sinh\sqrt{\dfrac{P}{EI}}\,x}{\cosh\sqrt{\dfrac{P}{EI}}\,L}$
   $+ \dfrac{w_0}{2P}x^2 + \dfrac{w_0 EI}{P^2}$

9. $\lambda = n^2$, $n = 1, 2, 3, \ldots$; $y = \sin nx$

11. $\lambda = \dfrac{(2n-1)^2\pi^2}{4L^2}$, $n = 1, 2, 3, \ldots$; $y = \cos\dfrac{(2n-1)\pi x}{2L}$

13. $\lambda = n^2$, $n = 0, 1, 2, \ldots$; $y = \cos nx$

15. $\lambda = \dfrac{n^2\pi^2}{25}$, $n = 1, 2, 3, \ldots$; $y = e^{-x}\sin\dfrac{n\pi x}{5}$

17. $\lambda = \dfrac{n\pi}{L}$, $n = 1, 2, 3, \ldots$; $y = \sin\dfrac{n\pi x}{L}$

19. $\lambda = n^2$, $n = 1, 2, 3, \ldots$; $y = \sin(n\ln x)$

21. $\lambda = 0$; $y = 1$
   $\lambda = \dfrac{n^2\pi^2}{4}$, $n = 1, 2, 3, \ldots$; $y = \cos\left(\dfrac{n\pi}{2}\ln x\right)$

23. $x = L/4$, $x = L/2$, $x = 3L/4$

25. $\omega_n = \dfrac{n\pi\sqrt{T}}{L\sqrt{\rho}}$, $n = 1, 2, 3, \ldots$; $y = \sin\dfrac{n\pi x}{L}$

27. $u(r) = \left(\dfrac{u_0 - u_1}{b - a}\right)\dfrac{ab}{r} + \dfrac{u_1 b - u_0 a}{b - a}$

EXERCISES 5.3 (PAGE 254)

7. $\dfrac{d^2x}{dt^2} + x = 0$

15. (a) 5 ft
   (b) $4\sqrt{10}$ ft/s
   (c) $0 \le t \le 15/4\sqrt{10}$; 7.5 ft

17. (a) $xv\dfrac{dv}{dx} + v^2 = 32x$
   (b) $\frac{1}{2}v^2x^2 - \frac{32}{3}x^3 = -288$ **(c)** approximately 0.66 s

19. (a) $xy'' = r\sqrt{1 + (y')^2}$.
   When $t = 0$, $x = a$, $y = 0$, $dy/dx = 0$.
   (b) When $r \ne 1$,
   $$y(x) = \dfrac{a}{2}\left[\dfrac{1}{1+r}\left(\dfrac{x}{a}\right)^{1+r} - \dfrac{1}{1-r}\left(\dfrac{x}{a}\right)^{1-r}\right] + \dfrac{ar}{1 - r^2}.$$
   When $r = 1$,
   $$y(x) = \dfrac{1}{2}\left[\dfrac{1}{2a}(x^2 - a^2) + \dfrac{1}{a}\ln\dfrac{a}{x}\right].$$
   (c) The paths intersect when $r < 1$.

CHAPTER 5 IN REVIEW (PAGE 259)

1. 8 ft **3.** $\frac{5}{4}$ m

5. False; there could be an impressed force driving the system.

7. overdamped **9.** 14.4 lb

11. $x(t) = -\frac{2}{3}e^{-2t} + \frac{1}{3}e^{-4t}$ **13.** $0 < m \le 2$

15. $\gamma = \dfrac{8\sqrt{3}}{3}$

17. $x(t) = e^{-4t}\left(\dfrac{26}{17}\cos 2\sqrt{2}t + \dfrac{28\sqrt{2}}{17}\sin 2\sqrt{2}t\right) + \dfrac{8}{17}e^{-t}$

19. (a) $q(t) = -\frac{1}{150}\sin 100t + \frac{1}{75}\sin 50t$

  (b) $i(t) = -\frac{2}{3}\cos 100t + \frac{2}{3}\cos 50t$

  (c) $t = \dfrac{n\pi}{50}, n = 0, 1, 2, \ldots$

EXERCISES 6.1 (PAGE 278)

1. $[-\frac{1}{2}, \frac{1}{2})$ **3.** $(-5, 15)$

5. $x - \frac{2}{3}x^3 + \frac{2}{15}x^5 - \frac{4}{315}x^7 + \cdots$

7. $1 + \frac{1}{2}x^2 + \frac{5}{24}x^4 + \frac{61}{720}x^6 + \cdots, (-\pi/2, \pi/2)$

9. $2c_1 + \sum_{k=1}^{\infty}[2(k+1)c_{k+1} + 6c_{k-1}]x^k$

13. $y_1(x) = c_0\left[1 + \dfrac{1}{3\cdot 2}x^3 + \dfrac{1}{6\cdot 5\cdot 3\cdot 2}x^6\right.$

$\left. + \dfrac{1}{9\cdot 8\cdot 6\cdot 5\cdot 3\cdot 2}x^9 + \cdots\right]$

$y_2(x) = c_1\left[x + \dfrac{1}{4\cdot 3}x^4 + \dfrac{1}{7\cdot 6\cdot 4\cdot 3}x^7\right.$

$\left. + \dfrac{1}{10\cdot 9\cdot 7\cdot 6\cdot 4\cdot 3}x^{10} + \cdots\right]$

15. $y_1(x) = c_0\left[1 - \dfrac{1}{2!}x^2 - \dfrac{3}{4!}x^4 - \dfrac{21}{6!}x^6 - \cdots\right]$

$y_2(x) = c_1\left[x + \dfrac{1}{3!}x^3 + \dfrac{5}{5!}x^5 + \dfrac{45}{7!}x^7 + \cdots\right]$

17. $y_1(x) = c_0\left[1 - \dfrac{1}{3!}x^3 + \dfrac{4^2}{6!}x^6 - \dfrac{7^2\cdot 4^2}{9!}x^9 + \cdots\right]$

$y_2(x) = c_1\left[x - \dfrac{2^2}{4!}x^4 + \dfrac{5^2\cdot 2^2}{7!}x^7 - \dfrac{8^2\cdot 5^2\cdot 2^2}{10!}x^{10} + \cdots\right]$

19. $y_1(x) = c_0; \; y_2(x) = c_1\sum_{n=1}^{\infty}\dfrac{1}{n}x^n$

21. $y_1(x) = c_0[1 + \frac{1}{2}x^2 + \frac{1}{6}x^3 + \frac{1}{6}x^4 + \cdots]$

$y_2(x) = c_1[x + \frac{1}{2}x^2 + \frac{1}{2}x^3 + \frac{1}{4}x^4 + \cdots]$

23. $y_1(x) = c_0\left[1 + \dfrac{1}{4}x^2 - \dfrac{7}{4\cdot 4!}x^4 + \dfrac{23\cdot 7}{8\cdot 6!}x^6 - \cdots\right]$

$y_2(x) = c_1\left[x - \dfrac{1}{6}x^3 + \dfrac{14}{2\cdot 5!}x^5 - \dfrac{34\cdot 14}{4\cdot 7!}x^7 - \cdots\right]$

25. $y(x) = -2\left[1 + \dfrac{1}{2!}x^2 + \dfrac{1}{3!}x^3 + \dfrac{1}{4!}x^4 + \cdots\right] + 6x$

$= 8x - 2e^x$

27. $y(x) = 3 - 12x^2 + 4x^4$

29. $y_1(x) = c_0[1 - \frac{1}{6}x^3 + \frac{1}{120}x^5 + \cdots]$

$y_2(x) = c_1[x - \frac{1}{12}x^4 + \frac{1}{180}x^6 + \cdots]$

EXERCISES 6.2 (PAGE 289)

1. $x = 0$, irregular singular point

3. $x = -3$, regular singular point; $x = 3$, irregular singular point

5. $x = 0, 2i, -2i$, regular singular points

7. $x = -3, 2$, regular singular points

9. $x = 0$, irregular singular point; $x = -5, 5, 2$, regular singular points

11. for $x = 1$: $p(x) = 5$, $q(x) = \dfrac{x(x-1)^2}{x+1}$

  for $x = -1$: $p(x) = \dfrac{5(x+1)}{x-1}$, $q(x) = x^2 + x$

13. $r_1 = \frac{1}{3}, r_2 = -1$

15. $r_1 = \frac{3}{2}, r_2 = 0$

$y(x) = C_1 x^{3/2}\left[1 - \dfrac{2}{5}x + \dfrac{2^2}{7\cdot 5\cdot 2}x^2\right.$

$\left. - \dfrac{2^3}{9\cdot 7\cdot 5\cdot 3!}x^3 + \cdots\right]$

$+ C_2\left[1 + 2x - 2x^2 + \dfrac{2^3}{3\cdot 3!}x^3 - \cdots\right]$

17. $r_1 = \frac{7}{8}, r_2 = 0$

$y(x) = C_1 x^{7/8}\left[1 - \dfrac{2}{15}x + \dfrac{2^2}{23\cdot 15\cdot 2}x^2\right.$

$\left. - \dfrac{2^3}{31\cdot 23\cdot 15\cdot 3!}x^3 + \cdots\right]$

$+ C_2\left[1 - 2x + \dfrac{2^2}{9\cdot 2}x^2 - \dfrac{2^3}{17\cdot 9\cdot 3!}x^3 + \cdots\right]$

19. $r_1 = \frac{1}{3}, r_2 = 0$

$y(x) = C_1 x^{1/3}\left[1 + \dfrac{1}{3}x + \dfrac{1}{3^2\cdot 2}x^2 + \dfrac{1}{3^3\cdot 3!}x^3 + \cdots\right]$

$+ C_2\left[1 + \dfrac{1}{2}x + \dfrac{1}{5\cdot 2}x^2 + \dfrac{1}{8\cdot 5\cdot 2}x^3 + \cdots\right]$

21. $r_1 = \frac{5}{2}, r_2 = 0$

$y(x) = C_1 x^{5/2}\left[1 + \dfrac{2\cdot 2}{7}x + \dfrac{2^2\cdot 3}{9\cdot 7}x^2\right.$

$\left. + \dfrac{2^3\cdot 4}{11\cdot 9\cdot 7}x^3 + \cdots\right]$

$+ C_2\left[1 + \dfrac{1}{3}x - \dfrac{1}{6}x^2 - \dfrac{1}{6}x^3 - \cdots\right]$

23. $r_1 = \frac{2}{3}, r_2 = \frac{1}{3}$

$y(x) = C_1 x^{2/3}[1 - \frac{1}{2}x + \frac{5}{28}x^2 - \frac{1}{21}x^3 + \cdots]$

$+ c_2 x^{1/3}[1 - \frac{1}{2}x + \frac{1}{5}x^2 - \frac{7}{120}x^3 + \cdots]$

25. $r_1 = 0, r_2 = -1$

$y(x) = C_1\sum_{n=0}^{\infty}\dfrac{1}{(2n+1)!}x^{2n} + C_2 x^{-1}\sum_{n=0}^{\infty}\dfrac{1}{(2n)!}x^{2n}$

$= C_1 x^{-1}\sum_{n=0}^{\infty}\dfrac{1}{(2n+1)!}x^{2n+1} + C_2 x^{-1}\sum_{n=0}^{\infty}\dfrac{1}{(2n)!}x^{2n}$

$= \dfrac{1}{x}[C_1\sinh x + C_2\cosh x]$

27. $r_1 = 1, r_2 = 0$

$y(x) = C_1 x + C_2[x\ln x - 1 + \frac{1}{2}x^2 + \frac{1}{12}x^3 + \frac{1}{72}x^4 + \cdots]$

29. $r_1 = r_2 = 0$

$y(x) = C_1 y(x)$

$+ C_2\left[y_1(x)\ln x + y_1(x)\left(-x + \dfrac{1}{4}x^2 - \dfrac{1}{3\cdot 3!}x^3 + \dfrac{1}{4\cdot 4!}x^4 - \cdots\right)\right]$,

where $y_1(x) = \sum_{n=0}^{\infty}\dfrac{1}{n!}x^n = e^x$

33. (b) $y_1(t) = \sum_{n=0}^{\infty} \frac{(-1)^n}{(2n+1)!} (\sqrt{\lambda} t)^{2n} = \frac{\sin(\sqrt{\lambda} t)}{\sqrt{\lambda} t}$

$y_2(t) = t^{-1} \sum_{n=0}^{\infty} \frac{(-1)^n}{(2n)!} (\sqrt{\lambda} t)^{2n} = \frac{\cos(\sqrt{\lambda} t)}{t}$

(c) $y = c_1 x \sin\left(\frac{\sqrt{\lambda} x}{x}\right) + c_2 x \cos\left(\frac{\sqrt{\lambda} x}{x}\right)$

EXERCISES 6.3 (PAGE 301)

1. $y = c_1 J_{1/3}(x) + c_2 J_{-1/3}(x)$ **3.** $y = c_1 J_{5/2}(x) + c_2 J_{-5/2}(x)$
5. $y = c_1 J_0(x) + c_2 Y_0(x)$ **7.** $y = c_1 J_2(3x) + c_2 Y_2(3x)$
9. $y = c_1 x^{-1/2} J_{1/2}(\lambda x) + c_2 x^{-1/2} J_{-1/2}(\lambda x)$
13. From Problem 10, $y = x^{1/2} J_{1/2}(x)$; from Problem 11,
$y = x^{1/2} J_{-1/2}(x)$.
15. From Problem 10, $y = x^{-1} J_{-1}(x)$; from Problem 11,
$y = x^{-1} J_1(x)$. Since $J_{-1}(x) = -J_1(x)$, no new solution
results.
17. From Problem 12 and $\lambda = 1$ and $\nu = \pm\frac{3}{2}$,
$y = \sqrt{x} J_{3/2}(x)$ and $y = \sqrt{x} J_{-3/2}(x)$.

23. $J_{-1/2}(x) = \sqrt{\frac{2}{\pi x}} \cos x$

27. $y = c_1 x^{1/2} J_{1/3}(\frac{2}{3}\alpha x^{3/2}) + c_2 x^{1/2} J_{-1/3}(\frac{2}{3}\alpha x^{3/2})$
29. (a) $P_6(x) = \frac{1}{16}(231x^6 - 315x^4 + 105x^2 - 5)$
$P_7(x) = \frac{1}{16}(429x^7 - 693x^5 + 315x^3 - 35x)$
(b) $P_6(x)$ satisfies $(1 - x^2)y'' - 2xy' + 42y = 0$.
$P_7(x)$ satisfies $(1 - x^2)y'' - 2xy' + 56y = 0$.

CHAPTER 6 IN REVIEW (PAGE 304)

3. $R = \sqrt{10}$
7. $r_1 = \frac{1}{2}, r_2 = 0$
$y_1(x) = C_1 x^{1/2}[1 - \frac{1}{3}x + \frac{1}{30}x^2 - \frac{1}{630}x^3 + \cdots]$
$y_2(x) = C_2[1 - x + \frac{1}{6}x^2 - \frac{1}{90}x^3 + \cdots]$
9. $y_1(x) = c_0[1 + \frac{3}{2}x^2 + \frac{1}{2}x^3 + \frac{5}{8}x^4 + \cdots]$
$y_2(x) = c_1[x + \frac{1}{2}x^3 + \frac{1}{4}x^4 + \cdots]$
11. $r_1 = 3, r_2 = 0$
$y_1(x) = C_1 x^3[1 + \frac{1}{4}x + \frac{1}{20}x^2 + \frac{1}{120}x^3 + \cdots]$
$y_2(x) = C_2[1 + x + \frac{1}{2}x^2]$
13. $y(x) = 3[1 - x^2 + \frac{1}{3}x^4 - \frac{1}{15}x^6 + \cdots]$
$- 2[x - \frac{1}{2}x^3 + \frac{1}{8}x^5 - \frac{1}{48}x^7 + \cdots]$
15. $x = 0$ is an ordinary point
17. $y(x) = c_0\left[1 - \frac{1}{3}x^3 + \frac{1}{3^2 \cdot 2!}x^6 - \frac{1}{3^3 \cdot 3!}x^9 + \cdots\right]$

$+ c_1\left[x - \frac{1}{4}x^4 + \frac{1}{4 \cdot 7}x^7 - \frac{1}{4 \cdot 7 \cdot 10}x^{10} + \cdots\right]$

$+ \left[\frac{5}{2}x^2 - \frac{1}{3}x^3 + \frac{1}{3^2 \cdot 2!}x^6 - \frac{1}{3^3 \cdot 3!}x^9 + \cdots\right]$

EXERCISES 7.1 (PAGE 312)

1. $\frac{2}{s}e^{-s} - \frac{1}{s}$ **3.** $\frac{1}{s^2} - \frac{1}{s^2}e^{-s}$ **5.** $\frac{1 + e^{-s\pi}}{s^2 + 1}$

7. $\frac{e^{-s}}{s} + \frac{e^{-s}}{s^2}$ **9.** $\frac{1}{s} - \frac{1}{s^2} + \frac{e^{-s}}{s^2}$ **11.** $\frac{e^7}{s - 1}$

13. $\frac{1}{(s - 4)^2}$ **15.** $\frac{1}{s^2 + 2s + 2}$ **17.** $\frac{s^2 - 1}{(s^2 + 1)^2}$

19. $\frac{48}{s^5}$ **21.** $\frac{4}{s^2} - \frac{10}{s}$ **23.** $\frac{2}{s^3} + \frac{6}{s^2} - \frac{3}{s}$

25. $\frac{6}{s^4} + \frac{6}{s^3} + \frac{3}{s^2} + \frac{1}{s}$ **27.** $\frac{1}{s} + \frac{1}{s - 4}$

29. $\frac{1}{s} + \frac{2}{s - 2} + \frac{1}{s - 4}$ **31.** $\frac{8}{s^3} - \frac{15}{s^2 + 9}$

33. Use $\sinh kt = \frac{e^{kt} - e^{-kt}}{2}$ to show that

$$\mathcal{L}\{\sinh kt\} = \frac{k}{s^2 - k^2}.$$

35. $\frac{1}{2(s - 2)} - \frac{1}{2s}$ **37.** $\frac{2}{s^2 + 16}$

EXERCISES 7.2 (PAGE 322)

1. $\frac{1}{2}t^2$ **3.** $t - 2t^4$ **5.** $1 + 3t + \frac{3}{2}t^2 + \frac{1}{6}t^3$
7. $t - 1 + e^{2t}$ **9.** $\frac{1}{4}e^{-t/4}$ **11.** $\frac{5}{7}\sin 7t$

13. $\cos\frac{t}{2}$ **15.** $2\cos 3t - 2\sin 3t$ **17.** $\frac{1}{3} - \frac{1}{3}e^{-3t}$

19. $\frac{3}{4}e^{-3t} + \frac{1}{4}e^t$ **21.** $0.3e^{0.1t} + 0.6e^{-0.2t}$
23. $\frac{1}{2}e^{2t} - e^{3t} + \frac{1}{2}e^{6t}$ **25.** $\frac{1}{5} - \frac{1}{5}\cos\sqrt{5}t$
27. $-4 + 3e^{-t} + \cos t + 3\sin t$
29. $\frac{1}{3}\sin t - \frac{1}{6}\sin 2t$ **31.** $y = -1 + e^t$ **33.** $y = \frac{1}{10}e^{4t} + \frac{19}{10}e^{-6t}$
35. $y = \frac{4}{3}e^{-t} - \frac{1}{3}e^{-4t}$
37. $y = 10\cos t + 2\sin t - \sqrt{2}\sin\sqrt{2}t$
39. $y = -\frac{8}{9}e^{-t/2} + \frac{1}{9}e^{-2t} + \frac{5}{18}e^t + \frac{1}{2}e^{-t}$

EXERCISES 7.3 (PAGE 333)

1. $\frac{1}{(s - 10)^2}$ **3.** $\frac{6}{(s + 2)^4}$

5. $\frac{1}{(s - 2)^2} + \frac{2}{(s - 3)^2} + \frac{1}{(s - 4)^2}$ **7.** $\frac{3}{(s - 1)^2 + 9}$

9. $\frac{s}{s^2 + 25} - \frac{s - 1}{(s - 1)^2 + 25} + 3\frac{s + 4}{(s - 1)^2 + 25}$ **11.** $\frac{1}{2}t^2 e^{-t}$

13. $e^{3t}\sin t$ **15.** $e^{-2t}\cos t - 2e^{-2t}\sin t$
17. $e^{-t} - te^{-t}$ **19.** $5 - t - 5e^{-t} - 4te^{-t} - \frac{3}{2}t^2 e^{-t}$
21. $y = te^{-4t} + 2e^{-4t}$ **23.** $y = e^{-t} + 2te^{-t}$
25. $y = \frac{1}{9}t + \frac{2}{27} - \frac{2}{27}e^{3t} + \frac{10}{9}te^{3t}$ **27.** $y = -\frac{3}{2}e^{3t}\sin 2t$
29. $y = \frac{1}{2} - \frac{1}{2}e^t\cos t + \frac{1}{2}e^t\sin t$
31. $y = (e + 1)te^{-t} + (e - 1)e^{-t}$
33. $x(t) = -\frac{3}{2}e^{-7t/2}\cos\frac{\sqrt{15}}{2}t - \frac{7\sqrt{15}}{10}e^{-7t/2}\sin\frac{\sqrt{15}}{2}t$

37. $\frac{e^{-s}}{s^2}$ **39.** $\frac{e^{-2s}}{s^2} + 2\frac{e^{-2s}}{s}$ **41.** $\frac{s}{s^2 + 4}e^{-\pi s}$
43. $\frac{1}{2}(t - 2)^2\mathcal{U}(t - 2)$ **45.** $-\sin t\,\mathcal{U}(t - \pi)$
47. $\mathcal{U}(t - 1) - e^{-(t-1)}\mathcal{U}(t - 1)$ **49. (c)**
51. (f) **53. (a)**

55. $f(t) = 2 - 4\mathcal{U}(t - 3)$; $\mathcal{L}\{f(t)\} = \frac{2}{s} - \frac{4}{s}e^{-3s}$

57. $f(t) = t^2\mathcal{U}(t - 1)$; $\mathcal{L}\{f(t)\} = 2\frac{e^{-s}}{s^3} + 2\frac{e^{-s}}{s^2} + \frac{e^{-s}}{s}$

59. $f(t) = t - t\mathcal{U}(t - 2)$; $\mathcal{L}\{f(t)\} = \frac{1}{s^2} - \frac{e^{-2s}}{s^2} - 2\frac{e^{-2s}}{s}$

61. $f(t) = \mathcal{U}(t - a) - \mathcal{U}(t - b)$; $\mathcal{L}\{f(t)\} = \frac{e^{-as}}{s} - \frac{e^{-bs}}{s}$

63. $y = [5 - 5e^{-(t-1)}]\mathcal{U}(t - 1)$

65. $y = -\frac{1}{4} + \frac{1}{2}t + \frac{1}{4}e^{-2t} - \frac{1}{4}\mathcal{U}(t-1) - \frac{1}{2}(t-1)\mathcal{U}(t-1)$
$\qquad + \frac{1}{4}e^{-2(t-1)}\mathcal{U}(t-1)$

67. $y = \cos 2t - \frac{1}{6}\sin 2(t-2\pi)\mathcal{U}(t-2\pi)$
$\qquad + \frac{1}{3}\sin(t-2\pi)\mathcal{U}(t-2\pi)$

69. $y = \sin t + [1 - \cos(t-\pi)]\mathcal{U}(t-\pi)$
$\qquad - [1 - \cos(t-2\pi)]\mathcal{U}(t-2\pi)$

71. $x(t) = \frac{5}{4}t - \frac{5}{16}\sin 4t - \frac{5}{4}(t-5)\mathcal{U}(t-5)$
$\qquad + \frac{5}{16}\sin 4(t-5)\mathcal{U}(t-5)$
$\qquad - \frac{25}{4}\mathcal{U}(t-5) + \frac{25}{4}\cos 4(t-5)\mathcal{U}(t-5)$

73. $q(t) = \frac{2}{5}\mathcal{U}(t-3) - \frac{2}{5}e^{-5(t-3)}\mathcal{U}(t-3)$

75. (a) $i(t) = \frac{1}{101}e^{-10t} - \frac{1}{101}\cos t + \frac{10}{101}\sin t$
$\qquad - \frac{10}{101}e^{-10(t-3\pi/2)}\mathcal{U}\left(t - \frac{3\pi}{2}\right)$
$\qquad + \frac{10}{101}\cos\left(t - \frac{3\pi}{2}\right)\mathcal{U}\left(t - \frac{3\pi}{2}\right)$
$\qquad + \frac{1}{101}\sin\left(t - \frac{3\pi}{2}\right)$

(b) $i_{max} \approx 0.1$ at $t \approx 1.6$, $i_{min} \approx -0.1$ at $t \approx 4.7$

77. $y(x) = \frac{w_0 L^2}{16EI}x^2 - \frac{w_0 L}{12EI}x^3 + \frac{w_0}{24EI}x^4$
$\qquad - \frac{w_0}{24EI}\left(x - \frac{L}{2}\right)^4 \mathcal{U}\left(x - \frac{L}{2}\right)$

79. $y(x) = \frac{w_0 L^2}{48EI}x^2 - \frac{w_0 L}{24EI}x^3$
$\qquad + \frac{w_0}{60EI}\left[\frac{5L}{2}x^4 - x^5 + \left(x - \frac{L}{2}\right)^5 \mathcal{U}\left(x - \frac{L}{2}\right)\right]$

EXERCISES 7.4 (PAGE 347)

1. $\dfrac{s^4 - 4}{(s^2 + 4)^2}$ **3.** $\dfrac{6s^2 + 2}{(s^2 - 1)^3}$ **5.** $\dfrac{12s - 24}{[(s-2)^2 + 36]^2}$

7. $\dfrac{6}{s^5}$ **9.** $\dfrac{s-1}{(s+1)[(s-1)^2 + 1]}$ **11.** $\dfrac{1}{s(s-1)}$

13. $\dfrac{s+1}{s[(s+1)^2 + 1]}$ **15.** $\dfrac{1}{s^2(s-1)}$ **17.** $\dfrac{3s^2 + 1}{s^2(s^2 + 1)^2}$

19. (a) $e^t - 1$ **(b)** $e^t - t - 1$ **(c)** $e^t - \frac{1}{2}t^2 - t - 1$

21. $\dfrac{1 - e^{-as}}{s(1 + e^{-as})}$ **23.** $\dfrac{a}{s}\left(\dfrac{1}{bs} - \dfrac{1}{e^{bs} - 1}\right)$

25. $\dfrac{\coth(\pi s/2)}{s^2 + 1}$

27. $y = -\frac{1}{2}e^{-t} + \frac{1}{2}\cos t - \frac{1}{2}t\cos t + \frac{1}{2}t\sin t$
29. $y = 2\cos 3t + \frac{5}{3}\sin 3t + \frac{1}{6}t\sin 3t$
31. $y = \frac{1}{4}\sin 4t + \frac{1}{8}t\sin 4t - \frac{1}{8}(t-\pi)\sin 4(t-\pi)\mathcal{U}(t-\pi)$
33. $y = \frac{1}{2}\sin t - \frac{1}{2}t\cos t + \frac{1}{4}t\sin t - \frac{1}{4}t^2\cos t$
35. $f(t) = \sin t$ **37.** $f(t) = -\frac{1}{8}e^{-t} + \frac{1}{8}e^t + \frac{3}{4}te^t + \frac{1}{4}t^2 e^t$
39. $f(t) = e^{-t}$ **41.** $f(t) = \frac{3}{8}e^{2t} + \frac{1}{8}e^{-2t} + \frac{1}{2}\cos 2t + \frac{1}{4}\sin 2t$
43. $y = \sin t - \frac{1}{2}t\sin t$
45. $i(t) = 100[e^{-10(t-1)} - e^{-20(t-1)}]\mathcal{U}(t-1)$
$\qquad - 100[e^{-10(t-2)} - e^{-20(t-2)}]\mathcal{U}(t-2)$

47. $i(t) = \dfrac{1}{R}\left(1 - e^{-Rt/L}\right)$
$\qquad + \dfrac{2}{R}\sum_{n=1}^{\infty}(-1)^n(1 - e^{-R(t-n)/L})\mathcal{U}(t-n)$

49. $x(t) = 2(1 - e^{-t}\cos 3t - \frac{1}{3}e^{-t}\sin 3t)$
$\qquad + 4\sum_{n=1}^{\infty}(-1)^n[1 - e^{-(t-n\pi)}\cos 3(t - n\pi)$
$\qquad\qquad - \frac{1}{3}e^{-(t-n\pi)}\sin 3(t - n\pi)]\mathcal{U}(t - n\pi)$

EXERCISES 7.5 (PAGE 354)

1. $y = e^{3(t-2)}\mathcal{U}(t-2)$ **3.** $y = \sin t + \sin t\,\mathcal{U}(t - 2\pi)$

5. $y = -\cos t\,\mathcal{U}\left(t - \dfrac{\pi}{2}\right) + \cos t\,\mathcal{U}\left(t - \dfrac{3\pi}{2}\right)$

7. $y = \frac{1}{2} - \frac{1}{2}e^{-2t} + [\frac{1}{2} - \frac{1}{2}e^{-2(t-1)}]\mathcal{U}(t-1)$
9. $y = e^{-2(t-2\pi)}\sin t\,\mathcal{U}(t - 2\pi)$
11. $y = e^{-2t}\cos 3t + \frac{2}{3}e^{-2t}\sin 3t$
$\qquad + \frac{1}{3}e^{-2(t-\pi)}\sin 3(t - \pi)\mathcal{U}(t - \pi)$
$\qquad + \frac{1}{3}e^{-2(t-3\pi)}\sin 3(t - 3\pi)\mathcal{U}(t - 3\pi)$

13. $y(x) = \begin{cases} \dfrac{P_0}{EI}\left(\dfrac{L}{4}x^2 - \dfrac{1}{6}x^3\right), & 0 \le x < \dfrac{L}{2} \\[2mm] \dfrac{P_0 L^2}{4EI}\left(\dfrac{1}{2}x - \dfrac{L}{12}\right), & \dfrac{L}{2} \le x \le L \end{cases}$

EXERCISES 7.6 (PAGE 358)

1. $x = -\frac{1}{3}e^{-2t} + \frac{1}{3}e^t$ **3.** $x = -\cos 3t - \frac{5}{3}\sin 3t$
$\quad\; y = \frac{1}{3}e^{-2t} + \frac{2}{3}e^t$ $\qquad\quad\; y = 2\cos 3t - \frac{7}{3}\sin 3t$

5. $x = -2e^{3t} + \frac{5}{2}e^{2t} - \frac{1}{2}$ **7.** $x = -\frac{1}{4}t - \frac{3}{4}\sqrt{2}\sin\sqrt{2}t$
$\quad\; y = \frac{8}{3}e^{3t} - \frac{5}{2}e^{2t} - \frac{1}{6}$ $\qquad y = -\frac{1}{4}t + \frac{3}{4}\sqrt{2}\sin\sqrt{2}t$

9. $x = 8 + \dfrac{2}{3!}t^3 + \dfrac{1}{4!}t^4$

$\qquad y = -\dfrac{2}{3!}t^3 + \dfrac{1}{4!}t^4$

11. $x = \frac{1}{2}t^2 + t + 1 - e^{-t}$
$\qquad y = -\frac{1}{3} + \frac{1}{3}e^{-t} + \frac{1}{3}te^{-t}$

13. $x_1 = \frac{1}{5}\sin t + \dfrac{2\sqrt{6}}{15}\sin\sqrt{6}t + \frac{2}{5}\cos t - \frac{2}{5}\cos\sqrt{6}t$

$\qquad x_2 = \frac{2}{5}\sin t - \dfrac{\sqrt{6}}{15}\sin\sqrt{6}t + \frac{4}{5}\cos t + \frac{1}{5}\cos\sqrt{6}t$

15. (b) $i_2 = \frac{100}{9} - \frac{100}{9}e^{-900t}$
$\qquad\;\; i_3 = \frac{80}{9} - \frac{80}{9}e^{-900t}$
(c) $i_1 = 20 - 20e^{-900t}$

17. $i_2 = -\frac{20}{13}e^{-2t} + \frac{375}{1469}e^{-15t} + \frac{145}{113}\cos t + \frac{85}{113}\sin t$
$\qquad i_3 = \frac{30}{13}e^{-2t} + \frac{250}{1469}e^{-15t} - \frac{280}{113}\cos t + \frac{810}{113}\sin t$

19. $i_1 = \frac{6}{5} - \frac{6}{5}e^{-100t}\cosh 50\sqrt{2}t - \dfrac{9\sqrt{2}}{10}e^{-100t}\sinh 50\sqrt{2}t$

$\qquad i_2 = \frac{6}{5} - \frac{6}{5}e^{-100t}\cosh 50\sqrt{2}t - \dfrac{6\sqrt{2}}{5}e^{-100t}\sinh 50\sqrt{2}t$

CHAPTER 7 IN REVIEW (PAGE 361)

1. $\dfrac{1}{s^2} - \dfrac{2}{s^2}e^{-s}$ **3.** false **5.** true **7.** $\dfrac{1}{s+7}$

9. $\dfrac{2}{s^2 + 4}$ **11.** $\dfrac{4s}{(s^2 + 4)^2}$ **13.** $\frac{1}{6}t^5$ **15.** $\frac{1}{2}t^2 e^{5t}$

17. $e^{5t}\cos 2t + \frac{5}{2}e^{5t}\sin 2t$
19. $\cos\pi(t-1)\mathcal{U}(t-1) + \sin\pi(t-1)\mathcal{U}(t-1)$
21. -5 **23.** $e^{-k(s-a)}F(s-a)$
25. $f(t)\mathcal{U}(t - t_0)$ **27.** $f(t - t_0)\mathcal{U}(t - t_0)$

29. $f(t) = t - (t - 1)\,\mathcal{U}(t - 1) - \mathcal{U}(t - 4);$

$$\mathcal{L}\{f(t)\} = \frac{1}{s^2} - \frac{1}{s^2}e^{-s} - \frac{1}{s}e^{-4s};$$

$$\mathcal{L}\{e^t f(t)\} = \frac{1}{(s-1)^2} - \frac{1}{(s-1)^2}e^{-(s-1)} - \frac{1}{s-1}e^{-4(s-1)}$$

31. $f(t) = 2 + (t - 2)\,\mathcal{U}(t - 2);$

$$\mathcal{L}\{f(t)\} = \frac{2}{s} + \frac{1}{s^2}e^{-2s};$$

$$\mathcal{L}\{e^t f(t)\} = \frac{2}{s-1} + \frac{1}{(s-1)^2}e^{-2(s-1)}$$

33. $y = 5te^t + \frac{1}{2}t^2 e^t$

35. $y = -\frac{6}{25} + \frac{1}{5}t^2 + \frac{3}{2}e^{-t} - \frac{13}{50}e^{-5t} - \frac{4}{25}\mathcal{U}(t - 2)$
$\qquad - \frac{1}{5}(t - 2)^2\,\mathcal{U}(t - 2) + \frac{1}{4}e^{-(t-2)}\,\mathcal{U}(t - 2)$
$\qquad - \frac{9}{100}e^{-5(t-2)}\,\mathcal{U}(t - 2)$

37. $y = 1 + t + \frac{1}{2}t^2$

39. $x = -\frac{1}{4} + \frac{9}{8}e^{-2t} + \frac{1}{8}e^{2t}$
$\qquad y = t + \frac{9}{4}e^{-2t} - \frac{1}{4}e^{2t}$

41. $i(t) = -9 + 2t + 9e^{-t/5}$

43. $y(x) = \dfrac{w_0}{12EIL}\left[-\dfrac{1}{5}x^5 + \dfrac{L}{2}x^4 - \dfrac{L^2}{2}x^3 + \dfrac{L^3}{4}x^2 \right.$
$\qquad\qquad \left. + \dfrac{1}{5}\left(x - \dfrac{L}{2}\right)^5 \mathcal{U}\left(x - \dfrac{L}{2}\right)\right]$

EXERCISES 8.1 (PAGE 373)

1. $\mathbf{X}' = \begin{pmatrix} 3 & -5 \\ 4 & 8 \end{pmatrix}\mathbf{X}$, where $\mathbf{X} = \begin{pmatrix} x \\ y \end{pmatrix}$

3. $\mathbf{X}' = \begin{pmatrix} -3 & 4 & -9 \\ 6 & -1 & 0 \\ 10 & 4 & 3 \end{pmatrix}\mathbf{X}$, where $\mathbf{X} = \begin{pmatrix} x \\ y \\ z \end{pmatrix}$

5. $\mathbf{X}' = \begin{pmatrix} 1 & -1 & 1 \\ 2 & 1 & -1 \\ 1 & 1 & 1 \end{pmatrix}\mathbf{X}$, where $\mathbf{X} = \begin{pmatrix} x \\ y \\ z \end{pmatrix}$

7. $\dfrac{dx}{dt} = 4x + 2y + e^t$

$\dfrac{dy}{dt} = -x + 3y - e^t$

9. $\dfrac{dx}{dt} = x - y + 2z + e^{-t} - 3t$

$\dfrac{dy}{dt} = 3x - 4y + z + 2e^{-t} + t$

$\dfrac{dz}{dt} = -2x + 5y + 6z + 2e^{-t} - t$

17. Yes; $W(\mathbf{X}_1, \mathbf{X}_2) = -2e^{-8t} \neq 0$ implies that $\mathbf{X}_1$ and $\mathbf{X}_2$ are linearly independent on $(-\infty, \infty)$.

19. No; $W(\mathbf{X}_1, \mathbf{X}_2, \mathbf{X}_3) = 0$ for every t. The solution vectors are linearly dependent on $(-\infty, \infty)$. Note that $\mathbf{X}_3 = 2\mathbf{X}_1 + \mathbf{X}_2$.

EXERCISES 8.2 (PAGE 389)

1. $\mathbf{X} = c_1\begin{pmatrix} 1 \\ 2 \end{pmatrix}e^{5t} + c_2\begin{pmatrix} 1 \\ -1 \end{pmatrix}e^{-t}$

3. $\mathbf{X} = c_1\begin{pmatrix} 2 \\ 1 \end{pmatrix}e^{-3t} + c_2\begin{pmatrix} 2 \\ 5 \end{pmatrix}e^{t}$

5. $\mathbf{X} = c_1\begin{pmatrix} 5 \\ 2 \end{pmatrix}e^{8t} + c_2\begin{pmatrix} 1 \\ 4 \end{pmatrix}e^{-10t}$

7. $\mathbf{X} = c_1\begin{pmatrix} 1 \\ 0 \\ 0 \end{pmatrix}e^{t} + c_2\begin{pmatrix} 2 \\ 3 \\ 1 \end{pmatrix}e^{2t} + c_3\begin{pmatrix} 1 \\ 0 \\ 2 \end{pmatrix}e^{-t}$

9. $\mathbf{X} = c_1\begin{pmatrix} -1 \\ 0 \\ 1 \end{pmatrix}e^{-t} + c_2\begin{pmatrix} 1 \\ 4 \\ 3 \end{pmatrix}e^{3t} + c_3\begin{pmatrix} 1 \\ -1 \\ 3 \end{pmatrix}e^{-2t}$

11. $\mathbf{X} = c_1\begin{pmatrix} 4 \\ 0 \\ -1 \end{pmatrix}e^{-t} + c_2\begin{pmatrix} -12 \\ 6 \\ 5 \end{pmatrix}e^{-t/2} + c_3\begin{pmatrix} 4 \\ 2 \\ -1 \end{pmatrix}e^{-3t/2}$

13. $\mathbf{X} = 3\begin{pmatrix} 1 \\ 1 \end{pmatrix}e^{t/2} + 2\begin{pmatrix} 0 \\ 1 \end{pmatrix}e^{-t/2}$

19. $\mathbf{X} = c_1\begin{pmatrix} 1 \\ 3 \end{pmatrix} + c_2\left[\begin{pmatrix} 1 \\ 3 \end{pmatrix}t + \begin{pmatrix} \frac{1}{4} \\ -\frac{1}{4} \end{pmatrix}\right]$

21. $\mathbf{X} = c_1\begin{pmatrix} 1 \\ 1 \end{pmatrix}e^{2t} + c_2\left[\begin{pmatrix} 1 \\ 1 \end{pmatrix}te^{2t} + \begin{pmatrix} -\frac{1}{3} \\ 0 \end{pmatrix}e^{2t}\right]$

23. $\mathbf{X} = c_1\begin{pmatrix} 1 \\ 1 \\ 1 \end{pmatrix}e^{t} + c_2\begin{pmatrix} 1 \\ 1 \\ 0 \end{pmatrix}e^{2t} + c_3\begin{pmatrix} 1 \\ 0 \\ 1 \end{pmatrix}e^{2t}$

25. $\mathbf{X} = c_1\begin{pmatrix} -4 \\ -5 \\ 2 \end{pmatrix} + c_2\begin{pmatrix} 2 \\ 0 \\ -1 \end{pmatrix}e^{5t}$
$\qquad + c_3\left[\begin{pmatrix} 2 \\ 0 \\ -1 \end{pmatrix}te^{5t} + \begin{pmatrix} -\frac{1}{2} \\ -\frac{1}{2} \\ -1 \end{pmatrix}e^{5t}\right]$

27. $\mathbf{X} = c_1\begin{pmatrix} 0 \\ 1 \\ 1 \end{pmatrix}e^{t} + c_2\left[\begin{pmatrix} 0 \\ 1 \\ 1 \end{pmatrix}te^{t} + \begin{pmatrix} 0 \\ 1 \\ 0 \end{pmatrix}e^{t}\right]$
$\qquad + c_3\left[\begin{pmatrix} 0 \\ 1 \\ 1 \end{pmatrix}\frac{t^2}{2}e^{t} + \begin{pmatrix} 0 \\ 1 \\ 0 \end{pmatrix}te^{t} + \begin{pmatrix} \frac{1}{2} \\ 0 \\ 0 \end{pmatrix}e^{t}\right]$

29. $\mathbf{X} = -7\begin{pmatrix} 2 \\ 1 \end{pmatrix}e^{4t} + 13\begin{pmatrix} 2t + 1 \\ t + 1 \end{pmatrix}e^{4t}$

31. Corresponding to the eigenvalue $\lambda_1 = 2$ of multiplicity five, the eigenvectors are

$$\mathbf{K}_1 = \begin{pmatrix} 1 \\ 0 \\ 0 \\ 0 \\ 0 \end{pmatrix}, \mathbf{K}_2 = \begin{pmatrix} 0 \\ 0 \\ 1 \\ 0 \\ 0 \end{pmatrix}, \mathbf{K}_3 = \begin{pmatrix} 0 \\ 0 \\ 0 \\ 1 \\ 0 \end{pmatrix}.$$

33. $\mathbf{X} = c_1\begin{pmatrix} \cos t \\ 2\cos t + \sin t \end{pmatrix}e^{4t} + c_2\begin{pmatrix} \sin t \\ 2\sin t - \cos t \end{pmatrix}e^{4t}$

35. $\mathbf{X} = c_1 \begin{pmatrix} \cos t \\ -\cos t - \sin t \end{pmatrix} e^{4t} + c_2 \begin{pmatrix} \sin t \\ -\sin t + \cos t \end{pmatrix} e^{4t}$

37. $\mathbf{X} = c_1 \begin{pmatrix} 5 \cos 3t \\ 4 \cos 3t + 3 \sin 3t \end{pmatrix} + c_2 \begin{pmatrix} 5 \sin 3t \\ 4 \sin 3t - 3 \cos 3t \end{pmatrix}$

39. $\mathbf{X} = c_1 \begin{pmatrix} 1 \\ 0 \\ 0 \end{pmatrix} + c_2 \begin{pmatrix} -\cos t \\ \cos t \\ \sin t \end{pmatrix} + c_3 \begin{pmatrix} \sin t \\ -\sin t \\ \cos t \end{pmatrix}$

41. $\mathbf{X} = c_1 \begin{pmatrix} 0 \\ 2 \\ 1 \end{pmatrix} e^t + c_2 \begin{pmatrix} \sin t \\ \cos t \\ \cos t \end{pmatrix} e^t + c_3 \begin{pmatrix} \cos t \\ -\sin t \\ -\sin t \end{pmatrix} e^t$

43. $\mathbf{X} = \begin{pmatrix} 28 \\ -5 \\ 25 \end{pmatrix} e^{2t} + c_2 \begin{pmatrix} 5 \cos 3t \\ -4 \cos 3t - 3 \sin 3t \\ 0 \end{pmatrix} e^{-2t}$
$\qquad + c_3 \begin{pmatrix} 5 \sin 3t \\ -4 \sin 3t + 3 \cos 3t \\ 0 \end{pmatrix} e^{-2t}$

45. $\mathbf{X} = -\begin{pmatrix} 25 \\ -7 \\ 6 \end{pmatrix} e^t - \begin{pmatrix} \cos 5t - 5 \sin 5t \\ \cos 5t \\ \cos 5t \end{pmatrix}$
$\qquad + 6 \begin{pmatrix} 5 \cos 5t + \sin 5t \\ \sin 5t \\ \sin 5t \end{pmatrix}$

EXERCISES 8.3 (PAGE 397)

1. $\mathbf{X} = c_1 \begin{pmatrix} 1 \\ 1 \end{pmatrix} + c_2 \begin{pmatrix} 3 \\ 2 \end{pmatrix} e^t - \begin{pmatrix} 11 \\ 11 \end{pmatrix} t - \begin{pmatrix} 15 \\ 10 \end{pmatrix}$

3. $\mathbf{X} = c_1 \begin{pmatrix} 2 \\ 1 \end{pmatrix} e^{t/2} + c_2 \begin{pmatrix} 10 \\ 3 \end{pmatrix} e^{3t/2} - \begin{pmatrix} \frac{13}{2} \\ \frac{13}{4} \end{pmatrix} t e^{t/2} - \begin{pmatrix} \frac{15}{2} \\ \frac{9}{4} \end{pmatrix} e^{t/2}$

5. $\mathbf{X} = c_1 \begin{pmatrix} 2 \\ 1 \end{pmatrix} e^t + c_2 \begin{pmatrix} 1 \\ 1 \end{pmatrix} e^{2t} + \begin{pmatrix} 3 \\ 3 \end{pmatrix} e^t + \begin{pmatrix} 4 \\ 2 \end{pmatrix} t e^t$

7. $\mathbf{X} = c_1 \begin{pmatrix} 4 \\ 1 \end{pmatrix} e^{3t} + c_2 \begin{pmatrix} -2 \\ 1 \end{pmatrix} e^{-3t} + \begin{pmatrix} -12 \\ 0 \end{pmatrix} t - \begin{pmatrix} \frac{4}{3} \\ \frac{4}{3} \end{pmatrix}$

9. $\mathbf{X} = c_1 \begin{pmatrix} 1 \\ -1 \end{pmatrix} e^t + c_2 \begin{pmatrix} -t \\ \frac{1}{2} - t \end{pmatrix} e^t + \begin{pmatrix} \frac{1}{2} \\ -2 \end{pmatrix} e^{-t}$

11. $\mathbf{X} = c_1 \begin{pmatrix} \cos t \\ \sin t \end{pmatrix} + c_2 \begin{pmatrix} \sin t \\ -\cos t \end{pmatrix}$
$\qquad + \begin{pmatrix} \cos t \\ \sin t \end{pmatrix} t + \begin{pmatrix} -\sin t \\ \cos t \end{pmatrix} \ln|\cos t|$

13. $\mathbf{X} = c_1 \begin{pmatrix} \cos t \\ \sin t \end{pmatrix} e^t + c_2 \begin{pmatrix} \sin t \\ -\cos t \end{pmatrix} e^t + \begin{pmatrix} \cos t \\ \sin t \end{pmatrix} t e^t$

15. $\mathbf{X} = c_1 \begin{pmatrix} \cos t \\ -\sin t \end{pmatrix} + c_2 \begin{pmatrix} \sin t \\ \cos t \end{pmatrix} + \begin{pmatrix} \cos t \\ -\sin t \end{pmatrix} t$
$\qquad + \begin{pmatrix} -\sin t \\ \sin t \tan t \end{pmatrix} - \begin{pmatrix} \sin t \\ \cos t \end{pmatrix} \ln|\cos t|$

17. $\mathbf{X} = c_1 \begin{pmatrix} 2 \sin t \\ \cos t \end{pmatrix} e^t + c_2 \begin{pmatrix} 2 \cos t \\ -\sin t \end{pmatrix} e^t + \begin{pmatrix} 3 \sin t \\ \frac{3}{2} \cos t \end{pmatrix} t e^t$
$\qquad + \begin{pmatrix} \cos t \\ -\frac{1}{2} \sin t \end{pmatrix} e^t \ln|\sin t| + \begin{pmatrix} 2 \cos t \\ -\sin t \end{pmatrix} e^t \ln|\cos t|$

19. $\mathbf{X} = c_1 \begin{pmatrix} 1 \\ -1 \\ 0 \end{pmatrix} + c_2 \begin{pmatrix} 1 \\ 1 \\ 0 \end{pmatrix} e^{2t} + c_3 \begin{pmatrix} 0 \\ 0 \\ 1 \end{pmatrix} e^{3t}$
$\qquad + \begin{pmatrix} -\frac{1}{4} e^{2t} + \frac{1}{2} t e^{2t} \\ -e^t + \frac{1}{4} e^{2t} + \frac{1}{2} t e^{2t} \\ \frac{1}{2} t^2 e^{3t} \end{pmatrix}$

21. $\mathbf{X} = \begin{pmatrix} 2 \\ 2 \end{pmatrix} t e^{2t} + \begin{pmatrix} -1 \\ 1 \end{pmatrix} e^{2t} + \begin{pmatrix} -2 \\ 2 \end{pmatrix} t e^{4t} + \begin{pmatrix} 2 \\ 0 \end{pmatrix} e^{4t}$

23. $\mathbf{X} = c_1 \begin{pmatrix} -1 \\ 1 \end{pmatrix} e^{-t} + c_2 \begin{pmatrix} -3 \\ 1 \end{pmatrix} e^t + \begin{pmatrix} -1 \\ 3 \end{pmatrix}$

25. $\mathbf{X} = c_1 \begin{pmatrix} 1 \\ -3 \end{pmatrix} e^{3t} + c_2 \begin{pmatrix} 1 \\ 9 \end{pmatrix} e^{7t} + \begin{pmatrix} \frac{55}{36} \\ -\frac{19}{4} \end{pmatrix} e^t$

27. $\begin{pmatrix} i_1 \\ i_2 \end{pmatrix} = 2 \begin{pmatrix} 1 \\ 3 \end{pmatrix} e^{-2t} + \frac{6}{29} \begin{pmatrix} 3 \\ -1 \end{pmatrix} e^{-12t}$
$\qquad - \frac{4}{29} \begin{pmatrix} 19 \\ 42 \end{pmatrix} \cos t + \frac{4}{29} \begin{pmatrix} 83 \\ 69 \end{pmatrix} \sin t$

EXERCISES 8.4 (PAGE 402)

1. $e^{\mathbf{A}t} = \begin{pmatrix} e^t & 0 \\ 0 & e^{2t} \end{pmatrix}$; $e^{-\mathbf{A}t} = \begin{pmatrix} e^{-t} & 0 \\ 0 & e^{-2t} \end{pmatrix}$

3. $e^{\mathbf{A}t} = \begin{pmatrix} t+1 & t & t \\ t & t+1 & t \\ -2t & -2t & -2t+1 \end{pmatrix}$

5. $\mathbf{X} = c_1 \begin{pmatrix} 1 \\ 0 \end{pmatrix} e^t + c_2 \begin{pmatrix} 0 \\ 1 \end{pmatrix} e^{2t}$

7. $\mathbf{X} = c_1 \begin{pmatrix} t+1 \\ t \\ -2t \end{pmatrix} + c_2 \begin{pmatrix} t \\ t+1 \\ -2t \end{pmatrix} + c_3 \begin{pmatrix} t \\ t \\ -2t+1 \end{pmatrix}$

9. $\mathbf{X} = c_3 \begin{pmatrix} 1 \\ 0 \end{pmatrix} e^t + c_4 \begin{pmatrix} 0 \\ 1 \end{pmatrix} e^{2t} + \begin{pmatrix} -3 \\ \frac{1}{2} \end{pmatrix}$

11. $\mathbf{X} = c_1 \begin{pmatrix} \cosh t \\ \sinh t \end{pmatrix} + c_2 \begin{pmatrix} \sinh t \\ \cosh t \end{pmatrix} - \begin{pmatrix} 1 \\ 1 \end{pmatrix}$

13. $\mathbf{X} = \begin{pmatrix} t+1 \\ t \\ -2t \end{pmatrix} - 4 \begin{pmatrix} t \\ t+1 \\ -2t \end{pmatrix} + 6 \begin{pmatrix} t \\ t \\ -2t+1 \end{pmatrix}$

15. $e^{\mathbf{A}t} = \begin{pmatrix} \frac{3}{2} e^{2t} - \frac{1}{2} e^{-2t} & \frac{3}{4} e^{2t} - \frac{3}{4} e^{-2t} \\ -e^{2t} + e^{-2t} & -\frac{1}{2} e^{2t} + \frac{3}{2} e^{-2t} \end{pmatrix}$;
$\qquad \mathbf{X} = c_1 \begin{pmatrix} \frac{3}{2} e^{2t} - \frac{1}{2} e^{-2t} \\ -e^{2t} + e^{-2t} \end{pmatrix} + c_2 \begin{pmatrix} \frac{3}{4} e^{2t} - \frac{3}{4} e^{-2t} \\ -\frac{1}{2} e^{2t} + \frac{3}{2} e^{-2t} \end{pmatrix}$ or
$\qquad \mathbf{X} = c_3 \begin{pmatrix} 3 \\ -2 \end{pmatrix} e^{2t} + c_4 \begin{pmatrix} 1 \\ -2 \end{pmatrix} e^{-2t}$

17. $e^{\mathbf{A}t} = \begin{pmatrix} e^{2t} + 3t e^{2t} & -9t e^{2t} \\ t e^{2t} & e^{2t} - 3t e^{2t} \end{pmatrix}$;
$\qquad \mathbf{X} = c_1 \begin{pmatrix} 1 + 3t \\ t \end{pmatrix} e^{2t} + c_2 \begin{pmatrix} -9t \\ 1 - 3t \end{pmatrix} e^{2t}$

23. $\mathbf{X} = c_1 \begin{pmatrix} \frac{3}{2}e^{3t} - \frac{1}{2}e^{5t} \\ \frac{3}{2}e^{3t} - \frac{3}{2}e^{5t} \end{pmatrix} + c_2 \begin{pmatrix} -\frac{1}{2}e^{3t} + \frac{1}{2}e^{5t} \\ -\frac{1}{2}e^{3t} + \frac{3}{2}e^{5t} \end{pmatrix}$ or

$\mathbf{X} = c_3 \begin{pmatrix} 1 \\ 1 \end{pmatrix} e^{3t} + c_4 \begin{pmatrix} 1 \\ 3 \end{pmatrix} e^{5t}$

CHAPTER 8 IN REVIEW (PAGE 404)

1. $k = \frac{1}{3}$

5. $\mathbf{X} = c_1 \begin{pmatrix} 1 \\ -1 \end{pmatrix} e^t + c_2 \left[\begin{pmatrix} 1 \\ -1 \end{pmatrix} te^t + \begin{pmatrix} 0 \\ 1 \end{pmatrix} e^t \right]$

7. $\mathbf{X} = c_1 \begin{pmatrix} \cos 2t \\ -\sin 2t \end{pmatrix} e^t + c_2 \begin{pmatrix} \sin 2t \\ \cos 2t \end{pmatrix} e^t$

9. $\mathbf{X} = c_1 \begin{pmatrix} -2 \\ 3 \\ 1 \end{pmatrix} e^{2t} + c_2 \begin{pmatrix} 0 \\ 1 \\ 1 \end{pmatrix} e^{4t} + c_1 \begin{pmatrix} 7 \\ 12 \\ -16 \end{pmatrix} e^{-3t}$

11. $\mathbf{X} = c_1 \begin{pmatrix} 1 \\ 0 \end{pmatrix} e^{2t} + c_2 \begin{pmatrix} 4 \\ 1 \end{pmatrix} e^{4t} + \begin{pmatrix} 16 \\ -4 \end{pmatrix} t + \begin{pmatrix} 11 \\ -1 \end{pmatrix}$

13. $\mathbf{X} = c_1 \begin{pmatrix} \cos t \\ \cos t - \sin t \end{pmatrix} + c_2 \begin{pmatrix} \sin t \\ \sin t + \cos t \end{pmatrix} - \begin{pmatrix} 1 \\ 1 \end{pmatrix}$

$+ \begin{pmatrix} \sin t \\ \sin t + \cos t \end{pmatrix} \ln|\csc t - \cot t|$

15. (b) $\mathbf{X} = c_1 \begin{pmatrix} -1 \\ 1 \\ 0 \end{pmatrix} + c_2 \begin{pmatrix} -1 \\ 0 \\ 1 \end{pmatrix} + c_1 \begin{pmatrix} 1 \\ 1 \\ 1 \end{pmatrix} e^{3t}$

EXERCISES 9.1 (PAGE 415)

1. $h = 0.1$

x_n	y_n
1.00	5.0000
1.10	3.9900
1.20	3.2545
1.30	2.7236
1.40	2.3451
1.50	2.0801

$h = 0.05$

x_n	y_n
1.00	5.0000
1.05	4.4475
1.10	3.9763
1.15	3.5751
1.20	3.2342
1.25	2.9452
1.30	2.7009
1.35	2.4952
1.40	2.3226
1.45	2.1786
1.50	2.0592

3. $h = 0.1$

x_n	y_n
0.00	0.0000
0.10	0.1005
0.20	0.2030
0.30	0.3098
0.40	0.4234
0.50	0.5470

$h = 0.05$

x_n	y_n
0.00	0.0000
0.05	0.0501
0.10	0.1004
0.15	0.1512
0.20	0.2028
0.25	0.2554
0.30	0.3095
0.35	0.3652
0.40	0.4230
0.45	0.4832
0.50	0.5465

5. $h = 0.1$

x_n	y_n
0.00	0.0000
0.10	0.0952
0.20	0.1822
0.30	0.2622
0.40	0.3363
0.50	0.4053

$h = 0.05$

x_n	y_n
0.00	0.0000
0.05	0.0488
0.10	0.0953
0.15	0.1397
0.20	0.1823
0.25	0.2231
0.30	0.2623
0.35	0.3001
0.40	0.3364
0.45	0.3715
0.50	0.4054

7. $h = 0.1$

x_n	y_n
0.00	5.0000
0.10	0.5215
0.20	0.5362
0.30	0.5449
0.40	0.5490
0.50	0.5503

$h = 0.05$

x_n	y_n
0.00	0.5000
0.05	0.5116
0.10	0.5214
0.15	0.5294
0.20	0.5359
0.25	0.5408
0.30	0.5444
0.35	0.5469
0.40	0.5484
0.45	0.5492
0.50	0.5495

9. $h = 0.1$

x_n	y_n
1.00	1.0000
1.10	1.0095
1.20	1.0404
1.30	1.0967
1.40	1.1866
1.50	1.3260

$h = 0.05$

x_n	y_n
1.00	1.0000
1.05	1.0024
1.10	1.0100
1.15	1.0228
1.20	1.0414
1.25	1.0663
1.30	1.0984
1.35	1.1389
1.40	1.1895
1.45	1.2526
1.50	1.3315

11. $h = 0.1$

x_n	y_n	Exact
0.00	2.0000	2.0000
0.10	2.1220	2.1230
0.20	2.3049	2.3085
0.30	2.5858	2.5958
0.40	3.0378	3.0650
0.50	3.8254	3.9082

$h = 0.5$

x_n	y_n	Exact
0.00	2.0000	2.0000
0.05	2.0553	2.0554
0.10	2.1228	2.1230
0.15	2.2056	2.2061
0.20	2.3075	2.3085
0.25	2.4342	2.4358
0.30	2.5931	2.5958
0.35	2.7953	2.7997
0.40	3.0574	3.0650
0.45	3.4057	3.4189
0.50	3.8840	3.9082

13. (a) $y_1 = 1.2$

(b) $y''(c)\dfrac{h^2}{2} = 4e^{2c}\dfrac{(0.1)^2}{2} = 0.02e^{2c} \le 0.02e^{0.2} = 0.0244$

(c) Actual value is $y(0.1) = 1.2214$. Error is 0.0214.

(d) If $h = 0.05$, $y_2 = 1.21$.

(e) Error with $h = 0.1$ is 0.0214. Error with $h = 0.05$ is 0.0114.

15. (a) $y_1 = 0.8$

(b) $y''(c)\dfrac{h^2}{2} = 5e^{-2c}\dfrac{(0.1)^2}{2} = 0.025e^{-2c} \le 0.025$ for

$0 \le c \le 0.1$.

(c) Actual value is $y(0.1) = 0.8234$. Error is 0.0234.

(d) If $h = 0.05$, $y_2 = 0.8125$.

(e) Error with $h = 0.1$ is 0.0234. Error with $h = 0.05$ is 0.0109.

17. (a) Error is $19h^2 e^{-3(c-1)}$.

(b) $y''(c)\dfrac{h^2}{2} \le 19(0.1)^2(1) = 0.19$

(c) If $h = 0.1$, $y_5 = 1.8207$. If $h = 0.05$, $y_{10} = 1.9424$.

(d) Error with $h = 0.1$ is 0.2325. Error with $h = 0.05$ is 0.1109.

19. (a) Error is $\dfrac{1}{(c+1)^2}\dfrac{h^2}{2}$.

(b) $\left| y''(c)\dfrac{h^2}{2} \right| \le (1)\dfrac{(0.1)^2}{2} = 0.005$

(c) If $h = 0.1$, $y_5 = 0.4198$. If $h = 0.05$, $y_{10} = 0.4124$.

(d) Error with $h = 0.1$ is 0.0143. Error with $h = 0.05$ is 0.0069.

EXERCISES 9.2 (PAGE 421)

1.

x_n	y_n	Exact
0.00	2.0000	2.0000
0.10	2.1230	2.1230
0.20	2.3085	2.3085
0.30	2.5958	2.5958
0.40	3.0649	3.0650
0.50	3.9078	3.9082

3.

x_n	y_n
1.00	5.0000
1.10	3.9724
1.20	3.2284
1.30	2.6945
1.40	2.3163
1.50	2.0533

5.

x_n	y_n
0.00	0.0000
0.10	0.1003
0.20	0.2027
0.30	0.3093
0.40	0.4228
0.50	0.5463

7.

x_n	y_n
0.00	0.0000
0.10	0.0953
0.20	0.1823
0.30	0.2624
0.40	0.3365
0.50	0.4055

9.

x_n	y_n
0.00	0.5000
0.10	0.5213
0.20	0.5358
0.30	0.5443
0.40	0.5482
0.50	0.5493

11.

x_n	y_n
1.00	1.0000
1.10	1.0101
1.20	1.0417
1.30	1.0989
1.40	1.1905
1.50	1.3333

13. (a) $v(5) = 35.7678$

(c) $v(t) = \sqrt{\dfrac{mg}{k}}\tanh\sqrt{\dfrac{kg}{m}}\,t$; $v(5) = 35.7678$

15. (a)

$h = 0.1$			$h = 0.05$	
x_n	y_n		x_n	y_n
1.00	1.0000		1.00	1.0000
1.10	1.2511		1.05	1.1112
1.20	1.6934		1.10	1.2511
1.30	2.9425		1.15	1.4348
1.40	903.0282		1.20	1.6934
			1.25	2.1047
			1.30	2.9560
			1.35	7.8981
			1.40	1.1E + 15

17. (a) $y_1 = 0.82341667$

(b) $y^{(5)}(c)\dfrac{h^5}{5!} = 40e^{-2c}\dfrac{h^5}{5!} \le 40e^{2(0)}\dfrac{(0.1)^5}{5!}$
$$= 3.333 \times 10^{-6}$$

(c) Actual value is $y(0.1) = 0.8234134413$. Error is $3.225 \times 10^{-6} \le 3.333 \times 10^{-6}$.

(d) If $h = 0.05$, $y_2 = 0.82341363$.

(e) Error with $h = 0.1$ is 3.225×10^{-6}. Error with $h = 0.05$ is 1.854×10^{-7}.

19. (a) $y^{(5)}(c)\dfrac{h^5}{5!} = \dfrac{24}{(c+1)^5}\dfrac{h^5}{5!}$

(b) $\dfrac{24}{(c+1)^5}\dfrac{h^5}{5!} \le 24\dfrac{(0.1)^5}{5!} = 2.0000 \times 10^{-6}$

(c) From calculation with $h = 0.1$, $y_5 = 0.40546517$. From calculation with $h = 0.05$, $y_{10} = 0.40546511$.

EXERCISES 9.3 (PAGE 426)

1. $y(x) = -x + e^x$; $y(0.2) = 1.0214$, $y(0.4) = 1.0918$, $y(0.6) = 1.2221$, $y(0.8) = 1.4255$

3.

x_n	y_n
0.00	1.0000
0.20	0.7328
0.40	0.6461
0.60	0.6585
0.80	0.7232

5.

x_n	y_n		x_n	y_n
0.00	0.0000		0.00	0.0000
0.20	0.2027		0.10	0.1003
0.40	0.4228		0.20	0.2027
0.60	0.6841		0.30	0.3093
0.80	1.0297		0.40	0.4228
1.00	1.5569		0.50	0.5463
			0.60	0.6842
			0.70	0.8423
			0.80	1.0297
			0.90	1.2603
			1.00	1.5576

7.

x_n	y_n		x_n	y_n
0.00	0.0000		0.00	0.0000
0.20	0.0026		0.10	0.0003
0.40	0.0201		0.20	0.0026
0.60	0.0630		0.30	0.0087
0.80	0.1360		0.40	0.0200
1.00	0.2385		0.50	0.0379
			0.60	0.0629
			0.70	0.0956
			0.80	0.1360
			0.90	0.1837
			1.00	0.2384

EXERCISES 9.4 (PAGE 432)

1. $y(x) = -2e^{2x} + 5xe^{2x}$; $y(0.2) = -1.4918$, $y_2 = -1.6800$

3. $y_1 = -1.4928$, $y_2 = -1.4919$

5. $y_1 = 1.4640$, $y_2 = 1.4640$

7. $x_1 = 8.3055$, $y_1 = 3.4199$;
$x_2 = 8.3055$, $y_2 = 3.4199$

9. $x_1 = -3.9123$, $y_1 = 4.2857$;
$x_2 = -3.9123$, $y_2 = 4.2857$

11. $x_1 = 0.4179$, $y_1 = -2.1824$;
$x_2 = 0.4173$, $y_2 = -2.1821$

EXERCISES 9.5 (PAGE 437)

1. $y_1 = -5.6774$, $y_2 = -2.5807$, $y_3 = 6.3226$

3. $y_1 = -0.2259$, $y_2 = -0.3356$, $y_3 = -0.3308$, $y_4 = -0.2167$

5. $y_1 = 3.3751$, $y_2 = 3.6306$, $y_3 = 3.6448$, $y_4 = 3.2355$, $y_5 = 2.1411$

7. $y_1 = 3.8842$, $y_2 = 2.9640$, $y_3 = 2.2064$, $y_4 = 1.5826$, $y_5 = 1.0681$, $y_6 = 0.6430$, $y_7 = 0.2913$

9. $y_1 = 0.2660$, $y_2 = 0.5097$, $y_3 = 0.7357$, $y_4 = 0.9471$, $y_5 = 1.1465$, $y_6 = 1.3353$, $y_7 = 1.5149$, $y_8 = 1.6855$, $y_9 = 1.8474$

11. $y_1 = 0.3492$, $y_2 = 0.7202$, $y_3 = 1.1363$, $y_4 = 1.6233$, $y_5 = 2.2118$, $y_6 = 2.9386$, $y_7 = 3.8490$

13. (c) $y_0 = -2.2755$, $y_1 = -2.0755$,
$y_2 = -1.8589$, $y_3 = -1.6126$, $y_4 = -1.3275$

CHAPTER 9 IN REVIEW (PAGE 438)

1. Comparison of Numerical Methods with $h = 0.1$

x_n	Euler	Improved Euler	Runge-Kutta
1.00	2.0000	2.0000	2.0000
1.10	2.1386	2.1549	2.1556
1.20	2.3097	2.3439	2.3454
1.30	2.5136	2.5672	2.5695
1.40	2.7504	2.8246	2.8278
1.50	3.0201	3.1157	3.1197

Comparison of Numerical Methods with $h = 0.05$

x_n	Euler	Improved Euler	Runge-Kutta
1.00	2.0000	2.0000	2.0000
1.05	2.0693	2.0735	2.0736
1.10	2.1469	2.1554	2.1556
1.15	2.2329	2.2459	2.2462
1.20	2.3272	2.3450	2.3454
1.25	2.4299	3.4527	2.4532
1.30	2.5410	2.5689	2.5695
1.35	2.6604	2.6937	2.6944
1.40	2.7883	2.8269	2.8278
1.45	2.9245	2.9686	2.9696
1.50	3.0690	3.1187	3.1197

3. Comparison of Numerical Methods with $h = 0.1$

x_n	Euler	Improved Euler	Runge-Kutta
0.50	0.5000	0.5000	0.5000
0.60	0.6000	0.6048	0.6049
0.70	0.7095	0.7191	0.7194
0.80	0.8283	0.8427	0.8431
0.90	0.9559	0.9752	0.9757
1.00	1.0921	1.1163	1.1169

Comparison of Numerical Methods with $h = 0.05$

x_n	Euler	Improved Euler	Runge-Kutta
0.50	0.5000	0.5000	0.5000
0.55	0.5500	0.5512	0.5512
0.60	0.6024	0.6049	0.6049
0.65	0.6573	0.6610	0.6610
0.70	0.7144	0.7194	0.7194
0.75	0.7739	0.7802	0.7801
0.80	0.8356	0.8431	0.8431
0.85	0.8996	0.9083	0.9083
0.90	0.9657	0.9757	0.9757
0.95	1.0340	1.0453	1.0452
1.00	1.1044	1.1170	1.1169

5. $h = 0.2$: $y(0.2) \approx 3.2$; $h = 0.1$: $y(0.2) \approx 3.23$

7. $x(0.2) \approx 1.62$, $y(0.2) \approx 1.84$

EXERCISES 10.1 (PAGE 446)

1. $x' = y$
$y' = -9 \sin x$; critical points at $(\pm n\pi, 0)$

3. $x' = y$
$y' = x^2 + y(x^3 - 1)$; critical point at $(0, 0)$

5. $x' = y$
$y' = \epsilon x^3 - x$;
critical points at $(0, 0)$, $\left(\dfrac{1}{\sqrt{\epsilon}}, 0\right)$, $\left(-\dfrac{1}{\sqrt{\epsilon}}, 0\right)$

7. $(0, 0)$ and $(-1, -1)$

9. $(0, 0)$ and $(\frac{4}{3}, \frac{4}{3})$

11. $(0, 0)$, $(10, 0)$, $(0, 16)$, and $(4, 12)$

13. $(0, y)$, y arbitrary

15. $(0, 0)$, $(0, 1)$, $(0, -1)$, $(1, 0)$, $(-1, 0)$

17. **(a)** $x = c_1 e^{5t} - c_2 e^{-t}$ **(b)** $x = -2e^{-t}$
$y = 2c_1 e^{5t} + c_2 e^{-t}$ $y = 2e^{-t}$

19. **(a)** $x = c_1(4 \cos 3t - 3 \sin 3t) + c_2(4 \sin 3t + 3 \cos 3t)$
$y = c_1(5 \cos 3t) + c_2(5 \sin 3t)$
(b) $x = 4 \cos 3t - 3 \sin 3t$
$y = 5 \cos 3t$

21. **(a)** $x = c_1(\sin t - \cos t)e^{4t} + c_2(-\sin t - \cos t)e^{4t}$
$y = 2c_1(\cos t)e^{4t} + 2c_2(\sin t)e^{4t}$
(b) $x = (\sin t - \cos t)e^{4t}$
$y = 2(\cos t)e^{4t}$

23. $r = \dfrac{1}{\sqrt[4]{4t + c_1}}, \theta = t + c_2; r = 4\dfrac{1}{\sqrt[4]{1024t + 1}}, \theta = t$; the solution spirals toward the origin as t increases.

25. $r = \dfrac{1}{\sqrt{1 + c_1 e^{-2t}}}, \theta = t + c_2; r = 1, \theta = t$ (or $x = \cos t$ and $y = \sin t$) is the solution that satisfies $\mathbf{X}(0) = (1, 0); r = \dfrac{1}{\sqrt{1 - \frac{3}{4}e^{-2t}}}, \theta = t$ is the solution that satisfies $\mathbf{X}(0) = (2, 0)$. This solution spirals toward the circle $r = 1$ as t increases.

27. There are no critical points and therefore no periodic solutions.

29. There appears to be a periodic solution enclosing the critical point $(0, 0)$.

EXERCISES 10.2 (PAGE 456)

1. **(a)** If $\mathbf{X}(0) = \mathbf{X}_0$ lies on the line $y = 2x$, then $\mathbf{X}(t)$ approaches $(0, 0)$ along this line. For all other initial conditions, $\mathbf{X}(t)$ approaches $(0, 0)$ from the direction determined by the line $y = -x/2$.

3. **(a)** All solutions are unstable spirals that become unbounded as t increases.

5. **(a)** All solutions approach $(0, 0)$ from the direction specified by the line $y = x$.

7. **(a)** If $\mathbf{X}(0) = \mathbf{X}_0$ lies on the line $y = 3x$, then $\mathbf{X}(t)$ approaches $(0, 0)$ along this line. For all other initial conditions, $\mathbf{X}(t)$ becomes unbounded and $y = x$ serves as the asymptote.

9. saddle point **11.** saddle point

13. degenerate stable node **15.** stable spiral **17.** $|\mu| < 1$

19. $\mu < -1$ for a saddle point; $-1 < \mu < 3$ for an unstable spiral point

23. **(a)** $(-3, 4)$
(b) unstable node or saddle point
(c) $(0, 0)$ is a saddle point.

25. **(a)** $(\frac{1}{2}, 2)$
(b) unstable spiral point
(c) $(0, 0)$ is an unstable spiral point.

EXERCISES 10.3 (PAGE 467)

1. $r = r_0 e^{\alpha t}$
3. $x = 0$ is unstable; $x = n + 1$ is asymptotically stable.
5. $T = T_0$ is unstable.
7. $x = \alpha$ is unstable; $x = \beta$ is asymptotically stable.
9. $P = a/b$ is asymptotically stable; $P = c$ is unstable.
11. $(\frac{1}{2}, 1)$ is a stable spiral point.
13. $(\sqrt{2}, 0)$ and $(-\sqrt{2}, 0)$ are saddle points; $(\frac{1}{2}, -\frac{7}{4})$ is a stable spiral point.
15. $(1, 1)$ is a stable node; $(1, -1)$ is a saddle point; $(2, 2)$ is a saddle point; $(2, -2)$ is an unstable spiral point.
17. $(0, -1)$ is a saddle point; $(0, 0)$ is unclassified; $(0, 1)$ is stable but we are unable to classify further.
19. $(0, 0)$ is an unstable node; $(10, 0)$ is a saddle point; $(0, 16)$ is a saddle point; $(4, 12)$ is a stable node.
21. $\theta = 0$ is a saddle point. It is not possible to classify either $\theta = \pi/3$ or $\theta = -\pi/3$.
23. It is not possible to classify $x = 0$.
25. It is not possible to classify $x = 0$, but $x = 1/\sqrt{\epsilon}$ and $x = -1/\sqrt{\epsilon}$ are each saddle points.
29. (a) $(0, 0)$ is a stable spiral point.
33. (a) $(1, 0), (-1, 0)$
35. $|v_0| < \frac{1}{2}\sqrt{2}$
37. If $\beta > 0$, $(0, 0)$ is the only critical point and is stable. If $\beta < 0$, $(0, 0)$, $(\hat{x}, 0)$, and $(-\hat{x}, 0)$, where $\hat{x}^2 = -\alpha/\beta$, are critical points. $(0, 0)$ is stable, while $(\hat{x}, 0)$ and $(-\hat{x}, 0)$ are each saddle points.
39. (b) $(5\pi/6, 0)$ is a saddle point.
   (c) $(\pi/6, 0)$ is a center.

EXERCISES 10.4 (PAGE 477)

1. $|\omega_0| < \sqrt{\dfrac{3g}{L}}$

5. (a) First show that $y^2 = v_0^2 + g \ln\left(\dfrac{1 + x^2}{1 + x_0^2}\right)$.

9. (a) The new critical point is $(d/c - \epsilon_2/c, a/b + \epsilon_1/b)$.
   (b) yes
11. $(0, 0)$ is an unstable node, $(0, 100)$ is a stable node, $(50, 0)$ is a stable node, and $(20, 40)$ is a saddle point.
17. (a) $(0, 0)$ is the only critical point.

CHAPTER 10 IN REVIEW (PAGE 480)

1. true   **3.** a center or a saddle point   **5.** false
7. false   **9.** $\alpha = -1$
11. $r = 1/\sqrt[3]{3t + 1}$, $\theta = t$. The solution curve spirals toward the origin.
13. (a) center   **(b)** degenerate stable node
15. $(0, 0)$ is a stable critical point for $\alpha \le 0$.
17. $x = 1$ is unstable; $x = -1$ is asymptotically stable.
19. The system is overdamped when $\beta^2 > 12kms^2$ and underdamped when $\beta^2 < 12kms^2$.

EXERCISES 11.1 (PAGE 488)

7. $\dfrac{\sqrt{\pi}}{2}$ **9.** $\sqrt{\dfrac{\pi}{2}}$ **11.** $\|1\| = \sqrt{p}$; $\left\|\cos\dfrac{n\pi}{p}x\right\| = \sqrt{\dfrac{p}{2}}$

21. (a) $T = 1$ **(b)** $T = \pi L/2$
   (c) $T = 2\pi$ **(d)** $T = \pi$
   (e) $T = 2\pi$ **(f)** $T = 2p$

EXERCISES 11.2 (PAGE 494)

1. $f(x) = \dfrac{1}{2} + \dfrac{1}{\pi}\displaystyle\sum_{n=1}^{\infty}\dfrac{1 - (-1)^n}{n}\sin nx$

3. $f(x) = \dfrac{3}{4} + \displaystyle\sum_{n=1}^{\infty}\left\{\dfrac{(-1)^n - 1}{n^2\pi^2}\cos n\pi x - \dfrac{1}{n\pi}\sin n\pi x\right\}$

5. $f(x) = \dfrac{\pi^2}{6} + \displaystyle\sum_{n=1}^{\infty}\left\{\dfrac{2(-1)^n}{n^2}\cos nx \right.$
$\left. + \left(\dfrac{(-1)^{n+1}\pi}{n} + \dfrac{2}{\pi n^3}[(-1)^n - 1]\right)\sin nx\right\}$

7. $f(x) = \pi + 2\displaystyle\sum_{n=1}^{\infty}\dfrac{(-1)^{n+1}}{n}\sin nx$

9. $f(x) = \dfrac{1}{\pi} + \dfrac{1}{2}\sin x + \dfrac{1}{\pi}\displaystyle\sum_{n=2}^{\infty}\dfrac{(-1)^n + 1}{1 - n^2}\cos nx$

11. $f(x) = -\dfrac{1}{4} + \dfrac{1}{\pi}\displaystyle\sum_{n=1}^{\infty}\left\{-\dfrac{1}{n}\sin\dfrac{n\pi}{2}\cos\dfrac{n\pi}{2}x \right.$
$\left. + \dfrac{3}{n}\left(1 - \cos\dfrac{n\pi}{2}\right)\sin\dfrac{n\pi}{2}x\right\}$

13. $f(x) = \dfrac{9}{4} + 5\displaystyle\sum_{n=1}^{\infty}\left\{\dfrac{(-1)^n - 1}{n^2\pi^2}\cos\dfrac{n\pi}{5}x \right.$
$\left. + \dfrac{(-1)^{n+1}}{n\pi}\sin\dfrac{n\pi}{5}x\right\}$

15. $f(x) = \dfrac{2\sinh\pi}{\pi}\left[\dfrac{1}{2} + \displaystyle\sum_{n=1}^{\infty}\dfrac{(-1)^n}{1 + n^2}(\cos nx - n\sin nx)\right]$

19. Set $x = \pi/2$.

EXERCISES 11.3 (PAGE 501)

1. odd   **3.** neither even nor odd   **5.** even   **7.** odd
9. neither even nor odd

11. $f(x) = \dfrac{2}{\pi}\displaystyle\sum_{n=1}^{\infty}\dfrac{1 - (-1)^n}{n}\sin nx$

13. $f(x) = \dfrac{\pi}{2} + \dfrac{2}{\pi}\displaystyle\sum_{n=1}^{\infty}\dfrac{(-1)^n - 1}{n^2}\cos nx$

15. $f(x) = \dfrac{1}{3} + \dfrac{4}{\pi^2}\displaystyle\sum_{n=1}^{\infty}\dfrac{(-1)^n}{n^2}\cos n\pi x$

17. $f(x) = \dfrac{2\pi^2}{3} + 4\displaystyle\sum_{n=1}^{\infty}\dfrac{(-1)^{n+1}}{n^2}\cos nx$

19. $f(x) = \dfrac{2}{\pi}\displaystyle\sum_{n=1}^{\infty}\dfrac{1 - (-1)^n(1 + \pi)}{n}\sin nx$

21. $f(x) = \dfrac{3}{4} + \dfrac{4}{\pi^2}\displaystyle\sum_{n=1}^{\infty}\dfrac{\cos\dfrac{n\pi}{2} - 1}{n^2}\cos\dfrac{n\pi}{2}x$

23. $f(x) = \dfrac{2}{\pi} + \dfrac{2}{\pi}\displaystyle\sum_{n=2}^{\infty}\dfrac{1 + (-1)^n}{1 - n^2}\cos nx$

25. $f(x) = \dfrac{1}{2} + \dfrac{2}{\pi}\displaystyle\sum_{n=1}^{\infty}\dfrac{\sin\dfrac{n\pi}{2}}{n}\cos n\pi x$

$f(x) = \dfrac{2}{\pi}\displaystyle\sum_{n=1}^{\infty}\dfrac{1 - \cos\dfrac{n\pi}{2}}{n}\sin n\pi x$

27. $f(x) = \dfrac{2}{\pi} + \dfrac{4}{\pi} \displaystyle\sum_{n=1}^{\infty} \dfrac{(-1)^n}{1 - 4n^2} \cos 2nx$

$f(x) = \dfrac{8}{\pi} \displaystyle\sum_{n=1}^{\infty} \dfrac{n}{4n^2 - 1} \sin 2nx$

29. $f(x) = \dfrac{\pi}{4} + \dfrac{2}{\pi} \displaystyle\sum_{n=1}^{\infty} \dfrac{2 \cos \frac{n\pi}{2} - (-1)^n - 1}{n^2} \cos nx$

$f(x) = \dfrac{4}{\pi} \displaystyle\sum_{n=1}^{\infty} \dfrac{\sin \frac{n\pi}{2}}{n^2} \sin nx$

31. $f(x) = \dfrac{3}{4} + \dfrac{4}{\pi^2} \displaystyle\sum_{n=1}^{\infty} \dfrac{\cos \frac{n\pi}{2} - 1}{n^2} \cos \dfrac{n\pi}{2} x$

$f(x) = \displaystyle\sum_{n=1}^{\infty} \left\{ \dfrac{4}{n^2\pi^2} \sin \dfrac{n\pi}{2} - \dfrac{2}{n\pi}(-1)^n \right\} \sin \dfrac{n\pi}{2} x$

33. $f(x) = \dfrac{5}{6} + \dfrac{2}{\pi^2} \displaystyle\sum_{n=1}^{\infty} \dfrac{3(-1)^n - 1}{n^2} \cos n\pi x$

$f(x) = 4 \displaystyle\sum_{n=1}^{\infty} \left\{ \dfrac{(-1)^{n+1}}{n\pi} + \dfrac{(-1)^n - 1}{n^3\pi^3} \right\} \sin n\pi x$

35. $f(x) = \dfrac{4\pi^2}{3} + 4 \displaystyle\sum_{n=1}^{\infty} \left\{ \dfrac{1}{n^2} \cos nx - \dfrac{\pi}{n} \sin nx \right\}$

37. $f(x) = \dfrac{3}{2} - \dfrac{1}{\pi} \displaystyle\sum_{n=1}^{\infty} \dfrac{1}{n} \sin 2n\pi x$

39. $x_p(t) = \dfrac{10}{\pi} \displaystyle\sum_{n=1}^{\infty} \dfrac{1 - (-1)^n}{n(10 - n^2)} \sin nt$

41. $x_p(t) = \dfrac{\pi^2}{18} + 16 \displaystyle\sum_{n=1}^{\infty} \dfrac{1}{n^2(n^2 - 48)} \cos nt$

43. $x(t) = \dfrac{10}{\pi} \displaystyle\sum_{n=1}^{\infty} \dfrac{1 - (-1)^n}{10 - n^2} \left[\dfrac{1}{n} \sin nt - \dfrac{1}{\sqrt{10}} \sin \sqrt{10}t \right]$

45. (b) $y_p(x) = \dfrac{2w_0 L^4}{EI\pi^5} \displaystyle\sum_{n=1}^{\infty} \dfrac{(-1)^{n+1}}{n^5} \sin \dfrac{n\pi}{L} x$

47. $y_p(x) = \dfrac{w_0}{2k} + \dfrac{2w_0}{\pi} \displaystyle\sum_{n=1}^{\infty} \dfrac{\sin(n\pi/2)}{n(EIn^4 + k)} \cos nx$

EXERCISES 11.4 (PAGE 510)

1. $y = \cos \sqrt{\lambda_n}\, x;\ \cot \sqrt{\lambda} = \sqrt{\lambda};\ 0.7402,\ 11.7349,\ 41.4388,\ 90.8082;\ \cos 0.8603x,\ \cos 3.4256x,\ \cos 6.4373x,\ \cos 9.5293x$

5. $\frac{1}{2}(1 + \sin^2 \sqrt{\lambda_n})$

7. (a) $\lambda = \left(\dfrac{n\pi}{\ln 5}\right)^2,\ y = \sin\left(\dfrac{n\pi}{\ln 5} \ln x\right),\ n = 1, 2, 3, \ldots$

(b) $\dfrac{d}{dx}[xy'] + \dfrac{\lambda}{x} y = 0$

(c) $\displaystyle\int_1^5 \dfrac{1}{x} \sin\left(\dfrac{m\pi}{\ln 5} \ln x\right) \sin\left(\dfrac{n\pi}{\ln 5} \ln x\right) dx = 0, m \neq n$

9. $\dfrac{d}{dx}[xe^{-x}y'] + ne^{-x}y = 0;\ \displaystyle\int_0^{\infty} e^{-x} L_m(x) L_n(x)\, dx = 0,$
$m \neq n$

11. (a) $\lambda = 16n^2,\ y = \sin(4n \tan^{-1}x),\ n = 1, 2, 3, \ldots$

(b) $\displaystyle\int_0^1 \dfrac{1}{1 + x^2} \sin(4m \tan^{-1}x) \sin(4n \tan^{-1}x)\, dx = 0,$
$m \neq n$

EXERCISES 11.5 (PAGE 517)

1. 1.277, 2.339, 3.391, 4.441

3. $f(x) = \displaystyle\sum_{i=1}^{\infty} \dfrac{1}{\lambda_i J_1(2\lambda_i)} J_0(\lambda_i x)$

5. $f(x) = 4 \displaystyle\sum_{i=1}^{\infty} \dfrac{\lambda_i J_1(2\lambda_i)}{(4\lambda_i^2 + 1) J_0^2(2\lambda_i)} J_0(\lambda_i x)$

7. $f(x) = 20 \displaystyle\sum_{i=1}^{\infty} \dfrac{\lambda_i J_2(4\lambda_i)}{(2\lambda_i^2 + 1) J_1^2(4\lambda_i)} J_1(\lambda_i x)$

9. $f(x) = \dfrac{9}{2} - 4 \displaystyle\sum_{i=1}^{\infty} \dfrac{J_2(3\lambda_i)}{\lambda_i^2 J_0^2(3\lambda_i)} J_0(\lambda_i x)$

13. $f(x) = \frac{1}{4}P_0(x) + \frac{1}{2}P_1(x) + \frac{5}{16}P_2(x) - \frac{3}{32}P_4(x) + \cdots$

19. $f(x) = \frac{1}{2}P_0(x) + \frac{5}{8}P_2(x) - \frac{3}{16}P_4(x) + \cdots,$
$f(x) = |x|$ on $(-1, 1)$

CHAPTER 11 IN REVIEW (PAGE 519)

1. true **3.** cosine **5.** false **7.** 5.5, 1, 0

9. $\dfrac{1}{\sqrt{1 - x^2}},\ -1 \leq x \leq 1,\ \displaystyle\int_{-1}^1 \dfrac{1}{\sqrt{1 - x^2}} T_m(x) T_n(x)\, dx = 0,$
$m \neq n$

13. $f(x) = \dfrac{1}{2} + \dfrac{2}{\pi} \displaystyle\sum_{n=1}^{\infty} \left\{ \dfrac{1}{n^2\pi}[(-1)^n - 1] \cos n\pi x \right.$
$\left. + \dfrac{2}{n}(-1)^n \sin n\pi x \right\}$

15. $f(x) = 1 - e^{-1} + 2 \displaystyle\sum_{n=1}^{\infty} \dfrac{1 - (-1)^n e^{-1}}{1 + n^2\pi^2} \cos n\pi x,$
$f(x) = \displaystyle\sum_{n=1}^{\infty} \dfrac{2n\pi[1 - (-1)^n e^{-1}]}{1 + n^2\pi^2} \sin n\pi x$

17. $\lambda = \dfrac{(2n - 1)^2\pi^2}{36},\ n = 1, 2, 3, \ldots,$
$y = \cos\left(\dfrac{2n - 1}{2} \pi \ln x\right)$

19. $f(x) = \dfrac{1}{4} \displaystyle\sum_{i=1}^{\infty} \dfrac{J_1(2\lambda_i)}{\lambda_i J_1^2(4\lambda_i)} J_0(\lambda_i x)$

EXERCISES 12.1 (PAGE 525)

1. The possible cases can be summarized in one form $u = c_1 e^{c_2(x+y)}$, where c_1 and c_2 are constants.

3. $u = c_1 e^{y + c_2(x-y)}$ **5.** $u = c_1(xy)^{c_2}$

7. not separable

9. $u = e^{-t}(A_1 e^{k\lambda^2 t} \cosh \lambda x + B_1 e^{k\lambda^2 t} \sinh \lambda x)$
$u = e^{-t}(A_2 e^{-k\lambda^2 t} \cos \lambda x + B_2 e^{-k\lambda^2 t} \sin \lambda x)$
$u = (c_7 x + c_8)c_9 e^{-t}$

11. $u = (c_1 \cosh \lambda x + c_2 \sinh \lambda x)$
$\quad \times (c_3 \cosh \lambda at + c_4 \sinh \lambda at)$
$u = (c_5 \cos \lambda x + c_6 \sin \lambda x)(c_7 \cos \lambda at + c_8 \sin \lambda at)$
$u = (c_9 x + c_{10})(c_{11}t + c_{12})$

13. $u = (c_1 \cosh \lambda x + c_2 \sinh \lambda x)(c_3 \cos \lambda y + c_4 \sin \lambda y)$
$u = (c_5 \cos \lambda x + c_6 \sin \lambda x)(c_7 \cosh \lambda y + c_8 \sinh \lambda y)$
$u = (c_9 x + c_{10})(c_{11}y + c_{12})$

15. For $\lambda^2 > 0$ there are three possibilities:
$\quad u = (c_1 \cosh \lambda x + c_2 \sinh \lambda x)$
$\qquad \times (c_3 \cosh \sqrt{1 - \lambda^2}y + c_4 \sinh \sqrt{1 - \lambda^2}y),$
$\quad \lambda^2 < 1$

$$u = (c_1 \cosh \lambda x + c_2 \sinh \lambda x)$$
$$\times (c_3 \cos \sqrt{\lambda^2 - 1}\, y + c_4 \sin \sqrt{\lambda^2 - 1}\, y),$$
$$\lambda^2 > 1$$
$$u = (c_1 \cosh x + c_2 \sinh x)(c_3 y + c_4),$$
$$\lambda^2 = 1$$

The results for the case $-\lambda^2 < 0$ are similar. For $\lambda^2 = 0$ we have

$$u = (c_1 x + c_2)(c_3 \cosh y + c_4 \sinh y).$$

17. elliptic   **19.** parabolic   **21.** hyperbolic

23. parabolic   **25.** hyperbolic

EXERCISES 12.2 (PAGE 532)

1. $k \dfrac{\partial^2 u}{\partial x^2} = \dfrac{\partial u}{\partial t}, 0 < x < L, t > 0$

$u(0, t) = 0, \dfrac{\partial u}{\partial x}\Big|_{x=L} = 0, t > 0$

$u(x, 0) = f(x), 0 < x < L$

3. $k \dfrac{\partial^2 u}{\partial x^2} = \dfrac{\partial u}{\partial t}, 0 < x < L, t > 0$

$u(0, t) = 100, \dfrac{\partial u}{\partial x}\Big|_{x=L} = -hu(L, t), t > 0$

$u(x, 0) = f(x), 0 < x < L$

5. $a^2 \dfrac{\partial^2 u}{\partial x^2} = \dfrac{\partial^2 u}{\partial t^2}, 0 < x < L, t > 0$

$u(0, t) = 0, u(L, t) = 0, t > 0$

$u(x, 0) = x(L - x), \dfrac{\partial u}{\partial t}\Big|_{t=0} = 0, 0 < x < L$

7. $a^2 \dfrac{\partial^2 u}{\partial x^2} - 2\beta \dfrac{\partial u}{\partial t} = \dfrac{\partial^2 u}{\partial t^2}, 0 < x < L, t > 0$

$u(0, t) = 0, u(L, t) = \sin \pi t, t > 0$

$u(x, 0) = f(x), \dfrac{\partial u}{\partial t}\Big|_{t=0} = 0, 0 < x < L$

9. $\dfrac{\partial^2 u}{\partial x^2} + \dfrac{\partial^2 u}{\partial y^2} = 0, 0 < x < 4, 0 < y < 2$

$\dfrac{\partial u}{\partial x}\Big|_{x=0} = 0, u(4, y) = f(y), 0 < y < 2$

$\dfrac{\partial u}{\partial y}\Big|_{y=0} = 0, u(x, 2) = 0, 0 < x < 4$

EXERCISES 12.3 (PAGE 535)

1. $u(x, t) = \dfrac{2}{\pi} \displaystyle\sum_{n=1}^{\infty} \left(\dfrac{-\cos \dfrac{n\pi}{2} + 1}{n} \right) e^{-k(n^2\pi^2/L^2)t} \sin \dfrac{n\pi}{L} x$

3. $u(x, t) = \dfrac{1}{L} \displaystyle\int_0^L f(x)\, dx$
$+ \dfrac{2}{L} \displaystyle\sum_{n=1}^{\infty} \left(\int_0^L f(x) \cos \dfrac{n\pi}{L} x\, dx \right) e^{-k(n^2\pi^2/L^2)t} \cos \dfrac{n\pi}{L} x$

5. $u(x, t) = e^{-ht} \left[\dfrac{1}{L} \displaystyle\int_0^L f(x)\, dx \right.$
$\left. + \dfrac{2}{L} \displaystyle\sum_{n=1}^{\infty} \left(\int_0^L f(x) \cos \dfrac{n\pi}{L} x\, dx \right) e^{-k(n^2\pi^2/L^2)t} \cos \dfrac{n\pi}{L} x \right]$

EXERCISES 12.4 (PAGE 539)

1. $u(x, t) = \dfrac{L^2}{\pi^3} \displaystyle\sum_{n=1}^{\infty} \dfrac{1 - (-1)^n}{n^3} \cos \dfrac{n\pi a}{L} t \sin \dfrac{n\pi}{L} x$

3. $u(x, t) = \dfrac{6\sqrt{3}}{\pi^2} \left(\cos \dfrac{\pi a}{L} t \sin \dfrac{\pi}{L} x \right.$
$- \dfrac{1}{5^2} \cos \dfrac{5\pi a}{L} t \sin \dfrac{5\pi}{L} x$
$\left. + \dfrac{1}{7^2} \cos \dfrac{7\pi a}{L} t \sin \dfrac{7\pi}{L} x - \cdots \right)$

5. $u(x, t) = \dfrac{1}{a} \sin at \sin x$

7. $u(x, t) = \dfrac{8h}{\pi^2} \displaystyle\sum_{n=1}^{\infty} \dfrac{\sin \dfrac{n\pi}{2}}{n^2} \cos \dfrac{n\pi a}{L} t \sin \dfrac{n\pi}{L} x$

9. $u(x, t) = e^{-\beta t} \displaystyle\sum_{n=1}^{\infty} A_n \left\{ \cos q_n t + \dfrac{\beta}{q_n} \sin q_n t \right\} \sin nx$,

where $A_n = \dfrac{2}{\pi} \displaystyle\int_0^\pi f(x) \sin nx\, dx$ and $q_n = \sqrt{n^2 - \beta^2}$

11. $u(x, t) = \displaystyle\sum_{n=1}^{\infty} \left(A_n \cos \dfrac{n^2\pi^2}{L^2} at + B_n \sin \dfrac{n^2\pi^2}{L^2} at \right)$
$\times \sin \dfrac{n\pi}{L} x$,

where $A_n = \dfrac{2}{L} \displaystyle\int_0^L f(x) \sin \dfrac{n\pi}{L} x\, dx$

$B_n = \dfrac{2L}{n^2\pi^2 a} \displaystyle\int_0^L g(x) \sin \dfrac{n\pi}{L} x\, dx$

15. $u(x, t) = \sin x \cos 2at + t$

17. $u(x, t) = \dfrac{1}{2a} \sin 2x \sin 2at$

EXERCISES 12.5 (PAGE 546)

1. $u(x, y) = \dfrac{2}{a} \displaystyle\sum_{n=1}^{\infty} \left(\dfrac{1}{\sinh \dfrac{n\pi}{a} b} \int_0^a f(x) \sin \dfrac{n\pi}{a} x\, dx \right)$
$\times \sinh \dfrac{n\pi}{a} y \sin \dfrac{n\pi}{a} x$

3. $u(x, y) = \dfrac{2}{a} \displaystyle\sum_{n=1}^{\infty} \left(\dfrac{1}{\sinh \dfrac{n\pi}{a} b} \int_0^a f(x) \sin \dfrac{n\pi}{a} x\, dx \right)$
$\times \sinh \dfrac{n\pi}{a} (b - y) \sin \dfrac{n\pi}{a} x$

5. $u(x, y) = \dfrac{1}{2} x + \dfrac{2}{\pi^2} \displaystyle\sum_{n=1}^{\infty} \dfrac{1 - (-1)^n}{n^2 \sinh n\pi} \sinh n\pi x \cos n\pi y$

7. $u(x, y) = \dfrac{2}{\pi} \displaystyle\sum_{n=1}^{\infty} \dfrac{[1 - (-1)^n]}{n}$
$\times \dfrac{n \cosh nx + \sinh nx}{n \cosh n\pi + \sinh n\pi} \sin ny$

9. $u(x, y) = \displaystyle\sum_{n=1}^{\infty} (A_n \cosh n\pi y + B_n \sinh n\pi y) \sin n\pi x$,

where $A_n = 200 \dfrac{[1 - (-1)^n]}{n\pi}$

$B_n = 200 \dfrac{[1 - (-1)^n]}{n\pi} \dfrac{[2 - \cosh n\pi]}{\sinh n\pi}$

11. $u(x, y) = \dfrac{2}{\pi} \displaystyle\sum_{n=1}^{\infty} \left(\int_0^\pi f(x) \sin nx \, dx \right) e^{-ny} \sin nx$

13. $u(x, y) = \displaystyle\sum_{n=1}^{\infty} \left(A_n \cosh \dfrac{n\pi}{a} y + B_n \sinh \dfrac{n\pi}{a} y \right) \sin \dfrac{n\pi}{a} x$,

where $A_n = \dfrac{2}{a} \displaystyle\int_0^a f(x) \sin \dfrac{n\pi}{a} x \, dx$

$B_n = \dfrac{1}{\sinh \dfrac{n\pi}{a} b} \left(\dfrac{2}{a} \displaystyle\int_0^a g(x) \sin \dfrac{n\pi}{a} x \, dx - A_n \cosh \dfrac{n\pi}{a} b \right)$

15. $u = u_1 + u_2$, where

$u_1(x, y) = \dfrac{2}{\pi} \displaystyle\sum_{n=1}^{\infty} \dfrac{1 - (-1)^n}{n \sinh n\pi} \sinh ny \sin nx$

$u_2(x, y) = \dfrac{2}{\pi} \displaystyle\sum_{n=1}^{\infty} \dfrac{[1 - (-1)^n]}{n}$

$\times \dfrac{\sinh nx + \sinh n(\pi - x)}{\sinh n\pi} \sin ny$

EXERCISES 12.6 (PAGE 549)

1. $u(x, t) = 100 + \dfrac{200}{\pi} \displaystyle\sum_{n=1}^{\infty} \dfrac{(-1)^n - 1}{n} e^{-kn^2\pi^2 t} \sin n\pi x$

3. $u(x, t) = u_0 - \dfrac{r}{2k} x(x - 1) + 2 \displaystyle\sum_{n=1}^{\infty} \left[\dfrac{u_0}{n\pi} + \dfrac{r}{kn^3\pi^3} \right]$

$\times [(-1)^n - 1] e^{-kn^2\pi^2 t} \sin n\pi x$

5. $u(x, t) = \psi(x) + \displaystyle\sum_{n=1}^{\infty} A_n e^{-kn^2\pi^2 t} \sin n\pi x$,

where $\psi(x) = \dfrac{A}{k\beta^2} [-e^{-\beta x} + (e^{-\beta} - 1)x + 1]$

and $A_n = 2 \displaystyle\int_0^1 [f(x) - \psi(x)] \sin n\pi x \, dx$

7. $\psi(x) = u_0 \left(1 - \dfrac{\sinh \sqrt{h/k} \, x}{\sinh \sqrt{h/k}} \right)$

9. $u(x, t) = \dfrac{A}{6a^2} (x - x^3)$

$+ \dfrac{2A}{a^2\pi^3} \displaystyle\sum_{n=1}^{\infty} \dfrac{(-1)^n}{n^3} \cos n\pi a t \sin n\pi x$

11. $u(x, y) = (u_0 - u_1)y + u_1$

$+ \dfrac{2}{\pi} \displaystyle\sum_{n=1}^{\infty} \dfrac{u_0(-1)^n - u_1}{n} e^{-n\pi x} \sin n\pi y$

EXERCISES 12.7 (PAGE 554)

1. $u(x, t) = 2h \displaystyle\sum_{n=1}^{\infty} \dfrac{\sin \lambda_n}{\lambda_n[h + \sin^2\lambda_n]} e^{-k\lambda_n^2 t} \cos \lambda_n x$,

where the λ_n are the consecutive positive roots of $\cot \lambda = \lambda/h$

3. $u(x, y) = \displaystyle\sum_{n=1}^{\infty} A_n \sinh \lambda_n y \sin \lambda_n x$,

where $A_n = \dfrac{2h}{\sinh \lambda_n b [ah + \cos^2\lambda_n a]} \displaystyle\int_0^a f(x) \sin \lambda_n x \, dx$ and the λ_n are the consecutive positive roots of $\tan \lambda a = -\lambda/h$

5. $u(x, t) = \displaystyle\sum_{n=1}^{\infty} A_n e^{-k(2n-1)^2\pi^2 t/4L^2} \sin\left(\dfrac{2n - 1}{2L} \right)\pi x$,

where $A_n = \dfrac{2}{L} \displaystyle\int_0^L f(x) \sin\left(\dfrac{2n - 1}{2L} \right)\pi x \, dx$

7. $u(x, y) = \dfrac{4u_0}{\pi} \displaystyle\sum_{n=1}^{\infty} \dfrac{1}{(2n - 1) \cosh\left(\dfrac{2n - 1}{2} \right)\pi}$

$\times \cosh\left(\dfrac{2n - 1}{2} \right)\pi x \sin\left(\dfrac{2n - 1}{2} \right)\pi y$

9. (b) 1.8751, 4.6941

EXERCISES 12.8 (PAGE 558)

1. $u(x, y, t) = \displaystyle\sum_{m=1}^{\infty} \sum_{n=1}^{\infty} A_{mn} e^{-k(m^2 + n^2)t} \sin mx \sin ny$,

where $A_{mn} = \dfrac{4u_0}{mn\pi^2} [1 - (-1)^m][1 - (-1)^n]$

3. $u(x, y, t) = \displaystyle\sum_{m=1}^{\infty} \sum_{n=1}^{\infty} A_{mn} \sin mx \sin ny \cos a\sqrt{m^2 + n^2} t$,

where $A_{mn} = \dfrac{16}{m^3 n^3 \pi^2} [(-1)^m - 1][(-1)^n - 1]$

5. $u(x, y, z) = \displaystyle\sum_{m=1}^{\infty} \sum_{n=1}^{\infty} A_{mn} \sinh \omega_{mn} z \sin \dfrac{m\pi}{a} x \sin \dfrac{n\pi}{b} y$,

where $\omega_{mn} = \sqrt{(m\pi/a)^2 + (n\pi/b)^2}$

$A_{mn} = \dfrac{4}{ab \sinh(c\omega_{mn})} \displaystyle\int_0^b \int_0^a f(x, y)$

$\times \sin \dfrac{m\pi}{a} x \sin \dfrac{n\pi}{b} y \, dx \, dy$

CHAPTER 12 IN REVIEW (PAGE 559)

1. $u = c_1 e^{(c_2 x + y/c_2)}$ **3.** $\psi(x) = u_0 + \dfrac{(u_1 - u_0)}{1 + \pi} x$

5. $u(x, t) = \dfrac{2h}{\pi^2 a} \displaystyle\sum_{n=1}^{\infty} \dfrac{\cos \dfrac{n\pi}{4} - \cos \dfrac{3n\pi}{4}}{n^2} \sin n\pi a t \sin n\pi x$

7. $u(x, y) = \dfrac{100}{\pi} \displaystyle\sum_{n=1}^{\infty} \dfrac{1 - (-1)^n}{n \sinh n\pi} \sinh nx \sin ny$

9. $u(x, y) = \dfrac{100}{\pi} \displaystyle\sum_{n=1}^{\infty} \dfrac{1 - (-1)^n}{n} e^{-nx} \sin ny$

11. $u(x, t) = e^{-t} \sin x$

13. $u(x, t) = e^{-(x+t)} \displaystyle\sum_{n=1}^{\infty} A_n [\sqrt{n^2 + 1} \cos \sqrt{n^2 + 1} t$

$+ \sin \sqrt{n^2 + 1} t] \sin nx$

EXERCISES 13.1 (PAGE 565)

1. $u(r, \theta) = \dfrac{u_0}{2} + \dfrac{u_0}{\pi} \displaystyle\sum_{n=1}^{\infty} \dfrac{1 - (-1)^n}{n} r^n \sin n\theta$

3. $u(r, \theta) = \dfrac{2\pi^2}{3} - 4 \displaystyle\sum_{n=1}^{\infty} \dfrac{r^n}{n^2} \cos n\theta$

5. $u(r, \theta) = A_0 + \displaystyle\sum_{n=1}^{\infty} r^{-n}(A_n \cos n\theta + B_n \sin n\theta)$,

where $A_0 = \dfrac{1}{2\pi} \displaystyle\int_0^{2\pi} f(\theta) \, d\theta$

$A_n = \dfrac{c^n}{\pi} \displaystyle\int_0^{2\pi} f(\theta) \cos n\theta \, d\theta$

$B_n = \dfrac{c^n}{\pi} \displaystyle\int_0^{2\pi} f(\theta) \sin n\theta \, d\theta$

7. $u(r, \theta) = \dfrac{1}{2} + \dfrac{2}{\pi} \displaystyle\sum_{n=1}^{\infty} \dfrac{\sin \frac{n\pi}{2}}{n} \left(\dfrac{r}{c}\right)^{2n} \cos 2n\theta$

9. $u(r, \theta) = A_0 \ln\left(\dfrac{r}{b}\right) + \displaystyle\sum_{n=1}^{\infty}\left[\left(\dfrac{b}{r}\right)^{n} - \left(\dfrac{r}{b}\right)^{n}\right]$

$\times\, [A_n \cos n\theta + B_n \sin n\theta],$

where $A_0 \ln\left(\dfrac{a}{b}\right) = \dfrac{1}{2\pi}\displaystyle\int_0^{2\pi} f(\theta)\, d\theta$

$\left[\left(\dfrac{b}{a}\right)^{n} - \left(\dfrac{a}{b}\right)^{n}\right] A_n = \dfrac{1}{\pi}\displaystyle\int_0^{2\pi} f(\theta)\cos n\theta\, d\theta$

$\left[\left(\dfrac{b}{a}\right)^{n} - \left(\dfrac{a}{b}\right)^{n}\right] B_n = \dfrac{1}{\pi}\displaystyle\int_0^{2\pi} f(\theta)\sin n\theta\, d\theta$

11. $u(r, \theta) = \dfrac{4}{\pi} \displaystyle\sum_{n=1}^{\infty} \dfrac{1 - (-1)^n}{n^3} \dfrac{r^{2n} - b^{2n}}{a^{2n} - b^{2n}} \left(\dfrac{a}{r}\right)^{n} \sin n\theta$

13. $u(r, \theta) = \dfrac{u_0}{2} + \dfrac{2u_0}{\pi} \displaystyle\sum_{n=1}^{\infty} \dfrac{\sin \frac{n\pi}{2}}{n} \left(\dfrac{r}{2}\right)^{n} \cos n\theta$

EXERCISES 13.2 (PAGE 571)

1. $u(r, t) = \dfrac{2}{ac} \displaystyle\sum_{n=1}^{\infty} \dfrac{\sin \lambda_n at\, J_0(\lambda_n r)}{\lambda_n^2 J_1(\lambda_n c)}$

3. $u(r, z) = u_0 \displaystyle\sum_{n=1}^{\infty} \dfrac{\sinh \lambda_n(4 - z)\, J_0(\lambda_n r)}{\lambda_n \sinh 4\lambda_n\, J_1(2\lambda_n)}$

5. $u(r, t) = \displaystyle\sum_{n=1}^{\infty} A_n J_0(\lambda_n r) e^{-k\lambda_n^2 t}$,

where $A_n = \dfrac{2}{c^2 J_1^2(\lambda_n c)} \displaystyle\int_0^c r J_0(\lambda_n r) f(r)\, dr$

7. $u(r, t) = \displaystyle\sum_{n=1}^{\infty} A_n J_0(\lambda_n r) e^{-k\lambda_n^2 t}$,

where $A_n = \dfrac{2\lambda_n^2}{(\lambda_n^2 + h^2) J_0^2(\lambda_n)} \displaystyle\int_0^1 r J_0(\lambda_n r) f(r)\, dr$

9. $u(r, t) = 100 + 50 \displaystyle\sum_{n=1}^{\infty} \dfrac{J_1(\lambda_n) J_0(\lambda_n r)}{\lambda_n J_1^2(2\lambda_n)} e^{-\lambda_n^2 t}$

11. (b) $u(x, t) = \displaystyle\sum_{n=1}^{\infty} A_n \cos(\lambda_n \sqrt{g t}) J_0(2\lambda_n \sqrt{x})$,

where $A_n = \dfrac{2}{L J_1^2(2\lambda_n \sqrt{L})} \displaystyle\int_0^{\sqrt{L}} v J_0(2\lambda_n v) f(v^2)\, dv$

EXERCISES 13.3 (PAGE 576)

1. $u(r, \theta) = 50 \left[\dfrac{1}{2} P_0(\cos \theta) + \dfrac{3}{4}\left(\dfrac{r}{c}\right) P_1(\cos \theta)\right.$

$\left. - \dfrac{7}{16}\left(\dfrac{r}{c}\right)^{3} P_3(\cos \theta) + \dfrac{11}{32}\left(\dfrac{r}{c}\right)^{5} P_5(\cos \theta) + \cdots\right]$

3. $u(r, \theta) = \dfrac{r}{c} \cos \theta$

5. $u(r, \theta) = \displaystyle\sum_{n=0}^{\infty} A_n \dfrac{b^{2n+1} - r^{2n+1}}{b^{2n+1} r^{n+1}} P_n(\cos \theta)$, where

$\dfrac{b^{2n+1} - a^{2n+1}}{b^{2n+1} a^{n+1}} A_n = \dfrac{2n+1}{2} \displaystyle\int_0^{\pi} f(\theta) P_n(\cos \theta) \sin \theta\, d\theta$

7. $u(r, \theta) = \displaystyle\sum_{n=0}^{\infty} A_n r^{2n} P_{2n}(\cos \theta)$,

where $A_n = \dfrac{(4n + 1)}{c^{2n}} \displaystyle\int_0^{\pi/2} f(\theta) P_{2n}(\cos \theta) \sin \theta\, d\theta$

9. $u(r, t) = 100 + \dfrac{200}{\pi r} \displaystyle\sum_{n=1}^{\infty} \dfrac{(-1)^n}{n} e^{-n^2 \pi^2 t} \sin n\pi r$

11. $u(r, t) = \displaystyle\sum_{n=1}^{\infty} \left(A_n \cos \dfrac{n\pi a}{c} t + B_n \sin \dfrac{n\pi a}{c} t\right) \sin \dfrac{n\pi}{c} r$,

where $A_n = \dfrac{2}{c} \displaystyle\int_0^c r f(r) \sin \dfrac{n\pi}{c} r\, dr$

$B_n = \dfrac{2}{n\pi a} \displaystyle\int_0^c r g(r) \sin \dfrac{n\pi}{c} r\, dr$

CHAPTER 13 IN REVIEW (PAGE 578)

1. $u(r, \theta) = \dfrac{2u_0}{\pi} \displaystyle\sum_{n=1}^{\infty} \dfrac{1 - (-1)^n}{n} \left(\dfrac{r}{c}\right)^{n} \sin n\theta$

3. $u(r, \theta) = \dfrac{4u_0}{\pi} \displaystyle\sum_{n=1}^{\infty} \dfrac{1 - (-1)^n}{n^3} \sin n\theta$

5. $u(r, \theta) = \dfrac{2u_0}{\pi} \displaystyle\sum_{n=1}^{\infty} \dfrac{r^{4n} + r^{-4n}}{2^{4n} + 2^{-4n}} \dfrac{1 - (-1)^n}{n} \sin 4n\theta$

7. $u(r, t) = 2e^{-ht} \displaystyle\sum_{n=1}^{\infty} \dfrac{J_0(\lambda_n r)}{\lambda_n J_1(\lambda_n)} e^{-\lambda_n^2 t}$

9. $u(r, z) = 50 \displaystyle\sum_{n=1}^{\infty} \dfrac{\cosh \lambda_n z\, J_0(\lambda_n r)}{\lambda_n \cosh 4\lambda_n\, J_1(2\lambda_n)}$

11. $u(r, \theta) = 100 \left[\dfrac{3}{2} r P_1(\cos \theta) - \dfrac{7}{8} r^3 P_3(\cos \theta)\right.$

$\left. + \dfrac{11}{16} r^5 P_5(\cos \theta) + \cdots\right]$

EXERCISES 14.1 (PAGE 583)

1. (a) Let $\tau = u^2$ in the integral $\text{erf}(\sqrt{t})$.

7. $y(t) = e^{\pi t} \text{erfc}(\sqrt{\pi t})$

9. Use the property $\displaystyle\int_0^b - \int_0^a = \int_0^b + \int_a^0$.

EXERCISES 14.2 (PAGE 590)

1. $u(x, t) = A \cos \dfrac{a\pi t}{L} \sin \dfrac{\pi x}{L}$

3. $u(x, t) = f\left(t - \dfrac{x}{a}\right) \mathcal{U}\left(t - \dfrac{x}{a}\right)$

5. $u(x, t) = \left[\dfrac{1}{2} g\left(t - \dfrac{x}{a}\right)^2 + A \sin \omega\left(t - \dfrac{x}{a}\right)\right]$

$\times\, \mathcal{U}\left(t - \dfrac{x}{a}\right) - \dfrac{1}{2} g t^2$

7. $u(x, t) = a\dfrac{F_0}{E} \displaystyle\sum_{n=0}^{\infty} (-1)^n \left\{\left(t - \dfrac{2nL + L - x}{a}\right)\right.$

$\times\, \mathcal{U}\left(t - \dfrac{2nL + L - x}{a}\right)$

$-\left(t - \dfrac{2nL + L + x}{a}\right)$

$\left. \times\, \mathcal{U}\left(t - \dfrac{2nL + L + x}{a}\right)\right\}$

9. $u(x, t) = (t - x) \sinh(t - x) \mathcal{U}(t - x)$
$+ xe^{-x} \cosh t - e^{-x} t \sinh t$

11. (a) $u(x, t) = u_0 \text{erfc}\left(\dfrac{x}{2\sqrt{kt}}\right)$

13. $u(x, t) = u_1 + (u_0 - u_1) \operatorname{erfc}\left(\dfrac{x}{2\sqrt{t}}\right)$

15. $u(x, t) = u_0\left[1 - \left\{\operatorname{erfc}\left(\dfrac{x}{2\sqrt{t}}\right) \right.\right.$
$\left.\left. - e^{x+t} \operatorname{erfc}\left(\sqrt{t} + \dfrac{x}{2\sqrt{t}}\right)\right\}\right]$

17. $u(x, t) = \dfrac{x}{2\sqrt{\pi}} \displaystyle\int_0^t \dfrac{f(t - \tau)}{\tau^{3/2}} e^{-x^2/4\tau} \, d\tau$

19. $u(x, t) = 60 + 40 \operatorname{erfc}\left(\dfrac{x}{2\sqrt{t - 2}}\right) \mathcal{U}(t - 2)$

21. $u(x, t) = 100\left[-e^{1-x+t} \operatorname{erfc}\left(\sqrt{t} + \dfrac{1-x}{2\sqrt{t}}\right)\right.$
$\left. + \operatorname{erfc}\left(\dfrac{1-x}{2\sqrt{t}}\right)\right]$

23. $u(x, t) = u_0 + u_0 e^{-(\pi^2/L^2)t} \sin\left(\dfrac{\pi}{L} x\right)$

25. $u(x, t) = u_0 - u_0 \displaystyle\sum_{n=0}^{\infty} (-1)^n \left[\operatorname{erfc}\left(\dfrac{2n + 1 - x}{2\sqrt{kt}}\right)\right.$
$\left. + \operatorname{erfc}\left(\dfrac{2n + 1 + x}{2\sqrt{kt}}\right)\right]$

27. $u(x, t) = u_0 e^{-Gt/C} \operatorname{erf}\left(\dfrac{x}{2}\sqrt{\dfrac{RC}{t}}\right)$

29. (a) $c(x, t) = A\sqrt{\dfrac{k}{\pi t}} e^{-x^2/(4kt)}$

EXERCISES 14.3 (PAGE 600)

1. $f(x) = \dfrac{1}{\pi} \displaystyle\int_0^\infty \dfrac{\sin \alpha \cos \alpha x + 3(1 - \cos \alpha) \sin \alpha x}{\alpha} \, d\alpha$

3. $f(x) = \dfrac{1}{\pi} \displaystyle\int_0^\infty [A(\alpha) \cos \alpha x + B(\alpha) \sin \alpha x] \, d\alpha$,

where $A(\alpha) = \dfrac{3\alpha \sin 3\alpha + \cos 3\alpha - 1}{\alpha^2}$

$B(\alpha) = \dfrac{\sin 3\alpha - 3\alpha \cos 3\alpha}{\alpha^2}$

5. $f(x) = \dfrac{1}{\pi} \displaystyle\int_0^\infty \dfrac{\cos \alpha x + \alpha \sin \alpha x}{1 + \alpha^2} \, d\alpha$

7. $f(x) = \dfrac{10}{\pi} \displaystyle\int_0^\infty \dfrac{(1 - \cos \alpha) \sin \alpha x}{\alpha} \, d\alpha$

9. $f(x) = \dfrac{2}{\pi} \displaystyle\int_0^\infty \dfrac{(\pi\alpha \sin \pi\alpha + \cos \pi\alpha - 1) \cos \alpha x}{\alpha^2} \, d\alpha$

11. $f(x) = \dfrac{4}{\pi} \displaystyle\int_0^\infty \dfrac{\alpha \sin \alpha x}{4 + \alpha^4} \, d\alpha$

13. $f(x) = \dfrac{2k}{\pi} \displaystyle\int_0^\infty \dfrac{\cos \alpha x}{k^2 + \alpha^2} \, d\alpha$

$f(x) = \dfrac{2}{\pi} \displaystyle\int_0^\infty \dfrac{\alpha \sin \alpha x}{k^2 + \alpha^2} \, d\alpha$

15. $f(x) = \dfrac{2}{\pi} \displaystyle\int_0^\infty \dfrac{(4 - \alpha^2) \cos \alpha x}{(4 + \alpha^2)^2} \, d\alpha$

$f(x) = \dfrac{8}{\pi} \displaystyle\int_0^\infty \dfrac{\alpha \sin \alpha x}{(4 + \alpha^2)^2} \, d\alpha$

17. $f(x) = \dfrac{2}{\pi} \dfrac{1}{1 + x^2}, \quad x > 0$

19. Let $x = 2$ in (7). Use a trigonometric identity and replace α by x. In part (b) make the change of variable $2x = kt$.

EXERCISES 14.4 (PAGE 605)

1. $u(x, t) = \dfrac{1}{\pi} \displaystyle\int_{-\infty}^{\infty} \dfrac{e^{-k\alpha^2 t}}{1 + \alpha^2} e^{-i\alpha x} \, d\alpha$
$= \dfrac{1}{\pi} \displaystyle\int_{-\infty}^{\infty} \dfrac{\cos \alpha x}{1 + \alpha^2} e^{-k\alpha^2 t} \, d\alpha$

3. $u(x, t) = \dfrac{1}{\sqrt{1 + 4kt}} e^{-x^2/(1+4kt)}$

7. $u(x, t) = \dfrac{2}{\pi} \displaystyle\int_0^\infty \dfrac{1 - \cos \alpha}{\alpha} e^{-k\alpha^2 t} \sin \alpha x \, d\alpha$

9. $u(x, t) = \dfrac{2}{\pi} \displaystyle\int_0^\infty \dfrac{\sin \alpha}{\alpha} e^{-k\alpha^2 t} \cos \alpha x \, d\alpha$

11. (a) $u(x, t) = \dfrac{1}{2\pi} \displaystyle\int_{-\infty}^{\infty} \left(F(\alpha) \cos \alpha a t\right.$
$\left. + G(\alpha) \dfrac{\sin \alpha a t}{\alpha a}\right) e^{-i\alpha x} \, d\alpha$

13. $u(x, y) = \dfrac{2}{\pi} \displaystyle\int_0^\infty \dfrac{\sinh \alpha(\pi - x)}{(1 + \alpha^2) \sinh \alpha\pi} \cos \alpha y \, d\alpha$

15. $u(x, y) = \dfrac{100}{\pi} \displaystyle\int_0^\infty \dfrac{\sin \alpha}{\alpha} e^{-\alpha y} \cos \alpha x \, d\alpha$

17. $u(x, y) = \dfrac{2}{\pi} \displaystyle\int_0^\infty F(\alpha) \dfrac{\sinh \alpha(2 - y)}{\sinh 2\alpha} \sin \alpha x \, d\alpha$

19. $u(x, y) = \dfrac{2}{\pi} \displaystyle\int_0^\infty \dfrac{\alpha}{1 + \alpha^2} [e^{-\alpha x} \sin \alpha y + e^{-\alpha y} \sin \alpha x] \, d\alpha$

21. $u(x, y) = \dfrac{1}{2\sqrt{\pi}} \displaystyle\int_{-\infty}^{\infty} \dfrac{e^{-\alpha^2/4} \cosh \alpha y}{\cosh \alpha} e^{-i\alpha x} \, d\alpha$
$= \dfrac{1}{2\sqrt{\pi}} \displaystyle\int_{-\infty}^{\infty} \dfrac{e^{-\alpha^2/4} \cosh \alpha y}{\cosh \alpha} \cos \alpha x \, d\alpha$

CHAPTER 14 IN REVIEW (PAGE 607)

1. $u(x, y) = \dfrac{2}{\pi} \displaystyle\int_0^\infty \dfrac{\sinh \alpha y}{\alpha(1 + \alpha^2) \cosh \alpha\pi} \cos \alpha x \, d\alpha$

3. $u(x, t) = u_0 e^{-ht} \operatorname{erf}\left(\dfrac{x}{2\sqrt{t}}\right)$

5. $u(x, t) = \displaystyle\int_0^t \operatorname{erfc}\left(\dfrac{x}{2\sqrt{\tau}}\right) d\tau$

7. $u(x, t) = \dfrac{u_0}{2\pi} \displaystyle\int_{-\infty}^{\infty} \dfrac{\sin \alpha(\pi - x) + \sin \alpha x}{\alpha} e^{-k\alpha^2 t} \, d\alpha$

9. $u(x, y) = \dfrac{100}{\pi} \displaystyle\int_0^\infty \left(\dfrac{1 - \cos \alpha}{\alpha}\right)$
$\times [e^{-\alpha x} \sin \alpha y + 2e^{-\alpha y} \sin \alpha x] \, d\alpha$

11. $u(x, y) = \dfrac{2}{\pi} \displaystyle\int_0^\infty \left(\dfrac{B \cosh \alpha y}{(1 + \alpha^2) \sinh \alpha\pi} + \dfrac{A}{\alpha}\right) \sin \alpha x \, d\alpha$

13. $u(x, y) = \dfrac{1}{2\pi} \displaystyle\int_{-\infty}^{\infty} \dfrac{\cos \alpha x + \alpha \sin \alpha x}{1 + \alpha^2} e^{-k\alpha^2 t} \, d\alpha$

EXERCISES 15.1 (PAGE 616)

1. $u_{11} = \frac{11}{15}, u_{21} = \frac{14}{15}$

3. $u_{11} = u_{21} = \sqrt{3}/16, u_{22} = u_{12} = 3\sqrt{3}/16$

5. $u_{21} = u_{12} = 12.50, u_{31} = u_{13} = 18.75, u_{32} = u_{23} = 37.50,$
$u_{11} = 6.25, u_{22} = 25.00, u_{33} = 56.25$

7. (b) $u_{14} = u_{41} = 0.5427, u_{24} = u_{42} = 0.6707,$
$u_{34} = u_{43} = 0.6402, u_{33} = 0.4451, u_{44} = 0.9451$

1.

Time	$x = 0.25$	$x = 0.50$	$x = 0.75$	$x = 1.00$	$x = 1.25$	$x = 1.50$	$x = 1.75$
0.000	1.0000	1.0000	1.0000	1.0000	0.0000	0.0000	0.0000
0.025	0.6000	1.0000	1.0000	0.6000	0.4000	0.0000	0.0000
0.050	0.5200	0.8400	0.8400	0.6800	0.3200	0.1600	0.0000
0.075	0.4400	0.7120	0.7760	0.6000	0.4000	0.1600	0.0640
0.100	0.3728	0.6288	0.6800	0.5904	0.3840	0.2176	0.0768
0.125	0.3261	0.5469	0.6237	0.5437	0.4000	0.2278	0.1024
0.150	0.2840	0.4893	0.5610	0.5182	0.3886	0.2465	0.1116
0.175	0.2525	0.4358	0.5152	0.4835	0.3836	0.2494	0.1209
0.200	0.2248	0.3942	0.4708	0.4562	0.3699	0.2517	0.1239
0.225	0.2027	0.3571	0.4343	0.4275	0.3571	0.2479	0.1255
0.250	0.1834	0.3262	0.4007	0.4021	0.3416	0.2426	0.1242
0.275	0.1672	0.2989	0.3715	0.3773	0.3262	0.2348	0.1219
0.300	0.1530	0.2752	0.3448	0.3545	0.3101	0.2262	0.1183
0.325	0.1407	0.2541	0.3209	0.3329	0.2943	0.2166	0.1141
0.350	0.1298	0.2354	0.2990	0.3126	0.2787	0.2067	0.1095
0.375	0.1201	0.2186	0.2790	0.2936	0.2635	0.1966	0.1046
0.400	0.1115	0.2034	0.2607	0.2757	0.2488	0.1865	0.0996
0.425	0.1036	0.1895	0.2438	0.2589	0.2347	0.1766	0.0945
0.450	0.0965	0.1769	0.2281	0.2432	0.2211	0.1670	0.0896
0.475	0.0901	0.1652	0.2136	0.2283	0.2083	0.1577	0.0847
0.500	0.0841	0.1545	0.2002	0.2144	0.1961	0.1487	0.0800
0.525	0.0786	0.1446	0.1876	0.2014	0.1845	0.1402	0.0755
0.550	0.0736	0.1354	0.1759	0.1891	0.1735	0.1320	0.0712
0.575	0.0689	0.1269	0.1650	0.1776	0.1632	0.1243	0.0670
0.600	0.0645	0.1189	0.1548	0.1668	0.1534	0.1169	0.0631
0.625	0.0605	0.1115	0.1452	0.1566	0.1442	0.1100	0.0594
0.650	0.0567	0.1046	0.1363	0.1471	0.1355	0.1034	0.0559
0.675	0.0532	0.0981	0.1279	0.1381	0.1273	0.0972	0.0525
0.700	0.0499	0.0921	0.1201	0.1297	0.1196	0.0914	0.0494
0.725	0.0468	0.0864	0.1127	0.1218	0.1124	0.0859	0.0464
0.750	0.0439	0.0811	0.1058	0.1144	0.1056	0.0807	0.0436
0.775	0.0412	0.0761	0.0994	0.1074	0.0992	0.0758	0.0410
0.800	0.0387	0.0715	0.0933	0.1009	0.0931	0.0712	0.0385
0.825	0.0363	0.0671	0.0876	0.0948	0.0875	0.0669	0.0362
0.850	0.0341	0.0630	0.0823	0.0890	0.0822	0.0628	0.0340
0.875	0.0320	0.0591	0.0772	0.0836	0.0772	0.0590	0.0319
0.900	0.0301	0.0555	0.0725	0.0785	0.0725	0.0554	0.0300
0.925	0.0282	0.0521	0.0681	0.0737	0.0681	0.0521	0.0282
0.950	0.0265	0.0490	0.0640	0.0692	0.0639	0.0489	0.0265
0.975	0.0249	0.0460	0.0601	0.0650	0.0600	0.0459	0.0249
1.000	0.0234	0.0432	0.0564	0.0610	0.0564	0.0431	0.0233

3.

Time	$x = 0.25$	$x = 0.50$	$x = 0.75$	$x = 1.00$	$x = 1.25$	$x = 1.50$	$x = 1.75$
0.000	1.0000	1.0000	1.0000	1.0000	0.0000	0.0000	0.0000
0.025	0.7074	0.9520	0.9566	0.7444	0.2545	0.0371	0.0053
0.050	0.5606	0.8499	0.8685	0.6633	0.3303	0.1034	0.0223
0.075	0.4684	0.7473	0.7836	0.6191	0.3614	0.1529	0.0462
0.100	0.4015	0.6577	0.7084	0.5837	0.3753	0.1871	0.0684
0.125	0.3492	0.5821	0.6428	0.5510	0.3797	0.2101	0.0861
0.150	0.3069	0.5187	0.5857	0.5199	0.3778	0.2247	0.0990
0.175	0.2721	0.4652	0.5359	0.4901	0.3716	0.2329	0.1078
0.200	0.2430	0.4198	0.4921	0.4617	0.3622	0.2362	0.1132
0.225	0.2186	0.3809	0.4533	0.4348	0.3507	0.2358	0.1160
0.250	0.1977	0.3473	0.4189	0.4093	0.3378	0.2327	0.1166
0.275	0.1798	0.3181	0.3881	0.3853	0.3240	0.2275	0.1157
0.300	0.1643	0.2924	0.3604	0.3626	0.3097	0.2208	0.1136
0.325	0.1507	0.2697	0.3353	0.3412	0.2953	0.2131	0.1107
0.350	0.1387	0.2495	0.3125	0.3211	0.2808	0.2047	0.1071
0.375	0.1281	0.2313	0.2916	0.3021	0.2666	0.1960	0.1032
0.400	0.1187	0.2150	0.2725	0.2843	0.2528	0.1871	0.0989
0.425	0.1102	0.2002	0.2549	0.2675	0.2393	0.1781	0.0946
0.450	0.1025	0.1867	0.2387	0.2517	0.2263	0.1692	0.0902
0.475	0.0955	0.1743	0.2236	0.2368	0.2139	0.1606	0.0858
0.500	0.0891	0.1630	0.2097	0.2228	0.2020	0.1521	0.0814
0.525	0.0833	0.1525	0.1967	0.2096	0.1906	0.1439	0.0772
0.550	0.0779	0.1429	0.1846	0.1973	0.1798	0.1361	0.0731
0.575	0.0729	0.1339	0.1734	0.1856	0.1696	0.1285	0.0691
0.600	0.0683	0.1256	0.1628	0.1746	0.1598	0.1214	0.0653
0.625	0.0641	0.1179	0.1530	0.1643	0.1506	0.1145	0.0617
0.650	0.0601	0.1106	0.1438	0.1546	0.1419	0.1080	0.0582
0.675	0.0564	0.1039	0.1351	0.1455	0.1336	0.1018	0.0549
0.700	0.0530	0.0976	0.1270	0.1369	0.1259	0.0959	0.0518
0.725	0.0497	0.0917	0.1194	0.1288	0.1185	0.0904	0.0488
0.750	0.0467	0.0862	0.1123	0.1212	0.1116	0.0852	0.0460
0.775	0.0439	0.0810	0.1056	0.1140	0.1050	0.0802	0.0433
0.800	0.0413	0.0762	0.0993	0.1073	0.0989	0.0755	0.0408
0.825	0.0388	0.0716	0.0934	0.1009	0.0931	0.0711	0.0384
0.850	0.0365	0.0674	0.0879	0.0950	0.0876	0.0669	0.0362
0.875	0.0343	0.0633	0.0827	0.0894	0.0824	0.0630	0.0341
0.900	0.0323	0.0596	0.0778	0.0841	0.0776	0.0593	0.0321
0.925	0.0303	0.0560	0.0732	0.0791	0.0730	0.0558	0.0302
0.950	0.0285	0.0527	0.0688	0.0744	0.0687	0.0526	0.0284
0.975	0.0268	0.0496	0.0647	0.0700	0.0647	0.0495	0.0268
1.000	0.0253	0.0466	0.0609	0.0659	0.0608	0.0465	0.0252

Absolute errors are approximately 2.2×10^{-2}, 3.7×10^{-2}, 1.3×10^{-2}.

5.

Time	$x = 0.25$	$x = 0.50$	$x = 0.75$	$x = 1.00$	$x = 1.25$	$x = 1.50$	$x = 1.75$
0.00	1.0000	1.0000	1.0000	1.0000	0.0000	0.0000	0.0000
0.05	0.5265	0.8693	0.8852	0.6141	0.3783	0.0884	0.0197
0.10	0.3972	0.6551	0.7043	0.5883	0.3723	0.1955	0.0653
0.15	0.3042	0.5150	0.5844	0.5192	0.3812	0.2261	0.1010
0.20	0.2409	0.4171	0.4901	0.4620	0.3636	0.2385	0.1145
0.25	0.1962	0.3452	0.4174	0.4092	0.3391	0.2343	0.1178
0.30	0.1631	0.2908	0.3592	0.3624	0.3105	0.2220	0.1145
0.35	0.1379	0.2482	0.3115	0.3208	0.2813	0.2056	0.1077
0.40	0.1181	0.2141	0.2718	0.2840	0.2530	0.1876	0.0993
0.45	0.1020	0.1860	0.2381	0.2514	0.2265	0.1696	0.0904
0.50	0.0888	0.1625	0.2092	0.2226	0.2020	0.1523	0.0816
0.55	0.0776	0.1425	0.1842	0.1970	0.1798	0.1361	0.0732
0.60	0.0681	0.1253	0.1625	0.1744	0.1597	0.1214	0.0654
0.65	0.0599	0.1104	0.1435	0.1544	0.1418	0.1079	0.0582
0.70	0.0528	0.0974	0.1268	0.1366	0.1257	0.0959	0.0518
0.75	0.0466	0.0860	0.1121	0.1210	0.1114	0.0851	0.0460
0.80	0.0412	0.0760	0.0991	0.1071	0.0987	0.0754	0.0408
0.85	0.0364	0.0672	0.0877	0.0948	0.0874	0.0668	0.0361
0.90	0.0322	0.0594	0.0776	0.0839	0.0774	0.0592	0.0320
0.95	0.0285	0.0526	0.0687	0.0743	0.0686	0.0524	0.0284
1.00	0.0252	0.0465	0.0608	0.0657	0.0607	0.0464	0.0251

Absolute errors are approximately 1.8×10^{-2}, 3.7×10^{-2}, 1.3×10^{-2}.

7. (a)

Time	$x = 2.00$	$x = 4.00$	$x = 6.00$	$x = 8.00$	$x = 10.00$	$x = 12.00$	$x = 14.00$	$x = 16.00$	$x = 18.00$
0.00	30.0000	30.0000	30.0000	30.0000	30.0000	30.0000	30.0000	30.0000	30.0000
1.00	28.7733	29.9749	29.9995	30.0000	30.0000	30.0000	29.9995	29.9749	28.7733
2.00	27.6450	29.9037	29.9970	29.9999	30.0000	29.9999	29.9970	29.9037	27.6450
3.00	26.6051	29.7938	29.9911	29.9997	30.0000	29.9997	29.9911	29.7938	26.6051
4.00	25.6452	29.6517	29.9805	29.9991	29.9999	29.9991	29.9805	29.6517	25.6452
5.00	24.7573	29.4829	29.9643	29.9981	29.9998	29.9981	29.9643	29.4829	24.7573
6.00	23.9347	29.2922	29.9421	29.9963	29.9996	29.9963	29.9421	29.2922	23.9347
7.00	23.1711	29.0836	29.9134	29.9936	29.9992	29.9936	29.9134	29.0836	23.1711
8.00	22.4612	28.8606	29.8782	29.9898	29.9986	29.9898	29.8782	28.8606	22.4612
9.00	21.7999	28.6263	29.8362	29.9848	29.9977	29.9848	29.8362	28.6263	21.7999
10.00	21.1829	28.3831	29.7878	29.9782	29.9964	29.9782	29.7878	28.3831	21.1829

(b)

Time	$x = 5.00$	$x = 10.00$	$x = 15.00$	$x = 20.00$	$x = 25.00$	$x = 30.00$	$x = 35.00$	$x = 40.00$	$x = 45.00$
0.00	30.0000	30.0000	30.0000	30.0000	30.0000	30.0000	30.0000	30.0000	30.0000
1.00	29.7968	29.9993	30.0000	30.0000	30.0000	30.0000	30.0000	29.9993	29.7968
2.00	29.5964	29.9973	30.0000	30.0000	30.0000	30.0000	30.0000	29.9973	29.5964
3.00	29.3987	29.9939	30.0000	30.0000	30.0000	30.0000	30.0000	29.9939	29.3987
4.00	29.2036	29.9893	29.9999	30.0000	30.0000	30.0000	29.9999	29.9893	29.2036
5.00	29.0112	29.9834	29.9998	30.0000	30.0000	30.0000	29.9998	29.9834	29.0112
6.00	28.8212	29.9762	29.9997	30.0000	30.0000	30.0000	29.9997	29.9762	28.8213
7.00	28.6339	29.9679	29.9995	30.0000	30.0000	30.0000	29.9995	29.9679	28.6339
8.00	28.4490	29.9585	29.9992	30.0000	30.0000	30.0000	29.9993	29.9585	28.4490
9.00	28.2665	29.9479	29.9989	30.0000	30.0000	30.0000	29.9989	29.9479	28.2665
10.00	28.0864	29.9363	29.9986	30.0000	30.0000	30.0000	29.9986	29.9363	28.0864

(c)

Time	$x = 2.00$	$x = 4.00$	$x = 6.00$	$x = 8.00$	$x = 10.00$	$x = 12.00$	$x = 14.00$	$x = 16.00$	$x = 18.00$
0.00	18.0000	32.0000	42.0000	48.0000	50.0000	48.0000	42.0000	32.0000	18.0000
1.00	16.4489	30.1970	40.1561	46.1495	48.1486	46.1495	40.1561	30.1970	16.4489
2.00	15.3312	28.5348	38.3465	44.3067	46.3001	44.3067	38.3465	28.5348	15.3312
3.00	14.4216	27.0416	36.6031	42.4847	44.4619	42.4847	36.6031	27.0416	14.4216
4.00	13.6371	25.6867	34.9416	40.6988	42.6453	40.6988	34.9416	25.6867	13.6371
5.00	12.9378	24.4419	33.3628	38.9611	40.8634	38.9611	33.3628	24.4419	12.9378
6.00	12.3012	23.2863	31.8624	37.2794	39.1273	37.2794	31.8624	23.2863	12.3012
7.00	11.7137	22.2051	30.4350	35.6578	37.4446	35.6578	30.4350	22.2051	11.7137
8.00	11.1659	21.1877	29.0757	34.0984	35.8202	34.0984	29.0757	21.1877	11.1659
9.00	10.6517	20.2261	27.7799	32.6014	34.2567	32.6014	27.7799	20.2261	10.6517
10.00	10.1665	19.3143	26.5439	31.1662	32.7549	31.1662	26.5439	19.3143	10.1665

(d)

Time	$x = 10.00$	$x = 20.00$	$x = 30.00$	$x = 40.00$	$x = 50.00$	$x = 60.00$	$x = 70.00$	$x = 80.00$	$x = 90.00$
0.00	8.0000	16.0000	24.0000	32.0000	40.0000	32.0000	24.0000	16.0000	8.0000
1.00	8.0000	16.0000	24.0000	31.9979	39.7425	31.9979	24.0000	16.0000	8.0000
2.00	8.0000	16.0000	23.9999	31.9918	39.4932	31.9918	23.9999	16.0000	8.0000
3.00	8.0000	16.0000	23.9997	31.9820	39.2517	31.9820	23.9997	16.0000	8.0000
4.00	8.0000	16.0000	23.9993	31.9686	39.0175	31.9686	23.9993	16.0000	8.0000
5.00	8.0000	16.0000	23.9987	31.9520	38.7905	31.9520	23.9987	16.0000	8.0000
6.00	8.0000	15.9999	23.9978	31.9323	38.5701	31.9323	23.9978	15.9999	8.0000
7.00	8.0000	15.9999	23.9966	31.9097	38.3561	31.9097	23.9966	15.9998	8.0000
8.00	8.0000	15.9998	23.9950	31.8844	38.1483	31.8844	23.9950	15.9998	8.0000
9.00	8.0000	15.9997	23.9931	31.8566	37.9463	31.8566	23.9931	15.9997	8.0000
10.00	8.0000	15.9996	23.9908	31.8265	37.7498	31.8265	23.9908	15.9996	8.0000

9. (a)

Time	$x = 2.00$	$x = 4.00$	$x = 6.00$	$x = 8.00$	$x = 10.00$	$x = 12.00$	$x = 14.00$	$x = 16.00$	$x = 18.00$
0.00	30.0000	30.0000	30.0000	30.0000	30.0000	30.0000	30.0000	30.0000	30.0000
1.00	28.7733	29.9749	29.9995	30.0000	30.0000	30.0000	29.9998	29.9916	29.5911
2.00	27.6450	29.9037	29.9970	29.9999	30.0000	30.0000	29.9990	29.9679	29.2150
3.00	26.6051	29.7938	29.9911	29.9997	30.0000	29.9999	29.9970	29.9313	28.8684
4.00	25.6452	29.6517	29.9805	29.9991	30.0000	29.9997	29.9935	29.8839	28.5484
5.00	24.7573	29.4829	29.9643	29.9981	29.9999	29.9994	29.9881	29.8276	28.2524
6.00	23.9347	29.2922	29.9421	29.9963	29.9997	29.9988	29.9807	29.7641	27.9782
7.00	23.1711	29.0836	29.9134	29.9936	29.9995	29.9979	29.9711	29.6945	27.7237
8.00	22.4612	28.8606	29.8782	29.9899	29.9991	29.9966	29.9594	29.6202	27.4870
9.00	21.7999	28.6263	29.8362	29.9848	29.9985	29.9949	29.9454	29.5421	27.2666
10.00	21.1829	28.3831	29.7878	29.9783	29.9976	29.9927	29.9293	29.4610	27.0610

(b)

Time	$x = 5.00$	$x = 10.00$	$x = 15.00$	$x = 20.00$	$x = 25.00$	$x = 30.00$	$x = 35.00$	$x = 40.00$	$x = 45.00$
0.00	30.0000	30.0000	30.0000	30.0000	30.0000	30.0000	30.0000	30.0000	30.0000
1.00	29.7968	29.9993	30.0000	30.0000	30.0000	30.0000	30.0000	29.9998	29.9323
2.00	29.5964	29.9973	30.0000	30.0000	30.0000	30.0000	30.0000	29.9991	29.8655
3.00	29.3987	29.9939	30.0000	30.0000	30.0000	30.0000	30.0000	29.9980	29.7996
4.00	29.2036	29.9893	29.9999	30.0000	30.0000	30.0000	30.0000	29.9964	29.7345
5.00	29.0112	29.9834	29.9998	30.0000	30.0000	30.0000	29.9999	29.9945	29.6704
6.00	28.8212	29.9762	29.9997	30.0000	30.0000	30.0000	29.9999	29.9921	29.6071
7.00	28.6339	29.9679	29.9995	30.0000	30.0000	30.0000	29.9998	29.9893	29.5446
8.00	28.4490	29.9585	29.9992	30.0000	30.0000	30.0000	29.9997	29.9862	29.4830
9.00	28.2665	29.9479	29.9989	30.0000	30.0000	30.0000	29.9996	29.9827	29.4222
10.00	28.0864	29.9363	29.9986	30.0000	30.0000	30.0000	29.9995	29.9788	29.3621

(c)

Time	$x = 2.00$	$x = 4.00$	$x = 6.00$	$x = 8.00$	$x = 10.00$	$x = 12.00$	$x = 14.00$	$x = 16.00$	$x = 18.00$
0.00	18.0000	32.0000	42.0000	48.0000	50.0000	48.0000	42.0000	32.0000	18.0000
1.00	16.4489	30.1970	40.1562	46.1502	48.1531	46.1773	40.3274	31.2520	22.9449
2.00	15.3312	28.5350	38.3477	44.3130	46.3327	44.4671	39.0872	31.5755	24.6930
3.00	14.4219	27.0429	36.6090	42.5113	44.5759	42.9362	38.1976	31.7478	25.4131
4.00	13.6381	25.6913	34.9606	40.7728	42.9127	41.5716	37.4340	31.7086	25.6986
5.00	12.9409	24.4545	33.4091	39.1182	41.3519	40.3240	36.7033	31.5136	25.7663
6.00	12.3088	23.3146	31.9546	37.5566	39.8880	39.1565	35.9745	31.2134	25.7128
7.00	11.7294	22.2589	30.5939	36.0884	38.5109	38.0470	35.2407	30.8434	25.5871
8.00	11.1946	21.2785	29.3217	34.7092	37.2109	36.9834	34.5032	30.4279	25.4167
9.00	10.6987	20.3660	28.1318	33.4130	35.9801	35.9591	33.7660	29.9836	25.2181
10.00	10.2377	19.5150	27.0178	32.1929	34.8117	34.9710	33.0338	29.5224	25.0019

(d)

Time	$x = 10.00$	$x = 20.00$	$x = 30.00$	$x = 40.00$	$x = 50.00$	$x = 60.00$	$x = 70.00$	$x = 80.00$	$x = 90.00$
0.00	8.0000	16.0000	24.0000	32.0000	40.0000	32.0000	24.0000	16.0000	8.0000
1.00	8.0000	16.0000	24.0000	31.9979	39.7425	31.9979	24.0000	16.0026	8.3218
2.00	8.0000	16.0000	24.0000	31.9918	39.4932	31.9918	24.0000	16.0102	8.6333
3.00	8.0000	16.0000	23.9997	31.9820	39.2517	31.9820	24.0001	16.0225	8.9350
4.00	8.0000	16.0000	23.9993	31.9686	39.0175	31.9687	24.0002	16.0391	9.2272
5.00	8.0000	16.0000	23.9987	31.9520	38.7905	31.9520	24.0003	16.0599	9.5103
6.00	8.0000	15.9999	23.9978	31.9323	38.5701	31.9324	24.0005	16.0845	9.7846
7.00	8.0000	15.9999	23.9966	31.9097	38.3561	31.9098	24.0008	16.1126	10.0506
8.00	8.0000	15.9998	23.9950	31.8844	38.1483	31.8846	24.0012	16.1441	10.3084
9.00	8.0000	15.9997	23.9931	31.8566	37.9463	31.8569	24.0017	16.1786	10.5585
10.00	8.0000	15.9996	23.9908	31.8265	37.7499	31.8269	24.0023	16.2160	10.8012

11. (a) $\psi(x) = \frac{1}{2}x + 20$

(b)

Time	x = 4.00	x = 8.00	x = 12.00	x = 16.00
0.00	50.0000	50.0000	50.0000	50.0000
10.00	32.7433	44.2679	45.4228	38.2971
20.00	29.9946	36.2354	38.3148	35.8160
30.00	26.9487	32.1409	34.0874	32.9644
40.00	25.2691	29.2562	31.2704	31.2580
50.00	24.1178	27.4348	29.4296	30.1207
60.00	23.3821	26.2339	28.2356	29.3810
70.00	22.8995	25.4560	27.4554	28.8998
80.00	22.5861	24.9481	26.9482	28.5859
90.00	22.3817	24.6176	26.6175	28.3817
100.00	22.2486	24.4022	26.4023	28.2486
110.00	22.1619	24.2620	26.2620	28.1619
120.00	22.1055	24.1707	26.1707	28.1055
130.00	22.0687	24.1112	26.1112	28.0687
140.00	22.0447	24.0724	26.0724	28.0447
150.00	22.0291	24.0472	26.0472	28.0291
160.00	22.0190	24.0307	26.0307	28.0190
170.00	22.0124	24.0200	26.0200	28.0124
180.00	22.0081	24.0130	26.0130	28.0081
190.00	22.0052	24.0085	26.0085	28.0052
200.00	22.0034	24.0055	26.0055	28.0034
210.00	22.0022	24.0036	26.0036	28.0022
220.00	22.0015	24.0023	26.0023	28.0015
230.00	22.0009	24.0015	26.0015	28.0009
240.00	22.0006	24.0010	26.0010	28.0006
250.00	22.0004	24.0007	26.0007	28.0004
260.00	22.0003	24.0004	26.0004	28.0003
270.00	22.0002	24.0003	26.0003	28.0002
280.00	22.0001	24.0002	26.0002	28.0001
290.00	22.0001	24.0001	26.0001	28.0001
300.00	22.0000	24.0001	26.0001	28.0000
310.00	22.0000	24.0001	26.0001	28.0000
320.00	22.0000	24.0000	26.0000	28.0000
330.00	22.0000	24.0000	26.0000	28.0000
340.00	22.0000	24.0000	26.0000	28.0000
350.00	22.0000	24.0000	26.0000	28.0000

1. (a)

Time	x = 0.25	x = 0.50	x = 0.75
0.00	0.1875	0.2500	0.1875
0.10	0.1775	0.2400	0.1775
0.20	0.1491	0.2100	0.1491
0.30	0.1066	0.1605	0.1066
0.40	0.0556	0.0938	0.0556
0.50	0.0019	0.0148	0.0019
0.60	−0.0501	−0.0682	−0.0501
0.70	−0.0970	−0.1455	−0.0970
0.80	−0.1361	−0.2072	−0.1361
0.90	−0.1648	−0.2462	−0.1648
1.00	−0.1802	−0.2591	−0.1802

(b)

Time	x = 0.4	x = 0.8	x = 1.2	x = 1.6
0.00	0.0032	0.5273	0.5273	0.0032
0.10	0.0194	0.5109	0.5109	0.0194
0.20	0.0652	0.4638	0.4638	0.0652
0.30	0.1318	0.3918	0.3918	0.1318
0.40	0.2065	0.3035	0.3035	0.2065
0.50	0.2743	0.2092	0.2092	0.2743
0.60	0.3208	0.1190	0.1190	0.3208
0.70	0.3348	0.0413	0.0413	0.3348
0.80	0.3094	−0.0180	−0.0180	0.3094
0.90	0.2443	−0.0568	−0.0568	0.2443
1.00	0.1450	−0.0768	−0.0768	0.1450

(c)

Time	$x = 0.1$	$x = 0.2$	$x = 0.3$	$x = 0.4$	$x = 0.5$	$x = 0.6$	$x = 0.7$	$x = 0.8$	$x = 0.9$
0.00	0.0000	0.0000	0.0000	0.0000	0.0000	0.5000	0.5000	0.5000	0.5000
0.04	0.0000	0.0000	0.0000	0.0000	0.0800	0.4200	0.5000	0.5000	0.4200
0.08	0.0000	0.0000	0.0000	0.0256	0.2432	0.2568	0.4744	0.4744	0.2312
0.12	0.0000	0.0000	0.0082	0.1126	0.3411	0.1589	0.3792	0.3710	0.0462
0.16	0.0000	0.0026	0.0472	0.2394	0.3076	0.1898	0.2108	0.1663	−0.0496
0.20	0.0008	0.0187	0.1334	0.3264	0.2146	0.2651	0.0215	−0.0933	−0.0605
0.24	0.0071	0.0657	0.2447	0.3159	0.1735	0.2463	−0.1266	−0.3056	−0.0625
0.28	0.0299	0.1513	0.3215	0.2371	0.2013	0.0849	−0.2127	−0.3829	−0.1223
0.32	0.0819	0.2525	0.3168	0.1737	0.2033	−0.1345	−0.2580	−0.3223	−0.2264
0.36	0.1623	0.3197	0.2458	0.1657	0.0877	−0.2853	−0.2843	−0.2104	−0.2887
0.40	0.2412	0.3129	0.1727	0.1583	−0.1223	−0.3164	−0.2874	−0.1473	−0.2336
0.44	0.2657	0.2383	0.1399	0.0658	−0.3046	−0.2761	−0.2549	−0.1565	−0.0761
0.48	0.1965	0.1410	0.1149	−0.1216	−0.3593	−0.2381	−0.1977	−0.1715	0.0800
0.52	0.0466	0.0531	0.0225	−0.3093	−0.2992	−0.2260	−0.1451	−0.1144	0.1300
0.56	−0.1161	−0.0466	−0.1662	−0.3876	−0.2188	−0.2114	−0.1085	0.0111	0.0602
0.60	−0.2194	−0.2069	−0.3875	−0.3411	−0.1901	−0.1662	−0.0666	0.1140	−0.0446
0.64	−0.2485	−0.4290	−0.5362	−0.2611	−0.2021	−0.0969	0.0012	0.1084	−0.0843
0.68	−0.2559	−0.6276	−0.5625	−0.2503	−0.1993	−0.0298	0.0720	0.0068	−0.0354
0.72	−0.3003	−0.6865	−0.5097	−0.3230	−0.1585	0.0156	0.0893	−0.0874	0.0384
0.76	−0.3722	−0.5652	−0.4538	−0.4029	−0.1147	0.0289	0.0265	−0.0849	0.0596
0.80	−0.3867	−0.3464	−0.4172	−0.4068	−0.1172	−0.0046	−0.0712	−0.0005	0.0155
0.84	−0.2647	−0.1633	−0.3546	−0.3214	−0.1763	−0.0954	−0.1249	0.0665	−0.0386
0.88	−0.0254	−0.0738	−0.2202	−0.2002	−0.2559	−0.2215	−0.1079	0.0385	−0.0468
0.92	0.2064	−0.0157	−0.0325	−0.1032	−0.3067	−0.3223	−0.0804	−0.0636	−0.0127
0.96	0.3012	0.1081	0.1380	−0.0487	−0.2974	−0.3407	−0.1250	−0.1548	0.0092
1.00	0.2378	0.3032	0.2392	−0.0141	−0.2223	−0.2762	−0.2481	−0.1840	−0.0244

3. (a)

Time	$x = 0.2$	$x = 0.4$	$x = 0.6$	$x = 0.8$
0.00	0.5878	0.9511	0.9511	0.5878
0.05	0.5808	0.9397	0.9397	0.5808
0.10	0.5599	0.9059	0.9059	0.5599
0.15	0.5256	0.8505	0.8505	0.5256
0.20	0.4788	0.7748	0.7748	0.4788
0.25	0.4206	0.6806	0.6806	0.4206
0.30	0.3524	0.5701	0.5701	0.3524
0.35	0.2757	0.4460	0.4460	0.2757
0.40	0.1924	0.3113	0.3113	0.1924
0.45	0.1046	0.1692	0.1692	0.1046
0.50	0.0142	0.0230	0.0230	0.0142

(b)

Time	$x = 0.2$	$x = 0.4$	$x = 0.6$	$x = 0.8$
0.00	0.5878	0.9511	0.9511	0.5878
0.03	0.5860	0.9482	0.9482	0.5860
0.05	0.5808	0.9397	0.9397	0.5808
0.08	0.5721	0.9256	0.9256	0.5721
0.10	0.5599	0.9060	0.9060	0.5599
0.13	0.5445	0.8809	0.8809	0.5445
0.15	0.5257	0.8507	0.8507	0.5257
0.18	0.5039	0.8153	0.8153	0.5039
0.20	0.4790	0.7750	0.7750	0.4790
0.23	0.4513	0.7302	0.7302	0.4513
0.25	0.4209	0.6810	0.6810	0.4209
0.28	0.3879	0.6277	0.6277	0.3879
0.30	0.3527	0.5706	0.5706	0.3527
0.33	0.3153	0.5102	0.5102	0.3153
0.35	0.2761	0.4467	0.4467	0.2761
0.38	0.2352	0.3806	0.3806	0.2352
0.40	0.1929	0.3122	0.3122	0.1929
0.43	0.1495	0.2419	0.2419	0.1495
0.45	0.1052	0.1701	0.1701	0.1052
0.48	0.0602	0.0974	0.0974	0.0602
0.50	0.0149	0.0241	0.0241	0.0149

5.

Time	x = 10	x = 20	x = 30	x = 40	x = 50
0.00000	0.1000	0.2000	0.3000	0.2000	0.1000
0.20045	0.1000	0.2000	0.2750	0.2000	0.1000
0.40089	0.1000	0.1938	0.2125	0.1938	0.1000
0.60134	0.0984	0.1688	0.1406	0.1688	0.0984
0.80178	0.0898	0.1191	0.0828	0.1191	0.0898
1.00223	0.0661	0.0531	0.0432	0.0531	0.0661
1.20268	0.0226	−0.0121	0.0085	−0.0121	0.0226
1.40312	−0.0352	−0.0635	−0.0365	−0.0635	−0.0352
1.60357	−0.0913	−0.1011	−0.0950	−0.1011	−0.0913
1.80401	−0.1271	−0.1347	−0.1566	−0.1347	−0.1271
2.00446	−0.1329	−0.1719	−0.2072	−0.1719	−0.1329
2.20491	−0.1153	−0.2081	−0.2402	−0.2081	−0.1153
2.40535	−0.0920	−0.2292	−0.2571	−0.2292	−0.0920
2.60580	−0.0801	−0.2230	−0.2601	−0.2230	−0.0801
2.80624	−0.0838	−0.1903	−0.2445	−0.1903	−0.0838
3.00669	−0.0932	−0.1445	−0.2018	−0.1445	−0.0932
3.20713	−0.0921	−0.1003	−0.1305	−0.1003	−0.0921
3.40758	−0.0701	−0.0615	−0.0440	−0.0615	−0.0701
3.60803	−0.0284	−0.0205	0.0336	−0.0205	−0.0284
3.80847	0.0224	0.0321	0.0842	0.0321	0.0224
4.00892	0.0700	0.0953	0.1087	0.0953	0.0700
4.20936	0.1064	0.1555	0.1265	0.1555	0.1064
4.40981	0.1285	0.1962	0.1588	0.1962	0.1285
4.61026	0.1354	0.2106	0.2098	0.2106	0.1354
4.81070	0.1273	0.2060	0.2612	0.2060	0.1273
5.01115	0.1070	0.1955	0.2851	0.1955	0.1070
5.21159	0.0821	0.1853	0.2641	0.1853	0.0821
5.41204	0.0625	0.1689	0.2038	0.1689	0.0625
5.61249	0.0539	0.1347	0.1260	0.1347	0.0539
5.81293	0.0520	0.0781	0.0526	0.0781	0.0520
6.01338	0.0436	0.0086	−0.0080	0.0086	0.0436
6.21382	0.0156	−0.0564	−0.0604	−0.0564	0.0156
6.41427	−0.0343	−0.1043	−0.1107	−0.1043	−0.0343
6.61472	−0.0931	−0.1364	−0.1578	−0.1364	−0.0931
6.81516	−0.1395	−0.1630	−0.1942	−0.1630	−0.1395
7.01561	−0.1568	−0.1915	−0.2150	−0.1915	−0.1568
7.21605	−0.1436	−0.2173	−0.2240	−0.2173	−0.1436
7.41650	−0.1129	−0.2263	−0.2297	−0.2263	−0.1129
7.61695	−0.0824	−0.2078	−0.2336	−0.2078	−0.0824
7.81739	−0.0625	−0.1644	−0.2247	−0.1644	−0.0625
8.01784	−0.0526	−0.1106	−0.1856	−0.1106	−0.0526
8.21828	−0.0440	−0.0611	−0.1091	−0.0611	−0.0440
8.41873	−0.0287	−0.0192	−0.0085	−0.0192	−0.0287
8.61918	−0.0038	0.0229	0.0867	0.0229	−0.0038
8.81962	0.0287	0.0743	0.1500	0.0743	0.0287
9.02007	0.0654	0.1332	0.1755	0.1332	0.0654
9.22051	0.1027	0.1858	0.1799	0.1858	0.1027
9.42096	0.1352	0.2160	0.1872	0.2160	0.1352
9.62140	0.1540	0.2189	0.2089	0.2189	0.1540
9.82185	0.1506	0.2030	0.2356	0.2030	0.1506
10.02230	0.1226	0.1822	0.2461	0.1822	0.1226

Note: Time is expressed in milliseconds.

CHAPTER 15 IN REVIEW (PAGE 630)

1. $u_{11} = 0.8929$, $u_{21} = 3.5714$, $u_{31} = 13.3929$

3. (a)

Time	$x = 0.00$	$x = 0.20$	$x = 0.40$	$x = 0.60$	$x = 0.80$	$x = 1.00$
0.00	0.0000	0.2000	0.4000	0.6000	0.8000	0.0000
0.01	0.0000	0.2000	0.4000	0.6000	0.5500	0.0000
0.02	0.0000	0.2000	0.4000	0.5375	0.4250	0.0000
0.03	0.0000	0.2000	0.3844	0.4750	0.3469	0.0000
0.04	0.0000	0.1961	0.3609	0.4203	0.2922	0.0000
0.05	0.0000	0.1883	0.3346	0.3734	0.2512	0.0000

(b)

Time	$x = 0.00$	$x = 0.20$	$x = 0.40$	$x = 0.60$	$x = 0.80$	$x = 1.00$
0.00	0.0000	0.2000	0.4000	0.6000	0.8000	1.0000
0.01	0.0000	0.2000	0.4000	0.6000	0.8000	0.0000
0.02	0.0000	0.2000	0.4000	0.6000	0.5500	0.0000
0.03	0.0000	0.2000	0.4000	0.5375	0.4250	0.0000
0.04	0.0000	0.2000	0.3844	0.4750	0.3469	0.0000
0.05	0.0000	0.1961	0.3609	0.4203	0.2922	0.0000

(c) Yes; the table in part (b) is the table in part (a) shifted downward.

EXERCISES FOR APPENDIX I (PAGE APP-2)

1. (a) 24   (b) 720   (c) $\dfrac{4\sqrt{\pi}}{3}$ (d) $-\dfrac{8\sqrt{\pi}}{15}$

3. 0.297

EXERCISES FOR APPENDIX II (PAGE APP-20)

1. (a) $\begin{pmatrix} 2 & 11 \\ 2 & -1 \end{pmatrix}$ (b) $\begin{pmatrix} -6 & 1 \\ 14 & -19 \end{pmatrix}$

(c) $\begin{pmatrix} 2 & 28 \\ 12 & -12 \end{pmatrix}$

3. (a) $\begin{pmatrix} -11 & 6 \\ 17 & -22 \end{pmatrix}$ (b) $\begin{pmatrix} -32 & 27 \\ -4 & -1 \end{pmatrix}$

(c) $\begin{pmatrix} 19 & -18 \\ -30 & 31 \end{pmatrix}$ (d) $\begin{pmatrix} 19 & 6 \\ 3 & 22 \end{pmatrix}$

5. (a) $\begin{pmatrix} 9 & 24 \\ 3 & 8 \end{pmatrix}$ (b) $\begin{pmatrix} 3 & 8 \\ -6 & -16 \end{pmatrix}$

(c) $\begin{pmatrix} 0 & 0 \\ 0 & 0 \end{pmatrix}$ (d) $\begin{pmatrix} -4 & -5 \\ 8 & 10 \end{pmatrix}$

7. (a) 180   (b) $\begin{pmatrix} 4 & 8 & 10 \\ 8 & 16 & 20 \\ 10 & 20 & 25 \end{pmatrix}$ (c) $\begin{pmatrix} 6 \\ 12 \\ -5 \end{pmatrix}$

9. (a) $\begin{pmatrix} 7 & 38 \\ 10 & 75 \end{pmatrix}$ (b) $\begin{pmatrix} 7 & 38 \\ 10 & 75 \end{pmatrix}$

11. $\begin{pmatrix} -14 \\ 1 \end{pmatrix}$ 13. $\begin{pmatrix} -38 \\ -2 \end{pmatrix}$ 15. singular

17. nonsingular; $\mathbf{A}^{-1} = \dfrac{1}{4}\begin{pmatrix} -5 & -8 \\ 3 & 4 \end{pmatrix}$

19. nonsingular; $\mathbf{A}^{-1} = \dfrac{1}{2}\begin{pmatrix} 0 & -1 & 1 \\ 2 & 2 & -2 \\ -4 & -3 & 5 \end{pmatrix}$

21. nonsingular; $\mathbf{A}^{-1} = -\dfrac{1}{9}\begin{pmatrix} -2 & -2 & -1 \\ -13 & 5 & 7 \\ 8 & -1 & -5 \end{pmatrix}$

23. $\mathbf{A}^{-1}(t) = \dfrac{1}{2e^{3t}}\begin{pmatrix} 3e^{4t} & -e^{4t} \\ -4e^{-t} & 2e^{-t} \end{pmatrix}$

25. $\dfrac{d\mathbf{X}}{dt} = \begin{pmatrix} -5e^{-t} \\ -2e^{-t} \\ 7e^{-t} \end{pmatrix}$

27. $\dfrac{d\mathbf{X}}{dt} = 4\begin{pmatrix} 1 \\ -1 \end{pmatrix}e^{2t} - 12\begin{pmatrix} 2 \\ 1 \end{pmatrix}e^{-3t}$

29. (a) $\begin{pmatrix} 4e^{4t} & -\pi \sin \pi t \\ 2 & 6t \end{pmatrix}$ (b) $\begin{pmatrix} \frac{1}{4}e^{8} - \frac{1}{4} & 0 \\ 4 & 6 \end{pmatrix}$

(c) $\begin{pmatrix} \frac{1}{4}e^{4t} - \frac{1}{4} & (1/\pi) \sin \pi t \\ t^2 & t^3 - t \end{pmatrix}$

31. $x = 3$, $y = 1$, $z = -5$

33. $x = 2 + 4t$, $y = -5 - t$, $z = t$

35. $x = -\frac{1}{2}$, $y = \frac{3}{2}$, $z = \frac{7}{2}$

37. $x_1 = 1$, $x_2 = 0$, $x_3 = 2$, $x_4 = 0$

41. $\mathbf{A}^{-1} = \begin{pmatrix} 0 & \frac{2}{3} & \frac{1}{3} \\ 0 & -\frac{1}{3} & -\frac{2}{3} \\ \frac{1}{3} & -\frac{2}{3} & 0 \end{pmatrix}$

43. $\mathbf{A}^{-1} = \begin{pmatrix} 5 & 6 & -3 \\ 2 & 2 & -1 \\ -1 & -1 & 1 \end{pmatrix}$

45. $\mathbf{A}^{-1} = \begin{pmatrix} -\frac{1}{2} & -\frac{2}{3} & -\frac{1}{6} & \frac{7}{6} \\ 1 & \frac{1}{3} & \frac{1}{3} & -\frac{4}{3} \\ 0 & -\frac{1}{3} & -\frac{1}{3} & \frac{1}{3} \\ -\frac{1}{2} & 1 & \frac{1}{2} & \frac{1}{2} \end{pmatrix}$

47. $\lambda_1 = 6, \lambda_2 = 1, \mathbf{K}_1 = \begin{pmatrix} 2 \\ 7 \end{pmatrix}, \mathbf{K}_2 = \begin{pmatrix} 1 \\ 1 \end{pmatrix}$

49. $\lambda_1 = \lambda_2 = -4, \mathbf{K}_1 = \begin{pmatrix} 1 \\ -4 \end{pmatrix}$

51. $\lambda_1 = 0, \lambda_2 = 4, \lambda_3 = -4,$
$\mathbf{K}_1 = \begin{pmatrix} 9 \\ 45 \\ 25 \end{pmatrix}, \mathbf{K}_2 = \begin{pmatrix} 1 \\ 1 \\ 1 \end{pmatrix}, \mathbf{K}_3 = \begin{pmatrix} 1 \\ 9 \\ 1 \end{pmatrix}$

53. $\lambda_1 = \lambda_2 = \lambda_3 = -2,$
$\mathbf{K}_1 = \begin{pmatrix} 2 \\ -1 \\ 0 \end{pmatrix}, \mathbf{K}_2 = \begin{pmatrix} 0 \\ 0 \\ 1 \end{pmatrix}$

55. $\lambda_1 = 3i, \lambda_2 = -3i,$
$\mathbf{K}_1 = \begin{pmatrix} 1 - 3i \\ 5 \end{pmatrix}, \mathbf{K}_2 = \begin{pmatrix} 1 + 3i \\ 5 \end{pmatrix}$

INDEX